T0383854

Advance Praise

RNA, the Epicenter of Genetic Information

The origin story and emergence of molecular biology is muddled. The early triumphs in bacterial genetics and the complexity of animal and plant genomes complicate an intricate history. This book documents the many advances, as well as the prejudices and founder fallacies. It highlights the premature relegation of RNA to simply an intermediate between gene and protein, the underestimation of the amount of information required to program the development of multicellular organisms, and the dawning realization that RNA is the cornerstone of cell biology, development, brain function and probably evolution itself. Key personalities and their hubris as well as prescient predictions are richly illustrated with quotes, archival material, photographs, diagrams and references to bring the people, ideas and discoveries to life, from the conceptual cradles of molecular biology to the current revolution in the understanding of genetic information.

Key Features

- Documents the confused early history of DNA, RNA and proteins – a transformative history of molecular biology like no other.

- Integrates the influences of biochemistry and genetics on the landscape of molecular biology.

- Chronicles the important discoveries, preconceptions and misconceptions that retarded or misdirected progress.

- Highlights the major pioneers and contributors to molecular biology, with a focus on RNA and non-coding DNA.

- Summarizes the mounting evidence for the central roles of non-protein-coding RNA in cell and developmental biology.

- Provides a thought-provoking retrospective and forward-looking perspective for advanced students and professional researchers.

RNA, the Epicenter of Genetic Information

A new understanding of molecular biology

John Mattick and Paulo Amaral

CRC Press
Taylor & Francis Group
Boca Raton London New York

CRC Press is an imprint of the
Taylor & Francis Group, an **informa** business

Cover Art by Allan Rodrigo Carvalho, Visual Artist at CSBL - University of São Paulo. RNA image created from the Holo L-16 ScaI Tetrahymena ribozyme structure at 3.1 angstrom resolution (Su Z. et al., Nature 596: 603–607, 2021; PDB DOI: 10.2210/pdb7EZ2/pdb, EMDataResource: EMD-31386). Special thanks to the Tree of Life Web Project for Tree of Life image © 2007, in which the background illustration was based.

First edition published 2023
by CRC Press
6000 Broken Sound Parkway NW, Suite 300, Boca Raton, FL 33487-2742

and by CRC Press
4 Park Square, Milton Park, Abingdon, Oxon, OX14 4RN

CRC Press is an imprint of Taylor & Francis Group, LLC

© 2023 John Mattick and Paulo Amaral

Library of Congress Cataloging-in-Publication Data
Names: Mattick, John, author.
Title: RNA, the epicenter of genetic information : a new understanding of molecular biology / by John Mattick, Paulo Amaral.
Description: First edition. |
Boca Raton : Taylor and Francis, 2023. | Includes bibliographical references and index.
Identifiers: LCCN 2022004098 (print) | LCCN 2022004099 (ebook) |
ISBN 9780367623920 (hardback) | ISBN 9780367567781 (paperback) |
ISBN 9781003109242 (ebook)
Subjects: LCSH: Molecular biology–History. | RNA. | Genomes.
Classification: LCC QH506 .M393 2023 (print) | LCC QH506 (ebook) | DDC 572.8–dc23/eng/20220203
LC record available at https://lccn.loc.gov/2022004098
LC ebook record available at https://lccn.loc.gov/2022004099

ISBN: 9780367623920 (hbk)
ISBN: 9780367567781 (pbk)
ISBN: 9781003109242 (ebk)

DOI: 10.1201/9781003109242

Typeset in Times
by codeMantra

Contents

Contents

Preface

RNA likely underpinned the emergence of life, yet it is arguably the least appreciated of all biological macromolecules. For most of the past century, RNA has been regarded principally as an intermediary between gene and protein. However, most of the human genome expresses RNAs that do not encode proteins, which begs the question: why?

The understanding of the functions of RNA and the answer to this question are bound up with the history of molecular biology. The term 'molecular biology' was coined by the mathematician Warren Weaver in 1938[1] and has become synonymous with the nature, transmission and manifestation of genetic information, and the structure of the molecules involved. The field had its roots in the discovery of DNA in 1869 and the identification of proteins and their enzymatic activity in the late 19th century, key events that gave birth to "the science of the chemistry of life". Since then, proteins and DNA have been the primary focus of studies of cellular and developmental processes and the conceptions of 'genes', 'gene expression' and 'gene regulation'.

While it was clear early on that chromosomes are the vehicles of inheritance, and contain DNA, RNA and proteins, for a long time it was thought that genetic information is held in the proteins; nucleic acids seemed too simple. In the 1940s, however, it was shown that DNA is the reservoir of genetic information, although it took some time for this finding to be accepted.

The connection between DNA and protein production was solved by the convergence of genetics and biochemistry, mainly in experimentally amenable bacteria and fungi, which led to the breathtaking advances that elucidated the role of 'messenger' RNA (mRNA) and the 'genetic code' in the 1950s and 1960s. The assumption that genetic information is mostly transacted by proteins ('one gene – one enzyme'), with RNA a transient intermediate, became entrenched, reflecting the mechanical zeitgeist of the age.

This assumption led to many subsidiary assumptions about the nature of genetic information, and the conclusion that most of the genomes of plants and animals are junk, based on theoretical considerations of mutational load and the finding that protein-coding sequences occupy only a small fraction of animal and plant genomes. The naivety of this conclusion and its superficial support by intrinsically circular assessments of the 'neutral evolution' of 'non-coding' sequences in genomes were rarely challenged.

There were other assumptions as well, including that heritable information is not transmitted from somatic cells to reproductive cells. This assertion, supported by a peculiar 1868 experiment involving amputation of mice tails, accompanied the so-called Modern Synthesis in the 1930s, which reconciled Mendelian genetics with Darwinian evolution and ruled out Lamarckian evolution, to buttress the belief that 'mutations' are random.

Undoubtedly the biggest surprises in the history of molecular biology were the discoveries in the 1960s and 1970s that plant and animal genomes are replete with 'repetitive elements' and that their genes are mosaics of fragmented protein-coding and flanking regulatory mRNA sequences ('exons') separated by extensive tracts of intervening sequences ('introns'), which are subsequently removed from the primary transcripts by splicing. It was immediately and almost universally assumed that introns are evolutionary relics colonized by 'selfish' genetic hobos and that the excised intronic RNA is simply degraded.

Also unexpected were the discoveries in the 1980s that RNA has catalytic capacity and, at the turn of the century, that the number of protein-coding genes in humans is similar to that in nematode worms that only have ~1,000 cells. By contrast, the extent of intronic and 'intergenic' non-protein-coding DNA sequences was found to increase with developmental complexity, rising to ~98% in humans and other mammals.

High-throughput expression studies revealed that these 'non-coding' sequences are transcribed in spatially restricted patterns to produce hundreds of thousands of RNAs that do not encode proteins. Many of these RNAs were subsequently shown to have regulatory and organizational functions during differentiation and development.

Here we provide an account of the development of molecular biology from the 19th century to the present. We pay particular attention to the history of the understanding of RNA, which has been neglected. We also discuss the founder fallacies – where initial interpretations of limited data were generalized, became orthodox explanations and then articles of faith. Our central theme is that the extrapolation of bacterial genetics to complex organisms, compounded by expectational, ascertainment and interpretative biases, has led to a linked series of false dichotomies and the misunderstanding of roles of RNA in the transmission of genetic information. The subsidiary theme is the clumsy progress of science.

This book focuses on RNA as the main player in cell and developmental biology, but also on chromatin composition and regulatory logic. While most educated in the pre-genomic era were taught that gene regulation is primarily carried out by proteins, this became hard to reconcile with the finding that genes encoding regulatory RNAs vastly outnumber protein-coding genes in humans, and the demonstrations of widespread sequence-specific guidance of effector proteins by RNAs. The simplicity and logic of base-pairing for sense-antisense target recognition and the ability of RNA to form complex three-dimensional structures are almost as old as the double helix itself. The existence of regulatory RNAs was hinted during the early period of molecular biology by genetic observations in fruit flies and maize, and by the appearance of unexplained bands in biochemical fractionations, but these were treated as oddities or interpreted through the lens of transcription factors, until the genome projects revealed the full extent of RNA expression in plants and animals.

We highlight the pioneers and controversies that accompanied the many unexpected observations, with particular attention to those that challenged the prevailing consensus, often ignored, at least at first. The book spans the early confusion about the functions of proteins and nucleic acids, the elucidation of the double helix and the 'genetic code', the premature relegation of RNA to intermediary between gene and protein, the strange genomes and genetics of plants and animals, and the misguided musings that underpinned the idea of junk DNA. We chronicle the spectacular advances brought by gene cloning and genome sequencing, the small and large regulatory RNA revolutions, and the slowly dawning realization of the central role of transposon-derived sequences, intrinsically disordered proteins, 'enhancers' and RNA-directed epigenetic processes in multicellular development, which we have tried to integrate into a new framework for understanding genetic programming.

We have cited original references where possible, to give credit to the work of others and to provide the evidence for our assertions and conclusions, especially in relation to the findings of the last two decades. We have also included extensive footnotes that add detail and can be skipped, as well as suggestions for further reading.

While the story is still unfolding, we conclude that the genomes of humans and other complex organisms are not full of junk but rather are highly compact information suites that are largely devoted to the specification of regulatory RNAs. These RNAs drive the trajectories of differentiation and development, underpin brain function and convey transgenerational memory of experience, much of it contrary to long-held conceptions of genetic programming and the dogmas of evolutionary theory.

Acknowledgments

We are grateful to colleagues who have worked with us over the years in the Mattick lab and beyond. So many have contributed to the ideas and findings that have constituted our body of work, and the background to this book, that it is hard to single out individuals, but we particularly thank Marcel Dinger, Tim Mercer, Mike Pheasant and Ryan Taft, as well as John Rinn, a pioneer of RNA biology with whom we have collaborated on many occasions.

We wish to thank several funding agencies and institutions for their support during the preparation and writing of this book, in particular (JM) Green Templeton College Oxford and UNSW Sydney; and (PA) the Brocher Foundation (which granted him a residency to conduct research for the book in the summer of 2019); the Royal Society Newton International Fellowship, Corpus Christi College and Borysiewicz Biomedical Sciences Fellowship (University of Cambridge); and Insper Institute of Education and Research in São Paulo, Brazil.

We posthumously thank Denise Barlow for encouraging us to prepare this book in the initial stages of our endeavor. We also thank the following people, who have been kind enough to read sections of the manuscript, provide advice and identify errors: Bill Amos, Mandy Ballinger, Andy Bannister, Tim Bredy, Piero Carninci, Lindy Castell, Don Chambers, Kay Davies, Caroline Dean, Marcel Dinger, Tommaso Leonardi, Gonçalo Castelo-Branco, Laurence Hurst, Thomas Mailund, Mick McManus, Michael Morgan, Helder Nakaya, John Rinn, Jeff Rosen, Klaus Scherrer, Phillip Sharp, John Shine, Joan Steitz, Grant Sutherland, Jai Tree, David Thomas, David Tremethick, Carl Walkley, Robert Weatheritt and Eric Westhof. Of course, any remaining errors or important omissions, which are almost inevitable in a book of such broad sweep, are ours alone, and we welcome feedback to correct them.

Finally, John thanks his wife Louise, his family, colleagues and friends, and likewise Paulo thanks Dóris P. Amaral, his family, colleagues and friends, for their steadfast support, encouragement and forbearance during this journey.

The authors, some time ago.

"The things that you're liable to read in the Bible ... ain't necessarily so."

(Ira Gershwin, *Porgy & Bess* 1935)

Authors

John Mattick is Professor of RNA Biology at UNSW Sydney. He earned his PhD in Biochemistry at Monash University in Melbourne and undertook his post-doctoral training at Baylor College of Medicine in Houston. Most recently, he was the Chief Executive of Genomics England, Executive Director of the Garvan Institute of Medical Research in Sydney and Director of the Institute for Molecular Bioscience, the Australian Genome Research Facility, the ARC Special Research Centre for Molecular and Cell Biology and the ARC Special Research Centre for Functional and Applied Genomics at the University of Queensland.

His research interests have been focused for the past three decades on the role of regulatory RNAs in the evolution and development of complex organisms. He pioneered the thesis that the majority of the genome of humans and other complex organisms, previously considered to be 'junk', is devoted to regulatory RNAs that direct the epigenetic trajectories of differentiation and development. He has published more than 300 research articles and reviews, which have been cited over 85,000 times. His work has received editorial coverage in *Nature*, *Science*, *Scientific American*, *New Scientist* and *The New York Times*. It has also been highlighted in two books: *The Deeper Genome* by John Parrington and *Promoting the Planck Club* by Don Braben.

He has received numerous awards including the International Union of Biochemistry and Molecular Biology Medal, the University of Texas MD Anderson Cancer Center Bertner Award for Distinguished Contributions to Cancer Research, and the Human Genome Organization Chen Medal for Distinguished Achievement in Human Genetics and Genomic Research. He is a Fellow of the Australian Academy of Science, the Royal Society of New South Wales, the Australian Academy of Technology and Engineering and the Australian Academy of Health and Medical Sciences. He is also an Associate Member of the European Molecular Biology Organization.

Paulo Amaral is Assistant Professor of Bioengineering at Insper in São Paulo, Brazil. He earned a BSc in Biological Sciences (University of Brasília), MSc in Biochemistry at the Institute of Chemistry, University of São Paulo, and PhD in Molecular Genetics at the Institute for Molecular Bioscience, University of Queensland, in Brisbane. He undertook his postdoctoral work at the CRUK/Wellcome Trust Gurdon Institute and Milner Therapeutics Institute, University of Cambridge, supported by a Royal Society and British Academy Newton International Fellowship, Corpus Christi College Research Fellowship and Borysiewicz Biomedical Sciences Fellowship.

He has also taught at different institutions as a visiting lecturer at both the undergraduate and graduate levels, including at the Karolinska Institutet and Uppsala University in Sweden, and Humanitas University and Università Campus Bio-Medico di Roma in Italy. Since 2004, his research has explored the roles of non-coding DNA, regulatory RNAs and RNA modifications in a variety of biological systems, in both academia and industry. He was one of the first to show that non-coding RNAs associate with chromatin and guide chromatin modifications for gene activation, and has been a longstanding student of the history of molecular biology, especially of the overlooked middleman, but almost certainly the key player, RNA.

1 Overview

THE GENETIC MATERIAL?

Proteins have been associated with all biological processes since the inception of biochemistry in the 19th century, given their abundance, enzymatic properties and versatility. On the other hand, although identified in 1869, the functions of nucleic acids remained obscure until the 1940s. In the years leading up to the turn of the 20th century, DNA was found to be localized in chromosomes, which were shown to be the vehicles for genetic inheritance. In 1909, it was proposed that nucleic acids form simple tetramers containing each of the four component nucleotides, following which it was generally thought, because of their presumed repetitive structure, that nucleic acids have only peripheral functions.

Accordingly, proteins, which are also found in chromosomes, were regarded as the repository of genetic information for the first four decades of the 20th century, with DNA functioning as a scaffold. However, in 1944, DNA was demonstrated to be the 'transforming principle' in bacteria, although this finding was only widely accepted after bacteriophage infection 'pulse-chase' experiments in 1952, the elucidation of the structure of DNA in 1953 and the demonstration of its semi-conservative replication in 1958.

While having a nucleotide composition similar to DNA, RNA did not appear to play a role in the intergenerational transmission of genetic information, although RNA viruses were later found to exist. It was regarded for decades as an uninteresting metabolic molecule in bacteria, yeast and plants, and only conclusively shown to exist in animal cells in the 1930s.

Microbial genetics from the 1920s established that (some) genes encode proteins, but the mechanism by which this occurred was unknown. Gradually it dawned that RNA might be involved, inferred from histochemical, ultracentrifugation and spectroscopic studies in the 1940s that showed that RNA is present in cytoplasmic microsomes (ribosomes), which were becoming recognized as the sites of protein synthesis.

Meanwhile, theoretical biologists had declared, in the so-called Modern Synthesis reconciling Darwinian evolution with Mendelian genetics, that mutations are random, and that Lamarckian inheritance of experience does not occur. Moreover, the emphasis on lethal protein-coding mutations muddled the interpretation of genetic variation, with ongoing debates between the 'Mendelians' and the quantitative geneticists.

HALCYON DAYS

The abundant ribosomal RNAs (rRNAs) were identified in the mid-1950s, but a specific function for RNA was demonstrated only in 1958, when small RNAs were shown to act as 'adaptors' for the incorporation of amino acids into microsomal proteins, named 'transfer RNAs' (tRNAs). In 1961, the radioactive labeling of 'messenger RNAs' (mRNAs) finally identified the 'unstable' intermediate between genes and proteins, establishing the connection.

In the following decade, the triplet 'genetic code' for protein synthesis was deciphered. Analysis of the lactose (*lac*) operon of *Escherichia coli* cemented the conclusion that genes are synonymous with proteins. The regulation of gene activity by protein 'transcription factors' was established and assumed to hold not just in bacteria but also in developmentally complex organisms. All that remained to do, it seemed, was to flesh out the details.

WORLDS APART

It was obvious by that time that plants and animals are orders of magnitude more complex than bacteria and have different cellular and genetic features, including much greater internal compartmentalization and far larger genomes. It was later shown that eukaryotic cells arose by fusion of bacterial and archaeal cells and that developmentally complex organisms burst onto the scene in spectacular adaptive radiations, most likely following regulatory innovations required to orchestrate organized cell division and differentiation.

Studies using newer techniques in the 1960s and 1970s showed that that eukaryotic DNA is packaged in a repeating structure ('nucleosomes') comprised

DOI: 10.1201/9781003109242-1

1

of basic proteins called histones, and that chromatin is compacted and remodeled during development. It was found that histones are dynamically modified by methylation and acetylation, which suggested that histone modifications act as a regulatory mechanism. It was also shown that RNA is associated with chromatin and that very high molecular weight 'heterogeneous' RNAs are synthesized in the nucleus, predicted to be precursors of mRNAs, but the function of the remainder of these transcripts was mysterious.

STRANGE GENOMES, STRANGE GENETICS

The use of the fruit fly *Drosophila melanogaster* as a model genetic system from the 1910s enabled the mapping of genes along chromosomes by measuring recombination distances (co-inheritance frequencies), which established the view of genes as discrete, 'particulate' entities. Analysis of naturally occurring and radiation-induced mutations identified 'homeotic' loci that caused bizarre segmental transformations along with other encoding epigenetic 'modifiers' that exhibited strange interactions.

Odd genetic phenomena were also reported in plants. 'Rogue' non-Mendelian patterns of inheritance were observed in peas in 1915 and characterized in other species from the 1950s, termed 'paramutation', later understood to be a feature of transgenerational epigenetic inheritance. Mobile 'controlling elements' were identified in maize in the 1940s and shown to be due to the transposition of regulatory cassettes. In the mid-1960s, large fractions of the genomes of plants and animals were found to be comprised of 'repetitive sequences', most of which derive from transposable elements. It was also found that the repetitive sequences are differentially transcribed.

In 1969, these disparate molecular observations were integrated into a schema of gene regulation in embryonic development, which included the concepts of 'structural' (protein-coding) and 'integrator' genes (most likely) expressing regulatory RNAs recognized by cognate receptor sequences, connected into networks by repetitive sequences. Processed nuclear RNAs were posited in other models to be global regulators of gene expression, but the problem was the lack of detail about the actual information in genomes, which rendered these models, as reasonable as they were, speculative and largely overlooked.

THE AGE OF AQUARIUS

The problem of lack of detail began to be solved by the gene cloning revolution and the development of DNA sequencing in the 1970s. These technologies led an explosion in knowledge, and by the mid-1990s shotgun cloning and sequencing was being used to characterize the many mRNAs that had eluded identification by biochemical and genetic assays. A myriad of protein-coding genes was discovered in organisms from bacteria to humans, including those that regulate development, cell division, cell differentiation, cell signaling, trafficking pathways and immunological responses, among many others, as well as mutated versions in cancer. These advances, however, diverted attention from the broader questions of genome regulation and reinforced the concept of genes as protein-coding.

ALL THAT JUNK

By the 1970s, it was evident, however, that most sequences in the genomes of complex organisms are not protein-coding (Figure 1.1). The amount of cellular DNA was found to broadly increase with developmental complexity, but there were incongruities, termed the C-value enigma. Theoretical considerations of population genetics, the lethality of protein-coding mutations, the presence of large numbers of repetitive sequences and seemingly defective 'pseudogenes' all suggested that some, and perhaps most, multicellular organisms carry substantial loads of non-functional DNA.

The corollary of 'neutral' evolution of non-functional sequences was widely accepted, although there was debate between the 'near-neutralists' and 'adaptationists' concerning the signatures of protein-coding genes (and, later, regulatory sequences) underpinning quantitative trait variation. Nonetheless, there was growing consensus that much if not most of the DNA in plant and animal genomes must be junk and that the many repetitive sequences are 'selfish' genetic hobos.

The discovery in 1977 that eukaryotic genes are mosaics of short fragments of mRNA protein-coding and flanking regulatory sequences ('exons') interspersed with non-coding sequences ('introns') that are removed by post-transcriptional splicing explained heterogeneous nuclear RNA and was proffered as further evidence of junk. Introns were rationalized as the remnants of the prebiotic assembly of genes, which had been purged from microbial

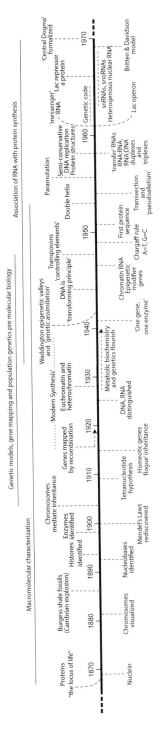
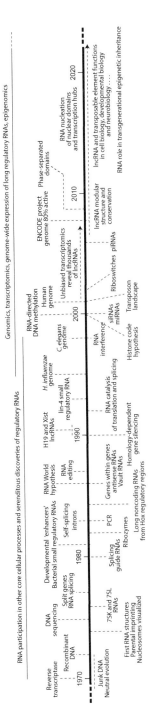

FIGURE 1.1   Historical overview.

genomes under selective pressure for rapid replication, even though the ancestors of complex organisms were also microbial. On the other hand, while small in unicellular eukaryotes, introns were found to increase in number and size with the developmental complexity of multicellular organisms, which suggested that these sequences had acquired important functions.

THE EXPANDING REPERTOIRE OF RNA

In parallel with the gene cloning revolution, the increasing sophistication of biochemical techniques identified relatively abundant RNA species beyond the canonical trio of tRNA, mRNA and rRNA. These included small nuclear RNAs (snRNAs) that guide splicing and other aspects of gene expression; small nucleolar RNAs (snoRNAs) that guide modifications of rRNAs, tRNAs and snRNAs; 7SK RNA, a negative regulator of transcription; 7SL RNA, an essential component of the 'signal recognition particle' that targets proteins to the endoplasmic reticulum and precursor of the ubiquitous Alu elements in the human genome; 'vault' RNAs of mysterious function but known to be involved in recycling of cellular components in lysosomes and neuronal synaptic plasticity; and rodent brain-specific transposon-derived RNAs that modulate behavior.

In the 1980s, RNAs were discovered to have self-splicing and cleavage activities, and that RNA catalyzes both translation and splicing, leading to the conclusion that RNA was the primordial molecule of life – the 'RNA World' hypothesis, whereby RNA subsequently outsourced its enzymatic functions to the more versatile proteins and its information functions to the more stable and easily replicable DNA. The early examples of the structural and functional capacities of RNA were, however, largely interpreted as relic infrastructural components rather than another dimension of molecular biology.

GLIMPSES OF A MODERN RNA WORLD

In the decades leading up to the turn of the century and shortly thereafter, as analytical sensitivity improved, many less abundant RNAs were identified. Small antisense RNAs ('riboregulators') and cis-acting RNA structures ('riboswitches') were found to control transcription and translation in bacteria, the latter by allosteric sensing of metabolites and environmental signals. Synthetic antisense oligonucleotides began to be used to artificially control gene expression in eukaryotic cells.

Overlapping 'antisense' transcription and 'nested' genes within genes were observed in animals and plants, hinting at intertwined genetic information and regulatory complexity. Differentially transcribed long 'untranslated' RNAs were reported to regulate ribosomal RNA transcription, and to be produced from the regulatory regions of homeotic and heat shock–induced genes in *Drosophila* and mammalian immunoglobulin class-switching, cancer-associated and parentally imprinted loci, among others.

Xist was identified as a long non-coding RNA that mediates female X-chromosome inactivation in mammals, and analogous RNAs mediating male X-chromosome activation were identified in *Drosophila*. 3' untranslated regions (3'UTRs) in mRNAs were found to be separately expressed and to transmit genetic information independently of their normally associated protein-coding sequences, and small RNAs antisense to 3'UTRs were found to control developmental timing in *C. elegans*. Although some speculated that these small and large RNAs may be the first examples of a more extensive RNA regulatory system in cell and developmental biology, they were generally regarded as oddities.

GENOME SEQUENCING AND TRANSPOSABLE ELEMENTS

By the mid-1990s, the extraordinary advances in DNA cloning, amplification and sequencing had made feasible the sequencing of whole genomes. The subsequent exponential growth of data led to progressively well-annotated genome databases and suites of computational tools for gene prediction, ortholog identification and the analysis of gene structure and expression. For the first time, the full complement of DNA sequence information in bacteria and archaea, protists, fungi, plants and animals began to be revealed, enabling comparative genomics to interrogate evolutionary relationships and functional indices at increasingly high resolution, including in complex microbial ecologies.

Prokaryote genomes were confirmed to be dominated by protein-coding genes, with phenotypic diversity achieved primarily by proteomic variation. On the other hand, animals differing by orders of magnitude in developmental complexity were unexpectedly found to have a similar number and repertoire of protein-coding genes – only about 20,000 in both nematodes and mammals – the 'G-value enigma'.

By contrast, increased developmental complexity correlated with the extent of non-protein-coding DNA, reaching over 98% in humans and other mammals, indicating that the developmental sophistication of multicellular organisms is achieved by the expansion of regulatory information. Moreover, transposable element and retroviral-derived repetitive sequences began to be recognized as major drivers of phenotypic innovation in a wide range of plants and animals.

THE HUMAN GENOME

The first draft of the human genome sequence was published in 2001, notwithstanding the controversies that surrounded the project. The number of identified human protein-coding genes was far lower than expected by most in the field. Comparison with the mouse genome suggested that ~95% of the human genome is non-functional, based on the assumption that ancient transposon-derived sequences can be used to measure the rate of neutral evolution. On the other hand, analyses of genomic features such as transcription, sequence accessibility, DNA and histone modifications and transcription factor binding led to the conclusion that most of the human genome exhibits biochemical indices of function.

Human 'Mendelian' disorders were mapped and confirmed to be largely due to disabling mutations in protein-coding sequences. By contrast, genome-wide association studies showed that variations affecting complex traits and disorders reside mainly in non-coding regions of the genome, although an appreciable fraction of the known genetic contribution to these traits appeared unaccounted, suggesting other factors at play.

SMALL RNAs WITH MIGHTY FUNCTIONS

Genetic observations in the 1980s and 1990s indicated that RNA may play a general role in gene regulation, when it was reported that sense and antisense RNAs could modulate endogenous gene expression transcriptionally and post-transcriptionally, referred to as 'co-suppression', 'gene silencing' and (ultimately) 'RNA interference' (RNAi). The finding that introducing sense and antisense RNAs together resulted in strong systemic repression of target genes led to the dissection of the RNAi pathways, showing that double-stranded RNAs are processed to form 'small interfering RNAs' (siRNAs) that guide DNA methylation and cleavage of orthologous sequences in mRNAs.

At the turn of the 21st century, it was discovered that the RNAi pathway is used extensively to control gene expression during animal and plant development, via 'microRNAs' (miRNAs) derived from introns and other non-protein-coding transcripts. Related small RNAs, 'piRNAs', many produced from repetitive sequences, were found to be required for fertility, germ and stem cell development in animals. Other classes of small regulatory RNAs were found to be derived from tRNAs, rRNAs, snoRNAs, snRNAs, gene promoters and splice junctions, and small RNAs were shown to have many functions, including intergenerational and interspecies communication.

A similar pathway, termed 'CRISPR', was later found in bacteria to use RNA guides to target cleavage of bacteriophage genomes, manipulation of which has revolutionized genetic analysis and genetic engineering. The common feature of the RNAi and CRISPR pathways is that they use small RNAs to guide generic effector proteins to target cognate sequences in RNAs and DNAs, a highly efficient and flexible system of gene control.

LARGE RNAs WITH MANY FUNCTIONS

The high-throughput RNA profiling projects that followed the genome projects revealed the existence in both animals and plants of large numbers of low abundance long, often multi-exonic, RNAs that have little or no protein-coding potential. These 'long non-coding RNAs' (lncRNAs) were found to be expressed 'intergenically', intronically and antisense to or overlapping protein-coding genes, as well as from thousands of 'pseudogenes' and 3'UTRs. The data also showed that most of the genome of eukaryotes is transcribed in highly complex overlapping patterns, substantially from both DNA strands.

Although initially suspected to be noise, lncRNAs were found to be dynamically expressed during differentiation and development, mostly in cell-type specific patterns. LncRNAs were also found be associated with membrane-less cellular organelles, chromatin-modifying proteins and/or chromatin domains. While the genetic signatures of lncRNAs are, in the main, subtler than protein-coding genes, many have been shown to be involved in cancer and developmental, autoimmune, neurodegenerative and neuropsychiatric disorders. Large numbers of lncRNAs – many of which are clade- or species-specific – were also discovered to have functions in cell fate determination and reprogramming, DNA damage repair, germ layer specification, hematopoietic, immunological and neuronal differentiation, retinal, skeletal, muscle and brain development, and memory and behavior, among many others.

THE EPIGENOME

It became increasingly evident during this period that the chromosomes of higher organisms are highly organized and epigenetically modified. Cytogenetic and molecular studies from the 1980s had shown the existence of chromosome territories, gene-rich and gene-poor regions and fine-scale 'topologically associated domains' with variable GC contents and non-random distributions of sequences derived from transposable elements. New genetic loci termed enhancers, hundreds of thousands of which exist in mammalian genomes, were identified and found to control plant and animal development by selective activation of protein-coding genes in their vicinity.

Nucleosomes were shown to contain canonical and specialist histones, some specific to mammalian germ and neuronal cells. The histones were found to be subject to a bewildering variety of post-translational modifications that are imposed, interpreted and erased by protein complexes that often have no intrinsic sequence specificity, including many essential for the developmental regulation of gene expression.

Histone modifications were shown to vary by gene expression and differentiation state. Exons were found to be preferentially located in nucleosomes, suggesting that epigenetic control of gene expression can be exon-specific. Vertebrate DNA was also found to be dynamically methylated during development, perturbed in cancer, and associated with gene repression. Little was known, however, of the pathways that determine the locus specificity of epigenetic modifications during development or in response to environmental influences.

THE PROGRAMMING OF DEVELOPMENT

The overarching question, rarely considered, is how much information is required to program development? The nematode worm has ~1,000 leaves in its developmental tree that are genetically hard-wired. Similarly, humans and other mammals must make trillions of divide or differentiate cell fate decisions with high accuracy, also hard-wired, as evidenced by the phenotypic congruency of monozygotic twins.

It had been widely assumed that Boolean combinatorics of transcription and other regulatory factors acting on cis-acting regulatory DNA sequences would suffice to direct developmental ontogeny, but this proposition was not rigorously justified theoretically, mathematically or mechanistically. By contrast, a decisional tree with N leaves requires an exponentially greater number of regulatory decisions, which is consistent with the quasi-quadratic increase in the number of regulatory genes with total gene number in bacteria. In all organisms, presumably, the proportion of the genome devoted to regulatory information increases with metabolic, developmental or cognitive complexity.

The fact that the genomes of plant and animals are transcribed in dynamic patterns during development and millions of different epigenetic marks are imposed at different positions in different cells across developmental stages suggests that RNA regulation has been enlisted as the most flexible and information efficient solution to the challenge of orchestrating multicellular ontogeny.

RNA RULES

Over the past two decades, RNA has been shown to regulate chromosome structure through interaction with transposon-derived sequences. DNA methylation, often differentially imposed at repetitive elements, had been known since the 1990s to be RNA-guided. Chromatin remodeling proteins, sometimes referred to as 'pioneer transcription factors', which have little or no sequence specificity and address different loci at different developmental stages, bind RNA. RNA-DNA hybrids and RNA-DNA-DNA triplexes were found to be common in eukaryotic chromatin. Histone-modifying proteins also have no intrinsic sequence specificity but some have been shown to associate with RNA 'promiscuously', i.e., bind to many different RNAs.

The largest class of sequence-specific transcription factors, containing zinc finger motifs, also addresses target loci differentially and binds RNA as well as DNA, with many having higher affinity for RNA-DNA hybrids than for double-stranded DNA. Half of the C_2H_2 zinc finger proteins in the human genome contain KRAB domains, many primate-specific, which wire them into regulatory networks by binding cognate transposon-derived sequences.

Enhancers were found to express non-coding RNAs that are required for enhancer action, which involves chromatin 'looping' to form transcriptional hubs. Enhancers have all of the signatures of genes, except that they do not encode proteins. The number of mapped enhancers is approximately the same as the number of lncRNAs expressed from the human genome, which resolves the G-value enigma.

It was also discovered that most proteins involved in regulating gene expression in plants and animals, including transcription factors and histone modifiers, contain 'intrinsically disordered regions' (IDRs), the fraction of which increases with developmental complexity. IDRs interact with RNAs to form phase-separated condensates, which are widely deployed to organize subnuclear and cytoplasmic domains, including topologically associated transcriptional hubs in chromatin. RNA interaction with primitive proteins containing IDRs to form phase-separated domains may also comprise the third dimension of the ancestral protocell.

LncRNAs have a modular and highly alternatively spliced structure, with many domains derived from 'repetitive' elements. LncRNAs also act as scaffolds and guides for ribonucleoprotein complexes, a highly efficient and flexible system that, like RNAi and CRISPR, uses RNA signals to regulate and direct generic protein effectors to their sites of action to program development and adaptive radiation.

PLASTICITY

Over 170 different modifications of nucleotides have been identified in RNA, some important for the structure, function or stability of rRNAs, tRNAs, snRNAs and snoRNAs, as well as mRNAs and other non-coding RNAs. These modifications have also been found to be, at least in some cases, reversible and to modulate the structure-function relationships of RNAs to control processes as diverse as chromatin organization, stem cell differentiation, development, brain function, stress responses, mRNA stability and miRNA processing, among others.

RNA modifications have been used to allow mRNA vaccines to evade the innate immune response.

RNA is also 'edited' by cytosine and adenosine deamination, to form uracil and inosine, respectively. Adenosine editing has expanded in vertebrate, mammalian and primate evolution, especially in the brain, and in humans occurs largely in Alu elements, which invaded the genome in three waves during primate evolution and occupy over 10% of the genome, with more than 1 million copies.

The APOBEC enzymes that deaminate cytosine to form thymine or uracil are vertebrate-specific, the first involved in somatic rearrangement and hypermutation of immunoglobulin domains. The ABOBECs have expanded under positive selection during mammalian and primate evolution, apparently to regulate transposable element and retroviral activity. Repetitive elements are mobilized in the brain, which is being shown to have many other unusual molecular dynamics associated with its ability to re-wire synaptic connections.

Transgenerational epigenetic inheritance (such as 'paramutation') was shown to involve small RNA signals and DNA methylation. Paramutation is associated with simple tandem sequence repeats (STRs), over 1 million of which are present in the human genome and are enriched in promoters of protein-coding genes and enhancers. STR variation has been linked with psychiatric disorders and cancer, as well as the modulation of physiological and neurological traits, suggesting that the extent of soft-wired inheritance of experience has been underestimated.

BEYOND THE JUNGLE OF DOGMAS

It seems that the nature of genetic information in complex organisms has been misunderstood since the inception of molecular biology, primarily because of the assumption that most genetic information is transacted by proteins. Other assumptions made during the formative years of genetics also appear to be incorrect, notably that mutations are random, and that epigenetic memory of experience is not inherited.

A transformation is taking place in the understanding of the role of RNA in evolution, inheritance, cell and developmental biology, brain function and disorders, ranging from basic science to a myriad of applications, including a new generation of RNA therapies.

Genomes contain biological software encompassing codes for components, self-assembly, differentiation and reproduction, supplemented by information transmitted by epigenetic memories. Not only has the data evolved, but also the data structures, implementation systems, evolutionary search algorithms and the interplay between hard- and soft-wired inheritance. Indeed, it is likely that evolution has learned how to learn, and that many primitive preconceptions will have to be reevaluated, with more surprises in store.

The details follow.

2 The Genetic Material?

For most of human history, the nature of matter, and life, was a subject of speculation. Famously, in the 4th century BCE, the Greek philosopher Aristotle cemented the ruminations of his predecessor Empedocles to assert that all matter – organic and inorganic – was composed of four elements – fire, air, water and earth – the ratio of which determined its properties. Aristotle also asserted that these elements are interchangeable but did not accept the deduction of his predecessors, Leucippus and Democritus, that all matter could be reduced to indivisible particles, 'atomos' or atoms. Aristotle's views held sway for two millennia.

In 1773, Joseph Priestley showed that heating of mercury 'calx' (mercuric oxide) not only produced liquid mercury, as had been known for thousands of years, but also a gas that caused candles to burn brightly. It was also known that treating metals with acids produced another, highly flammable, gas, termed 'phlogiston'. In 1778, Antoine Lavoisier mixed these two gases, added a lighted match and observed that they combined to form water. He named Priestley's gas 'oxygen' (from the Greek, meaning acid-maker) and re-named phlogiston 'hydrogen' (water-maker). He also burned other substances such as phosphorus and sulfur and showed that they combined with air to make new materials, 'compounds', gaining weight in the process, then in 1789 published a list of 33 chemical elements, grouping them into gases, metals, nonmetals and earths.[1]

In 1794, Joseph Proust proposed that compounds have defined chemical formulas,[2] and in 1803 John Dalton proposed the atomic theory of matter,[3] leading in the following decades to the identification and prediction of many other atomic 'elements', the development of simple nomenclature (H for hydrogen, O for oxygen, C for carbon, N for nitrogen, S for sulfur, P for phosphorus and so on), the distinction by Amedeo Avogadro in 1811 of atoms and combinations thereof ('molecules'),[4,5] and the development of the Periodic Table in 1869 independently by Dmitri Mendeleev and Julius Meyer.[6]

Such was the time that chemists started to explore the nature of matter in biology. The geneticists and physicists came later.

SUGARS AND FATS

Sugars, starches, oils and fats derived from plants and animals had of course been known for eons and used for nutrition, cooking and other practical applications.

In 1789, François Poulletier de La Salle and Michel Chevreul described a substance that could be extracted by alcohol from bile stones and named it 'cholesterine' (cholesterol) from the Greek 'chole', meaning bile and 'stereos', meaning solid.[7] In 1815, Henri Braconnot classified fats into 'suifs' (greases) and 'huiles' (fluid oils).[8] In the next decade, Chevreul developed a more detailed classification, encompassing greases, tallow, waxes, resins, oils and volatile oils, among others.[9] In 1847, Theodore Gobley isolated phospholipids from brains and egg yolks.[10] The identification of more complex forms, such as glycolipids and sphingolipids, came later as did the generic term 'lipid', coined in the 1920s by Gabriel Bertrand from the Greek 'lipos' (fat).[11]

Also in 1789, Lavoisier determined that sugar is composed of carbon, hydrogen and oxygen and that the fermentation of sugar by yeast produced ethanol and carbon dioxide, long exploited in brewing and baking.[1,12] In 1833, Anselme Payen and Jean-François Persoz discovered the first enzyme activity ('distase'),[13] and in 1839 Payen coined the term 'cellulose' from the French word 'cellule' for cell.[14,15]

In 1857, Claude Bernard isolated and introduced the term 'glycogen' for the starch-like substance stored in the livers of mammals.[16] The term 'carbohydrate' (French 'hydrate de carbone') also originated around this time to describe high molecular weight chains of simple sugars such as glucose, whose composition could be expressed generally as $C_n(H_2O)_n$.

DOI: 10.1201/9781003109242-2

PROTEINS: 'THE LOCUS OF LIFE'

In 1828, Friedrich Wöhler produced urea from ammonium cyanide, which showed that inorganic molecules can be converted into organic compounds[17] and demolished the widely held belief that the latter could only be produced through some sort of 'vitalism'.[18] In 1839, Gerhardus Mulder described substances (fibrin, egg albumin and gelatin) that contain large amounts of C, H, O and N, and also S and P, which he considered 'the most essential substances of the animal kingdom'.[19,20] He shared the results with the chemist Jons Berzelius, who suggested that these substances be called 'proteins', from the Greek πρωτειος ('primarium' or first).[19,21,22]

Mulder's work also showed that proteins from animals and plants (which played a 'principal role in their economy') shared a similar, but varied, atomic composition.[20] Their molecular nature, however, remained nebulous for decades.[23,24] Although proteins initially represented more a concept than a defined chemical entity, it became gradually accepted that, as major constituents of all organisms, they were central to the processes of life. Indeed, the 'colloidal nature'[a] of the substances of living tissues, at the time popularly described as the 'protoplasm',[20] exhibited many of the properties that were associated with proteins, and was in fact explicitly regarded a "proteinaceous substance".[25]

This idea was prevalent among the early proponents of Darwinian evolutionary theory. One of the most prominent, Thomas Huxley, defined the protoplasm in 1868 as the "locus of life" and postulated that the physical basis of life (including heredity) lay in this universal biological substance.[20,26] Huxley remarked: "It may be truly said, that the acts of all living things are fundamentally one" and that "all protoplasm is proteinaceous."[27]

Similarly, in 1871, Charles Darwin cautiously speculated about the role of proteins in the origin of life, while stressing the common ancestry of species:

> It is often said that all the conditions for the first production of a living organism are now present, which could ever have been present. But if (and oh! what a big if!) we could conceive in some warm little pond, with all sorts of ammonia and phosphoric salts, light, heat, electricity, &c., present, that a **proteine** compound [emphasis added] was chemically formed ready to undergo still more complex

changes, at the present day such matter would be instantly devoured or absorbed, which would not have been the case before living creatures were formed.[28]

During the second half of the 19th century, the diversity of proteins became apparent and the empirical observations made were key to framing the concepts of enzymes and 'biological specificity'.[20,29] During this time, Louis Pasteur and others demonstrated the nexus between enzymatic activity and 'life' in fermentation,[12] by then a well-established paradigm of physiological chemistry.

Pasteur also showed, using swan-necked flasks, which allowed air exchange but limited microbial contamination, that life did not arise spontaneously, and noted that organic molecules are chirally left-handed,[b] a fundamental step forward in biochemistry, underpinning, among other things, modern drug development. Also studying fermentation, in 1894 Emil Fischer promoted the importance of stereo-chemical rules (popularized as the 'Lock and Key Model') to explain the interaction of substrate with enzyme, highlighting the central role and exquisite specificity of the "so-called enzymes" in biological processes.[31] In 1897, Eduard Buchner demonstrated that yeast extracts ferment alcohol from sugar, showing that biochemical processes do not necessarily require living cells, but are catalyzed by the enzymes formed in cells.[32]

Between 1899 and 1908, Fischer and others, notably Ernest Fourneau, Franz Hofmeister and Albrecht Kossel, made important advances in the understanding of the chemistry of proteins, sugars and nucleic acids, including the description of the peptide bond. The latter was a crucial shift in understanding how animal substances arise, from the traditional view that proteins are acquired from plants to the realization that they are synthesized from a set of constituent parts (amino acids[c]) incorporated into peptide chains: the 'peptide theory', largely promoted by the work of Kossel.[20,23,34,35]

---

[a] We now know that RNA can nucleate colloidal domains and does so in many contexts (Chapter 16).

[b] Possibly due to the chiral mutation bias of cosmic radiation.[30]

[c] Interestingly, some amino acids were named after the plant and animal source from which they were initially isolated: the first discovered amino acid (in 1806) was asparagine, isolated from asparagus; later glutamate from gluten; serine from silk (from the Latin for silk, 'sericum'); tyrosine (the crystals in aged cheese) from the Greek for cheese 'tyrós'; valine from the roots of the valerian plant; glycine from sugarcane and gelatin (the Greek 'γλυκύς', sweet tasting); etc.[33]

It was not until 1926 that James Sumner first purified an enzyme (urease),[36] but proteins were already regarded as the molecules that underlie life processes, the 'Protein World'.[37] Aleksandr Oparin proposed in 1924 that life originated on Earth through gradual chemical evolution of carbon-based molecules in a "primordial soup",[38] with John ('J. B. S.') Haldane independently advancing a similar theory 5 years later.[20,38–40] Both suggested in the 1920s that polypeptides were the initial particles of 'colloidal size', whereas nucleic acids, which had been discovered 50 years earlier and known to be major components of chromosomes (see below), were not mentioned.[d] The idea that proteins were the primordial molecules was reinforced by the famous experiment by Stanley Miller and Harold Urey in 1953, which demonstrated the formation of amino acids[e] from inorganic molecules in the simulated reducing and highly electrified atmospheric conditions presumed to exist in the primitive Earth.[45,46]

Biochemistry became a recognized scientific discipline in the 1930s and 1940s with the advent of a better understanding of the structure-function relationships of proteins, driven by physicists using techniques such as X-ray crystallography and isotopic labeling.[20] These advances coincided with the emergence and the coining of the term 'molecular biology',[47] whose focus is to understand the molecular basis of genetic material and how it determines cellular and organismal phenotypes.

In this sense, molecular biology is distinct from biochemistry, although there is considerable overlap and many biochemists, but perhaps not molecular geneticists or cell and developmental biologists, would assert that the terms are synonymous.[f] In any case both new disciplines were centered on the study of proteins, which heavily influenced the conceptions of genetic information and the mechanisms of heredity and development.

The period from 1930 to 1970 also saw great progress in the characterization of intermediary metabolism, the other heart of biochemistry.[20] The achievements included specification of the glycolytic (fermentation) pathway, the urea/ornithine, citric acid and glyoxylate cycles,[49] the pathways for lipid synthesis and degradation, the synthesis of amino acids and complex carbohydrates, and the nucleoside and pentose phosphate pathways for the synthesis of nucleotides and enzymatic cofactors, notably by Hans Krebs, Gustav Embden, Otto Meyerhof (who also discovered the universal energy currency, adenosine triphosphate, ATP), Jakub Karol Parnas, Otto Warburg, Horace Barker, David Green, Peter Mitchell, Ephraim Racker and Salih Wakil, among many others.[20,50–52]

NUCLEIC ACIDS AND CHROMOSOMES

The history of nucleic acids[g] began with Friedrich Miescher's discovery in 1869 of "nuclein". This was a substance distinct from proteins, isolated from the nuclei of pus cells and characterized by resistance to protease digestion, a high content of phosphorus and the absence of sulfur.[54–56] Miescher made this discovery in the laboratory of Felix Hoppe-Seyler, who had the experiments repeated and 2 years later published Miescher's paper together with others describing the isolation of nuclein from various sources. Miescher floated the idea that nuclein might be the genetic material, presumably because it was enriched in sperm, but vacillated on the issue.[57] Nuclein was later found to have an acidic nature and was named 'nucleic acid' (German 'nucleïnsäure') by his student Richard Altmann in 1889.[56,58,59]

Although Miescher's nuclein was later shown to be deoxyribonucleic acid (DNA), the 'nuclein' isolated from yeast by Hoppe-Seyler was not (mainly) DNA but the first description of what would later become known as ribonucleic acid (RNA).[55]

---

[d] In 1954, following Avery's demonstration that DNA was the genetic material, Haldane speculated that nucleic acids may be more primitive than proteins.[40] Oparin also later agreed that RNA likely preceded proteins in the origin of life (Chapter 8).

[e] It was recently shown that similar conditions (involving hydrogen cyanide, hydrogen sulfide as the reductant, ultraviolet light as the energy source and copper photoredox and wet-dry cycling) can also produce ribonucleosides and lipid precursors.[41–44]

[f] Alan Turing, the breaker of the Nazi Enigma Code and the father of modern computing, posited in his influential paper in 1952, 'The Chemical Basis of Morphogenesis', that "a system of chemical substances, called morphogens … is adequate to account for the main phenomena of morphogenesis", noting that (in contrast) "the function of genes is presumed to be purely catalytic".[48]

[g] The first component of nucleic acids described was in fact the RNA nucleoside, inosine, by Justus von Liebig in 1847, which he isolated from beef broth and named 'inosinic acid'. The 'umami' savory ('brothy' or 'meaty') taste (one of the five basic culinary tastes, the others being sweet, salty, sour and bitter) derives from a combination of the amino acid L-glutamate and ribonucleotides such as guanosine monophosphate and inosine monophosphate.[53]

In 1882, Walther Flemming described fibrous structures in the nucleus and their separation in mitosis, which he visualized by staining and termed 'chromatin' (stainable material).[60–62] Along with Edouard Van Beneden, he also described centrosomes[63,64] (see Chapter 15), a term introduced by Theodor Boveri in 1888.[65,66] Later that year, Heinrich Waldeyer termed the nuclear fibers 'chromosomes' (stainable bodies) (Figure 2.1).[62,67]

In the years leading up to the turn of the 20th century, Albrecht Kossel showed that nucleic acids are an inherent component of chromatin and identified

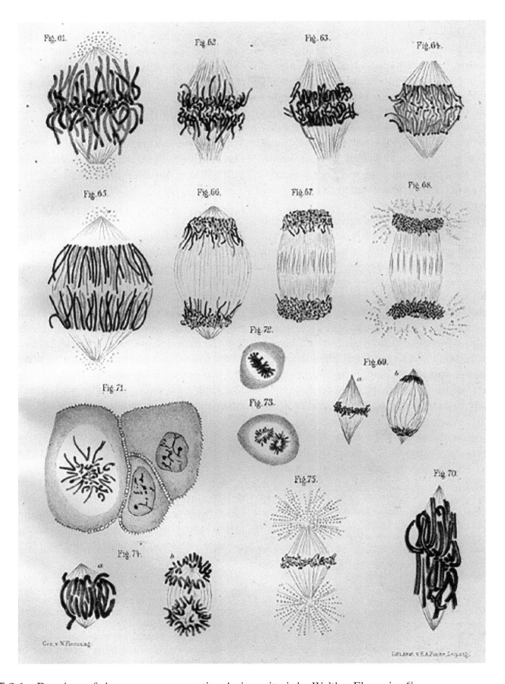

FIGURE 2.1   Drawings of chromosome segregation during mitosis by Walther Flemming.[61]

their constituent nucleoside bases: the 'purines'[h] guanine (G) and adenine (A), which have a double ring structure; and the 'pyrimidines' cytosine (C), thymine (T) and (what would be later recognized as its RNA counterpart) uracil (U), which have a single ring structure.[54,56,68,69]

CHROMOSOMES AS THE MEDIATORS OF GENETIC INHERITANCE

Gregor Mendel's principles of inheritance involving binary combinations of simple genetic traits were re-discovered around 1900 by the botanists Hugo DeVries, Carl Correns, and brothers Armin and Erich von Tschermak-Seysenegg,[70–73] and promulgated by William Bateson, who translated Mendel's 1865 paper in 1901.[74] Bateson also coined the word 'genetics' (from the Greek gennō, γεννώ; 'to give birth') and many other genetic terms that are still in use, such as 'homozygote' and 'heterozygote'.[75]

Around the turn of the century, Carl Rabi and Boveri[i] described chromosome territories, nuclear transplantation in sea urchins and abnormal chromosomes in cancer, leading them to conclude that developmental differentiation is a consequence of regulatory structures within the hereditary material.[76–79]

Based on these and other observations, William Sutton proposed in 1903 that chromosomes "constitute the physical basis of the Mendelian law of heredity" and that "one of the most characteristic features of chromatin is a large percentage content of highly complex and variable chemical compounds, the nucleo-proteids",[80] which had been first described by William Halliburton in 1895.[81] Shortly thereafter, Wilhelm Johannsen coined the terms 'gene', 'genotype' and 'phenotype', although he did not speculate about the nature of the gene.[82,83]

Direct evidence for the Chromosome Theory of Inheritance was first provided by the discovery of 'sex chromosomes' independently by Nettie Stevens and Edmund Wilson in 1905.[84,85] In 1914, chromosomes were conclusively demonstrated to be the entities that carry genes[j] by Calvin Bridges[86,87] working in the laboratory of Thomas Hunt Morgan, alongside Alfred Sturtevant and Hermann Muller, using the powerful fruitfly (*Drosophila melanogaster*) genetic system[k] that they had developed.[88]

The previously conceptual genes had found a physical home, a 'locus'. Morgan, Bridges, Muller and Sturtevant proposed that genes are linearly arranged on chromosomes like beads on a string and suggested that new combinations arise by 'crossing-over' (which was observed microscopically) and exchange of genetic material between pairs of chromosomes at meiosis, termed 'recombination'.[88–91] On the other hand, Muller's studies identified many loci ('allelomorphs')[92] that later turned to encode regulatory elements, including transposon insertions (Chapter 10) (Figure 2.2).

Analysis of the physical relationships between genetic markers (such as eye color or developmental mutations) relied on measuring the frequency of their co-inheritance in genetic crosses. Co-segregation of markers occurs in 50% of the progeny if the markers are on different chromosomes (or far apart on the same chromosome), but at a higher frequency if the loci are located near each other and separated only by occasional recombination events.

These inheritance patterns gave rise to the concepts of gene 'linkage'[l] and 'linkage groups' (i.e., genes connected on the same chromosome), and 'genetic distance' (the frequency of recombination between linked genes, measured in 'centimorgans', or cM[m]). It also led to the description of phenotypic

---

[h] The first purines discovered were caffeine and theobromine (found in chocolate). Purines, pyrimidines and related metabolites are remarkably versatile compounds, used widely in biology, not just as nucleic acid constituents but also as energy currency, such as ATP, GTP (guanosine triphosphate) and NAD (nicotinamide adenine dinucleotide), regulatory/signaling molecules (such as cyclic AMP) and protein modifier (such as ADP-ribose).

[i] Theodor and Marcella Boveri also observed that the cytoplasm played a role in hereditary processes and proposed that it was the interaction of cytoplasm and chromosomes that determined the development of an organism, although this interaction was not subjected to in-depth genetic analysis for nearly a century.[76]

[j] Initially, an X chromosome-linked recessive white eye mutation that affected males.

[k] *Drosophila* provided an ideal experimental system for genetic analysis, as it is easily maintained in the laboratory and has a short generation time. Its importance as a model for animal development throughout the 20th century cannot be overstated, despite the initial skepticism of medical researchers about its relevance to human biology, which largely evaporated later when gene sequencing revealed that the genes controlling development and neural function in *Drosophila* have equivalents in humans.[88] Many genes involved in *Drosophila* development are also involved in cancers (Chapters 6 and 14).

[l] Linkage between genes ('partial coupling') was first observed by Bateson and Reginald Punnet in sweet peas in 1904 (see[93]).

[m] A unit corresponding to a recombination frequency between linked loci of 1% per generation.

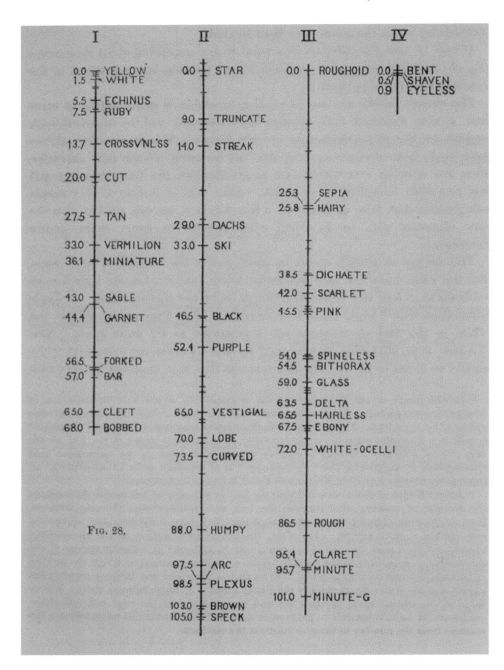

FIGURE 2.2 Morgan's diagram of *Drosophila melanogaster* chromosome linkage groups and genetic maps determined by recombination frequency, which "cannot be represented in space in any other way than by a series of points arranged in a line like beads on a string". Figure 28 in his 1922 Croonian Lecture on the mechanism of heredity.[91] (Reprinted with permission from The Royal Society.)

differences due to 'cis' interactions between genes located nearby on one chromosome; 'trans' interactions between genes located distally or on different chromosomes; homozygous and heterozygous variants, where the same or different variants ('alleles') are present on the parental chromosomes; 'recessive' mutations, where one copy of the 'wild-type' allele is sufficient for function and masks the presence of the damaged variant; 'dominant' mutations, where the mutant allele overrides the 'normal' gene; and partial 'penetrance' where the frequency of a phenotype differs from normal Mendelian ratios, due to 'haploinsufficiency' or the influence of 'modifier' genes.

The genetic maps resulting from such crosses led to the conclusion that genes are discrete objects with exclusive borders – a perception that reflects the low resolution of these early studies (and many others to this day). It also led to the idea of the 'gene for …', not just physical traits like eye color or, later, human genetic disorders such as cystic fibrosis (Chapter 11), but ultimately also for psychosocial traits, as if all genetic influences could be viewed in binary terms (genes -> traits), which underpinned the subsequent 'one gene – one enzyme (protein) – one function' assumption in biochemistry.[94,95]

The view of genes as particulate entities was reinforced by the work of Nikolay Timoféeff-Ressovsky, who attempted to measure the "radius" of genes,[96] which heavily influenced Max Delbrück, who co-authored their 'Classical Green Pamphlet' in 1935,[97] which was "the starting point" for Erwin Schrödinger's subsequent ruminations on the physical nature of genetic information (see below) and the "keystone in the formation of molecular genetics".[98]

However, Muller had shown by X-ray mutagenesis in 1930 that rearrangements that moved active genes near to heterochromatic regions (Chapter 4) of *Drosophila* chromosomes resulted in changes in the pattern of expression of these genes,[92] called 'position effect variegation' (PEV).[99] This was also observed by Barbara McClintock and others in maize, in that case with transposition underlying PEV (Chapter 5).

These observations led Muller and Richard Goldschmidt to challenge the conception of genes as distinct entities.[100–102] Nonetheless, the view of the gene as a discrete unit persisted even in the face of later discoveries that many genes (i.e., DNA segments that express RNA products) overlap or reside within other genes,[103–110] which has led to difficulties in genome annotation and a barrier to understanding the genome as an information continuum (Chapters 13–16).

Goldschmidt also coined the term 'phenocopy' to describe morphological changes in *Drosophila* that could be induced by imposition of stress during development, insisting on this basis, to little avail, that gene function had to be considered in

developmental context[111] (Chapter 5). Goldschmidt was considered a heretic by the neo-Darwinists.[102,112]

THE MODERN SYNTHESIS

During the first four decades of the 20th century, evolutionary biologists and geneticists, notably Bateson, Haldane, R. A. (Ronald) Fisher, Sewall Wright, Ernst Mayr, G. Ledyard Stebbins, Theodosius Dobzhansky[n] and Julian Huxley, among other theoreticians of the time – collectively referred to as the 'neo-Darwinists' – brought together Mendelian inheritance, Darwinian gradualism and selection, and statistical genetics into what is known as the 'Modern Synthesis', after the title of Huxley's 1942 book.[115] These pioneers of theoretical population genetics introduced important concepts such as the relevance of population size, genetic drift and the strength of selection, in parallel developing statistical methods and models that have found widespread applications to this day.[116,117]

A tenet was that all evolutionary phenomena and species diversity can be explained in a way consistent with known genetic mechanisms,[o] with the unifying theme that natural and artificial selection operates on heritable variation arising by random mutation.[p] Biometric population approaches (which showed continuous trait variation) and Mendelian inheritance were reconciled by Fisher and others by invoking polymorphic multi-factorial traits and the 'infinitesimal model' of variation and selection.[113,115,122–124]

A corollary of the supposition that mutation occurs randomly is that experience cannot influence the characteristics of the next generation, contrary to the proposal of Jean-Baptiste Lamarck a century

---

[n] Dobzhansky also worked with Morgan. He showed that natural populations had much greater genetic variability than assumed in previous models[113] and coined the famous aphorism "Nothing in biology makes sense except in the light of evolution".[114]

[o] Some, notably Cyril Darlington, recognized the limitations of the classical gene and argued that the conception of genotypes as the sum of genes could not explain variation in animal and plant populations, which must be dependent on interactions among genes and between the genotype, the cellular machinery, the reproductive habit and environment of the organism.[118,119]

[p] In the 1920s, Haldane analyzed the famous textbook example of the appearance of dark pigmentation in peppered moths during the Industrial Revolution and established that evolution could occur even faster than contemporaries such as Fisher had assumed.[120] It was much later shown that this occurred by a transposon insertion that altered gene expression[121] (see Chapters 5 and 10).

before Darwin.[q] As evidence against 'Lamarckian evolution', evolutionary biologists recalled the 19th-century work of August Weismann,[127] who asserted that somatic cells only received a subset of the information contained in the germline, and that information does not flow in reverse from somatic to germ cells and (therefore) cannot be transmitted to the next generation.[r] The latter came to be known as the 'Weismann Barrier',[128] but was challenged in the 1940s and 1950s by Conrad Waddington (the father of epigenetics, Chapter 14), who provided evidence of the inheritance ('genetic assimilation') of characteristics acquired in response to environmental perturbation[129–131] (Chapter 5).

The ideas introduced by the Modern Synthesis had a lasting influence on the conceptual landscape of molecular biology and the interpretations of experimental observations. These included the later gene-centric models of genome variation, 'fitness' and evolution, and the invocation of 'junk' DNA (Chapter 7), as well as the concept that the sequences of important genes are 'conserved' during evolution, which has frequently been assumed in efforts to discriminate functional and non-functional regions of genomes (Chapter 11).

There was considerable cross-fertilization in the 1940s and 1950s between theoretical evolutionary biologists and microbial geneticists, notably Delbrück[s] and Joshua and Esther Lederberg,

assessing lethal/competence phenotypes in bacteria and their viruses, which would lend support to the spontaneous and random nature of mutations as opposed to 'pre-adaptations' or 'directed mutations'[134] – a debate at the time.

Equally if not more importantly for the understanding of genetic programming in the following decades, the emphasis on lethal loss-of-function mutations overlooked the differences between essential genes and variations in regulatory (non-protein-coding) sequences that may have no effect on the viability of plants and animals but are the major drivers of quantitative trait variation, adaptive evolution, survival and reproductive success (Chapters 7 and 11).

These assumptions – that inheritance occurs entirely through Mendelian genes and mutations occur randomly – became entrenched before virtually anything was known about the nature of genetic variations and their phenotypic impact, especially in relation to the 'complex traits' that have large environmental components, and resulted, in part, in the lack of appreciation of Barbara McClintock's later work on mobile genetic elements (Chapter 5). They also underpinned the dichotomy of the classical and historical 'Nature versus Nurture'[t] debates that raged for decades, not realizing or even considering that such complex phenotypes may be the integrated outcomes of overlapping genetic and epigenetic processes.

The concept of genetic determinism overflowed into sociological and political arenas, notably the common and at the time fashionable idea[u] (notably, extending from observations of selective breeding in agriculture and of domesticated animals by Darwin and others) that there are superior and inferior human characteristics (especially intelligence) that vary between individuals and 'races'[137] and might be improved by selective breeding (promoted by Francis Galton, who coined the term 'eugenics') or, worse, by sterilization, as occurred in the USA and by genocide in Nazi Germany. Genetics was also politicized in Stalinist Russia by Trofim Lysenko, who favored

---

[q] Darwin himself was agnostic about the origin of the variations upon which he posited selection to act and was not antagonistic to Lamarck's ideas. In Chapter V of *The Origin of Species*, he remarked (making an interesting distinction between artificial selection and natural evolution): "I have hitherto sometimes spoken as if the variations—so common and multiform in organic beings under domestication, and to a lesser degree in those in a state of nature—had been due to chance. This, of course, is a wholly incorrect expression, but it serves to acknowledge plainly our ignorance of the cause of particular variation."[125] As pointed out by Devon Fitzgerald and Susan Rosenberg: "He also described multiple instances in which the degree and types of observable variation change in response to environmental exposures ... Darwinian evolution, however, requires only two things: heritable variation (usually genetic changes) and selection imposed by the environment. Any of many possible modes of mutation—purely 'chance' or highly biased, regulated mechanisms—are compatible with evolution by variation and selection."[126]

[r] Part of the evidence was Weismann's peculiar experiment in 1868 wherein he severed the tails of mice for five generations and showed that this experience had no effect on the presence or length of tails in the descendants.[127]

[s] Delbrück insisted that the primary problem in biology was to discover the physical structure of genes.[132] At one point, he estimated that the size of a gene was 1,000 atoms.[133]

[t] An expression also coined by Galton, based on studies of twins,[135] a common approach used to this day to assess the relative contributions of inheritance and environment to complex traits.

[u] Although scientific racism and eugenics was fashionable in many societies and intellectual circles, important thinkers and scientists opposed those ideas, including Dobzhansky and Alfred Russel Wallace, the co-discoverer of evolution by natural selection, who stated that Galton's eugenics was 'impractical, ineffective or immoral'.[136]

Lamarckian evolution in line with communist ideology, which led to the abolishment of population genetics and the persecution of geneticists and other scientists in the USSR.[138,139]

DISTINGUISHING DNA AND RNA

During the same period, research on nucleic acids was surprisingly limited, despite the finding that they are a major component of chromosomes, the repositories of genetic information.[140] Indeed, nucleic acids barely feature in the history of biochemistry before 1940,[20] a consequence of the prevailing view that proteins have greater chemical diversity than the assumed monotony of nucleic acids, thought to be merely structural or metabolic entities.

This view was strengthened by the 'tetranucleotide hypothesis' put forward in 1909 by the chemist Phoebus Levene, who identified ribose sugars in 'yeast nucleic acid'[141,142] and deoxyribose sugars in calf thymus gland.[143–145] Levene proposed that phosphate-sugar-base units formed circular tetramers containing each of the four 'nucleotides', based on the perception that they are present in equimolar proportions.[54,142,145–147]

Levene's work in identifying the five common nucleotides (A, G, C, T/U) and the phosphodiester bonds between them was a major advance,[145,147] but the tetranucleotide hypothesis implied that nucleic acids could not carry complex information. However, during the 1920s and 1930s, Einar Hammarsten showed that the molecule later recognized as DNA has a very high molecular weight, using a gentler method for its isolation that avoided harsh treatment with alkali.[148] Robert Feulgen (who discovered the eponymous DNA stain, see below), as well as Hammarsten and his student Erik Jorpes, then demonstrated that nucleic acid (RNA) from pancreas contained more guanine than other bases.[54,149]

These observations did not fit with a simple equimolar tetranucleotide structure (although Jorpes tried to rationalize the guanine excess within a pentanucleotide structure[149]), but were not recognized and Levene's hypothesis held sway for decades.[140,145,147] In addition, there was a perception that the different types of nucleic acids do not occur in all organisms, which argued against universal biological roles. One form, the 'thymonucleic acid' or 'zoonucleic acid' had been isolated from animal glands, pus cell nuclei and sperm and was initially thought to be present

exclusively in animals; the other, known as 'pentose nucleic acid', 'zymonucleic acid', 'yeast nucleic acid' or (plant) 'phytonucleic acid', was thought to be absent from animal cells until the early 1940s.[54,150,151] Thus, there was confusion about the distribution and functions of what later became known as DNA and RNA.

It was also shown during the same period that DNA and RNA have differences in their nucleoside base composition (with DNA containing adenine, guanine, cytosine, thymine, whereas RNA contains adenine, guanine, cytosine and uracil, the demethylated analog of thymine) and in the sugars in their sugar-phosphate backbone, deoxyribose in DNA and ribose in RNA, hence 'deoxyribonucleic acid' and 'ribonucleic acid'.[152]

Erwin Chargaff, who later observed the purine-pyrimidine equivalences that underpinned the base pairing critical to the elucidation of the structure of DNA and the templating of messenger RNA, pointed out in 1950 that

> Although only two nucleic acids, the deoxyribose nucleic acid of calf thymus and the ribose nucleic acid of yeast, had been examined analytically in some detail, all conclusions derived from the study of these substances were immediately extended to the entire realm of nature; a jump of a boldness that should astound a circus acrobat.[153]

In 1924, Feulgen and Heinrich Rossenbeck developed a sensitive histochemical reaction that intensely stained DNA.[154] It demonstrated that DNA is present in the nucleus, but not the cytoplasm,[v] of plants and animals, and constituted important evidence for DNA as the possible genetic material.[155] Jean Brachet, perhaps the most important of the early RNA biochemists, developed new methods to differentiate DNA and RNA in the 1930s and early 1940s, using basic dyes and cytochemical staining, later combined with ribonuclease treatments (Figure 2.3).[156–160] His work showed that both DNA and RNA are universal constituents of animal and plant cells and that RNA, unlike DNA, is mainly located in the cytoplasm. Later, it was shown in bacteria that, while the DNA composition varies between different species, both DNA and RNA are always present.[161,162]

Although DNA was linked with chromosomes, the function of RNA was for a long time a mystery.

---

[v] Leaving aside the small amount of DNA later shown to be present in mitochondria and chloroplasts.

EXPERIMENTAL MODIFICATIONS   75

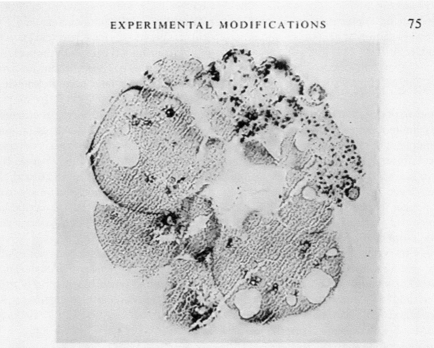

Fig. 29. Formation of atypical ectoderm in an amphibian egg treated with ribonuclease at the morula stage (Brachet and Ledoux, 1955).

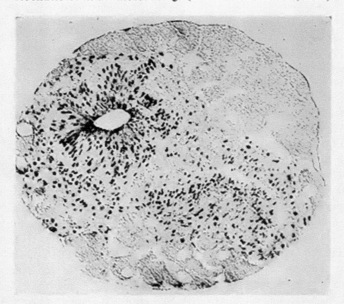

Fig. 30. Formation of a nervous system in an amphibian egg treated with a mixture of RNA and ribonuclease at the morula stage (Brachet and Ledoux, 1955).

FIGURE 2.3 Jean Brachet's photographs of the perturbation of amphibian morphogenesis by RNase treatment. (Reproduced from Brachet,[160] with permission from Elsevier under Creative Commons 4.0 license.)

It was speculated that RNA might serve as an energy repository or a cytoplasmic precursor converted into DNA during cell division (the 'conversion hypothesis').[160] It was also unclear what form RNA might take, since the 2′ hydroxyl group of ribose represented a potential bifurcation point, such that RNAs could exist as branched molecules. Only later, in the 1950s, was this idea dispensed with, when studies by Alexander Todd and others demonstrated that RNAs are linear polymers[w] whose nucleotides, as in DNA, are linked by 3′-5′ phosphodiester bonds.[165,166]

Confirmation of the role of RNA as an information molecule emerged from the study of plant and animal viruses that contained RNA but not DNA. In 1937, Frederick Bawden and Norman Pirie identified RNA in tobacco mosaic virus (TMV),[167] despite the previous suggestion by Wendell Stanley, who had crystallized TMV, that protein was the active 'autocatalytic' agent of the virus.[168] Later experiments revealed that the TMV and other plant viruses, as well as foot-and-mouth disease virus and other animal viruses, contained RNA as the sole nucleic acid, which indicated that RNA must act the template for viral replication and protein synthesis.[169,170]

ONE GENE-ONE PROTEIN AND THE 'NATURE OF MUTATIONS'

The association between 'genes' and proteins dates back to 1902, when Archibald Garrod provided evidence for the inheritance of human disorders and phenotypes that reflected enzymatic deficiencies, such as alkaptonuria.[171] The causes of the enzyme deficiencies in such 'inborn errors of metabolism' (which included other congenital disorders, such as albinism) were still only vaguely defined,[172] as they preceded the conceptual terms 'gene' and 'genome'.[x]

Indeed, even George Beadle and Edward Tatum's influential 1941 'one gene – one enzyme' principle,[174] based on studies of mutants of the filamentous fungus

Neurospora crassa (red bread mold) (Figure 2.4),[y] was not well accepted before the nature of hereditary material was understood.[178] The phrase subsequently morphed into 'one gene – one protein' when it was realized that not all proteins are enzymes, and even this changed when it was found much later that, in the higher eukaryotes, one gene can produce variations of the same protein by alternative splicing (Chapter 7).

Beadle and Tatums' work, nevertheless, united genetics and biochemistry.[94,181] It also popularized the use of microbes as model organisms.[182] In 1946, Tatum and one of his students, Joshua Lederberg (with whom he and Beadle later shared the Nobel Prize)[z] demonstrated genetic recombination in the enteric bacterium *Escherichia coli*,[184] which consequently became widely used as a model organism. Moreover, most genetic studies were focused on lethal, conditionally lethal (such as in auxotrophy[aa] or temperature-sensitivity) and/or phenotypically severe mutations, especially in haploid cells (bacteria and haploid *Neurospora* and yeast), which are overwhelmingly biased to protein-coding sequences (Chapter 7).

At the time, proteins were thought to comprise genes, rather than be the products of them. DNA was considered to have only peripheral functions, such as serving as an "intra-nuclear buffer",[160] or acting as a scaffold during gene replication. The latter was championed by the small community of scientists then engaged in nucleic acids research, such as Hammarsten and Torbjörn Caspersson.[140] Caspersson, whose work was also instrumental in establishing the involvement of RNA in protein synthesis, proposed a metabolic relationship between nucleic acids and "gene reproduction", noting that "synthesis of nucleic acid is closely connected with gene reproduction", and that "it may be that the

---

w On the other hand, the 2′ hydroxyl of RNA increases its ability to form three-dimensional structures through hydrogen bonding.[163,164]

x The word "Genom" was coined before the nature of the genetic material was known, by the botanist Hans Winkler in 1920: "I propose the expression Genom for the haploid chromosome set, which, together with the pertinent protoplasm, specifies the material foundations of the species …" (in his book Verbreitung und Ursache der Parthenogenesis; see [173]).

y The concept had earlier roots. Lucien Cuénot showed at the turn of the 20th century by his studies of mouse coat color variation that Mendelian inheritance occurred in animals as well as plants. He concluded that "mnemons" (genes) are responsible for the production of enzymes and is credited with the first enunciation of the gene = enzyme concept.[175–178] Among other important findings, he also described the first alleles and lethal mutation in mice, the recessive yellow agouti allele. This same locus was used a century later by Emma Whitelaw and David Martin to show the epigenetic inheritance of metastable alleles determined by transposable elements[179,180] (Chapters 5, 14 and 17).

z Apparently overlooking the contributions of his wife, Esther.[183]

aa The inability of an organism to synthesize a particular organic compound required for its growth in minimal media.

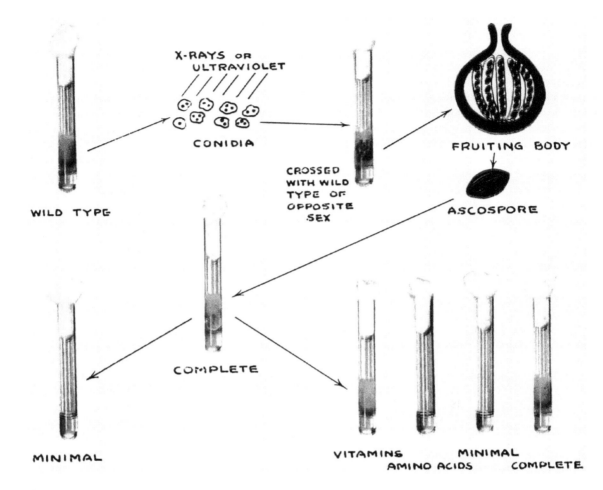

FIGURE 2.4 George Beadles' 'lantern slide' explaining the procedure he and Edward Tatum used to isolate metabolic mutants of the filamentous fungus *N. crassa*, showing complementation and Mendelian segregation of the affected gene. (Reproduced from Horowitz[181] with permission from Oxford University Press.)

property of a protein which allows it to reproduce itself is its ability to synthesize nucleic acid".[185] However, his conception was that the "structure-forming properties" of DNA (alluding to the high molecular weight DNA polymers earlier demonstrated by Hammarsten) was simply auxiliary to the basic proteins of the nucleus.[140,185]

Research on genetic material up until World War II, therefore, concentrated on proteins as the prime candidate for the genetic material, the "protein version of the central dogma".[54] Even the first reported infectious agents of bacteria, characterized as "filter-passing viruses" (later termed bacteriophages or 'phages' for short, soon to play a major role in the understanding of genes, Chapter 3), were considered to be "enzymes with power of growth".[186]

DNA IS THE GENETIC MATERIAL

In 1944, the physicist Erwin Schrödinger wrote a book entitled *What Is Life?*, in which he made the logical deduction that the genetic material would be comprised of an "aperiodic crystal" – that is, a molecule of regular structure with information embedded in its fine-scale variations – a "miniature code", the first use of the term 'code' in relation to biology.[187]

Schrödinger's prediction was borne out in the same year[bb] by Oswald Avery, Colin MacLeod and Maclyn McCarty, who built on the studies of bacterial 'transformation' by Fred Griffith in the late 1920s[189] and showed that the change from a benign

---
[bb] And later by the structure of DNA, with Schrödinger's prediction explicitly recognized by Francis Crick (see [188]).

to a virulent form of the bacterium *Streptococcus pneumoniae* could be effected by DNA but not by protein.[54,190] Their finding was confirmed in *E. coli* a year later by André Boivin.[191]

However, such was the entrenchment of prior expectations[140] that it took almost 10 years for the conclusion to be widely accepted,[192] and "even Avery himself was reluctant to accept it until he had buttressed his experiments with the most rigorous controls".[144] Arguing that a small amount of contaminating proteins could be present in Avery's preparations, not only Hammarsten, but also other eminent scientists, especially Alfred Mirsky, were unconvinced that DNA was the genetic material.[140,144,193] James Watson said, in retrospect, that (unfortunately) "most people didn't take him [Avery] seriously".[194] Moreover, as Gunther Stent observed later,[192] Avery's finding was "premature".[cc] General acceptance only came after the 1952 experiments by Alfred Hershey and Martha Chase that showed the uptake of ^{32}P-labeled DNA, but not ^{35}S-labeled proteins, into bacteria after infection with bacteriophage T4,[190,195] and the elucidation of the structure of DNA in 1953.

Levene's tetranucleotide structure was finally and convincingly refuted by Chargaff's demonstrations in the late 1940s that DNA "formed extremely viscous solutions in water" (confirming Hammarsten's earlier observations), implying a structure much larger than a tetranucleotide, and by paper chromatography[dd] that its constituent bases were not present in equimolar proportions.[153] Instead, Chargaff showed that the pyrimidine (T, C) and purine bases (A, G) are present in equal amount in DNA, and that A and T, as well as G and C, occurred in the same proportions.[ee] That is, %G=%C and %A=%T, and that the ratios of these pairs of nucleotides (%G+C/%A+T) were the same in different tissues of the same organism, but varied between organisms.[207–209]

THE DOUBLE HELIX – ICON OF THE COMING AGE

Chargaff's data was crucial for Watson and Crick's interpretation of the X-ray diffraction patterns of DNA fibers obtained by Rosalind Franklin[ff] and Raymond Gosling, building on work by others,[gg] notably Bill Astbury and Florence Bell who more than a decade earlier had determined the planar structure of, and the ~3.4 angstrom spacing between, the bases (stacked like "a pile of plates"),[210–213] which led to the elucidation of its double-helical structure.[214–216]

As is well known, and was enormously compelling at the time,[217] although it took a few years for its significance to be widely appreciated,[218] this structure is governed by nucleotide base (purine-pyrimidine) pairing rules, which immediately suggested a mechanism for the duplication of genetic information.[214] This was subsequently demonstrated in 1958 by Matthew Meselson and Franklin Stahl using labeling with heavy isotopes of nitrogen to distinguish the template from newly synthesized DNA strands in buoyant density gradients.[219] An enzyme mediating the synthesis of new, complementary strands was discovered by Arthur Kornberg and colleagues in 1956 and termed DNA-dependent DNA polymerase.[220–222]

Not so well known is that John Masson Gulland, Denis Jordan and colleagues had shown in 1947 that DNA is held together by hydrogen bonds,[223] and that their PhD student James Creeth had, in his 1948 PhD thesis, proposed a model for the structure of DNA comprising two chains with a sugar-phosphate backbone on the exterior and hydrogen-bonded bases between the nucleotide bases of opposite chains in the interior (Figure 2.5).[224,225]

---

cc Stent said "A discovery is premature if its implications cannot be connected by a series of simple logical steps to canonical, or generally accepted, knowledge."[192]

dd While first reported in 1925,[196] paper and ion-exchange chromatography was also detecting modified bases in DNA and RNA,[197,198] although the import of these modifications was not appreciated until much later (Chapters 14 and 17).

ee Chargaff later showed that the %G=%C and %A=%T in single-stranded bacterial DNA.[199] Chargaff's Second Parity Rule has since been shown to hold for all double-stranded genomes, except mitochondrial DNA, over a scale of kilobases (*E. coli*) and megabases (human), due to abundant inverse symmetries thought to reflect the distribution of repetitive elements,[200–206] but may reflect the preservation of RNA secondary structure in the encoded transcripts (including in repetitive elements) (Chapter 16).

ff Franklin was initially skeptical that DNA had a helical structure, as evidenced by her comments in a satirical note, in the style of an *in-memoriam* card, sent to Maurice Wilkins in July 1952: "It is with great regret that we have to announce the death, on Friday 18th July 1952 of DNA helix … A memorial service will be held next Monday or Tuesday." The card is reproduced in Brenda Maddox's book 'Rosalind Franklin: The Dark Lady of DNA', p. 185.[210]

gg The considerable back story of the application of X-ray crystallography to understanding the structure of DNA (and proteins) is laid out in Gareth Williams' book 'Unravelling the Double Helix: The Lost Heroes of DNA'.[211]

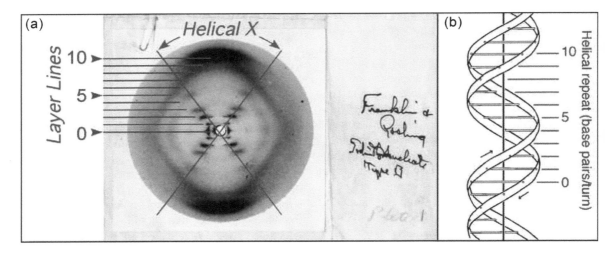

FIGURE 2.5   (a) Photograph 51 of B-DNA. X-ray diffraction photograph of a DNA fiber at high humidity (Franklin and Gosling, 1953).[215] Interpretation of the helical-X and layer lines added in blue. (b) Watson-Crick model of B-DNA. (Adopted from Watson and Crick,[214] with the helical repeat associated with the layer lines labeled. Reproduced from Shing Ho and Carter[226] with permission from the authors under Creative Commons 4.0 license.)

Acceptance of the DNA as the genetic material and the significance of its double-helical structure in genetic inheritance and gene expression came only after concerns about the ability of its strands to unwind had been resolved by the Meselson and Stahl experiment, and a plausible mechanism for the expression of genetic information (i.e., RNA-templated protein synthesis) had been established[218,227] (Chapter 3).

A subtle but important feature of the structure of DNA[hh] is that its strands are antiparallel, with both strands going from 5′ to 3′ (with respect to the phosphate linkages that connect the sugar-phosphate backbone) but in the opposite direction to each other, with DNA replication, RNA transcription and protein synthesis all proceeding from 5′ to 3′ in relation to the sugar-phosphate linkages. The linear arrangement of the bases along the backbone of the helix was fundamental to the logical deductions and experimental approaches to deciphering the genetic code.

The principles of copying/reading by base pairing and the directionality of the information were critical for understanding the synthesis and roles of RNA. Similarly, the subsequent demonstration by Julius Marmur, Paul Doty and colleagues that complementary strands of DNA (and RNA transcribed from the DNA) could recognize each other by base pairing[233,234] played an important part in the identification of messenger RNA and in the first analyses of genomic sequence composition and complexity (Chapter 3).

FURTHER READING

Cairns J. (2008) The foundations of molecular biology: A 50th anniversary. *Current Biology* 18: R234–36.

Cobb M. (2014) Oswald Avery, DNA, and the transformation of biology. *Current Biology* 24: R55–R60.

Crow J.F. (2005) Hermann Joseph Muller, evolutionist. *Nature Reviews Genetics* 6: 941–5.

Harman O.S. (2005) Cyril Dean Darlington: The man who 'invented' the chromosome. *Nature Reviews Genetics* 6: 79–85.

Horder T.J. (2006) Gavin Rylands de Beer: How embryology foreshadowed the dilemmas of the genome. *Nature Reviews Genetics* 7: 892–8.

Hunter G.K. (2000) *Vital Forces: The Discovery of the Molecular Basis of Life* (Academic Press, Cambridge, MA).

Huxley J. (1942) *Evolution: The Modern Synthesis* (George Allen & Unwin Ltd, Crows Nest).

[hh] Another feature of DNA, often overlooked, is that there are alternative forms of base pairing, notably Hoogsteen pairing, which delayed acceptance of the Watson-Crick model.[228] It can exist in different forms, specifically the A- and B-forms as shown by Franklin (the B-form being the classic double helix), the Z-form discovered by Alex Rich, and others such as G-quadruplexes and I-motifs, which exist naturally *in vivo*.[229–231] It was also later shown that the base stacking and helical dimensions vary according to nucleotide sequence.[232]

Kohler R.E. (1975) The history of biochemistry: A survey. *Journal of the History of Biology* 8: 275–318.

Morgan T.H. (1922) Croonian Lecture:—On the mechanism of heredity. *Proceedings of the Royal Society B: Biological Sciences* 94: 162–97.

Nature conference: Thirty years of DNA. *Nature* 302: 651–4 (1983). https://doi.org/10.1038/302651a0.

Nilsson E.E., Maamar M.B. and Skinner M.K. (2020) Environmentally induced epigenetic transgenerational inheritance and the Weismann Barrier: The dawn of Neo-Lamarckian theory. *Journal of Developmental Biology* 8: 28.

Olby R. (1994) *The Path to the Double Helix: The Discovery of DNA* (Dover Publications, Mineola, NY).

Satzinger H. (2008) Theodor and Marcella Boveri: Chromosomes and cytoplasm in heredity and development. *Nature Reviews Genetics* 9: 231–8.

Stent G.S. (1972) Prematurity and uniqueness in scientific discovery. *Scientific American* 227: 84–93.

Strauss B.S. (2016) Biochemical genetics and molecular biology: The contributions of George Beadle and Edward Tatum. *Genetics* 203: 13–20.

Teich M. and Needham D. (1992) *A Documentary History of Biochemistry 1770–1940* (Leicester University Press, Leicester).

Williams G. (2019) *Unravelling the Double Helix: The Lost Heroes of DNA* (Weidenfeld & Nicolson, London).

3 Halcyon Days

THE BIG QUESTION

Following the double helix, the overarching question was how the information in DNA is transduced into proteins. It was by then established that, in eukaryotic cells, DNA resided in the chromosomes and doubled at cell division. These observations were consistent with its role as the carrier of genetic information, but not with an involvement in protein synthesis.

In the early 1950s, Jean Brachet showed that enucleated cells could temporarily maintain protein synthesis,[1] and based on a series of grafting experiments with the unicellular green algae *Acetabularia*, Joachim Hammerling showed the existence of morphogenetic substances produced "under the influence" of the nucleus and transported to the cytoplasm that are "products of gene action, which stand between gene and character".[2]

RNA gradually emerged as the intermediate. It was found to be present in high levels in the cytoplasm, particularly in association with the "ergastoplasm" (endoplasmic reticulum) structure[3–5] (Chapter 4), and that its levels vary in different tissues and metabolic states. The correlation between the amount of RNA and the rate of protein synthesis, independently observed in the early 1940s by Brachet and Caspersson, led them to propose that RNA was involved in protein synthesis.[5,6]

DISCOVERY OF THE RIBOSOME

Around the same time, using ultracentrifugation to fractionate mammalian liver cells infected with the cancer-causing Rous sarcoma virus, Albert Claude,[a] who was the first to isolate the mitochondrion, the chloroplast, the Golgi apparatus and the lysosome, observed cytoplasmic granules associated with membranes, initially called microsomes. He found that the granules contained large quantities of nucleic acids of the "ribose type",[8,9] which Brachet postulated to be the sites of protein synthesis.[10] Later, microsomes were found to correspond to microvesicles, arising from fragments of membranes from the endoplasmic reticulum, with which ribosomes are commonly associated. The granules were visualized in 1955 using electron microscopy by George Palade and Philip Siekevitz,[11] and re-named 'ribosomes' in 1958 by Richard Roberts,[12] in view of their abundant ribonucleic acid component (Figure 3.1).

Ribosomes were initially characterized by their physical sedimentation rates, with bacterial ribosomes designated as '70S' and eukaryotic ribosomes as '80S', terminology that is still in use today. Subsequently it was found that ribosomes could be separated into two main components, a large '50S' subunit and a small '30S' subunit in bacteria, and equivalently a large '60S' subunit and a small '40S' subunit in eukaryotes.[13] These studies were aided by the use of detergent solubilization, chaotropic agents and phenol extraction techniques to isolate intact RNAs,[14–16] prior to which most preparations contained mainly degradation products due to the ubiquity of RNases released in lysed cells and present on skin.[17]

Small ribosomal subunits were found to be composed of a number of proteins complexed with an RNA termed 16S and 18S rRNA in bacteria and eukaryotes, respectively. The large subunit in bacteria contains two RNAs (28S and 5S) whereas the large subunit in eukaryotes contains three RNAs (30S, 5.8S and 5S), all complexed with proteins. The ribosomal RNAs are transcribed as a single large precursor, which is processed to form the individual rRNAs, from one operon in bacteria and many in eukaryotes, the latter shown later to be tissue- and developmental stage-specific,[18] as are many ribosomal proteins.[19,20]

---

[a] Claude was also the first to show, controversially, that the active agent in the cancer causing Rous sarcoma virus was not "thymonucleic acid" but "strongly positive … for pentoses" (i.e., RNA), 6 years before Avery.[7] Rous sarcoma virus became famous later for its role in the discovery of the first 'oncogene' (Chapter 6).

DOI: 10.1201/9781003109242-3

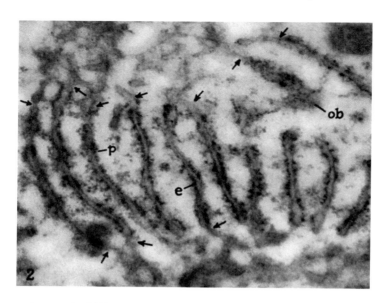

FIGURE 3.1   Electron micrograph of 'ribosomes' associated with the endoplasmic reticulum and free in the cytoplasm of liver cells. (Reproduced from Palade and Siekevitz[11] with permission from Rockefeller Institute Press.)

It was also shown by radioactive tracing that ribosomes are the sites – the cellular factories[b] – where amino acids are assembled into proteins,[21] although the mechanism was yet to be defined.

Thus, the original association of RNA with protein synthesis was a result of the detection of the most abundant RNAs (i.e., rRNAs) allowed by the techniques of the time. This abundance obscured the far more complex population of other RNAs that are expressed from the genome, as later the relatively high abundance of messenger RNAs also obscured the presence of equally if not more complex populations of cell-specific regulatory RNAs in plants and animals (Chapters 12 and 13).

THE MESSENGER AND THE ADAPTOR

The discovery of messenger RNA (mRNA), and its templating of protein synthesis in ribosomes by interaction with adaptor molecules, was science at its best, involving the interplay of observation, logical deductions, discussions and ingenious experiments by many individuals, notably François Jacob,

Jacques Monod,[c] Crick, Watson, Sydney Brenner, Marshall Nirenberg and their collaborators.[24]

The dominant hypothesis that emanated from the 1940s to explain protein biosynthesis was summarized in 1950 by Peter Caldwell and Cyril Hinshelwood: "In the synthesis of protein, the nucleic acid, by a process analogous to crystallization, guides the order by which the various amino acids are laid down."[25]

The importance of protein sequence was reinforced by Linus Pauling's team's discovery in 1949 that the electrophoretic mobility, and therefore the amino acid composition, of hemoglobin is altered in sickle cell anemia, for the first time showing the molecular basis of a genetic disease,[26] with the specific amino acid changes in this and other mutant hemoglobins subsequently identified.[27–30]

It was clear by the early 1950s that the nucleus is the source of RNA[10,31] and that RNA could serve as the template for protein assembly, first proposed

---

[b] As both subunits contain many proteins, RNA was simply thought to be the framework for the machine, until it was shown much later that RNA in fact lies at the catalytic heart of peptide bond formation (Chapter 9).

[c] Both Jacob and Monod were eclectic individuals with interesting histories, including participation in the French Resistance in World War II. Between them, in addition to the 1965 Nobel Prize in Physiology or Medicine, awards included France's World War II highest decoration for valour, the Cross of Liberation, as well as the Croix de Guerre, the Légion d'Honneur and the American Bronze Star Medal. Monod also shared a deep postwar friendship with the writer-philosopher Albert Camus.[22,23]

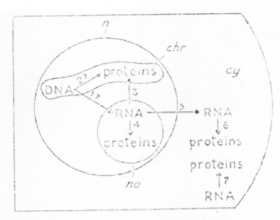

Fig. 41. Scheme proposing relationship between DNA, RNA and proteins in the different parts of the cell. chr: chromosomes; cy: cytoplasm; n: nucleus; no: nucleolus.

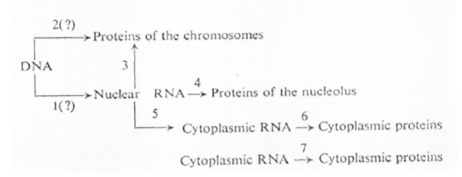

FIGURE 3.2   Jean Brachet's speculations on the flow of genetic information. Figure 41 in his 1959 paper on the biological role of ribonucleic acids. (Reproduced from Brachet,[37] with permission from Elsevier under Creative Commons 4.0 license.)

by André Boivin[d] and Roger Vendrely in 1947[32] and elaborated by Alexander Dounce and Brachet in the early 1950s[33–37] (Figure 3.2). It was also becoming evident that protein synthesis requires "the ordered interaction of three classes of RNA – ribosomal, soluble, and messenger"[38] (see below).

In 1958, as a key part of the theoretical considerations of the process of protein encoding by DNA (the "coding problem"[39]), Crick proposed the 'Adaptor Hypothesis', by which some molecule must serve as the carrier for amino acid incorporation into peptide chains during protein synthesis. Crick postulated that the adaptor was RNA, given that "base

pairing made RNA uniquely suited for a role as a small, specific RNA recognition molecule".[40]

Such RNAs had just been identified by Mahlon Hoagland, Paul Zamecnik and colleagues, who showed that small soluble 'sRNAs' could be conjugated to amino acids (labeled with a radioactive carbon isotope,[14]C) and transfer the labeled amino acids to proteins in microsomal preparations. This reaction required GTP (guanosine triphosphate), later shown to be the energy source for peptide bond formation. From this they concluded that such RNAs, later named transfer RNAs (tRNAs), function as the intermediate carrier of amino acids in protein synthesis.[41]

The factory and the adaptor had been found, but the template and the 'code' remained undefined. With the association of ribosomes with protein synthesis increasingly accepted, one hypothesis was that

---

[d] André Boivin was one of the earliest and most visionary supporters of Avery's claim that DNA was the hereditary material.[24]

distinct ribosomes served as templates for different proteins, leading to the new aphorism "one gene – one ribosome – one enzyme".[40]

However, given the rapid rates of protein synthesis that were observed, for example, after infection of bacteria with bacteriophages (phages[e]) and that the two known RNA species (rRNAs and tRNAs) were essentially homogeneous, stable and similar in different species, this hypothesis seemed implausible: these RNA species did not fulfill the requirements of dynamic templates for protein synthesis.[22,38,45–47]

Early clues for the intermediate candidate had been obtained with the use of nucleotides labeled with a radioactive isotope of phosphorus (^{32}P). In 1953, Hershey and colleagues found that, unlike DNA, a small fraction of RNA is synthesized "extremely rapidly" following T2 phage infection.[48] In 1956, Elliot Volkin and Lazarus Astrachan reported that while T2 phage infection arrested bacterial protein synthesis, it triggered massive phage protein synthesis. They also noticed that while cellular RNA remained essentially unchanged, short-lived RNA with the same base composition as the viral DNA was contemporaneously produced. Volkin and Astrachan called this variant RNA "DNA-like-RNA" and remarked that "such RNA molecules may be an entire new species, possibly related to phage growth".[49] In 1959, Arthur Pardee, Jacob and Monod showed the same rapid inducibility of short-lived RNA from the *lac* operon following lactose exposure[50] (the 'Pajama' experiment[51]).[f]

Others, notably Sol Spiegelman, Benjamin Hall and Masayasu Nomura, confirmed and extended these observations,[52–54] which, although not widely recognized at the time, were crucial for the discovery of the messenger.[22,51] Consequently, Jacob and Monod, in their famous 1961 paper describing the operon model (see below), postulated the existence of an "unstable" RNA that conveyed the genetic information for protein production to the cytosol.

The "candidate" (which they first called "X") was named "messenger RNA" (mRNA).[46]

Brenner, Jacob and Meselson were already working on this hypothesis[22] and in the same year proved the existence of mRNA using the phage system and incorporation of labeled RNA into previously existing ribosomes.[47] At the same time, François Gros, Walter Gilbert and colleagues in Watson's laboratory demonstrated rapid turnover of mRNAs and their "DNA-like" base composition in bacteria,[55] and three groups demonstrated DNA-dependent RNA synthesis in bacteria and in isolated nuclei of mammalian cells.[56–58]

THE 'GENETIC CODE'

In parallel, attention was turning to the nature of the information that instructed the sequence of amino acids in proteins. Theoretical considerations of a 'genetic code' was a major theme in the post-World War II period, especially in the so-called 'RNA Tie Club', which included physicists such as George Gamov and Richard Feynman, as well as Crick, Brenner, Martynas Yčas and others, founded in 1954 to "solve the riddle of the RNA structure and to understand how it built proteins".[59–61]

An intellectual world away from biochemistry but emergent in the background[62] was the revolutionary work in the 1940s on information theory and computational control systems connecting machines with biology by Claude Shannon[g] and Norbert Wiener,[64–66] which ushered in the digital age and, *inter alia*, the concepts of genetic coding and genetic programming.[67]

In 1951, Fred Sanger and colleagues established that proteins are linear polymers of amino acids, and produced the first protein sequence, that of human insulin, by partial hydrolysis of the two peptide chains,[68–71] an approach whose principles he later applied to RNA sequencing (Chapter 6).

Of central importance was Seymour Benzer's use of genetic recombination and temperature-sensitive mutants of bacteriophage T4 at the turn of the decade to map the fine structure of genes.[72–76] Benzer,[h] who later went on to become a pioneer of behavioral genetics,[78] showed that genes are linear but not indivisible, using the resolving power of his system to identify deletions and nucleotide changes, some of

---

[e] The term given to bacterial viruses, from the Greek meaning 'bacteria eater', discovered by Frederick Twort and Félix d'Hérelle in 1915–1917,[42,43] coined by the latter and often shortened to 'phages'. The use of bacteriophages was instrumental in the analysis and elucidation of gene structure, replication and expression, as they comprised an extremely powerful system that could introduce genetic changes and poll millions of genetic events in overnight bacterial culture.[44]

[f] The Pajama experiment also revealed that the induction of beta-galactosidase from the *lac* operon is regulated by a repressor,[50] which ushered in the concept of the regulation of gene expression.[51]

---

[g] Shannon's PhD thesis was entitled 'An algebra for theoretical genetics'.[63]

[h] Benzer was later described as the researcher who "more than any other single individual, enabled geneticists adapt to the molecular age".[77]

TABLE 3. NUCLEOTIDE SEQUENCES OF RNA CODONS

1st Base	2nd Base				3rd Base
	U	C	A	G	
U	PHE*	SER*	TYR*	CYS*	U
	PHE*	SER*	TYR*	CYS	C
	leu*?	SER	TERM?	cys?	A
	leu*, f-met	SER*	TERM?	TRP*	G
C	leu*	pro*	HIS*	ARG*	U
	leu*	pro*	HIS*	ARG*	C
	leu	PRO*	GLN*	ARG*	A
	LEU	PRO	gln*	arg	G
A	ILE*	THR*	ASN*	SER	U
	ILE*	THR*	ASN*	SER*	C
	ile*	THR*	LYS*	arg*	A
	MET*, F-MET	THR	lys	arg	G
G	VAL*	ALA*	ASP*	GLY*	U
	VAL	ALA*	ASP*	GLY*	C
	VAL*	ALA*	GLU*	GLY*	A
	VAL	ALA	glu	GLY	G

FIGURE 3.3 The genetic code for amino acids presented by Nirenberg et al.[89] (Reproduced with permission of Cold Spring Harbor Laboratory Press.)

which specify a different amino acid and others that corrupt or terminate protein synthesis.[79]

Reasonably, then, genes and proteins were presumed to be co-linear,[80] that is, the order of nucleotides is the same as that of their specified amino acids, but it was unknown whether the code is overlapping or non-overlapping.[i] The former was considered unlikely on logical grounds by Brenner[39] and experimentally by Akira Tsugita and Heinz Fraenkel-Conrat, who showed in 1960 that a point mutation resulted in just one amino acid change.[82]

The matter came to a climax in 1961. Crick and others reasoned that the length of the coding units ('codons') must be at least three to be able to specify all 20 amino acids that standardly occur in proteins, which in turn implied that, if so, there may be more than one ('redundant') codon for each amino acid,[j]

or at least some of them ($4^2 = 16$; $4^3 = 64$), with corresponding 'adaptor RNAs' (tRNAs) linked to cognate amino acids.[40,84]

In 1961, Crick, Brenner and colleagues used Benzer's high-resolution bacteriophage gene system and some of his mutants to show that insertion of one or two nucleotides in the coding sequence resulted in a non-functional protein, because it threw the subsequent codons out of kilter ('frame-shift' mutations), whereas the insertion or deletion of three nucleotides had more subtle effects, thereby demonstrating that the coding unit was indeed a triplet.[81]

In the same year, experiments with RNA homopolymers in cell-free extracts by Marshall Nirenberg and Heinrich Matthaei demonstrated that polyuridine can direct the incorporation of the amino acid phenylalanine into proteins.[85] This not only proved that messenger RNA directs protein synthesis, but also provided the platform for working out the entire triplet-based genetic code by the mid-1960s using combinations of nucleotides in synthetic RNAs[62,86–91](Figure 3.3).

As André Lwoff put it, "the messenger ceased to be an *être de raison* and became a molecule",[92] and the aphorism 'one gene – one enzyme' had found the intermediate.

---

[i] This also led to the common one-dimensional conception of RNAs, which have complex three-dimensional structures, which also transmit information (Chapters 8 and 16). However, only tRNAs were explicitly considered to have a "protein-like structure".[81]

[j] In 1966, following the first determination of the sequence and secondary structure of a tRNA (see below), Crick published 'The Wobble Hypothesis', which provided a structural explanation for the degeneracy of the genetic code.[83]

THE *lac* OPERON AND GENE REGULATION

The year 1961 also saw the publication of Jacob and Monod's classic model of gene regulation in the same paper that proposed the existence of mRNAs,[46] based on studies of lambda phage infection and the genetic dissection of the *lac* 'operon' of *E. coli*. This had a decisive impact on the conceptual framework of the regulation of gene expression and the protein-centric paradigm of genetic information that has dominated molecular biology for most of its history. Every undergraduate student of molecular biology is taught the *lac* operon as the exemplar of gene regulation.

The *lac* operon consists of three 'structural' genes (transcribed as one 'polycistronic' mRNA containing three open reading frames) that specify three proteins involved in the uptake and metabolic utilization of the milk sugar lactose by the bacteria in the gut, including the enzyme beta-galactosidase,[k] together with a nearby 'repressor' gene, whose product keeps the *lac* genes silent until and unless lactose is present – there is no point in producing the enzymes to utilize lactose if none is present.

Jacob and Monod articulated the notion that genomes contained both "structural genes", which encoded enzymes and other proteins, such as hemoglobin and insulin, etc., and "regulator genes",[46] which specified regulatory systems that control the expression of the former.[l] In this model, structural genes obeyed the 'one-gene, one-protein' principle, and regulator genes encoded a trans-acting "repressor" (of unknown composition) that would interact with other DNA sequences ("operators") linked in *cis* (that is, adjacent to the promoter) to block the initiation of transcription, which in turn implied that they would generally lie upstream of the target genes.[m] There was no consideration of the converse, that there may also be activators that operate similarly.

It was discovered later that the product of another gene with a more universal activator function makes it easier for specialized sugar utilization genes to be induced if energy levels are low (Figure 3.4).

In any case, the concept that differential gene activity underlies cell differentiation was obvious and had already been proposed in the 1950s (e.g.,[93]), but whether the regulation of gene expression could be explained simply in terms of the action of regulatory proteins encoded by other genes was uncertain, although beginning to be widely assumed. The cytogeneticist Barbara McClintock intuited that there was a distinction between (protein-coding) "gene elements" and (mobile) "controlling elements", based on genetic studies of transposon mobilization in maize (Chapter 5). She proposed that these mobilized elements, despite not being part of the "gene" nor (likely) enzymes, would act as modifiers, suppressors or inhibitors of gene activity, and predicted their general occurrence in other organisms.[94]

Accordingly, McClintock warned against settling too quickly on a protein-centric definition of genes (and mutations[n]) based on studies in bacteria before the structure of DNA or the nature of the controlling factors she had found – or any 'gene' – were defined. In 1950, she wrote in a letter to a colleague:

> Are we letting a [protein-coding] philosophy of the gene, control [our] reasoning? What then is the philosophy of the gene? Is it a valid philosophy? … When one starts to question the reasoning behind the present notion of the gene (held by most geneticists) the opportunity for questioning its validity becomes apparent.[95]

Moreover, early evidence had indicated that RNAs might have other properties, beyond their roles in protein translation, of potential importance in genetic transactions. Soon after the publication of the double-helical structure of DNA, Alexander Rich (a founding member of the RNA Tie Club) and David Davies showed that RNA molecules could base pair to form double-stranded RNAs (dsRNAs),[96,97] a discovery that was met with some skepticism or disregard,[98]

---

[k] Beta-galactosidase became a favorite target of assays for gene expression by linking its coding sequences to presumed regulatory elements and assaying its activity using an artificial ('chromogenic') substrate that produced a blue color in response to enzyme activity.

[l] And perhaps other regulatory genes, in a chicken-and-egg hierarchy, especially during the complex suites of gene expression during multicellular differentiation and development – see Chapter 15.

[m] What was initially called the 'operator' is now referred to as the gene 'promoter', which encompasses the regulatory sequences upstream of the transcribed regions of genes recognized by regulatory factors and RNA polymerase. Operator may be the better term.

[n] McClintock told Charles Burnham in January 1950: "Even though the details are manifold, obviously, there is a consistency that does not fail. You can see why I have not dared publish an account of this story. There is so much that is completely new and the implications are so suggestive of an altered concept of gene mutation that I have not wanted to make any statements until the evidence was conclusive enough to make me confident of the validity of the concepts."[95]

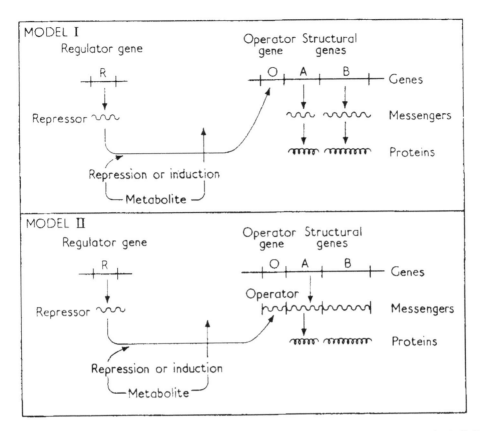

FIGURE 3.4 Jacob and Monod's 1961 models of the *lac* operon and the regulation of protein synthesis.[46] (Reproduced with permission from Elsevier.) Note that in both models the *lac* repressor is drawn as an RNA.

although later shown to be a feature of cellular RNA interactions[99–101] and regulatory systems (Chapters 8 and 12). In 1957, Rich, Davies and Gary Felsenfeld showed that RNA can also interact sequence-specifically with double-stranded RNA or DNA to form three-stranded (triplex) structures,[102] via non-canonical 'Hoogsteen' base pairing[103] in the major groove of the helix,[o] which "may have significance as a prototype for a biologically important three-stranded complex, as, for example, a single ribonucleic acid chain wrapped around a two-stranded DNA".[102] In 1961, Spiegelman and Hall showed that RNA-DNA hybrids exist naturally in cells.[53]

A few years later, Robert Holley, Ada Zamir and colleagues showed by the first sequencing of a tRNA (alanine tRNA, using partial ribonuclease digestion and two-dimensional fractionation[p]) that RNAs form secondary structures via internal base pairing,

forming a 'cloverleaf' structure with double-helical base-paired regions when displayed in two dimensions.[106] These analyses, which took 9 years, also identified ten chemical modifications of its nucleotides[107] (Chapter 17).

The tRNA structure was confirmed and its canonical L-shape revealed almost a decade later by Rich and colleagues using X-ray crystallography, the first determination of the 3D structure of a natural RNA.[108,109] Later studies showed that all tRNAs have four hairpin helices and three variable loop structures inserted between two hairpin structural elements and that the 3′ end of all tRNA molecules contain a conserved CCA sequence, to which the relevant amino acid is attached by specific enzymes (Figure 3.5).[110,111]

These and subsequent structural studies revealed unusual structural motifs, non-canonical base pairing, tertiary interactions, intercalated strands, pseudoknots, coaxial stacking and bound metals, pointing to the structural complexity and versatility of RNA

---

[o] Hoogsteen base pairing occurs in tRNAs.[104]

[p] Fred Sanger and colleagues developed a similar method to sequence RNAs at the same time.[105]

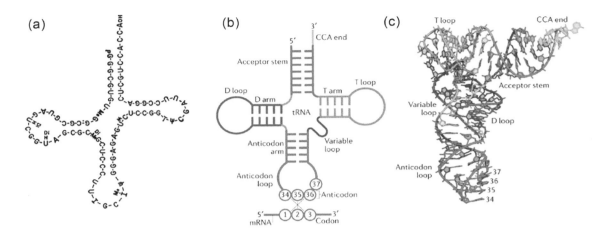

FIGURE 3.5 (a) The original two-dimensional representation of the cloverleaf structure of the 'adaptor' alanine tRNA, drawn by Holley et al.[106] (Reproduced with permission of the American Association for the Advancement of Science.) (b) The general secondary and (c) spatial L-shaped structure (c) of tRNA. (Reproduced from Suzuki[112] with permission from Springer Nature.)

molecules.[108,109,113–116] Although the significance of these properties of RNAs was cryptic,[97] the possible regulatory implications of DNA-RNA and RNA-RNA interactions did not go entirely unnoticed.

In 1959, Arthur Pardee and Louise Prestidge, and a year later Leo Szilard, suggested that RNA would make a good candidate for the *lac* repressor.[117,118] Jacob and Monod subsequently, in their *magnum opus* on the *lac* operon,[46] also proposed that RNA may be the agent produced by the regulatory gene, emphasizing its sequence specificity. In their words: "the operator tends to combine (by virtue of possessing a particular base sequence) specifically and reversibly with a certain (RNA) fraction possessing the proper (complementary) sequence." Because RNA can base pair with DNA and RNA, they proposed two models by which RNA could act as repressors, either at the RNA transcriptional ("genetic operator model") or post-transcriptional levels ("cytoplasmic operator model", where the operator is present in the "polycistronic" transcript). Jacob and Monod favored the former.[46] This was a special moment for RNA in the history of molecular biology, with great conceptual implications, but was short-lived, for reasons explained below.

Rich proposed in 1961 that both strands of DNA in the cells could potentially template complementary ('antisense') copies of RNAs, which was shown much later to be a widespread occurrence, especially in animal and plant cells (Chapter 13). He speculated that "it does not seem likely that both of these [DNA strands] go on to manufacture a protein molecule" and that there is "an interesting possibility in that this may be part of the control apparatus for turning on or off the synthesis of a given class of proteins", suggesting that this could involve the formation of double-stranded RNA,[119] which also turned out (much later) to be correct (Chapter 12). Again using similar principles, Kenneth Paigen elaborated on the operon model in 1962, and speculated that diffusible (trans-acting) RNAs produced by regulator genes could base pair with the non-template DNA strand of structural genes, reversibly regulating the "release" of messenger RNAs produced from the template DNA strand.[120]

Paul Sypherd and Norman Strauss offered the possibility that the repressor system could involve complexes with both RNA and protein, wherein proteins had specificity for small molecule ligands (in this case, lactose) and the RNA provided specificity for the operator.[121] The latter (RNA guidance of transcription factors and chromatin-modifying proteins to specific genomic locations) was a prescient prediction (Chapter 16). RNA was also later shown to be capable, like proteins, of binding small molecules and responding allosterically ('riboswitches') to regulate gene expression (Chapter 9).[q]

Although it is clear that these models were proposed in the absence of knowledge of many of the

---

[q] Even ribosomal RNA was later shown to regulate the expression of genes that control development.[122]

enzymatic components (such as helicases) and mechanisms involved in DNA replication and transcription, the recurrent theme was that they invoked the "simplicity" and "logic" of RNA regulation via base pairing, which only required an RNA size of 10–12 bases to "provide the necessary specificity",[120] foreshadowing the action of microRNAs and other small RNAs discovered at the turn of the next century (Chapter 12).

Nevertheless, in the years following the *lac* operon, the proposition of regulatory RNAs was disfavored, because the emerging models required that the repressor interact with small molecules (metabolic effectors), for which proteins with three-dimensional structures (such as allosteric enzymes, which alter their shape and activity upon binding small molecules[r]) seemed more suitable, even though these views were "more doctrinal than empirical" and "the proteinaceous nature of the repressor was taken for granted".[126]

This expectation was confirmed and its generality assumed when Walter Gilbert and Benno Muller-Hill found in 1966 that the *lac* repressor is a protein and Mark Ptashne subsequently isolated the bacteriophage lambda repressor protein, both shown to specifically bind to regulatory ('operator') DNA sequences upstream of the protein-coding genes.[127–130] Moreover, in 1965, Ellis Englesberg and collaborators had demonstrated the existence of protein 'activators' in the control of gene expression in bacteria, expanding the dominant 'repressor' model (negative control),[131] although, oddly, Monod was unconvinced.[s] Then, with the discovery of the 'sigma

factors' controlling RNA polymerase transcription initiation,[133,134] the basic mechanism of regulating gene expression in bacteria, and presumptively in higher organisms, seemed generally understood, despite RNA players emerging in the background (Chapter 9).[t]

These findings consolidated the conclusion that proteins comprise not only the 'enzymes' but also the 'regulators' of gene expression, and the suggestions by Jacob and Monod, Rich and others that some genes might specify regulatory RNAs were relegated to history.

PROTEIN STRUCTURE

As aforementioned, an important development during this period was protein sequencing, first achieved in the late 1940s and early 1950s by Fred Sanger using partial acid hydrolysis to sequence insulin,[143] soon supplanted by cyclic cleavage of terminal amino acids called Edman degradation, after its developer, Pehr Edman in 1950,[144] which was later automated.[145]

The determination of the amino acid sequence of proteins (later to be much more efficiently and accurately deduced from gene sequences) was an essential prerequisite to the determination of their three-dimensional structures using X-ray crystallography.[u] It was developed and applied by John Kendrew and Max Perutz and colleagues in 1958 to hemoglobin and muscle myoglobin, which showed

---

[r] The important concept of allostery ('allosteric inhibition') was advanced by Monod and Jacob in 1961 to describe binding of a ligand to one site in a protein causing a structural change that hampers the binding of a second ligand (a DNA sequence or an enzyme substrate) at another site.[123] Monod and colleagues correctly predicted that allostery might be a general form of cellular regulation, and that allosteric sites might be useful drug targets.[123–125] RNAs can also act as allosteric 'riboswitches' that respond to small molecules and other cues, especially in bacteria, not discovered until 2002 (Chapter 9).

[s] During this period, Englesberg gave several seminars at the Pasteur Institute. As told by Jacob: "After each seminar, however, [Englesberg] received a severe lesson in regulatory genetics from Monod, who always insisted on a notion 'that even a schoolboy cannot ignore: negative × negative equals positive!' Englesberg said that 'whenever I spoke with Jacob and Monod, they would say that they were 33.3% convinced, and then 50% convinced, about positive control. When I gave a seminar at the Pasteur … in 1972, they said 'Well, we are 66.6% convinced'."[132]

[t] The identification of the *lac* repressor protein occurred in the same year as the first RNAs not associated with translation were detected in human cells,[135] and just 1 year before the first abundant non-tRNA small RNA was discovered in *E. coli*. The latter was a 6S ubiquitous small (180–200nt) regulatory RNA,[136] whose function as repressor of sigma factor-dependent gene transcription and regulator of RNA polymerase promoter use was not determined until 30 years later.[137] A second small (109nt) RNA found in *E. coli* in 1973[138,139] (a transcript named *Spot 42*, encoded by the *spf* gene) also had an unknown function until 2002: it is also a trans-acting antisense RNA, which represses the galactose operon (and indeed many other operons) at the post-transcriptional level by base pairing with the *galK* mRNA[140–142] (Chapter 9).

[u] The entry of physicists into biology in the 1940s and 1950s revolutionized macromolecular analysis. There is a deeper history, with the development of the principles of X-ray diffraction by crystals and the mathematics involved to probe their atomic structure, dating back to the early 1900s, pioneered by Max von Laue and the father and son team of William Henry and William Lawrence Bragg.[146,147] X-ray fiber diffraction data was, of course, also central to the elucidation of the double-helical structure of DNA.

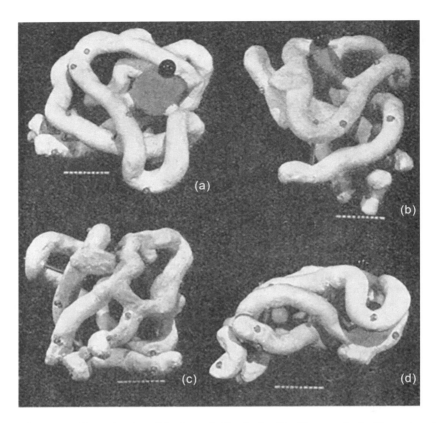

FIGURE 3.6   The first three-dimensional model of myoglobin obtained by X-ray analysis. (Reproduced from Kendrew et al.[148] with permission of Springer Nature.)

that these proteins folded into three-dimensional globular structures[148–152] (Figure 3.6). These studies also revealed the major structural elements of proteins, initially encompassing α-helices, β-sheets, turns and later transmembrane domains and enigmatic 'intrinsically disordered regions' (Chapter 16).

The structural analysis of proteins was initially restricted by their ability to form crystals – creating these was and is an art in itself. Later, nuclear magnetic resonance imaging, first described by Isidor Rabi in 1938 and developed in 1946 by Felix Bloch and Edward Purcell, allowed the determination of relatively small proteins in solution by Kurt Wüthrich, Richard Ernst, Ad Bax, Marius Clore, Angela Gronenborn and Gerhard Wagner, among others, in the 1970s and 1980s,[153,154] and has continued to be refined. More recently, the development of improved methods of cryo-electron microscopy by Jacques Dubochet, Joachim Frank, Richard Henderson[155] and others has allowed structural characterization of much larger proteins and protein complexes.[156–159]

Atomic resolution of protein structure fueled enduring discovery, accelerated by high-throughput methods and many technical innovations that revealed the structure-function relationships, fine chemistry and dynamics in the enzymes, molecular machines and macromolecular components of cells.

THE CENTRAL DOGMA

In the late 1950s, before the identification of mRNA, Crick publicly articulated what he termed the "Central Dogma" of the directional flow of genetic information,[40,160] reflecting earlier considerations by Boivin and Vendrely, Brachet, Watson[v] and Dounce. In this paradigm, proteins were the final destination of the information contained in DNA and conveyed by RNA, as once "information has passed into protein it cannot get out again".[40]

---

[v] Watson sketched the Central Dogma in his lab notebook in 1952.[161,162]

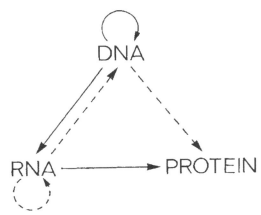

NATURE VOL. 227 AUGUST 8 1970

Fig. 3. A tentative classification for the present day. Solid arrows show general transfers; dotted arrows show special transfers. Again, the absent arrows are the undetected transfers specified by the central dogma.

FIGURE 3.7   Crick's 1970 formulation of the 'Central Dogma'.[163] (Reprinted by permission from Springer Nature.)

In a subsequent formalization in 1970,[163] Crick also included – by way of a dashed line – that RNA may be itself copied, along with a dashed line indicating that information could flow in reverse from RNA to DNA,[w] but not from protein to RNA (Figure 3.7). These modifications were presumably made in the wake of the finding in the 1960s that RNA viruses replicate[164–171] and the discovery of virally encoded reverse transcriptase[x] independently, and with dogged determination in the face of skepticism, by David Baltimore and Howard Temin in 1970.[175–177] Reverse transcriptase was critical to enable the coming gene cloning revolution (Chapter 6) and the understanding of retroviral biology, the full implications of which have yet to be realized (Chapter 10).

The Central Dogma has held true to this day (except for the speculative transfer of information directly from DNA to protein), but became widely interpreted, including by Watson,[162] as 'DNA makes RNA makes proteins', with its implicit assumption, not necessarily intended by Crick, that RNA functions only as an intermediate.

IT'S ALL OVER NOW

The two decades from 1953 to 1972 were exhilarating and the new crop of molecular biologists were rightly pleased with what had been achieved, but their self-satisfaction and hubris were palpable. The *lac* operon and the Central Dogma consolidated the notion that (with exceptions like the few rRNA and tRNA types) genes are synonymous with proteins, and that all genetic information, including regulatory information, is transacted by proteins, not only in bacteria but also in developmentally complex plants and animals.

Consequently, the hegemony of proteins as both structural and regulatory molecules was established, prematurely, within the first two decades of molecular biology, despite the odd molecular and genetic observations in plants and animals (Chapters 4 and 5) and a looming surprise that should have given pause for thought (Chapter 7), with more to come (Chapters 8–13).

As Crick opined in 1958: "Biologists should not deceive themselves with the thought that some new class of biological molecules, of comparable importance to the proteins, remains to be discovered".[40]

---

[w] Crick's original diagram of the flow of genetic information included a dotted line from RNA back to DNA.[40]

[x] RNA-dependent RNA polymerases are also found in eukaryotic cells.[172] There are a number of 'DNA repair' enzymes with reverse transcriptase activity in the brain, with obvious implications.[173] Information also moves laterally from DNA via RNA to DNA by retrotransposition,[174] the outcomes of which dominate the genome and genetics of complex organisms (Chapters 4, 10 and 16).

And Brenner in 1963:

> It is now widely realized that nearly all the 'classical' problems of molecular biology have either been solved or will be solved in the next decade. … Molecular biology succeeded in its analysis of genetic mechanisms partly because geneticists had generated the idea of one gene-one enzyme. … Molecular biology succeeded also because there were simple model systems such as phages which exhibited all the essential features of higher organisms so far as replication and expression of the genetic material were concerned.[y]
> … It is probably true to say that no major discovery comparable in importance to that of, say, messenger RNA, now lies ahead in this field.[z]

Gunther Stent proclaimed in his 1968 article entitled 'That Was the Molecular Biology That Was':

> All hope that paradoxes would still turn up in the study of heredity had been abandoned long ago, and what remained now was the need to iron out the details … [and that there remained] … only one major frontier of biological enquiry for which reasonable molecular mechanisms cannot be envisaged: the higher nervous system.[179]

And Brenner added [Stent's point is]

> that once we knew both the structure of DNA and that nucleotide sequences encoded amino acid sequences of proteins, and that once the principle of gene regulation had been found by Jacob and Monod, there was nothing left to do. Thus embryology could be accounted for by simply turning on the right genes in the right place at the right time and that was the solution to the problems of development. Not only did we not have to bother investigating the developmental biology of the millions of different species of animals and plants, but there would be no motivation for scientists to pursue those fields because the mystery had vanished.[180]

The belief that genes are synonymous with proteins reflected the mechanical zeitgeist of the age. Bicycles and cars have parts, and so do organisms – proteins

that carry oxygen (hemoglobin), form skin (keratins), signal energy levels (insulin) or control the activity of other genes ('transcription factors'), etc. It was just assumed that these 'conserved' components, whose expression is regulated by *trans*-acting transcription factors acting on malleable adjacent promoter-operator sequences, were enough to explain all of biology.

Little thought was given at the time to the enormous differences between bacteria and developmentally complex organisms. The 'biochemical unity of life' was just taken as given.[181] As Monod said in 1954, in a recapitulation of a 1926 assertion by the microbiologist Albert Jan Kluyver:[aa] "Anything found to be true of *E. coli* must also be true of elephants."[181]

That may be the case, but the logical trap was that the reciprocal might not be. No one knew.

FURTHER READING

Ball P. (2018) Schrödinger's cat among biology's pigeons: 75 years of what is life? *Nature* 560: 548–50.

Cairns J. and Watson J.D. (2007) *Phage and the Origins of Molecular Biology* (Cold Spring Harbor Laboratory Press, Cold Spring Harbor, NY).

Carroll S.B. (2013) *Brave Genius: A Scientist, a Philosopher, and Their Daring Adventures from the French Resistance to the Nobel Prize* (Crown Publishers, New York).

Cobb M. (2015) Who discovered messenger RNA? *Current Biology* 25: R526–32.

Cobb M. (2016) *Life's Greatest Secret: The Race to Crack the Genetic Code* (Profile Books Ltd., London).

Darnell J.E. (2011) *RNA: Life's Indispensable Molecule* (Cold Spring Harbor Laboratory Press, Cold Spring Harbor, NY).

Kay L. (2000) *Who Wrote the Book of Life? A History of the Genetic Code* (Stanford University Press, Redwood City, CA).

Morange M. (2020) *The Black Box of Biology: A History of the Molecular Revolution (translated by Matthew Cobb)* (Harvard University Press, Cambridge, MA).

Olby R. (1994) *The Path to the Double Helix: The Discovery of DNA* (Dover Publications, Mineola, NY).

Rich A. (2009) The era of RNA awakening: Structural biology of RNA in the early years. *Quarterly Reviews of Biophysics* 42: 117–37.

Siekevitz P. and Zamecnik P.C. (1981) Ribosomes and protein synthesis. *Journal of Cell Biology* 91: 53s–65s.

Watson J.D., Gann A. and Witkowski J. (2012) *Double Helix, Annotated and Illustrated* (Simon & Schuster, New York).

---

[y] Sydney Brenner, Excerpts from Letter to Max Perutz, June 1963; reproduced in Wood (1988).[178]

[z] Sydney Brenner, excerpts from Proposal to the Medical Research Council, October, 1963; reproduced in Wood (1988).[178]

[aa] "From the elephant to butyric acid bacterium—it is all the same!"[181]

4 Worlds Apart

It was already evident that multicellular eukaryotes are orders of magnitude more complex than bacteria. Humans, for example, have ~30–40 trillion cells[1,2] that are precisely assembled during embryonic and post-natal development into a myriad of different and precisely sculpted muscles, bone and other organs, and a brain with over 85 billion neurons and a trillion synaptic connections.[3–5]

Eukaryotic cells are also generally much larger[a] and have more complex organization than bacterial cells. They also have larger genomes,[b] especially in multicellular organisms, split between linear chromosomes, with variable but usually substantial amounts of 'repetitive' sequences (Chapters 5 and 10).

Eukaryotic cells have a membrane-bound nucleus where the chromosomes are located, an important consequence of which is the separation of transcription from translation[c] (Chapter 7). They also contain other internal membranous structures and membrane-bound 'organelles'[11] including caveolae (surface pits for endocytosis of external material);[12] endosomes (which traffic proteins, lipids and other components within the cell);[13] peroxisomes (where specialized oxidative metabolism takes place);[14] lysosomes (which degrade engulfed particles and intracellular components);[15–17] the endoplasmic reticulum (ER, the 'rough' form of which is studded with ribosomes);[18,19] the Golgi apparatus (a distribution network wherein proteins are imported from the ER, tagged with carbohydrates, sorted and packaged into endosomal vesicles destined for lysosomes, the cell surface or export,[20,21] named after their discoverer Camillo Golgi in 1898[22]); mitochondria (which generate energy by oxidation of carbohydrates and fatty acids);[23] and (in plants and algae) chloroplasts (photosynthetic energy capturing factories that produce sugars, sometimes called 'plastids')[24] (Figure 4.1). These organelles display an intricate degree of interaction and coordination.[25–30]

Eukaryotic cells also have many non-membrane-bound compartments, such as the nucleolus (the site of ribosomal biogenesis within the nucleus), which was first observed in 1835.[31] These compartments are phase-separated domains nucleated by RNAs and proteins containing intrinsically disordered regions (Chapter 16). Prokaryotic (bacterial and archaeal) cells have no internal structures as obvious as those in eukaryotes, although there is spatial organization and compartmentalization.[32–34] Phase-separated domains occur in prokaryotes and may predate cellular life (Chapter 16).

Recent advances in scanning electron microscopy and cryo-electron tomography have also enabled high-resolution imaging of cellular organelles and subcellular structures.[35–39]

THE ORIGIN OF CELLS

The origin of life involved the evolution of macromolecules capable of transmitting information and catalyzing biosynthetic reactions, as well as their encapsulation and the harnessing of energy – the reversal of entropy to create ordered systems, as first argued by Schrödinger in 1944.[40,41] There are two leading hypotheses about where and how this might have occurred: in deep ocean hydrothermal vents where proton gradients could form, proposed by Michael Russell, Nick Lane, Crispin Little and colleagues;[42–46] or in terrestrial hot springs where hydrothermal pools undergo wet-dry cycles (reprising Darwin's "warm little ponds"[47]) that favor the synthesis of organic polymers, lipids, peptides and nucleic acids, put forward by David Deamer, Martin Van Kranendonk, Armen Mulkidjanian, Eugene Koonin, Steve Benner and others,[48–55] with evidence favoring the latter.[47,49,50,56] There is also evidence that

---

[a] There are exceptions.[6]

[b] Bacterial genomes have a maximum size of around 10 Mb,[7] are usually circular and replicated bidirectionally from a single origin of replication, first shown by John Cairns in 1963.[8] Bacteria can also contain additional circular DNAs called plasmids, a term introduced in 1952 by Joshua Lederberg to refer to "any extrachromosomal hereditary determinant".[9] Plasmids often carry antibiotic resistance genes or others that confer selective advantage and may replicate autonomously or become integrated into the chromosome.

[c] Transcription and translation are coupled processes in bacteria. Translational stalling can result in transcription termination, as in the *trp* operon, where it is used to attenuate the production of tryptophan biosynthetic enzymes, shown by Charles Yanofsky in the late 1970s.[10]

DOI: 10.1201/9781003109242-4

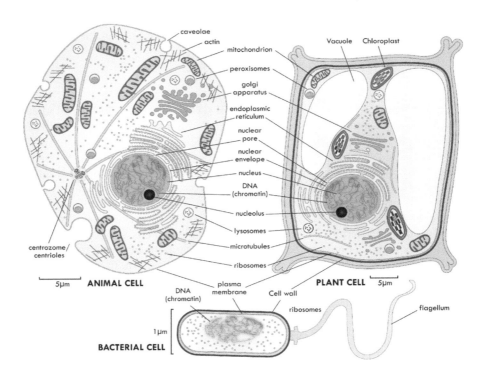

FIGURE 4.1   Comparison of prokaryotic and eukaryotic cell structures, original illustration by Heidi Cartwright.

the ubiquitous presence of ATP is due, in part, to its ability to aid protein solubility.[57]

In 1977, Carl Woese[d] and George Fox[59] made the unexpected discovery by ribosomal RNA gene sequencing that there are not two but three distinctive domains of life on Earth:[60] the unicellular Bacteria and the superficially similar Archaea, collectively called Prokarya, and the unicellular and multicellular Eukarya – protozoa, algae, fungi, plants and animals.[e]

The eukaryotes – the last common ancestor of which is estimated to have emerged around two billion years ago[63] – appear to have arisen from an archaeal progenitor that fused with a bacterium, as they possess many nuclear genes derived from each, but those involved in the core processes of DNA replication, transcription and translation, including the histones that are used to package eukaryotic

chromatin (Chapter 14), are clearly archaeal in origin.[64–74] Whether there are two (original) or (now) three primary branches of life[f] is a moot and almost semantic point.[76]

Mitochondria and chloroplasts are also descended from bacterial ancestors[g] captured by endosymbiosis,[24,67] proposed controversially by Lynn Margulis (Sagan) in 1967[81] but later confirmed by Robert Schwartz and Margaret Dayhoff.[82] Mitochondria and chloroplasts contain remnant small circular genomes[80,83–86] and bacterial-like translation systems for key hydrophobic proteins that must be made *in situ*.[87,88]

A plausible theory advanced by Tom Cavalier-Smith is that eukaryotes initially made their living as cellular scavengers and predators (think amoebae), which required the development of a flexible external membrane for phagocytosis.[89,90] It also required internal membranes for the protection of the genome and compartmentalization of lysosomes and other

---

[d] Woese was also one of the originators of the RNA World Hypothesis in 1967 (Chapter 9), although not by that name.[58] The 3-domain theory took many years to be accepted, although the data was clear.

[e] The terms 'eukaryote' and 'prokaryote' were first coined by Édouard Chatton in 1925, and the distinction formalized in 1962 by Roger Stanier and C. B. (Cornelius) van Niel, based on the presence or absence of a nucleus.[61,62]

[f] It has also been suggested that the giant viruses (Megavirales), discovered in amoebae in 2002, comprise a fourth super-kingdom of life.[75]

[g] Mitochondria are descended from a hydrogen-producing α-proteobacterium;[77,78] chloroplasts from a cyanobacterium.[79,80]

organelles, which is consistent with the flexible membranes and microvesicles observed in the extant lineage of the proposed archaeal ancestor of the eukaryotes.[69,72]

GENETIC RECOMBINATION

Whereas prokaryotes only have one genome copy (in addition to self-replicating extrachromosomal plasmids), eukaryotic cells are (usually) 'diploid',[h] having obtained one nuclear genome copy from each parent,[i] produced by the process of meiosis to form 'haploid' 'gametes' (sperm and ova). This and the elaboration of two sexes in eukaryotes may have arisen (although there are many possible explanations[92–95]) to allow recombinational exchange of larger genomes in complex cells and especially between multicellular organisms.[j]

In (most) eukaryotes, having two genome copies means having two copies of each of the genes therein, which are referred to as 'alleles' if they differ in any demonstrable way. These alleles may be dominant or recessive with respect to the other in their impact on phenotype, as demonstrated by Mendel. This arrangement provides the advantage that defective alleles can be tolerated, as having one functional version is usually enough to compensate,[k] which enables flexibility. Regulatory variation may be co-dominant, which allows more complex dynamics in the evolution and expression of quantitative traits.

DNA exchange occurs *ad hoc* in prokaryotes and at meiosis in eukaryotes, where it involves chromosomal pairing, formation of a 4-stranded cruciform structure between homologous DNA duplexes[l] ('crossing-over') and strand exchange.[100] Homologous recombination was the basis of genetic mapping by Benzer, Morgan and in the first half of the 20th century (Chapter 2), and later used to track down the protein-coding genes that are damaged in disorders such as cystic fibrosis (Chapter 11).

Recombinational exchange is an evolvability strategy that appears to have arisen at the dawn of life to enable genetic variations to be separated and discriminated by selection.[101] Gene assortment and fault tolerance were major considerations in the development of evolutionary theory and mathematical models of population genetics.

The evolution of the pathways and infrastructure for genetic recombination is an example of second-order Darwinian selection, where an accidental innovation has no particular or immediate phenotypic consequence but confers long-term advantages. There are almost certainly other evolutionary search optimization strategies that have not yet been recognized, because of the emphasis on phenotypic selection and the belief that mutation occurs randomly (Chapter 18).

It is also worth noting the growing appreciation of transposons (Chapters 5 and 10) and viruses,[m] the most abundant biological entities on Earth (which may have predated cellular life), as wider currencies for genetic exchange and dissemination, with central roles in the early evolution of cells, the 'invention' of DNA and DNA replication, the formation of the three domains of life and the diversification of multicellular organisms.[105–110]

THE EMERGENCE OF COMPLEX ORGANISMS

Multicellular plants appeared around 1–1.2 billion years ago, initially in the oceans, then in freshwater environments.[111,112] They colonized the land around 850 million years (Myr) ago[113] and diversified into more complex vascular forms with roots, leaves

---

[h] Some are 'polyploid', such as wheat, which has six copies of each chromosome (hexaploid), and whose gametes have three.

[i] Mitochondrial genomes are maternally transmitted in ova, the source in part of the cytoplasmic inheritance of some characteristics first observed by the Boveris around 1900 (Chapter 2). Chloroplast genomes are usually also maternally inherited.[91]

[j] Linear chromosomes may be required for meiotic segregation.[96] They are copied via multiple replication origins and an overlapping series of bidirectional replication bubbles, a highly complex process that is tightly controlled during differentiation and development[97] (Chapter 15).

[k] Some defective genes can be dominant for mechanistic reasons, such as those encoding proteins involved in multi-component complexes, because only one copy is expressed (as in parental imprinting; Chapter 5) or because they only occur on sex chromosomes, as exemplified by the higher frequency of color blindness in males (the genes are located on the X chromosome). Some 'heterozygous' combinations of functional and defective alleles produce intermediate effects, called 'haploinsufficiency', which may be more common than appreciated and have benefits in some circumstances.

[l] Called a 'Holliday junction', after Robin Holliday who proposed it in 1964.[98,99]

[m] Norton Zinder and Joshua Lederberg showed in 1952 that bacterial viruses can integrate into the bacterial genome,[102] providing an explanation for the lysogeny phenomenon earlier described by Eugene Wollman and André Lwoff, which became an important genetic tool for gene mapping.[103] Animal retroviruses were described by multiple groups in the 1960s and 1970s.[104]

and seeds between 500 and 200 Myr ago, when the angiosperms (flowering plants) emerged following an ancient genome duplication.[114,115] The phylogenetic tree from algae to angiosperms is now being constructed from DNA and RNA sequence data.[116] Interestingly, improvements in the efficiency of CO_2 fixation, known as the C4 pathway, appeared just 25–32 Myr ago, likely as an adaptation to declining CO_2 concentrations due to biological sequestration into calcium carbonate and carbonaceous deposits.[117]

Sponges, the most primitive of the animal phyla, existed around 890 Myr.[118] The first animals with complex body plans, the Ediacaran fauna,[n] appear in the fossil record between 620 and 550 Myr ago in an evolutionary radiation called the Avalon explosion,[120] with antecedents up to 800 Myr ago.[111] The Ediacaran fauna were soft-bodied organisms, ranging in size from 1 cm to over 1 m, likely making a living as scavengers – like fungi, with which animals share a common ancestor.[121–124] Ediacarans had radial or bilateral symmetry and segmented tube-, quilt- or frond-like structures, some with similarities to modern worms and jellyfish,[125–129] which may be their descendants.[130–133]

The Ediacarans were largely supplanted around 540–500 Myr ago by a second and more spectacular large-scale radiation. It is known as the Cambrian explosion, first revealed by fossils in the Burgess Shale of the Canadian Rocky Mountains, discovered by Richard McConnell in 1886 and characterized in detail in the early 20th century by Charles Walcott and others since, notably Simon Conway Morris,[134] in many Cambrian 'Lagerstätten'. In one strata of rock, and an estimated time window of ~10–20 Myr, recognizable ancestors of all extant metazoan phyla appear, including arthropods and chordates, with hard skeletons, advanced predation and locomotive capacity, along with other bizarre forms[135–139] (Figure 4.2), likely driven by the evolution of macrophagy.[140]

Soon after, fish were swimming in the oceans.[141] Similar rapid phenotypic diversifications also occurred after later mass extinction events,[142] including that following the meteorite strike in the Gulf of Mexico 66 Myr ago, which wiped out the non-avian dinosaurs and allowed the rise of mammals into vacated ecological niches.[143,144]

The initial appearance and rapid evolution of animals has commonly been thought to have been potentiated by the increase in atmospheric oxygen from photosynthesis and the advantages of aerobic energy generation by mitochondrial electron transport[o] in eukaryotes.[132,146–148] However, there was substantial atmospheric oxygen long before the evolution of animals,[149,150] and it seems that sufficient oxygen may have been an enabler but not a direct cause of their emergence.[151,152] Rather, the transition from unicellular to developmentally complex organisms with highly organized assemblages of specialized cell types was likely achieved by advances in genome organization and regulatory systems (Chapters 14–16). Later transitions, undoubtedly also requiring genetic innovations, occurred in the colonization of the land, to enable physiological adaptability to a more variable environment and more complex structures for terrestrial mobility.

CHROMATIN

The most obvious molecular genetic difference between prokaryotes and eukaryotes is that the much larger genomes of the latter are not only sequestered in a nucleus but also segmented into chromosomes and packaged into complex chromatin structures.

Eukaryotic chromatin is not homogeneous, and contains regions with different properties, broadly divided into open 'euchromatin' (gene rich, transcriptionally active, lightly stained) and compacted 'heterochromatin' (transcriptionally quieter, densely stained), first documented in the late 1920s by Emil Heitz,[153] a pioneer of cytogenetics,[154] notably in the giant 'polyploid' or 'polytene' chromosomes[p] in the salivary glands of insects.[157–159]

Dynamic changes in chromatin were observed in the appearance and disappearance at different developmental stages of 'facultative' heterochromatin by Heitz and others,[153] and 'puffs' in polytene chromosomes formed by localized decondensation of small

---

[n] Named after the Australian site where they were found in abundance by Reginald Sprigg in 1947.[119]

[o] There are conflicting views.[145]

[p] Polythene chromosomes were first observed by Édouard-Gérard Balbiani in 1881, who is remembered in the term 'Balbiani ring' which refers to the large chromosomal puffs where transcription occurs. They contain multiple DNA molecules in parallel, generated by multiple rounds of DNA replication without an intervening cell division.[155] Their banded structure corresponds to topologically associated domains (Chapter 14), which are preserved between polytene and diploid cells.[156]

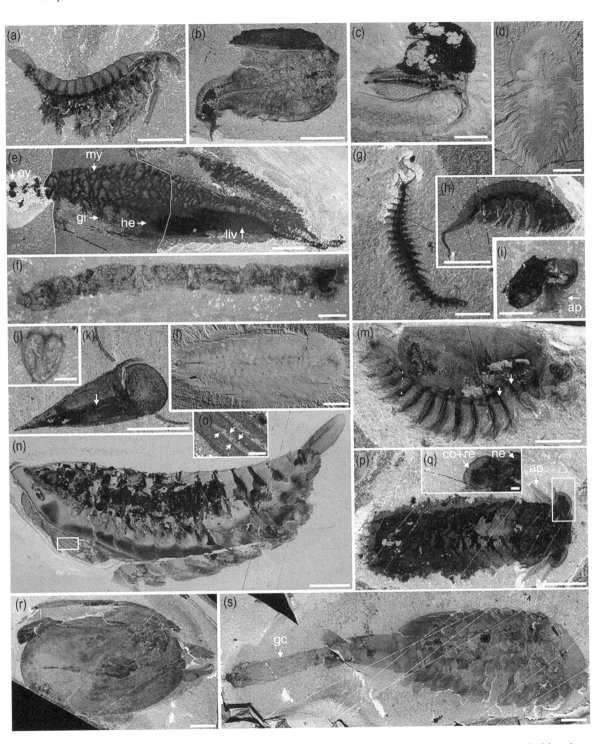

FIGURE 4.2 Cambrian fossils from the Burgess Shales, including ancestral arthropods (a–d, h–j, l–s), primitive chordates (e, f), annelid (g) and mollusk (k). (Reprinted from Caron et al.[139] by permission of Springer Nature.)

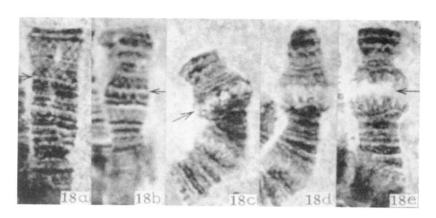

FIGURE 4.3 Micrographs of banded polytene chromosomes, arrows indicate 'puffs'. (Reprinted from Poulson and Metz [160] by permission of John Wiley and Sons.)

chromosomal segments, described by Donald Poulson and Charles Metz in 1938 (Figure 4.3), and by others in the 1950s and 1960s.[160–162] Puffs exhibit developmental stage- and tissue-specific patterns,[163,164] can be induced by heat shock[165] and hormones such as ecdysone,[166] and are sites of RNA synthesis.[167–172] In 1973, it was shown by Adolf and Monika Graessman that puff induction involves RNA,[173] and later by Subhash Lakhotia and colleagues that the product of a puff induced by heat shock is an RNA that does not encode a protein[174,175] (Chapter 9).

Specific heterochromatic regions of eukaryotic chromosomes form centromeres,[176,177] which act as organizing centers of cell division, described first by Edouard van Beneden and then by Boveri (who coined the name) in the 1870s and 1880s.[178,179] Centromeres contain internal granules (called 'centrioles' by Boveri) and attach to kinetochores for spindle formation and chromatid pairing and separation to daughter cells during mitosis and meiosis[180] (see Chapter 15).

In 1959, Susumu Ohno showed that one of the two X-chromosomes in female mammals is heterochromatic[181] (called 'nucleolar satellite' and later the 'Barr body' after its discoverer, Murray Barr[182]). In 1961, Mary Lyon demonstrated that X-chromosome inactivation occurs randomly in early embryogenesis:[183,184] females are mosaics of active X-chromosomes inherited from either parent,[q] a 'dosage compensation'

mechanism to equalize with males, who only have one X-chromosome – a traditional system in genetic and cytological studies[r] (Chapter 2) later shown to be controlled by RNA[s] (Chapter 9). It was also known that in some insects one entire set of chromosomes becomes heterochromatic during male early embryonic development,[189] and that chromosomes in embryonic cells often have a different morphology than those in adult cells.[190]

The existence of 'facultative' heterochromatin, position effect variegation, chromosomal puffs and 'lampbrush' chromosomes[t] (described by Alexander Flemming in 1882[193]), which occur in the oocytes of all animals except, curiously, mammals,[194] suggested that there are higher-order genomic arrangements and additional modes of gene regulation during plant and animal development.[153]

It was also found that eukaryotic DNA is wrapped

---

[q] The mosaic pattern of X-chromosome inactivation can be observed in variegated coat colors, such as in 'tortoise shell' cats, which are almost invariably female (except when chromosomal aberrations such as XXY occur[185]), and in the mosaic pattern of sweat glands in women.

[r] X-linked (sex-linked) traits have also been of great value to human genetics (Chapter 11), given that recessive mutations are exposed in males, classic examples being red-green color-blindness and Duchene muscular dystrophy,[186] with variable intermediate phenotypes (severity of effect) in females because of mosaic expression.[187]

[s] In *Drosophila*, dosage compensation is achieved not by inactivation of one of the X-chromosomes in females, but by global upregulation of the activity of the single X-chromosome in males,[188] which also centrally involves non-coding RNAs (Chapter 9).

[t] In the late 1950s, electron microscopic visualization of elongating transcripts on lampbrush chromosomes first suggested rapid packaging of nascent RNAs with proteins.[191] In the 1970s, several laboratories began focusing on biochemical purification and compositional/structural analysis of non-ribosomal ribonucleoproteins, which led to identification of mRNA cap–binding proteins, polyA-binding protein, pre-mRNA splicing proteins (Chapter 8) and mRNA transport proteins, among others.[192]

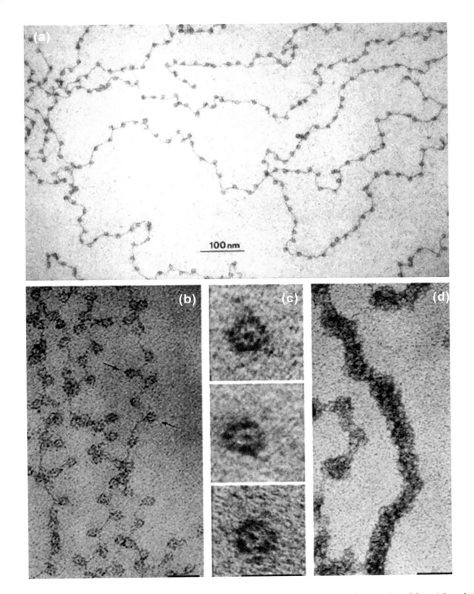

FIGURE 4.4 Electron micrographs taken by Donald and Ada Olins of (a) and (b, size marker 30 nm) low ionic strength chromatin spreads showing the 'beads on a string'; (c) nucleosomes derived from nuclease-digested chromatin (size marker 10 nm); (d) chromatin spread at a moderate ionic strength showing a 30 nm higher-order structure (size marker 50 nm). (Reproduced from Olins and Olins[201,202] with permission of American Scientist (a) and Springer Nature (b–d).)

around proteins called histones,[u] like cotton around a spool in a repeating structure, called 'nucleosomes'.[198] Histones were identified by Kossel in 1894,[199] but it was another 80 years before nucleosomes were visualized in the electron microscope by Ada and Donald Olins[200–202] (Figure 4.4) and their octameric histone complement defined by Roger Kornberg and colleagues.[203,204] It was even longer before it became evident that histones are the major repositories of epigenetic information (Chapter 14), although hints of a role in gene regulation were emerging.

[u] Eukaryotic histones have ancestral orthologs in archaea,[70,71,74] where they regulate gene activity in response to environmental circumstances.[195] Histones have been shown to have copper reductase activity,[196] suggesting their original role was to protect against oxygen toxicity.[197]

In 1950, Ellen and Edgar Stedman[v] proposed that histones are repressors that could inactivate genes in a tissue-specific manner, based on the quantities of histones in growing and non-growing tissues.[205,206] Their proposal was supported by work of Ru-chih Huang and James Bonner, and Vincent Allfrey, Alfred Mirsky and colleagues, who found that histones inhibit the transcription of DNA *in vitro*.[207,208]

However, histones, superficially at least, displayed uniformity between tissues and species, as well as between 'repressed' and active chromatin (see below), which led John Frenster, Allfrey and Mirsky to conclude in 1963 that they were unlikely to be gene-specific regulators.[209] For decades thereafter, nucleosomes were considered primarily a mechanism for compacting large genomes,[210–212] given the widespread conviction that transcription factors are the primary means of gene regulation.[213]

In the mid-1960s, Allfrey and Mirsky proposed that post-translational modifications of histones (acetylation and methylation) have regulatory functions.[214,215] They showed that lymphocyte activation triggers massive acetylation of chromatin[216] and that histone acetylation also occurs in insects.[217] A decade on DNA methylation was also suggested as a mechanism to regulate gene activity,[218,219] although these ideas would only be tested and confirmed much later[220–223] (Chapter 14).

Beyond this, how the structure of chromatin was organized and how it affected gene expression in eukaryotes was unknown; progress was slow because of the sheer size and complexity of the genomes and chromosomes, and the difficulties of working with all but unicellular models such as yeast.

CHROMATIN-ASSOCIATED RNAs

By this time, it was also clear that RNA is the third component of chromatin. In his early histological experiments on the distribution of DNA and RNA, Brachet showed that RNA is present not only in the cytoplasm but also in chromatin,[224,225] subsequently found by Mirsky and Hans Ris to reside in a NaCl insoluble fraction, which comprised only ~10% of the total but retained all the usual features

of chromatin.[226] RNA was also reported to remain attached to chromosomes during cell division.[227]

Indeed, while largely overlooked during the heady days of the genetic code, a number of publications in the 1960s and early 1970s reported the presence of RNA in chromatin fractions,[228] some of which were proposed to be structural and regulatory agents. These included Frenster's 1965 model of 'De-repressor RNAs', based on the observation that RNA added to heterochromatin fractions increased the level of transcription, which was most pronounced when nuclear RNAs were added (compared to cytoplasmic and non-specific RNAs such as rRNA or yeast RNA), posited to involve RNA hybridization to complementary sequences in the repressed DNA.[229,230]

In 1965 also, Huang and Bonner reported the presence of low molecular weight RNAs in chromatin. These "chromosomal RNAs" (cRNAs) were protected from RNase degradation and corresponded to ~8% of the total nucleic acid mass present in nucleohistones.[231] This was the first in a series of reports of the existence of tissue-specific short RNAs in chromatin in plants and animals that associate with non-histone chromatin proteins and can hybridize to homologous DNA,[232–236] leading to the hypothesis that cRNAs had a role in regulating gene expression.[235,237–239]

These short RNAs are not precursors for any cytoplasmic product[240] and some were distinguished by a high content of methylated or dihydropyrimidine nucleotides,[233,240,241] a signature of small nucleolar and small spliceosomal RNAs (Chapter 8).

Interestingly, cRNAs were found to hybridize extensively to "middle repetitive" DNA sequences, which Bonner proffered as evidence that repetitive sequences may be regulatory elements.[237,242,243] These observations would have impact on models of genome regulation in the higher organisms (Chapter 5), but were sidelined later by the widespread assumption that much of the genomes of higher organisms is junk, partly and ironically because they contained so many 'repetitive' sequences (Chapters 7 and 10).

Soon after the Frenster and Bonner publications, William Benjamin and colleagues reported that RNA isolated from a rat liver nucleoprotein fraction co-sedimented with histones.[244] This RNA had high adenine and uridine content and had heterogeneous sizes by sucrose gradient analysis, adding to the complexity of the types of RNAs of unknown functions found in the eukaryotic nucleus. Although

---

[v] The Stedmans had earlier suggested that non-histone chromosomal proteins, which they called "chromosomins", represent the "basis of inheritance" and are also involved in gene regulation, predicting that the physical association of chromosomins and nucleic acids was required for synthesis of specific proteins.[205]

recognizing that these RNAs might represent an intermediate in the synthesis of mRNA, Benjamin et al. also speculated that these RNAs could play a role in the control of gene expression, invoking Paul Sypherd and Norman Strauss' 1963 suggestion that regulatory systems involve both protein and RNA,[245] such that associated RNAs might confer specificity to repressive histones by base-pairing with the target gene[244] (Chapter 16).

During the 1970s, nuclear RNAs began to be better characterized. Some were visualized with chromatin at specific stages of the cell cycle and proposed to act as "programmers" of "chromosomal information" and gene regulation.[246] In vitro experiments by Takeharu Kanehisa and colleagues with purified chromatin indicated that specific short chromatin-associated RNAs could "modify" chromatin structure and stimulate RNA synthesis, particularly in chromatin isolated from the same tissue, suggesting a tissue-specific effect.[247–249]

In 1973, Isaac Bekhor showed that "chromosomal RNA-protein complexes" can interact with DNA in vitro and increase its melting temperature, indicating a stabilizing effect, leading him to postulate that cRNAs present in chromosomal RNA-protein complexes constituted a structural component of chromatin, rather than regulatory molecules,[250] an idea favored by other studies reporting RNAs associated with heterochromatin.[251] In 1978 Sheldon Penman's group showed that stable species of high molecular weight RNAs also associate with nuclear complexes, from which it was again hypothesized that "RNA networks" had structural roles in the nucleus.[252]

Thoru Pederson and Jaswant Bhorjee reported in 1979 that three short RNAs are associated with chromatin and favored the hypothesis that these "DNA-linked RNAs" are involved in the control of the tertiary structure of chromatin.[253] These RNAs were ~130–200-nt long, highly abundant, relatively stable, and were designated small nuclear RNAs D, C and G'[253] (later small nuclear RNAs U1, U2 and U5, respectively, Chapter 8). They showed that only a fraction (<10%) of these RNAs is associated with chromatin, supporting the earlier results of Mirsky and Ris, while the remainder was nucleoplasmic.[253] Others reported that separated fractions contained between 6 and 11 size classes of small RNAs, most of which seemed to be reversibly bound to chromatin proteins,[233,254,255] but some of which did not dissociate in high salt concentrations, possibly reflecting RNA-DNA hybrids.[256]

Later studies using hybridization and cytogenetic techniques showed that cRNAs from human placenta displayed a widespread pattern of hybridization to metaphase chromosomes, preferentially in telomeric regions and heterochromatic short arms of acrocentric chromosomes, as well as regions with a high content of repetitive DNA.[257] Chromatin fractionation studies using Frenster's method had also indicated that heterochromatin contains a more stable fraction of chromatin-associated RNAs compared to euchromatin.[251]

Other studies focused on the effects of high molecular weight RNAs and nascent transcripts in chromatin and in other nuclear structures. Jean-Pierre Bachellerie and colleagues used subnuclear fractionations, autoradiographic and ultrastructural techniques to show that "perichromatin fibrils represent the morphological state of newly formed heterogeneous nuclear RNA (hnRNA)".[258]

Much of this would make sense later (Chapters 7, 8 and 16), but at the time there was a fierce debate over the existence, biological relevance and specificity of chromatin-associated RNAs. The concerns ranged from the reproducibility of the findings, the questionable purity of cellular and chromatin fractions (complicated by a plethora of fractionation procedures), the presence of nucleases that could lead to contamination with degradation products of tRNAs, rRNAs and heterogeneous nuclear RNAs (see below), and the feeling that the models proposed were too speculative.[259–264]

On the other hand, a number of lines of evidence were proffered against the criticisms, including that cRNAs have characteristic elution properties, their complexity and hybridization kinetics differed from common RNAs such as tRNA and rRNA, they had a different stability, and chromatin preparations using stringent methods contained a reproducible RNA fraction.[233,236,239,243,256,265,266]

By the end of the 1970s, different groups that had analyzed the composition of total chromatin estimated that the RNA/DNA ratio of the chromatin was approximately ~0.05–0.2 in different eukaryotes, and that distinct classes of RNAs with specific properties were chromatin-associated, including species of varying stability and molecular weight,[253,256,267–269] which would include new classes of infrastructural RNAs involved in rRNA biogenesis and splicing (Chapter 8).

Nevertheless, uncertainty about chromatin-associated RNA remained because the characteristics of the reported RNA profiles varied by tissue and the method of chromatin isolation. The exact nature of these RNAs was unknown, given that the methods of identification were rudimentary at that time. It was a complicated muddle.

EARLY MODELS OF RNAs IN NUCLEAR ARCHITECTURE

There was also emerging evidence that RNA is involved in the organization of the nuclear 'matrix', a fibrous structure first reported in 1948,[270] although its composition, stability and function has been the subject of ongoing conjecture.[271]

Penman recognized that chromosomes are not randomly distributed in the nucleus, much later described in detail (Chapter 14), and that the nuclear matrix played a central role in its three-dimensional architecture. His group used a chromatin-depletion strategy to show that ribonucleoprotein networks extend throughout a nuclear structural lattice and that the integrity of nuclear and chromatin architecture is dependent on RNA, as indicated by the treatment of cells with transcription inhibitors or extracted nuclei with RNase.[w] Penman concluded that RNA is not only a structural component of the nuclear matrix (an "RNA-dependent nuclear matrix") but also organizer of higher-order structures of chromatin ("architectural RNA"),[273,274] an idea that would reemerge powerfully much later when it was discovered that RNA and transcription modulate chromatin territories and nucleate subcellular domains (Chapter 16).

Similar ideas were elaborated by others, including the 'Unified Matrix Hypothesis' postulated in 1989 by Klaus Scherrer, in which the transcribed and non-transcribed part of non-coding DNA would have a direct morphogenic function.[275] According to this hypothesis, these regions have an intrinsic role in "the tridimensional network of chromatin and nuclear topological organization".[275] This was posited to explain phenomena such as the 'Chromosome Fields' with co-localization of linked genetic loci within chromosome regions and the specificity of sites of chromosome recombination in cancer.[275] Scherrer also suggested that RNA processing may play a role in nuclear architecture by organizing the selective transport and control of individual

transcripts, which would then act as signals for specific proteins and in combination define the nuclear matrix.[276]

HETEROGENEOUS NUCLEAR RNA

Radioactive labeling kinetics, sucrose gradient sedimentation and hybridization studies in the early 1960s by Scherrer, James Darnell, Georgii Georgiev and colleagues indicated that rapidly labeled transcripts were formed in the nucleus of mammalian cells, which exhibited high molecular weight and heterogeneous sedimentation profiles (Figure 4.5), AU-rich composition and unstable character.[277–282] These unexpected 'giant' nuclear transcripts were only broadly defined and described variously as DNA-like RNA ('dRNA'),[278] nascent messenger-like RNA (nascent 'mlRNAs')[281] and heterogeneous (or heterodisperse) nuclear RNA ('hnRNA').[282]

In 1963, Scherrer and Darnell showed that ribosomal RNAs in human cells are initially produced as large precursor molecules and subsequently processed into mature rRNAs,[279,283] confirmed by Penman and Guiseppe and Barbara Attardi,[284,285] and shown to take place in the nucleolus.[286,287] This may have been an idiosyncratic feature of the ribosomal operons, but they also reported that other "giant" RNAs with "messenger RNA properties" also exist in the nucleus.[283,284,288]

The biological significance of the large heterogeneous transcripts was puzzling. The difference in base composition between hnRNA and cytoplasmic mRNA, as well as differences in their half-lives, did not suggest a simple relationship.[289] Henry Harris noted in 1965 that "only a small proportion of the RNA made in the nucleus of animal and higher plant cells serves as a template for the synthesis of protein", and considered that "most of the nuclear RNA, however, is made on parts of the DNA which do not contain information for the synthesis of specific proteins. This RNA does not assume the configuration necessary for protection from degradation and is eliminated" (quoted in [290]). Harris later noted that "pulse-labeled RNA was almost universally misdiagnosed as messenger RNA" and that other suggestions were considered profoundly heretical at the time.[291]

In 1966, Scherrer was the first to propose that hnRNAs are precursors of mRNAs[288] but, despite his intense efforts, his proposal was not widely entertained.[276] In 1968, Penman showed that

---

w RNase treatment had been used previously (in 1974) to show that RNA is involved in the maintenance of condensation of dinoflagellate chromosomes.[272]

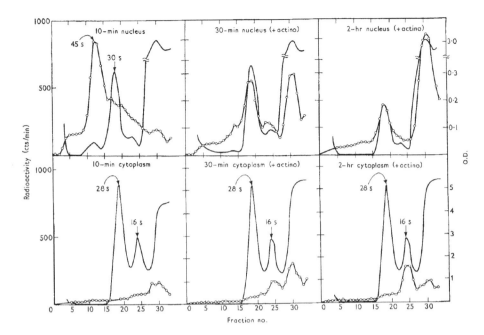

FIGURE 4.5 Sucrose gradient profiles of pulse-labeled DNase-treated nuclear and cytoplasmic preparations before and following actinomycin D treatment (which blocks RNA synthesis) showing the presence of high molecular weight RNAs in the nucleus but not the cytoplasm. Open circles, radioactivity; continuous line optical density at 260nm, which detects nucleic acids. (Reproduced from Penman[286] with permission of Elsevier.)

"heterogeneous nucleoplasmic RNA" is produced in the absence of ribosomal RNA synthesis (blocked by specific concentrations of actinomycin D) and is turned over rapidly, with a mean life of approximately 1 hour, although he did not think that this might be a precursor to mRNA.[292,293] Penman also showed that the size of hnRNA increased with increased genome size,[294] all of which is consistent with the later discovery of introns and large pre-mRNA primary transcripts that are spliced to produce mRNAs (Chapter 7).

Hybridization studies by Allfrey and Mirsky indicated that while ~80% of the DNA was not accessible for transcription, transcription products covered up to 20% of the DNA in a given mammalian cell.[295] This and similar observations were potentially relevant for gene regulation because, given that the majority of the genome DNA was inactive in a particular cell type, it implied a mechanism of genome repression and allowed the suggestion of specialization of chromosomal regions for the control of growth and development of differentiated tissues.[208,295]

In 1967, Ruth Shearer and Brian McCarthy showed that while all the sequences of cytoplasmic RNA were present in the nucleus, the latter contains a much greater fraction of sequences that are not exported to the cytoplasm, but rather are rapidly turned over,[296] confirmed by others.[297,298] They went on to say: "The existence of RNA molecules specific to the nucleus suggests a role as mediators of the regulation of gene transcription ... although these functions are entirely speculative, the finding that the majority of the active genome codes for short-lived RNA molecules which are restricted to the nucleus opens up exciting possibilities for the study of the regulation of gene action in mammals."[296]

HEROES OR FOOLS?

Some working in animal genetics at the time, such as Ed Lewis (Chapter 5), drew the obvious conclusion that that "the 'genome' of phage and bacteria may be structurally organized in manner different from the chromosomes of higher forms".[299] However, not much notice was taken. The *lac* operon dominated the models of gene organization and the regulation of gene expression in eukaryotes; it was the basis of the ruminations by Jacob and Monod,[300] and others such as Ernst Mayr[301] on "genetic programmes" and gene networks (or "nets of interacting genes"), in what developed into the belief that the combinatorial action of 'transcription factors' is sufficient to execute complex developmental programs[302–304] (Chapter 15).

In a 1969 paper entitled 'On the Structural Organization of Operon and the Regulation of RNA Synthesis in Animal Cells', Georgiev assumed that the principles of regulation of transcription defined in bacteria are retained in multicellular organisms.[304] He defined an operon as an "elementary unit of transcription" and proposed that operons in higher organisms consisted of a promoter-proximal regulatory "zone" and a structural zone that contained the coding sequences of several mRNAs with related functions, as in the *lac* operon. In his view, the entire operon would be transcribed as a "giant D-RNA", with regulatory sequences at the 5' end degraded in the nucleus and the mRNA transferred into the cytoplasm. Being aware of Roy Britten's findings regarding the abundance of repeat regions in the genome,[305] he also posited that repetitive sequences could be present in many operons and be targets for regulatory proteins, thereby assigning a functional role for the repetitive sequences scattered throughout the genome, similar to that proposed in the same year by Britten and Eric Davidson (Chapter 5).

Scherrer also proposed (in 1968) that hnRNAs are polycistronic precursors of mRNAs and be subject to both transcriptional and post-transcriptional regulation, the "cascade regulation" model.[281] Accordingly, a polycistronic hnRNA may be cleaved sequentially in the nucleus or in the cytoplasm to generate multiple mRNAs in a regulated fashion. Indeed, this logic assumed that "as a consequence of the central dogma, postulating that gene activity leads to phenotypic expression of genetic information through the mediation of mRNA, localized genes related to particular phenotypic characteristics should produce mRNA".[281]

In the following years many observations pointed to a relationship between hnRNA and mRNA.[306] In 1971, Darnell, George Brawerman, Mary Edmonds and colleagues found that eukaryotic mRNAs[x] and hnRNAs both contain an extended sequence of adenines ('polyA tails')[y] at their 3' end,[314–316] confirmed by others,[316–318] and that eukaryotic mRNAs are derived from longer precursors.[319–321] In 1974, Kin-Ichiro Miura, Yasuhiro Furuichi, Aaron Shatkin, Fritz Rottman and colleagues showed that eukaryotic mRNAs and pre-mRNAs both also contain an inverted modified (methylated) nucleotide ('cap') structure[z] at their 5' end[324–329] (m7G, later shown to play a role in RNA splicing and translational control[330,331]), also confirmed,[332–334] along with a contemporary report of widespread methylation of mRNA[335] (Chapter 17).

These findings supported the conclusion that hnRNAs are, in fact, precursors to mRNAs,[316] with a suggestion that mRNA might be comprised of fragments from each end of the hnRNA,[306,318] but the idea that hnRNAs are mRNA precursors was only accepted and understood after the discovery of introns and splicing in 1977[276] (Chapter 7).

On the other hand, it was gradually recognized by the late 1960s that the multigenic operon was an unlikely mode of eukaryotic genome organization and regulation since, for example, related genes such as those encoding alpha and beta hemoglobins or the enzymes for galactose metabolism were not co-localized in the genome, and the length of hnRNAs was much larger than that which would be expected for reasonably sized polycistronic mRNAs.[281]

Because the relation of hnRNAs and mRNA was not yet established, Scherrer and Lise Marcaud contemplated that hnRNAs might contain sequences "other than those carrying structural cistrons", such as regions for interaction with proteins through secondary and tertiary structures, thus conferring specificity to the "functional RNA".[281] Alternatively,

---

[x] Prokaryotic mRNAs can also contain a shorter polyA sequence at their 3' end.[307]

[y] The existence of 3' polyA tails in mRNAs was exploited and remains widely used in cloning and sequencing protocols (Chapter 6), to separate mRNA from the large amounts of ribosomal, transfer and other RNAs in cells by its hybridization to oligomers of U or T ('oligo dT') affixed to solid or colloidal surfaces.[308]

Later the 5' cap would also be exploited for mRNA purification,[309] although it turned out that many other transcripts that do not encode proteins are also capped and polyadenylated (Chapter 13). Moreover, while widely overlooked, a large fraction of the RNAs detected in human cells are not polyadenylated, although they are often capped.[310–313]

[z] In fact, there are at least 25 different types of 5' caps in eukaryotic cells, at least some of which are cell- and tissue-specific, with roles in the initiation of protein synthesis, protection from exonuclease cleavage and as identifiers for recruiting protein factors for pre-mRNA splicing, polyadenylation and nuclear export.[322,323]

it was suggested that these RNAs might have independent regulatory roles, such as interacting with other regulatory molecules (inducers and repressors) or regulating allosteric proteins.[281]

In one way or another, many of these speculations would prove true, but the technology of the time was still too limited to test them, and animal and plant systems were much too complicated. At that time only a handful of laboratories tried, valiantly, to understand genetic information and gene expression in eukaryotes. As recalled by Scherrer,

> only a few investigators were interested in the molecular biology of animal cells; the 'serious' research was with *E. coli* and bacteriophages. James Watson visited MIT frequently, and would discuss our strange results. One day, he told me 'To work with animal cells, you've got to be a hero or a fool!'[276]

FURTHER READING

Brown S.W. (1966) Heterochromatin. *Science* 151: 417–25.

Cooper G.M. (2000) *The Cell, A Molecular Approach* (Sinauer Associates, Sunderland, MA).

Darnell J.E. (2011) *RNA: Life's Indispensable Molecule* (Cold Spring Harbor Laboratory Press, Cold Spring Harbor, NY).

Lane N. (2015) *The Vital Question: Why is Life the Way It Is?* (Profile Books, London).

Pederson T. (2009) The discovery of eukaryotic genome design and its forgotten corollary--the postulate of gene regulation by nuclear RNA. *FASEB Journal* 23: 2019–21.

Quammen D. (2018) *The Tangled Tree: A Radical New History of Life* (Simon Schuster, New York).

Smith J.M. (1978) *The Evolution of Sex* (Cambridge University Press, Cambridge).

Valentine J.W. (1978) The evolution of multicellular plants and animals. *Scientific American* 239: 140–58

Van Kranendonk M.J., Deamer D.W. and Djokic T. (2017) Life springs. *Scientific American* 317: 28–35.

5 Strange Genomes, Strange Genetics

The genomes of 'higher organisms' are thousands of times larger than that of *E. coli*.[1] Viral genomes range from ~1 kb to 2.5 Mb.[2,3] Bacterial and archaeal genomes range from ~300kb to 11 Mb (the upper limit, apparently, see Chapter 15), protozoan and fungal genomes from ~8 to 100 Mb (most less than 50 Mb),[4] animal genomes from ~100 Mb to 5 Gb (average 1.3 Gb) and plant genomes[a] from ~100 Mb to 20 Gb (average 1.5 Gb).[8] By comparison, the combined length of the two largest human genes, dystrophin (2.4 Mb; required for muscle integrity)[9,10] and the neurexin CNTNAP2 (2.3 Mb; which functions in the nervous system)[11] is approximately the same as the entire *E. coli* genome.

In 1964, Friedrich Vogel estimated that the human genome (~3.3 Gb) encodes around 7 million genes, based on the size of hemoglobins and the assumption that "most of the DNA works as genetic material".[12] Similar numbers were obtained using similar logic by others – in 1969 the theoretical biologist Stuart Kauffman extrapolated James Watson's estimate of 2,000 genes in *E. coli* in the first (1965) edition of his book 'Molecular Biology of the Gene'[13] to predict that there are 2 million genes in humans.[14]

Vogel recognized that his estimate was disturbingly high and speculated that "the systems of higher order which are connected with structural genes in operons and regulate their activity might occupy a much larger part of the genetic material than the structural genes".[12]

On the other hand, the existence of upward exceptions, notably in some amoebae, plants and arthropods, to an otherwise consistent increase in the amount of DNA in cells of eukaryotes of differing developmental complexity (Figure 5.1) led to the so-called C-value enigma,[8] which was invoked to support the growing idea that the genomes of higher organisms carry a lot of superfluous DNA (Chapter 7).

REPETITIVE DNA

In the late 1960s, Roy Britten and colleagues analyzed the patterns of DNA renaturation[16,17] and concluded that, unlike bacteria, "in general, more than one-third of the DNA of higher organisms is made up of sequences which recur anywhere from a thousand to a million times per cell" likely to have "arisen from large-scale precise duplication of selected sequences, with subsequent divergence caused by mutation and the translocation of segments of certain member sequences" (Figure 5.2).

They suggested that these sequences and events have important evolutionary implications:

> The range of frequency of repetition is very wide, and there are many degrees of precision of repetition in the DNA of individual organisms. During evolution the repeated DNA sequences apparently change slowly and thus diverge from each other. There appears to be some mechanism which, from time to time, extensively reduplicates certain segments of DNA, replenishing the redundancy.[17]

Britten and colleagues also cited many studies that observed changes in the pattern of types of hybridizable RNA synthesized during embryonic development and liver regeneration (noting that the hybridization conditions used selectively favored repetitive sequences), which showed "direct evidence for the genetic function of at least some of the repeated DNA sequences" and "that during the course of differentiation different families of repeated sequences are expressed at different stages".[17,18] Many other studies have since confirmed that 'repeat sequences' are a prevalent component of the genomes of multicellular eukaryotes generally,[19] and that they are transcribed in a highly regulated manner, especially in the early stages of embryonic development (Chapter 16).

These results also explained the earlier observation that eukaryotic DNA often exhibited the presence of 'satellite' bands in equilibrium density

---

[a] The smallest genome known is that of the parasitic microsporidian *Encephalitozoon intestinalis* (haploid 2.3 Mb);[5] the largest known are the marbled lungfish *Protopterus aethiopicus* (ca. 130 Gb)[6] and the monocot plant *Paris japonica* (ca. 150 Gb).[7]

DOI: 10.1201/9781003109242-5

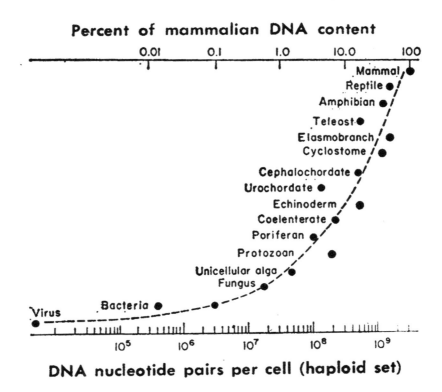

FIGURE 5.1 Britten and Davidson's 1969 graph of the increase in the minimum amount of DNA that had been recorded for "species at various grades of organization". (Reproduced from Britten and Davidson[15] with permission of the American Association for the Advancement of Science.)

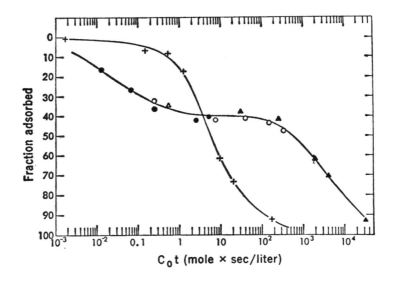

FIGURE 5.2 Britten and Kohne's 1968 graph of renaturation hybridization kinetics of denatured calf thymus DNA (circles and triangles) showing ~40% rapidly renaturing DNA, indicating the existence of high copy numbers (estimated to be an average of 100,000) of repetitive sequences (left side of graph) and 50%–60% single copy sequences (right side). The curve with + indicates the renaturation kinetics of radiolabeled *E. coli* DNA spiked into the same assay. (Reproduced from Britten and Kohne[17] with permission of the American Association for the Advancement of Science.)

gradients, indicating a substantial fraction with different nucleotide composition. Heterogeneity of sequence composition was subsequently found to occur throughout the mammalian genome, due to regional variation in G+C content, termed 'isochores' by Giorgio Bernardi.[20,21]

However, because they did not fit the conventional conception of a 'gene', the presence of large amounts of 'repetitive' DNA, a term that became pejorative, reinforced the idea that complex eukaryotes could tolerate a great deal of nonfunctional DNA, rather than stimulate new thinking, with the notable exception of Britten himself and a few others (see below).

The repetitive sequences discovered by Britten turned out to be largely derived from transposons, short (SINE) and long (LINE) elements mobilized by reverse transcription, and DNA segments mobilized by a 'cut-and-paste' mechanism, which range in length from a few hundred to 30 thousand base pairs and can be excised or transcribed, copied and reinserted into genomes at other places, conveying cassettes of genetic information.[19,22]

CONTROLLING ELEMENTS

Transposons were discovered in 1948 by Barbara McClintock,[b] who observed unexpected position-dependent effects of mobile DNA segments in maize, which changed the colors and color patterns of corn kernels (among other things), specifically the mobile 'Dissociator-Activator' (Ds-Ac) elements, and another that she called 'Suppressor-mutator' (Spm), published well before Jacob and Monod's work on the regulation of the lac operon in E. coli,[25,26] with which she later drew parallels[27] (Figure 5.3).

McClintock's work was revolutionary. Rather than focusing on mutations that affect viability, she developed and used cytogenetic analysis of X-ray-induced chromosomal breakage-fusion-bridge cycles for creating and mapping mutations.[28,29] She was the first to show that genetic recombination involved physical exchange of chromosomal segments,[30] the first to observe 'jumping genes' and the first to show that the position of their insertion alters the expression of nearby genes (for example at the bronze locus, which is involved in the biosynthetic pathway for the pigment anthocyanin)[31] – quite different from normal conceptions of regulatory sequences.

McClintock reported her data on transposition in the formal scientific literature in 1950,[33] and in person in 1951 at the influential Cold Spring Harbor Symposium on Quantitative Biology,[25] at the conclusion of which geneticist Evelyn Witkin recalls:

> In her own [McClintock's] words, the response was puzzlement, and in some instances hostility. Certainly, as I remember it, there was baffled silence after her talk and little or no discussion of her densely documented evidence and argument for transposable elements and their effects on gene expression. The audience seemed embarrassed by the lack of response, understandably, because McClintock was respected and admired as a great geneticist.[c] Here, her conclusions were too radically in conflict with the entrenched genetic concept of a stable genome, and her data too complex, to allow for rapid or easy acceptance, although a small number of geneticists who had come to know her work well believed it to be profoundly important.[35]

Highlighting the parallels with position-effect variegation in Drosophila and heterochromatin (and also citing the work of Ed Lewis, see below), McClintock said:

> The behavior of these new mutable loci in maize cannot be considered peculiar to this organism. The author believes that the mechanism underlying the phenomenon of variegation is basically the same in all organisms … Is it usually an orderly mechanism, which is related to the control of the processes of differentiation?[33]

McClintock's observations were confirmed by Royal Alexander Brink and Robert Nilan studying another mobile element called 'Modulator', who concluded that their findings "cannot be explained in conventional genetic terms" and speculated that Modulator "may belong to the obscure category of germinal substances, termed heterochromatin, which has been associated … with different variegated phenotypes in Drosophila".[36]

---

[b] McClintock also made many cytological observations of subcellular structures, most notably the "nucleolar organizing region",[23] now known to contain tandem arrays of ribosomal RNA genes.[24]

[c] On more conventional grounds, McClintock was elected to the National Academy of Sciences in 1944 at the relatively young age of 42, and in 1945 she was elected the first female president of the Genetics Society of America.[34]

FIGURE 5.3   Ears of corn kept by McClintock to illustrate the transposition of 'controlling elements' – the 'standard' position of the transposable element *Ds* (*Dissociation*) located after *C* (*colorless*), *Sh* (*shrunken*) and *wx* (*waxy*) on chromosome 9 (upper ear, from 1947) and the transposed position of *Ds* now positioned before *Sh*, *Bz* (*bronze*) and *wx* (lower ear, from 1949). (Courtesy of Cold Spring Harbor Laboratory Archives (photo: Sarah Vermylen). Reproduced from Chomet and Martienssen [32] with permission from Elsevier.)

The genetic behavior of these families of elements is complex but, in general, they are comprised of some small, transposition-competent sequences and a larger number of derived transposition-defective sequences[d] whose mobility is dependent on a factor supplied by an active member of the family.[39] Indeed, *Ac* and *Ds* are autonomous and non-autonomous transposable elements,[40] respectively, explaining the dependence of the latter on the former. Other elements, such as the *Suppressor-mutator* element, are more complicated.[41,42]

McClintock eventually received the Nobel Prize in 1983 for having discovered transposition, coinciding with demonstrations by Nina Fedoroff and others that *Ac* encodes a transposase responsible for the transposition of itself as well as *Ds*,[39,43] the identification of TEs (transposable elements, see below) in *Drosophila*[44–46] and bacteria,[47,48] and the growing use of TEs as mutagenic tools, especially 'P-elements' in *Drosophila*.[49,50]

McClintock insisted that her most important finding was that such 'controlling elements' - which others such as Brink preferred to call 'transposable elements', as the term was "less interpretative"[51] – played a role in normal development, as both the timing and frequency of transposition and associated chromosomal rearrangements are developmentally regulated and have developmental

---

[d] The human genome, for example, has over 1 million copies of the Alu element (comprising over 10% of the genome), only ~5% of which are active.[37] The name 'Alu' derives from the fact that these elements were first recognized as the short (~300nt) repeated DNA sequences commonly contain a cleavage site for the *Arthrobacter luteus* (*Alu*I) restriction enzyme.[38]

consequences.[25,26,52–54] She also showed that some of these elements exhibited reversible mutations.[55]

McClintock conceptualized the signatures of epigenetic control of gene expression decades before it was formally recognized and studied. She identified genetic elements that were either active or inactive only in the crowns of corn kernels, as well as elements that were active only in lower side branches (called 'tillers') or exhibited a different pattern of activity in cobs from the tillers and primary stalk.[52,54] This implied the existence of an additional regulatory mechanism that determines the pattern of TE expression/activity during development. McClintock correctly speculated that this mechanism involved heritable modifications of the gene that are not caused by changes to its DNA sequence, later shown by Rob Martienssen and colleagues to be due to DNA methylation and histone modifications that regulate heterochromatin formation and are imposed by enzymes directed by RNA[56–60] (Chapters 12, 14 and 16).

McClintock intuited that her controlling elements did not encode regulatory proteins but rather functioned as regulatory modules. However, when transposons carrying protein-coding antibiotic resistance genes were discovered in bacteria by James Shapiro in 1969,[47] her theory that TEs are involved in regulating gene expression during development was much ignored, to her ongoing frustration.[51] The fact that the genetic phenomena involved were complex and not easily understood, as opposed to the simple *lac* operon/transcription factor operator-repressor model, did not help.[32]

Nonetheless, she was the pioneer of the concept of the dynamic genome, including in response to stress,[61] and her ideas were well ahead of her contemporaries. We now know that transposon-derived sequences are major sites of epigenetic modification, regulate many aspects of gene expression during plant and animal development, and are mobilized in evolutionary time for phenotypic diversification and in real time in the brain (Chapters 10, 16 and 17).

PARAMUTATION, IMPRINTING AND TRANSINDUCTION

In 1915, Bateson and Caroline Pellew reported unusual "rogue" non-Mendelian patterns of inheritance in peas,[62,63] which were occasionally observed thereafter in other species, but were not studied in a systematic

way until Brink coined the term 'paramutation' in the 1950s to describe the atypical inheritance of traits displayed by some genes in maize, tomato and other species.[64–68] While initially thought to be restricted to plants, later studies showed that paramutation and related phenomena also occur in animals, including mammals, likely to a much greater extent than has been appreciated, and that it involves the intergenerational transmission of epigenetic information (Chapter 17).

Waddington, who pioneered the concept that epigenetic mechanisms control the trajectories of development,[69] reported, also in the 1950s, that exposure of *Drosophila* eggs to ether for a few generations resulted in the inheritance of a bithorax-like phenotype (see below) in subsequent (untreated) generations.[70] He also showed that another phenotype, 'crossveinless' wings, induced by heat shock of pupae, showed similar inheritance in later generations.[71] Waddington concluded that:

> All phenotypes are modified, to a greater or lesser extent, by the environment. All genotypes under natural conditions will be subject to selection pressures relating to the manner in which their development is modified by the environment. The phenotypic effect of any new gene mutation must therefore be to some extent influenced by the kind of developmental flexibility which has been built into the rest of the genotype by selection for its response to the environment.[70]

Such 'genetic assimilation' has profound implications (Chapter 18), but was waved away by most theoretical and molecular biologists of the time.[72] It is difficult, in any case, to disentangle the interplay of genetic variation with epigenetic inheritance,[73,74] and recent evidence suggests that, at least in some instances, the cause is stress-induced mutations.[75,76]

In 1974, D. R. Johnson reported that an allele of the *brachyury* gene (which encodes a transcription factor required for mammalian gastrulation and early organogenesis[77]) has a lethal phenotype if inherited from the mother, but not if inherited from the father.[78] A similar phenomenon was described at a translocated locus affecting the relative number of male and female offspring by Mary Lyon in 1977.[79] This 'parental imprinting' was characterized independently by the groups of Azim Surani and Davor Solter in 1984 as a requirement for both female and male genomes for development, due to 'marked' loci ('epialleles') with differential patterns of expression depending on parent of origin.[80–83]

Parental imprinting only occurs in mammals and is thought to be an adaptation to placental biology driven by sexual antagonism, maternal-offspring coadaptation and/or kinship interest.[84–87] A high proportion of the imprinted genes have arisen through retrotransposition[88–90] (although they are depleted of SINE elements[91]) and imprinting appears to have evolved with the invasion of particular classes of repeats at the marsupial-eutherian interface.[86,92] Interestingly, non-coding RNAs are at the heart of the regulation of genomic imprinting and X-chromosome inactivation in mammals (Chapters 9, 13 and 16).

Dysregulated imprinting has since been associated with disorders such as Angelman and Prader-Willi Syndromes in humans,[93–95] as well as with the callipyge ('beautiful buttocks') 'polar overdominance' phenotype in sheep involving cis- and trans-epiallelic interactions[96–98] and non-coding RNAs.[99] Similar epigenetic mechanisms that regulate transposon activity for genome defense and gene regulation were found to underlie the phenomena (discovered in the 1980s and 1990s) of 'repeat-induced' and 'homology-dependent' gene silencing[e] in fungi, plants and animals[100–105] (Chapter 12).

Another odd phenomenon, termed 'transinduction', was described in 1997 by Alison Ashe, Nick Proudfoot and colleagues, who found that transient cell transfections with a beta-globin gene construct induced the expression of non-coding RNAs from the regulatory 'locus control region' and intergenic regions of the globin locus (but not the endogenous beta-globin mRNA), which required transcription of the transgene but not its translation.[106] That is, ectopic expression of an mRNA has regional effects on the gene expression profile of its locus in the absence of the encoded protein, which indicates that transcription itself and/or mRNAs also convey regulatory signals.

THE BITHORAX 'COMPLEX LOCUS'

Clues pointing to epigenetic control of development and the involvement of regulatory RNAs were also emerging in the 1940s and 1950s from Ed Lewis' studies on the genetics of Drosophila body segment specification, based on mutant and recombination

phenotypes, particularly in 'homeotic' genes, which produce transformations of segment identity.[107–109]

'Homeosis' was initially described and the term coined in 1894 by Bateson to describe phenotypic variations in which "something is changed into the likeness of something else".[110,111] The first homeotic gene was identified in 1915 by Calvin Bridges,[107] who identified a mutation ('bithorax') that converted the third thoracic segment into the second, producing an additional pair of wings,[f] and mapped it to a genomic region later named by Lewis the 'bithorax complex' (Figure 5.4).[113–119]

Mutations in the bithorax complex showed spectacular perversions of development. The bithorax phenotype is caused by loss-of-function mutations in the protein-coding gene ultrabithorax (Ubx). Other mutations convert antennae into legs (Antennapedia, Antp) or genitalia into legs or antennae (Abdominal-A, abd-A, and Abdominal-B, Abd-B), all also caused by loss-of-function of 'homeotic' proteins.[113–119] It was subsequently realized, after the gene cloning revolution (Chapter 6), that duplicated paralogous clusters of homeotic genes, with the same spatial order of paralogs, occur in vertebrates.[121]

Lewis found that the spatial expression patterns of the homeotic genes and their phenotypes are modulated by mutations in nearby 'cis'-regulatory loci called anterobithorax (abx), bithorax (bx), bithoraxoid (bxd), contrabithorax (Cbx) and postbithorax (pbx), as well as infra-abdominal (iab1–4, modulating abd-A; and iab5–8, modulating Abd-B), which he termed 'pseudoalleles'.[122] Some of these loci turned out to reside in the introns of the protein-coding genes, but most could be easily separated by recombination, indicating that they are discrete, separate genetic elements.[123]

These genes also exhibited intriguing position effects, trans-heterozygotes being more abnormal than cis-heterozygotes, indicating a local interaction, with trans-heterozygotes having defective regulation on one chromosome and a defective protein on the other, whereas cis-heterozygotes had a normal pattern of expression of a functional protein from one chromosome.[122,124]

---

[e] These phenomena have also been referred to as 'quelling' and 'co-suppression'.

[f] The evolutionary significance of this transformation was not lost on Lewis, who noted the similarity of bithorax mutations to (and therefore the ease of generating) the additional set of wings on a dragonfly.[112]

FIGURE 5.4   (a) Wild-type and (b) a *bithorax* mutant *Drosophila melanogaster.* (Reproduced from Akbari et al.[120] with permission from Elsevier.)

TRANSVECTION

In 1954, Lewis discovered that a structural rearrangement that moved *Ubx* to a different chromosomal location resulted in a mutant phenotype in flies heterozygous but not those homozygous for the rearrangement, which he attributed to disruption of the pairing of the two alleles.[125]

Based on the '*cis-trans* position effect', Lewis proposed that pseudoallelic regions control the hierarchical expression relations (or 'polarity') observed in the cluster by either "cis-vection", in which the substances generated by the pseudoalleles acted on adjacent genes, or by "trans-vection" based on the proximity (somatic pairing) of homologous sequences to explain the influence of alleles in one chromosome over the alleles in another.[122,124]

Transvection, or 'allelic cross-talk', has been observed at many other loci in *Drosophila* and, among other things, regulates the sexually dimorphic expression of X-linked genes.[126–128] The phenomenon appears to be proximity-dependent, but not always.[129,130] Some interactions can occur at a distance and may not be strictly dependent on homolog

pairing as translocations and mutations in the gene *zeste*, which normally disrupt transvection, have limited or no effects on pairing.[131]

To explain transvection, Lewis initially invoked the operon model, postulating that pseudoallelic loci contained genes responsible for the synthesis of substances that are coordinately regulated by an operator element[108,132] and that a locally produced diffusible substance would activate a cascade of reaction steps involving the other substances (the "sequential reaction model").[122,124] However, if these substances were proteins their production had to occur close to their sites of action because of the proximity effect, i.e., in the nucleus, whereas translation occurred in the cytoplasm.

Consequently, like transposition, it was problematic to explain the 'pairing-dependent' transvection phenomenon in conventional terms, unless "homologous chromosomes cooperate with one another in the transcription process".[108] Lewis then floated the idea that substances "used only briefly in development" are synthesized "at the chromosomal level".[108]

While the obvious candidate was RNA, Lewis suggested that the substances involved might be

RNAs, polypeptides, or products of the enzymatic activity of such polypeptides,[132,133] and was careful not to make assumptions,[g] famously saying that "The laws of genetics had never depended upon knowing what the genes were chemically and would hold true even if they were made of green cheese", perhaps a frustrated reaction and pointed riposte to colleagues who told him that he "was simply dealing with mis-sense and nonsense mutants within a protein and that all we were doing was mapping sites within a single protein coding unit!".[132,134]

Decades later, in the 1980s, David Hogness (who cloned the first eukaryotic transposon), Mike Akam and colleagues found that the *bithorax* complex produces multiple overlapping and intergenic RNAs, many of which do not encode proteins,[135–138] including those from the (presumed) *cis*-regulatory sequences encompassing Polycomb and Trithorax 'response elements'[126,139] (see below). For example, the *bxd* locus produces a 27 kb transcript whose expression is highly regulated during embryogenesis, in a pattern that is partially reflexive of the expression pattern of *Ubx*.[135–138] The regulatory elements of *abd-A* and *Abd-B* are also transcribed.[140–142]

With this evidence in mind, others also suggested that transvection involves RNA transactions,[143–145] concluding in one case that "the genetic analysis of these transvection effects suggests that the transcription of the CbxIRM and Cbx2 alleles depends on RNAs of short radius of action from the homologous Ubx gene".[143] Subsequent studies showed that trans-vection occurs by sequences on one homolog direct-ing the expression of the cognate transcription unit on the other,[146] and that these sequences are 'enhanc-ers',[147–150] genes that control the spatial expression patterns of nearby and distal protein-coding genes during development and produce non-coding RNAs in the cells in which they are active (Chapters 14 and 16).

Like other genes and genetic phenomena affect-ing development that were first discovered in maize or *Drosophila* and thought to be idiosyncratic, trans-vection was subsequently found to occur in fungi, plants and mammals, and to be a general feature of multicellular eukaryotes, although still poorly understood.[151,152]

EPIGENETIC MODIFIERS

The other major finding of this period, whose importance was also not broadly appreciated until much later, was the identification of genes that had a global effect on the expression of homeotic genes, termed Polycomb Group (PcG) and Trithorax group (TrxG). The first PcG gene, *Polycomb* (*Pc*) was discovered in 1947 by Pamela Lewis,[153] so named because the most common effect of PcG mutants is the appearance of extra pairs of sex combs on the legs of the second and third thoracic segment in males, a feature normally found only on the legs of the first thoracic segment. These mutants are lethal when homozygous but cause mild homeotic trans-formations when one copy is incapacitated, a semi-dominant 'haplo-insufficient' phenotype.

Related genes with similar effects, such as *Extra sex combs*, *Sex combs on midleg* and *Additional sex combs*, were subsequently also identified.[154–156] In 1978, Lewis observed that *Pc* "seems to code for a repressor of the complex", as it repressed expres-sion of *Ubx* in anterior segments, causing them to be transformed into more posterior ones.[113]

Mutants that had the opposite effect were also identified, notably by Phil Ingham in the 1980s. These mutants, including *Trithorax* and others such as *Enhancer of zeste*, cause embryonic segments to be transformed into anterior ones by antagoniz-ing PcG proteins.[123,157–161] Many *TrxG* genes were subsequently identified through genetic screens for mutations that suppress the phenotype of *PcG* genes.[162–164]

The observation that PcG and TrxG factors are required to maintain homeotic gene expression gave rise to the hypothesis that PcG (repressive) and TrxG (activating) proteins act as a "cellular memory system".[165,166] Orthologs of *PcG* and *TrxG* genes were later found to be ubiquitous in plants and animals[167] and to encode histone-modifying proteins and ATP-dependent chromatin-remodel-ing complexes, or repressors thereof, for the epigen-etic control of gene expression during development (Chapter 14).

Indeed, if all this sounds complicated, it is, and was too much for the operon crowd to digest. Max Delbruck, the physicist turned bacteriophage

---

[g] Lewis later capitulated, but not completely, to the conventional view of gene regulation, stating in 1994 that: "The remain-ing regions are thought to include enhancer-like sequences to which regulatory proteins bind, thereby conferring spatial- and temporal specific production of the proteins encoded by the complexes."[111]

geneticist, wrote in a 1963 letter to a former member of Lewis' laboratory: "I then plunged into the bithorax saga for which Lewis very kindly sent me his latest manuscript ... I must say I am puzzled ... [and] ... strongly suspect that there is something wrong here in the analysis."[132]

THE BRITTEN AND DAVIDSON MODEL

In 1969, Britten and Eric Davidson[h] published a paper entitled 'Gene regulation for higher cells: a theory',[15] which attempted to integrate the findings of the preceding decades, including the prevalence of repetitive sequences in higher organisms, Davidson's observations on 'informational RNAs' in amphibian and sea urchin embryos,[170] and the data on embryo development that Davidson had assembled the year before in his book 'Gene activity in early development'.[171] This paper devoted special attention to the enormous size of the genomes of higher organisms, the diversity of transcripts in the nucleus and the abundance of repetitive sequences that are transcribed in a cell-specific fashion. It was the first serious consideration of gene networks ("gene batteries") and regulatory circuits in the evolution and development of higher eukaryotes.[172]

Britten and Davidson's theory was partly but not entirely influenced by Jacob and Monod's principles (with elements analogous to operators, regulator and structural genes), but was distinct in that it encompassed the difference in genomic organization and transcriptional output between eukaryotes and prokaryotes, and emphasized the importance of genomic structure in the cascade of gene expression during development. An important proposition was that "the (normal) state of the higher cell genome is histone-mediated repression and that regulation is accomplished by specific activation".[15]

Britten and Davidson proposed that genes in eukaryotes were differentiated into functional classes: structural (or "producer") genes (which encode proteins) and "integrator genes" (which encode regulatory

molecules), the latter of which was influenced by McClintock's model of gene regulation, developed before it was known that transposable elements were the main sources of repeated sequences.

As in Jacob and Monod's original model, Britten and Davidson posited that regulatory genes produce RNAs. They extended this notion to suggest that such RNAs would connect regulatory networks to activate the gene batteries that specified the phenotypes of different cell types. Focusing on regulation at the level of transcription, the proposed advantage of RNAs as regulatory molecules was again its base pairing ability, which allowed genomic regulation by recognition of specific "receptor sequences".[15]

A major factor in their proposal was the literature describing chromatin-associated RNAs and the huge heterogeneous nuclear RNAs that contain repetitive sequences. According to the model, RNAs bind in a sequence-specific manner to DNA to provide the basis of specific gene regulation, explaining the complexity of RNAs in the nucleus compared to the cytoplasm. It would also explain the different spectrum of RNAs transcribed from repeated sequences during early development and cell differentiation.[170,173]

Britten and Davidson thought that, while it was possible either RNA or proteins were the diffusible regulatory molecules, RNA represented the "simpler alternative".[15] In addition, according to them, the potential of formation of new gene batteries (acting in a specific cellular program) appeared to differ greatly between the two alternatives, because RNA specificity would only depend on target sequence complementarity, with implications for the modularity and co-evolution of these sequences (Chapter 16).

They were also motivated by early measurements of the cellular DNA content in different species, which showed that the amount of DNA per cell was greater in "higher" than in "primitive" forms of invertebrates.[174] Britten and Davidson plotted the amount of DNA of a vast range of organisms (from viruses and bacteria to mammals) and were able to show that the *minimum* genome size greatly increased concomitant with increased complexity of their organization. They then reasoned that most of the known biosynthetic pathways were already present in unicellular organisms, and that a significant increase in the number of 'structural genes' in higher organisms was unlikely.[15] Therefore, they posited that the difference between sponges and mammals would lie in the increased complexity of regulation, which corresponded to

---

[h] Davidson was a colorful character. He played American football at a senior level, rode a Harley-Davidson, and was lead singer for an Appalachian folk music ensemble and played banjo in the Iron Mountain String Band.[168,169] He had worked with Mirsky and he and Roy Britten were part of the Caltech group that included Delbrück and Leroy Hood (who later invented the first automated DNA sequencer, Chapter 10), one of a number of overlapping schools of thought that influenced research and concepts throughout the early period of molecular biology. McClintock was an outsider.

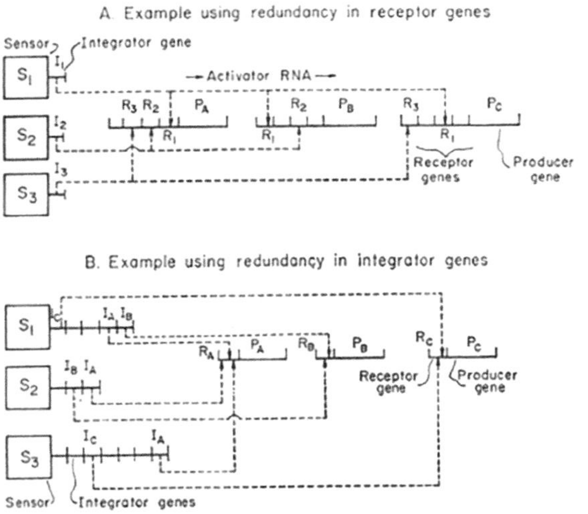

Fig. 1. Types of integrative system within the model. (A) Integrative system depending on redundancy among the regulator genes. (B) Integrative system depending on redundancy among the integrator genes. These diagrams schematize the events that occur after the three sensor genes have initiated transcription of their integrator genes. Activator RNA's diffuse (symbolized by dotted line) from their sites of synthesis—the integrator genes—to receptor genes. The formation of a complex between them leads to active transcription of the producer genes P_A, P_B, and P_C.

Gene Regulation for Higher Cells: A Theory
Roy J. Britten and Eric H. Davidson
Science 165, 349-357 (1969)

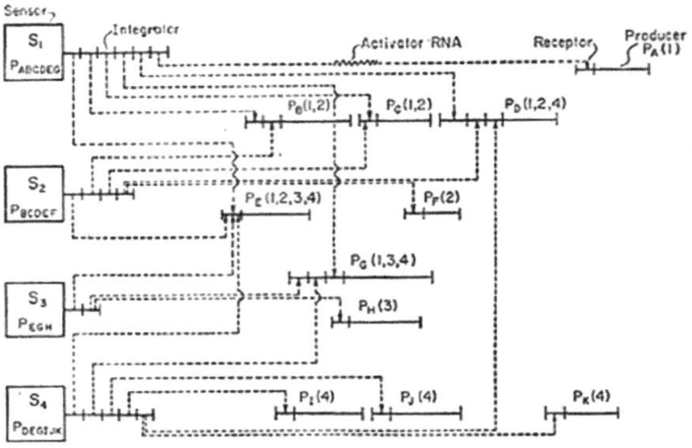

Fig. 2. This diagram is intended to suggest the existence of overlapping batteries of genes and to show how, according to the model, control of their transcription might occur. The dotted lines symbolize the diffusion of activator RNA from its sites of synthesis, the integrator genes, to the receptor genes. The numbers in parentheses show which sensor genes control the transcription of the producer genes. At each sensor the battery of producer genes activated by that sensor is listed. In reality many batteries will be much larger than those shown and some genes will be part of hundreds of batteries.

FIGURE 5.5 The 1969 Britten and Davidson gene regulatory network model. (Reproduced from Britten and Davidson[15] with permission from the American Association for the Advancement of Science.)

the "integrator genes" and "receptor sequences". The expansion of non-protein-coding sequences, but not coding sequences, in animal genomes was later confirmed by genome sequence data (Chapter 10).[175–177]

Finally, Britten and Davidson proposed that the repetitive sequences played a role in coordinating the expression of gene networks during development and differentiation. In fact, Britten and David Kohne in their original paper published in the previous year, perhaps aware of the growing consensus that repetitive and other non-protein-coding sequences in the huge genomes of higher organisms is junk (Chapter 7), stated: "A concept that is repugnant to us is that about half of the DNA of higher organisms is trivial or permanently inert (on an evolutionary time scale). Furthermore, at least some of the members of DNA families find expression as RNA. We therefore believe that the organization of DNA into families of related sequences will ultimately be found important to the phenotype."[17] In their model, the repeat elements comprise "receptor genes" adjacent to "producer genes", functioning as target sequences for regulatory RNAs (Figure 5.5).

These ideas were extended in subsequent publications during the 1970s, including proposed molecular mechanisms for gene-by-gene rewiring in development and evolution.[178–182] They were also supported by Nina Fedoroff's observation that "large RNAs in the mammalian cell nucleus contain elements complementary to ones in other large nuclear RNAs",[183] although it was later noted by Pederson that "this intriguing finding was never pursued by this or any other group, at least as the extant literature speaks".[184]

Importantly, Britten and Davidson developed the idea that multicellular development was dependent on an extensive, interwoven regulatory architecture embedded in the structural organization of the genome, and that developmental 'novelty' – phenotypic diversification – is achieved mainly by variation in the regulatory architecture,[172,185–187] not the repertoire of encoded proteins (although there are lineage-specific expansions of protein families, such as homeotic genes in vertebrates and olfactory receptors in mice, and some novel proteins that appear from time to time in evolution). The concept that phenotypic diversity in plants and animals is largely achieved through mutations altering regulatory circuits[i] was extended by Mary-Claire King and Allan Wilson in their 1975 paper 'Evolution

[i] A feature especially amenable to transposon insertion and associated epigenetic control.

at two levels in humans and chimpanzees'[188] and Jacob's 1977 paper 'Evolution and tinkering',[189] and is now well accepted in the field referred to as 'Evo-Devo',[190,191] even extending to the evolution of enzyme activity.[192]

Waddington commented that the Britten-Davidson model of the mechanisms controlling embryogenesis was "the first … to make sense".[193]

BOOLEAN MODELS OF COMBINATORIAL CONTROL

In parallel with Britten and Davidson, a Boolean network function for the regulation of gene activity was advanced by Kauffman.[14,194–196] Kauffman invoked bacterial operon promoter-repressor systems as the model[195] and suggested that combinations of binary ('on-off') DNA-protein interactions could produce stable gene activity patterns from small numbers of variables, later embraced by Davidson.[185,197–199] Such approaches (which did not consider but does not prohibit regulatory RNAs) have been used to correctly predict some gene expression profiles;[198,200,201] they also gave comfort to the generally accepted idea that combinatorics of protein interactions are sufficient to account for ontogeny, although it has otherwise had little impact on the consciousness of the molecular biologists traditionally concerned with nuts and bolts.

In any case, by the beginning of the 1970s, two major problems identified in the 1960s were becoming generally recognized: for what purpose is so much of the DNA in a cell transcribed, and why does only a proportion of nuclear hnRNA contain the poly(A) sequence typical of mRNAs? A commentary in the journal *Nature* at the time remarked that "all the attention has focused on proving the existence of the messenger proportion of the HnRNAs", and then predicted that "In the future it may be the other sequences, among which controlling elements might be found, that will command more interest".[202]

PROCESSED RNAs AS GLOBAL REGULATORS

Explicitly influenced by the ideas of Britten and Davidson, as well as of Vogel,[12] Crick (later a proponent of selfish and junk DNA, Chapter 7) ventured in 1971 that most DNA of 'higher organisms' does not encode proteins and proposed a general model for the chromosomes of higher organisms.[203] Crick's model was based primarily on cytogenetic studies of *Drosophila* polytene chromosomes, in which dense bands contained high concentrations of DNA and histones separated by less dense 'interband' areas (see Chapter 14 for modern views on chromatin and genome organization). He suggested that the protein-coding sequences were found in the small fraction of fibrous DNA characterizing the interbands. On the other hand, the dense chromosome bands corresponded to "globular structures", which comprised most of the genome and which, he proposed, would contain "unpaired DNA" available for gene control, the 'Unpairing Postulate'. Crick speculated that "this postulate has even more force if single-stranded RNA is also used to recognize the control sequences on DNA elements", and that repetitive sequences may be the specific interaction sites regulated by interactions with histones.[203]

Based on the properties of RNA, Gerald Kolodny proposed in the early 1970s that regulatory RNAs originating from the breakdown of hnRNAs constituted major drivers of cell differentiation during development. Kolodny hypothesized that the derived "short activator RNAs" could base pair with unique single-stranded DNA sequences in control regions of the target genes (such as in areas of the chromatin where the DNA is exposed and able to be experimentally cleaved by DNases[j]) and promote regulation at the transcriptional level.[204,205] According to this model, hnRNA processing could occur either in the nucleus or in the cytoplasm, as there was evidence from Lester Goldstein and colleagues that short RNAs are shuttled back to the nucleus, including RNAs that then would become associated with chromatin;[206–209] the short RNAs would act as "primers" for transcription of one or several mRNA species containing matching sequences.[204]

In addition, Kolodny and colleagues showed that RNAs are secreted by mammalian cells, confirming Pierre Mandel and Pierre Métais' 1948 report of RNAs circulating in plasma and transmitted between cells.[210–212] In Kolodny's model, activator RNAs are derived from stored maternal RNAs (thus having a role in inheritance) and

---

[j] 'DNase hypersensitivity sites' occur at different chromosomal positions in different cell types and were later exploited to map the positions of transcription initiation and transcription factor binding sites in genomes (Chapters 11 and 14).

initiate an unfolding developmental program during embryogenesis.[204,205,211,212]

Burke Judd and Michael Young suggested in 1974 that individual chromosomal subunits (or 'chromomeres') corresponded to cistrons and were modules of gene regulation in eukaryotes.[213] This idea was based on evidence that large proportions of chromomeres are transcribed into very large hnRNAs[214] and likely processed into one or several mRNAs. Judd and Young hypothesized that these segments contained much more information than protein-coding sequences, and yet acted as a single operational unit. This theory of 'one cistron − one chromomere' considered that the large proportion of DNA that was transcribed, but apparently not translated, might be "coding for regulatory functions", conveying information for regulating transcription and post-transcriptional maturation and translation of that unit. Finally, they proposed that some of the "extra RNA" released during the processing of the large transcripts could activate other cistrons of a (related) biosynthetic or developmental pathway, in which mutations might manifest in a pleiotropic (i.e., multilateral) fashion.[213]

In 1975, Stuart Heywood and colleagues proposed mRNA regulation by "translational-control RNA", short RNAs hypothesized to be generated by processing of hnRNAs in the nucleus,[215,216] conceptually presaging the biogenesis and action of microRNAs discovered 30 years later (Chapter 12).

In 1976, George Brawerman developed an RNA-based model for the control of transcription during multicellular development "distinct from those operating in bacteria", mediated by "primer RNAs" that could bind by complementary base pairing to target sites in the genome, including repetitive sequences, and be abnormally expressed in cancer. He hedged his bets however, noting that recent studies had also indicated "that unique proteins associated with the DNA in chromatin appear to be directly responsible for the specificity of transcription ... [which does] not seem to leave any room for a specific role of primer RNA".[217]

In the same year, Elizabeth Dickson and Hugh Robertson published an article entitled 'Potential regulatory roles for RNA in cellular development'.[218,219] They proposed that RNAs may represent "informed signals" that regulate gene expression by interacting either "directly or indirectly" with DNA.[218] They

used examples of emerging cellular phenomena involving RNAs (such as the existence of 'non-coding' infectious RNA molecules, called 'viroids', in plants; Chapter 8) to highlight the biological capacities of RNAs, and canvassed the features that made RNAs "prime candidates" as regulatory molecules: rapid turnover by nuclease degradation; the ability to fold in complex ways, with globular regions "reminiscent of protein structure" that are ideal for interaction with proteins; base pairing specificity for interaction with nucleic acids; and, finally, high (informational) "coding" capacity.

They highlighted that an RNA sequence of just 17 bases is sufficient to specify a unique region in the human genome, compared to the information required to produce a regulatory protein such as the *lac* repressor (~1,000 nucleotides). Thus, they suggested, besides the protein regulatory systems, "the use of RNA as an additional control element would add flexibility, efficiency, and elegance to a logical system of gene control".[218]

Dickson and Robertson speculated about the possible sources of regulatory RNAs and their mechanisms of action, including the regulation of transcription, translation, target degradation and even DNA replication. In short, like previous suggestions, they indicated that hnRNAs could be processed in the nucleus in a cell-type specific manner into mRNAs and "extra RNAs", generating a "stable RNA molecule from a previously unstable RNA (precursor) species" that could act, for example, as trans-acting transcriptional primers that coordinately promoted expression of specific genes. Finally, they also proposed that RNAs could be utilized as signals to transfer external information to the genome during cellular development, resulting in a change in the state of differentiation.[218]

This model not only explored the potential of RNA regulators, but integrated it with the broader spectrum of regulatory options in eukaryotic cells, involving both proteins and RNAs. It was remarkably prescient.

OUT ON A LIMB

Receptive reactions, however, were not common, and most contemporaries preferred the perceived simplicity of a protein-based regulatory schema, believing that RNA regulation was unnecessary, and that such models were flights of fancy.

As Ellen Rothenberg, who worked with Davidson, wrote in her obituary of him:[220]

> Now, by the 1970s, most regulatory biologists in my own molecular biology orbit (at Harvard, MIT, University of California San Francisco, and the Salk Institute) had been massively influenced by Jacob and Monod's work, by models of bacterial operon regulation, and by the precedents for elegant λ phage regulation of lytic vs. lysogenic growth by a mini network of mutually antagonistic activator/repressor proteins.[221-226] How could these be skimmed over so lightly in a book about differential gene regulation as the foundation for development?
>
> It was not just this particular work of Eric's that failed to draw upon Jacob and Monod. Interestingly, one of the most controversial predictions in the 1969 Britten and Davidson paper was that regulatory RNAs rather than regulatory proteins might be responsible for complex gene regulation.[15] Yet this was presented without regard for the clear evidence already in hand at the time that gene regulatory molecules were proteins in these bacterial systems. Why? Asked about this many years later, Eric often explained that for him in the 1960's, the evident differences between bacterial gene regulation and complex eukaryotic gene regulation in development completely dwarfed the similarities. Hybridization kinetic analyses of bacterial and multicellular eukaryotic genomes had already showed these to have vastly different kinds of sequence organization, with a severe paucity of repeat sequences in the bacterial genomes compared to the multicellular eukaryotes. If these were regulatory sites, then bacteria were missing this kind of regulation.
>
> Also, Eric's view of development was that this irreversible, hierarchical process of increasing complexity that he was interested in was so different from the reversible, physiological nutrient responses of bacteria that there was no reason to posit the same kinds of molecular mechanisms. In this way, Eric and Roy were indeed charting their own course. But were they actually solving developmental mechanisms?[220]

It seems that many felt the same way, and the thoughtful models of Britten and Davidson, and Dickson and Robertson, and their ideas of regulatory RNA, although they attracted some attention at the time, were ultimately sidelined and overrun in the excitement of the emerging gene cloning revolution.

Brow-beaten by the orthodoxy,[k] Davidson spent the rest of his highly successful and influential career studying the role of transcription factors and gene regulatory networks in sea urchin development[172,185-187,197,198,220,228-233] (Chapter 15). Davidson did however report that "nontranslatable transcripts containing interspersed repetitive sequence elements constitute a major fraction of the poly(A) RNA stored in the cytoplasm of both the sea urchin egg and the amphibian oocyte",[234] one of the first explicit descriptions of long 'non-coding' RNAs beyond those in the ribosome (Chapter 9).

Roy Britten remained studying repetitive DNA mainly from an evolutionary perspective but not specifically as regulatory cassettes, although he did acknowledge their importance as sources of variation,[235-237] a theme that would be picked up later by others.

FURTHER READING

Britten R.J. and Davidson E.H. (1969) Gene regulation for higher cells: A theory. *Science* 165: 349–57.

Chomet P. and Martienssen R. (2017) Barbara McClintock's final years as nobelist and mentor: A memoir. *Cell* 170: 1049–54.

---

k As Davidson recalled later: "I read it [Jacob and Monod's 1961 lac operon paper], but it didn't penetrate ... As I recall, everything they did was about repression ... We knew that there were all kinds of RNAs that bacteria didn't make. And nuclear RNA had been discovered, but nobody knew what it did. So the relevance of the Jacob & Monod ideas was definitely not obvious to me... We built [our] model on RNA reading the sequences because it was simpler to deal with complementarity than with what was completely unknown, namely how proteins can read DNA sequences. ... It was known that some proteins bind to DNA, and we knew about viral proteins that read DNA sequences. In the '71 application of that model to evolution we said it could be either RNA or protein, but the logic is going to be the same. And the logic is the same, but the first model was basically built on RNA recognition. Well, now it turns out that there are a number of regulatory functions that do use complementary sequence recognition by regulatory RNAs. But the main heavy lifting of regulation is done by protein–DNA interaction, of the kind that had already been found in bacteria. I suppose if I had paid more attention to Jacob and Monod we probably would not have built the model using regulatory RNA. We discussed whether we were going to talk about protein or RNA, and RNA was neater and easier to deal with because of natural complementarity. But that was wrong, from the standpoint of how it works."[227]

Comfort N.C. (2003) *The Tangled Field: Barbara McClintock's Search for the Patterns of Genetic Control* (Harvard University Press, Cambridge, MA).

Crow J.F. and Bender W. (2004) Edward B. Lewis, 1918–2004. *Genetics* 168: 1773.

Davidson E.H. (2006) *The Regulatory Genome: Gene Regulatory Networks In Development And Evolution* (Academic Press, Cambridge, MA).

Deichmann U. (2016) Interview with Eric Davidson. *Developmental Biology* 412: S20–S29.

Kauffman S. (1971) *Current Topics in Developmental Biology*, Vol. 6 (A. A. Moscona & Alberto Monroy) 5, pp. 145–82 (Academic Press, Cambridge, MA).

Lewis E.B. (2007) *Genes, Development and Cancer, The Life and Work of Edward B. Lewis* (Springer, Berlin).

6 The Age of Aquarius

While the nature of gene regulation in higher organisms was being mooted, hamstrung by lack of molecular data, elsewhere a technological revolution was taking place, which would provide the toolkit to determine the repertoire of genes, their structures and products, and ultimately the composition of entire genomes.

The pace of molecular biology research and discovery was accelerating because of the expansion of the university and research sectors after World War II, especially in the 1970s when a new 'baby boomer' generation entered the workforce, eager to embrace new ways. The focus on bacterial molecular biology dissipated and there was a "mass migration in biomedicine" into "higher organisms" with the advent of gene cloning and the use of animal viruses to understand cancer.[1]

RECOMBINANT DNA AND 'GENE CLONING'

In 1972, Paul Berg, Stanley Cohen and Herbert Boyer[a] ushered in the biotechnology era by demonstrating that DNA could be cut and joined *in vitro* to make a 'recombinant' DNA molecule that could be propagated in a host cell to generate large numbers of copies by clonal amplification.

The roots of this advance lay in bacterial genetics, specifically in another strange phenomenon called 'host restriction-modification', discovered in the early 1950s by Salvador Luria, Mary Human, Giuseppe Bertani and Jean Weigle, whereby a bacteriophage infects a different strain of a bacterium (initially of *E. coli* and the closely related *Shigella dysenteriae*) orders of magnitude less efficiently than it infects the host strain from which it was derived, and reciprocally in reverse.[2–4]

A decade later, Werner Arber and Daisy Dussoix, and Mathew Meselson and colleagues, showed that this odd behavior was a manifestation of a bacterial defense system comprised of an enzyme (a 'restriction endonuclease') that cleaves foreign DNA at or near a specific nucleotide sequence, and a complementary enzyme that insulates the same sequences in the host genome, typically by methylation of one of the nucleotides. Most copies of invading viral genomes are cut by the nuclease, but a few become protected by the modification and escape cleavage. These survivors are then able to replicate efficiently, be insulated and infect the same bacterial strain with high efficiency, but not the original strain or others with different host restriction-modification systems.[5–13]

Different types of restriction endonucleases were defined:[14] Type I enzymes cleave DNA at a random distance, up to one kilobase, away from a complex recognition site;[b] Type II enzymes, of which there are several subtypes, recognize and cleave palindromic recognition sites (such as GGATCC), usually 4–8 base pairs in length, to produce either blunt or overhanging (complementary or 'sticky') ends by staggered cleavage; Type III enzymes recognize two separate non-palindromic sequences that are inversely oriented, and cleave 20–30 base pairs from the recognition site; and Type IV enzymes recognize and cleave specific modified, usually methylated, DNA sequences.[c]

The nomenclature follows a convention of an abbreviation of the name of the species and the order in which the enzyme was isolated, for example, *E. coli* strain RY13 enzyme 1, is called *Eco*R1. Thousands of different restriction enzymes are now known, including artificial enzymes produced by engineering, many of which are available commercially, fostering a molecular biology service industry to supply enzymes, cloning vectors and other tools (see below).

The enzymes studied by Arber and Meselson were Type I, which had limited utility. In 1970, however, Hamilton Smith and colleagues isolated the first Type II restriction enzyme (from *Haemophilus influenzae*, *Hind*II), which enabled the reproducible cleavage of DNA molecules at specific sequences.[17,18] This

---

a Berg, Cohen and Boyer were all part of the south San Francisco UCSF/Stanford community, the epicenter of the early development of recombinant DNA technology.

b Type I restriction enzymes restrict the influx of foreign DNA via horizontal gene transfer while maintaining sequence-specific methylation of host DNA and have the ability to change sequence specificity by domain shuffling and rearrangements.[15,16]

c And more recently, Type V, whose cleavage sites are determined by guide RNAs (the CRISPR systems; Chapter 12).

DOI: 10.1201/9781003109242-6

was used by Kathleen Danna and Daniel Nathans to construct the first physical map of a genome, that of simian virus 40 (SV40), using size separation by gel electrophoresis of the resulting fragments, ushering in the era of 'restriction mapping'.[19]

It also brought about the era of recombinant DNA technology, as restriction fragments from different genomes could be mixed and joined together. This required another innovation – the purification and use of DNA 'ligases' capable of joining complementary or blunt DNA ends by the formation of a phosphodiester bond, isolated in 1967 by Bernard Weiss and Charles Richardson.[20] The first recombinant DNA molecule was produced by Paul Berg, Bob Symons and colleagues in 1972, who mixed *EcoR*I-cleaved SV40 DNA with a DNA segment containing lambda phage genes and the galactose operon of *E. coli*,[21] but fearing the dangers that might be created, declined to introduce these molecules into living cells.[22]

That was left to Cohen, who in 1973–1974, together with Annie Chang, Robert Helling, Boyer[d] and other colleagues, ligated *EcoR*I restriction fragments from *Staphylococcus aureus* and the frog *Xenopus laevis* with a plasmid that contained a replication origin, an antibiotic resistance gene (for selection) and was cleaved just once (i.e., linearized) by *EcoR*I. They reintroduced the recombined molecules into *E. coli* using a $CaCl_2$ ('transformation') procedure developed by Cohen.[24–26] These experiments showed that, to first approximation, genes could be successfully exchanged between species by human intervention.[e]

In 1977, Boyer's laboratory developed the first plasmid vector specifically designed for gene cloning, called pBR322, which was small, ~4 kb, and had two antibiotic resistance genes, one for selection of transformants and the other with unique restriction enzyme sites for DNA insertion to enable identification of recombinant plasmids (Figure 6.1).[29]

More sophisticated cloning vectors were developed, notably by Joachim Messing and colleagues from bacteriophage M13, containing multiple clustered sites (MCS) for restriction endonuclease cleavage, whereby double digestion prevents self-ligation and only allows re-circularization with a compatible insert.[31] Later versions contained an MCS within the *lacZ* gene, which allowed identification of colonies containing recombinant plasmids based on colorimetric detection of the encoded enzyme (beta-galactosidase) activity (blue colonies) or lack thereof (white, disrupted by an insert), and the ability to isolate single-stranded forms to aid DNA sequencing.[30,32] Cloning sites were also added into other genes that enabled direct (positive) selection of recombinant clones (e.g., [33]).

Many other variations and elaborations were then, and still are, being developed on the core requirements of a 'vector' (plasmid or virus) capable of being replicated in a desired host,[f] a selectable marker (usually an antibiotic resistance gene or metabolic enzyme to complement a deficiency in the host) to discriminate transformants from non-transformants,[g] a restriction ('cloning') site (or battery thereof) to insert foreign DNA, a means of favoring[h] or discriminating recombinant clones from those containing vector alone and a means of identifying the desired insert (the target gene to be cloned)[i] among the many others that may be produced from restriction endonuclease digestion of the input DNA.

Because the production of an encoded protein was an important scientific and commercial objective, many host-vector 'expression' systems were developed in the following decades to enable the high-level transcription, translation and purification of the encoded protein (see, e.g., [34,35]), often assembled in

---

[d] Boyer was well aware of the potential, having written a review on DNA restriction and modification systems the year before.[23]

[e] These advances led to the famous Asilomar Conference on Recombinant DNA technology in 1975, which "placed scientific research more into the public domain, and can be seen as applying a version of the "precautionary principle' via an initial voluntary moratorium and then strict controls on recombinant DNA construction and the release of genetically modified organisms into the environment", with "one felicitous outcome [being] the increased public interest in biomedical research and molecular genetics .. [and stimulation of] knowledgeable public discussion some of the social, political, and environmental issues that are and will be emerging from genetic medicine and the use of genetically modified plants in agriculture".[27,28] The participation of the public in the implications, applications, 'ethical' considerations and prescribed limits of genetic technologies was to be revived again with the later advent of techniques for precise engineering of animal and plant genomes (Chapter 12). Many genome research programs, notably those funded by Genome Canada, have required a proportion of the funding to be allocated to the social, ethical, economic, environmental and legal aspects of the work.

---

[f] That is, having an origin of replication that is recognized by the host cell.

[g] DNA transformation by $CaCl_2$ treatment of cells is inefficient, $~10^{-4}$ at best; more efficient methods were developed later.

[h] By using two different restriction enzymes so that the vector could not be re-joined without an insert.

[i] Cloning genes that would complement a deficiency in the host cell, usually bacteria or yeast, was relatively straightforward.

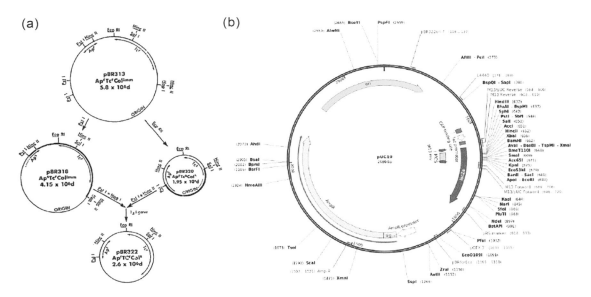

FIGURE 6.1 (a) The procedure used by Herb Boyer and colleagues to produce the plasmid cloning vector pPR322 containing unique restriction endonuclease sites with its two antibiotic resistance genes, such that recombinant plasmids may be identified by the insertional loss of one or the other. (Reproduced from Bolivar et al.[29] with permission of Elsevier.) (b) pBR322 was the precursor to more sophisticated cloning vectors such as pUC19[30] containing multiple clustered unique sites for restriction endonuclease cleavage and strategies to physically favor or genetically select for recombinant molecules. pUC19 image (https://www.addgene.org/50005/) generated by SnapGene software (snapgene.com). (Courtesy of Addgene.)

gene cloning protocol manuals that became ubiquitous in this period.[j]

ENABLING TECHNOLOGIES

The practical potential of the technology was realized immediately by Cohen and Boyer, who patented their method and started the first of the new generation of biotech companies, DNAX and Genentech, respectively, while Berg was a major proponent of strict regulation.

The initial targets were genes encoding medically important hormones, such as insulin, growth hormone and erythropoietin, among others. The difficulty was to identify the rare bacterial clone that contained the desired gene, a needle-in-a-haystack problem, especially when dealing with large genomes. This was approached via RNA, as it was reasoned that the tissues in which the proteins are

highly expressed would be an enriched source of the corresponding (mRNA) coding sequences – a fortuitous approach given the (at that time) unknown problem that the protein-coding sequences of most genes are not contiguous in complex organisms (Chapter 7). To do this, additional technologies had to be developed.

The first was complementary DNA ('cDNA') synthesis – conversion of mRNAs to a DNA equivalent. This was achieved using reverse transcriptase, discovered a couple of years earlier by David Baltimore[42] and Howard Temin,[43] with an oligo dT primer that would anneal to the polyA tail of mRNAs to initiate synthesis of complementary DNA strands, then conversion of the single-stranded copies into double-stranded DNA with DNA polymerase.

The second was the synthesis of specific DNA (and later RNA) sequences, based on phosphonate, phosphodiester, phosphite triester and phosphoramidite anhydrous chemical synthesis methods pioneered in the 1950s, 1960s and 1970s by Alexander Todd,[44–46] Har Gobind Khorana,[47] Robert Letsinger[48,49] and Colin Reese,[50,51] among

---

[j] Two of the most popular have been 'Molecular Cloning: A Laboratory Manual'[36] and 'Current Protocols in Molecular Biology',[37] both regularly updated in new editions or volumes (see, e.g., [38–41]).

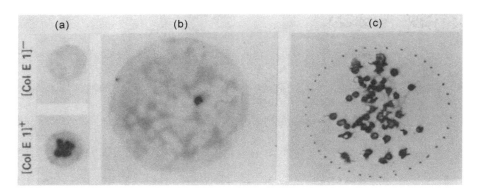

FIGURE 6.2   Autoradiographs of radiolabeled ColE1 plasmid RNA hybridized to lysed *E. coli* on nitrocellulose filters containing or lacking the plasmid at different ratios (B) 1:100 and (C) 1:1. (Reproduced from Grunstein and Hogness[70] with permission.)

others, and adapted to solid phase synthesis by Marvin Caruthers and colleagues in 1981.[52,53] These developments enabled the automation of oligonucleotide synthesis, as well as the incorporation of both natural and non-natural bases, novel linkages such as phosphorothioate or peptidyl bonds to improve biological stability or interaction strength, and other additions such as biotin for oligonucleotide capture (for reviews and recent developments see [51,54–57]), a vibrant domain of the biotechnology industry.

Synthetic designed oligonucleotides have become an indispensable part of the toolkit for molecular biology research and genetic engineering. Their uses encompass not only the detection of corresponding DNA or RNA sequences by hybridization, including highly parallel microarrays and bead arrays for target quantification and capture,[k] but also primers for DNA sequencing and amplification, introduction of restriction sites and mutations for genetic and protein engineering, construction of hybrid and other forms of artificial genes, mutagenic screening, production of antisense sequences to block gene expression, gene therapy,[58] large scale genome engineering[59] and many others. Much later enzymatic methods would be developed, using engineered terminal transferase,[60,61] which make possible the production of longer DNA sequences for synthetic biology and even the prospect of using DNA for data storage.[62]

The third technology was DNA, RNA and protein 'blotting', first developed by Ed Southern in 1975 using radioactively (and later biotin) labeled probes (usually cDNAs) to detect the location of corresponding genomic sequences in a restriction digest displayed by electrophoresis (the eponymous 'Southern blot'),[63] which played an important role in the discovery of 'genes-in-pieces' (Chapter 7). The RNA equivalent ('Northern blot') was developed by James Alwine, David Kemp and George Stark in 1977,[64] and the protein equivalent ('Western blot')[65] by Harry Towbin and colleagues in 1979 using labeled antibodies or other ligands to detect specific proteins in electrophoretic displays of cellular contents or fractions.[65–67] Subsequent variations were 'Southwestern blots'[68] and 'Northwestern blots'[69] to detect DNAs and RNAs bound by specific proteins, respectively.

These were early days, the technology was in its infancy and the cloning of specific genes was a major challenge, taking months and sometimes years. A typical strategy was to construct a cDNA 'library' from a tissue known to express the gene of interest, often involving size fractionation to enrich the desired mRNA (monitored by *in vitro* translation and Western blotting of the products), insertion of the cDNAs into a phage or plasmid vector and transformation into a bacterial host, usually *E. coli*, then screening for the desired clones among the tens of thousands of transformants by colony hybridization using radiolabeled oligonucleotide probes, developed by Michael Grunstein and Hogness in 1975[70] (Figure 6.2), commonly designed to be specific for a subsequence of the encoded protein with minimal codon redundancy. Those involved at the time can attest to the considerable celebrations that followed the successful cloning of a desired gene.

---

[k] Including sequence capture for 'exome' sequencing (Chapter 11) and targeted RNA sequencing (Chapter 13).

The revolutionary advance was that individual genes and genomic segments could now be isolated, amplified and characterized.

DNA SEQUENCING

The other technology was DNA sequencing, required to verify the identity and understand the details of the cloned gene. The first methods were developed in the late 1960s by George Brownlee, Fred Sanger and Bart Barrell, who used a paper fractionation method to sequence the 120nt 5S rRNA from *E. coli*,[71,72] and by Ray Wu and colleagues who used a primer extension approach (copying the sequence *in vitro* using DNA polymerase) to sequence the ends of phage lambda.[73–75] The first complete gene (encoding the MS2 RNA phage coat protein) and complete genome sequences (of phage MS2) were in fact RNA sequences, achieved by Walter Fiers and colleagues in 1972[76] and 1976,[77] respectively, using two-dimensional electrophoresis after partial nuclease digestion of the phage RNA.

In 1977, two new and more generalizable methods for DNA sequencing were published, made possible and widely applicable by the large amounts of cloned DNA. The first, by Gilbert and Allan Maxam, used terminal radiolabeling followed by base-specific partial chemical cleavage and size separation of the resulting set of fragments by (one-dimensional) electrophoresis, visualized by radiography.[78]

The second, developed by Sanger[i] and colleagues, extended Wu's primer extension method to produce the first sequence of a DNA genome (that of bacteriophage φX174) using chain terminating dideoxynucleotide analogs, terminal radiolabeling and size separation of the resulting set of fragments by electrophoresis.[79,80] Sanger sequencing (as it became known) quickly overtook the Maxam-Gilbert cleavage method, as it was easier to implement and more scalable (Figure 6.3).

Incremental technical improvements were made, which increased the length of the sequence reads. The next big leap forward was the introduction of fluorescently labeled primers by Leroy Hood

and colleagues in 1986[81] and chain terminators by James Prober and colleagues in 1987,[82] which led to the development of the first automated DNA sequencer by Lloyd Smith in the same year,[83] using a repertoire of labels that allowed all four base-specific chain termination events to be identified in single reaction and read by continuous electrophoresis past a photodetector, with the data directly analyzed by a linked computer (Figure 6.4). The later development of 'sequencing by synthesis' (SBS)[84] using reversible chain terminators in high density on solid phase surfaces, resulted in another step change in the volume of data and a reciprocal decrease in cost and enabled the industrialization and massive parallelization of DNA sequencing that led to the genome projects at the turn of the century and ultimately to the feasibility of personal genome sequencing for precision healthcare[85] (Chapters 10 and 11).

THE GOLD RUSHES

These new technologies led to a stampede in the late 1970s and following years to clone and sequence genes or cDNAs encoding proteins of interest from bacteria, archaea, fungi, plants and animals, the ease of which depended on the availability of suitable genetic complementation (for genes in microorganisms) or tissues in multicellular organisms where the gene was highly expressed. For the latter reason, the first vertebrate cDNAs to be isolated were those encoding hemoglobin,[87] immunoglobulins, the chicken egg protein ovalbumin, and highly expressed muscle and milk proteins. They were followed by many others as the technology developed and became adopted across a wider spectrum of biological and biomedical disciplines, not just biochemistry and genetics, but also botany, zoology, microbiology, developmental biology, physiology, pharmacology, cell biology, pathology, anthropology, evolutionary biology, cancer biology, etc.

Importantly, molecular biology connected plant and animal developmental and behavioral genetics with biochemistry. Many of the genes that affect phenotype in model organisms and others in other species were cloned and sequenced, leading to an explosion of discovery and characterization of whole new families of proteins involved in body plan specification, cell differentiation and cell biology. Many of these genes turned out to be similar from yeast and invertebrates to humans.

---

[i] Sanger was one of the few people to be awarded two science Nobel Prizes in the same category (Chemistry), for protein sequencing and DNA sequencing, the other being John Bardeen (Physics) for developing the transistor and superconductor theory. The only other dual winners were Marie Skłodowska-Curie (Physics and Chemistry) and Linus Pauling (Chemistry and Peace).

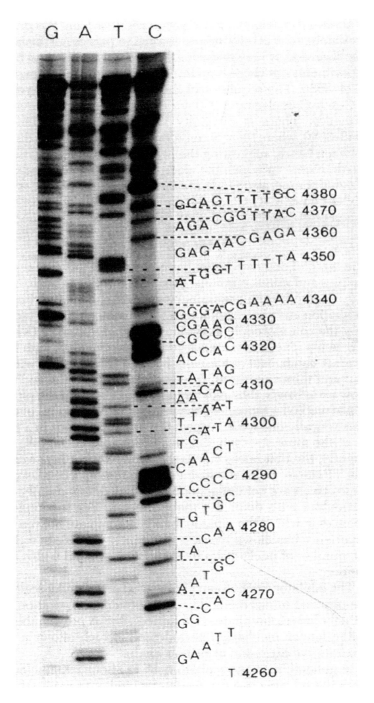

FIGURE 6.3   One of the autoradiographs presented by Sanger and colleagues in their 1977 paper on DNA sequencing,[80] showing electrophoretic size separation of X174 DNA sequences copied *in vitro* from specific primers using radiolabeled nucleotide triphosphates, with each of the four separate reactions containing specific A, G, C or T chain terminating nucleotide analogs (ddATP, ddGTP, ddCTP and ddTTP). The nucleotide sequence is read bottom to top (5'>3') from the ascending fragments in the different tracks. Later refinements optimized the reaction conditions, including the ratios of ddNTPs to dNTPs and the use of radiolabeled primers to yield even labeling. (Reproduced with author permission from Sanger et al.[80])

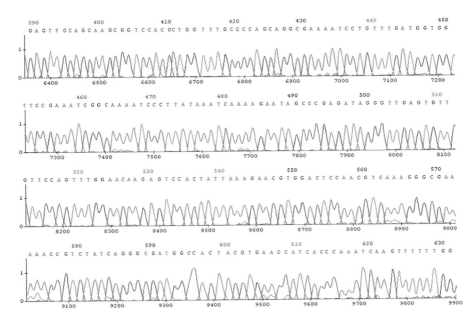

FIGURE 6.4 An example of the output of an automated fluorescent DNA sequencer. (Reproduced from Foret et al.[86] with permission of Elsevier.)

There was scientific gold to be unearthed everywhere. Investigators built their careers on the discovery of an important gene and study of its associated biology, all the better if the encoded protein may have medical significance.[m] This in turn reinforced the orthodox view of genetic information and fostered a generation of molecular biologists occupied with the "brutal reductionism" of identifying and characterizing genes encoding proteins.[92]

Nonetheless, there were wonderful discoveries in the decades that led up to the genome sequencing projects at the turn of the century. One could – as many have – fill a book on these alone: genes controlling cell division or enabling host colonization by bacteria; genes controlling flowering in plants; genes encoding molecular machines, all the way from ribosomal proteins to chloroplasts to flagella and muscle fibers; genes forming the cytoskeleton, and those encoding histones and histone modifiers, etc.

The avalanche of protein sequences (deduced from cloned genes) also led to the recognition of similar functional modules in different proteins, such as protease, phosphorylation, methylation, nucleotide binding and DNA binding domains, nuclear and mitochondrial localization signals, secretion signals, etc., information about which is now housed in databases such as Pfam.[93,94]

HOX GENES

The cloning in the 1980s by Walter Gehring and others of the genes in the Drosophila *bithorax* complex - studied by Lewis, Hogness[n] and others - identified the 'homeotic' proteins and the core 'homeobox' domain,[98–100] as well as many others identified by genetic screens, which enabled their expression patterns during development to be monitored.

---

[m] This is a source of unconscious bias by investigators and may explain in part why many biomedical studies have proven difficult to reproduce.[88–91]

[n] Anticipating the first successful cloning of eukaryotic DNA, in 1972 Hogness proposed using large insert clones to enable the detailed study of chromosome structure. His laboratory generated the first random clones from any organism in 1974, mapped a cloned DNA segment to a specific chromosomal location a few months later, and by early 1975, had generated clone libraries encompassing the entire *Drosophila* genome. In the late 1970s and early 1980s, Hogness, Lewis, Wellcome Bender and colleagues achieved the first 'positional cloning' of a gene, *Utrabithorax* (*Ubx*), and then others, using chromosomal 'walking' and 'jumping' aided by inversions. Many of the mutant alleles in the loci studied turned out to be the result of chromosomal breakage or transposon insertions rather than alterations to protein-coding sequences,[95–97] contrasting with the spectrum of chemical mutagen-induced single base changes used widely in mammalian genetic studies.

Unexpectedly, work from Mike Akam and colleagues revealed that the regulatory loci in the *bithorax* complex did not encode proteins, but rather expressed non-protein-coding RNAs,[101–103] but these were overlooked in favor of the homeotic proteins and the preconception that regulatory regions functioned in *cis* by binding regulatory proteins.

There was also a strong emphasis on finding equivalents of genes identified in model organisms (including, for example, neurological and transporter proteins) in other species by sequence homology, using initially a Southern blot variation dubbed 'zoo blots' and later sequence similarity.[99,104–106] Such approaches led to the discovery that not only do homeotic gene clusters occur in vertebrates in multiple copies but also that their introns (Chapter 7), relative orientation and temporal expression patterns (including antisense transcripts) are conserved between *Drosophila* and mammals[107–111](Figure 6.5).

Many other genes involved in *Drosophila* development were found to have human orthologs, a great surprise at the time, including that encoding the homeobox-containing protein Pax6, which is required for eye morphogenesis in both insects and mammals, indicating a common evolutionary origin[o] despite the differences between compound and camera-type eyes.[106] They also included many mutated in human diseases, like the homolog of the fly gene *patched* in the etiology of the skin cancer basal cell carcinoma[p] and the brain tumor medulloblastoma.[112,113]

ONCOGENES AND TUMOR SUPPRESSORS

Many genes that play a role in the etiology of cancer[q] – 'oncogenes' (in which mutations and activation drive cancer) and 'tumor suppressors' (whose

inactivation facilitates cancer) – were unearthed in those years and are still being identified today.

The first oncogene was identified by a brilliant experiment which sought to understand how the Rous sarcoma[r] virus (RSV) transforms avian cells into a cancerous state. RSV was discovered in 1911 by Francis Peyton Rous, who showed that this retrovirus, from which Baltimore and Temin later isolated reverse transcriptase, was the infectious agent present in cell-free extracts of chicken tumors that could transmit cancer to other birds,[116] consistent with observations of others in leukemia[117] and sarcoma.[118] The surprising finding that cancer could be caused by a virus was, as so often the case, not believed and was "met with reactions ranging from indifference to skepticism and outright hostility",[119] although Rous ultimately received the Nobel Prize (55 years later) for his discovery. Analysis of viral mutants that could replicate but not 'transform' cells in culture led to the isolation of the *v-src* gene, encoding a protein tyrosine kinase (which phosphorylates other proteins). Just as importantly, *v-src* was shown to be a constitutively activated version of a normal human gene, the 'proto-oncogene' *c-SRC*, which is mutated in many cancers.[119,120] Subsequent studies, initially of Burkitt's Lymphoma in the early 1980s, showed that somatic chromosomal translocations involving the *c-myc* gene could create oncogenic hybrids,[121] which also occurs in the *bcr-abl* fusion characteristic of chronic myeloid leukemia.[122]

In 1969, Henry Harris showed that fusion of normal cells with tumor cells suppressed their tumorigenicity, indicating that cells express genes that control cell growth, which are lost in cancers. In 1971, Alfred Knudson and others studying rare cases of familial retinoblastoma hypothesized that the heritability was due to a loss-of-function mutation in one copy of a germline gene, followed by a later *de novo* (somatic) mutation in the other allele: the 'two-hit hypothesis'.[123] Nearly 15 years later this led to the identification of the first tumor suppressor gene, encoding the retinoblastoma tumor suppressor protein, *RB1*.[124,125] The two-hit hypothesis explained the relationship between inherited and acquired mutations in cancer predisposition genes, including *TP53*, also referred to as 'the guardian of the genome', which was discovered in 1979 by several groups and is mutated in about 50% of all cancers.[126,127]

---

[o] The evolution of the eye has been a popular and controversial topic in evolutionary biology and often cited as an example of 'intelligent design'.

[p] Tracked down by what is termed 'positional cloning' (also referred to as 'forward genetics'), whereby genetic and physical mapping techniques are combined to home in on the chromosomal locations of mutations causing serious genetic disorders (Chapter 11).

[q] Cancer is fundamentally a life-threatening disease of metazoans, resulting from the reversion of individual cells in complex differentiated organisms to a primitive, atavistic state,[114] wherein mutations disrupt the interactions between ancestral genes that promote cellular growth and those that control cell division and differentiation in multicellular development[115] (Chapter 15).

---

[r] Sarcoma is the collective name given to cancers of connective tissue.

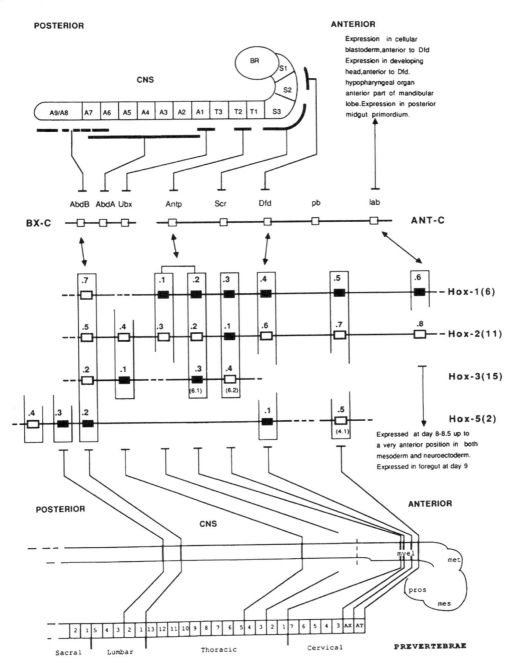

FIGURE 6.5 Schematic representation of the correlation between the *Drosophila* homeotic gene complexes and the murine (vertebrate) *Hox* gene network. (Reproduced from Duboule and Dollé[107] with permission from John Wiley and Sons.) The upper part represents the domains of expression of *Drosophila* homeotic genes in the embryonic central nervous system (CNS). In the central part all the genes belonging to the same subfamily are indicated by the vertical open or closed rectangles, the latter being the *Hox* loci that had been studied by comparative *in situ* hybridization experiments and whose expression domains had been defined at that time. The bottom part schematically represents the antero-posterior boundaries of expression of these genes along the fetal CNS and pre-vertebral column.

Such experiments resolved the controversy about the common causes of cancer – it is due to mutations that result in the ectopic activation of genes that promote cell division[121] and the loss of function of other genes that constrain cell division and migration.[128] These mutations can be inherited, occur spontaneously or be induced by DNA-damaging carcinogens and radiation, a complex landscape that is still being mapped by sequencing of tumor genomes in thousands of human cancers (Chapter 11).

Driven by medical need, the most intensely studied genes in the human genome[s] are those involved in cancer and other diseases.[134] Many altered cancer-causing genes, such as the breast cancer predisposition genes *BRCA1* and *BRCA2*, maintain genome integrity,[135,136] with mutations leading to the chaotic genomes seen in many cancers. Other genes, such as the mismatch repair genes associated with Lynch syndrome,[t] cause a high tumor mutation burden, creating 'neo-antigens'[138] that can be recognized by the immune system.[138–142]

IMMUNOLOGY AND MONOCLONAL ANTIBODIES

Gene cloning also provided molecular insight into the previously arcane world of immunology, showing that antibody genes undergo rearrangement and hypermutation to generate a wide arsenal of antigen-recognition molecules. It was found that cells that express antibodies to foreign antigens undergo secondary changes to improve the binding of the antibody, as well as clonal selection,[143,144] as had been predicted by Macfarlane Burnett.[145] It also led to the identification of inflammatory molecules ('cytokines') that excite immune responses and drive autoimmune disorders,[146] as well as tangentially to the development and production of mouse monoclonal antibodies in culture, pioneered by Georges Köhler and César Milstein in 1975[147] (Figure 6.6) and later humanized by Greg Winter and colleagues,[148] which have proved so efficacious as therapeutics for cancer

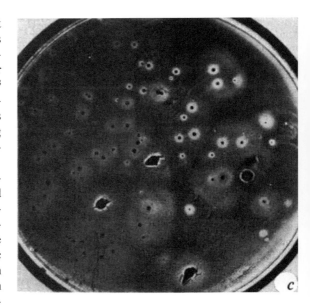

FIGURE 6.6 Photograph of immortalized cells created by a fusion between a myeloma cell line and spleen cells from a mouse immunized with sheep red blood cells, showing individual clones that secrete 'monoclonal' antibodies that lyse sheep red cells, indicated by the halos. (Reproduced from Köhler and Milstein[147] with permission from Springer Nature.)

and autoimmune diseases, among other conditions and applications.

BIOTECHNOLOGICAL EXPLOITATION

Not only did the gene cloning revolution create a vibrant technology support industry, it also transformed the pharmaceutical industry. The cloning, engineering and high-level expression of genes encoding human hormones such as insulin (which had previously been isolated from pig and cattle pancreas), erythropoietin (used to stimulate red cell production after bone marrow transplantation[u]) and growth hormone, among others, spawned multibillion-dollar products and companies, such as Genentech and Amgen.

Valuable tools were also developed by gene cloning and manipulation, notably the green fluorescent protein from jellyfish and its variants with different emission wavelengths by Osamu Shimomura, Douglas Prasher, Martin Chalfie and Roger Tsien,[149]

---

[s] Second only to TP53, and the most popular non-human gene, is the mouse *Rosa26*, which was identified in 1991 by Philippe Soriano and Glenn Friedrich as a locus that is ubiquitously active in mammals,[129] and subsequently widely used for the construction of transgenic mice and other species, as well as transgenic human cells.[130–132] The *Rosa* locus encodes two overlapping non-protein-coding RNAs,[133] whose functions are presently unknown.

[t] Which result in dinucleotide repeat instability.[137]

---

[u] And used illegally by athletes, as is growth hormone, to boost oxygen carrying capacity and muscle mass, respectively.

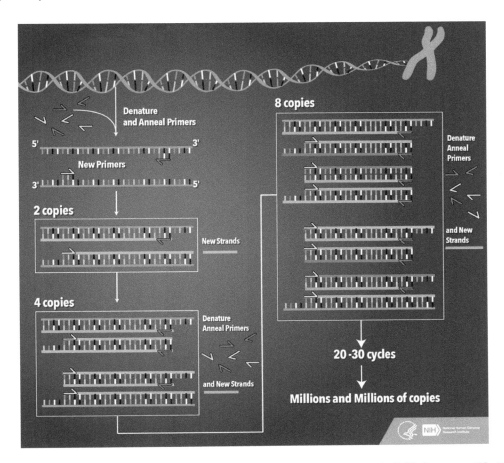

FIGURE 6.7   The principle of exponential amplification of targeted DNA sequences by PCR. Image modified from the US National Human Genome Research Institute fact sheet.[157]

and firefly luciferase by Marlene DeLuca and colleagues,[150,151] which have been widely used in cell and developmental biology to track the expression of genes fused to these visual 'reporters'.

Many of the genes discovered and characterized during this period were patented for medical or industrial use, largely on the basis of the inventiveness of the technology and the novel uses claimed for the products of these genes, a practice that was later circumscribed.[152] However, and despite criticism of gene patenting, it had the beneficial effect of allowing the development of new pharmaceuticals, which require (limited) monopoly rights to protect and recover the required massive investments in clinical safety and efficacy trials, following the tragic teratogenic effects of the anti-nausea drug thalidomide on limb development in embryos.[153]

CELL-FREE DNA AMPLIFICATION AND SHOTGUN CLONING

In 1983, Kary Mullis conceived a brilliantly simple strategy to amplify defined segments of DNA *in vitro*, using flanking oligonucleotide primers and DNA polymerases for cyclic, exponential replication of the targeted sequence, termed 'Polymerase Chain Reaction' or PCR[154] (Figure 6.7). The crucial technical advance was the use of thermostable DNA polymerases isolated from thermophilic archaea,[155] originally identified by Thomas Brock in hot springs in Yellowstone National Park in 1964.[156] PCR allowed ultra-sensitive detection and amplification of known DNA segments and transformed gene cloning, genetic engineering and diagnostic assays for mutations and infectious agents, especially viruses.

However, most genes that had been cloned in those days encoded proteins that had been identified biochemically or genetically. There were many more, as Craig Venter, Mark Adams and colleagues demonstrated in 1992 when they introduced the concept of 'shotgun' mRNA cloning and sequencing ('expressed sequence tags') to double the number of known human proteins in a single publication.[158] There was also great controversy when the US National Institutes of Health attempted to patent these genes *en masse*.[159] Similar agnostic approaches identified thousands of new genes in other organisms including plants.[160] Importantly, the shotgun strategy along with advances in the technology for high-throughput DNA sequencing allowed gene discovery on an industrial scale[161–164] and set the foundations for the genome projects (Chapter 10).

A WORLD OF PROTEINS

Throughout this period gene cloning and characterization was almost exclusively focused on protein-coding sequences,[v] due to a number of intrinsic and mutually reinforcing biases: expectational bias that most genes encode proteins; perceptual bias due to the strong phenotypes of disabling protein-coding mutations that are readily observed and genomically mapped; a sampling bias, as protein-coding genes are generally highly expressed;[w] technical bias due to the use of oligo(dT) priming of cDNA synthesis, which favors mRNAs; the difficulty of sequencing vast tracts of non-protein-coding DNA, and reticence to do so; and the problem of identifying causative mutations among the many variations in introns and 'intergenic' sequences.

The concept of a gene became synonymous with 'open reading frames', reinforcing the presumed equivalence of gene and protein, which in turn had a major influence on the interpretation of the discoveries of the mosaic structure of eukaryotic genes and the vast tracts of non-protein-coding sequences in animal and plant genomes (Chapter 7).

As observed by Ed Rubin and Lewis:

> Ironically, the success in cloning and studying individual genes dampened enthusiasm for an organized genome project, which was seen as unnecessary. Over 1300 genetically characterized genes—nearly 10% of all the genes in Drosophila—have been cloned and sequenced by individual labs. This is over twice the percentage of genes in any other animal for which both the loss-of-function phenotype and sequence have been determined. Nevertheless, for flies as well as other animals, less than a third of genes have obvious phenotypes when mutated, emphasizing the critical importance of genome sequencing as a gene discovery method.[95]

A very large fraction of discovered proteins in all kingdoms of life have no known function.[172]

On the other hand, not only did the genetic and biochemical approaches used to identify and characterize proteins reveal surprising cases of regulatory RNAs (Chapters 8, 9, 12 and 13), genome sequencing and high-throughput assays later showed that the largest class of proteins in the human genome is RNA binding proteins.[173,174]

FURTHER READING

Cohen S.N. (2013) DNA cloning: A personal view after 40 years. *Proceedings of the National Academy of Sciences* 110: 15521.

Roberts R.J. (2005) How restriction enzymes became the workhorses of molecular biology. *Proceedings of the National Academy of Sciences* 102: 5905.

Rubin G.M. and Lewis E.B. (2000) A brief history of *Drosophila*'s contributions to genome research. *Science* 287: 2216–8.

Russo E. (2003) Special report: The birth of biotechnology. *Nature* 421: 456–7.

Watson J.D. and Tooze J. (1981) *The DNA Story. A Documentary History of Gene Cloning*. (W.H. Freeman and Company, New York).

---

[v] Some early work indicated that polymorphisms in the 5' region of human protein-coding genes are associated with variations in gene expression and disorders, such as hemoglobinopathies and hypertriglyceridemia.[165–168]

[w] In general, protein-coding genes are more highly and broadly expressed than genes that express regulatory RNAs, which show high cell specificity,[169–171] although there are exceptions (Chapter 13).

7 All That Junk

THE C-VALUE ENIGMA

Following Avery's demonstration that DNA is the genetic material, cytological and biochemical measurements of the amount of cellular DNA showed that species have a characteristic DNA content[a] (termed the 'C-value' by Hewson Swift[1]) and that the amount of cellular DNA broadly increases with developmental complexity if taxa are compared on the basis of their minimal DNA content.[2,3] Related studies during this period used the drug colchicine to block DNA replication at metaphase, enabling the complement and size distribution of chromosomes also to be determined.[b]

However, anomalies were found. In many taxa, the cellular DNA content of species varies over a wide range:[6,7] some simple protists and plants such as green algae and mosses have more DNA than flowering plants; and many plants (including onions, a popular example[8]) and some other protozoans such as amoebae have more DNA per cell than mammals[3,9,10] (Figure 7.1). Since it was assumed that more complex organisms[c] require more genetic information (and the understanding of gene structure and regulation was derived from microbial genetics and biochemistry studies), these anomalies led to the coining of the term 'C-value paradox'[9] or 'C-value enigma'.[7]

There has been, and remains, considerable speculation about the significance of the spectrum of DNA content, which is often interpreted as evidence of the ability of eukaryotes to maintain superfluous DNA.[7] Correlations were sought and sometimes found with

cell size, involving an increase in the number of nuclei and/or copies of the genome, possibly to support the metabolism of larger cells.[11–15]

The inordinately large amounts of DNA in some species transpired to be due to two factors: polyploidy, i.e., multiple copies of the genome, which occurs commonly in plants and sporadically in animals, especially insects;[17–21] and lineage-specific expansions of transposon-derived sequences, notably in some fishes and amphibians (especially lungfish[22] and salamanders[23]),[d] some clades of arthropods[24,25] and cnidarians (hydra),[26] and many plants, where they play a major role in adaptive evolution (Chapter 10).[e]

The G-value enigma emerged later, when the genome projects showed that there is no correlation between the number of protein-coding genes and developmental complexity (Chapter 10). Genome sequencing also showed that the ratio of non-protein-coding to protein-coding DNA (which intrinsically corrects for ploidy) increases with morphological complexity,[21,27] suggesting that, whatever else may be at play, increased complexity is associated with the expansion of regulatory information. This imputation can only be falsified by a downward exception, i.e., the identification of developmentally complex organisms that have little non-coding DNA, of which none have been found to date.[f]

For decades, however, the notion that "the number of distinct protein-coding genes that an organism made use of was a valid measure of its complexity" was deeply rooted and well accepted.[28]

DUPLICATION AND TRANSPOSITION

The mechanisms by which genomes can be enlarged are gene, segmental or whole genome duplication – first

---

[a] Measured in picograms / cell, converted to base pairs on the basis that 1 base pair = 660 daltons, assuming an equimolar amount of the bases, i.e., 1 pg = 10^9 base pairs, or 1,000 Mb (1 Gb).

[b] Only in 1956 was it reported that the correct diploid number of human chromosomes is 2n = 46,[4] until which time it had been thought to be 48, based on studies in the 1920s by Theophilus Painter.[5]

[c] The definition of biological complexity is controversial and susceptible to pedantry. We define three types of complexity: metabolic complexity, which is collectively high in microorganisms, and lower in plants and animals; developmental complexity, the numbers of positionally and functionally distinct cells and structures (Chapter 15); and cognitive complexity, the ability to process information and learn, which is highest in mammals (Chapter 17).

[d] It should be noted, however, that the smallest amphibian genome is half the size of the smallest mammalian genome. See http://www.genomesize.com.

[e] Some species and cell types increase chromosomal/chromatid copy number during development. The giant polytene chromosomes of salivary glands in *Drosophila* is one example.

[f] Upward exceptions do not negate the possibility that large amounts of regulatory DNA are needed to program the ontogeny of complex organisms.

DOI: 10.1201/9781003109242-7

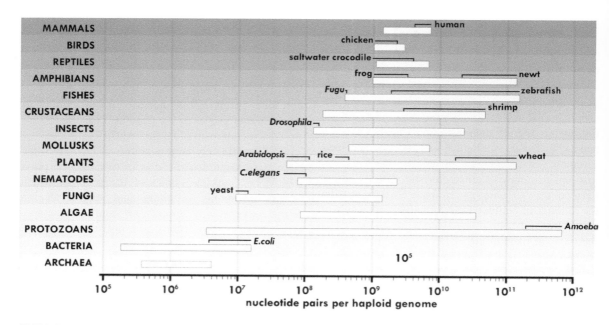

FIGURE 7.1   The range of haploid genome sizes for the groups of organisms listed. (Adapted from an image by Steven Carr, Memorial University of Newfoundland, published in Fedoroff,[16] with permission from the American Association for the Advancement of Science.)

proposed by Susumo Ohno in 1970[29] – and copy-and-paste insertion of sequences from external sources or elsewhere in the genome by transposition. That is, the raw material for evolutionary innovation is sequence duplication and transposition. The former has been documented in many species, for example, in yeast and at the origin of the vertebrates, where it is evident that whole genome duplication has occurred at some point in their evolutionary history, with some duplicated genes having acquired new functions and been retained, whereas those that remained redundant were largely lost.[30–32] Partial genome (segmental) duplication is also well documented.[33]

The work of Leslie Gottlieb, Donald Levin and others has shown that genome duplication ('auto-polyploidy') and fusion of genomes between related species (termed 'allopolyploidy')[g] creates phenotypic novelty and speciation – altering patterns of gene expression, physiological responses, growth rates, developmental features, reproductive outputs, mating systems and ecological tolerances,[34–36] including in Darwin's finches.[37] This led Levin to suggest that such nucleotypic effects may "'propel' a population into a new adaptive sphere, perhaps accounting

for the distribution of polyploids, both auto- and allopolyploids, in areas beyond those of the diploid parents".[36]

Transposition is a specialized and highly flexible form of sequence relocation or (more commonly) 'duplication' (multiplication) that mobilizes protein-coding and/or regulatory cassettes,[38–42] which explains its evolutionary value and distribution (see below). Transposases are, in fact, the most abundant and ubiquitous genes in nature.[43] Large numbers of various classes of retrotransposed sequences occur in multicellular eukaryotes, especially plants and animals[44,45] and the colonization of genomes by transposons appears to have occurred in bursts[45] likely associated with major evolutionary adaptations (see, e.g.,[22,46]) (Chapter 10). Transposable elements (TEs) are diverse and have been widely incorporated into regulatory networks in different clades,[47] as predicted by Britten and Davidson (Chapter 5), with, for example, most primate-specific regulatory sequences having been derived from these elements.[38,39]

It is reasonable to assume that genomes contain some duplicated or transposed sequences that are in suspension between functional exaptation on the one hand and degradation or deletion on the other, i.e., have not (yet) acquired a useful (new) function nor been lost. It is currently difficult, if not impossible,

---

[g]   Wheat is a familiar example.[34]

to determine the extent of such limbo sequences in any given lineage. One might speculate, however, that the more ancient the duplication or TE, the more likely it is to have acquired, or already have, a useful function that has contributed to its retention. One might also speculate that recently acquired transposable elements have played a role on phenotypic diversification, which has now been well documented[39,48] (Chapter 10).

MUTATIONAL LOAD, NONSENSE DNA, NONSENSE RNA

The problem was that the large genomes of protists, plants and animals, and their large numbers of 'repetitive' sequences, could not be reconciled with the protein-centric conception of genetic information.

The population geneticists and evolutionary theorists at the time, notably Müller[h] and Ohno, suggested that, since increases in genome sizes in eukaryotes occurred by polyploidization, much of the duplicated DNA is redundant. They also argued that, if the unique sequences (~50% of the genome) in mammals specified structural (i.e., protein-coding) genes, there would be ~1 million such genes, which would, by comparison with bacteria, impose an unbearable mutational load, the escape from which was the prime function of recombination.[49,51–53]

Ohno extended this logic to regulatory information, speculating that "in order not to be burdened with an unbearable mutation load, the necessary increase in the number of regulatory systems had to be compensated by simplification of each regulatory system. It would not be surprising if each mammalian regulatory system is shown to have fewer components than the lac-operon system of *Escherichia coli*".[53]

Based on these considerations and Haldane's 1957 'cost of selection' principle, which stated that the number of gene loci in a genome is a key determinant of the rate of evolution,[54] Masatoshi Nei

concluded in 1969 that, given the "high probability of accumulating ... lethal mutations in duplicated genomes... it is to be expected that higher organisms carry a considerable number of nonfunctional genes (nonsense DNA) in their genome" and that "higher organisms, including man ... are using only a small fraction of the maximum amount of genetic information their DNA molecules are able to store".[55,56] This logic has persisted to the present,[8,57] underpinning the recent claim, for example, that "the functional fraction within the human genome cannot exceed 15%".[58]

Such early musings were based on the analysis of easily discernible simple traits, which constituted the majority of genetic studies up until that time and indeed until the end of the 20th century. These traits included metabolic defects,[i] flower and eye color[j] and severe genetic disorders, which usually result from high-impact loss-of-function mutations in protein-coding sequences. By and large, they did not take into account that variations in regulatory sequences that control quantitative traits in complex organisms may be more subtle, although they may have a strong influence on complex traits and reproductive fitness: this was a huge blind spot.

In this context, it should be noted that the mathematical foundations of quantitative genetics were laid down with a very different set of problems in mind – such as the prediction of short-term responses to artificial selection – which went on to focus on genetic diversity based on enzyme polymorphisms,[59] again before crucial details of the variation in genome sequences and of genome regulation in complex organisms were known.[60]

Incorporating molecular considerations, John Paul (1972) stated the alternatives that, considering the existence of hnRNAs and the size of mammalian genomes, "either that the mutational load argument does not hold for eukaryotes or [as concluded by others] that much of the DNA in eukaryotes is not informational".[61]

He speculated that the more and less compact regions of chromosomes differ chemically, in that "modified histones, modified DNA or extra substances"

[h] Interestingly, based on mutational load arguments at the time, Müller estimated that there would be ~30,000 genes in mammals, repeated by Ohno, which turned out (much later) to be surprisingly accurate for protein-coding genes.[49] King and Jukes used similar calculations to predict an upper limit of 40,000 essential genes.[50] Such considerations do not apply to regulatory sequences if variations within them lead to complex trait variation, shown later by genome-wide association studies (Chapter 11).

[i] The work of Garrod, Cuénot, Beadle and Tatum, Luria and Delbruck, Lederberg, Benzer, Müller and others on 'biochemical mutations' that led to the 'one gene – one enzyme' hypothesis (Chapter 2).
[j] Which highly influenced early geneticists, including R. A. Fisher (a founder of the field of Population Genetics) and the Modern Synthesis.

determined the conformation of 'nucleohistone' (chromatin). He reasoned that non-histone proteins would perform this function, with auxiliary participation of nascent RNAs. In his model, "address sites" in the interbands would be targets for "polyanionic" regulators, allowing relaxation and transcription of nascent RNA that would not only contain an mRNA sequence but would also accumulate in these regions, recruiting RNA binding proteins and inducing further unwinding of chromatin.

Paul also used this possible role for nascent transcripts to explain the existence of the very large transcriptional units (hnRNAs) in animals and plants, which would contain the sequence of mRNAs together with redundant sequences producing 'nonsense RNAs' that perform an 'unwinding role'. Although vaguely defined, this was one of the first models that posited RNAs and histone modifications acting together to regulate gene expression. This model also predicted, as did others (Chapter 5), that these nascent RNAs are processed to generate mRNAs, and even suggested the existence of sequence signals in the hnRNAs that guide the processing into the RNA parts to be degraded in the nucleus or to be exported to the cytoplasm.[61]

The issue was summarized by Ed Southern in 1974: "The outstanding problem presented by eukaryotic DNA is that of finding a role for these large fractions not used in coding for proteins or cytoplasmic RNAs."[62]

Speculation was rife. The evolutionary biologist Tom Cavalier-Smith wrote in 1978:

> Eukaryote DNA can be divided into genic DNA (G-DNA), which codes for proteins (or serves as recognition sites for proteins involved in transcription, replication and recombination), and nucleoskeletal DNA (S-DNA) which exists only because of its nucleoskeletal role in determining the nuclear volume …[11,12]

Others suggested the excess non-coding DNA might be retained for "genome balance",[63] have some value as a mutational sponge[49] or buffering,[64] or be a reservoir for evolutionary innovation.[65–68]

'NEUTRAL' EVOLUTION

A natural corollary of the idea that much of the genomes of plants and animals is not functional is that these sequences are evolving 'neutrally'. In parallel, the growing availability of amino acid sequence data revealed that protein sequences have been diverging between lineages at a relatively constant rate,[k] referred to as the "molecular clock" by Emile Zuckerkandl and Linus Pauling,[71] or "genetic equidistance" by Emanuel Margoliash[72] in the early 1960s. In 1971, Richard Dickerson showed that the clock runs at different rates for different proteins[73] (Figure 7.2), later shown to vary by orders of magnitude, useful to measure evolutionary relationships over different genetic distances and evolutionary timescales.[74–76]

The divergence of the sequences of homologous proteins over time was surprising and, after taking into account the frequencies of deleterious mutations and mutational load, led Motoo Kimura to propose in 1968 the neutral theory of molecular evolution, or 'genetic drift', which posited that "an appreciable fraction" of the genome was evolving independently of natural selection.[77–81] Like Nei, Kimura was also motivated by Haldane's argument and the 1970s finding that the numbers of nucleotide changes observed between humans and chimpanzees could not be explained by selection, called "Haldane's Dilemma",[82] tacitly assuming that mutation is random and not influenced by other mechanisms (see Chapter 18).

A similar proposal was made by Jack King and Thomas Jukes in their 1969 article entitled 'Non-Darwinian Evolution,' which extolled the importance of random genetic changes and genetic drift in evolution.[50] The theory was refined in 1973 by Kimura's student, Tomoko Ohta, and later by others, notably Michael Lynch, who emphasized the importance of nearly neutral ("slightly deleterious") mutations, whose exposure to selection is dependent on the size of the interbreeding population.[83–88]

The extension of this logic, posited as the 'null' hypothesis,[89] is that highly complex organisms with small effective population sizes such as mammals accumulate greater loads of transposable elements, larger introns (see below) and larger intergenic regions,

---

[k] These were the first manifestations of the nascent field of bioinformatics, pioneered also by Margaret Dayhoff and Richard Eck,[69] who introduced the concept of molecular phylogeny, reflecting a prescient prediction by Crick a few years earlier, when he stated: "Biologists should realise that before long we shall have a subject which might be called 'protein taxonomy'— the study of the amino acid sequences of the proteins of an organism and the comparison of them between species. It can be argued that these sequences are the most delicate expression possible of the phenotype of an organism and that vast amounts of evolutionary information may be hidden away within them."[70]

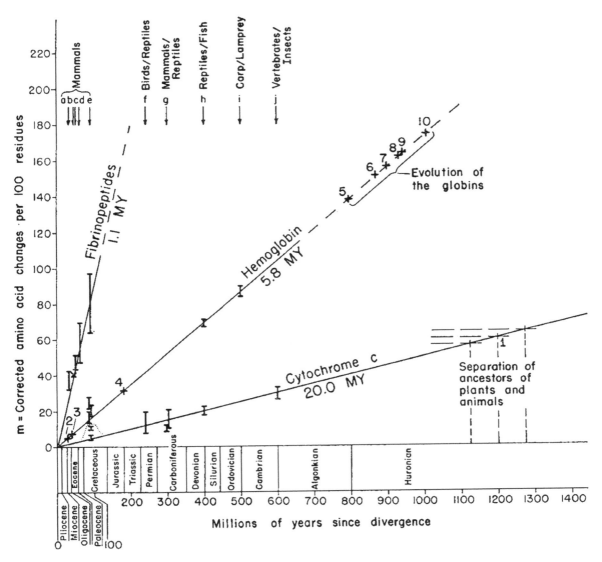

FIGURE 7.2 The molecular clock. Dickerson's graph of the rates of molecular evolution in fibrinopeptides, hemoglobin and cytochrome c. (Reproduced from Dickerson[73] with permission from Springer Nature.)

all of which co-vary inversely with population size, such that especially large bodied species with low population sizes have bloated genomes and difficulty in purging even slightly deleterious mutations.[85-88,90] Later theoretical studies also concluded, mainly based on 'non-conservation', that alternative transcription, polyadenylation, RNA modification and RNA editing[1] sites in complex organisms are non-adaptive.[91-95]

Neutral evolution was controversial in evolutionary circles, reflecting a long-standing disagreement between 'classical' and 'quantitative' geneticists that simmered for decades, although thought to have been resolved by Fisher's infinitesimal model[96] (Chapter 2). The classical geneticists viewed the normal state to be a wildtype (protein-coding) gene with a low frequency of deleterious (usually recessive) mutants in the population, influenced by Mendel's simple trait segregation in peas and by genetic ('Mendelian') disorders in humans. On the other hand, the quantitative geneticists, mainly working in agriculture, citing

---

[1] Despite the fact that RNA editing has expanded greatly and the enzymes involved have been subject to strong positive selection in the primate lineage (Chapter 17).

the abundant variation in quantitative traits in crop plants and livestock and the 'concealed variability' revealed by inbreeding experiments, proposed that many genes have two or more alleles maintained at intermediate frequencies in populations by 'balancing' selection, perhaps influenced by environmental factors.[59]

A renewed debate between the 'near-neutralists' and 'adaptationists' ensued following Kimura's and Ohta's papers.[97] The former maintained that genetic drift accounts for most differences within populations or between species, whereas the latter credited them to positive selection for adaptive traits,[98,99] although as Laurence Hurst later observed "the two positions are often hard to discriminate as they make many similar predictions".[97]

These debates did not often consider that there might be an important distinction between the genetic signatures of protein and regulatory variation and were mostly thought of in terms of binary (wildtype and 'defective') alleles rather than interconnected networks.[99,100] Nor did they take into account the role of transposons in phenotypic variation (see below; Chapters 5 and 10), positive selection for reproductive success,[99,100] or the amount of information that might be required to organize the four-dimensional development of multicellular organisms[101] (Chapter 15). Moreover, nearly neutral genetic drift does not account for the rapid evolution of animal phyla and species, such as observed in the Cambrian explosion, Darwin's finches[102] and primates,[103–109] and is at odds with the whole genome biochemical indices of function that were revealed later (Chapter 13).[m]

As Mayr observed in 1970:

> The day will come when much of population genetics will have to be rewritten in terms of interaction between regulator and structural genes. This will be one more nail in the coffin of beanbag genetics. It will lead to a strong reinforcement of the concept that the genotype of the individual is a whole and that the genes of a gene pool form a unit.[110]

And Jacob in 1977:

> It seems likely that divergence and specialization of mammals, for instance, resulted from mutations altering regulatory circuits rather than chemical structures. Small

changes modifying the distribution in time and space of the same structures are sufficient to affect deeply the form, the functioning, and the behavior of the final product – the adult animal.[111]

The situation was summarized in 2014 by Karl Niklas:

> Beginning with a series of papers in the early 20th century and culminating with his book The Genetical Theory of Natural Selection, Ronald A. Fisher (1930) founded the field of population genetics and designated the gene as the unit of stable hereditary transmission between successive generations. This genocentric view of inheritance asserted the preeminent importance of allele frequency distributions and differential reproductive success in evolutionary processes. However, it failed to explore alternative origins of phenotypic variation. It simply assumed that all phenotypic variants result from [protein-coding] gene mutations … Perhaps even more restrictive was the additional assumption that the phenotype could be mapped directly onto the genotype and thus described simply by changes exclusively at the level of individual genes or sets of genes.[112]

Niklas continued:

> This outlook was challenged in the 1970s and 1980s within a field of study soon to be called evolutionary-developmental biology, or simply evo-devo, which asserted that evolutionary phenotypic transformations are the result of changes in gene expression patterns rather than the immediate products of mutations of individual genes … Arguably … this perspective can be traced back to a seminal paper by Britten and Davidson.[112]

Richard Lewontin, who developed some of the statistical tools for assessing genetic drift and selection (largely from studies of electrophoretic variation in proteins in natural populations of *Drosophila*),[59] observed in 1974 that

> For many years population genetics was an immensely rich and powerful theory with virtually no suitable facts on which to operate. It was like a complex and exquisite machine, designed to process a raw material that no one had succeeded in mining… Quite suddenly the situation has changed. The mother-lode has been tapped and facts in profusion have been poured into the hoppers

---

[m] That is not to say, however, that genetic drift is not an important evolutionary process, and there are likely many passenger or hitchhiker sequence variations of subtle effect.[97]

of this theory machine. And from the other end has issued – nothing … The entire relationship between the theory and the facts needs to be reconsidered.[98]

In 1996, Ohta admitted, with respect to nucleotide substitution patterns, that "all current theoretical models suffer either from assumptions that are not quite realistic or from an inability to account readily for all phenomena".[113]

CONSERVATION AND SELECTION

The concept of neutral evolution led to attempts to define a subset of sequences that are evolving neutrally, to measure the unconstrained rate of sequence drift, and thereby determine which (other) sequences in the genome might be evolving more rapidly or slowly under positive or negative selection,[n] and therefore be functional.

One obvious candidate was the 'redundant' (usually third) base of synonymous codons, first exposed by the pioneering sequencing in 1983 of 11 cloned alcohol dehydrogenase (*Adh*) genes in natural populations of *Drosophila*, which revealed 43 previously hidden polymorphisms. Only one of these polymorphisms altered a codon specificity (and resulted in the known electrophoretic variant of the protein), implying that nonsynonymous changes have phenotypic consequences and are deleterious, whereas the others were possibly neutral.[114] However, later analyses showed that amino acid codon sequences are not evolving neutrally (as is also the case for many non-coding sequences), possibly reflecting selection pressures on translational efficiency or RNA structure,[115–119] with others showing that the genetic code is optimal for encoding additional information.[120–122] Later studies showed that non-coding polymorphisms affect *Adh* expression[123] and that *Adh* variants are selected indirectly.[124]

The field also looked to other sequences, notably 'pseudogenes' (see below) and ancient retrotransposons, to estimate the rate of neutral evolution, on the questionable and likely incorrect assumption that they are non-functional (Chapter 10), leading to a vast underestimation of the amount of the human genome that is under selection (Chapter 11).

The debate continues,[125,126] but the concept of neutral evolution is coming under siege. As concluded recently by Andrew Kern and Matthew Hahn:

> The neutral theory was supported by unreliable theoretical and empirical evidence from the beginning, and … we argue that, with modern data in hand, each of the original lines of evidence for the neutral theory are now falsified, and that genomes are shaped in prominent ways by the direct and indirect consequences of natural selection.[127]

The adherents begged to differ.[128]

Of course, different types of sequences have different structure-function constraints and different selection pressures, as is seen within protein-coding sequences where the amino acid sequences of active sites are highly conserved but associated scaffolding and domain linker sequences are quite plastic.[129,130] Regulatory sequences are even more plastic;[131–135] orthologous promoters that have no obvious sequence homology direct similar expression patterns in fish and humans,[136] and less than 5% of human embryonic stem cell developmental 'enhancers' (Chapter 14) are 'conserved' in mouse.[135] These regulatory sequences also encompass small regulatory RNAs and vast numbers of tissue- and cell-type specific long non-coding RNAs, which seem to be even more evolutionarily flexible, with different sequence-structure-function constraints and including increasing numbers of functionally validated species- and clade-restricted RNAs (Chapters 12, 13 and 16).

It is now well established that adaptive radiation in complex organisms, including primates, is mostly due to regulatory variation,[137–142] which may be co-dominant and therefore immediately visible to selection.[143] Regulatory sequences evolve rapidly,[o] mostly (initially at least) under positive selection for changes in morphological and physiological phenotypes.

It has also been known for some time, confirmed later by the genome projects (Chapters 10 and 11), that the mutation spectrum varies enormously across the genome,[144,145] which has been rationalized for example as local variation in the underlying mutation rate (due to regional differences in

---

[n] These contraforces are hard to disentangle over evolutionary time. By definition, any useful variation is subject to positive selection until it becomes fixed in the population (appearing initially to have evolved rapidly by supplanting the previous sequence). It is then subject to negative selection as its loss is disadvantageous, and thereafter evolves slowly (Chapters 10 and 11).

[o] In general, there are exceptions, such as the ultraconserved elements whose sequences evolved more rapidly than those of proteins during tetrapod evolution but are evolving far more slowly in the amniotes (birds and mammals; Chapter 10).

nucleotide composition) or the activity of DNA repair enzymes.[144–147] The alternative explanation that the vast non-coding regions of plant and animal, especially mammalian, genomes are under selection could not be countenanced, both because it was assumed to be junk and because it appeared impossible due to the mathematical models of selection operating on random mutation in small populations.

A later analysis showed that there are at least seven different rate classes of sequence evolution in the mammalian genome,[148] and different rates of sequence evolution in gene promoters.[149] Others concluded that ~95% of the human genome is influenced by background selection and biased gene conversion,[150] commensurate with the proportion of the genome that is dynamically transcribed into (mainly non-protein-coding) RNA (Chapter 13), while observations in natural *Arabidopsis* accessions show that epigenome-associated mutation bias occurs differentially across the genome and gene regions, with essential genes (in particular gene bodies) subject to stronger purifying selection having a lower mutation rate.[151,152]

JUNK DNA

It did not seem to occur to most at the time, apart from McClintock, Britten and Davidson and a few others (Chapter 5), that the enormous numbers of 'repetitive' sequences and nuclear-localized RNAs might play a role in plant and animal differentiation and development.

And since no one could countenance gigabases of regulatory protein-binding sites,[p] and for all of the other reasons cited above, Ohno summed up the growing consensus when in 1972 he wrote about "all that 'junk' DNA in our genome"[q] arguing that only a fraction of the human DNA functions as 'genes' and that there is "more than 90% degeneracy contained within our genome".[49] Ohno's conclusions were reinforced by the existence of seemingly defective 'pseudogenes'[161] (first identified in 1977 and described as 'relics of evolution'[162]), the 'gene-poor' and transposable element-rich heterochromatin, and the extensive intergenic

regions in intensively studied loci, thought to be genetically and transcriptionally silent but later shown to be the sites of 'enhancer' and other regulatory elements that control gene expression patterns in development (Chapters 14 and 16).

Duplications of globin genes were highlighted by Ohno and others not only as the source of new functional genes, but also of defective pseudogenes with untranslatable sequences ("recent degenerates"),[49,163] many of which have since been shown to have regulatory functions[164] (Chapters 10 and 13).[r] Notably, the pseudogene in the human hemoglobin cluster on chromosome 11, hemoglobin subunit beta pseudogene 1 (*HBBP1*, or *η-globin* pseudogene in primates), was later found to be subject to strong selection, tissue-specifically expressed, essential for erythropoiesis, mutated in a form of thalassemia and to regulate the switches of globin gene expression during development.[170–173]

Some did take issue. Herb Boyer questioned the calculations of the extent of functionality in the genome based on the assumption that lethal mutation rates apply to the whole genome, noting in the discussion of Ohno's paper that "we can only measure what we see".[49]

Stephen O'Brien wrote shortly afterwards that the

> conclusions (that) indicate that more than 90% of the eukaryotic genome may be composed of nonfunctional or noninformational 'junk' DNA … have not been fundamentally proven; rather they are based on simplifying assumptions of questionable validity, in some cases contradictory to experimental data.[174]

He challenged the notion that lethal mutation frequency was a good metric for gene number and genome functionality, citing several lines of evidence suggesting that mutations only result in lethality in a minority of genes.[s] Noting that hybridization stud-

---

[p] Recent high-resolution data suggest that transcription factor binding sites occupy just 0.2% of the genome.[153]

[q] Non-protein-coding DNA has been called by many names. These include 'excess DNA',[154,155] 'surplus, nonessential, degenerate or silent DNA',[156,157] 'garbage DNA',[29] 'non-informational or nonsense DNA',[49] 'vestigial DNA',[158] 'supplementary DNA'[159] and 'incidental DNA'.[160]

[r] In 1986, John McCarrey and Art Riggs proposed that pseudogenes might have roles as regulatory switches or 'determinator-inhibitor pairs' during development based on antisense relationships,[165] a prediction validated, at least in part[166–169] (Chapters 9 and 13).

[s] Later high-throughput studies showed that a large fraction of protein-coding genes in *E. coli* and yeast do not result in lethality or in easily discernible phenotypes when deleted, presumably because laboratory conditions do not recapitulate natural selection for more subtle functions or variations in gene expression.[175,176] The same problem, of limited phenotypic screens, applies also to animals, especially in relation to inter- and intra-species competition, and behavioral and cognitive characteristics.

ies indicate the presence of a minimum of 300,000 different transcripts of 1 kb or more in mouse brain alone, he made the very reasonable point that "RNA does not have to be translated to have a function" and that their existence and their tissue- and developmental stage-specific expression transcription support their functionality.[174]

Heterochromatin was also widely thought to be inert, despite the fact that it is dynamic, and there was, even then, considerable evidence of its importance in developmental processes.[177] As observed by Spencer Brown in his incisive 1966 paper: "Our present picture of gene action comes almost exclusively from microorganisms. It is a verbally simply one ... the systems controlling gene regulation in higher organisms probably involve highly complex mechanisms necessary for developmental integration." Among several considerations on the potential roles of heterochromatin, he noted the reports of chromatin associated RNAs and pondered regarding the abundant RNAs present exclusively in the nucleus that "Such observations would make sense if the genes in higher organisms were required to build complex machinery for their own control".[177]

Jim Peacock and colleagues pointed out in 1978:

> In recent years it has become clear that specific genetic properties are attributable to heterochromatic regions of chromosomes and that the different segments of heterochromatin in a genome may have different properties ... we present data, primarily from *Drosophila*, to show that the heterochromatin of each chromosome has a unique, segmental identity, and that DNA sequences in heterochromatin have, as do DNA sequences in euchromatin, defined patterns of conservation and change during evolution. We show that the properties discovered in Drosophila apply to other eukaryotes, including plants and mammals.[178]

Put simply, the use of the frequency of lethal mutations by evolutionary theorists, the emphasis on negative selection to assess the extent of functionality of the genomes of complex organisms, and the assumptions that repetitive sequences and pseudogenes are non-functional were based on only rudimentary knowledge of molecular genetic information and were biased by emphasizing deleterious mutations over quantitative trait variations. It was conceptually primitive, but unfortunately influential.

Cloning and more advanced genetic mapping techniques[i] would later show the majority of mutations that cause severe phenotypic consequences in mammals map to protein-coding sequences – what might be called 'catastrophic component damage'. However, the vast majority (~95%) of variations affecting complex traits – with few or only subtle effects on viability – occur outside of protein-coding sequences (Chapter 11). This is largely invisible to high-fitness-impact and lethality-based measures of genetic load but constitutes the more important component of phenotypic variability in natural populations.

However, Mayr, O'Brien and others were swimming against the tide. The phrase 'junk DNA' entered the popular lexicon, uncritically embraced by those – including Brenner[65] – who were convinced of the primacy of proteins in the specification of cell and developmental biology, seemingly incurious about what all that non-protein-coding DNA might be doing. In fact, as seen below, proponents of junk DNA explicitly discouraged research into the possible roles of non-coding regions of genomes.

SELFISH DNA

A logical extension of the junk DNA view was the proposal promulgated and popularized by Richard Dawkins in 1976, following earlier theorizing by George Williams[179] and William Hamilton,[180] that DNA sequences have a propensity to select for their survival, which he termed "the selfish gene".[181] Dawkins argued that the selfish gene hypothesis can explain the fact that "a large fraction of the DNA is never translated into protein", stating "The simplest way to explain the surplus DNA is to suppose that it is a parasite, or at best a harmless but useless passenger, hitching a ride in the survival machines created by the other DNA".[181]

The concept was extended in 1980 by back-to-back papers by Ford Doolittle and Carmen Sapienza, and Leslie Orgel and Crick, entitled 'Selfish genes, the phenotype paradigm and genome evolution' and 'Selfish DNA: the ultimate parasite', respectively.[155,182,183] As put by the latter:

> In summary, then, there is a large amount of evidence which suggests, but does not prove, that much DNA in higher organisms is little better than junk. We shall assume, for the

---

[i] Such as the exome sequencing and genome-wide association studies, Chapter 11.

CLASS 1 – RETROTRANSPOSONS

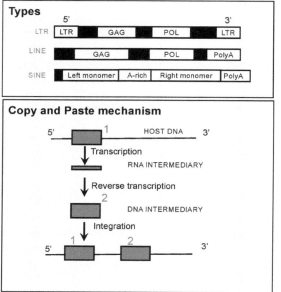

CLASS 2 – DNA TRANSPOSONS

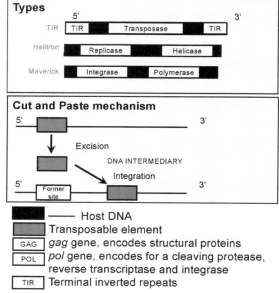

FIGURE 7.3 Classification of transposable elements and mechanisms of transposition. Class I retrotransposons mobilize via an RNA intermediate. Class II DNA transposons utilize a DNA intermediate. Autonomous elements encode the enzymatic machinery necessary for their transposition. Non-autonomous elements typically do not encode proteins but are capable of being mobilized using the machinery produced by their autonomous counterparts. (Reproduced from Serrato-Capuchina and Matute,[44] under Creative Common CC BY license.)

rest of this article, that this hypothesis is true … What we would stress is that not all selfish DNA is likely to become useful. Much of it may have no specific function at all. It would be folly in such cases to hunt obsessively for one.[183]

The exemplars of selfish DNA were sequences derived from (endogenous) retroviruses, transposons and other types of repetitive elements (Figure 7.3), which reinforced the view that these elements are (mainly) genetic hobos,[u] notwithstanding McClintock's demonstrations that transposon mobilization changes developmental phenotype and responses to the environment.[184] The view of transposons as functionless and/or deleterious parasites[v] has endured,[189–191] reinforced by the discovery that they are restrained by

methylation (Chapter 14) and other 'silencing' mechanisms. Although their discovery earned McClintock a Nobel Prize in 1983, it seems that her emphasis and insistence on them as controlling elements in differentiation and development, and the finding that TEs cause insertional mutations in bacteria, delayed the award.[192,193]

The role of transposons as mobile cassettes of genetic (especially regulatory) information in evolutionary and biological processes, and the proportion of transposon-derived sequences that may have contributed or acquired useful functions in genomes[194] was then and is still not known, but it suited the zeitgeist to assume that it is low. The selfish fattening out by transposons of intergenic sequences and introns within genes (see below) then became the widely accepted explanation for all that junk, and the C-value enigma.[182,195,196]

GENES-IN-PIECES!

Perhaps the most unexpected discovery in the history of molecular biology was that genes in eukaryotes, especially developmentally complex eukaryotes, are

---

[u] They "acquired the anthropomorphic labels of 'selfish' and 'parasitic' because of their replicative autonomy and potential for genetic disruption".[16]

[v] It is clear that retroviral and retrotransposon insertions in some instances disrupt protein-coding genes,[185] but it is also clear that some, if not many or most, have far from random genomic- and clade distributions and have been exapted to function, 'nature's tools for genetic engineering'[186–188] (Chapters 10 and 16).

not co-linear with their encoded proteins, but rather are fragmented and separated by non-protein-coding 'intervening' sequences or 'intragenic regions', dubbed 'introns'.[157,197,198] The fragments of protein-coding sequences (and flanking regulatory sequences in mRNAs) were reciprocally called 'exons'.[w]

In 1975, Darnell and colleagues showed that adenovirus mRNA is derived from a high molecular weight precursor.[199] In 1977, Phillip Sharp and Rich Roberts and their colleagues observed under the electron microscope that adenovirus mRNAs do not hybridize contiguously to the adenovirus genome, but rather loop out in segments (Figure 7.4), indicating that the mRNA is derived from regions of the genome that are not adjacent,[200–202] confirmed by others.[203]

The same phenomenon was soon reported in vertebrate genes encoding β-globin,[204,205] chicken ovalbumin (Figure 7.5) and lysozyme,[206–208] and immunoglobulin light chain,[209] and in ribosomal RNA genes in *Drosophila*,[210] whose cloned mRNA (cDNA or complementary DNA) sequences hybridized to multiple larger sized fragments of restriction endonuclease-digested genomic DNA in Southern blots. This is impossible to explain unless the cDNA sequences were spread over a large section of genomic real estate, which was confirmed by transcript mapping and sequence analysis.[211–213]

It transpired that the intervening sequences are 'spliced' out from the primary transcripts[197,202,214,215] – now called pre-mRNAs – in the nucleus, by a complex RNA-guided and catalyzed process (see following chapter), which explained the previously observed hnRNAs. The re-assembled 'mature' mRNAs are then exported to the cytoplasm for translation into proteins, so all was right with the gene=protein worldview, even if it is stranger than could possibly have been imagined.

The discovery of split genes (or "genes-in-pieces" as phrased by Walter Gilbert[157]) and mRNA splicing in eukaryotic cells was "a complete shock to the scientific world", as it broke another fundamental tenet of gene expression – the concept of collinearity – as "everyone assumed that the structure of a gene was a contiguous string of base pairs, from

which information was transferred for synthesis of a protein".[215]

The presence of introns interrupting the mRNA sequences of eukaryotic genes was immediately and universally assumed to be another manifestation of, and proffered as further evidence for, junk DNA[155,182,183,216] – notwithstanding and not considering the obvious alternative that other information may be transmitted by the excised non-protein-coding RNA,[217] and contemporary reports of "intron-mediated enhancement" of gene expression.[218] That the possibility that introns or intronic RNAs contain functional signals was not canvassed at the time is testimony to the strength of the belief that genetic information is (only) transduced through proteins, entrenched just 16 years after the *lac* operon.

Nonetheless the discovery of introns meant that the mystery of mammalian mRNA biogenesis had been solved.[219] It also helped to explain the vast amount of non-coding DNA in the genomes of higher organisms, and the existence of hnRNAs and the excess of RNA in the nucleus, reconciling the Central Dogma with these unusual features of eukaryotic gene expression. Crick described introns as "'nonsense' stretches of DNA interspersed within the sense DNA".[214] As put by Ohno, while bacterial genomes are "small and tidy", filled with polycistronic genes, the genomes of vertebrates are "untidy to the extreme", with genes spaced very apart from each other in such a way that "translation through so long spacer is out of question … [and] there was no choice but to achieve the fusion of adjacent coding sequences at the post-transcriptional level".[220]

It was also assumed that the excised intronic RNA is quickly degraded and the ribonucleotides recycled, although the technology of the time was too primitive to draw this conclusion. Northern blots, which may have been able to track the fate of the excised RNAs, had just been introduced (Chapter 7) and often relied on polyA-based purification protocols that neither capture nor detect spliced out RNAs.[x] Sharp and colleagues stated (with supporting references) that "introns are excised from pre-mRNAs with a half-life of 3 seconds to 30 seconds"[224] but a year later asserted (without supporting references) that excised introns "are rapidly degraded… (with) a half-life of … the order of a few seconds",[225] an entirely different statement, indicative of the logical

---

[w] The etymology is exon = EXpressed regiON, coined by Gilbert in 1978: "The notion of the cistron… must be replaced by that of a transcription unit containing regions which will be lost from the mature messenger – which I suggest we call introns (for intragenic regions) – alternating with regions which will be expressed – exons."[157]

[x] Other studies showed that at least 40% of all RNAs in human cells are not polyadenylated (Chapter 13).[221–223]

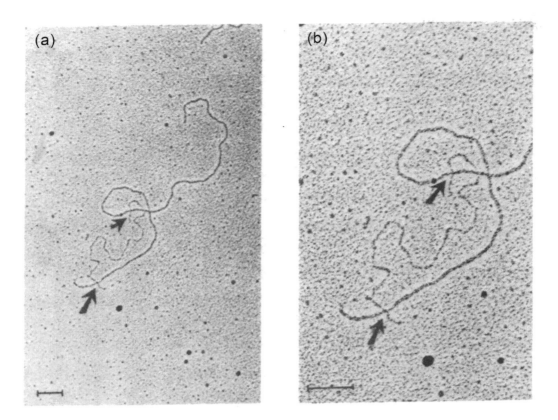

FIGURE 7.4 Electron micrographs of a hybrid between and adenovirus-derived mRNA and adenovirus DNA, with arrows showing boundaries of the R-loop of single-stranded DNA that is not present in the mRNA, the first demonstration of the presence of introns. (Reproduced with author permission from Berget et al.[200])

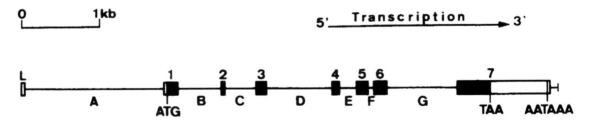

FIGURE 7.5 Exon intron structure of the chicken ovalbumin Y gene. Filled boxes indicate protein-coding sequences, with unfilled areas indicating 5' and 3' untranslated regions in the mature mRNA. (Reproduced from Heilig et al.[228] with permission from Oxford University Press.)

slip. In fact, intron-specific *in situ* hybridization showed that excised intronic RNAs can be relatively stable and easily detectable in the nucleus.[226] Later studies showed that many functional RNAs are derived from introns (Chapter 8) and that intronic RNAs – including retained introns and intron-derived RNAs – constitute the major fraction of the non-coding RNAs in mammalian cells.[227]

Sweeping introns under the intellectual carpet as junk (like the transposon-derived sequences within them) still left the question of how the split gene arrangement came to be in the first place. Their existence was subsequently rationalized by Gilbert as a hangover of the primordial assembly of genes from fragments of protein-coding information (the 'introns-early hypothesis').[157]

Gilbert also predicted that the presence of introns would enable 'alternative' splicing and thereby the evolution of modularity in protein-coding genes, expanding the repertoire of protein isoforms in complex organisms[157] (Figure 7.6). This proved to be correct,[229–232] and was recently shown to include the exonic capture of fragments of transposable elements to allow the protein to act as a genome-wide transcriptional regulator, leading to the conclusion that "TEs interacting within their host genome provide the raw material to generate new combinations of functional domains that can be selected upon and incorporated within the hierarchical cellular network".[41]

Gilbert's hypothesis was elaborated independently by Darnell,[233] Doolittle[234] and Colin Blake,[235] with the sequitur that exons would be predicted to encode protein functional units or "smaller, supersecondary structures".[235] While there was evidence that

some exons corresponded to protein domains,[236–238] it was difficult to show that most protein-coding exons comprised modular elements of protein structure,[217,239] and later studies showed that alternative splicing is more common in regulatory sequences in mRNAs and non-coding RNAs than in protein-coding exons.[240]

Developmentally complex organisms have a greater number and larger size of introns,[241,242] comprising at least 40% of the human genome – and likely much more, given that there are many distal alternative promoters and 5' exons expressed in early development, introns in genes encoding non-coding RNAs, and many genes enclosed within introns of other genes (Chapter 13).

By contrast, it was argued, the genomes of fast-growing microorganisms had been streamlined under pressure for rapid replication, overlooking the fact that developmentally complex eukaryotes had microbial ancestors for at least a billion years, which would have been subject to the same pressures. As Gilbert expressed it: "… introns were lost in the course of evolution … [and] only genes in slowly replicating cells of complex organisms still retain the full stigmata of their birth".[243] This is, in evolutionary terms, a *non sequitur*, but nonetheless was repeated by others. Brenner in 1990:

> There is a view that *E. coli* is primitive and we are advanced. That is true from the point of view of function and action. But it is not true from the point of view of genome structure. Here it is *E. coli* that is streamlined and sophisticated, whereas it is our genome that has preserved a far more primitive condition.[244]

Introns were later found to reside in out-of-the-way places (non-translated RNA genes) in bacteria,[245] to have self-splicing capability and to be able to act as mobile genetic elements[246] (Chapter 8). Cavalier-Smith, Norman Palmer and John Logsdon Jr. argued that it was more likely that introns entered (by reverse splicing, Chapter 8) and expanded in complex organisms late in evolution,[216,241,247,248] while not challenging the assumption that they are devoid of information. This view persisted despite the examples of conserved sequences within them, which if removed or mutated have phenotypic effects (and, as seen later, also encoding distinct and stable RNA species).[249–252]

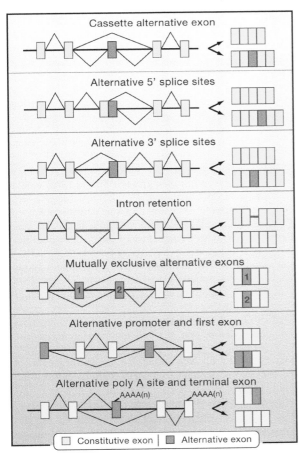

FIGURE 7.6 Types of alternative splicing. (Reproduced from Blencowe [232] with permission of Elsevier.)

The prevailing view was summarized by Matt Ridley in his 1999 book *Genome: The Autobiography of a Species in 23 Chapters*:

> Each gene is far more complicated than it needs to be, it is broken up into many different 'paragraphs' (called exons) and in between lie long stretches (called introns) of random nonsense and repetitive bursts of wholly irrelevant sense, some of which contain real genes of a completely different (and sinister) kind ... But ninety-seven per cent of our genome does not consist of true genes at all. It consists of a menagerie of strange entities called pseudogenes, retropseudogenes, satellites, minisatellites, microsatellites, transposons and retrotransposons: all collectively known as 'junk DNA', or sometimes, probably more accurately, as 'selfish DNA'. Some of these are genes of a special kind, but most are just chunks of DNA that are never transcribed into the language of protein.[253]

The presumed irrelevance of the vast tracts of transcribed non-protein-coding RNAs – "Mother Nature's dirty little secret"[253] or "junk in the attic"[65] – became accepted as such.

NOT JUNK?

There was an alternative, equally if not more plausible, and far more interesting, possibility canvassed by John Mattick in 1994,[217] i.e., that the separation of transcription from translation in eukaryotes allowed the invasion of protein-coding genes[y] by introns.[246,254] He posited that the evolution of the spliceosome then allowed these sequences to explore new genetic space and to acquire functions as RNA regulatory signals (or 'informational RNA', iRNA) expressed in parallel, akin to efference signals in neurobiology,[255] accounting for the expansion of these sequences. He also predicted that some, and perhaps most, genes in complex organisms express regulatory RNAs and that the evolution of RNA regulatory networks was the enabler of the appearance and radiation of developmentally complex animals.[217] That is, plant and animal genomes are not full of non-functional remnants of early evolution colonized by parasitic genetic hobos but are largely devoted to the specification of regulatory RNAs required for multicellular development[256–266] (Chapters 12–14 and 16). Later studies showed, *inter alia*, that 'enhancers' with tissue-specific activity (Chapters 14 and 16) are enriched in introns[267] and that many small regulatory 'microRNAs' are derived from introns (Chapter 12).[268]

FURTHER READING

Barton N.H., Briggs D.E.G., Eisen J.A., Goldstein D.B. and Patel N.H. (2007) *Evolution* (Cold Spring Harbor Laboratory Press, Cold Spring Harbor, NY).

Berk A.J. (2016) Discovery of RNA splicing and genes in pieces. *Proceedings of the National Academy of Sciences USA* 113: 801.

Brown S.W. (1966) Heterochromatin. *Science* 151: 417–25.

Callier, V. (2018) Theorists debate how 'neutral' evolution really Is. Quantamagazine, November 18, 2018. https://www.quantamagazine.org/neutral-theory-of-evolution-challenged-by-evidence-for-dna-selection-20181108/#.

Darnell J.E. (2011) *RNA: Life's Indispensable Molecule* (Cold Spring Harbor Laboratory Press, Cold Spring Harbor, NY).

Darnell J.E. (2013) Reflections on the history of pre-mRNA processing and highlights of current knowledge: A unified picture. *RNA* 19: 443–60.

Dover G. (2000) *Dear Mr Darwin: Letters on the Evolution of Life and Human Nature* (Weidenfeld & Nicolson, London).

Hurst L.D. (2009) Genetics and the understanding of selection. *Nature Reviews Genetics* 10: 83–93.

Kumar S. (2005) Molecular clocks: Four decades of evolution. *Nature Reviews Genetics* 6: 654–62.

Chambon P. (1978) Summary: The molecular biology of the eukaryotic genome is coming of age. *Cold Spring Harbor Symposia on Quantitative Biology* 42: 1209–34.

Mattick J.S. (1994) Introns: Evolution and function. *Current Opinion in Genetics and Development* 4: 823–31.

Mattick J.S. (2001) Non-coding RNAs: The architects of eukaryotic complexity. *EMBO Reports* 2: 986–91.

Mattick J.S. (2004a) RNA regulation: A new genetics? *Nature Reviews Genetics* 5: 316–23.

Mattick J.S. (2004b) The hidden genetic program of complex organisms. *Scientific American* 291: 60–7.

Scherrer K. (2003) Historical review: The discovery of 'giant' RNA and RNA processing: 40 years of enigma. *Trends in Biochemical Sciences* 28: 566–71.

---

[y] Transcription and translation are tightly coupled in prokaryotes, meaning that introns in protein-coding genes are disruptive, whereas the separation of transcription from translation by the nuclear membrane in eukaryotes allows intron excision before translation. The only introns found in prokaryotes to date are located in rRNA and tRNA genes, which would not be subject to such strong negative selection.

8 The Expanding Repertoire of RNA

The biochemical analyses of RNA in the 1960s detected many short RNAs in the nucleus and cytoplasm of eukaryotic cells[a] using new techniques of radioactive labeling, differential sedimentation and gel electrophoresis, and better procedures for isolating intact RNAs with detergents and chaotropic agents[1] to overcome degradation by RNases.

Initially there were concerns that the small RNAs are by-products of the biogenesis or degradation of larger RNAs. On the other hand, contrary to the generalization that nuclear RNAs are transient, destined for processing and export to the cytoplasm, the few groups studying these newly identified low molecular weight RNAs reported that they are highly expressed and account for ~20% of the nuclear RNA in mammalian cells.[2] They were also found to differ in size and sequence composition from tRNAs and rRNAs, and to be metabolically stable.[3]

At least 10 discrete RNA species were identified, some of which contained methylated nucleotides, localized in specific subnuclear fractions (the nucleoplasm, chromatin or the nucleolus), with others in the cytoplasm.[4–9] Many of those RNAs in the nucleus were uridine-rich, leading to their designation as 'U RNAs', numbered in the order of their discovery as U1, U2, U3 and so on.[2,6] Robert Weinberg and Sheldon Penman named them "small nuclear RNAs" (snRNAs).[6,10]

It was also found that RNA polymerase III is responsible for the transcription of many of these small RNAs, not RNA polymerase II, which transcribes mRNAs (and long non-protein-coding RNAs, Chapter 13), indicating that different RNA polymerases synthesize different classes of RNA.[11–13] RNA polymerase III products also include RNAs originating from repetitive sequences, only later characterized, such as those transcribed in human cells from Alu elements.[13,14]

The characterization of snRNAs in the following decades revealed that they had 'housekeeping' functions in the modification and maturation of rRNAs, tRNAs and mRNAs, as well as other functions in gene regulation and cellular processes such as protein export. These years also saw RNAs encroach on the traditional domain of proteins, catalysis, which in turn led to a plausible explanation of the molecular origin of genetic information, with RNA at its core.

SPLICEOSOMAL RNAs

Although their roles were unknown, snRNAs were too small to function as mRNAs and, being mainly nuclear, did not seem to be directly involved in protein synthesis. On the other hand, some snRNAs contained sequences complementary to hnRNAs. This led Michael Lerner and Joan Steitz, and independently John Rogers and Randolph Wall, to propose in 1980 that these RNAs play a role in RNA splicing,[15–17] based on earlier work on sense-antisense interactions between mRNA and rRNAs sequences in translation initiation.[18,19] In particular, the complementarity of the U1 snRNA sequence to both the 5' and 3' splice site sequences of hnRNAs led to the hypothesis that splice sites are recognized and aligned through RNA–RNA interactions between the splice sites and U1 snRNA.[15–17]

The characterization of the functions of these RNAs in the 1980s was aided by the fortuitous discovery that antibodies in the serum of individuals suffering autoimmune disorders, such as lupus erythematosus, precipitated ribonucleoprotein complexes (RNPs) containing snRNAs.[15,20–23] Several of the snRNAs interacted with a common antigen, the Sm (Smith) antigen,[b] named after the first lupus patient in whom such antibodies were detected.[20,23] Other autoantigens were associated with other RNP complexes.[25]

These antibodies were used not only to purify the complexes but also to block the function of the corresponding snRNPs *in vitro*, which showed that splicing of pre-mRNA is inhibited by targeting the U1 RNP and therefore that snRNAs are required.[26,27] Chemical cross-linking confirmed that U1 and U2 RNAs do, in fact, base pair to hnRNAs in the nucleus,[28,29] and genetic complementation experiments confirmed that U1, U2, U4, U5 and U6

---

[a] Also in bacteria (Chapter 9).

[b] Sm proteins participate in a wide variety of RNA transactions and predate the Archaea-Eukarya split.[24]

DOI: 10.1201/9781003109242-8

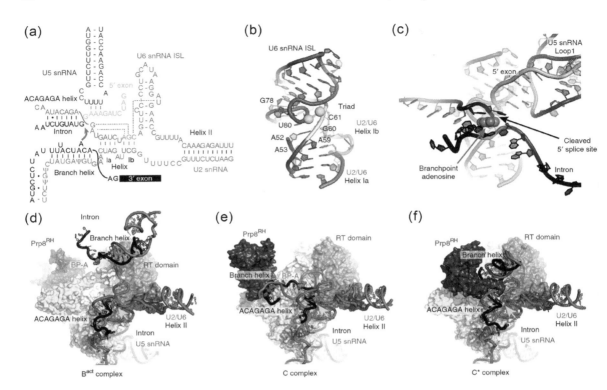

FIGURE 8.1 The mechanism of splicing and the complexity of the RNA interactions and structures (a) The RNA interaction network before the first trans-esterification reaction. The dotted lines indicate the triplex interactions at the catalytic core. (b) Three-dimensional structure of the active site RNA in the C complex. Magnesium ions are represented by two yellow spheres located between the backbone of the catalytic triad and the highly twisted backbone at the bulge in the internal stem loop. (c) Structure of the 5′ exon and branched intron bound to the active site (overlaid on the structure in b). (d–f) Interaction of the catalytic core of the spliceosome and movement of the branch helix, RNAs are color-coded as in a. The yellow arrows indicate the active site metals. (Reproduced from Fica and Nagai[31] with permission from Springer Nature.)

snRNAs function in splicing.[25,30] Characterization of the process revealed that large RNPs, which came to be known as 'spliceosomes', incorporate these snRNAs, wherein they interact with each other and with target pre-mRNAs to guide the process of splicing (Figure 8.1).[25,30–33]

Much later, it was found that U1 and U4 snRNAs also regulate transcriptional initiation, transcript structure and chromatin architecture,[34–40] and that U2 snRNA is required for RNA polymerase II pausing,[41] coupling transcription to splicing, which are intertwined processes (Chapters 14 and 16). It was also discovered that there is a minor class of spliceosome, which contains specific snRNAs (U11, U12, U4atac and U6atac) equivalent to but distinct from their counterparts in the major U2-type spliceosome (U1, U2, U4 and U6), and which recognizes a rare class of introns

initially referred to as AT–AC introns, now called U12-type introns.[42–45] The minor spliceosome is required for development and may have particular functions in the brain.[45–48]

SMALL NUCLEOLAR RNAs

Other small RNAs had other functions. The highly conserved U3 RNA was found to associate with 28S rRNA[8,49] and to be localized in the nucleolus,[11,50] where ribosome biogenesis occurs, which led Jean-Pierre Bachellerie to suggest in 1983 that U3 and other "small nucleolar RNAs" (snoRNAs)[c] participate in this process.[50] Subsequently it was also found

---

[c] Although they were originally named based on their nucleolar localization (in contrast to snRNAs), snoRNAs are also found beyond the nucleolus[51] and are secreted from cells.[52]

that autoimmune antibodies recognizing a nucleolar protein, fibrillarin,[53] would co-precipitate not only U3 but also the less abundant U8 and U13 RNAs.[54] It was later found also to bind U16.[55]

Similar to the observations that led to the elucidation of the roles of snRNAs in pre-mRNA splicing, snoRNAs were found to have short sequence motifs complementary to rRNA sequences, which indicated that snoRNAs were involved via base pairing in rRNA processing, modification or other aspects of ribosome biogenesis.[50]

Nevertheless, it was only in 1990 that U3 was shown to be essential for rRNA processing,[56] and in 1996 that snoRNA U24 directs site-specific methylation of rRNAs[57] with subsequent studies showing that other snoRNAs perform similar functions via base pairing with target sequences adjacent to modification sites.[51,58] Thus, although first identified in the 1960s, it was three decades before snoRNAs were defined as a new "class of RNAs" with demonstrated functions.[51,59]

SnoRNAs are ~60–300nt in length and are classified into two families (based on typical sequence motifs and structural features) that guide enzyme complexes to perform 2′-O-ribose methylation ('C/D box' RNAs) or pseudouridylation ('H/ACA box' RNAs) respectively of target nucleotides (Figure 8.2), not only in rRNAs and tRNAs but also in snRNAs.[60] Homologs of C/D box and H/ACA box snoRNAs occur in archaea, where they also guide modifications of tRNAs, indicating that they first evolved over 3 billion years ago.[61–63]

There are also many snoRNAs that show tissue-specific expression and whose targets are unknown, described as "orphan" snoRNAs.[60] Some were later found also to be involved in RNA processing, including the C/D box snoRNAs U8, U14, and U22, as well as H/ACA box snoRNAs snR10, snR30, E2 and E3, which direct site-specific cleavage of pre-rRNAs.[65] Yet, other snoRNAs were shown to regulate alternative splicing by base pair recognition[66] and to guide other modifications such as acetylation of specific cytosine residues in 18S rRNA.[67]

More was to come. In 2001, Beáta Jády and Tamás Kiss identified a sno-like RNA, U85, containing

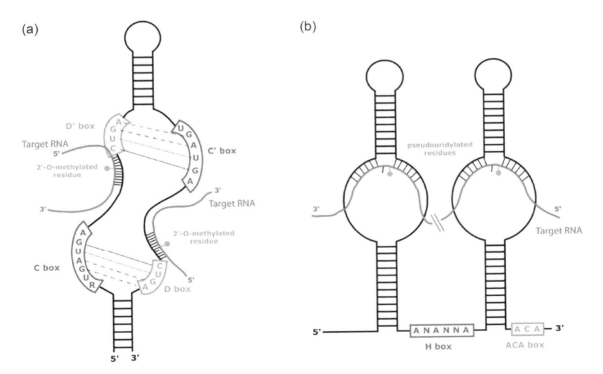

(a)   (b)

FIGURE 8.2 Schematic representation of C/D (a) and H/ACA (b) box snoRNAs. Consensus box sequences are highlighted in green. The conserved structures within these snoRNAs guide effector protein complexes that catalyze 2′-O-methylation or pseudouridylation respectively. (Reproduced from Abel and Rederstorff[64] with permission from Elsevier.)

both H/ACA and C/D box motifs, which is localized in 'Cajal bodies' (a subnuclear domain associated with nucleoli, discovered by Santiago Ramón y Cajal in 1903[68]) and guides 2′-O-ribose methylation and pseudouridylation of the U5 spliceosomal snRNA.[69] The snRNA U7 is also localized in Cajal bodies and participates in histone pre-mRNA 3′ end formation.[70,71] Other small Cajal body-specific ('sca') RNAs have since been identified, some with a composite structure similar to U85, while others have only H/ACA box or C/D box domains, which guide modifications of spliceosomal snRNAs[72] and tRNAs, the latter as part of a stress response.[73,74] The RNA component of the human telomerase complex also contains a characteristic H/ACA box scaRNA-like structure and also localizes in Cajal bodies.[75–77] Primate-specific H/ACA snoRNA-like RNAs, called AluACA RNAs, are derived from intronic Alu-repeat RNAs[78] and new functions of snoRNAs continue to be discovered, such as the maintenance of chromatin accessibility.[79]

There are over 700 known snoRNA and snoRNA-like RNAs encoded in the human genome.[80] Most snoRNAs are produced by processing of intronic RNAs excised from host transcripts,[81] commonly those of genes encoding proteins involved in translation or ribosome biogenesis, including ribosomal proteins, translation factors and nucleolar proteins such as fibrillarin,[60,82] the first evidence of parallel genetic output. Many other snoRNAs are derived from the introns of transcripts that do not encode proteins,[83–88] some involving species-specific alternative splicing,[89] and whose primary function in many cases is uncertain, although others, like *Gas5*,[90] have demonstrated roles as long regulatory RNAs (Chapters 9 and 13). Some snoRNAs are expressed exclusively in the brain.[51,91–93]

The complexity of the relationships between small RNAs is illustrated by the later discovery that snoRNAs, from yeast to humans, are processed to produce three subspecies, one of which functions as a microRNA in the RNA interference pathway[64,94–96] (Chapter 12). The complexity of the networks, and an indication of how much is yet to be understood about them, is highlighted, for instance, by the observation that a human-specific snoRNA is attached to the end of a longer non-coding RNA that regulates rRNA biogenesis and nucleolar structure (via phase separation, Chapter 16).[97–99] In addition, many snoR-NAs accumulate in the form of stable lariats instead of fully processed snoRNP particles, with as yet unknown functions.[100]

Aberrant expression of snoRNAs has been linked to human disease. There is a large cluster of C/D box snoRNAs in a parentally imprinted gene that is normally expressed in the brain from the maternally derived allele, perturbations of which are associated with Angelman and Prader-Willi syndromes.[88,91,101,102] One of the snoRNAs in this region, HBII-52, contains an 18 nucleotide sequence that is complementary to an exon in the serotonin receptor 5-HT(2C) mRNA and mediates its alternative splicing.[103]

OTHER SMALL GUIDE, SCAFFOLDING AND REGULATORY RNAs

7SL and 7SK RNA species, their names reflecting their sedimentation coefficient, were identified by Penman in 1976.[104] These RNAs were initially thought to have a viral origin[3] but were shown to be present in uninfected cells.[104,105]

7SL RNA is ~300nt in length. It is ubiquitous in eukaryotic cells and was found, accidently,[d] to be an essential component of the protein export 'signal recognition particle' (SRP),[106] characterized in the 1970s and 1980s by Günter Blobel and colleagues.[107] The SRP associates with the ribosome and targets nascent proteins to the endoplasmic reticulum via an N-terminal 'signal' or 'leader' sequence for membrane insertion or secretion into the extracellular milieu,[106,108,109] with 7SL RNA acting as a scaffold upon which the six proteins of the SRP assemble.[109,110] Similar RNAs were later shown to be involved in protein export in bacteria and archaea.[111,112]

7SL RNA was subsequently found to be required for the selective packaging of the RNA/DNA modifying enzymes APOBEC3G and 3F into retroviral particles[113–117] and to repress the translation of the tumor suppressor TP53.[118] It is also a precursor of the Alu elements in the human genome (Figure 8.3).[119–121]

7SK RNA is highly expressed in vertebrates and, like most RNAs, has a complex structure.[122] Early evidence indicated that 7SK regulates transcription and transcription termination in a tissue- and species-specific manner,[123,124] but mechanistic insights would not emerge until the early 2000s.

---

d Blobel and colleagues were expecting only proteins to be components of the SRP. The presence of RNA was revealed by a strong UV absorbance signal at a wavelength typical of nucleic acids in the SRP preparation, detected as a result of an incorrect setting on the detector.[106]

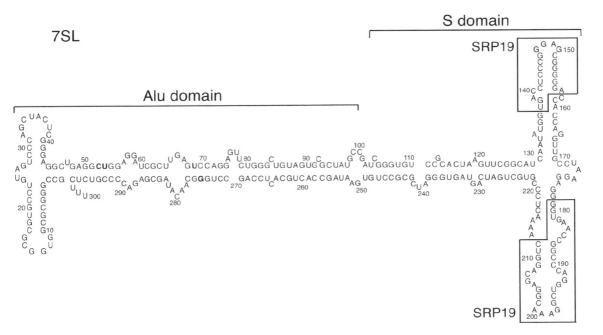

FIGURE 8.3 The structure of 7SL RNA, showing the part coopted into Alu transposable elements in primates. SPR19 binding sites are shown. (Reproduced with minor modifications from Itano et al.[117] with permission from John Wiley and Sons.)

These studies showed that 7SK RNA acts as a trans-acting negative regulator, uncovered serendipitously in biochemical assays to identify factors that regulate RNA polymerase II, similar to the "general transcription factor" role identified for the small 6S RNA in bacteria[125] (Chapter 9), but having additional roles in gene expression regulation in animals[126–134] (Chapter 13). 7SK is also present in invertebrates, and, despite the fact that it is little primary sequence similarity, its identification was possible due to its conserved secondary structural motifs and domains.[135,136]

In one of the earliest demonstrated regulatory RNA roles, in 1980, Hugh Pelham, Robert Roeder and colleagues discovered that the transcription factor TFIIIA[e] not only binds the 5S rRNA gene[f]

but also its transcript, which results in a feedback loop that titrates the transcription factor away from the gene, inhibiting further transcription and stabilizing the transcript until required for ribosome assembly.[140,141]

Y RNAs were identified in 1981 as components, like snRNAs, of autoantigens in systemic lupus patients.[21,142] There are four distinct and highly conserved Y RNAs (in humans ranging from ~80 to ~110nt),[143] which are structural components of the Ro autoantigen.[144–146] The Ro protein, lack of which causes a lupus-like syndrome in mice, appears to prevent autoimmunity by recognizing misfolded RNAs, with Y RNAs regulating the process.[147–149] Y RNAs also occur in bacteria where they associate with orthologs of mammalian Ro and are involved in rRNA maturation and stress responses,[150–152] with a modular structure that includes a domain that mimics tRNAs,[153] indicating an ancient function in cellular RNA biology.

Vault RNAs, short 80–150nt RNAs transcribed by RNA polymerase III, so named because of their presence in large ovoid ribonucleoprotein particles in the cytoplasm of eukaryotic cells that resemble the arches of cathedral vaults, were discovered in

---

[e] TFIIIA was shown by Aaron Klug and colleagues to interact with RNA and DNA via repetitive domains stabilized by zinc, called 'zinc fingers'.[137] Zinc finger transcription factors were later found to be the largest class of transcription factors in plants and animals, comprising 3% of the genes in the human genome[138] (Chapter 16).

[f] A non-coding RNA, termed 5S-OT, is transcribed from 5S rDNA loci in eukaryotes and has been shown to regulate transcription of 5S rRNA in mammals. An antisense Alu element has inserted at the 5S-OT locus in monkeys, apes and humans and regulates alternative splicing of other genes via Alu/anti-Alu pairing.[139]

1986 by Nancy Kedersha and Leonard Rome.[154,155] Vault RNAs are considered essential for eukaryotic cell biology because of their high conservation and near ubiquitous presence.[156] Their function is not well understood, but recent evidence indicates that they play a role in regulating autophagy,[157] i.e., the degradation and recycling of cellular components in lysosomes,[158] as well as apoptosis[159,160] and signaling pathways involved in neuronal synapse formation and plasticity.[161]

Other RNAs discovered in these years were the highly abundant viral RNAs in infected cells, some of which interact with autoantigens. The ~160nt VA RNAs (virus-associated RNAs) present in cells infected with adenovirus[163] (Figure 8.4) were in fact the first non-coding RNAs described after the 'canonical' RNAs (rRNA, tRNA and mRNA) and the first non-coding RNAs shown to be expressed from mammalian viruses, reported in the same year as the identification of the *lac* repressor.

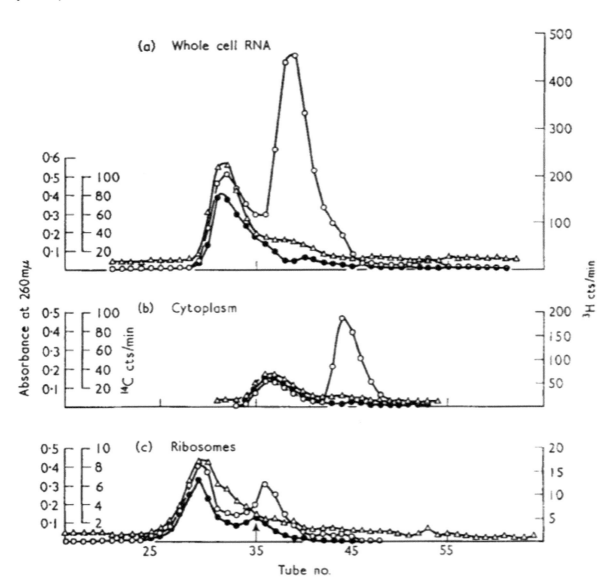

FIGURE 8.4   The discovery of low molecular weight virus-associated RNA (VA RNA): size chromatography of newly synthesized (radiolabeled) RNA isolated from infected (open circles) and uninfected (closed circles) cells infected with adenovirus (UV absorbance shown by triangles). (Reproduced from Reich et al.[163] with permission from Elsevier.)

VA RNAs were later shown (among other roles) to inhibit protein kinase R (PKR)[g] to curb the innate immune response and enhance the translation of viral RNAs.[163–169]

RNAs that are not highly conserved across large evolutionary distances but have specific expression patterns were also identified during the 1980s. Examples include neuronal RNAs that are transported into dendrites, such as B2, a ~180nt RNA transcribed by RNA polymerase III from repeated sequences in mice, which shows higher expression in some tumor cells and heat-shocked cells,[170–172] and BC1 (brain cytoplasmic RNA 1), a ~150nt RNA transcribed in rats from repeated sequences derived from a tRNA,[173–175] with a human equivalent, BC200.[176,177] Transgenic mice lacking BC1 have no obvious developmental deficiency, but display reduced exploration activity and increased anxiety, a phenotype that is invisible in the cage but likely lethal in the wild.[178]

Small RNAs are also required as primers for DNA replication[179,180] and for the maintenance of telomeres,[h] shown by Elizabeth Blackburn, Carol Greider and Jack Szostak to be accomplished by retrotransposon-derived RNAs operating on repeat sequences to replicate chromosome ends.[182–189]

Small common RNAs such as snRNAs and snoRNAs were relatively easy to find. Others were identified later in genomic datasets by characteristic sequence motifs and secondary structures, using a growing suite of bioinformatic tools[190–193] and databases such as *Rfam*[194,195] and *RNAcentral*.[196] On the other hand, lower copy number transcripts that are less conserved and expressed only in particular circumstances or cells were difficult to detect, a problem compounded by the lack of anticipation of the existence of cell-specific transcripts beyond mRNAs and their nuclear precursors (Chapters 12 and 13).

CATALYTIC RNAs AND THE ANCIENT RNA WORLD HYPOTHESIS

The participation of RNAs in a variety of cellular processes and the existence of many RNPs with different functions signaled that RNAs are versatile. As put by Crick in 1966, referring to the ability of RNA to form complex secondary structures: "tRNA looks like Nature's attempt to make RNA do the job of a protein."[197]

In 1957, an influential symposium in Moscow on the origin of life speculated that RNA most likely preceded proteins in the origin of life, a view supported there by Brachet, Mirsky, Oparin[i] and others.[199] In 1962, Alex Rich proposed that RNAs had a central role in the origin of life.[200] In the late 1960s, Orgel, Woese and Crick also hypothesized that RNAs might have preceded proteins in a pre-cellular world, predicting that RNAs possessed the required enzymatic activities.[197,201,202]

Nonetheless, the existence of catalytic RNAs in extant organisms was completely unexpected.[203] In 1982, Tom Cech and colleagues discovered that RNA can perform autocatalytic 'self-splicing' rearrangements, removing the intervening sequence of rRNA precursors by excision and cyclization in the ciliate protozoan *Tetrahymena*. They named these catalytic RNAs "ribozymes",[204] later categorized as 'self-splicing group I introns'.

Group II self-splicing introns were recognized in 1983 in organelle genomes by François Michel and Bernard Dujon.[205] They consist of a catalytically active intron RNA and an intron-encoded reverse transcriptase, enabling intron proliferation within genomes. Group II intron RNA catalyzes its own splicing via transesterification reactions that are the same as those of spliceosomal introns, yielding spliced exons and an excised intron lariat RNA.[206–208] Thus, group II introns appear to be the ancestors of modern spliceosomal introns[209–213] and likely entered the eukaryotic lineage through the bacterial ancestor of the mitochondrion.[214,215]

Also in 1983, Sidney Altman, Norman Pace and colleagues showed that RNA is the catalytic component of the bacterial RNase P complex, which produces mature tRNAs by cleaving a 5' end sequence in a process analogous to splicing.[216] The RNA in RNaseP is one of the only two ribozymes found in all domains of life:[217] a closely related eukaryotic RNA was also described in the early 1980s[218] and later shown to be the catalytic center of a complex involved in sequence-specific processing of mitochondrial and other RNAs,[219] hence called RMRP

---

[g] PKR is in fact variously inhibited or activated by trans-acting RNAs.[162]

[h] The existence of telomeres to protect chromosome ends was inferred by McClintock and Muller in the 1930s. Muller coined the term 'telomere' from the Greek 'telos' (end) and 'meros' (part).[181]

[i] Oparin previously had a long running debate with Hermann Muller, with Oparin maintaining that life was the outcome of a step-wise process of pre-cellular evolution of membrane-bound polymolecular systems, whereas Muller argued that life started with the appearance of the first nucleic acid molecule.[198]

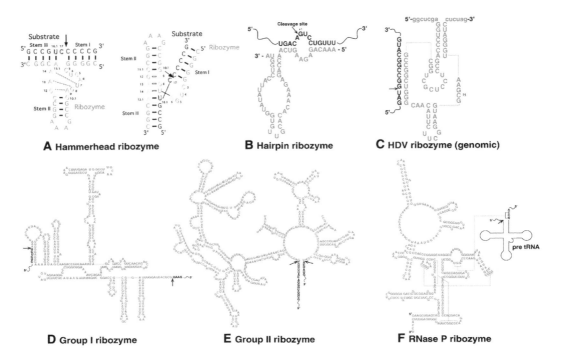

FIGURE 8.5 Secondary structures of six types of ribozymes. (Reproduced from Takagi et al.[222] with permission of Oxford University Press. The ribozyme or intron portion is printed in green. The substrate or exon portion is printed in black.)

(RNase MRP), although it is mainly located in the nucleolus and has other important regulatory functions (Chapter 13) (Figure 8.5).[220,221]

The demonstration that RNA molecules are able to cleave and join themselves or other RNAs, and capable of the phosphodiester bond transfers needed for RNA synthesis, prompted Walter Gilbert to formalize the "RNA World" hypothesis.[223] In this view, RNA molecules, not proteins, were the precursors of existing life, having performed the catalytic and information storage functions[j] in the pre-cellular world.[k]

Since then self-cleaving ribozymes have been found in bacteria, protists, fungi, plants, nematodes,

arthropods, insects and vertebrates, including human,[222,227–234] one of which has been shown to methylate other RNAs.[235] More surprisingly for its implications, B2 and Alu elements – 'repetitive' sequences that occur in ~350,000 copies and over 1 million copies in the mouse and human genomes respectively – have been shown to harbor self-cleaving ribozyme activity that is induced upon stress and T-cell activation by binding to the Polycomb histone methyltransferase protein EZH2[233] (Chapter 16).

It has been reasonably postulated that modified nucleotides may have enhanced the early catalytic capacities of RNAs and facilitated their path to self-replication.[236] Ribosomal RNAs, small nucleolar RNAs and spliceosomal RNAs are all heavily modified, and RNA modification has been widely deployed as a mechanism to introduce plasticity into RNA regulatory circuits (Chapter 17).

THE CATALYTIC HEART OF SPLICING AND TRANSLATION

In 1992, Harry Noller and colleagues showed that rRNA is not just a structural scaffold for the

---

[j] This view is supported by many observations, including that an RNA polymerase ribozyme obtained by *in vitro* evolution can copy complex RNA templates, including itself, albeit at low fidelity.[224] There are also plausible scenarios for the prebiotic synthesis of the pyrimidine and purine building blocks of RNA.[225]

[k] RNA can also nucleate 'liquid crystal' phase-separated domains, explored early on by Oparin (Chapter 2), who worked extensively on the role of RNA as a polyanion in the formation of 'coacervates',[226] a property that would come to the fore as central to both modern cell biology and prebiotic evolution as RNAs would have been able to sequester organic molecules in a proto-cell (Chapter 16).

(a)

(b)

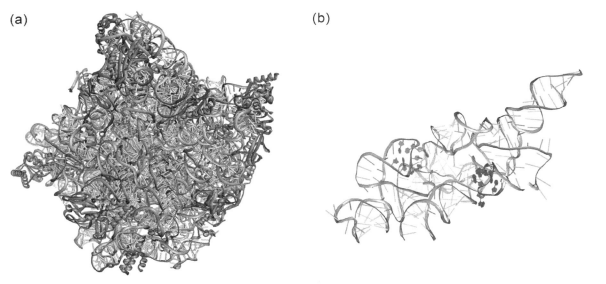

FIGURE 8.6 The active site of the ribosome, the peptidyl transferase center, is located on the large ribosomal subunit within a highly conserved region of the ribosomal RNA (red). (a) Model of RNA structure. (b) detail. (Image courtesy of Marina Rodnina and Wolfgang Wintermeyer (Max Planck Institute for Biophysical Chemistry, Göttingen).)

ribosome, as had been widely assumed, but harbors the central peptidyl transferase activity for protein chain extension in translation, making the ribosome a complex and conserved ribozyme[237,238] (Figure 8.6). RNAs must therefore have pre-existed proteins.

As noted above, RNA splicing is also an RNA catalyzed reaction,[1] with sequence and mechanistic similarities to group II self-splicing introns in bacteria, capable of inserting into new locations by reversal of the splicing reaction.[209,239]

Many small molecules, including antibiotics, target ribosomal and other RNAs: these include tetracyclins, aminoglycosides such as streptomycin and gentamycin, chloramphenicol, carbomycin A, blasticidin S, puromycin and hygromycin B.[240–242] They also include poisons such as ricin, an N-glycosidase RNA-modifying enzyme that depurinates a conserved loop of 28S rRNA and leads to irreversible arrest of protein synthesis.[243] Indeed, therapeutic targeting of diverse RNA types by small molecules is an area of rapidly growing interest.

THE DIGITAL AND ANALOG FACES OF RNA

This period also saw the expansion of RNA structural biology, led by Noller, Eric Westhof, Tom Steitz, Robin Gutell, Jennifer Doudna and others, who showed that RNAs have extraordinarily complex structures, capable of binding proteins, as a consequence of being able to form hydrogen bonds on all three faces – the Watson-Crick face, the Hoogsteen face (within the double-stranded groove) and the ribose, because of its 2' hydroxyl, which is lacking in DNA.[244–246] They also have exposed sequences that can base pair with other RNAs and DNA, through RNA:RNA duplexes, R-loops (RNA:DNA duplexes with displaced single stranded DNA) and RNA:DNA:DNA triplexes, which are common in eukaryotic chromatin[247,248] (Chapter 16).

The structural versatility of RNA has been explored and exploited *in vitro*. In the 1990s, the groups of Larry Gold and Jack Szostak developed SELEX ('Systematic Evolution of Ligands by Exponential Enrichment') to evolve artificial RNAs that bind specific ligands ('aptamers') or have other activities,[249–252] speculating that the same will have occurred *in vivo*.[253]

In addition, "free" low molecular weight circular RNA molecules lacking protein-coding capacity but

---

[1] U2 and U6 snRNAs interact to form the conserved structure of the catalytic triplex, coordinating two magnesium ions to form the active site of the spliceosome.[212]

having "peculiar" secondary structures and ribozyme activity were discovered in the 1970s to infect and autonomously replicate in plant cells (baptized as 'viroids'),[254,255] and postulated to represent living fossils[256] subject to Darwinian evolution in a prebiotic world.[257] Other curious virally associated RNAs were also described in the late 1970s and 1980s.[258–260]

CANDLES IN THE DARK

By 1985, it was becoming apparent that RNAs are multifaceted molecules that have specific subcellular locations, form complex structures, interact with (many) proteins and perform a vast array of functions beyond protein synthesis, from gene regulation to acting as components of cellular complexes, catalytic molecules and antisense guides. These observations did little to challenge the assumption that the destination of genetic information is (nearly always) the production of a protein and were regarded as interesting but idiosyncratic additions to the tapestry of molecular biology, rather than the first indications of a wider role for RNA in cell and developmental biology.

In his 1986 article describing the RNA World hypothesis, Gilbert proposed that, after the emergence of DNA as the carrier of genetic information, RNA was then "relegated to the intermediate role that it has today – no longer the centre of the stage, displaced by DNA and the more effective protein enzymes".[223] Others concurred.[261]

Penman lamented in 1991:

> If genes just make proteins and our proteins are the same, then why are we so different? … we have the bizarre proposal dominating biology that the incredibly complex living systems are described entirely by component

proteins and their coding sequences. Where is the genetic information that executes the design of an organism? We do not have to look far for a candidate. There is plenty of information in the more than 95% of the genome that is devoid of open reading frames. These sequences, heavily transcribed in all cells, appear to have little, if anything, to do with making proteins.[262]

FURTHER READING

Darnell J.E. (2011) *RNA: Life's Indispensable Molecule* (Cold Spring Harbor Laboratory Press, Cold Spring Harbor, NY).

Gesteland R.F., Cech T. and Atkins J.F. (2006) *The RNA World: The Nature of Modern RNA Suggests a Prebiotic RNA World* (Cold Spring Harbor Laboratory Press, Cold Spring Harbor, NY).

Joyce G.F. (1991) The rise and fall of the RNA world. *The New Biologist* 3: 399–407.

Lazcano A. (2012) The biochemical roots of the RNA world: From zymonucleic acid to ribozymes. *History & Philosophy of the Life Sciences* 34: 407–23.

Mount S.M. and Wolin S.L. (2015) Recognizing the 35th anniversary of the proposal that snRNPs are involved in splicing. *Molecular Biology of the Cell* 26: 3557–60.

Müller, F., et al. (2022) A prebiotically plausible scenario of an RNA–peptide world. *Nature* 605: 279-284.

Robertson M.P. and Joyce G.F. (2012) The origins of the RNA world. *Cold Spring Harbor Perspectives in Biology* 4: a003608.

Vicens Q. and Kieft J.S. (2022) Thoughts on how to think (and talk) about RNA structure. *Proceedings of the National Academy of Sciences USA* 119: e2112677119.

Wilkinson M.E., Charenton C. and Nagai K. (2020) RNA splicing by the spliceosome. *Annual Review of Biochemistry* 89: 359–88.

9 Glimpses of a Modern RNA World

The discovery of abundant small RNAs with functions beyond translation and the recognition of their target specificity by base-pairing prompted exploration of the regulatory potential of 'antisense' molecules.[1] In the late 1970s, Paul Zamecnik (whose group discovered tRNA, Chapter 3) and colleagues demonstrated that binding of short synthetic oligonucleotides to complementary sequences in Rous sarcoma virus and human T-cell lymphotropic virus blocked the replication and translation of viral RNA and the oncogenic transformation of cells.[2,3]

Such findings suggested that short antisense RNAs might exist naturally in cells, not as common species involved in core processes but as (individually rarer) regulators of specific genes or transcripts, but difficult to identify and characterize by the analytical techniques of the time.

RIBOREGULATORS

The clues were already there. Studies in the 1960s had revealed a number of small RNAs (sRNAs)[a] of unknown function in bacteria.[4,8,9] The existence of 'antisense' RNAs and 'bidirectional transcription' was first reported in 1972 in phage lambda, where it was proposed to control expression of the lambda repressor.[10] The subsequent sequencing of the genomes of bacteriophages and eukaryotic viruses showed that the occurrence of overlapping genes and transcripts was a general phenomenon.[11–14]

Conserved bacterial RNAs such as the 10Sa[b] and 10Sb[c] RNAs,[20,21] as well as regulatory motifs and structures, had also been reported around that time, for example, in feedback mechanisms controlling rRNA and ribosomal protein levels, akin to RNA structure-dependent regulatory mechanisms identified in bacteriophages.[22–26]

In 1975, Stuart Heywood and colleagues demonstrated that short RNA sequences from chicken muscle ribonucleoprotein fractions could control translation of the mRNA encoding myosin. They called the RNAs "translation control RNAs" (tcRNAs) and proposed that tcRNAs act by binding their mRNA targets in a sequence-specific manner.[27,28] In follow-up work a decade later, Heywood demonstrated that one of these tcRNAs, tcRNA102, recognizes a sequence in the 5′UTR of the myosin mRNA.[29,30]

In the 1980s, a number of studies identified bacterial plasmid-encoded small 'untranslatable' antisense RNAs that formed stable secondary structures and regulated plasmid replication, plasmid incompatibility, transposition and translation, among others.[31–35] For example, mutation analysis revealed that the ~108nt antisense transcript 'RNA I' blocked replication of the ColE1 plasmid (Chapter 6) by base pairing with the RNA that forms the replication primer[31] – one of the first regulatory roles demonstrated for any RNA (see Chapter 8). Soon after, a ~70nt RNA transcribed from a promoter of the Tn10 transposon was shown to repress transposition by preventing translation of the transposase mRNA, representing the first example of transposon regulation by antisense RNAs.[34] A ~170nt antisense RNA (micRNA) expressed in *E. coli* was found to inhibit translation of OmpF mRNA, which encodes a major outer membrane protein.[35] Packaging of phage DNA during infection was found to be directed by the phage-encoded ~120nt phi29 RNA, as part of the DNA-packaging machine.[36]

Before the end of the decade, enough examples had accumulated to allow generalizations around the theme of antisense RNA control of gene expression and the potential of fine-tuning interactions in a way not readily achieved by proteins.[37] Masayori Inouye speculated at the time that this "regulatory system may be a general regulatory phenomenon in *E. coli* and in other organisms, including eukaryotes"[38] and that "RNA species may have additional roles in the regulation of various cellular activities".[35]

---

[a] One of these sRNAs was discovered in 1967,[4] 9 months after the identification of the *lac* repressor. It was initially dubbed 6S (also known as SsrS) and later shown to be a structurally conserved molecule that regulates RNA polymerase promoter use.[5–7] One can only speculate what the impact on the conceptual framework of RNA and protein function in molecular biology might have been if this had come to light earlier.

[b] 10Sa RNA is also known as tmRNA (a 'transfer-messenger RNA' with properties of a tRNA and an mRNA) or SsrA.[15–17] It was later shown to play a key role in the symbiosis between the bacterium *Vibrio fischeri* and squid[18] (Chapter 12).

[c] 10Sb RNA was later found to be the bacterial homolog of the RNase RMP ribozyme[19] (Chapter 8).

DOI: 10.1201/9781003109242-9

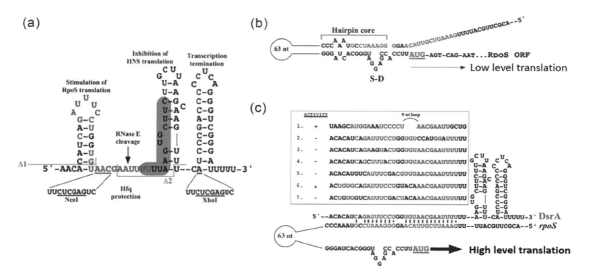

FIGURE 9.1   Structure (a) and mode of action of DsrA in controlling RpoS translation. The Shine-Dalgarno (S-D) ribosome binding sequence and the AUG start codon are sequestered by the 5' sequence of the RpoS mRNA (b) but released by DsrA binding (c). (Reproduced from Majdalani et al.[59])

In subsequent decades, it was shown that sRNAs regulate many bacterial processes,[d] including virulence, quorum (community) sensing, symbiosis, stress responses, the physiological transition from growth to stationary phase, other aspects of metabolism and environmental responses, bacteriophage packaging, DNA exchange, transcription and translation, among others.[18,38,42,43] Thousands of bacterial sRNAs that regulate gene expression at both transcriptional and post-transcriptional levels[42] have now been described,[44–46] aided by new high-throughput RNA-protein interaction technologies.[47–50]

The following examples are illustrative. The *E. coli* DsrA RNA (Figure 9.1), which is induced at low temperatures, inhibits transcriptional silencing by the nucleoid-associated H-NS protein and stimulates translation of the stress sigma factor RpoS, both depending on association with the RNA-binding protein Hfq.[51–53]

Another ~109nt sRNA, oxyS, was found to repress translation of RpoS by interacting with Hfq and altering its activity, acting as a global regulator to activate or repress the expression of approximately 40 genes involved in stress responses.[54,55] In fact, many sRNAs, such as the Spot 42 sRNA that regulates the galactose operon (Chapter 3), also require Hfq[e] for their stability and function.[57,58]

The common involvement of Hfq, which acts as a general cofactor for stabilizing small antisense RNAs, facilitating RNA-RNA interactions and gene expression control in many bacteria,[57,60–64] including the regulation of utilization of the intestinal metabolite ethanolamine by a Hfq-dependent sRNA,[65] indicated the existence of broader RNA-regulated networks.[42,60] This was also an early example of the use of a generic protein infrastructure to execute RNA-directed regulatory events, a theme that would later be writ large in eukaryotes (Chapters 12 and 16).

Other RNAs that control global processes in bacteria were also discovered, such as the inducible CsrB and CsrC RNAs of *E. coli*, which bind (via conserved sequences and hairpin structures) and inhibit the RNA-binding protein CsrA, a translational regulator, by outcompeting mRNA targets.[59,66] Homologs of this system have been implicated in the regulation

---

[d] As explained by Kai Papenfort and Jörg Vogel: "The late appreciation of regulatory RNA might be attributed to the fact that loci encoding such regulators were rarely selected in genetic screens for virulence factors, likely owing to a usually smaller gene size, missing annotations in genome sequences, and typically subtle phenotypes, as compared to virulence-associated proteins."[39] For example, the ~514nt RNAIII was originally described as the δ-hemolysin mRNA of *Staphylococcus aureus* but subsequent molecular analysis revealed that, in addition to expressing hemolysin from its 5' region, RNAIII acts as an antisense regulator of virulence and surface protein synthesis through its 3' region, a dual coding and regulatory RNA.[40,41]

[e] Hfq was originally described in 1968 as an *E. coli* host factor required for the synthesis of bacteriophage Qβ RNA and the replication of the bacteriophage Qβ RNA genome.[56]

of gluconeogenesis, biofilm formation and virulence factor expression in a variety of bacterial pathogens,[59] and represent some of the first examples of mimicry, protein-sequestration or sponging by regulatory RNAs at a post-transcriptional level.

In 2020, the late promoter of the Shiga toxin-encoding bacteriophage in enterohemorrhagic *E. coli* was found to produce an abundant regulatory RNA to silence the expression of the toxin during lysogeny (the Shiga toxins cause renal failure and neurological damage), which had been hiding in plain sight despite decades of research on Shiga toxin.[67]

Synthetic riboregulators have been constructed for eukaryotic translational control.[68] A common principle underlying the functions of these small regulatory RNAs is the ability to combine secondary structures that can bind proteins or small ligands (as exemplified by some ribozymes[69,70] and SELEX) with exposed nucleotide stretches that can recognize other RNAs or DNA in a sequence-specific manner.

RIBOSWITCHES

In 2002 and following years, Ron Breaker, Wade Winkler, Alexander Mironov, Evgeny Nudler and others showed that the ability of RNA to sense ligands, previously thought to be the sole province of proteins, is widely used by bacteria to connect the regulation of transcription and translation to metabolic and environmental signals, including thiamin (vitamin B1), riboflavin-5′-phosphate (vitamin B2), biotin (vitamin B7), cobalamin (vitamin B12), fluoride, various amino acids, S-adenosyl methionine (SAM) and glucosamine-6-phosphate, among many others, and even temperature (RNA 'thermometers').[71-77] These RNA ligand-sensing modules have become known as 'riboswitches',[78-81] with more evident in genomic analyses,[82] and high-resolution studies revealing the molecular dynamics involved.[83]

For example, the SAM riboswitch (the 'S-box leader') is a highly conserved RNA domain that responds to the coenzyme SAM with high affinity and specificity. In *Bacillus subtilis*, it occurs in the 5′ region of dozens of genes encoding proteins involved in methionine or cysteine biosynthesis, where it allosterically regulates their expression at the level of transcription termination. When SAM is unbound to the RNA aptamer, the anti-terminator sequence sequesters the terminator, which is then unable to form, whereas when SAM is bound, the anti-terminator is sequestered and transcription is terminated (Figure 9.2).[73,84,85]

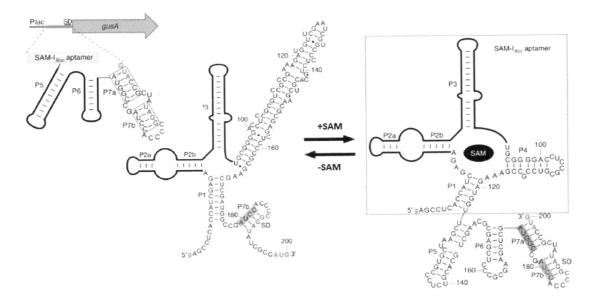

FIGURE 9.2 Structure and function of the S-adenosyl methionine (SAM) riboswitch in the 5′ untranslated region of the mRNA in the polycistronic *met* operon of *Xanthomonas campestris*, which encodes three enzymes for the biosynthesis of methionine, replaced here by the reporter gene *gusA*. Binding of SAM to the RNA aptamer in the riboswitch (boxed) causes an allosteric structural rearrangement that sequesters the Shine-Dalgarno sequence (purple) and AUG start codon (cyan) to inhibit translation. (Adapted from Tang et al.[85] under license from Creative Commons.)

Thiamin riboswitches also occur in plants, fungi and protists,[86–88] and shown, *inter alia*, to regulate RNA splicing.[89] Riboswitches likely also exist in animals, although their repertoire is not so well explored, possibly because of the difficulty of their characterization in complex organisms. Artificial riboswitches have been constructed to respond to pH[90] and light[91] and to control RNA splicing.[92]

Ligand-induced allosteric changes in RNA structure are similar to those observed in proteins upon binding of small molecules, such as nucleotides (ATP, AMP, GTP, etc.) to sense energy status or transduce extracellular signals, and even the *lac* repressor's recognition of lactose, which causes a conformational change in the repressor so that it can no longer bind DNA to block transcription of lactose metabolizing enzymes.

Riboswitches may have predated proteins and have been suggested to be the oldest mechanism for the regulation of gene expression.[93] As Breaker speculated: "The characteristics of some riboswitches suggest they could be modern descendants of an ancient sensory and regulatory system that likely functioned before the emergence of enzymes and genetic factors made of protein."[94]

ANTISENSE RNAs AND COMPLEX TRANSCRIPTION IN EUKARYOTES

The first evidence of regulatory antisense RNAs in eukaryotes was obtained in 1987, also by Zamecnik's group, who reported endogenous small (<30nt) RNA oligonucleotides in mammalian cells using radioactive labeling, proposing that they "may play a regulatory role in intracellular metabolism and may conceivably travel from one cell to another in a similar role".[95] It was another 13 years before the ubiquity and power of such regulatory 'microRNAs' in eukaryotes started to be revealed and appreciated (Chapter 12).

Nonetheless, based on the principles of the activity of antisense RNAs in bacteria, as well as experiments by Zamenick's group with exogenous antisense oligonucleotides, "anti-message" RNAs began to be used from 1984 as a tool for suppressing the expression of specific genes[f] in eukaryotes, including globin, at the level of transcription, translation and/or RNA

stability[1,99–102] before the discovery of natural antisense transcripts in eukaryotes.[103,104]

The use of synthetic antisense oligonucleotides was quickly adopted and is still widely employed to study gene function in a wide range of eukaryotes, including frogs, insects, plants and mammalian cells, as specificity of inhibition is easy to achieve independently of any knowledge of the function of the gene under investigation.[103]

Antisense oligonucleotides also form a vital component of the toolkits for genetic engineering and gene therapy,[g] aided by artificial chemistries, such as peptide or phosphorothioate linkages and methylene bridges ('locked nucleic acids'), to increase the target affinity and half-life of the molecules *in vivo*.[107–111] The use of synthetic nucleic acids has since been given new impetus by the discovery of another natural antisense RNA regulatory pathway, the small RNA-guided 'CRISPR' systems that have revolutionized genetic engineering (Chapter 12).

At that time, however, despite the emerging examples of regulatory sRNAs in bacteria and the ability of antisense molecules to artificially modulate gene expression in eukaryotes, "the extent to which this novel form of regulation of gene expression is utilized in prokaryotes and eukaryotes … [remains] … to be established".[112]

The first discoveries of natural regulatory RNAs in eukaryotic cells were serendipitous[h]–a pattern repeated over the next ~15 years, until the genome projects revealed the full extent of RNA expression (Chapter 13). Nevertheless, an unexpected by-product of genetic screens and conventional gene cloning and mapping approaches were many early observations that hinted

---

f Antisense interactions between cDNAs and mRNAs ('hybrid-arrested translation') was developed in the late 1970s for gene mapping and identification.[96–98]

g The first RNA therapeutics company, Isis, now Ionis, was established in 1989 by Stanley Crooke.[105] As of 2021, eight antisense oligonucleotide drugs had been approved for commercial use.[106]

h One of the relevant discoveries during this period was the HIV TAR (trans-activating response element), an RNA stem-loop structure located at the 5′ ends of nascent HIV-1 transcripts, which was proposed to be a "novel type of regulatory element" for transcriptional activation.[113–115] In addition to the viral regulatory RNAs mentioned in the previous chapter, several other non-coding RNAs were subsequently characterized from DNA and RNA viruses that infect eukaryotic cells. These include highly abundant small RNAs and lncRNAs discovered in the late 1970s and 1980s, such as the EBV-encoded RNAs found to have different roles including recruitment of transcription factors to control expression[116] (Chapter 16) and the 2.7 kb repeat-derived RNA that comprises ~20% of the early transcription from the human cytomegalovirus Beta2.7 gene[117] and whose function only started to be identified decades later.[118]

at the existence of longer (>200nt) non-protein-coding RNAs in eukaryotes.

These early studies also revealed the existence of 'nested' genes and non-coding transcripts in intensely studied genomic regions, such as developmental and cancer-related loci, as well as in studies using differential cDNA cloning and hybridization strategies[i] to identify transcripts from genes that are active or repressed in specific tissues and/or developmental stages.

In 1986, Steven Henikoff and colleagues reported the first case of a "gene within a gene" in *Drosophila*, showing that a pupal cuticle protein is encoded within the intron of an unrelated gene, on the opposite strand and independently expressed. They described their findings as "an unambiguous exception to the classical linear model of gene organization" and, considering the possible commonality of genes nested within large introns and extended loci, remarked that "it is interesting to consider the genetic complexity that could result".[120]

In the same year, Trevor Williams and Mike Fried showed that a region of the mouse genome encodes two RNAs that are transcribed in opposite direction and overlap at their 3' ends, contemplating the implications in the light of the findings of experimentally introduced antisense RNAs inhibiting gene activity.[121]

Similarly, in the same issue of *Nature*, Charlotte Spencer and colleagues reported that a transcript of unknown function overlaps that of the dopa decarboxylase (*Ddc*) gene on the opposite strand in *Drosophila*.[122] Given that the transcripts showed differences in temporal and spatial expression, they proposed that the antisense transcript could have regulatory function based on either RNA-RNA base pairing or via transcriptional interference and that "such arrangements in eukaryotes may be more common than previously supposed",[122,123] a prediction that was confirmed 20 years later when high-throughput transcriptome analyses were undertaken in the wake of the genome projects[124–126] (Chapter 13).

Also in 1986, Alain Nepveu and Kenneth Marcu showed that the protein-coding and opposite complementary strands of the *c-Myc* locus in mice are transcribed and regulated independently,[127] confirmed the following year by Gail Sonenshein and colleagues,[128] suggesting a role of the antisense RNAs in c-Myc processing or transcriptional interference.[127] The *TP53*

tumor suppressor locus was shown in 1989 to also express a long antisense RNA, 'inRNA', speculated to be involved in the maturation of p53 mRNAs.[129]

Other examples in different organisms followed. An antisense RNA expressed in the silk moth *Bombyx mori* was found to display extensive complementarity to the chorion gene *Hcb.12* and to be co-expressed in follicular cells during development.[130] An RNA antisense (aHIF) to the 3'UTR of the human Hypoxia-inducible factor-1 alpha (HIF1α) mRNA was found to be co-expressed in cancer and hypoxia.[131] The expression of other antisense RNAs was found to negatively correlate with that of their complementary protein-coding mRNAs, such as the intronic antisense RNA from the human *eIF2A* locus,[132] transcripts antisense to the chicken alpha-I collagen gene[133] and transcripts antisense to the *EB4* locus in the slime mold *Dictyostelium*, whose functional analysis suggested a role in regulating the stability of EB4 mRNA.[134]

In some cases, such as the tcRNA identified by Heywood, antisense RNAs had only limited sequence complementarity to their targets, suggesting the potential existence of 'trans-acting' RNAs that originated from different loci.[30,135] Other cases involved large regions of overlap between antisense RNA and mRNA, sometimes spanning most of the length of the transcription units.[136]

A large number of similar but disparate observations followed at the end of the decade, including demonstrations of other nested genes and sense-antisense pairs in plants, insects, birds and mammals,[136–142] including in well-studied loci such as the globin clusters, which also showed evidence of transcription of non-coding regions encompassing enhancers and *cis*-regulatory elements,[143,144] as did loci exhibiting parental imprinting (see below).

Some of these early reports considered the conceptual and practical implications of the interleaved organization of genes and transcripts, challenging orthodox concepts, especially that the exons of protein-coding transcripts are the only biologically relevant portions of genes. As put by Adelman and colleagues in 1987:

> this situation may be a significant form of molecular evolution. By using both strands of the same DNA, the information content (regulatory and/or structural) of a particular genetic segment becomes amplified, adding a new complexity to the concept of a eukaryotic gene.[145]

---

[i] Subtractive cDNA hybridization and differential display.[119]

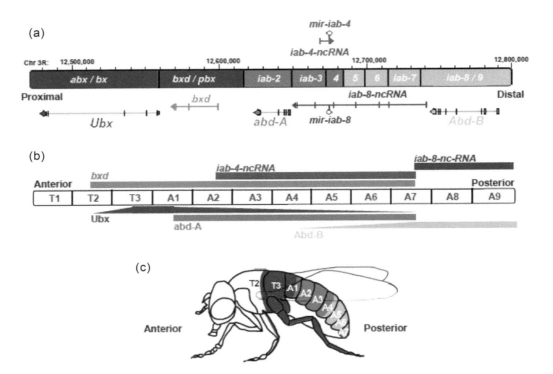

FIGURE 9.3 Map of the *bithorax* complex in *Drosophila*, showing the coding and non-coding transcripts produced from each locus – the protein-coding genes *Ubx*, *abd-*A and *Abd-b*, and the regulatory genes *bxd*, *iab-4* and *iab-8* – and their expression in different segments of the fly. The non-coding RNAs expressed from *iab-4* to *iab-8* are also regulated by microRNAs (Chapter 12). (Reproduced from Garaulet and Lai[150] with permission from Elsevier.)

LONG UNTRANSLATED RNAs

In addition to antisense RNAs, a number of other "unconventional" RNAs were found to be transcribed from 'intergenic'[j] regions of eukaryotic genomes and associated with genetic effects, notably in the well-studied regulatory regions of the *bithorax* complex studied by Lewis in *Drosophila*.[146]

As noted in Chapter 5, David Hogness, Michael Akam and colleagues discovered in 1985 that only one of the five mapped mutations ('pseudoalleles') in the *bithorax* complex corresponded to a protein-coding gene (*Ultrabithorax* or *Ubx*), while the others are derived from a much larger region containing regulatory elements. The pseudoallelic mutations were located in introns or in the upstream *bxd* region: the latter was found to be transcribed into a ~27 kb RNA that has a number of large introns and is subjected to differential splicing to produce various smaller (~1.2 kb) polyadenylated non-coding RNAs, none of

which has protein-coding potential.[147,148] The expression of these transcripts was also shown to be highly regulated during embryogenesis, in a pattern that is partially reflective of *Ubx*.[147,149]

Moreover, while the extended *BX-C* cluster contained three protein-coding genes (*Ubx*, *abd-A* and *Abd-B*), it produced at least seven distinct RNAs that are co-linearly transcribed during development[149,151] (Figure 9.3). Other non-protein-coding transcripts were also reported in the nearby *iab-4* locus.[150,152] As their relevance was unclear, it was mooted that these transcripts are "functionless"[149] or "might function in cis by some unprecedented mechanism".[151] Some suggested that transcription of these loci is a passive by-product of the recruitment of transcription factors to enhancer sites that act distally by looping to contact promoters of protein-coding genes, or that it is simply the act of transcription (not the transcribed RNA) that is relevant by remodeling and/or exposing chromatin and underlying DNA sequences to transcription factors.[153,154] Many (presumed) cis-regulatory elements encompassing different developmental 'enhancers' and 'response elements' for the epigenetic regulators

---

[j] That is, transcribed from genomic sequences between annotated protein-coding genes, a description reflecting the deep bias that genes = proteins.

Polycomb and Trithorax[155,156] have been characterized in the loci that express these RNAs,[157] but their mode of action is only now being elucidated, especially in the context of RNA-directed chromatin modifications that control the expression of the clusters (Chapters 14 and 16).

Another early example is the *93D* locus, one of the largest of the genomic regions in *Drosophila* that 'puff' (i.e., become transcriptionally active) after heat shock. In the early 1980s, the groups of Subhash Lakhotia and Mary Lou Pardue found that the *93D* locus does not specify a protein, but rather a set of rapidly evolving non-coding transcripts (although an intron is highly conserved[158]), known as *hsr omega* RNAs: long repeat-containing transcripts with at least three different isoforms of ~1–10 kb that are differentially expressed in different tissues and developmental stages.[159–164]

Transcription of repeat-containing spliced and polyadenylated long non-coding RNAs was also reported in loci involved in immunoglobulin class switching and recombination in the 1980s, with evidence that such transcripts are associated with 'enhancer' action and alteration of chromatin architecture, including V(D)J recombination at antigen receptor loci,[165–168] whose roles in these processes are now being understood.[169–172]

Many long 'nontranslatable' antisense, sense and intergenic RNAs from higher eukaryotes were also cloned during the 1980s and 1990s. One of the first was an interspersed maternally derived 3.7 kb non-coding RNA called ISp1, characterized by Britten and Davidson's group in 1988, studying the sea urchin *Strongylocentrotus purpuratus*. This transcript, whose function is unknown, is polyadenylated and apparently processed into shorter (400–600 nt) RNAs stored in the cytoplasm of sea urchin eggs.[173]

Large 'transcription units' of unknown function that appeared to lack protein-coding potential were being reported in humans as early as 1985, notably from the *PVT-1* (plasmacytoma variant translocation) locus that activates expression of the *MYC* oncogene and is heavily implicated in cancers through amplifications, retroviral insertions and translocations. The *PVT-1* locus spans at least 200 kb and expresses large multi-exonic non-coding RNAs that initiate 57 kb downstream of *MYC*.[174–181]

Other cancer-associated long non-coding RNAs (lncRNAs) identified in following years, such as BIC,[182–185] TP53TG1 ('TP53 target gene 1')[186,187] and DD3

(Differential Display Code 3, also known as PCA3, prostate cancer antigen 3),[188,189] turned out, like PVT1,[190] to be, at least in part, microRNA precursors (Chapter 12), as did, for example, the developmentally regulated and highly conserved 7H4 RNAs, found by subtractive hybridization to be highly enriched in synaptic nuclei of rat skeletal neuromuscular junctions.[191,192]

However, the first mammalian long non-coding RNA to be well recognized was H19. It was cloned in 1990 by Shirley Tilghman and colleagues by differential hybridization, and corresponded to a transcript that was originally identified by the same group in 1984 as an abundant RNA in a screen of a mouse fetal liver cDNA library (named after the cDNA clone designated pH19).[193] It was also found to be expressed in rat skeletal muscle (where it was called ASM).[194] H19 is transcribed by RNA polymerase II, spliced and polyadenylated, with a number of short open reading frames that are not conserved between mouse and human.[k] It also did not seem to be associated with ribosomes. *H19* was, therefore, proposed to represent an "unusual gene" whose product differs from a "classical mRNA" in that it may act as an RNA, not an intermediate to a protein.[193] It was subsequently found to be part of an imprinted locus that encompasses the *Igf2* (insulin-like growth factor) gene, to have tumor suppressor capacity[196–201] and to be lethal when ectopically expressed.[202]

In the following year, Carolyn Brown, Hunt Willard and colleagues cloned an RNA expressed exclusively from the 'X-inactivation center' in the inactive X chromosome in females. This alternatively spliced transcript was a candidate for the factor responsible for dosage compensation and was named Xist ('X-inactive specific transcript').[203] They found no evidence of an encoded protein, and characterized human XIST[l] as a 17-kb RNA containing conserved tandem repeats, localized within the nucleus and coating the inactive X chromosome in female cells in a position "indistinguishable from the X inactivation-associated Barr body", leading them to propose that Xist functions as a "structural RNA".[204]

Xist functions partially by recruiting chromatin-repressive complexes to promote heterochromatin formation and transcriptional silencing of the chromosome[205–209] (Figure 9.4), although how it selects

---

[k] Later proteomic analysis indicated that H19 produces a short protein in humans but not mice,[195] suggesting that it was either gained or lost in one or other lineage.

[l] The gene naming convention uses all capitals for human genes, and only the first letter in uppercase (e.g., *Xist*) in mouse.

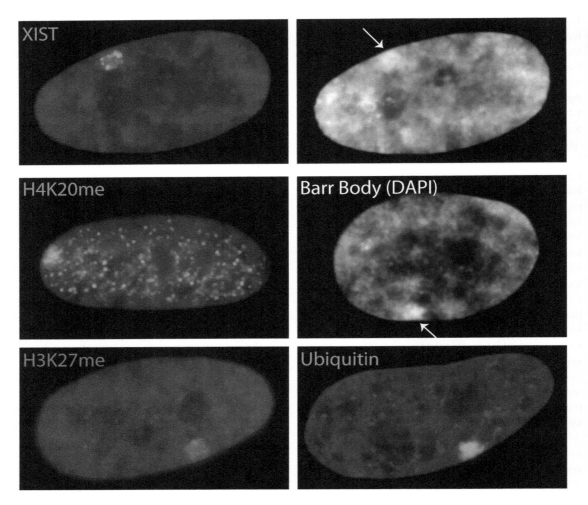

FIGURE 9.4 Localization of Xist on the transcriptionally inactive condensed X chromosome in interphase female human cells (the Barr body, with intense DAPI DNA staining), which is accompanied by repressive histone modifications such as methylation of H4K20 and H3K27, and H2AK119 ubiquitination (Chapter 14). (Image courtesy of Jeanne Lawrence,[205,206] UMass Chan School of Medicine).

just one of the two X-chromosomes[m] is not fully understood. While thought of as a special case at the time, Xist has become emblematic of the extraordinary complexity of non-coding RNA control of chromatin architecture including the nucleation of phase-separated domains (Chapter 16). It emerged later that the *Xist* locus originated by the fusion of a 'pseudogenized' protein-coding gene with a set of

transposable elements that are essential to its function.[212,213] And it does explain how the heterochromatic Barr body described by Ohno (Chapters 4 and 7) is formed.

Analogous non-coding RNAs balancing X-chromosome dosage in *Drosophila*, roX1 and roX2 (RNA on the X chromosome), were identified by use of enhancer traps and male-specific hybridization a few years later by Victoria Meller, Richard Axel, Mitzi Kuroda, Ron Davis, Richard Kelley, Asifa Akhtar and colleagues. The roX RNAs (whose activity is modulated by alternative splicing) act not to repress one of the two X-chromosomes in females but to globally upregulate gene expression from the single X chromosome in

---

[m] There are differences between rodents and primates, associated with distinctions in early development. The paternal allele of *Xist* is silenced in the trophectoderm (placenta) in mice, but not in humans. Seemingly random inactivation of parental alleles occurs in human trophectoderm and in both human and mouse embryos.[210,211]

males via tandem stem-loop structures that bind effector proteins to remodel chromatin and compartmentalize the X chromosome (also involving Phase Separation, Chapter 16).[214–222]

By the end of the 1990s, dozens of lncRNAs with regulatory functions had been identified in a wide variety of eukaryotes.[223,224] Early examples that hinted at the diversity of these non-coding RNAs included: 'meiRNA' in fission yeast, essential for the pairing of homologous chromosomes in meiosis,[225,226] involving phase separation[227] (Chapter 16); the dutA RNA in *Dictyostelium*, induced in development during mold aggregation;[228–230] an "unusual family" of 1.8–2.4 kb transcripts in the malarial protozoan parasite *Plasmodium falciparum* involved in the expression or rearrangements of virulence genes;[231] the bifunctional enod40 RNA in lucerne, expressed during nodule organogenesis, wherein the RNA structures are more highly conserved than the encoded peptides;[232–234] the pseudogene-derived antisense transcript[n] pseudoNOS (or antiNOS) suppressing the expression of the cognate nitric oxide synthase (*NOS*) gene in neurons of the snail *Lymnaea stagnalis*;[236,237] Xlsirt RNAs in frogs, "interspersed repeat transcripts" localized and playing structural roles in the vegetal pole cytoskeleton of *Xenopus* oocytes;[238,239] the 'yellow crescent RNA' in the ascidian *Styela clava*, a maternal transcript localized in the zygotic myoplasm;[240,241] the mammalian non-coding multi-exonic alternatively spliced 'growth arrest-specific' gas5 gene that hosts several snoRNAs (Chapter 8)[242–244] and is also active as a mitochondrially localized long non-coding RNA;[245] and SRA, a highly conserved steroid receptor coactivator (Figure 9.5), which was found accidentally in a protein-binding screen and later described as being "different from eukaryotic transcriptional coactivators in its ability to function as an RNA transcript to selectively regulate the activity of a family of transcriptional activators".[246–249] Such screens also detected many other RNAs that act as transcriptional activators "when tethered to DNA".[250–252]

As in *Drosophila*, several lncRNAs in vertebrates identified during the 1990s were found to originate from developmental loci, often showing coordinated expression and functional relationships with their associated protein-coding genes. These included RNAs antisense to homeobox-containing genes, such as *HoxA11*,[253,254] *HoxD3*,[255] and *Dlx1* and *Dlx6*.[256,257] *Xist* was also found to be overlapped by a gene specifying a 40 kb unstable antisense lncRNA, Tsix, which was identified by RNA fluorescence *in situ* hybridization and negatively regulates *Xist* expression during the early steps of X inactivation,[258,259] an early indication of the highly intricate regulatory networks involving lncRNAs (Chapter 16).

From the end of the 1990s, it emerged that the imprinted *Igf2/H19* locus and many other imprinted loci differentially express overlapping sense and antisense transcripts,[198,260–264] including the *Igf2 receptor* (*Igf2r*) locus, where an astoundingly long (~108 kb) antisense RNA was serendipitously found to be transcribed from a promoter located in an intron of the *Igf2r* gene.[265,266] In contrast to the protein-coding gene *Igf2r*, which is expressed from the maternal allele, the non-coding RNA, termed Air (Antisense Igf2r RNA), is expressed only from the paternal allele.[265–267]

Similar phenomena were observed at other loci,[268–270] including *Xist*,[271] *Meg3*[272] (also known as *Gtl2*, first isolated by a gene trap approach in mice[273,274]) and the *Kncq1* locus. In the latter, transcription of a 91 kb lncRNA named *KvLQT1-AS* (*KvLQT1 antisense*, later known as *Kcnq1ot1*), like Air, initiates in an intron of the paternal allele of the protein-coding *Kncq1* gene from a 'CpG island' (a region of high GC dinucleotide content that is a target for DNA methylation, normally associated with gene repression; Chapter 14) called the imprinting control region (IC2)[275–279] (Figure 9.6). Given their reciprocal expression, these antisense RNAs were proposed to be involved in the silencing of the associated protein-coding genes,[280] demonstrated for Air in 2002.[281]

Both *Air* and *Kcnq1ot1* RNAs were later shown by the groups of Peter Fraser and Chandrasekhar Kanduri respectively, to silence transcription by binding to and targeting the histone methylase G9a and DNA methyltransferases to chromatin to alter the epigenetic state of the locus[282,283] (Chapters 14 and 16).

UTR-DERIVED RNAs

The protein-coding portion (the open reading frame) of mRNAs is flanked by 'upstream' and 'downstream' untranslated regulatory regions, referred to as 5′UTRs and 3′UTRs, respectively. 3′UTRs have increased in size with increasing morphological

---

[n] The first report of an antisense transcript from a pseudogene was that from human topoisomerase I by Bing-Sen Zhou and colleagues in 1992.[235]

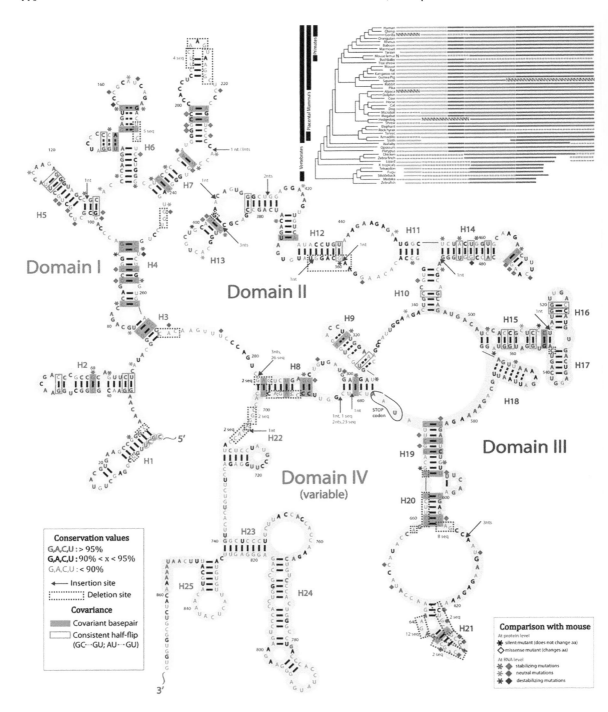

FIGURE 9.5 The structure and evolutionary conservation from fish to mammals of the steroid receptor coactivator RNA (SRA). (Reproduced from Novikova et al.[249] with permission from Oxford University Press.)

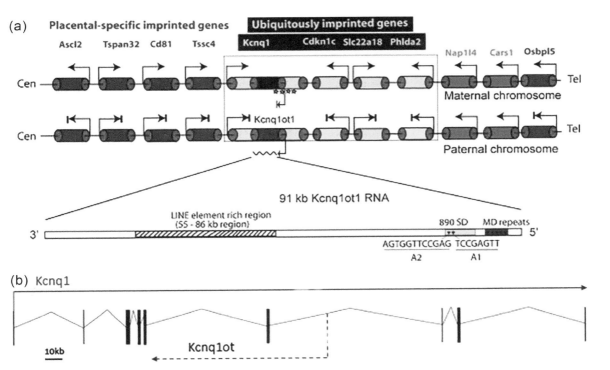

FIGURE 9.6 A composite figure showing (a) the genomic arrangement of the *Kcnq1* imprinted locus in mouse with (b) the exon-intron structure of the *Kncq1* gene and the antisense Kcnq1ot transcript initiated from its 6th intron. (Adapted from Kanduri[279] (a) and Pandey et al.[278] (b) with permission from Elsevier.)

complexity during animal evolution,[o] especially in vertebrates, where they usually occupy as much or more of the mRNA as the coding sequence in mammals and are often highly conserved.[285–287] 3′UTRs contain modules that bind regulatory proteins and small RNAs (Chapter 12) to control the translation, localization and stability of the mRNA.[288,289]

In 1993, Helen Blau and colleagues discovered that the 3′UTRs of three muscle associated genes (troponin I, tropomyosin and α-cardiac actin) could inhibit cell division and suppress malignancy in a myogenic cell line independently (i.e., in the absence) of the normally associated protein-coding sequence.[290,291] Other trans-acting 3′UTR-derived RNAs with similar properties were reported from the genes encoding ribonucleotide reductase[292] and prohibitin.[293] It was also shown that the loss of oogenesis caused by the lack of the *Drosophila* gene *oskar* can be rescued by its 3′UTR

alone, indicating that this RNA "acts as a scaffold or regulatory RNA essential for oocyte development".[294]

Later studies showed that independent expression of 3′UTR sequences is widespread (Chapter 13), occurring in as many as half of all mammalian genes[289,295–299] as well as commonly in plants.[300] These "UTR-associated RNAs" (uaRNAs) or "downstream of genes" (DoGs) are, at least in some cases, nuclear-localized and induce differentiation separately from their usually associated protein-coding sequences (Chapter 13).[290–295,297,301–303] In the testis, for example, the coding sequences of the *Myadm* gene are expressed in the cytoplasm of the interstitial cells, whereas the 3′UTR is not expressed in these cells but highly expressed in the nuclei of germ and Sertoli cells.[295] This phenomenon is particularly pronounced in the brain[295–297] where, for example, the 3′UTR of the *Klhl31* gene but not the coding region is highly expressed in the cerebellum and the hippocampus[295] (Figure 9.7), and cytoplasmic cleavage of the IMPA1 3′UTR is necessary to maintain axon integrity.[303]

---

[o] Exons specifying 5′UTRs have expanded in humans and are highly alternatively spliced.[284]

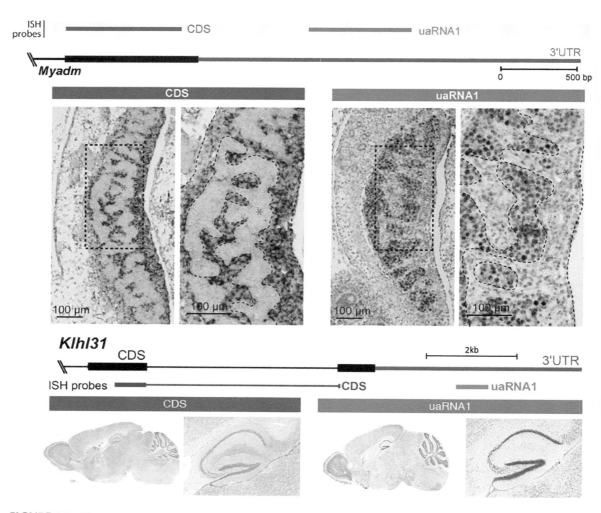

FIGURE 9.7 Top panel: Expression of *Myadm* coding sequences in the interstitial cells of the developing mouse testis and the 3'UTR in the nuclei of Sertoli and germ cells in the testis cords. Bottom panel. Expression of the Klhl31 3'UTR but not coding sequences in the cerebellum and hippocampus, with close-up of the hippocampus showing especially strong expression in the dentate gyrus. (Reproduced from Mercer et al.[295] with permission from Oxford University Press.)

The regulatory and evolutionary logic of having a covalently linked RNA sequence that regulates mRNA activity *in cis* but also acts independently *in trans* is an astounding observation, whose biological *raison d'être* is yet to be satisfactorily explained, but is emblematic of the complexity and mysteries of the emerging world of RNA regulation. It also presaged later findings that non-coding RNAs can act as 'decoys' or 'sponges' for bacterial[47,59] and eukaryotic small RNAs[301] and RNA-binding proteins (Chapters 12, 13 and 16). It is also clear, and in retrospect unsurprising, that the terms messenger and regulatory RNA are not mutually exclusive and

that individual RNAs can have multiple functions (Chapter 13).

FIRST EXAMPLES OF SMALL REGULATORY RNAs IN ANIMALS

Also in 1993, two articles published by the groups of Gary Ruvkun and Victor Ambros described a small RNA that played a role in developmental regulation in the nematode worm *Caenorhabiditis elegans*.[304–307] Previous genetic screens by their groups had shown that the product of the *lin-4* gene regulates the expression of *lin-14*, a heterochronic gene

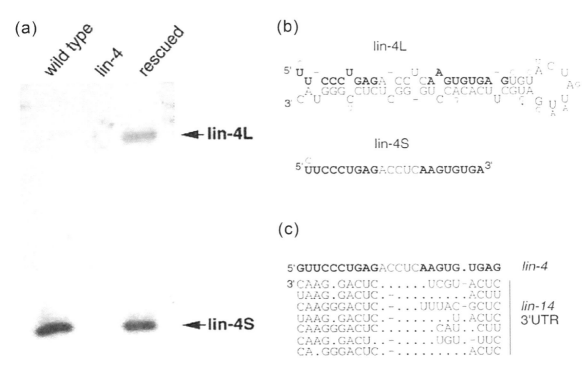

FIGURE 9.8 (a) Northern blot showing the small ~22nt RNA (lin-4S) produced by the *lin-4* gene and its precursor (lin4-L) in wildtype *C. elegans*, absent in a deletion mutant. (b) The sequences lin-4S and lin4-L, the latter showing the predicted secondary stem-loop structure. Sequences complementary to the lin-14 3'UTR are bold. (c) The complementarity between lin-4 and seven copies of a repeated element in the 3'UTR of lin-14 RNA that is conserved in *C. elegans* and *C. briggsae*. (Reproduced from Lee et al.[304] with permission from Elsevier.)

encoding a nuclear protein involved in the temporal control of post-embryonic development, by a mechanism targeting the 3'UTR of *lin-14*.

Both groups were expecting a regulatory protein,[306,307] but the *lin-4* locus mapped to the intron of a long spliced non-coding RNA.[304] The Ambros' group found that the primary *lin-4* transcript was processed into two overlapping RNAs of 61nt and 22nt in length.[304] Similar to the elucidation of the roles of spliceosomal snRNAs and snoRNAs, both groups found that the small RNAs produced from the *lin-4* locus had partial complementary to a number of sequences in the 3'UTR of the *lin-14* mRNA (Figure 9.8). Curiously, while noticing that lin-14 protein levels are reduced in development without changes in the transcript abundance, they proposed that these "small temporal RNAs" formed multiple RNA duplexes that inhibited the translation of *lin-14* mRNA. This inhibition depended on the (partial)

base pair complementarity, which was conserved in the homologous sequences of the related species *C. briggsae*.[304,305,308,309] Thus, it was speculated that there may be a "novel kind of antisense translational control mechanism" and that "lin-4 may represent a class of developmental regulatory genes that encode small antisense RNA products".[304]

They were not to realize how prophetic these words would be (Chapter 12).

CURIOSITIES OR EMISSARIES?

The general significance of this finding was, once again, not recognized at the time. tRNAs were for a long time considered the "smallest biologically active nucleic acids known".[310] Given the "incredible" small size of these RNAs and the lack of obvious homologs outside of worms, even the groups of Ruvkun and Ambros saw them as a "curiosity" of worms, comparable to the "gene regulatory vignettes" of small bacterial regulatory RNAs and the few known eukaryotic non-coding RNAs.[304,306,307]

Not only was it novel that such tiny RNAs could have regulatory properties, it was also surprising that they originated from an intron and targeted non-coding regions in mRNAs, at a time when the regulatory relevance of 3'UTRs was still being established.[311] However, this did not disturb the prevailing conceptual framework. According to Ambros, "there was no theoretical need to explain existing phenomena in terms of new mechanisms or new classes of molecules. Transcription factors-mediated regulation of cell fate was a successful model to account for developmental biology".[307]

A 1994 editorial in *Science* highlighted these emerging findings, the various interpretations and Mattick's hypothesis, remarking on the existence of "too many cases of odd RNAs" and speculating that "there might be a whole family of regulatory RNAs".[312] Similarly, an editorial in *Nature* posed the question: "Are these RNAs all grotesque deviants, one-of-a-kind aberrations, like characters in a Fellini film?" and admonished "But pay close attention to them. [They may] instead have been the first emissaries from an unexplored and vast RNA world".[311]

FURTHER READING

Breaker R.R. (2012) Riboswitches and the RNA world. *Cold Spring Harbor Perspectives in Biology* 4: a003566.

Darnell J.E. (2011) *RNA: Life's Indispensable Molecule* (Cold Spring Harbor Laboratory Press, Cold Spring Harbor, NY).

Lee R., Feinbaum R. and Ambros V. (2004) A short history of a short RNA. *Cell* 116: S89–S92.

Mattick J.S. (2004) RNA regulation: A new genetics? *Nature Reviews Genetics* 5: 316–23.

Morange M. (1998) *A History of Molecular Biology* (Harvard University Press, Cambridge, MA).

Morris K.V. and Mattick J.S. (2014) The rise of regulatory RNA. *Nature Reviews Genetics* 15: 423–37.

Ruvkun G., Wightman B. and Ha I. (2004) The 20 years it took to recognize the importance of tiny RNAs. *Cell* 116: S93–6.

10 Genome Sequences and Transposable Elements

GENOME MAPPING

The prelude to genome sequencing was genome mapping, physically for microbial genomes, which generally range from 400 kb to ~10 Mb, and, initially, genetically for animal and plant genomes, which are orders of magnitude larger (Chapter 7).

Physical mapping was performed by cleavage of genomic DNA with restriction endonucleases with rare recognition sites and electrophoretic size separation of the resulting fragments, which were then ordered into linear or circular maps by partial or sequential digestion and DNA cross-hybridization. Genetic markers were integrated into these maps in well-studied bacterial species, such as *E. coli* (4.7 Mb)[1] and *Pseudomonas aeruginosa* (5.9 Mb),[2] as well as in the brewing and baking yeast *Saccharomyces cerevisiae*, which also has a relatively small genome enabling relatively accurate estimation of its size.[3] Maps were also developed for other well-studied fungi, notably the 'fission' yeast *Schizosaccharomyces pombe* (which is also used in brewing[a]) and *Neurospora crassa*.

Genetic mapping of large genomes based on the frequency of co-inheritance of linked markers was pioneered in *Drosophila* in the early part of the 20th century (Chapter 2) and well advanced by its end. Genome maps based on linkage analysis were also constructed for other widely studied species, such as maize, rodents, cattle and the model flowering plant *Arabidopsis thaliana*, in which mutants were first described in 1873 (see[4]) and which gained wide currency from the 1950s and 1960s, especially when it became obvious that, by plant standards, it has an unusually compact genome.

However, physical mapping of genomes of complex organisms at high resolution was a monumental task beyond the capability of any laboratory or consortium before the 1980s. The restriction fragment pattern was far too complex for any individual segment to be resolved and identified – a blur on electrophoretic display – until these genomes could be partitioned by cloning.

GENETICS AT GENOME SCALE

In the 1970s, Hogness, Welcome Bender and colleagues constructed libraries of randomly cloned large inserts that encompassed the entire *Drosophila* genome (Chapter 6). This allowed physical restriction enzyme maps to be developed for individual segments and genomes to be virtually assembled by 'chromosomal walking' across overlapping segments. It also allowed the screening of these libraries for specific sequences by 'colony hybridization' and the first 'positional cloning' of a gene, *Ultrabithorax*, by mapping a disruptive inversion, an approach that was then extended to many other alleles and genes in which mutants had arisen by chromosomal breakages or transposon insertions.[5-9]

In 1980, Christiane Nüsslein-Volhard and Eric Wieschaus undertook systematic genome-wide screens for genes involved in *Drosophila* development,[10] which led to the discovery of the components of the major signaling pathways, many of which then turned out, like *Ultrabithorax*, *Polycomb* and *Trithorax*, to have homologs in other animals, including mammals. These new genes were then isolated and characterized by positional cloning and transposon tagging,[11,12] an approach extended to other species including *Arabidopsis* and mice.

Similar large insert libraries and physical maps of chromosomes were developed for many other organisms and used extensively for the mapping of mutations, especially those causing human genetic disorders, using restriction site polymorphisms as guideposts to track alleles in affected families (Chapter 11). They were also used as the platforms for whole genome sequencing projects prior to the introduction of highly parallel random sequencing and assembly approaches.

---

[a] Pombe is the Swahili word for beer.

In the 1980s, Martin Evans, Oliver Smithies and Mario Capecchi developed methods for constructing transgenic mice using retroviral vectors and homologous recombination in embryonal stem cells,[13–15] which permitted the introduction of specific mutations to examine their consequences and the rescue of mutant phenotypes by gene transfer.[16–21] These approaches were further extended by 'enhancer traps' to screen for genes based on their pattern of expression[22] and systems for ectopic expression.[23]

WHOLE GENOME SEQUENCING OF BACTERIA AND ARCHAEA

While some viral and organelle genomes had been sequenced,[b] the sequencing of organismal genomes was made feasible by the development in the mid-1980s of fluorescently labeled oligonucleotide sequencing primers by Leroy Hood and colleagues, which enabled optical (laser) reading of electrophoretic displays of fragments generated by the Sanger chain termination method and the consequent development of highly parallel automated DNA sequencers[26] (Chapter 6).

Using this technology, the first whole genome from an organism to be sequenced was that of the bacterium *Haemophilus influenza* by Craig Venter, Hamilton Smith and colleagues in 1995,[27] who devised a strategy of 'shotgun' cloning and sequencing to avoid the tedious work of mapping a genome and the difficulty of coordinating laboratories working on different parts of it. Venter and Smith rationalized that it was easier, at least for small genomes, to sequence random fragments *en masse*, and then assemble a continuous sequence *in silico* by matching overlaps, called 'contigs',[c] a logical extension of their shotgun sequencing of human cDNAs.[28]

And so it proved, and the sequence of 1.83 Mb *H. influenzae* genome was completed 2 years before that of *E. coli* (albeit having a larger genome, 4.6 Mb[29]), whose sequencing was begun earlier. This was the first time that molecular biologists were able read the entire genome sequence of a living cell, identify all

of the protein-coding genes (Figure 10.1) and start to understand its genetic programming and evolutionary history holistically, exemplified by the insights gained from sequencing the intracellular parasite *Rickettsia prowazekii*, the causative agent of epidemic typhus.[30]

These studies revealed many previously unknown features of bacterial genomes, including remnants of bacteriophages – which proved the signpost to a spectacular new technology for genetic engineering (Chapter 12) – and other sequences that suggested genome evolution and plasticity through transposition and horizontal DNA exchange, often using bacteriophages as the vehicle.

Thousands of bacterial and archaeal genomes have since been sequenced and deposited in the public databases. The data reveal that prokaryotic genomes, while encoding some short regulatory RNAs (Chapter 9), are comprised, in the main, of protein-coding genes, separated by short regions that contain cis-acting transcriptional and translational control sequences.[d] Nonetheless, their genomes collectively encode extraordinary proteomic diversity and fluidity of gene content, reflecting their range of ecologies from commensal pathogens to deep ocean volcanic vents and industrial waste.[35,36]

For example, most *E. coli* strains contain between 4,000 and 5,000 genes,[e] but only 20% of the genes in a typical *E. coli* genome are shared among all strains,[37] which is the core proteome that defines the species, whereas the total number of different protein-coding genes observed in different strains exceeds 16,000.[34,38] A recent analysis of 303 million bacterial genes from 13,174 publicly available metagenomes showed that most genes are specific to a single habitat and that the majority of species-level genes and protein families are rare.[39] That is, phenotypic diversity in prokaryotes, primarily metabolic and ecological versatility, is achieved by varying the proteome.

[b] The human mitochondrial genome is only ~17 kb (sequenced in 1981 by Sanger and colleagues[24]); the tobacco chloroplast genome is ~156 kb, and its sequencing by two Japanese research teams[25] in 1986 was a tour-de-force at the time.
[c] This approach is easier in bacteria because of the low frequency of repetitive sequences whose locations are ambiguous.

[d] Prokaryotic genomes range in size from just 160 kb in the insect symbiotic bacteria *Carsonella ruddii*[31] and *Nasuia deltocephalinicola*[32] (and are generally small in endosymbiotic and obligate intracellular parasitic species, such as *Mycoplasma*) to 14.8 Mb in the free-living soil bacterium *Sorangium cellulosum*,[33] with ~11,500 protein-coding genes, which appears to be close to the upper limit[34] (Chapter 15).
[e] The standard laboratory strain of *E. coli* ('K-12') has ~4,300 protein-coding genes.[29]

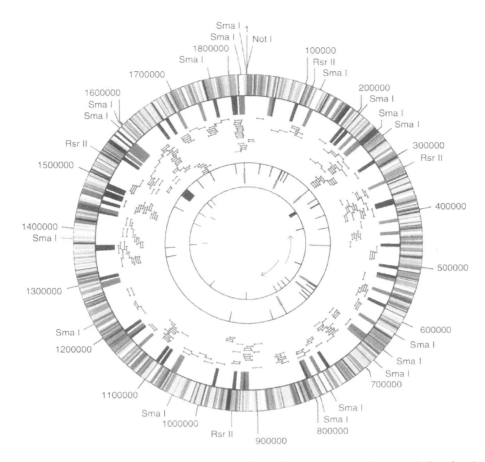

FIGURE 10.1 Circular map of the *H. influenzae* genome illustrating the location of key restriction sites (outer perimeter), color-coded predicted coding regions (outer concentric circle), regions of high and low G+C content (inner concentric circle), coverage of clones used to generate the sequence (third concentric circle), locations of the six ribosomal operons (green), tRNAs (black) and the cryptic mu-like prophage (blue) (fourth concentric circle), simple tandem repeats (fifth concentric circle), the putative origin of replication at the outward pointing green arrows and putative termination signals (red). (Reproduced from Fleischmann et al.[27] with permission of the American Association for the Advancement of Science.)

GENOME SEQUENCING OF UNICELLULAR EUKARYOTES

The first eukaryotic genome to be sequenced was the 12.4 Mb genome[f] of *S. cerevisiae*, achieved chromosome-by-chromosome by an international consortium led by André Goffeau in 1996, which identified almost 6,000 protein-coding genes and 140 rRNAs, 40 snRNAs and 275 tRNAs.[41,42]

---

[f] At the time of writing, the smallest known eukaryotic genome is that the microsporidian parasite *Encephalitozoon intestinalis* (2.3 Mb). The largest known plant genome is that of the monocot *Paris japonica* (~150 Gb). The largest known animal genome is that of the genome of the marbled lung fish, *Protopterus aethiopicus* (~130 Gb) (Chapter 7).[40]

The *S. cerevisiae* genome contains only a few (~270) short introns (average 247bp), located in just 4.5% of protein-coding genes,[43] deletion of which was later shown to have physiological and growth effects,[44–47] i.e., even these small introns contain information. Interestingly, the number of introns in *S. cerevisiae* is substantially lower than in the superficially similar *S. pombe*, whose 13.8 Mb genome has fewer protein-coding genes (~4,900) but many more introns (~4,700, in 40% of protein-coding genes), albeit still mainly small (30–800bp, modal length 48bp).[48]

Both *S. cerevisiae* and *S. pombe* have proven invaluable in the genetic dissection of the control and mechanics of cell division and other basic processes such as protein trafficking, and the identification of homologous genes in plants and animals, including human.[49,50] Indeed, genome analyses showed that the protein components of core cellular processes have been conserved throughout eukaryotic evolution.

Large numbers of yeasts (including such pathogens as *Candida albicans*, which causes thrush) have now been sequenced, showing, for example, that many brewing yeasts are hybrids of *S. cerevisiae* and *S. eubayanus*, and that both originated in East Asia, with different strains being domesticated in different places.[51,52]

The genome sequence of *N. crassa* was published in 2003 and found to contain ~10,000 protein-coding genes, with introns and intergenic regions occupying ~56%. About 10% of the genome is comprised of repetitive sequences.[53]

And, of course, the genomes of important parasites, such as *Plasmodium falciparum*, which causes malaria, were also soon sequenced,[54] along with that of its mosquito host in 2002.[55] The richness of the information in these genome sequences is extraordinary, and still being investigated. To do so has required the construction and maintenance of databases and the acquisition of bioinformatic tools and skills, a major change to the investigative landscape of all biological domains, from evolution to neurobiology.

GENOME SEQUENCING OF MODEL PLANTS AND ANIMALS

The first (nearly) complete sequence of the genome of any multicellular organism was that of *C. elegans*, accomplished in 1998 by a consortium led by John Sulston,[56] who (with Sydney Brenner and H. Robert Horvitz) pioneered it as an experimental organism.[57,58] *C. elegans* has only ~ 1,000 somatic cells, whose ontogeny had been determined, and has been a useful model for many processes including cell differentiation (Chapter 15), RNA interference (Chapter 12), transgenerational inheritance (Chapter 17), drug responses such as nicotine withdrawal[59,60] and aging,[61] among others. The *C. elegans* genome was found to be 97 Mb in size and to contain just over 20,000 protein-coding genes, "one-fifth to one-third the number predicted for humans".[56] Seventy-three percent of the *C. elegans* genome is comprised of introns (26%) and 'intergenic' (47%) sequences (Figure 10.2).

Two years later, the sequence of the *Drosophila melanogaster* genome was completed by a consortium led by Venter and Jerry Rubin.[62] This time the approach was different, not sequencing of previously mapped cloned segments, as was the case with *C. elegans*, but mainly sequencing of random fragments, as had been done with *H. influenzae*. This achievement put paid to the skepticism that many had expressed of this approach because of the problem of repetitive sequences in genome assembly; shotgun sequencing is now the standard method.

The *Drosophila* genome is ~120 Mb in length and encodes only ~13,600 protein-coding genes, many of which have equivalents in humans,[63] only twice the number of protein-coding genes in yeast and fewer than in *C. elegans*, which is developmentally far simpler than an insect.[62] This anomaly was noted in a commentary at the time: "… there is little relationship between total gene number, neuron number, morphology and behavioral capacities of diverse organisms in different phyla … (which) merely highlight our ignorance of biological complexity and how it is instantiated."[64]

By contrast, the *Drosophila* genome contains a greater proportion than *C. elegans* of introns (>41,000 ranging up to 70 kb in length, for example, in the *bithorax* and *DMD*[g] genes) and 'intergenic' sequences, which collectively comprise ~80% of the genome, one of the first hints from genome sequencing that increased developmental complexity is not a function of the number of protein-coding genes, but rather of information in non-coding regions.

In the same year (2000), the first plant genome sequence was also published, that of *Arabidopsis thaliana*, which has one of the most compact plant genomes known (125 Mb), similar in size to that of *Drosophila*, but contains almost twice as many protein-coding genes, ~25,500.[68] The genome sequences of two rice cultivars (~450 Mb; 30–50,000 protein-coding genes) were published in 2002.[69,70]

A 'first draft' of human genome sequence was published in 2001,[71,72] and a more complete compendium in 2004[73] (Chapter 11), revealing the full extent of the complement of sequences derived from transposable elements, other repeats, introns and 'intergenic' regions. The mouse genome was published

---

[g] Due to its exceptionally large introns, *DMD* is one of the largest protein-coding genes not only in *Drosophila*[65] but also in *Fugu*[66] and human,[67] suggesting that both the exons and the introns of this gene have been conserved.

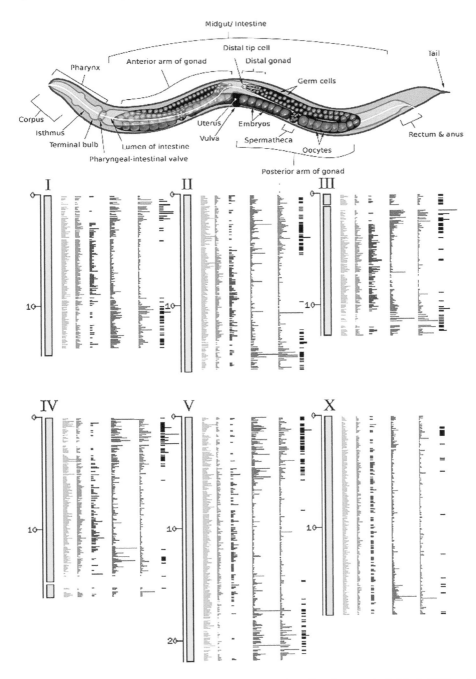

FIGURE 10.2 *C. elegans* and map of its genome: Distributions of predicted genes (pale blue); EST matches (green); yeast protein similarities (dark blue); and inverted (purple), tandem (red), and TTAGGC repeats (black) along each chromosome. Numbers are Mb. (Reproduced from The *C. elegans* Sequencing Consortium[56] with permission of the American Association for the Advancement of Science. *C. elegans* image by K. D. Schroeder made available under Creative Commons Attribution-Share Alike 3.0 Unported license.)

in 2002,[74] followed soon by the sequences of rat,[75] dog,[76] cow,[77] chimpanzee,[78] chicken[79] pufferfish,[80] the ascidian *Ciona* (a primitive chordate)[81] and many others.

These studies revealed high conservation of the protein-coding gene complement among vertebrates (~20,000 protein-coding genes, 75% orthologous between fish and human), and especially mammals (~90% orthologous),[82] with lineage-specific expansion or contraction of some gene families such as those encoding cytokines and olfactory receptors.[83–85]

The genome of the pufferfish (*Takifugu ruprides*, often referred to simply as *Fugu*) was sequenced because it is unusually compact (just 365 Mb, an order of magnitude smaller than in human, but three times bigger than in *Arabidopsis*), and held up as a model of a streamlined vertebrate genome with minimal 'junk'. The *Fugu* genome contains 11% protein-coding, 22% intronic and 67% intergenic non-coding DNA, 17% comprised of repetitive sequences.[80]

THE G-VALUE ENIGMA

The unexpected finding from the genome projects was the lack of correlation between the number of protein-coding genes and developmental complexity.[86] Up until this point, gene number had been widely proffered to be a valid measure of biological complexity[87] – and may still be, if the definition of a 'gene' is extended to those encoding regulatory RNAs (Chapters 12, 13 and 16).

To recap, *C. elegans*, a simple nematode with only ~1000 somatic cells has ~20,000 protein-coding genes, as has its sister species *C. briggsiae*.[56,88–90] Sponges, the most basal metazoans, have ~30,000 protein-coding genes.[91] The far more complex insect *Drosophila* has ~13,600 protein-coding genes, mosquitos have ~16,000,[55,92] whereas the water flea *Daphnia* has ~30,000, the increase in the latter apparently related to ecological flexibility rather than developmental complexity.[93]

Humans have ~40 trillion cells sculpted into a myriad of different muscles, bones and organs with complex architectures,[94] as well as a brain with approximately 85 billion neurons (Chapter 15),[94,95] but just ~20,000 protein-coding genes, similar to *C. elegans* and other mammals.[96–102] Indeed, despite fluctuations, the number of protein-coding genes remains remarkably static across the animal kingdom, despite enormous differences in developmental

complexity and cognitive capacity.[103,104] Moreover, the majority of protein-coding genes in animals are orthologous, including most of those involved in multicellular development and brain function.[91,105] That is, all animals have a similar protein toolkit.

On the other hand, in contrast to lack of scaling of protein-coding genes, the fraction of the genome that is intronic and 'intergenic' increases with developmental complexity, crudely defined as the number of different 'cell types' (Chapter 7),[106–108] although this definition underestimates the different spatial identities, architectures and ontogenies of functionally similar (e.g., muscle or bone) cells (Chapter 15). Prokaryotes have ~10%–15% non-protein-coding sequences, mainly specifying cis-regulatory elements controlling transcription and translation. The non-coding fraction of the genomes of unicellular eukaryotes (protists) generally lies in the range of 40%–50%, fungi 50%–60%, plants 70%–90%, and animals mostly in excess of 90%, with the human genome having 98.8% non-coding DNA (Figure 10.3).[103,104]

Clearly the majority of the information that orchestrates developmental programs and phenotypic diversity lies in the non-protein-coding regions of the genome, which raises the questions of what form the information takes and how is it transduced? The conventional view has been that it involves the combinatorics of *cis*-regulatory protein-binding sites, more complex post-translational modifications, and expansion of the range of protein isoforms by alternate splicing,[h] all of which requires additional regulatory information[86,113,114] (Chapter 15). However, *cis*-regulatory elements cannot conceivably occupy more than a small fraction of gigabase-sized vertebrate genomes (recently estimated to be ~7%[115]). On the other hand, the high-throughput RNA sequencing that followed on the heels of genome sequencing revealed that the non-coding regions of animal and plant genomes express thousands of regulatory

---

[h] The extent to which alternative splicing expands the proteome may be limited. While it is clear that alternative splicing can increase greatly the number of isoforms of particular proteins, such as in the classic example of the *Drosophila Dscam* (*Down syndrome cell adhesion molecule*) gene, which expresses over 38 thousand distinct mRNAs,[109,110] much of the alternative splicing in mRNAs, especially in humans, occurs among 5' non-coding regulatory exons, not within the body of the protein-coding exons,[111] and a large proportion generates non-coding transcripts (Chapter 13). Most protein-coding genes express a single dominant splice isoform.[112]

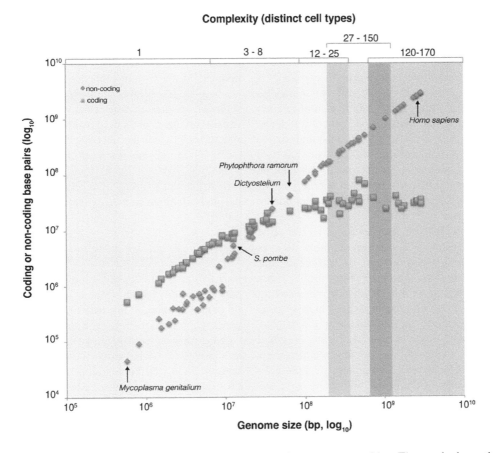

FIGURE 10.3   The relationship between biological complexity and genome composition. The y-axis shows the amount of protein-coding sequence (red) and non-protein-coding sequence (blue), which together comprise the total genome size (x-axis) in 76 organisms across the phylogenetic spectrum encompassing 23 species of bacteria, 7 protozoa, 9 simple and complex fungi, 14 plants (including *Chalmydomonas*, the green alga *Volvox carteri*, *Arabidopsis*, rice, maize and grape), 9 invertebrates (including sponge, *C. elegans, Drosophila melanogaster* and the ascidian *Ciona intestinalis*) and 14 vertebrates (including the pufferfish *Takafugu rubripes*, zebrafish, frog, the lizard *Anolis carolinensis*, chicken, mouse, cow, dog, and human). The number of different cell types in each organism is taken from[108] as indication of developmental complexity. In unicellular organisms, protein-coding sequences dominate but that the proportion of non-coding sequences increases relative to protein-coding sequences, intersecting in simple multicellular organisms, following which the protein-coding sequences remain relatively constant, whereas the extent of non-protein-coding sequences increases exponentially. (Reproduced from Liu et al.[104])

RNAs in different cells and tissues at different developmental stages (Chapters 12 and 13).

COMPARATIVE GENOMICS AT NUCLEOTIDE RESOLUTION

Since that time, there has been a myriad of comparative analyses of genomes. An early example was the comparison of 12 genomes in *Drosophila* phylogeny, which identified, among other things, many putatively non-neutral changes in protein-coding genes, non-coding RNA genes and cis-regulatory regions, high conservation of shared microRNA sequences, including target mismatches, and adaptive evolution of lineage-restricted microRNAs[116] (Chapter 12).[i]

---

[i] The sequences of 101 Drosophilid genomes have recently been published.[117]

Another unexpected finding of the comparison of vertebrate genomes was the discovery of several hundred "ultraconserved" elements >200bp (UCEs) that are identical between the human, mouse and rat genomes, none of which are protein-coding.[118] A subsequent analysis requiring identity of sequences >100bp between any three of five mammalian genomes (human, rat, mouse, dog and cattle) identified almost 14,000 such UCEs,[119] the vast majority of which are not protein-coding, and showed that they evolved rapidly, presumably under positive selection, between fish and amniotes, but then became essentially frozen, subject to fierce negative selection in birds and mammals.[119,120]

Each UCE is different but has followed the same evolutionary trajectory and so presumably there is some commonality of function. At least some are derived from retrotransposons.[121] They are far more conserved than those specifying protein-coding sequences and rRNAs, which are highly constrained by structure and multilateral RNA-RNA and RNA-protein interactions. Many are enriched in the vicinity of developmental genes and appear to overlap developmental 'enhancers' (Chapters 14 and 16) especially in the brain, and many are transcribed into non-protein-coding RNAs, with highly specific expression patterns that are perturbed in cancers and other diseases.[121–131] UCEs are also dosage-sensitive.[132,133]

However, in contrast to their extraordinary pan-amniote conservation, deletion of four UCEs that function as enhancers in transgenic assays showed no overt developmental perturbation[134] and insertion of sequences into UCEs made no change to enhancer activity,[123] although cognitive phenotypes were not examined. Subsequent deletion of UCEs in the vicinity of the neuronal transcription factor Arx also resulted in viable and fertile mice, but showed subtle neurological or growth abnormalities.[135] Recent results show that the ultraconservation of enhancers is not necessary for their function,[136] and the reason for the fierce conservation of UCEs in birds and mammals remains a mystery.[137]

Reciprocally, comparative analyses also identified many RNA genes that have been subject to positive selection in hominid evolution.[138,139] One of the most rapidly evolving sequences in the human genome lies within a gene (*HAR1F*) specifying a highly structured non-protein-coding RNA expressed in Cajal–Retzius neurons during embryonic development of the neocortex,[138] a six-layered structure that is far larger and more complex in humans than in other mammals, including Old World monkeys.[140,141] Only two nucleotide changes have occurred in the 118bp *HAR1* sequence between chickens and chimpanzees, but there have been 18 changes in the human sequence since our split from the latter.[138] A number of such "human accelerated regions" regulate dosage-sensitive neural genes, acting as enhancers and/or expressing regulatory RNAs, mutations in which disrupt cognition and social behavior.[142–145] Moreover, many primate-specific RNAs, including 'repeat-derived' long non-coding RNAs, are involved in a variety of developmental, physiological and cognitive processes (see below and Chapter 13).[146–156]

PSEUDOGENES AND RETROGENES

Large numbers of 'pseudogenes', rivaling the number of protein-coding genes, were also identified in genomic data – almost 20,000 in the human genome.[157] Pseudogenes are fragments of duplicated protein-coding genes and 'processed' intronless copies of mRNAs that (presumably) have been reverse transcribed and retroposed into the genome ('retrogenes'), which have been interpreted as nonfunctional 'molecular fossils', because they contain incomplete open reading frames or disabling mutations[158–160] (Chapter 7). Curiously, retrogenes are found mainly in mammals.[158,161] At least some have been subject to evolutionary selection.[162–164] Many are transcribed in specific cells, and several have been shown to regulate the expression of their protein-coding counterparts, with medical implications (Chapter 13).[165–175]

TRANSPOSABLE ELEMENTS

The genome sequencing projects also revealed the repertoire, distribution, age, activity and features of sequences derived from transposons and retroviruses, collectively referred to as TEs (transposable elements), the dominant components of most plant and animal genomes. TEs comprise only a small fraction of yeast, slime mold and *Drosophila* genomes, can be almost absent or occupy a large fraction of protozoan parasite genomes, and are highly variable both in extent and type in vertebrate and plant genomes (Figure 10.4).[176,177]

In agreement with Britten's early estimates, nearly half, and perhaps as much as two-thirds, of the human genome is derived from DNA transposons

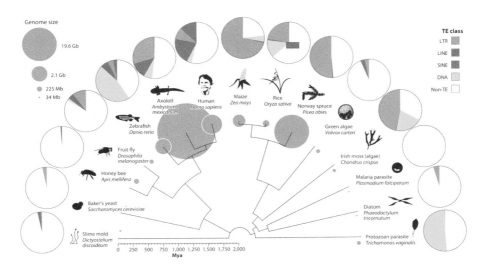

FIGURE 10.4 Distribution of TEs across eukaryote phylogeny. Reference genome size (sea green circles) varies dramatically across eukaryotes and is loosely correlated with TE content. Abbreviations: LINE, long interspersed nuclear element; LTR, long terminal repeat; SINE, short interspersed nuclear element; DNA, class II transposons. (Figure reproduced from Wells and Feschotte[177] with permission from Annual Reviews.)

and from short and long interspersed retrotransposable elements ('SINEs' and 'LINEs') and endogenous retroviruses ('ERVs') that replicate and invade genomic sites via RNA intermediates,[j] although most are now quiescent.[71,74,179-184] A quarter of these TEs correspond to ~ 1.2 million highly similar, but not identical, copies of Alu SINE elements[67] (derived from 7SL RNA, Chapter 8) that entered the human lineage in three waves during primate evolution,[179,180] and were the substrate for a massive expansion of RNA editing, especially in the brain,[185-190] (Chapter 17), with evidence that at least some have been exapted as cell type specific enhancers[191] (Chapters 14 and 15).

Similar numbers and distributions of SINEs (some also descended from 7SL RNA) occur in the mouse genome, although they are distinct from Alu elements and entered the rodent lineage independently.[74] In both species SINEs are clustered in gene-rich regions, especially near promoters, while LINEs (17% of the genome[183]) are concentrated in "gene-poor" regions and depleted from promoters,[192] indicating different roles (examples in Chapter 16). There are hundreds of thousands of LINE elements in mammalian genomes, but much lower numbers in most non-mammalian vertebrates,[193] although there are exceptions (see below).

The Consortium human genome paper concluded "the organization of Alu elements … suggests that there may be strong selection in favour of preferential retention of Alu elements in GC-rich regions and that these 'selfish' elements may benefit their human hosts",[71] a conclusion confirmed by a later study that showed "that Alu and B1 elements have been selectively retained in the upstream and intronic regions of genes belonging to specific functional classes … (with) no evidence for selective loss of these elements in any functional class".[194]

Indeed, while sequences derived from and distributed by transposable elements have been thought to be largely non-functional (Chapter 7), there is a wealth of evidence, dating back to McClintock's studies showing transposition altering phenotype in maize and the regulated expression of TEs in development observed by Britten, Davidson and others (Chapter 5), as well as logic,[195,196] that TEs are major sources of genetic innovation.[177,197-201] They contain and mobilize modular cassettes of (mainly) regulatory information to influence phenotype in evolutionary historical[202-216] and real time.[206,217-221] They are perhaps the most important mediators of genetic fluidity, often called 'jumping genes' although most do not fit the traditional concept of a 'gene', as McClintock intuited by referring to them

---

[j] Retroviruses and TEs are thought to share an evolutionary relationship. Similar to retroviruses, ERVs and LINEs encode a reverse transcriptase and mobilize via an RNA intermediate.[178]

as 'controlling elements' (Chapters 2 and 5). The electronic analogy is control packets.

TRANSPOSABLE ELEMENTS AS FUNCTIONAL MODULES

Thousands of human TEs appear to have undergone positive selection in the vicinity of developmental genes.[222] Other genomic regions, mainly non-coding but also associated with developmental regulation, have been refractory to transposon insertions.[223] A substantial fraction of regulatory sequences in humans, including 25% of promoters and many developmental enhancers, contain sequences derived from TEs.[224] 30%–40% of mouse and human RNA transcripts initiate within repetitive elements,[225,226] and analysis of approximately 250,000 retrotransposon-derived transcription start sites showed that the derived transcripts are generally tissue-specific, coincide with gene-dense regions and often function as alternative promoters and/or express non-coding RNAs (Chapter 13).[225] Some ancient TEs in the vertebrate lineage contain subsequences that have been retained over huge evolutionary distances.[121,227–230]

TEs have been shown to be the source of protein-coding and non-coding genes or exons,[121,214,231–237] centromeres,[238,239] transcription factors, their binding sites and networks,[201,240–242] lineage-specific regulatory RNAs and tissue-specific developmental enhancers (Chapters 14 and 16),[121,153,155,200,224,237,243–246] promoters and transcription start sites,[200,208,218,225,237,246–251] epigenetic control modules,[252–259] neocentromeres,[258] targets for parental imprinting,[260] splice sites,[204,261] translational controls,[262] microRNAs and microRNA targets, RNA nuclear localization signals[263,264] and behavioral modifiers.[265]

TEs are the building blocks for epigenetic regulation and chromatin organization,[192,255,266–269] the senior level of the control of gene expression and cell fate decisions during development in complex organisms (Chapter 14). Many 'repeats' are involved in the formation of heterochromatin, the importance of which was historically downplayed, albeit with exceptions[270] (Chapter 7) but now known to be regulated, *inter alia*, by KRAB zinc finger proteins that bind to TEs[271,272] and other transcription factors[273] that have evolved to regulate TE-derived regulatory sequences during embryogenesis and neuronal differentiation[271,274] (Chapters 14 and 17).

TEs are also a common source of functional domains in regulatory RNAs,[214] for example, as modules for protein-binding and interaction partners for enhancer action (Chapter 16). They are also prevalent in mRNAs of rapidly evolving mammalian-specific genes.[240] Retrotransposon-derived sequences are widely incorporated into coding and non-coding transcripts in human pluripotent stem cells.[275] Primate-specific retroviral 'enhancers' (Chapter 14) and associated TE-containing non-coding RNAs are required for maintenance of stem cell identity and the pluripotency network in humans,[153,245,276,277] and the majority of primate-specific regulatory sequences are derived from transposable elements.[278] They also occur in the most abundant transcripts in the mouse oocyte and regulate gene expression during early embryogenesis.[217,279] Numerous retrotransposons act as preimplantation-specific gene regulatory elements and a mouse-specific retrotransposon is essential for mouse preimplantation development.[280] Developmental transitions and cellular stresses increase the expression of both human and mouse SINE transcripts, suggesting a role in both development and physiology.[281–291] LINE1 elements are spliced into non-canonical transcript variants to regulate T cell quiescence and exhaustion.[292] A retrotransposon is also required for small-RNA-induced pathogen avoidance memory in *C. elegans* and horizontal transfer of that memory to naïve animals.[293]

TRANSPOSABLE ELEMENTS AS DRIVERS OF PHENOTYPIC INNOVATION

TEs underpin many aspects of quantitative trait variation, due to their capacity to alter gene expression patterns in differentiation and development, and thereby to act as drivers of adaptive/regulatory evolution.[211] Bursts of retrotransposition have been linked with major diversification and speciation events.[213,294] Transposon insertions have been associated with developmental innovations and transitions in vertebrates,[200,295] including tetrapod evolution,[296] tail loss in the apes,[297] human-specific hippocampal development,[298] the derivation of small breeds of dogs from gray wolves[299] and the differences between Poodles, Boxers and Great Danes.[300,301] The 'calico' white coat color with spotting in cats arose through a retroviral insertion in an intron that regulates the spatial expression of the *c-kit* gene, which in turn controls melanocyte differentiation.[302] Similarly, a

FIGURE 10.5 Adaptive evolution by transposon insertion into the first intron of the *cortex* gene of the British Peppered Moth (*Biston betularia*; panel A) in the early Industrial Revolution, which increases the expression the gene to create the sooty black form (*Biston betularia f. carbonaria*; panel B), presumably to improve camouflage and reduce bird predation. Panel C shows the gene structure (insertion in yellow) and detail (panel D) of the class II DNA transposon containing three repeated units, flanked by direct repeats resulting from target site duplication (black nucleotides) next to inverted repeats (red nucleotides). Moth photographs A,B by Olaf Leillinger (Creative Commons Attribution-Share Alike 2.5 Generic license). (Gene structure C, D reproduced from v'ant Hof et al.[310] with permission from Springer Nature.)

transposon-derived inverted repeat in an intron of a gene, '*goldentouch*', is commonly associated with color polymorphism in Midas cichlid fishes.[303]

The Rag recombinase proteins involved in V(D) J recombination and the signal sequences therein in the adaptive immune system of vertebrates are also derived from transposons,[304,305] as are the regulatory networks underlying MHC (major histocompatibility complex) expression.[306] The regulation of innate immunity has also occurred through the co-option of endogenous retroviruses.[307]

The classic textbook example of adaptive micro-evolution of the British peppered moth into a black form during the industrial evolution,[k] which was widely cited during the development of mathematical evolutionary theory and the Modern Synthesis (Chapter 2),[308,309] proved to be due to an intronic TE insertion that increases the expression of the gene

cortex,[310] a member of a conserved family of cell cycle regulators that controls pigmentation pattern,[311] estimated to have occurred in or around 1819, when Charles Darwin was 10 years old (Figure 10.5).[312]

Transposon insertions also underlie morphological variations of tomatoes[313] and the changes in the branching structure[l] that marked the domestication of maize from its wild teosinte ancestor,[314] as well as subsequent flowering time adaptations that allowed cultivars to be grown at higher latitudes.[315] Transposable elements change the color of grapes[316] and apples[317] by insertions in the promoters of genes encoding transcriptional activators of pigment production. Analogous insertions occurred independently in Sicilian and Chinese strains of 'blood' oranges, where the cold dependency of the pigmentation reflects the induction of the retroelement by

---

[k] The dark form is long thought to be positively selected because it provided better camouflage from bird predation in a sooty environment (Chapter 2).

[l] Altering the pattern of expression of action of a distal regulatory 'enhancer'.[314]

stress.[318] TE mobilization also appears to be a major generator of genetic variation in *Arabidopsis*.[319]

The huge genome sizes of many native and cultivated plants, including wheat, maize, apples and onions that have been selected in recent history, may reflect their greater flexibility to use TE insertions and polyploidy to generate phenotypic plasticity,[215,317,320–323] in the face of being rooted to the spot, unlike animals, which can move and (consequently) have more precise developmental requirements and limited re-wiring options. This may also explain the extraordinary diversity of phenotypes among closely related plants, such as the varieties produced by artificial selection of the wild mustard plant *Brassica oleracea*, including broccoli, broccolini, Brussels sprouts, white and red cabbage, cauliflower, kale, and kohlrabi, all of which are the same species. Some animal lineages too – including salamanders and other chordates – may have life history (and specific niche/environmental factors) affecting and being affected by changes in TE content and genome size, with potentially significant impact in adaptations and diversity within the lineage.[324–326]

TE insertions are linked intimately to 'epigenetic' control of gene activity, notably by methylation[259,327] (Chapter 14). The cycling of transposable elements between active and inactive states in maize is determined by the methylation state of the element[328–330] and genome sequencing has revealed the spontaneous insertion of a methylation-insensitive TE-derived

'epiallele' in an inbred strain of mouse.[331] Indeed, careful analysis of the features of TEs, which comprise greater than 85% of the maize genome, their insertion sites, expression and methylation profiles, etc., "reveal a diversity of survival strategies … with each TE family representing the evolution of a distinct ecological niche … (and whose impact) is highly family- and context-dependent".[323]

As Nina Fedoroff said in her 2012 Presidential Address to the American Association for the Advancement of Science:

> I contend that it is precisely the elaboration of epigenetic mechanisms from their prokaryotic origins as suppressors of genetic exchanges that underlies both the genome expansion and the proliferation of TEs characteristic of higher eukaryotes. This is the inverse of the prevailing view that epigenetic mechanisms evolved to control the disruptive potential of TEs. The evidence that TEs shape eukaryotic genomes is by now incontrovertible. My thesis, then, is that TEs and the transposases they encode underlie the evolvability of higher eukaryotes' massive, messy genomes…[181]

As the genomes of more and more species, including those representing key phylogenetic transitions, are sequenced, and awareness grows, the focus has changed from analyzing the repertoire of protein-coding genes to the nature and distribution of

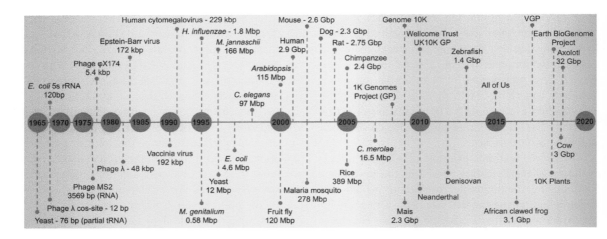

FIGURE 10.6 Timeline illustrating the major genome sequencing achievements from the mid-1960s to 2019, placed in a color-coded background according to the sequencing approach. Orange, early sequencing methods; yellow: Sanger-based shotgun sequencing; green: 'next generation sequencing' (NGS) technologies based on sequencing by synthesis;[356] blue, NGS plus long-read sequencing for whole genome assembly. (Reproduced from Giani et al.[3512] with permission of Elsevier.)

TEs. As a prime example, the publication reporting the marsupial opossum (*Monodelphis domestica*) genome sequence emphasized innovations in TEs and other non-coding sequences in the mammalian lineage, in sharp contrast with the stasis in protein-coding sequences, as "an important creative force in mammalian evolution".[332]

The 5 Gb genome of *Tuatara*, the only remaining member of an archaic order that last shared a common ancestor with other reptiles about 250 million years ago and is a link to the now-extinct stem reptiles from which dinosaurs, modern reptiles, birds and mammals evolved, is 64% composed of an amalgam of TEs with both reptilian and mammalian features.[333] In 2021, the complete sequence of the 43 Gb genome of lungfish, the closest living relative of the tetrapods, which is 14 times larger than the human genome, showed that it is 90% composed of intergenic and intronic TEs, mainly LINE elements, that resemble those of tetrapods more than those of ray-finned fish.[296,334]

Even 'simple' repeats (dinucleotide and trinucleotide 'microsatellites') or 'short tandem repeats', used as markers in gene mapping and DNA fingerprinting,[335,336] have been shown to play a role in adaptive radiation,[337] be flexible[338] and function in modulating gene expression.[339,340] Simple repeats are also associated with quantitative trait variation,[341] environmental adaptation[342] and human neurodegenerative and neuropsychiatric conditions,[339,343,344] likely with intergenerational consequences (Chapter 17). The naïve idea that 'repetitive' sequences can be *a priori* and collectively dismissed as junk (with a few 'exceptions') is unsustainable in the face of these observations.

A blind spot in genome analysis and comparative genomics, particularly with short-read sequencing, is the difficulty of mapping repetitive sequences and segmental duplications. A related issue has been the widespread use of the 'RepeatMasker' program, which masks repeats and low complexity DNA sequences, hiding over 50% of the human genomic sequence.[345,346] These problems are being relieved by advent of long-read technologies (see below and Chapter 11), such as nanopore sequencing, which drags single molecules of DNA (or RNA) through engineered protein pores embedded in membranes and measures the disturbance in the electrical current as nucleotides pass through, enabling sequencing of much longer fragments than SBS (over 1 Mb) and direct sequencing of RNA.[322,347–355]

THE GREAT EXPLORATION – THE DIVERSITY OF LIFE

The pace of genomic exploration was given a huge boost with new technologies allowing massive parallelization of the sequencing process. The most successful to date has been the 'sequencing by synthesis' (SBS) method, invented by Shankar Subramanian and David Klenerman,[356] and later commercialized. SBS uses fluorescently labeled nucleotides containing reversible terminators to optically sequence high density clusters of PCR-amplified fragments on solid surfaces. SBS, along with other technologies, permitted a hyper-exponential increase in the volume of DNA sequence data produced and reciprocal reduction in cost – at a much faster rate than the so-called Moore's Law of computing (at one point a ~2-fold increase in capacity / processing speed and reciprocal halving of cost every 18 months) – the fastest technology revolution in human history.[357]

Over the past decade, there has been an explosion of genome sequencing across the entire phylogenetic spectrum (Figure 10.6). Not only have the genomes of tens of thousands of bacterial and archaeal species been sequenced,[358] but sequencing has become so sensitive and efficient that it became possible to sequence and deconvolute complex microbial communities (termed 'metagenomics'), such as those in soil, sea- and fresh-water, mining and industrial sites, extreme environments, and the digestive tracts of ruminants and humans.[359] Indeed this is the only way to characterize the vast majority of prokaryotic life on earth, which cannot be cultured as (single) colonies on artificial media such as agar plates,[360] although this may be changing,[361] not to mention the estimated billion viruses[m] in every cubic meter of the ocean.[366]

A subset of metagenomics is the human 'microbiome' – the bacteria, archaea, protists and fungi, and their own viruses, such as bacteriophages, that inhabit our gut and other places (skin, mouth etc.), and which vastly outnumber our own (human) cells – termed a human "supra-organism".[367]

The human microbiome appears to have a large influence on health, including metabolic activity, autoimmune and inflammatory disorders,

---

[m] Viruses may be the universal genetic currency, trading information across species and kingdom boundaries. They have co-evolved with cellular life[362–364] and may have been instrumental in the formation of the eukaryotic nucleus.[365]

atherosclerosis and cancer,[368–374] neurodegenerative and neurodevelopmental disorders,[375–378] neurotransmitter biosynthesis,[379,380] social development,[381,382] depression,[383] sensory and locomotor behavior,[384,385] stress responses,[386] obesity[387] and immunity,[388] all of which are associated with particular types of gut bacteria and bacteriophages. The Human Microbiome Project was initiated in 2007.[367,389]

A similar exploration of the hugely varied world of protists and fungi is also underway and is reshaping the eukaryotic tree of life.[390] There are far too many projects to catalog here, except to say that it is now or soon will be unacceptable to study any species or ecosystem without sequencing the genomes involved.

And so it is with plants and animals: for example, the 1,000 Plant Genomes Project initiated in 2008[391] – over 200 angiosperm (flowering plant) genomes and over 1000 plant transcriptomes had been sequenced by 2019;[392,393] the 10K Vertebrate Genomes Project initiated in 2009;[394,395] the 'Earth BioGenome Project' initiated in 2018 to characterize the genomes of all of Earth's eukaryotic biodiversity;[396] the genomes of hundreds of butterfly species;[397,398] and the 'Zoonomia Project' to characterize the genomes of eutherian mammals, with 131 assemblies reported in 2020.[399] A comparative analysis of 363 bird genomes in 2020 more than doubled the fraction of bases that are predicted to be conserved between species and revealed extensive patterns of selection in non-coding DNA.[400]

Despite the numerous (now mainly computational) challenges, genome databases are moving beyond simple gene catalogs to encompass the diversity of variations (nucleotide substitutions, insertions and deletions, as well as structural changes, rearrangements and transposon insertions), and the presence or absence of particular genomic regions in individuals, populations and clades (the 'pangenome' of a species), to allow a greater exploration of genome dynamics and the basis of phenotypic diversity.[38,321,401–403]

The examination and comparison of the evolution and divergence of genomes and their sequence elements[404–407] is an enterprise that will continue for the foreseeable future. Analysis of the genomes of extinct hominids such as Neanderthals and Denisovans by Svante Pääbo and colleagues and others is revealing the details of recent human

evolution,[408–412] indicating that there have been multiple bursts of adaptive changes specific to modern humans during the past 600,000 years involving genomic regions related to brain development and function.[413] Others are documenting the diversity in the human population[414] and the details of the migrations out of Africa,[415–420] the provenance of the biblical Dead Sea scrolls,[421] and the genomes of extinct megafauna, such as the mammoth[422] and cave bear.[423]

FROM GENOME SEQUENCE TO GENOME BIOLOGY

In the years following the completion of the pioneering projects, many studies were described as "genome-wide" or "global", even though they were limited to protein-coding genes or the 'exonic' component of the genome, again on the assumption that most of the relevant information resides therein.

Fortunately, other studies extended to the whole genome, allowing the discovery of many dynamic features outside of coding sequences. The dramatic improvement in sequencing technologies enabled the implementation of unbiased methodologies to globally study the dynamic properties of genomes, including the progressive identification of all transcribed sequences (the 'transcriptome') and the positional modifications of histones and DNA (the 'epigenome'), as well as protein-binding sites, chromatin structure and other features in different cell types at single cell resolution (Chapters 13–14).

FURTHER READING

Feschotte C. and Pritham E.J. (2007) DNA transposons and the evolution of eukaryotic genomes. *Annual Review of Genetics* 41: 331–68.

Genomic sequencing, https://www.nature.com/articles/d42859-020-00099-0 (2021).

Heather J.M. and Chain B. (2016) The sequence of sequencers: The history of sequencing DNA. *Genomics* 107: 1–8.

Sasidharan R. and Gerstein M. (2008) Protein fossils live on as RNA. *Nature* 453: 729–31.

Shendure J., et al. (2017) DNA sequencing at 40: Past, present and future. *Nature* 550: 345–53.

Slotkin R.K. and Martienssen R. (2007) Transposable elements and the epigenetic regulation of the genome. *Nature Reviews Genetics* 8: 272–85.

11 The Human Genome

THE PROJECT

The flagship project of the age was, of course, the Human Genome Project (HGP). We devote a chapter to it, primarily in the light of its controversies and controversial findings, the interpretation of which bears heavily on the understanding of genetic programming and the use of genomic information in healthcare.

The HGP was first mooted at a conference organized at the University of California Santa Cruz in 1985 by the biophysical molecular biologist Robert Sinsheimer,[1] attended by David Botstein, John Sulston, Bob Waterston, Leroy Hood, Walter Gilbert and George Church, among others. It was formally proposed in 1986 by the cancer virologist Renato Dulbecco[2] at a meeting in Santa Fe attended by, among others, Sinsheimer, Watson and Charles DeLisi from the US Department of Energy (DOE).[3–5] The HGP was then recommended for funding in 1987 by a subcommittee of the Office of Health and Environmental Research of the DOE (including Sinsheimer, Dulbecco and Hood),[6] supported by many luminaries of the time, albeit with reservations.[7] The first human genome sequencing conference was held in 1989 at Wolf Trap Farm near Washington.[4]

The project captured the imagination and ambition of the US government – the biomedical equivalent of the Apollo Space Program – which provided most of the funding through the National Institutes of Health (NIH) and the DOE, with a large contribution from the Wellcome Trust in the UK and support from the governments of Japan, France, Germany and China – the 'public' project. There was also a parallel project undertaken by a private company, Celera Genomics Corporation, headed by the *bête noir* of the human genetic establishment, Craig Venter. The public project was officially launched in 1990, but there was a lot of civil and not-so-civil toing-and-froing before it got seriously underway.

There are three aspects worth recalling. The first is the debate about whether to sequence just mRNAs (cDNAs, as an extension of Venter's 1995 study) or

to sequence the entire genome. Why spend all that money of sequencing acres of junk?[8] Moreover, the view of "a surprisingly vocal group" was that the project (in any case) was a waste of money that would be better allocated to other areas of research or to healthcare, exacerbated by a fear of, or antagonism to, 'big science'.[7,8]

For example:

> It is doubtful that much of the resulting information will provide insights into human diseases or fundamental biological processes … (repeated sequences and introns) serve mainly to space exons or represent junk DNA. Obtaining the sequence of these genomic regions is, in my view, simply a waste of money and effort … Genome projects should be severely curtailed or, better still, abandoned.[9]

And Brenner, astride the fence: "If something like 98% of the genome is junk, then the best strategy would be to find the important 2%, and sequence it first".[10]

By contrast from Sinsheimer:

> There is currently a facile assumption that only 1 or 2 or 5 percent of the genome is 'of interest.' I am not convinced we know that. Surely, in an evolutionary sense, much more will be of interest. Knowledge of the variability among the genomes of individuals will surely shed light on variations in physiology and susceptibility to disease, as well as on questions of human origin.[11]

Others, Watson[a] in particular (who was made initial director of the project, and whose genome was the second to be sequenced[12]), agreed and maintained that the human genome could not be understood unless it was sequenced in its entirety, including its non-coding elements, whatever their extent and form might be.[13]

The second aspect was the speculation at the time about the numbers of 'genes' in the human genome,

---

[a] Watson's recollections may be found at https://wellcomecollection.org/works/m4kr8fz5.

which had declined from early and seemingly ludicrous estimates of millions (based on genome size and bacterial-like gene density; Chapter 5) to somewhere in the range of 30,000–150,000.[14–19] As always the underlying assumption was that, apart from those specifying infrastructural RNAs involved in mRNA splicing and translation, and a few others, the gene complement would be mainly protein-coding.

The third was the effort to assemble a coordinated international consortium to undertake the project, mainly at three meetings in Bermuda, co-chaired by the NIH, DOE and the UK's Wellcome Trust, in 1996, 1997 and 1998. They set out the so-called 'Bermuda principles',[20] which held that the human genome sequence data should be made public immediately, promulgated by the Wellcome Trust (and its senior investigators, notably John Sulston), which had no external stakeholders to satisfy, and the NIH, which likely realized that US interests would benefit most because of their capacity for fast adoption. This initially made life difficult for those from other countries, notably Germany, France and Japan, whose governments wanted to capture commercial value from their investment, but eventually the main players prevailed.[20–22]

The Bermuda conferences also discussed which group(s) would take responsibility for, and have provenance over, the sequencing of specific chromosomes, or parts thereof, based on their historical work on mapping of genetic disorders, and the resources they had developed along the way – based on the clone and map then sequence strategy proposed by Gilbert (see [8]). Venter announced that he would just sequence the whole lot, shotgun-style, and assemble the genome from overlapping 'contigs', which was met with a mixed reaction.[b]

The competition between the 'public' and 'private' genome sequencing initiatives had two interesting consequences: it spurred the funding agencies, notably the Wellcome Trust, to increase their investment in the project;[c] reciprocally Celera used the flow of data releases from the public project to accelerate the development of its draft of the human genome sequence.[23]

In any event, as is so often the case, competition was a good thing, and the project was completed

FIGURE 11.1 Industrial scale sequencing for the human genome. (Reproduced from the Human Genome Sequencing Consortium[24] with permission of Springer Nature.)

ahead of time and under budget, with an estimated total cost around $USD 3 billion. Rapprochement was achieved and in 2001 'first drafts' of the sequence (totaling ~2.9 gigabases, or ~90%) of a composite genome amalgamated from a number of anonymous individuals by the public consortium and of Craig Venter's genome by Celera were published contemporaneously in *Nature*[24] and *Science*,[25] respectively. These publications were accompanied by fanfare announcements on both sides of the Atlantic by then US President Clinton (flanked by Venter and Francis Collins, who coordinated the public project as the then director of the National Human Genome Research Institute) and UK Prime Minister Blair. A more complete sequence was published in 2004 (Figure 11.1).[26]

Analysis of the assembled sequences showed that just ~1% of the genome is protein-coding, with ~2% of the total represented in mRNAs (including the 5' and 3'UTRs that control mRNA localization, translation and turnover),[d] whereas 24% is intronic and 74% is 'intergenic' DNA. The genome was found to contain fewer protein-coding genes than expected, the initial counts by the two camps being 30,000–40,000[24] and 26,588 with "an additional approximately 12,000 computationally derived genes with mouse matches or other weak supporting evidence".[25]

---

[b] Personal recollection of JSM.

[c] https://www.sanger.ac.uk/news_item/1998-05-13-wellcome-trust-announces-major-increase-in-human-genome-sequencing/.

[d] More recent estimates indicate that only 0.77% of the human genome contains protein-coding information, and that exons in mature mRNAs occupy 1.74% of the genome.[27]

Even these surprisingly low estimates also turned out to be inflated, likely biased by prior expectations. The actual number of human protein-coding genes has since been revised downward to ~20,000,[27–29] although increasingly offset by growing numbers of genes found to express small and large non-protein-coding RNAs[30] (Chapters 12 and 13).

ASSESSMENT OF FUNCTIONALITY

The subsequent publication and comparative analysis of the mouse genome sequence in 2002 (almost half of which can be aligned to the human genome, with 99% protein orthology[e]) included the estimate that only ~5% of the sequences in mammalian genomes has been 'conserved' during evolution, and by imputation is functional.[34]

The estimate assumed that ancient 'repeats' (i.e., transposon-derived sequences) that have persisted in both genomes since their divergence over 100 million years ago are non-functional and can be used to determine the rate and distribution of 'neutral' evolution of unconstrained sequences over time. Applying this estimate to the remainder of the alignable sequences showed that 95% had diverged to similar extent, with only 5% diverging more slowly, under evolutionary pressure for preservation of particular sequences, termed 'purifying selection',[34] despite dramatic variations in the 'neutral' substitution rates across the genome (Figure 11.2).[35–38]

The conclusion that most of the human genome is not under evolutionary selection and is therefore not functional was widely accepted. It supported the orthodox view and has remained a central plank of the argument of that most of the genome is junk,[39,40] and is therefore important to address.

There are several logical problems with the analysis upon which this conclusion relies. First, it is entirely circular: the assumption that ancient transposon-derived sequences that are orthologous in both genomes are non-functional was used to justify the conclusion that most of the rest of the genome is also non-functional. If the assumption is correct (although there was no evidence to support it), the conclusion is reasonable. If the assumption is wrong, then the conclusion is also wrong.[41]

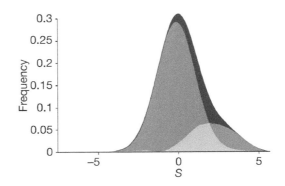

FIGURE 11.2 Comparison of the distribution of sequence divergence between the alignable fraction of the human and mouse genomes (dark blue), decomposed into a mixture of two scaled component distributions: neutrally evolving recognizable common ancient repeats (red) and sequences imputed to be under selection after subtraction of the red distribution from the blue distribution (light blue and gray), corresponding to approximately 5% of the total, which contains most of the orthologous protein-coding sequences (estimated to be about 1.5%). The remainder is assumed to be conserved regulatory elements. Note that if the red curve comprises only the recognizable highly conserved end of the original distribution, the presumed neutral rate of sequence divergence will have been underestimated (the red distribution will be shifted left), and the proportion of the genome imputed to be under selection will be higher. (Reproduced from Waterston et al.[34] with permission of Springer Nature.)

Indeed, this was a questionable assumption given that the reference sequences have been retained independently in the mouse and human genomes for over 100 million years, especially in view of the considerable evidence of the biological functions of TEs, the known cases of which, however, were regarded as exceptions rather than examples of a general phenomenon. This increasingly appears to be incorrect (Chapter 10).

Second, even if the assumption that most ancient retrotransposon-derived sequences that date back to the common ancestor are non-functional is correct, the analysis had an inherent flaw: many of the 'ancient repeats' used in the comparison are barely recognizable as being orthologous, because their sequences have drifted apart, which means there may be, and likely are, an unknown number that have diverged further, to the point of being unrecognizable.[42–45] Indeed the mouse genome analysis stated:

> The ability ... to detect (common ancestral) repeats was found to fall off rapidly for

---

[e] Most protein-coding genes are conserved in vertebrates, with a large proportion of proteins shared between human, birds and fish,[31,32] many in all metazoans[33] (Chapter 10).

divergence levels above about 37%. If we simulate the events … the proportion of the genome that would still be recognizable as ancestral repeats falls to only 6%.[34]

Consequently, the rate of (supposed) neutral evolution in mammalian genomes, and therefore the extent of their functionality, was underestimated to an unknown extent, and even a small increase in the true neutral evolution rate results in a large increase in the proportion of the human genome that is under 'purifying' selection.[41]

Third, while sequence conservation imputes function – highly structured RNAs like rRNAs and proteins are constrained by their physicochemical structure-function relationships – lack of sequence conservation imputes nothing.[46] Not only do non-conserved, lineage-specific sequences underlie evolutionary novelties, regulatory sequences (including gene promoters and some enhancers) can and indeed do evolve quickly,[36,47–50] like language,[f] under different sequence-function constraints and positive selection for adaptive radiation.[40,41] A high proportion of non-coding RNAs and other regulatory sequences, including many that have been functionally validated, show little sequence conservation, use TE-derived modular elements and are lineage-restricted, sometimes with only short conserved sequence and structural 'motifs' embedded in large RNA molecules (Chapters 13 and 16).

Subsequent studies showed that there are at least seven different rate classes of sequence evolution in the human genome,[51] at least 18% of the human genome is conserved at the level of predicted RNA structure,[52] there is strong negative selection across both coding and non-coding sequences,[53] and the vast majority of sequence variations influencing complex traits and diseases occurs in the non-coding regions of the genome (see below).

A pairwise comparison of eight mammalian species concluded that "there is a high rate of turnover of functional non-coding elements in the mammalian genome, so measures of functional constraint based on human-mouse comparisons may seriously underestimate the true value",[36] later reaffirmed by analyzing broader genomic datasets in avian lineages.[54] This

challenges the use of primary sequence conservation and *a priori* dismissal of repetitive sequences in the assessment of genome functionality.

THE MAJORITY OF THE GENOME IS ACTIVE

The possibility that the widely held belief that the vast majority of the human genome is non-functional may be wrong soon became evident in other ways.

The large-scale transcriptome sequencing projects that followed the genome projects revealed that most of the mammalian genome is differentially transcribed, producing an extraordinarily complex interlacing suite of coding and non-protein-coding RNAs, the latter exhibiting exquisitely precise expression patterns (Chapter 13).

The subsequent ENCODE ('Encyclopedia of DNA Elements') project, a large international study which aimed to identify functional elements in the human genome, encompassing RNA expression, the distribution of chromatin modifications (Chapter 14), transcription factor binding sites, DNase hypersensitive (exposed) regions, promoters, etc.[g] in different cell types (Figure 11.3), concluded, in its 2007 'pilot' publication covering 1% of the genome, that most of the studied regions exhibited (these) biochemical indices of function.[55,56]

This figure, however, was at odds with the estimate in the same paper (reiterating that from the earlier human-mouse genome comparison) that only "5% of the bases in the genome can be confidently identified as being under evolutionary constraint in mammals" … and therefore that "Surprisingly, many functional elements are seemingly unconstrained across mammalian evolution".[55]

After some internal (pre-submission) debate among the authors, a decision was made not to canvas the alternative possibility that the estimate of the extent of 'conservation' of the genome- and that only clearly conserved sequences are functional-might be incorrect.[36,47,48] Instead, the incongruity was rationalized as the existence of "a large pool of neutral elements that are biochemically active but provide no specific benefit to the organism".[55] This somewhat contradictory statement became a key talking point and led to a wager publicized in *Nature* as to whether more or less than 20% of the human genome

---

[f] The evolution of language is a useful comparison. The English word 'brother' and the French word 'frère' have no obvious homology, but not only do both have meaning, they have the same meaning and are derived from a common antecedent, having diverged under loose sequence-function constraints (sender-receiver recognition, as in regulatory circuits).[40]

[g] Unfortunately, the project did not include an examination of the incidence or distribution of alternative DNA structures.

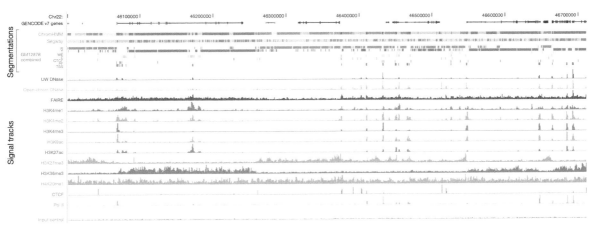

FIGURE 11.3 Representative compilation of ENCODE genomic features cataloged on part of human chromosome 22 in the GM12878 lymphoblastoid cell line. Annotated (protein-coding) genes and their exon-intron structures are shown at top. Chromosome segmentation refers to blocks sharing similar features. Other tracks show predicted enhancers (E, Chapter 14), transcription start sites (TSS), RNA polymerase II binding sites (Pol II), open chromatin (DNase accessibility), nucleosome depleted sequences (FAIRE[56]) and the positions of nucleosomes marked with various histone modifications (Chapter 14). (Reproduced from Dunham et al.[60] with permission of Springer Nature.)

is functional,[57] which at the time of writing had still not been settled.

The more comprehensive 2012 genome-wide ENCODE paper[h] confirmed that at least 80% of the human genome "participates in at least one biochemical RNA- and/or chromatin-associated event in at least one cell type", and addressed the conservation conundrum by stating that

> an appreciable proportion of the unconstrained elements are lineage-specific elements required for organismal function … and the remainder are probably 'neutral' elements that are not currently under selection but may still affect cellular or larger scale phenotypes without an effect on fitness.[60]

This paper spawned another round of controversy,[61,62] with some apoplectic at the suggestion that a large fraction of the genome may be functional, invoking the C-value paradox, mutational load and circular conservation arguments (Chapter 7), while rejecting any suggestion that dynamic transcription or differential chromatin modifications in non-protein-coding regions might be valid indices of genetic

function, including in a species- and clade-restricted fashion.[39,63,64] It is clear that some of the antagonism was related to the invocation of junk in genomes as a line of argument against proponents of intelligent design,[40] who seize and misuse scientific ideas and observations to try to justify non-scientific, untestable beliefs.

DAMAGED GENES

Naturally, the human genome was a major focus of genetic mapping by medical geneticists to identify genes responsible for serious inherited 'Mendelian' metabolic, physiological, developmental and/or cognitive disorders.[i] These diseases are the result of "catastrophic component damage",[68] i.e., disruptive mutations (mainly) in protein-coding sequences, which are generally lethal or severely disabling in the homozygous state, and many deleterious in the heterozygous state.

---

[h] The data included identification of ~2.9 million DNase hypersensitivity sites, ~580,000 of which could be connected to promoters,[58] and evidence that over 75% of the genome is differentially transcribed,[59] identifying many more transcripts than just those encoding 20,000 proteins and their splice variants (Chapter 13).

[i] Most such mutations are recessive, meaning that two damaged copies are required for the disorder to manifest, and reciprocally that there may be high frequency of heterozygous carriers, especially for mutations that may, like cystic fibrosis (1 in 25 carrier frequency in Caucasian populations)[65] and sickle cell anemia (common in tropical and subtropical regions), have provided protection in the heterozygous state against tuberculosis[66] and malaria,[67] respectively, a positive evolutionary trade-off.

However, of course, controlled breeding was not possible to construct conventional genetic maps of the human genome to locate and identity damaged genes and, in any case, the number of known genes with trackable allelic variants that segregated in large families was limited. A different approach was needed.

The solution, proposed by Ellen Solomon and Walter Bodmer in 1979[69] and again by David Botstein, Ray White, Mark Skolnick and Ron Davis in 1980,[70] was to take advantage of single nucleotide polymorphisms ('SNPs') in genomes that resulted in gain or loss of restriction endonuclease sites and a resulting change in the size of the corresponding fragments. Such restriction fragment length polymorphisms, or RFLPs[j] could be tracked as surrogate genetic markers by hybridization of Southern blots with cloned sequence probes, and linked to the inheritance of a condition in extended families.[72]

The search was assisted by the construction of chromosome-specific cloned libraries by flow cytometry sorting of metaphase chromosomes in the early 1980s by Kay Davies, Bryan Young, Rob Krumlauf and colleagues,[73,74] by the use of somatic cell hybrids developed by Bodmer and colleagues and Stephen Goss and Henry Harris in the 1970s,[75,76] and 'radiation hybrid' mapping developed in 1990 by David Cox, Richard Myers and colleagues, whereby individual human chromosomes or parts thereof can be separated and maintained in mouse cell lines.[77–79]

These approaches were made feasible by high penetrance disorders, which are easy to trace in affected pedigrees, especially if dominant or located on the X-chromosome, i.e., commonly exposed in males. On the other hand, the difficulty in identifying the causative gene was increased by the relatively large genomic regions identified by genetic mapping, a needle in a haystack problem.

One of the complications was that meiotic recombination rates across the human genome (and indeed across mammalian genomes in general) are not uniform, but rather occur at hotspots,[k] between which there is little recombinational exchange,[81] referred to as 'linkage disequilibrium'. The recombination-poor regions between hotspots are termed 'haplotype blocks',[82] which parse the genome into 'HapMaps'[83,84] that subsequently formed the analytical platform for population-scale mapping of genetic variations influencing complex traits and multifactorial diseases[85] (see below).

The identification of damaged genes by cloning and mapping approaches was, at the time, a *tour-de-force*, achieved by high-resolution mapping of the chromosomes carrying affected genes, and searching for markers (i.e., sequence variants) that are co-inherited with the condition, aided by homozygosity mapping, since most damaged genes are recessive.[86] Coarse mapping to haplotype blocks was relatively easy, but fine mapping to locate the affected gene within the region, especially in the absence of obvious candidates, relied on rare recombinational or deletion events in particular families, which were hard to find.

Eventually the hard grind paid off, culminating in the identification in 1986 by Tony Monaco, Lou Kunkel and colleagues of the protein-coding gene on the X-chromosome that is damaged in Duchenne's Muscular Dystrophy (*dystrophin*).[87–89] Dystrophin is required for the maintenance of muscle integrity and, as noted previously, is one of the largest genes and proteins[l] in vertebrates.[90,91] In 1989, Lap-Chee Tsui and colleagues identified the gene responsible for cystic fibrosis on chromosome 7, which encodes a chloride ion transporter ('Cystic Fibrosis Transmembrane Regulator' or 'CFTR')[92,93] and explained the symptoms of the disease, including salty sweat, in both cases providing targets for diagnosis and gene therapy.[65,93–99]

In 1991, a CGG trinucleotide repeat expansion was identified in the 5'UTR of the *FMR1* gene (which encodes a synaptic protein) in Fragile X Syndrome, the most common form of inherited intellectual disability[100,101] (which is also associated with autism).[102] Similar repeat expansions were subsequently identified in other genes causing X-linked or autosomal dominant neurological disorders such as Kennedy's Disease, Myotonic Dystrophy, Huntington's Disease and Spinocerebellar Ataxia,[103–113] which were initially thought to result in defective proteins or translation (since many lie in the introns or UTRs) but may also be RNA toxicity disorders, an increasingly prominent theme in neurodegenerative diseases (Chapter 16) (Figure 11.4).

---

[j] Sequence polymorphisms in intronic sequences that altered restriction sites were later patented as a means of diagnosing tightly linked genetic disorders,[71] a surprising decision by patent offices in view of the long history of linkage mapping.

[k] Human genetic data suggests that 60% of recombination events happen in 6% of the genome.[80]

[l] Human dystrophin is composed of 79 exons (encoding 3,684 amino acids) that account for 0.6% of its 2.4Mb sequence.[90]

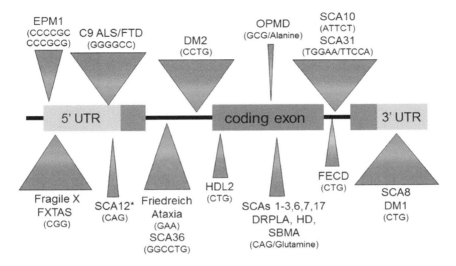

FIGURE 11.4 Schematic of gene showing repeat expansions that cause neurologic diseases. The differing sizes of the associated triangles roughly reflect the range of repeat expansion sizes in each disease. The SCA12* repeat is in an intron. (Reproduced from Paulson[113] with permission of Elsevier.)

Others followed, and it became easier as the technology improved.

A PLETHORA OF 'RARE DISEASES'

About 3%–5% of all children are born with a serious physical or intellectual disability due to a mutation in a protein-coding gene or a chromosomal abnormality,[114] which are also major causes of miscarriage.[115,116] While some genetic disorders, like cystic fibrosis and thalassemia, are relatively common, most are individually rare, due to damage to any one of thousands of protein-coding genes. Collectively, however, they account for a high proportion of all infant deaths and pediatric hospital admissions, as well as a lifetime burden on survivors, their families and health systems.[m]

Such damaged genes, because of their low allele frequency in the population and mostly recessive nature, often lie silent in family histories, as the incidence of homozygosity is low.[n] The high collective frequency of defective alleles,[121–123] however, means that at least 1 in 10 couples are at serious risk of bearing a disabled child with every pregnancy, due to the 1 in 4 chance of each transmitting to their child a damaged gene that they unknowingly have in common. There is also a surprisingly high frequency of new ('de novo') mutations that result in intellectual disability.[124,125]

The identification of damaged genes in individuals suffering severe disabilities is now done not by genetic mapping (impossible due to their rarity) but by whole 'exome' or genome sequencing, comparing their genome (and usually those of their parents, called a 'trio') with a reference, to identify mainly variations in protein-coding sequences (or proximal non-coding regions, such as splicing signals) that introduce a frame shift or stop codon that result in a truncated protein, or codon changes that disrupt protein structure and function.[o]

Exome sequencing[p] has been favored by many clinical geneticists and others because of the

---

[m] Protein-coding mutations, whose presence may not be evident in early life, account for up to 50% of pediatric hospital conditions and 10%–20% of all hospital admissions, as well morbidity and premature death in later life.[117–119] Examples of such (damaged) genes are those causing familial hypercholesterolemia and cardiac defects, which result in catastrophic heart failure in otherwise healthy adults. Many adults carry protein-coding mutations that have yet to become pathogenic.[120]

[n] Higher in communities that have consanguineous, i.e., mostly first cousin, marriages, which increases the odds of homozygosity of mutated genes in the children.

[o] There are many more possible amino acid changes (allelic variants) with more subtle effects.

[p] Exome sequencing is accomplished by oligonucleotide-based hybridization capture of known protein-coding sequences, thereby removing ~99% of the genome prior to sequencing.

emphasis on protein-coding mutations and because it is cheaper and easier to analyze than whole genome sequencing. Its diagnostic yield is, however, lower than whole genome sequencing because it is limited to annotated exons in annotated genes, has technical biases, and is generally unable to detect other types of damage such as translocations and copy number variations, which can be found using other means.[126–128]

Nonetheless the process is becoming increasingly efficient, supported by databases that have cataloged thousands of genetic disorders, and sophisticated software that can sift through millions of individual sequence variations. It is also being aided by increasing detection of 'expressed regions' in transcriptome studies, leading to better annotation of both coding and non-coding genes.[127,129]

Currently, several genetic conditions are polled at birth by the so-called 'Guthrie' heel prick blood test, which uses biochemical and genetic tests to screen for genetic disorders that can be treated by early intervention. The prototype, and good example, is the test for phenylketonuria, a rare recessive disorder whereby infants cannot metabolize the aromatic amino acid phenylalanine, leading to mental retardation, which can be avoided by dietary modification.[130] In the near future, it is likely that the Guthrie test will be replaced, conditional on parental consent, with whole genome sequencing,[q] which will provide a much more comprehensive view of incipient genetic problems and allow early intervention to prevent or mitigate their effects.

Moreover, the surprise finding that a significant proportion (~6%) of the DNA circulating in the blood of pregnant women comes from the fetus[131,132] has led to the rapid rise of non-invasive prenatal testing (NIPT), which can detect chromosomal trisomies (such as trisomy 21, Down's Syndrome) more accurately and with no threat to the embryo (unlike the preexisting amniocentesis and chorionic villi sampling tests).[133] This has also contributed to the progressive demise of medical cytogeneticists, whose other main activity is detecting chromosomal translocations in cancer and balanced translocations in reproductive failure, also soon to be replaced by genome sequencing.

The identification of mutations causing serious disorders will inevitably become fully automated, as computers match the spectrum of patient sequence variation and clinical features with recorded cases, reducing and eventually obviating the need for *ad hoc* sleuthing by clinical geneticists in hospital laboratories.[134] Computerization will also allow such information, and recommended evidence-based actions based on the latest publications and national guidelines, to be delivered to the desktop of health professionals, including general practitioners on the front line.

COMPLEX TRAITS AND DISORDERS

Human genetic analyses have progressively moved from protein-centric (such as the classic blood-group and HLA allele frequencies) to variable microsatellite loci[135–137] and more recently to genome-wide approaches involving very large numbers of individuals.[138,139] Human genomes vary by ~0.1%, i.e., unrelated individuals have 4–5 million sequence differences, although the total number of differences that occur among humans is many times greater, with no absolute differences yet found between populations, although allele frequencies vary.[140–142] Studies comparing the congruence or difference between identical and non-identical twins showed that genetic factors play a substantial part in susceptibility to almost all human traits and disorders, including to infectious diseases.[143,144]

However, the identification of genetic loci contributing to complex traits and diseases is not amenable to the approaches used in mapping severe genetic deficiencies because each causal locus often only makes a small contribution to overall heritability.[145,146] The problem was to a significant extent solved by the development of haplotype maps and oligonucleotide arrays (originally developed by Patrick Brown and colleagues for transcriptome analysis[147]) that poll common sequence variants.[141] SNP arrays are cheap to produce and enabled large population-scale surveys, termed 'genome-wide association studies' (GWAS), which compare the distribution of sentinel SNPs (and by imputation other variants that co-segregate in the same haplotype block) to identify variants that are statistically over- or under-represented with respect to the trait or disease under study.[146] The statistical probabilities are then graphed across the genome to produce so-called 'Manhattan' plots, with

---

[q] Whole genome sequencing can also allow detection of mutations in non-coding regulatory RNAs, such that occur in some cases of phenylketonuria – see Chapter 13.

a P-value of ~10^{-8} commonly used as a significance threshold.[148,149]

The first GWAS was conducted in 2002 on 94 Japanese individuals who had suffered myocardial infarction and 658 controls using (protein-coding) gene-centric SNPs, identifying, among others, an intronic SNP that enhanced the transcriptional level of the lymphotoxin-alpha gene, confirmed in a more focused analysis of over 1,000 affected individuals and controls.[150] This was followed by a study in 2005 on 96 individuals suffering macular degeneration with 50 controls, which identified two significantly associated SNPs in an intron of a gene encoding a blood complementation factor.[151] Two years later the Wellcome Trust Case Control Consortium published a multilateral GWAS involving 14,000 cases of seven common diseases (~2,000 individuals for each of coronary heart disease, type 1 diabetes, type 2 diabetes, rheumatoid arthritis, Crohn's disease, bipolar disorder and hypertension) with 3,000 shared controls.[152] Other studies at the time led to the discovery of many variants in non-coding regions regulating the developmental expression of human fetal hemoglobin.[153-155]

Since then, study sizes have grown to millions of individuals and have encompassed over 1,000 different conditions and traits, usually using 'biobank' samples integrated by international consortia.[156]

Despite the still-present difficulties in teasing out the interplay of genetic variations and heterogeneous social-environmental factors in complex traits, the phenotypes examined include psychological traits such as temperament,[157] neuropsychiatric disorders (such as autism[158,159]), schizophrenia and bipolar disorder,[160-162] ADHD (attention deficit hyperactivity disorder), panic disorder and depression,[163-167] vertigo,[168] neurodegenerative diseases such as Alzheimer's and Parkinson's Disease,[161,169-171] as well as various types of cancer,[172] immunological disorders (such as ankylosing spondylitis, ectopic dermatitis, asthma and inflammatory bowel disease),[173-176] hypertension,[177] height and body mass index,[178] bone density and osteoporosis,[179] alcoholism and other drug dependences,[180-184] caffeine consumption,[185] handedness,[186] insomnia,[187] aging,[188] and even cognitive performance,[189] intelligence[190-193] and correlated (and environmentally contingent) educational attainment (Figure 11.5).[194]

Two general findings emerged from this fleet of GWAS, apart from the identification of tens of thousands of SNPs/haplotype blocks associated with various conditions and traits.

The first is that GWAS does not appear to identify, quantitatively, all of the genetic contribution to complex traits, traditionally determined by pedigree estimates and twin studies, although these are

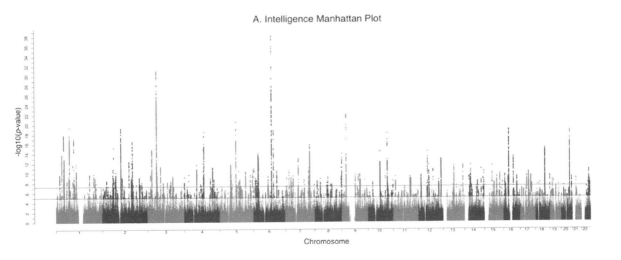

FIGURE 11.5 Combined Manhattan plot of two large genome-wide association studies of education and intelligence. (Reproduced from Hill et al.[193] under Creative Commons Attribution 4.0 International License.) The red line indicates threshold for genome-wide significance and the black line the threshold for suggestive associations. The data suggest that genes involved in neurogenesis, myelination, expressed in the synapse and involved in the regulation of the nervous system play a role in the variation in intelligence.

limited by confounding environmental and methodological factors.[143,195–197] The emblematic example is height, a deceptively simple trait that is known to be highly (80%–90%) genetically determined after controlling for environmental variables such as nutrition,[198] implicit in twin studies, but where only ~25% of the variance could be accounted by GWAS-identified loci, of which there at least 180.[178,199,200] More extensive studies identified several thousand "near-independent" DNA markers and rare variants that appear to account for 60%–70% of the genetic contributions to height,[178,201] possibly overestimated due to uncorrected stratification.[197] Another confounding factor is 'hidden epistasis' (i.e., synergistic interactions between loci involving regulatory networks).[195,202,203] There is similar complexity of polygenic contributions to other traits such as urate, insulin-like growth factor 1 and testosterone levels.[204]

In addition to the plethora of environmental factors and life histories that can interact with genotypes differently, haplotype analysis masks 'private' mutations and tandem repeat variations that have occurred in individual lineages since the divergence of the common versions of haplotype blocks, which occurred hundreds of generations ago.[205] For example, whole genome sequencing of families found that

> Variation in height in our sample arises from a combination of a small number of QTLs[r] with large effects - which are not tagging previously identified common variants, and so cannot be imputed from them - and a large number of common variants with small effects.[206]

Tandem repeat variations make a significant contribution to autism,[207–209] as also do rare mutations that are only a few generations old.[210] There is also an unknown contribution of transgenerational epigenetic inheritance (Chapter 17), which cannot be polled by DNA variants.

The second general finding from GWAS is that (unsurprisingly) the vast majority of genetic variations associated with complex traits and diseases, including cancer predisposition,[172] occur outside of protein-coding sequences, in intronic and intergenic sequences.[211–219]

Although some pleiotropic SNPs (affecting multiple traits) occur in coding sequences, UTRs and promoters, loci containing multiple-trait associated variants cover the majority of the genome.[220] A high proportion of the imputed loci exhibit the signatures of being (real) genes, including promoters characterized by DNase hypersensitivity, typical chromatin modification signatures and transcription.[68,213,221–225] Variation between individuals also occurs by recombination between endogenous proviral sequences,[226] as well as in tandem repeat sequences.[227]

There is enrichment of variations in enhancers (Chapters 14 and 16) that are active in disease-relevant cell types associated with developmental abnormalities, cancers, Alzheimer's Disease, schizophrenia, autoimmune diseases including diabetes, rheumatoid arthritis and multiple sclerosis, and cardiovascular disorders. Functional analyses are uncovering increasing numbers of causal variations, many linked to non-coding RNAs transcribed from these loci.[160,228–238] Indeed, while most haplotype blocks identified as being associated with complex traits and diseases in GWAS studies are devoid of protein-coding genes ('gene deserts'),[216,239,240] most produce multi-exonic non-protein-coding RNAs,[223–225,241–243] at least some of which comprise or are candidates for the molecular basis of trait association[234,235,244–251] (Chapter 13).

Identifying the relevant variations within haplotype blocks among the many differences between individuals is a huge challenge but will likely be achieved by analysis of large datasets of genome sequences, as recently in the case of autism.[252] These analyses will be informed by model organism[s] studies,[254] RNA expression and predicted structural variants, DNA and histone modifications in affected tissues (epigenome-wide association scans/studies or EWAS) and transcription factor binding profiles, to link genomic variants with molecular and phenotypic indices.[53,255–262]

THE TRANSFORMATION OF MEDICAL RESEARCH AND HEALTHCARE

Frustration has often been expressed at the delay in the delivery of health benefits from the HGP, many based on promises and expectations that arguably had their roots in a century-old 'genes for'

---

[r] QTL = Quantitative trait loci.

[s] High-throughput genetic screening of *C. elegans* orthologs of human obesity-candidate genes reported in GWAS identified 17 protein-coding loci that are causally linked to obesity across phylogeny.[253]

mentality, which was boosted by successes in the study of monogenic disorders but does not reflect the complexity of most human diseases and traits. Nonetheless, by 2011 it was estimated that there had been a 140-fold economic return on investment by the US government in the HGP,[263] and, fueled by the major scientific advances that the genome sequences have made possible, there is renewed interest in harnessing the information in whole genome sequences for healthcare at individual and population scales.

The $1,000 human genome sequence cost barrier was breached in 2014 and is likely to decline further. New competing approaches and technologies are emerging and reaching the market, such as long-read sequencing and the combination of chromosome conformation capture and deep sequencing for chromosome-length assembly of large genomes.[264] Other technologies using solid-state devices and high-resolution microscopy may not be far away, with the $100 human genome sequence in sight.

The declining cost of sequencing prompted the establishment of projects to explore human genetic diversity and the etiology of cancer, beginning with the 1,000 Genomes[140,141] and the International Cancer Genomes Consortium projects,[265,266] followed quickly by the UK 100,000 Genomes Project, the first to apply population-scale genomic sequencing to the diagnosis of genetic disorders and cancer,[267] completed in 2018. Larger projects are underway, with the UK announcing a minimum of 1 million genomes (and an "ambition" of 5 million genomes) to be sequenced by the UK Biobank and the National Health Service[268] and 1 million genomes to be sequenced by the US 'All of US' program, along with accompanying clinical and lifestyle data,[269] with similar projects under way in China and many other places. In fact, with the accumulation of such studies (with high-coverage WGS of very large number of individuals with diverse ancestral and admixed backgrounds, deep phenotyping, longitudinal assessment and improved imputation methods[270]), the focus is shifting to the dissection of the contribution of rare non-coding variants to human phenotypic variation,[215,271,272] including polygenic risk variant calling for complex diseases[273–275] (an approach that is still controversial[276]) and pharmacogenomic indices to guide drug selection and dose.[277]

Sequencing of tumor DNA is also revolutionizing the understanding and treatment of cancer,[278] showing that cancers that arise in different tissues are caused by a similar spectrum of mutations.[279,280]

Most of the main 'driver' mutations occur in protein-coding genes, such as TP53, whereas there are many other non-coding variants that also contribute,[t] including previously undetected 'weak drivers' with aggregated effects on cancer phenotypes.[266,283,284] Increasing numbers of the protein mutations can be treated with targeted drugs[285,286] and, in the case of tumors with high mutational load, with immunotherapies, which are proving extraordinarily successful in increasing survival.[287–289]

Genetic screening is allowing the identification of cancer-susceptibility genes and monogenic diseases at population level, finding a large number of previously unsuspected carriers and individuals with latent disease risk.[290–292] Within a decade or so it is likely, depending on the pace of reduction in sequencing and storage/analysis costs, that genomic analysis will become routine in the early detection, identification, treatment and prevention or mitigation of genetically linked disorders and risks, including those usually manifested later in life, such as familial hypercholesterolemia, arthritis and cancer. Although the evidence framework and underlying databases are still evolving, identification of underlying mutations is already leading to improvements in outcomes, through either the selection of targeted drugs or the likely response to immunotherapies, leading to substantial increases in life expectancy and, sometimes, to permanent remission.[285,293]

Twenty years after the publication of the human genome drafts, there is virtually complete coverage of entire chromosomes[294,295] and a full picture of the diversity of human genomes (as 'the genome' is a misnomer) is emerging,[296] including the secrets of the highly repetitive and heterochromatic regions and the functional impact of their variations.

The acquisition of human whole genome sequences at scale will continue to illuminate human biology and transform medical research, drug discovery and healthcare over the coming decades. Millions of genomes, accompanied by billions of data points from clinical records, self-phenotyping and smart sensors (which record real-time physiological and environmental parameters), will create a multidimensional information ecology that can be mined for new genotype-phenotype correlations using machine learning and other

---

t Some non-coding mutations in regulatory regions, such as – prominently – in the promoter of the *TERT* telomerase component, also have driver malignant effects.[281,282]

methods of artificial intelligence, consequently refining patient stratification and treatment.[u] Once the infrastructure is in place to analyze and report the consequences of genomic variants to clinicians (and patients), medicine will change from the art of crisis management to the science of good health, and radically improve the quality, efficiency and sustainability of healthcare, arguably the most important and fastest growing industry in the world.

---

[u] Machine learning on transcriptome and genomic data in relation to cell differentiation stage will also transform understanding of the genes and genetic variants controlling development.

FURTHER READING

Davies K. (2010) *$1,000 Genome: The Revolution in DNA Sequencing and the New Era of Personalized Medicine* (Free Press, New York).

Tam V., et al. (2019) Benefits and limitations of genome-wide association studies. *Nature Reviews Genetics* 20: 467–84.

Watson J.D. (1990) The human genome project: past, present, and future. *Science* 248: 44.

Watson J.D. and Cook-Deegan R.M. (1991) Origins of the human genome project. *The FASEB Journal* 5: 8–11.

12 Small RNAs with Mighty Functions

At the same time as the genome projects were getting into full swing, another revolution was underway, which brought into light the natural capacity of RNA to bind other RNAs and DNA sequence-specifically to combat viruses and to regulate gene expression,[1,2] as first proposed decades earlier (Chapters 3, 5 and 9).

Such logical capacity expands enormously the range of regulatory options by using RNA to address different targets and execute specific actions using a generic infrastructure of guidable protein effectors. This is a highly efficient and flexible system of gene control, which has opened up new avenues for gene modulation and genetic engineering. It also lies at the heart of the epigenetic control of gene expression during multicellular differentiation and development (Chapter 16).

UNUSUAL GENETIC PHENOMENA INVOLVING RNA

The demonstration that RNAs can use base pairing to form double-stranded structures dates back to 1956.[3] In the 1960s and 1970s, numerous reports showed the existence of double-stranded RNAs (dsRNAs) derived from viral RNA genomes replicating in eukaryotic cells. It had also been known since the 1960s that synthetic and naturally occurring viral, bacterial and fungal dsRNAs can induce antiviral cytokines called 'interferons'[a] and inhibit protein synthesis in animal cells at low concentrations, an important feature of innate immunity.[8–15] The first candidate RNA drug (reported in 1967) was a fungal extract that contained "an anti-viral agent … [which] is an RNA of viral origin".[10,15,16]

The discovery of interferons and their induction by dsRNAs explained, at least in part, a phenomenon known since the 1920s as 'virus interference',[4] whereby viral infection results in resistance to subsequent infections by the same or related viruses.[17,18] The mechanism was subsequently shown to involve two interferon-induced, dsRNA-dependent enzymes that inhibit translation and activate an endoribonuclease.[19,20] In 1985, Masayori Inouye and colleagues developed an antisense platform ('micRNA') to regulate host gene expression, which has "great potential as a novel cellular immune system for blocking bacteriophage or virus infection",[21] with Sergio Giunta and Giuseppe Groppa suggesting that

> Upon virus infection and exposure to viral nucleic acids, the cell synthesizes micRNAs that … constitute some of the dsRNAs that are involved both in interferon induction and in the cellular metabolic pathways activated by interferon … and may also be preferentially cleaved by cellular enzymes which would account for the still unexplained discrimination between cleavage of cellular and viral mRNAs in interferon treated cells.[22]

presaging both the mechanism and the sensing of foreign nucleic acids by the Toll pathway.[b]

In the late 1980s, Andrew Fire, Craig Mello and others showed that the introduction of DNA constructs containing fragments of genes in *C. elegans* caused suppression of the endogenous homologous gene.[32] As expected, this silencing was observed when constructs were designed to drive antisense RNA transcription.[33] Surprisingly, however, silencing was also observed when designed to drive 'sense' gene expression, which was expected to do the opposite, i.e., increase the mRNA levels.[33] Adding to this puzzle, in 1995, Su Guo and Ken Kemphues obtained the same result by directly introducing *in vitro* transcribed RNAs in

---

[a] Interferons were first described and named in 1957 by Alick Isaacs and Jean Lindenmann, who showed inhibition of influenza virus growth by prior exposure of cells to active or heat-inactivated virus.[4] They are now known to comprise three classes of proteins, two of which (interferon types I and III) bind to cell surface receptors and induce expression of proteins that prevent the virus from replicating.[5,6] Interferon gamma (type II) enhances specialized subsets of immune responses and is involved in the development of autoimmune disorders such as multiple sclerosis.[7]

[b] Animal cells also discriminate foreign nucleic acids by their lack of nucleotide modifications, in DNA by lack of CpG methylation and in RNA by lack of base modifications, sensed by the so-called Toll-like receptors,[23–26] which also play a role in development and brain synaptic architecture.[27–30] The modification of synthetic mRNAs to circumvent the innate immune response was a critical step in the development of mRNA vaccines.[31]

DOI: 10.1201/9781003109242-12

FIGURE 12.1 Homology-dependent gene silencing. The 'Cossack Dancer' variegated pattern in transgenic petunia expressing a chimeric chalcone synthase pigmentation gene resulting in co-suppression of the endogenous genes. (Reproduced from Napoli et al.[44] with permission of Oxford University Press.)

either the sense or antisense orientation, which caused 'silencing' of the corresponding gene in *C. elegans*.[34]

Similar silencing effects in plants and fungi were also reported in the 1990s,[35] when it was established that the presence of multiple copies of introduced transgenes resulted in repression not only of the transgenes, but also of the homologous endogenous genes.[c]

The iconic example is flower color in petunia, where the effect of silencing of endogenous loci by antisense RNAs was already well established, as famously illustrated a decade earlier by the changes of pigmentation in transgenic petunia flowers expressing antisense RNAs targeting genes specifying the anthocyanin biogenesis pathway,[39–42] which appeared to result from "increased RNA turnover".[43]

In 1990, two groups did the opposite experiment but unexpectedly obtained the same result – they overexpressed a pigment synthesis gene hoping to produce deep purple petunia flowers, but instead obtained predominantly white flowers, with remarkable mosaic patterns, including one dubbed "The Cossack Dancer" (Figure 12.1).[41,44,45]

These phenomena were variously termed 'co-suppression', 'homology-dependent gene silencing', 'transgene silencing' ('quelling' in fungi)[35,46–49] or 'RNA-mediated interference' (RNAi),[50] and later shown also to occur in *Drosophila*.[51,52]

Despite being initially regarded as a peculiar phenomenon, as early as in 1986 'co-suppression' began to be used as a biotechnological tool to modulate plant phenotypes,[36,53–55] only a few years after the first transgenic plants had been constructed by the introduction of foreign genes.[56] Similar to the puzzling RNA interference observed in worms, it was suggested that the silencing caused by introduced copies of 'sense' transgenes in plants[44,57,58] might be triggered by "unexpected" promoter activity in the antisense orientation.[59,60] Nevertheless, this remained a matter of discussion, as the mechanisms behind these phenomena were unknown and appeared to be heterogeneous.[60]

Several models involving RNA-RNA, RNA-DNA and DNA-DNA pairing interactions (or RNA-DNA triplexes[45]) were proposed and investigated by the pioneering groups, those of Michael Wassenegger, Marjori and Antonius Matzke, David Baulcombe, Jan Kooter and others.[48,61–65] In some cases, silencing appeared to be caused by epigenetic mechanisms; in particular sequence-specific RNA-directed DNA methylation[48,65–69] (see below and Chapter 14), which was termed 'transcriptional gene silencing' (TGS).

Other evidence suggested that antisense RNAs interfered with translation or directed the degradation of the target mRNA via dsRNA formation,[43,54,58,70] termed 'post-transcriptional gene silencing' (PTGS), without a supported idea of how 'sense' RNAs might result in the same outcome, or whether there may be more than one RNA-mediated pathway, although there was evidence that TGS and PTGS are mechanistically related.[71,72]

The finding that the silencing of specific genes was able to spread systemically in plants indicated the participation of a diffusible molecule,[73–76] likely RNA itself,[47,54,64,77,78] with the prediction that there existed a "sequence-specific signaling mechanism in plants that may have roles in developmental control as well as in protection against transposons and viruses".[76]

In 1999, Baulcombe and Andrew Hamilton, considering the lack of evidence for long antisense RNAs, hypothesized that low molecular weight RNAs might be involved but had escaped detection due to their small size. They showed that short

---

[c] Sequence-specific transgene-induced virus resistance had also been observed[36] and appeared to be a natural plant defense mechanism.[37,38]

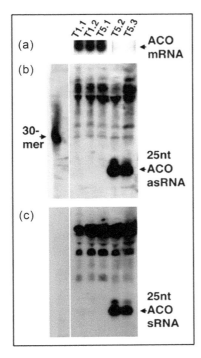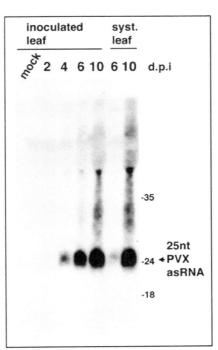

FIGURE 12.2 The identification of small sense and antisense RNAs in post-transcriptional gene silencing in plants. (Reproduced from Hamilton and Baulcombe[79] with permission of the American Association for the Advancement of Science.) The left panel shows that transgenic lines of tomato mRNA containing *ACO* transgenes, in which the endogenous *ACO* gene (involved in fruit ripening[80]) has been silenced, express 25nt small sense and antisense RNAs. The right panel shows the accumulation of a 25nt RNA following infection with potato virus X.

antisense RNAs (estimated to be ~25nt) are produced in transgenic or virus-infected plants undergoing PTGS when both antisense and target mRNAs (cellular or viral) are expressed. They proposed that these RNAs acted as guides for PTGS, stressing that small RNAs "are long enough to convey sequence specificity yet small enough to move through plasmodesmata", and are thus able to act as "the systemic signal and specificity determinants of PTGS" (Figure 12.2).[79]

THE RNA INTERFERENCE PATHWAY

In 1998, Andrew Fire, Craig Mello and co-workers resolved the sense-antisense conundrum by showing (inadvertently, as part of an experimental control) that addition of sense and antisense RNAs together potently triggered systemic and heritable gene silencing in *C. elegans*, which "was at least two orders of magnitude more effective than either single strand alone".[81] Although the mechanism of silencing was still unknown, these observations explained

the RNAi phenomenon, suggesting that the previous silencing observed with 'sense' RNAs was due to contaminating dsRNAs formed during *in vitro* RNA synthesis and antisense transcription.

Independently, also in 1998, Peter Waterhouse and colleagues showed that gene silencing and transgene-mediated virus resistance in plants were induced and targeted by dsRNAs,[82] and Elisabetta Ullu and colleagues showed that dsRNA induced mRNA degradation in trypanosomes.[83] It was soon also demonstrated that dsRNAs trigger gene silencing in *Drosophila*[84–87] and mammalian cells.[88–90]

These findings were followed by genetic dissection of the biochemical pathways and proteins involved.[91,92] The first identified was an RNA-dependent RNA polymerase (RdRP) named *qde-1* (for *quelling defective mutant 1*), which had been predicted to participate in RNA silencing (as, by definition, they would produce dsRNAs).[93,94] Although RdRPs had been detected in prokaryotes and eukaryotes decades before, mostly associated with RNA virus replication, RdRP activity had also been described in tissues of "apparently healthy plants" whose product seemed to be "double-helical RNA",[95] with other reports showing the existence of non-viral RdRPs.[96,97] In 1999, Carlo Cogoni and Giuseppe Macino showed that RdRP was involved in gene silencing,[94] with others subsequently showing that homologs are present not only in fungi, but also in many plants and nematodes,[d] indicating a common pathway.[101–104]

Many other components of the 'RNAi machinery' were soon identified and shown to occur in plants, invertebrates and vertebrates.[92,105] To cut another long story short, work from the laboratories of David Bartel, Greg Hannon, Thomas Tuschl, Phil Zamore and others showed that long dsRNA is recognized by dsRNA-binding proteins (of which there are many versions) that interact with and activate an enzyme named Dicer[e] to cleave the bound RNA into short dsRNA fragments of ~20–23nt in length (which can be amplified by RdRP, hence making RNAi auto-catalytic),[91,101,104,107–109] as had been proposed previously,[110] explaining the powerful and relatively stable systemic and transgenerational silencing obtained with small initial amounts of RNAs.

The short dsRNA fragments, called siRNAs (small interfering RNAs),[89] are then unwound and one of the two strands (the 'guide strand', determined by the strength of the hydrogen bond interaction at the 5′-ends of the RNA duplex, known as the 'asymmetry rule') is loaded into an effector complex, the 'RNA-induced silencing complex' (RISC). The guide strand pairs with a complementary sequence in an mRNA (or other target RNA molecule) and induces its cleavage by an 'Argonaute'[f] protein, the catalytic component of the RISC.[89,107–109,113]

There is a large family of Argonaute proteins[114,115] and a range of small RNA signaling pathways, especially in plants.[116–118] In eukaryotes, Argonaute proteins are found both in the cytoplasm, where they are involved in target mRNA degradation, whereas others are nuclear.[119] Argonaute orthologs also occur in prokaryotes,[114] where they chop foreign plasmids or phage DNA (or transcribed sequences from foreign DNA)[g] into small (13–25 nt) guide sequences, which then target and cleave complementary DNA,[120–124] indicating that RNA-guided targeting of genes for viral defense and gene regulation dates back to the dawn of biology.[h] This system is similar to but distinct from bacterial and archaeal CRISPR-Cas systems that also derive guides from foreign DNA (see below), although some CRISPR-Cas loci encode Argonaute proteins that associate with RNA guides, suggesting that the two mechanisms work synergistically in some organisms.[120,124]

Moreover, the identification of highly conserved protein complexes that mediate small regulatory RNA functions is consistent with principles found in RNA control of other processes, such as the actions of snRNAs and snoRNAs in eukaryotes (Chapter 8), and sRNAs in bacteria, where Hfq acts as a general cofactor for antisense RNAs that rely on short stretches of base pairing[127–132] (Chapter 9).

Genetic studies showed that membrane-bound dsRNA-binding proteins are required for systemic transmission of RNAi.[133] There are two such proteins in C. elegans,[134,135] and two orthologs in vertebrates, the latter of which transport internalized extracellular dsRNAs from endocytic compartments into the cytoplasm for immune activation.[136,137] However, while systemic RNAi transmission is well understood in plants,[138] the extent of and mechanisms for exocrine RNA transport and signaling in mammals[i] are unclear; in part it appears to involve packaging of small RNAs (including microRNAs – see below) into exosomes and other circulating vesicles,[141–144] reported to alter gene expression in immune and other cells,[145–150] and to influence aging,[151] as well as tumor invasion and metastasis in cancer,[152–154] although there are conflicting reports.[155,156]

---

d Some animals, including mammals, have lost RdRPs but have instead evolved a 'ping-pong' amplification system to amplify piRNAs (see below).[98–100]

e Mammalian stem cells protect themselves from some RNA viruses (including Zika virus and SARS-CoV-2) by expressing an alternatively spliced isoform of Dicer, which potentiates antiviral RNAi, a process similar to that in insects and worms, which also use Dicer-dependent RNA interference in antiviral defense, in contrast to mammalian differentiated cells, which generally rely on the interferon system.[106]

f So named because the first reported Argonaute mutant – in plants – yielded a phenotype that resembled the octopus Argonauta argo.[111] Different classes of small RNAs bind to specific members of the Argonaute protein family.[112]

g It is not yet known how the Argonaute proteins distinguish invading DNA from host sequences,[120] although one possibility is that it involves base modification.

h The cooption of endogenous dsRNA 'killer viruses' by the yeast S. cerevisiae to gain competitive advantage over related strains may be a least a partial reason for the loss of the RNAi pathway in this and some other fungal species.[125,126]

i First mooted by Dickson and Robertson in 1976[139] (Chapter 5) and by Stephen Benner in 1988.[140]

TRANSCRIPTIONAL GENE SILENCING: RNA-DIRECTED DNA METHYLATION

Consistent with a nuclear role, there is now considerable evidence that TGS (as distinct from PTGS) imposed by dsRNAs involves methylation of cognate sequences in the genome,[69,157-160] which, although at that time best documented in plants, led Wassenegger to predict in 2000 that "RNA-directed DNA methylation is involved in epigenetic gene regulation throughout eukaryotes", including mammalian X-chromosome inactivation and parental imprinting.[161] This prediction was confirmed in 2004 when Manika Pal-Bhadra, Tatsuo Fukagawa and colleagues showed that RNAi is required for heterochromatin formation in *Drosophila* and vertebrate cells,[162,163] and Kevin Morris and colleagues showed that siRNAs can induce methylation and TGS of homologous loci in mammalian cells,[164] since confirmed by many others[165] and shown to also involve Argonaute proteins.[166-168]

The groups of Rob Martienssen, Shiv Grewal, Danesh Moazed, Robin Allshire, Caroline Dean and others showed that RNAi pathways also control many aspects of chromosome structure and dynamics.[169-179] They demonstrated that transposon-dense heterochromatic regions are not transcriptionally inert, but express RNAs with key functions, many of which involve targeting of nascent transcripts and chromosome-associated RNAs by small RNA species.[180,181] These include regulation of centromere function,[182-185] meiotic and mitotic chromosome segregation,[169] repression of meiotic genes by targeting nearby retrotransposon-derived repeats,[186] other aspects of meiotic progression including chromosome condensation,[169] rectification of copy number imbalances,[169] genome rearrangements[187,188] and chromosome reassembly,[189,190] plant flowering time[191] and DNA damage responses from fission yeasts to mammals,[169,192,193] with interconnections to transcription and splicing.[180,181] RNAi pathways are required for transposon silencing, thought to protect the genome against uncontrolled transposition of mobile elements,[194-198] but have been adapted to regulate developmental processes.[186,199,200]

In plants, RNA-directed DNA methylation induces TGS through the production by specialized plant RNA polymerases of 24nt RNAs that are loaded into AGO4,[201-203] which then recruits DNA methyltransferase and directs DNA methylation to regulate plant-pathogen interactions, stress responses and development.[159,204-207] Similar pathways operate in animals, although less is known about them.[167,169,196-198,205] The notable common feature is the use of embedded transposon-derived sequences as the targets of locus-specific regulation by RNAs, which may explain, in large part, the widespread presence of such sequences in genomes of higher organisms (Chapters 10 and 16).

The observations in *C. elegans* and other organisms showed that RNA signals also mediate transgenerational epigenetic inheritance,[81,208-211] requiring, *inter alia*, a nuclear Argonaute protein[212] and RNA modification[213] (Chapter 17).

RESEARCH AND BIOTECHNOLOGY APPLICATIONS

The ability of natural or modified small siRNAs, delivered directly or via a vector that produces a 'small/short hairpin RNA' (shRNA), to inhibit gene expression specifically and potently in plants and animals has been widely adopted to investigate and modulate gene function *in vivo* and *in vitro*.

In mammals, siRNA was first used to 'knock down' protein-coding genes in oocytes and newly fertilized zygotes by Florence Wianny and Magdalena Zernicka-Goetz, where the phenotypes observed were similar to those observed with null mutants of the endogenous genes.[88] siRNA-directed gene knockdown was later shown to be generally applicable to mammalian cells[89] and applied to systematically target protein-coding genes in culture to identify those involved in various aspects of cell biology (e.g.,[214]), viral replication (e.g.,[215]) and drug responses or sensitivity (e.g.,[216]).

The attraction of RNAi (or any nucleic acid-based system) for human therapy is substantial, as it flexibly employs a relatively common chemistry to inhibit the expression of target genes, notwithstanding the possibility of off-target effects, which are progressively being addressed.[217] The longstanding problem is *in vivo* delivery, especially tissue specificity and transport across the blood-brain barrier to treat neurological disorders,[218] as dsRNAs are not easily absorbed except by the liver.[219] The delivery problem has been solved in part by utilizing viruses, notably AAV (adenovirus-associated virus, because of its low immunogenicity, many tissue-specific serotypes, and low rate of chromosomal integration), as well as other approaches and vehicles such as artificial exosomes, liposomes, nanoparticles and peptide-conjugates that have been engineered to deliver stably expressed shRNAs and small and large RNAs *in vivo*.[220–225]

There are less technical problems in plants, and shRNA-based transgenes have been widely used to engineer or to exogenously deliver, for instance, viral and fungal resistance in horticultural crops and ornamental plants.[226,227]

Viral vectors may also be used to deliver normal copies of defective genes, although their payload is limited by the amount of DNA that can be packaged in the virus, and there may be residual immunogenicity and genotoxicity problems.[218,228,229] Nonetheless, AAV vectors carrying replacement genes or antisense oligonucleotides to rescue or bypass splicing defects have been used successfully to treat different conditions, such as spinal muscular atrophy,[230,231] muscular dystrophy[232] and retinal dystrophy.[233]

However, the use of siRNA and shRNA in research and gene therapy is being rapidly supplanted by a far more efficient and precise genome manipulation system, based on RNA targeting of engineered nucleases, called CRISPR, discussed later in this chapter.

MICRORNAs

Closely related endogenous RNA regulatory pathways that regulate differentiation and development were waiting in the wings to be discovered. As described in Chapter 9, a 22nt small RNA, lin-4, had been described by Victor Ambros and colleagues in 1993 to control developmental timing in *C. elegans*.[234] In 2000, a second small (~21-nt) regulatory RNA species, let-7, was identified by Gary Ruvkun and colleagues, produced from a locus that, like *lin-4*, controls developmental timing by negatively regulating the *lin-41* (protein-coding) gene through partial complementarity to target sequences in its 3'UTR.[235] The similarity to the *lin-4* system led the authors to conclude that

"the mechanism by which they [lin-4 and let-7] regulate the expression of their target could be related" and that they "may constitute elements of a cascade of stage-specific regulatory RNAs that control the temporal sequence of events in *C. elegans* development".[235]

Shortly thereafter, Amy Pasquinelli, Ruvkun and colleagues found that the let-7 sequence, size, and 'hairpin' structure of its precursor are highly

conserved in invertebrates and vertebrates,[236] with the latter having multiple paralogs.[237] The timing of *let-7* expression during development and its target sites in the lin-41 mRNA are also conserved.[236,238] These "small temporal RNAs"[236] became known as 'microRNAs' (miRNAs).

The similarity of size of miRNAs and siRNAs (and the small RNAs described in plant PTGS), suggested a relationship between them,[236] confirmed when Alla Grishok and colleagues found that inactivation of Dicer and other components of the RNAi pathway impaired miRNA processing and activity.[239]

After the realization of the possible existence of other small RNA "classes", and despite "lingering skepticism" and wariness "of what we imagined was a vast detritus of partially degraded RNAs in cells",[240] the development of strategies for large-scale isolation and sequencing of small RNAs with "probable regulatory roles",[237,241,242] as well as bioinformatic searches using the available genome sequences,[243] led to the cloning and characterization of large numbers of small ~21nt RNAs from a variety of organisms, including mammals,[237,241,242,244–249] "deviants no longer".[250]

As Ruvkun observed at the time:

why have more of the miRNAs not been revealed by genetic analysis? …the number of genes in the tiny RNA world may turn out to be very large, numbering in the hundreds or even thousands in each genome. Tiny RNA genes may be the biological equivalent of dark matter — all around us but almost escaping detection, until first revealed by *C. elegans* genetics.[251]

Subsequent studies showed that miRNAs are also produced from processed introns and other non-coding transcripts,[j] including RNAs derived from repetitive sequences, that form an internally folded dsRNA stem-loop structure with an imperfectly paired double-stranded stem.[254,259–265] This structure is recognized and cleaved from precursor host RNAs by a 'Microprocessor' complex comprised of the RNaseIII endoribonuclease Drosha and its partner dsRNA-binding protein Pasha/DGCR8,[266–270] or generated directly from the debranching of

---

[j] Examples, among many others, include the non-coding RNA BIC (B-cell integration cluster) originally identified as a common site for proviral insertion in lymphomas[252–255] and classic lncRNAs such as H19,[256,257] similar to small nucleolar RNA host genes. At least some (like H19) have a dual function as regulatory lncRNAs, such as the miR-205 host transcript that regulates pituitary hormone production.[258]

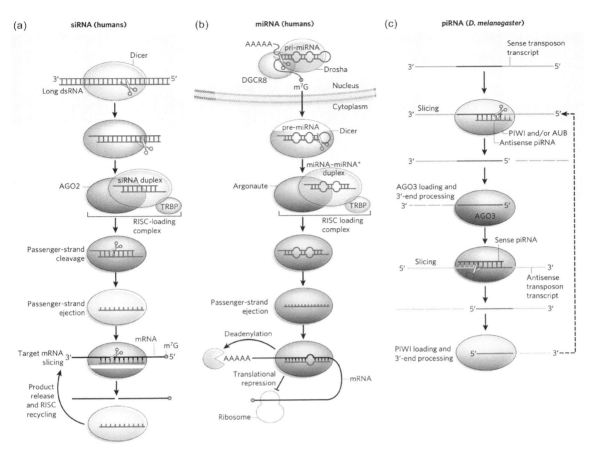

FIGURE 12.3 Biogenesis and mechanism of action of the main classes of small regulatory RNA. (Reproduced from Jinek and Doudna[275] with permission of Springer Nature.)

lariat intermediates formed during intron excision ('mirtrons')[271–273] to yield a hairpin-shaped RNA of ~65–70 nucleotides in length. This 'pre-miRNA' is then cut by Dicer in the cytoplasm, loaded into the RISC complex, which (after ejection of the 'passenger' strand) targets cognate mRNAs for deadenylation of their polyA tails and degradation and/or translational inhibition by binding to complementary sequences primarily in 3'UTRs (Figure 12.3).[91,239,274–276]

Thousands of tissue-specific miRNAs[244,277–280] – some of which are conserved over long evolutionary distances,[281] whereas others are more lineage-restricted[247–249,282] – have been identified and shown to regulate a wide range of differentiation and cellular processes.[274,277,283,284] These include maintenance of pluripotency and stem cell states,[277,285] cell proliferation,[286] epithelial to mesenchymal transition,[287] brain morphogenesis,[288] hematopoietic

differentiation,[289] neuronal asymmetry,[290] dendritic spine development,[291] muscle development,[292] programmed cell death,[293] innate immune responses,[294] metabolism,[295] leaf development[296,297] and flowering time,[298] among many others.[k] MicroRNAs also undergo developmentally regulated post-transcriptional modifications,[299,300] and exhibit altered expression in developmental disorders and cancers[301–303] (dubbed 'oncoMiRs'[304]).

A feature of miRNAs, which generally distinguishes them from siRNAs, is that the precursor dsRNA structure that is recognized by Drosha does not have perfect base complementarity, and either strand may be utilized. Importantly, both miRNAs

---

[k] Despite their ubiquity, miRNAs have rarely shown up in genetic screens (see Chapter 13). Indeed, prior to their biochemical discovery, only one miRNA locus had originally been picked up in *Drosophila* – a gene called *bantam* – and was not initially recognized as a microRNA.[286]

and siRNAs can repress gene expression without activating the interferon pathway,[89] which is triggered by longer dsRNAs.[305] MicroRNAs appear to have evolved before multicellularity, as they are also found in unicellular algae,[306] although their repertoire expanded greatly during metazoan, vertebrate and mammalian evolution.[307,308]

Curiously, it appears that an individual miRNA may recognize sites in many mRNAs,[309-311] and that many, if not most, mRNAs, also contain *bona fide* recognition sites for multiple miRNAs.[312,313] The regulatory logic for this reciprocal multiplexing is unclear and, while miRNAs appear to locate their targets through an 8-nucleotide 'seed' sequence,[297,310] the rules for target recognition are still being teased out,[314-316] as also is the place of miRNAs in decisional hierarchies (Chapter 15).[317]

Although miRNAs are thought to operate primarily on mRNAs in the cytoplasm, they also operate in the nucleus to target pre-mRNAs,[318] as well as enhancer RNAs[319] (to modulate chromatin architecture; Chapters 14 and 16) and nuclear RNAs such as 7SK.[320] They also bind to long non-coding RNAs, sometimes thought to function as miRNA 'sponges' or 'decoys',[321-323] although they are likely legitimate and common targets in their own right. Some miRNAs also trigger the production of 21–22nt siRNAs from transposons,[324] referred to as 'easiRNAs' (epigenetically activated small interfering RNAs), to control genome dosage response and hybridization barriers in plants.[325] MicroRNAs also influence nuclear DNA methylation,[326,327] but the intersecting small RNA pathways and networks have yet to be understood.

Nonetheless, a sort of consensus has emerged. In general, siRNAs are perfectly complementary to their targets and promote their degradation, and in plants are used for defense against viruses and transposons, as well as adaptive responses to environmental stress.[328] siRNAs are also produced in animals - in oocytes, embryonic and somatic cells, from pseudogenes[l] and TEs and other naturally formed dsRNAs[331,332] - where they regulate transposons, genome dosage response and developmental processes, likely interrelated.[94,195,198,265,329,330,333-337] On the other hand, miRNAs are imperfectly complementary to mRNA (usually) 3'UTR sequences and often direct translation inhibition, as well as less well characterized tar-

gets, for the regulation of developmental processes as described above.

PIWI-ASSOCIATED RNAs

In 2006, a related general class of small RNAs was described,[338-342] although they had been detected a few years earlier by Alexei Aravin and colleagues in analyses of tandem repeat and retrotransposon silencing[333] and small RNA sequencing datasets in the testis.[343] A subgroup of Argonaute proteins, known as the 'Piwi family',[m] was known to be required for germ- and stem-cell development in animals, three homologs (Mili, Miwi and Miwi2) being essential for spermatogenesis in mammals[345,346] (Figure 12.4). These proteins bind 'Piwi-interacting RNAs', or piRNAs, which are longer than miRNAs, generally ranging from 25 to 33nt, and have a characteristic 2'O-methylation[n] of the 3' terminal ribose,[348] which is recognized by the PAZ domain of Piwi proteins,[349,350] with those interacting with different Miwi/Mili proteins having a different mean length.[346,351,352]

Thousands of genomic loci have been found to produce piRNAs.[o] These RNAs are present in millions of copies in testes and are so abundant that, like the discovery of small nuclear RNAs in the 1960s, they were discovered by visualization in gel electrophoretic displays. PiRNAs are also expressed in ovaries[353-358] and somatic cells,[354,359] where they regulate gene expression in stem cells and embryonic development,[357,358] and in the brain, where they are required for neurogenesis and memory formation[360-363] (Chapter 17). piRNA expression is also dysregulated in cancer.[364] They are not produced by Drosha processing of dsRNA intermediates but the best understood mechanism for piRNAs derived from repeat sequences and present in diverse organisms is through a 'ping-pong' amplification system from single-stranded RNA templates.[98-100]

---

[l] Processed from duplexes between antisense pseudogene (noncoding) RNAs and corresponding mRNAs.[329-331]

[m] The name given to the original *Drosophila* mutant: P-element-induced wimpy testis (Piwi).[344]

[n] All miRNAs and siRNAs in plants and siRNAs in *Drosophila* are 2'-*O*-methylated at their 3' ends, apparently to improve their stability.[347]

[o] There are over 2,500 miRNA and over 23,000 piRNA candidate loci documented in humans, rivaling the number of protein-coding genes.[351]

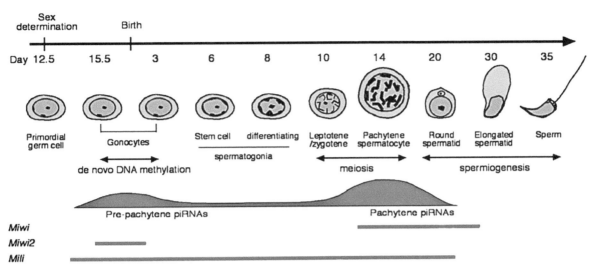

FIGURE 12.4 Biphasic production of piRNAs and Miwi expression during mouse sperm development, (Adapted from Zheng and Wang,[382] under Creative Commons Attribution License.)

piRNAs are thought to be (mainly) involved in the repression of transposons,[p] likely their ancestral function.[196,346,352,367–371] They are also involved in parental imprinting.[372] However, their range is much greater. For instance, male-specific chromosome Y-derived long non-coding transcripts, named Pirmy and Pirmy-like RNAs, which exhibit a large number of splice variants in testis, act as templates for piRNAs that regulate autosomal genes.[373] piRNAs are also derived from other sequences, including intergenic transcripts, pseudogenes and the 3'UTRs of mRNAs,[112,374,375] the latter targeting the introns of other genes.[376] Along with their roles in development and brain function, these observations lead to the conclusion that genomes have co-opted the TE-derived piRNA system for regulating transposon activity and other dimensions of genome regulation.[375–379] These include the modulation of nuclear architecture at target loci containing TEs, shown in a *Drosophila* ovarian somatic cell line and involving directing these genomic regions to the nuclear periphery, heterochromatin formation and chromatin conformational changes to promote transcriptional silencing; thus providing a possible mechanism for the proposed role of TEs as "functional motifs" for the dynamic regulation of chromatin state and nuclear architecture[380,381] (see Chapters 14 and 16).

In sperm development in *Drosophila* and mammals, there are two bursts and classes of piRNAs and two corresponding rounds of Piwi protein expression. The first class is expressed before the meiotic 'pachytene' stage of spermatogenesis, is derived from transposon-rich clusters, and associates with Mili and Miwi2. The second class is expressed during the pachytene stage from thousands of genomic loci, is extremely abundant and associates with Mili and Miwi1.[382–384] Their function is unknown,[385] although an intriguing possibility is that this dual system evolved to enhance evolvability (Chapter 18), with quality control, when progeny numbers are small.[386] There is widespread formation of dsRNAs in the testis, different in profile from that observed in somatic cells.[387] Moreover, the loss of fertility in Piwi mutants lacking piRNAs is not due to de-repression of repetitive elements: pachytene piRNAs repress gene expression by directing the cleavage of meiotic transcripts required for sperm function.[388,389]

It may also be that the phenotypes of the loss of piRNA function are mediated by epigenetic silencing of histone genes,[390] implicating small RNAs in the transmission of epigenetic information across generations (Chapter 17). With many more piRNA loci yet to be studied, there is much to be learned about this large class of small RNAs in animals.

---

[p] Involving the long-described but mysterious, germ-cell-specific 'nuage' organelle,[365] which forms a phase-separated domain (Chapter 16) that concentrates single-stranded DNA but largely excludes double-stranded DNA.[366]

OTHER CLASSES OF SMALL RNAs

The landscape is even more complicated, as most RNAs form double-stranded structures that can be and frequently are processed into small RNAs.

A variety of small RNA species are produced from rRNAs ('rRFs') in a developmental stage-specific manner in fungi, plants and animals.[391–393] Most derive from 28S rRNA, and have various sizes, some shown to be siRNAs ('phasiRNAs'), miRNAs and piRNAs, as well as 'qiRNAs'[394] that function in DNA damage response.[q] The 18S rRNA is also the source of smaller RNAs, including miRNAs, as well as a longer species of 130nt that may be an miRNA precursor. miRNAs are also processed from the 5.8S rRNA and the 45S rRNA precursors.[391] The piRNA pathway is required to control the abundance of rRNA-derived siRNAs and the ribosomal RNA pool in *C. elegans*.[398]

In all branches of life, tRNAs are processed into highly abundant smaller species ('tRFs' for tRNA-derived fragments, or 'tsRNAs' for tRNA-derived small RNAs),[399–403] initially thought to be degradation products,[404] despite their production being cell-state and tissue-specific[399,405,406] and (at least in some cases) Dicer-dependent,[400,407] possibly playing a role in the global regulation of RNA silencing.[408] Some tRNA fragments function as miRNAs.[409–411] The production of tRFs is influenced by methylation of the tRNA by the RNA methyltransferases Dnmt2[412] and Nsun2, which may be involved in transgenerational inheritance (see below) and loss of which causes neurodevelopmental disorders.[413] tRNAs and tRNA fragments are also modified and regulated by TET2, which oxidizes 5-methylcytosine in DNA[414] (Chapter 14). tRFs and modifications thereof are stress-induced and inhibit translation,[415–419] control retrotransposons,[420,421] promote cell proliferation,[401] and are depleted in immortalized cell lines.[406]

tRFs also convey long distance regulatory signals. Their levels are increased in *Arabidopsis* roots upon phosphate starvation,[422] and rhizobial (symbiotic bacterial) tRFs modulate host nodulation in legumes via the RNAi pathway.[423] They are present, along with miRNAs, in secreted exosomes[144] from stem cells[424] and activated T-cells.[425] They are also abundant in semen and sperm,[420,426,427] and have been found to contribute to intergenerational inheritance of acquired metabolic disorders, with evidence that this involves their methylation[428–430] (see Chapter 17). tRFs have also been shown to affect the biogenesis of snoRNAs, scaRNAs and snRNAs, as well as histone levels and chromatin organization[431] (Chapter 14).

tRNA-like small RNAs are also produced from the processing of long non-protein-coding RNAs, such as MALAT1,[432,433] one of the most highly expressed genes in the vertebrate genome, which also produces piRNAs.[434]

All known snoRNAs (in fungi, plants and animals, from yeast to human) are processed to produce smaller species: those derived from C/D snoRNAs show a bimodal size distribution of 17–19nt and >27nt, the latter similar in size to piRNAs.[436] Those derived from H/ACA snoRNAs are predominantly 20–24nt in length[436] (Figure 12.5), at least some of which have been shown to be Dicer-dependent, bind to Argonaute, function as miRNAs and have developmental effects.[265,435–439]

Likewise, spliceosomal snRNAs are processed into smaller fragments,[409] although their significance is unknown. Vault RNAs (Chapter 8) are processed as well, dependent on cytosine methylation, into small RNAs that bind to Argonaute proteins and exhibit miRNA-like functions.[440] Other short (80–100nt) intronically derived RNAs, some of which are highly conserved, also associate with Argonaute proteins and are capable of repressing mRNAs via target sequences in the 3'UTRs.[441] All of these observations indicate deep evolutionary and complex functional connections between regulatory RNA pathways.

Finally, another class of small RNAs with a modal length of 17–18nt is associated with transcription start sites ('tiRNAs') and exonic splice sites ('spliRNAs') in animals, but not in plants, with evidence that they are associated with binding sites for epigenetic regulators and mark nucleosome positions,[442–444] which may be subject to fine control in animals because of the required precision of ontogeny (Chapter 15).

RNA COMMUNICATION BETWEEN SPECIES

Some of the early experimental RNAi methods in *C. elegans* indicated the potential of interspecies communication by small RNAs: "dsRNAs expressed

---

[q] Similar 20–35nt small 'ddRNAs' have been reported elsewhere to control the DNA damage response via interaction with damage-induced long non-coding RNAs (Chapter 13) and the formation of phase-separated domains (Chapter 16) at DNA double strand breaks.[395–397]

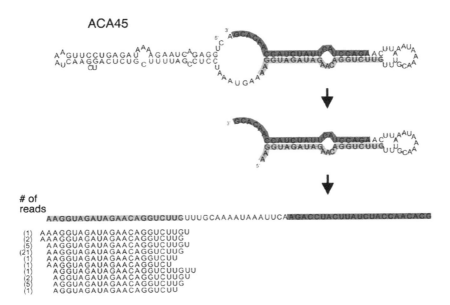

FIGURE 12.5 Processing of an miRNA from snoRNA ACA45. The blue bar shows the sequences of the putative miRNA products and the number of reads in Argonaute-associated small RNA sequencing data, while the red bar shows putative 'star' product from the stem-loop precursor. (Reproduced from Ender et al.[435] with permission of Elsevier.)

in *Escherichia coli* could induce gene silencing in nematode larvae that feed on them … raising the possibility that dsRNAs may be transferred from prokaryotic pathogens to their hosts".[445]

Small RNAs have subsequently been shown to mediate communication between species, for pathogenicity, defense or symbiosis. For example, the small RNA *SsrA* produced by the bacterium *Vibrio fischeri* is loaded into outer membrane vesicles that are transported into the epithelial cells in the light organ of the squid and suppress antimicrobial responses and permit light-organ symbiosis between the bacterium and its host.[446]

The soil bacterium *Pseudomonas aeruginosa*, which is pathogenic in many species, releases a small RNA termed P11 into *C. elegans* that is processed by the RNAi and piRNA machinery, and specifically silences the expression of the protein maco-1, which functions in chemotaxis and neuronal excitability, leading to avoidance behavior that is heritable for four generations, of value to both.[447] *C. elegans* transmits this memory of learned avoidance to naive animals and four generations of progeny through virus-like particles encoded by the Cer1 retrotransposon.[448]

Similarly, the bacterial pathogen *Listeria monocytogenes* secretes a small RNA-binding protein named Zea that binds a subset of *L. monocytogenes* RNAs

and RIG-I, the non-self-RNA innate immunity sensor in mammalian cells. In mice, this was shown to dampen the immune response and suggested to allow sustainable host-bacterium symbiosis.[449] Zea orthologs occur in other bacteria that are rarely associated with disease, suggesting that this may be a general mechanism to assist co-existence between bacteria and multicellular hosts,[449] especially in view of the growing awareness of the importance of the enteric microbiome in ruminant biology and mammalian physiology.

These may occur between eukaryotic kingdoms, as shown between the mycorrhizal fungus *Pisolithus microcarpus* and its host *Eucalyptus grandis*, in which the fungus miRNA *Pmic_miR-8* is transported into the plant roots, targeting host genes to promote symbiosis.[450] Such interactions may influence pathogenicity and aggressiveness of parasites, as exemplified by the interaction between the fungal pathogen *Botrytis cinerea* and host plants, proposed to involve retrotransposon-derived small RNAs derived from the fungus and delivered into plant cells and cross-kingdom RNA interference (ckRNAi) to undermine host defense.[451]

Even complex ecological interactions may be affected by regulatory RNAs. This is suggested by the reported effect of short satellite RNAs (satRNAs, a type of subviral agent associated with some viruses, Chapter 8) that accompanies cucumber mosaic virus (CMV) and are processed into small RNA (sRNA) species in infected tobacco leaves. The production of these Y-sat sRNAs affects host gene expression and leads to a change of the color of the leaves from green to bright yellow, which preferentially attracts aphid vectors. Surprisingly, the Y-sat RNAs are also processed in the aphids feeding on the leaves, leading to the production of small RNAs that affect aphid physiology, turning them red and subsequently promoting wing formation. This in turn facilitates CMV and Y-sat RNA spread without apparent detrimental effects on the plant, thus suggesting a potential intricate 'survival strategy' by a subviral non-coding RNA, which depends on interaction with (and affects the interactions between) the CMV helper virus, the host plant and the insect vector.[452]

It is early days in the exploration and understanding of the extent of communication by "social RNA" (a term coined by Eric Miska)[453,454] in the competition and cooperation within and between species in complex environments, but there are further documented examples of RNA exchange, including between parasitic nematodes and mammalian cells,[455] plants and fungal pathogens,[456,457] parasitic and host plants,[458] and even between queen and worker bees via royal and worker jellies,[459,460] a pathway which has been exploited to artificially introduce virus resistance to commercial hives.[461]

CRISPR

The exponential growth and analysis of DNA and RNA sequence data transformed molecular biology, no better demonstration being the serendipitous discovery[r] of the CRISPR RNA-based immune system in prokaryotes.[466–470]

In 1987, Yoshizumi Ishino and colleagues reported the existence of mysterious, partly palindromic ~30nt tandemly repeated sequences, separated by ~30 nt 'spacers' in *E. coli*.[471] A similar peculiar array of interspersed repeats was described in the salt-tolerant archaeon *Haloferax*

mediterranei by Francisco Mojica and colleagues in 1993, who went on to collate instances of such sequences in many different species, which they called 'Short Regularly Spaced Repeats'.[472,473] Others doing related bioinformatic analyses showed that a common set of protein-coding '*cas*'s genes lay adjacent to these repeats, which they termed 'Clustered Regularly Interspaced Short Palindromic Repeats',[474] whose acronym, CRISPR, being more euphonic, became adopted.

The function of CRISPR was revealed in 2005, when three groups, including Mojica's,[t] discovered that the spacers separating the repeats corresponded to fragments of bacteriophages and plasmids, indicating that these sequences recorded past invasions and constituted a memory-defense system,[477–479] reinforced by the bioinformatically inferred nuclease function of the CRISPR-associated *cas* genes and analogies with RNAi.[479,480]

This conclusion was confirmed when Philippe Horvath, Rodolphe Barrangou, Sylvain Moineau and colleagues working at the French company Danisco[u] showed in 2007 that recently selected phage-resistant mutants of the lactic acid-producing bacterium *Streptococcus thermophilus*, used in yogurt and cheese production, had acquired phage-derived sequences at their CRISPR loci and that phage isolates that overcame the immunity had single nucleotide changes that altered the sequences corresponding to the spacers.[481] These and other investigators, notably Luciano Marraffini, John van der Oost and colleagues, also showed that the CRISPR array is transcribed and cleaved within the repeats to produce 61 nt 'crRNAs' wherein the spacer (target) sequence is flanked by the palindromic sequences of the repeats. The crRNAs are then loaded into a *cas*-encoded

---

[r] Another serendipitous discovery was bacterial 'retrons' which were first discovered in 1989 in myxobacteria and *E. coli*.[462,463] Retrons are chimeric covalently linked RNA/DNA molecules produced by reverse transcriptase, prior to which reverse transcriptases had not been known to exist in bacteria.[464] Later studies showed that retrons form a second line of defense against phages (abortive infection by cell death) that is triggered if the first lines of defense have collapsed.[465]

[s] 'CRISPR-associated'.
[t] Mojica's paper was first submitted for publication in 2003, but not published until early 2005, due to multiple rejections. In fact all three papers were rebuffed by leading journals, being considered by their editors to lack sufficient novelty and importance to send out for peer review.[469] A similar story of editorial rejection and reviewing delays by leading journals robbed Giedrius Gasiunas and colleagues of recognition for the co-discovery of RNA programmable Cas9–crRNA target cleavage.[475,476]
[u] As in many other cases, like antisense RNAs and RNAi (to generate virus-resistant transgenic plants), some practical applications of CRISPR preceded mechanistic understanding. The company (Danisco), which was acquired by DuPont, announced in 2012 the commercial availability of several *S. thermophilus* strains immune to various bacteriophages for pizza cheese-making.

nuclease,[481–484] which uses the spacer guide sequence to target and introduce a double strand break in cognate DNA via two distinct nuclease domains.[v] The Cas proteins are essentially RNA-programmable restriction enzymes[487,488] (Chapter 6), some of which are also used for transposon homing.[489,490]

A vital missing link was discovered when Emmanuelle Charpentier and colleagues, who had earlier shown that streptococcal virulence factors are controlled by regulatory RNAs,[491] found that the third most highly expressed sequence (after rRNA and tRNA) in *Streptococcus pyogenes* is a small RNA that is transcribed from a sequence adjacent to the CRISPR locus and has 25 bases of near-perfect complementary to the repeats. This RNA, termed 'tracrRNA' (trans-activating CRISPR RNA), hybridizes to the transcribed CRISPR array to guide cleavage by host-encoded RNaseIII,[492] is loaded into the cas nuclease along with a crRNA, and is essential for its function[485] (Figure 12.6). An endogenous guide RNA autoregulates CRISPR-*Cas* expression to mitigate autoimmune toxicity.[493]

There are a number of variants of CRISPR systems in bacterial and archaeal genomes.[496,497] Type I and type III systems employ a large multi-Cas protein complex for crRNA binding and target cleavage.[466,467] The simplest, Type II, consists of the CRISPR array, the gene encoding the tracrRNA, and four protein-coding *cas* genes, three of which are involved in generating new spacers and repeats and the other, *cas9*, encodes the targeting nuclease.[486]

Virginijus Siksnys, Horvath, Barrangou and colleagues showed that the CRISPR system from *S. thermophilus* could be functionally transferred into *E. coli* and that all that was required to target and cleave a specific DNA sequence is the tracrRNA, the Cas9 nuclease and the corresponding crRNA,[498] which could be as short as 20 nt as the flanking repeat sequences are not required, paving "the way for engineering of universal programmable RNA-guided DNA endonucleases".[475]

Shortly thereafter, Charpentier, Martin Jinek, Jennifer Doudna and colleagues showed that the normally base-paired cRNA and tracrRNAs could be covalently linked and produced as a single chimeric guide RNA (sgRNA), simplifying delivery of the system.[485,486]

CRISPR systems have also been coopted naturally for various structural and regulatory functions by repurposing of diverged repeats encoded outside of CRISPR arrays, including by transposons for RNA-guided transposition, by plasmids for interplasmid competition, and by viruses for antidefense and interviral conflicts. There are also multiple highly derived CRISPR variants of yet unknown functions.[499]

RNA-DIRECTED GENOME EDITING

In 2013, the groups of Feng Zhang[500] and George Church[501] demonstrated that the CRISPR system (Cas9 plus sgRNAs) could be used in human and mouse cells to introduce sequence-specific double strand breaks,[w] which, when repaired *in vivo*, lead to incorporation of guided mutations, the latest in a long line of 'homing endonucleases' with wide practical applications.[504,505] Zhang and colleagues showed that they could engineer the Cas9 sequence to disable one of its two nuclease domains, converting it into a 'nickase' enzyme that only introduces a single-strand break, facilitating the insertion of new sequences into genomes by homology-directed repair. They also showed that multiple guide sequences can be encoded into a single CRISPR array to enable the simultaneous alteration of several sites.[500]

CRISPR technology has proven far more versatile and efficient than the previous methods developed for targeted genome modification, which employed nucleases linked to the DNA-binding domains of transcription factors whose sequence specificity could be engineered by altering amino acids in these domains (called ZFNs[x] and TALENs), an effective but relatively cumbersome methodology.[506–508]

Moreover, CRISPR technology has been adapted by engineering the Cas9 and other Cas nucleases to inactivate, manipulate, image and annotate specific DNA and RNA sequences in a wide variety of organisms,[495,509–514] including high-throughput mutagenesis to identify the roles of new genes and genetic pathways in cell and developmental biology.[515] CRISPR has also been widely used in industrial biotechnology[516,517] and in agriculture to construct transgenic plants and animals with desirable properties.[518,519]

---

[v] Cas9 contains a HNH-like nuclease domain that cleaves the DNA strand complementary to the guide RNA sequence, and an RuvC-like nuclease domain that cleaves the opposite DNA strand.[485,486]

[w] Group II introns were the first RNAs to be used for gene/genome targeting.[502–504]

[x] Involving 'zinc fingers', see Chapter 16.

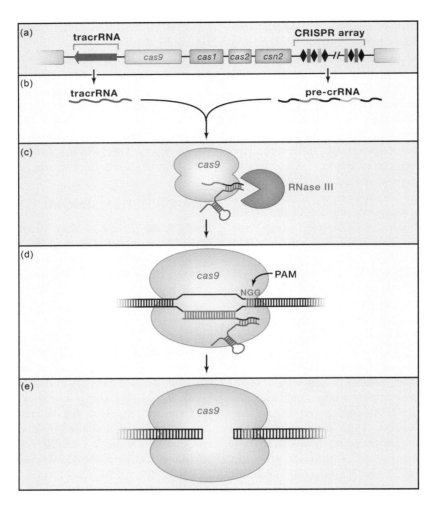

FIGURE 12.6   The type II CRISPR-Cas9 System from *Streptococcus thermophilus*. (Reproduced from Lander[469] with permission of Elsevier.) Like RNAi, CRISPR-Cas systems rely on RNA guidance for target specificity. (a) The locus contains a CRISPR array, four protein-coding genes (*cas9*, *cas1*, *cas2* and *cns2*) and the tracrRNA. The CRISPR array contains repeat regions (black diamonds) separated by spacer regions (colored rectangles) derived from phage and other invading genetic elements. The *cas9* gene encodes a nuclease that confers immunity by cutting invading DNA that matches existing spacers, while the *cas1*, *cas2* and *cns2* genes encode proteins that function in the acquisition of new spacers from invading DNA. (b) The CRISPR array and the tracrRNA are transcribed, giving rise to a long pre-crRNA and a tracrRNA. (c) These two RNAs hybridize via complementary sequences and are processed to shorter forms by Cas9 and RNase III. (d) The resulting complex (Cas9 + tracrRNA + crRNA) then searches for DNA sequences that match the spacer sequence (shown in red). Binding to the target site also requires the presence of the protospacer adjacent motif (PAM), which functions as a molecular handle for Cas9. (e) Upon binding to a target site matching the crRNA sequence, Cas9 cleaves the DNA three bases upstream of the PAM site. Cas9 contains two endonuclease domains, HNH and RuvC, which cleave, respectively, the complementary and non-complementary strands of the target DNA, creating blunt ends. Other natural or artificial Cas systems employ different nucleases.[494,495]

The toolkit and applications have been advancing at breathtaking pace.[520] For example, nuclease-inactive and RNA-targeting Cas proteins have been fused to a plethora of different effector proteins to alter gene expression, epigenetic modifications and chromatin interactions.[495,521–526] These include the fusion of transcriptional repressors or activators to catalytically inactive Cas9 to repress ('CRISPRi') or enhance ('CRISPRa') expression of specific protein-coding and non-coding genes (Figure 12.7).[527–531]

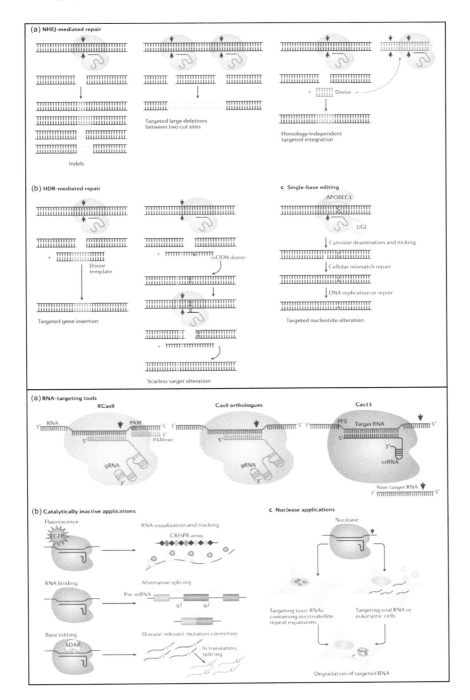

FIGURE 12.7 CRISPR applications. Top panel, genome editing; bottom panel, RNA-targeting, tagging and editing. (Reproduced from Pickar-Oliver and Gersbach[495] with permission of Springer Nature.)

They also include the fusion of cytidine deaminase or reverse transcriptase to Cas9 by to enable single base replacement or the insertion of any desired sequence (programmed into the guide RNA) at any specific position in the genome, termed 'base editing' and 'prime editing', respectively. These have been demonstrated to reverse damaging mutations at high efficiency in cell culture[494,532–535] and (to a limited extent, but with increasing success) *in vivo* including amelioration of deafness and amyloidosis,[536–538] which "in principle could correct up to 89% of known genetic variants associated with human diseases".[534] CRISPR systems have also been adapted to edit RNA with high specificity[539] to correct a range of genetic defects,[540–543] and to develop multiplexed assays for high sensitivity detection of viral RNAs.[544–546]

CRISPR is also being deployed to propagate the ingression of desirable genes into natural populations, first mooted by CF Curtis in 1968 (using translocations)[547,548] and again by Austin Burt in 2003 (using "site-specific selfish genes" such as homing endonucleases, group II introns and some types of transposable elements).[505] To date, effective introduction of 'gene-drive' systems based on CRISPR-Cas9 has been achieved in yeast,[549,550] fruitfly[551] and mosquito,[552–554] among others, with applications including eliminating vector-borne and parasitic diseases, promoting sustainable agriculture and curtailing invasive species, albeit with biosafety and 'ethical' concerns,[555,556] which in turn have motivated the design of tunable and reversible systems.[557–560]

Advanced nanoparticle delivery systems are being developed in parallel to enable high-efficiency tissue-specific delivery *in vivo* for gene therapy, vaccination, cancer treatment and investigative genetic manipulation.[561–563] The efficiency and specificity of genome editing continues to improve, with constant factor being the flexibility of RNA guides to unlock the potential of targeting generic effector proteins to specific DNA or RNA locations.

A billion years ago, RNA guides also solved the challenges of regulating cell fate decisions and tissue architecture in multicellular organisms (Chapters 15 and 16).

FURTHER READING

Abbott A. (2016) The quiet revolutionary: How the co-discovery of CRISPR explosively changed Emmanuelle Charpentier's life. *Nature* 532: 432–4.

Castel S.E. and Martienssen R.A. (2013) RNA interference in the nucleus: Roles for small RNAs in transcription, epigenetics and beyond. *Nature Reviews Genetics* 14: 100–12.

Gutbrod M.J. and Martienssen R.A. (2020) Conserved chromosomal functions of RNA interference. *Nature Reviews Genetics* 21: 311–31.

Lander E.S. (2016) The heroes of CRISPR. *Cell* 164: 18–28.

Ledford H. (2016) The unsung heroes of CRISPR. *Nature* 535: 342–4.

Maraganore J. (2022) Reflections on Alnylam. *Nature Biotechnology* 40: 641-50.

Özata D.M., Gainetdinov I., Zoch A., O'Carroll D. and Zamore P.D. (2019) PIWI-interacting RNAs: Small RNAs with big functions. *Nature Reviews Genetics* 20: 89–108.

Pasquinelli A.E. (2015) MicroRNAs: Heralds of the noncoding RNA revolution. *RNA* 21: 709–10.

Veigl S.J. (2021) Small RNA research and the scientific repertoire: A tale about biochemistry and genetics, crops and worms, development and disease. *History and Philosophy of the Life Sciences* 43: 30.

13 Large RNAs with Many Functions

Following fast on the heels of the genome projects came high-throughput RNA sequencing projects, which, notwithstanding the excitement around the RNA interference pathway and small RNA control of gene expression, marked the turning point from regarding non-coding RNAs as ancillary contributors to mainstream players in cell and developmental biology.[1,2]

The technique of cloning end fragments of RNAs as 'Expressed Sequence Tags' (ESTs) was developed and popularized in the 1990s to identify protein-coding genes[a] and their spliced isoforms.[7,8] Because mRNAs typically comprise only ~3% of the total RNA in cells (rRNAs comprise ~95%) and it was assumed that polyadenylated RNAs are mRNAs,[b] oligo(dT) hybridization was used to purify these RNAs and to prime reverse transcription from their 3' ends.

Unexpectedly, the large-scale cDNA cloning efforts yielded many sequences that lacked protein-coding capacity,[13,14] which were initially suspected to be degradation products or DNA contamination during library preparation. This led to alternative methods to 'filter' for protein-coding genes, such as by evolutionary conservation.[15] The existence of long 3'UTRs in mRNAs also presented difficulties, as the cloned ESTs often did not extend into the upstream protein-coding sequences.[c]

To circumvent the latter problem strategies were developed to increase the representation of internal exons of transcripts by priming reverse transcription with random oligonucleotides.[18] This approach was used in the Human Cancer Genome Project to generate nearly one million 'Open Reading Frame' ESTs ('ORESTES') from cancers and normal tissues.[18–21] Many novel RNAs were identified, leading to the suggestion in 2000 that 36,000 human genes was likely a "significant underestimate", even after sequences that did not correspond to predicted (protein-coding) genes were excluded.[19] It was later shown that approximately half of the ORESTES were differentially expressed intronic or intergenic RNAs.[13]

The confounding problem was the wide dynamic range of gene expression, both within and between cells in heterogeneous tissues.[22] Of the ~500,000 polyadenylated RNA molecules estimated to exist in a human cell, at least half are accounted by a small number of highly expressed mRNAs.[23] Over 95% of the RNA species are expressed at low levels: in a tissue such as rat liver, for example, ~10 species are present at ~10,000 copies, 500 at ~200 copies, and 15,000 at ~10 copies or less per cell.[24] The expression of the small fraction (3%) of 'housekeeping' genes is ~30-fold higher than all other transcripts,[25] so the latter can only be reliably observed at high sequencing depth or with enrichment strategies.[22]

Consequently, studies aiming to comprehensively survey the transcriptome used hybridization subtraction methods to deplete abundant RNAs and improve the representation of cell- and tissue-specific transcripts.[26,27] To reduce interpretation problems, a method termed 'Cap Analysis of Gene Expression' (CAGE) was developed by Piero Carninci, Yoshihide Hayashizaki and colleagues using biotinylation of terminal nucleotides to capture 'full-length' RNAP II transcripts[d] containing both the 5' end and the 3' polyA tail.[14,27,29–32]

---

[a] High-throughput mRNA sequencing provided the necessary information not only for better annotation of gene (exon-intron) structures and alternative splicing,[3,4] but also for 'proteomics', which matches amino acid sequences predicted by mass spectrometry of peptides generated (usually) by proteolytic digestion with those in mRNA open reading frames. Although having a high false-positive rate, and complicated by post-translational modifications, proteomics has proved useful for identifying the protein constituents of subcellular organelles and complexes.[5,6]

[b] This assumption ignored historical evidence that some mRNAs such as histone mRNAs are not polyadenylated[9] and that polyA-RNAs are abundant in human cells.[10–12]

[c] The 'Mammalian Gene Collection' initiative (also) sequenced cDNAs from their 5' ends but discarded the clone if an AUG start codon and a following open reading frame were not identified,[16,17] thereby excluding noncoding transcripts.

[d] A large fraction of the RNAs in human cells are not polyadenylated, although they are often capped.[10–12,28]

DOI: 10.1201/9781003109242-13

PERVASIVE TRANSCRIPTION

In the early 2000s, Hayashizaki, Carninci and colleagues established the influential FANTOM[e] projects,[33,34] which undertook large-scale sequencing of normalized full-length cDNA libraries constructed from a wide variety of mouse cell types, tissues and developmental stages, with the objective of characterizing the proteome. However, the unexpected and ultimately headline result was the identification of over 34,000 long "mRNA-like" (5' capped and 3' polyadenylated) transcripts often emanating from 'intergenic' regions, many of which were spliced and differentially expressed, but did not appear to code for proteins.[14,34–37] At least 70% of protein-coding loci were also found to express overlapping antisense RNAs, some of which were shown to have regulatory function[38,39] and to be conserved over large evolutionary distances in the vertebrates.[40] Widespread tissue-specific sense-antisense and intronic transcription was also observed across the spectrum from yeast to humans.[41–52]

Similar findings were reported using the orthogonal method of high-density genome tiling arrays[f] independently by Tom Gingeras, Mike Snyder and colleagues, who showed that transcribed sequences covered far more of the human genome[g] than predicted from protein-coding gene annotations,[28,54–57] with pervasive transcription of coding and non-coding regions in embryonic stem cells, becoming more restricted as differentiation proceeds.[58] They also showed that almost half of the transcripts in human cells are not polyadenylated,[28] confirming largely forgotten reports from the 1970s and 1980s.[10–12] The non-polyadenylated RNAs are derived from repeats and introns,[28,59] often transcribed at high levels by

RNA polymerase III[h],[69–71] or from processing of RNAPII transcripts.[72–75] Moreover, the total mass of the non-ribosomal, non-protein-coding RNAs in human cells and brains was found to exceed that of the mRNAs.[76,77]

While controversial at first, these findings were confirmed by other studies in animals, plants and fungi using a variety of techniques, including additional large-scale cloning and sequencing of cDNAs,[57,72,78–81] serial analysis of gene expression and massively parallel signature sequencing[82–85] and microarrays and genome-wide tiling arrays that probe the expression of non-coding regions.[13,37,46,52,55,56,78,86–89] Over 85% of the *Drosophila* genome was found to be dynamically expressed during the first 24 h of embryonic development,[49,90–95] over 70% of the *C. elegans* genome to be transcribed in mixed-stage populations[88] and over 85% of the yeast genome to be expressed in rich media.[87] A subsequent intensive survey of 1% of the human genome by the ENCODE[i] Consortium showed that at least 93% of the nucleotides in the studied regions are transcribed in one or more of the 11 cell lines analyzed and that most of the unannotated transcripts are expressed in just one or a few.[96]

At the same time, we and others showed that the expression of unannotated mammalian long

---

[e] FANTOM: Functional Annotation of the Mouse (later Mammalian) Genome. The successive FANTOM projects introduced many technical innovations and have produced a wealth of data, including well-annotated transcription start and termination atlases, full-length cDNA clones and other valuable resources for the international research community.[33,34]

[f] Hybridized to cDNAs randomly primed from polyA+ or polyA- RNAs.

[g] Again described as the "dark matter" in the genome.[53]

[h] RNA polymerase III (RNAPIII) produces various types of regulatory RNAs from repeat sequences, some of which are clade-specific, such as B2 SINE RNAs in mice and Alu RNAs in humans, in both cases with modular structures that repress RNA polymerase II during stresses like heat shock at specific loci, a striking example of convergent evolution.[60–62] Large numbers of human- or primate-specific Alu-derived short RNAs transcribed by RNAPIII have been identified by bioinformatic and biochemical strategies, including a class of structured small (<120nt) RNAs (snaRs) complexed with the dsRNA-binding nuclear factor 90 family. These are mostly genomically clustered and differentially expressed in regions of the brain, other tissues and cancer cells.[63–65] Recently it has been shown that transcription of *snaR-A*, which produces an miRNA that targets a metastasis inhibitor,[66] is driven by an embryonic isoform of RNAPIII that is also upregulated in cancer cells, whereas the other isoform is expressed in specialized tissues.[67] Other TE-derived RNAs are involved in other aspects of genome expression and organization, discussed in Chapter 16. For example, bidirectional transcription by RNAPII and RNAPIII of the B2 SINE sequence in mice was found to restructure the growth hormone locus into nuclear compartments and define the heterochromatin-euchromatin boundary to regulate the expression of the gene during organogenesis, suggesting a role of these abundant elements in the topological organization of the genome.[68]

[i] ENCODE: Encyclopedia of DNA Elements.

non-coding RNAs (lncRNAs)[j] is highly dynamic and tissue-specific (details below), being most extensive in brain and testis.[37,98] LncRNAs are also differentially expressed during development in other organisms.[99] Some of the non-coding transcripts were found to be huge, tens and sometimes hundreds of kilobases in length ('macroRNAs'),[100–102] a well-characterized example being Air (108 kb) from the imprinted *Igf2r* locus (Chapter 9).[100,103,104] A more recent pan-transcriptome analysis reported the discovery of thousands of novel RNAs, including a previously poorly cataloged class of non-polyadenylated single-exon lncRNAs.[105]

Widespread antisense transcription is also observed in bacteria[106] and viruses.[107,108] Indeed, global transcriptome analyses showed that the vast majority of all nucleotides in all genomes from viruses to humans are transcribed from one or both strands at some point in their life cycle.[109]

THE AMAZING COMPLEXITY OF THE TRANSCRIPTOME

Sequencing of expressed RNAs and RACE[k]-tiling arrays also revealed that most transcripts in mammals and insects have alternative transcription start and termination sites, the former often initiating hundreds of kilobases upstream of the previously annotated gene starting point(s) and spanning other genes in between.[57,91,110,111] Approximately 250,000 transcriptional start sites in mammals reside within transposon or retroviral[l] derived sequences, which account for up to 30% of the transcribed loci, produce 5' capped RNAs that are generally tissue-specific, and frequently function as alternative promoters of protein-coding genes and/or express non-coding RNAs.[113]

Extraordinary complexity of tissue- and lineage-specific alternative splicing was also observed,[3,114–118] particularly in lncRNAs.[4,119] LncRNAs appear to be less efficiently spliced than mRNAs, a feature that is correlated with heightened alternative over constitutive splicing,[120–122] possibly related to chromatin retention.[123] LncRNAs are enriched in transposon-derived elements and other repeat sequences,[99,124–127] which is likely related to their functional modularity (Chapter 16). Thousands of 'pseudogenes' are also transcribed,[128–131] as are developmental 'enhancers', whose numbers far outweigh those of protein-coding genes[132–142] (Chapter 14).

Transcriptome analyses progressively revealed the existence (and, in most cases, drastically expanded the repertoire) of other types of transcripts[143] various 'classes' of promoter-associated RNAs,[144–148] 3'UTRs (Chapter 9) and regulatory RNAs originating from intergenic spacers between rRNA genes (in which promoters and transcripts had been known for decades – see references in[149–151]), as well as other classes such as circular lncRNAs (circRNAs)[152–154] (see below) and intron-derived lncRNAs with "snoRNA-ends" (sno-lncRNAs).[155] Intron retention was also found to be common in plants and animals, where it is used to control cell differentiation.[46,59,156–163] Surprisingly, even RNAs modified with N-glycans and a range of other RNAs are displayed on cell surfaces, apparently cell type specifically.[164,165]

The picture that emerged is that eukaryotic genomes express a semi-continuum of interlacing and overlapping coding and non-coding transcripts from both DNA strands,[m] especially in animals[14,45,57,171–175] (Figure 13.1). There are genes encoding proteins, snoRNAs, miRNAs and other small non-coding RNAs located within other genes encoding proteins and lncRNAs, with unclear boundaries, often in nested chains where three or more transcripts overlap in "complex loci", and the landscape of expressed transcripts, as well as their promoters, splicing patterns and termination points, are different in different cells and tissues.[45] Indeed, the true extent of the repertoire of RNA expression is still unknown, given that most analyses to date have been carried out in cultured cells and do not capture the fine scale transcriptomes of diverse cells during the ontogeny of

---

[j] LncRNAs are defined as non-protein-coding RNAs >200 nt, an arbitrary classification partly based on a size cutoff in biochemical/biophysical commercial RNA purification kits and protocols that exclude most infrastructural RNAs, such as tRNAs, snoRNAs and snRNAs, as well as miRNAs, siRNAs and piRNAs.[97]

[k] RACE: Rapid amplification of cDNA ends.

[l] Later studies showed that the number of long noncoding RNAs expressed from endogenous retroviral promoters correlates with pluripotency or the degree of malignant transformation.[112]

[m] A similar albeit less complex genomic organization pertains in prokaryotes, where hundreds of transcriptional start sites are located within operons, as well as opposite to annotated genes, indicating that the complexity of gene expression is increased by uncoupling polycistronic linkages and the genome-wide use of antisense transcription.[166] Apart from the thousands of short regulatory RNAs (Chapter 9) there are other, often still mysterious, longer (>200nt) highly structured and conserved noncoding RNAs that have been discovered in bacteria.[167–170]

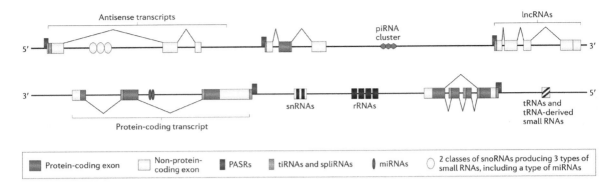

FIGURE 13.1 Graphical representation of the complexity of the transcriptional landscape in mammals. (Reproduced from Morris and Mattick.[176])

differentiation and development, although this is rapidly changing with the ubiquity of RNA sequencing, including increasingly powerful single-cell sequencing analyses of a variety of organisms, tissues and conditions (see below and following chapters).

Collectively, these observations were revolutionary in their implications. They challenged both the equivalence of genes with proteins[n] and the notion of 'genes' as discrete entities.[72,174,178] They also suggested the opposite of what had been long thought, i.e., that the genomes of humans and other complex organisms are information dense, not information sparse[179] (Chapter 7). The genome could no longer be envisaged as a linear array of protein-coding genes and associated cis-regulatory sequences, with some infrastructural and idiosyncratic non-coding RNAs. Rather, genome biology had to be reimagined as a highly dynamic continuum of coding and non-coding transcription,[174,178,180] the latter becoming more extensive as developmental complexity increases.[109] Moreover, almost every gene is overlapped by an exon or intron of another gene expressed in some cell type, and any given sequence can be intronic, exonic or intergenic, depending on the expression state of the cells.

PROTEIN-CODING OR NOISE?

Unsurprisingly, these findings were initially met with skepticism. They indicated a massive hidden layer of RNAs of unknown function that had not been countenanced by the existing models of gene

regulation, although the precedents had been there for decades, especially in the homeotic loci controlling organism development (Chapters 5 and 9). The protein-centric conception of genetic information and gene regulation could accommodate a few idiosyncratic regulatory RNAs, and post-transcriptional control of gene expression by miRNAs, but not tens of thousands of non-coding RNAs. No wonder there were reservations.

One difficulty was discriminating coding from non-coding transcripts,[181] leading to the suggestion that some or many of the newly identified transcripts might contain short open reading frames that had fallen below the radar of the genome annotations, which had generally used a minimum open reading frame of 100 codons.[o] Subsequent studies using ORF conservation, proteomic analyses and ribosomal profiling showed that, while there may be hundreds of unrecognized short proteins, including hormones and small peptides encoded in some RNAs annotated as non-coding,[182–197] the vast majority of lncRNAs exhibit no evidence of protein-coding capacity.[185,186,198]

The dichotomy may itself be false: at least some RNAs have dual function as both coding and regulatory RNAs[133,181,189,194,199–210] and mRNAs appear to play a role in cellular organization.[211,212]

---

[n] This created problems for genome annotations, which had been traditionally organized around protein-coding genes, although they are still used as landmarks.[177]

---

[o] The initial annotation of the human genome generally used the presence of a conserved open reading frame of 100 codons in RNAs (exonic fragmentation of protein-coding sequences made this assessment impossible at genome sequence level) as an arbitrary cutoff on the basis that this was unlikely to occur by chance. The sequencing of large numbers of vertebrate genomes will allow more accurate assessment of conserved open reading frames, ultimately with near statistical certainty at the codon level.

Some protein-coding loci also generate miRNAs or lncRNAs[p] by alternative splicing,[209,213–221] and there is "a non-negligible fraction of protein-coding genes (where) the major transcript does not code a protein".[222] Some regulatory lncRNAs have evolved from protein-coding ancestors and pseudogenes.[128,131,223–225] Some genes that express lncRNAs with enhancer activity also produce miRNAs[226,227] and micropeptides,[189,190,208,228,229] further examples of parallel outputs from complex loci. Moreover, lncRNAs have many alternative splice isoforms[4] (Chapter 16), some of which have been shown to have different functions.[230–235]

Skepticism of the significance of non-coding transcription took many forms, including speculation that the unannotated RNAs are technical artifacts, genomic DNA contamination or have dispensable functions.[179,236,237] The most common reaction to these findings, however, was to assert that the bulk of the observed transcription, although dynamic and often cell- or tissue-specific, is 'noise': most non-coding transcripts were detected at low levels, many are or appeared to be comprised of just one exon, and many seemingly 'random' fragments abounded in initial RNA sequencing datasets, with differential expression sometimes attributed to variations in chromatin accessibility.[238–241]

The concept of transcriptional noise was introduced in the 1990s by the observation of the cellular heterogeneity and stochastic fluctuations in the firing of known promoters in bacteria and yeast, not spurious transcription from illegitimate initiation sites.[242–246] Nonetheless it was seized upon, and conflated with 'neutral evolution',[247] leading to debates about the functionality or otherwise of the plethora of non-coding transcripts in eukaryotic cells,[239,248–251] reprising previous discussions about the functionality of introns, transposon-derived sequences and pseudogenes.

THE RESTRICTED EXPRESSION OF LONG NON-CODING RNAs

It transpired that the low-level and fragmentary signals from lncRNAs in sequencing datasets is mainly a consequence of their highly developmental stage-specific expression, exacerbated by insufficient sequencing depth,[q] especially in complex tissues.[22,248,252,253]

The expectation that high expression levels reflect functionality is based on the prevalence of protein-coding RNAs, which are, on average, more highly expressed than regulatory RNAs, although there are exceptions.[254,255] Indeed, mRNAs encoding regulatory proteins such as transcription factors are usually expressed at lower levels than those encoding structural or metabolic proteins, and have shorter half-lives.[256,257]

Regulatory RNAs would likewise require relatively low average expression levels, i.e., more localized expression, and more dynamic control.[254,258] Examples, among many others, include functionally validated chromatin-associated lncRNAs detected on average in less than ten copies per cell in populations.[259–265] Single-molecule RNA FISH revealed that the localized TERT (telomerase reverse transcriptase) pre-mRNA occurs in 9–10 copies per cell and is only spliced during mitosis.[163,266] Similar low expression levels are also observed for a number of functionally well validated regulatory RNAs,[179] including XIST (Chapter 16).

Many smaller "cryptic unstable transcripts" expressed antisense from promoters and from intra- and intergenic regions are rapidly degraded by RNA turnover and surveillance[r] pathways,[47,286–288] which were thought to be a quality control mechanism to limit "inappropriate expression",[286] but were

---

[p] For example, the *PNUTS* gene encodes both PNUTS mRNA and lncRNA-PNUTS by alternative splicing of the primary transcript, each eliciting distinct biological functions; PNUTS mRNA is ubiquitously expressed, whereas the production of lncRNA-PNUTS is tightly regulated.[213]

[q] This problem also confounds the attempts to construct 'gene networks' from transcriptomic data, especially when different cells in a population are expressing different genes and responding differently to stimuli, and most data is derived from the 3' end (UTRs) of transcripts, which may or may not be part of a corresponding protein-coding mRNA.[75]

[r] Via RNA-degrading 'exosomes'[267–269] (not to be confused with the extracellular vesicles that have the same name) and 'nonsense-mediated RNA decay' (NMD). NMD is known as a quality control mechanism to ensure only mRNAs with complete open reading frames are exported for translation, by degrading "aberrant RNAs" that have "exon junction complexes" 3' to stop codons (because the stop codon of protein-coding genes is located primarily in the last exon).[270–276] However, recent evidence suggests that it also distinguishes short-lived regulatory RNAs from mRNAs and controls their steady-state levels in different contexts,[277–280] including stress responses.[281] The NMD pathway is developmentally regulated,[279,282] and required for embryonic stem cell fate determination[283,284] and neuronal architecture.[285] Loss of NMD components leads to developmental abnormalities and neurological disorders.[276,285]

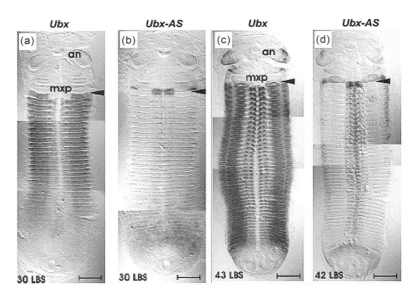

FIGURE 13.2 Reflective *in situ* hybridization patterns of the expression of the protein-coding gene *ultrabithorax* (*Ubx*) and its overlapping antisense RNA in embryos of the centipede *Strigamia maritima* at mid- and late-segmentation stages. (Reproduced from Brena et al.[319] with permission of John Wiley and Sons.) The arrow marks the anterior margin of the first leg bearing segment (LBS).

subsequently found to regulate promoter and enhancer function (Chapter 16).[289–297]

Similarly, other transcripts associated with promoters were detected only after depletion of components of RNA-degrading 'exosomes', including 'promoter-associated RNAs' (PASRs) of ~250–500nt identified in yeast and human cells,[111,144] 'promoter upstream transcripts' (PROMPTs) of ~0.5–2.5 kb, identified in human cells in both sense and antisense orientations upstream of the transcription start sites of expressed genes.[145] Exosomes have also been shown to control the levels of "a vast number" of lncRNAs with enhancer activity in B cells and pluripotent embryonic stem cells.[298]

Analysis of *in situ* hybridization patterns of over 1,000 lncRNAs showed that a high proportion exhibit precise expression patterns in the brain/central nervous system, easily detected in highly specific and localized populations of cells in the striatum, retina, hippocampus, cerebellum, olfactory bulb and cortical layers, among others.[299–302] Complex region- or cell-specific and developmentally transient expression patterns of 'intergenic' and antisense lncRNAs have also been observed in fish,[303] *Drosophila*,[304,305] honey bees[306–308] and other multicellular and unicellular eukaryotes[99] (Figure 13.2). It has also been observed in globin[133,309–311] and interleukin loci,[312,313] now being extended to others

by detailed examination of more tissues and developmental time points, aided by the advent of single-cell sequencing.[314–318]

To get around the problem of low sequencing depth, John Rinn, Tim Mercer and colleagues developed a method called 'RNA CaptureSeq',[s] akin to exome sequencing, to enrich transcripts expressed from specific genomic locations.[252,253,326] This approach showed that, even in a relatively homogenous population of cultured fibroblasts, regions that appeared devoid of transcripts – sometimes referred to as 'gene deserts' – expressed lncRNAs in a subset of the cells.[252] It detected previously unknown isoforms of intensively studied protein-coding genes, such as *TP53*, and lncRNAs expressed from homeotic and other developmental loci.[252,253,326–328] It also revealed that most GWAS regions, including those associated with neuropsychiatric functions, are transcribed into lncRNAs.[329–331] Other studies showed that most intergenic lncRNAs originate from enhancers and are specifically expressed in cell types relevant to the associated GWAS trait[332,333] (see below and Chapters 14 and 16).

RNA CaptureSeq also revealed that many of the existing annotations of lncRNA (and some mRNA)

---

[s] Along with sophisticated internal standards to properly measure the sensitivity of DNA and RNA sequencing analysis.[320–325]

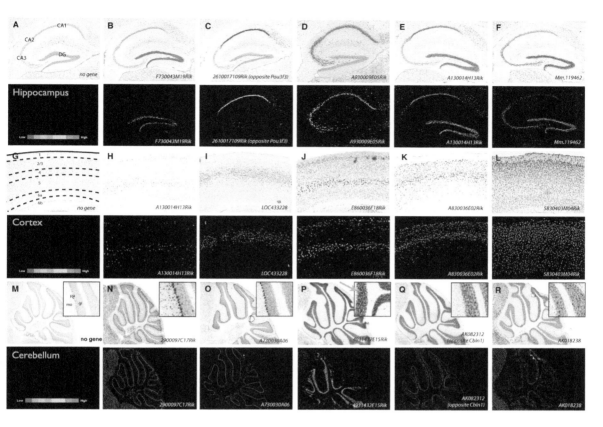

FIGURE 13.3 *In situ* hybridization of lncRNAs in mouse brain. (Original images from the Allen Brain Atlas,[338,339] reproduced from Mercer et al.[300])

structures were incomplete,[t] that many lncRNAs are multi-exonic, and that the internal exons of lncRNAs (but not mRNAs) are almost universally alternatively spliced.[4] These high resolution and recent single-cell RNA sequencing studies also indicate that the number of lncRNAs and lncRNA isoforms that are expressed is far greater than cataloged in current databases.[105,334]

Thus, lncRNAs generally show more tissue-restricted and transient expression patterns than mRNAs,[46,185,254,300,318,335–337] helping to explain their low representation in RNA sequencing datasets (Figure 13.3). This in turn suggests that lncRNAs are more specific markers and regulators of cell state, including disease state, than proteins with generic functions in, e.g., muscle, bone or neuronal cells.

OTHER INDICES OF FUNCTIONALITY

LncRNAs are dynamically expressed during all aspects of animal differentiation and development[u] along developmental axes,[89,132,260,303] in embryonic stem cells,[58,124,341,342] neuronal cells,[258,343–345] muscle cells,[346] mammary gland,[347] hematopoietic and immune cells,[37,348,349] among many others.[99,350] They are also differentially expressed in neurological responses, for example, in the songbird zebra-finch where "40% of transcripts in the unstimulated auditory forebrain are non-coding and derive from intronic or intergenic loci... Among the RNAs that are rapidly suppressed in response to new vocal signals ... two-thirds are ncRNAs".[351] LncRNAs show altered expression in cancer and other diseases[352] (see below) and are also dynamically expressed during plant and fungal development.[51,353–355] LncRNAs are also trafficked to specific subcellular locations

---

[t] These studies also showed that there are far more regulatory 5′ exons in human than in mouse mRNAs, which are also highly alternatively spliced, suggesting that humans have evolved a more complex cis-regulatory architecture of mRNAs,[4] possibly related to brain function.

[u] Including in sponge.[340]

in the nucleus and cytoplasm, and specific domains within them (Chapter 16).[28,300,303,307,336,356–372]

The half-lives of lncRNAs are broadly similar to those of mRNAs, over an equally wide range, many being highly stable.[257,373,374] In fact, the loci expressing lncRNAs exhibit most of the characteristics of *bona fide* genes:[375] their expression is regulated by conventional hormones, morphogens and transcription factors;[344,376–382] many have polyadenylation sites;[383] they show non-neutral mutational patterns;[384] their promoters and exons have chromatin marks similar to those of protein-coding genes[385] (Chapter 14); and their splice junctions and structures are conserved,[4,14,384–386] allowing the identification of orthologs in other species.[387–389]

Surprisingly, the promoters of lncRNAs are, on average, more conserved than those of protein-coding genes,[14,384,390] suggesting higher cell specificity, consistent with their restricted expression and the conserved expression patterns of syntenic RNAs.[391–393] Although the proportion of lncRNAs with primary sequence similarity is low among vertebrates,[384,389] thousands of RNAs in mammals, *Drosophila*, plants and yeast have conserved secondary structures and sequence motifs, and a minimum of 20% of the mammalian genome has been shown to be under evolutionary selection at the level of predicted RNA structure.[386,394–401]

Lack of conservation does not mean lack of function.[402] There are well-described examples of lncRNAs (including *Drosophila* roX RNAs) that evolve rapidly while maintaining functional interactions.[402–404] Xist shows only patchy primary sequence conservation among mammals, notably in its 'repeat' sequences, and its adjacent lncRNA, Jpx, which activates Xist, while sharing no obvious sequence or structural homology between human and mouse, is functionally interchangeable between them.[404,405].

Many other well-studied lncRNAs involved in developmental processes, including Air,[406] DISC2,[407] NTT,[408] BORG[409] and UM 9(5), show only short stretches of conserved sequences.[410] In addition, many lncRNAs have conserved functions in vertebrate development despite rapid sequence divergence[411] and orthologous lncRNAs that are developmentally regulated in different species have been identified solely on the basis of the conservation of splice sites and associated introns.[387] This indicates that lncRNAs have greater orthology than is evident from conventional sequence comparisons,

reflecting positive selection for phenotypic variation[412] and more plastic structure-function relationships[413,414] than protein-coding sequences.[91,392]

Furthermore, bearing in mind the difficulty of identifying orthologs that are evolving to alter the fine control of developmental processes, many lncRNAs are clade-specific and, consequently, largely unstudied. One example is the lncRNA Sphinx, which regulates courtship behavior in *Drosophila* and is expressed from a chimeric gene that arose by the retrotransposition of a sequence from an ATP synthase gene and capture of an adjacent exon and intron, fixed by positive selection.[415] There are many other examples of species-specific lncRNAs, with evidence of recent birth and selective sweeps, controlling cell differentiation (e.g., [416]) and brain functions (see below and Chapter 17).

GENETIC SIGNATURES

Another reservation was that few lncRNAs had, at the time, been identified in genetic screens, which intrinsically favored protein-coding mutations.[179,375]

Protein-coding mutations are frequently disastrous, including those affecting enzymes, motor proteins, transporters, signaling proteins, etc., as well as transcription factors, epigenetic modifiers and other regulatory proteins, which cause system-wide malfunctions. The same holds for some highly expressed non-coding RNAs with generic functions,[417] such as RMRP (the RNA component of RNase MRP, Chapter 8), mutations in which cause a pleiotropic human disease, cartilage-hair hypoplasia, first identified by linkage analysis[418,419] and later shown also to produce miRNAs,[420] to perturb helper T-cell epigenetic regulation[421] and to inactivate the tumor suppressor P53.[422]

A major blind spot was phenotypic bias: the severe and pleiotropic effects of damage to proteins or 'housekeeping' RNAs contrast with damage to regulatory sequences, which may only affect a part of the networks that control differentiation and development or environmental responses, with more subtle context- and/or cell type-specific consequences, often referred to as quantitative trait variation. Indeed, the use of the word 'mutation', as opposed to 'variation', reflects an inherent bias in the identification of genetic factors that affect phenotypes in animals and plants, with those exhibiting strong negative effects (being easier to identify and map) understandably having taken precedence over those that do not (Chapters 7 and 11).

The related blind spots were expectational, technical and interpretative bias: historically, most genetic screens used experimental and informatic approaches that prioritized protein-coding genes and exons. Many now known important mutations in lncRNAs were consequently missed by exome sequencing and chromosomal microarray analyses.[423] Mutations that could not be tracked to and shown to introduce stop codons or different amino acids in a protein-coding sequence were rarely pursued, given the large number of variations in non-coding sequences, which were mostly invisible and untraceable before the availability of genome and transcriptome sequences.[v] Even those that were confidently mapped outside of protein-coding sequences were routinely interpreted as affecting cis-acting protein-binding sites that regulate nearby coding genes. Put simply, it was assumed that most disease-causing mutations occur in protein-coding sequences, where it was easy to identify them, or in cis-regulatory sequences that bind regulatory proteins, with scant knowledge of regulatory RNAs that might be expressed from the locus, some of which are now being identified.[333,425–432]

In *Drosophila*, where careful genetic analysis identified many enhancers and other regulatory regions affecting development,[w] the same interpretative bias occurred, despite the abundant evidence of differential expression of lncRNAs from these regions (Chapter 5), although there were exceptions, such as the roX RNAs identified by careful genetic and expression mapping (Chapter 9).[433,434]

There were other exceptions, especially in farm animals, where controlled breeding permitted accurate dissection of the genetic causes of quantitative trait variation. The mutation underpinning the 'callipyge' polar overdominance phenotype in sheep (Chapter 5) was mapped to a non-coding RNA expressed from the complex *Dlk1-Dio3* imprinted region,[435–438] as were others affecting quantitative trait variation.[439] Similar strong effects are observed for mutations in other lncRNAs (see below) and non-coding regulatory regions, as exemplified by the *Crest* mutation in chickens, which causes a spectacular phenotype in which the small feathers normally present on the head are replaced by much larger feathers normally present in dorsal skin. It was shown to be caused by a 197 bp duplication of an evolutionarily conserved sequence in the intron of *HoxC10*, which causes the ectopic expression of *HoxC10* and other *Hox* genes, altering cell regional identity (Figure 13.4).[440]

Indeed, many lncRNAs that are now known to be important went undetected or were overlooked in genetic screens. These included nearly all miRNAs,[x] many of which are not individually essential for viability or development;[442] a conserved lncRNA ('yar') that lies within *Drosophila Achaete-Scute* complex locus studied by Muller and others many decades ago, which was recently discovered to regulate sleep behavior;[443] and 3'UTR of the *oskar* gene that was unexpectedly found to function as a lncRNA controlling *Drosophila* oogenesis (Chapter 9).[444,445]

Moreover, other lncRNAs initially thought to be non-essential, many of which display little 'conservation' and are lineage-restricted (including human-specific RNAs), have been implicated in disease,[446–449] with more coming to light with the growing awareness of their relevance to complex traits.[423,450] These RNAs have been identified by chromosome breakpoints, fusions and translocations, deletions, copy number variations, point mutations, insertions and deletions, aberrant imprinting, other epigenetic defects and haploinsufficiency, among others.[450]

Examples, some exhibiting Mendelian inheritance, but most involved in complex disorders, include Di George Syndrome (a range of symptoms including congenital heart problems, unusual facial features, frequent infections, developmental delay, learning problems and cleft palate);[452] other neurodevelopmental and

---

[v] There are few promoters in the catalogs of mutations associated with genetic disorders,[424] although no one disputes their functionality.

[w] An important subtlety, and difference between genetic screens in *Drosophila* and mammals is that many if not most naturally occurring mutations and those experimentally induced in the latter (by the mutagen ENS) are single nucleotide mutations, which can have serious consequences on a protein-coding sequence, but often subtle consequences on regulatory sequences. By contrast, most experimental mutagenesis in *Drosophila* involved transposable element insertion or large deletions, which have more serious phenotypic effects on both coding and regulatory sequences, and hence it is no surprise that so many regulatory loci were unearthed in *bithorax* and other intensively studied gene regions in *Drosophila*.

[x] It has been proposed that miRNAs operate in a hierarchical and canalized series of regulatory networks (see Chapter 15), a fraction of miRNAs acting at the top of this hierarchy, with their loss resulting in broad developmental defects, whereas most miRNAs are expressed with high cellular specificity and play roles at the periphery of development, affecting the terminal features of specialized cells.[441] It is likely that the same applies to lncRNAs.

FIGURE 13.4 The 'crest' phenotype resulting from a 197bp duplication in the intron of *HoxC10*, which alters cell identity. (Reproduced from Li et al.,[440] under Creative Commons 4.0 license.)

craniofacial disorders;[453–455] developmental defects, e.g., involving the lncRNA Chaserr, which regulates the expression of a chromatin remodeler implicated in neurological disease (Figure 13.5);[451] limb malformations, brachydactyly and other skeletal abnormalities (Figure 13.6);[430,456–458] Angelman[y] and Prader-Willi Syndromes;[460–463] schizophrenia;[407,464–466] Kallmann Syndrome (a subtype of gonadotropin-releasing hormone deficiency with a loss of smell),[467] pseudo-hypoparathyroidism;[468] alcohol use disorders;[469] non-alcoholic fatty liver disease;[470] diabetes;[471] multiple sclerosis;[472] autoimmune thyroid disease;[473] Sjögren syndrome;[474] celiac disease;[475] Hypereosinophilic Syndrome;[476] Kawasaki Disease;[261] psoriasis;[477,478] inflammatory bowel disease;[479] atherosclerosis;[231,429,480,481] cardiac hypertrophy,[482] Alzheimer's

Disease;[483] ataxias;[484,485] myocardial infarction;[486] and some types of thalassemia.[131,487,488]

It has also been shown that phenylketonuria, one of the first documented human genetic disorders, which is mostly due to mutations in the enzyme phenylalanine hydroxylase, can also be caused by perturbations in a regulatory lncRNA and be modulated by administration of modified RNA mimics in mouse models.[489]

As noted already, genomic regions associated with a wide variety of complex disorders and characteristics, including psychiatric traits and disorders and neurodegenerative diseases, are replete with lncRNAs,[329,330,332,333] which are therefore candidates for the mechanistic basis of the association. Sensibly, studies have started to focus not only on non-coding regions, but also on mutations/variations that affect RNA structure.[490]

Many lncRNAs have been also associated with the etiology, progression, genomic instability and therapy resistance of cancers,[390,491–496] through altered

---

[y] Therapies are being developed for Angelman's Syndrome by knocking down the regulatory *Ube3a-ATS* non-coding RNA with antisense oligonucleotides to restore expression of the normally silent (imprinted) paternal *Ube3a* allele in patients lacking the maternal allele.[459]

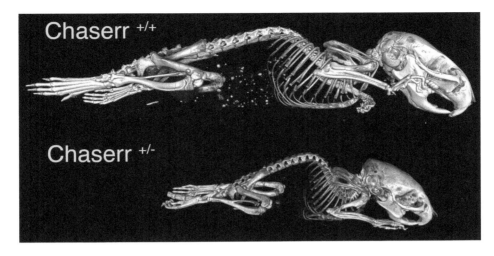

FIGURE 13.5 The severe phenotype resulting from haploinsufficiency of the lncRNA Chaserr. (Reproduced from Rom et al.[451] under Creative Commons CC BY license.)

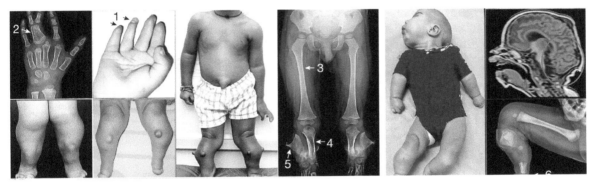

FIGURE 13.6 Skeletal malformations due to the loss of lncRNA Maenli. (Reproduced from Allou et al.[430] with permission from Springer Nature.)

expression, insertional mutagenesis and/or naturally occurring mutations, and functional validation of previously known and novel RNAs that act as oncogenes[z] or tumor suppressors, many of which are enriched in repetitive elements.[390] In some cases (such as H19, PVT1, MIAT/Gomafu, OIP5-AS1/Cyrano, TUG1,[aa] HOTAIR, MEG3, XIST, TSIX, MALAT1 and NEAT1), perturbations in lncRNAs are associated with multiple cancers;[235,390,494,499–519] and in other cases with particular types of cancers, including leukemias and lymphomas,[135,520–525] melanoma,[526,527]

osteosarcoma,[528] gastric,[529] lung,[530] breast,[531,532] prostate,[13,533–535] bladder[536] and many others, including bone metastasis,[537] with increasing understanding of the mechanisms involved.[505,537–539]

A major problem is that many if not most mutations in lncRNAs are cell type- and context-dependent,[540] and not evident in fish tanks or mouse cages[296] unless subjected to specific challenges or behavioral assays. Indeed, in *Drosophila* and *C. elegans*, both intensively studied, less than a third of protein-coding genes have obvious phenotypes when mutated,[541,542] and many apparently disruptive mutations in human genes are common in the population, indicating more subtle interactions.[543,544] Targeted knockout of the rodent-specific and highly expressed brain lncRNA, BC1 (Chapter 8), yielded no obvious developmental

---

[z] The lncRNA Cherub is required for the transformation of stem cells into malignant cells.[497]

[aa] TUG1 is required for mouse retinal differentiation[498] and male fertility.[194]

phenotype, but causes behavioral changes, and impaired experience-dependent plasticity and learning in mice.[545,546] Deletion of the highly expressed and relatively highly conserved lncRNAs, Neat1 and Malat1, similar to that observed with ultraconserved elements (Chapter 10), did not result in dramatic developmental deficiencies,[547,548] but later analyses showed changes in behavior and placental biology, as well as involvement in synapse formation, myogenesis, cancers and responses to pathogen infections.[549–553] The loss of another lncRNA, FosDT, which is highly expressed in the cerebral cortex and interacts with chromatin-modifying proteins associated with the neuronal transcription factor REST, causes no developmental anomalies but reduces brain damage from strokes.[554,555] The lncRNA Pnky regulates neuronal differentiation and its deletion affects postnatal cortical development, although the mice do not exhibit superficial defects.[556,557]

On the other hand, deletion of other lncRNAs in mice by Rinn and colleagues, with visual markers that revealed exquisite expression patterns, resulted in a range of more obvious phenotypes including homeotic transformations, skeletal and neuronal abnormalities, heart and gastrointestinal defects, muscle wasting, abnormal lung morphology and aging.[558,559] Many more have since been reported using various *in vivo* approaches.[540]

High-throughput siRNA and CRISPR reverse genetic screens combined with molecular phenotyping[560] are now increasing the search speed, identifying, for example, lncRNAs that are involved in chromatin interactions,[561] required for heart development,[562] regulate nuclear factor trafficking,[563] activate resistance to BRAF inhibitors in melanoma,[564] respond to Wnt signaling,[565] sensitize glioma cells to radiation and essential to cancer cell viability,[566] involved in cell growth and migration,[560,567] spermatogenesis,[568] lung cancer[569] or have various fitness effects.[570] A recent large-scale study using CRISPR mutagenesis of over 16,000 lncRNAs in seven cell lines identified almost 500 required for normal cellular proliferation, 89% of which were expressed in only one cell type.[571]

A systematic high-throughput loss-of-function analysis of 248 protein-coding genes and 141 lncRNAs using the fission yeast *S. pombe*, assessing mutant growth and viability in "benign" and 145 variable conditions, showed that phenotypes are found much more frequently for the former compared to the latter (47.5% for the lncRNAs and 96%

for the protein-coding genes). However, on more careful inspection (also evaluating the effect on cell-size and/or cell-cycle control), 59.6% of lncRNAs yielded phenotypes and, upon overexpression of the lncRNAs under 47 different conditions, 90.3% led to altered growth under certain conditions. These results reinforce the notion that most of the lncRNAs exert cellular functions in specific environmental or physiological contexts.[572]

Intriguingly, it appears that some gene deletions may be masked by compensatory mechanisms, whereas acutely disturbing transcript levels may have more severe effects.[573] There is also evidence that 'shadow' enhancers (which express and likely operate through lncRNAs, Chapter 16) provide redundancy and robustness to developmental programs (Chapter 15),[574–578] which makes sense given the criticality of the process for survival and reproductive success. On the other hand, knockdown of lncRNA expression in culture often has visible effects, in terms of changes in cell shape, behavior and gene expression profiles,[375,560] which may have stronger manifestations in artificial *in vivo* settings such as xenograft models.[565]

AN AVALANCHE OF LONG NON-CODING RNAs

With the popularization of unbiased transcriptomic studies, growing numbers of non-coding RNAs have been identified and studied in model organisms, human cell lines and disease systems, mainly *ad hoc* by the differential expression of intronic, intergenic, pseudogene-derived and antisense lncRNAs, but also increasingly by functional screens. The ENCODE project alone found over 850 pseudogenes that are "transcribed and associated with active chromatin"[140] with many since been shown to function as regulatory RNAs.[130,522,579–584]

In addition to viroids that have small circular genomes (Chapter 8), circular RNAs (circRNAs) occur in plant and animal cells, one of the first discovered being a variant of the *Sry* mammalian male sex-determining gene transcript,[585] as ever considered an interesting oddity at the time. CircRNAs remained under the radar because traditional cloning and sequencing protocols and informatic methods mitigated against their detection. They are predominantly produced by back-splicing facilitated by reverse-complementary (often recently acquired transposon-derived sequences) that promote pre-mRNA

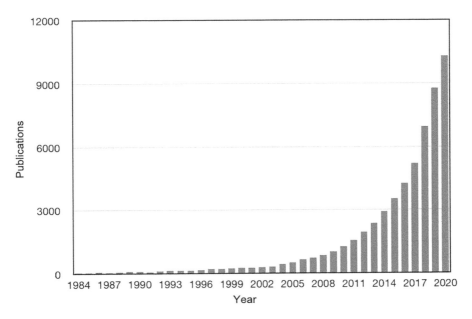

FIGURE 13.7   The increase in the number of publications that have the terms 'long/large non(-)coding RNA' or variations thereof (lncRNA or lincRNA) in their PubMed entry.

folding.[586] Since their rediscovery, circRNAs have become recognized as a *bona fide* class of functional RNAs, in many cases comprising the dominant transcript isoform.[152,154,587,588]

CircRNAs are predominantly nuclear[ab] and act through several mechanisms, having been shown to regulate, *inter alia*, transcription, immune responses, behavior, neural cell function and pluripotency,[589–594] *Drosophila* lifespan[595] and centromeric chromatin organization in maize.[596] Many circRNAs have regulatory interactions with the cognate protein-coding gene, as illustrated by the neuronal-enriched psychiatric-disease associated circHomer1 RNA and its host gene *Homer1*. Based on *in vitro* and *in vivo* studies with mouse models, they were found to be functionally antagonistic in synapses of the orbitofrontal cortex, with this opposing interplay regulating synaptic gene expression, cognitive flexibility and behavioral performance, with potential relevance for brain function and psychiatric diseases.[594]

Over the past decade, there have been ~50,000 publications with long non-coding RNA as a key term (Figure 13.7) and over 2,000 publications reporting validated long non-coding RNA functions.[597] These

studies have been assisted by the systematic cataloging and annotation of lncRNAs by the GENCODE consortium,[ac] The Cancer Genome Atlas (TCGA)[516] and the extension of the FANTOM projects to transcriptomic atlases, associated Web resources and functional annotation of lncRNAs.[34]

There are now hundreds of thousands of cataloged lncRNAs and dozens of databases (and databases of databases) with curated information.[600] Well over 100,000 human lncRNAs have been recorded,[601] many of which are specific to the primate lineage,[389,602] including retrovirus-derived lncRNAs,[603] a vastly incomplete catalog due to the still limited analysis of different cells at different developmental stages and physiological conditions.

A PLETHORA OF FUNCTIONS

LncRNAs have been shown to regulate many aspects of mammalian development, cell differentiation, (Figure 13.8) physiology and brain function[605,606] (see Table and Chapter 17), as well as many other roles in other organisms,[354,607–610] including the translation of

---

[ab] Containing exons,[589] or a mixture of intronic and exonic sequences.[590] Some are derived entirely from introns, and have been shown to regulate their parent protein-coding genes.[591]

[ac] Initially formed as part of the pilot phase of the ENCODE project[598] but expanded to annotate "human and mouse genes and transcripts supported by experimental data with high accuracy, providing a foundational resource that supports genome biology and clinical genomics".[177,336,599]

Examples of Functions of lncRNAs in Mammalian Biology

Stem cell pluripotency, self-renewal, lineage commitment and reprogramming[262,315,617–629]

Cell cycle,[635] proliferation and migration[560,567,636–640]

Brain evolution,[557,602,642] neocortex,[643] forebrain,[644,645] and retinal[498,646,647] development

Synaptic plasticity and function[655–660]

Mammary,[347,664] sclerotome,[665] heart,[562,666–669] lung,[670] skeletal,[457,458] limb[430] and intestinal[671] development

Liver regeneration[691,692]

Angiogenesis[694,695] and fibrogenesis[696]

Hematopoiesis,[699] granulocyte,[349] megakaryocyte,[416] T-cell[471,472,700] and keratinocyte[701–703] differentiation

Innate and adaptive immune responses[708–715]

Microbial susceptibility, endotoxic shock and immunity[474,717–721]

Growth hormone and prolactin production[729]

Testis development and spermatogenesis[194,568,731–733]

Thermogenic adipocyte regulation[736]

Epithelial-mesenchymal transition[213,536,630,631]

mesoderm and endoderm differentiation[632–634]

Cellular senescence[234] and apoptosis[370,476,641]

Maintenance of neural progenitor cells,[593] neuronal differentiation,[556,607,648–650] outgrowth and regeneration,[637,651,652] axon integrity,[653] myelination[654]

Memory,[661,662] sex-specific depression[663] and social hierarchy[658]

Muscle differentiation and function,[209,672–684] myogenesis and muscle fiber type switching[631,685–690]

Cholesterol biosynthesis and homeostasis[470,693]

Formation of vascular endothelial cell junctions[697,698]

Erythropoiesis and developmental regulation of globin gene expression[311,704–707]

Inhibition of viral replication[716]

V(D)J and Ig class switch recombination[722,723]

inflammation and neuropathic pain[428,475,479,724–728]

Glucocorticoid resistance[730]

DNA damage repair[517,734,735]

Mitochondrial function[737]

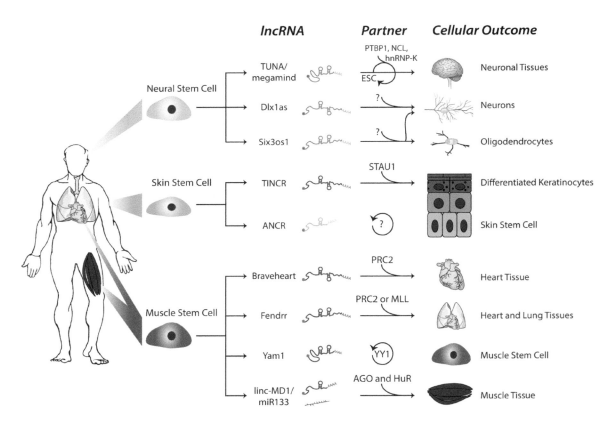

FIGURE 13.8 Control of cell differentiation and self-renewal by lncRNAs. (Reproduced from Flynn and Chang[607] with permission of Elsevier.)

Doublesex in *Drosophila*,[175] female honeybee development,[308] plant vernalization (see below), strawberry fruit ripening,[611] DNA elimination and genome rearrangements in ciliate life cycle and reproduction;[612] the mitotic to meiotic switch[613] and meiotic chromosomal pairing in yeast,[614,615] and carotenoid biosynthesis in filamentous fungi.[616]

Mechanistic details have started to emerge, some serendipitously. As with the discovery of the 7SL RNA in signal recognition particles in the early 1980s (Chapter 8), biochemical assays used to identify protein interactions detected other regulatory RNAs, such as the identification by yeast two- and three-hybrid screens of mammalian *SRA* RNA as a transcriptional coactivator, *Gas5* RNA as a repressor of glucocorticoid receptor activity, and the plant *ENOD40* RNA as a regulator of the localization of an RNA-binding protein in cytoplasmic granules.[738–741] Other lncRNAs have been found to associate with cell membranes to alter their permeability and dynamics, thereby modulating signal transduction and transport pathways,[742–746] including the reprogramming of glucose metabolism.[747]

Functionally characterized examples have established that lncRNAs participate in virtually all levels of genome organization and gene expression, via RNA-RNA, RNA-DNA and RNA-protein interactions, often involving repeat elements within them, including SINEs in 3'UTRs.[748] These encompass the regulation of transcription, chromatin architecture and the organization of subcellular domains (see Chapter 16), control of protein translation and localization,[361,563,661,748–750] splicing[381,655,751–762] and other forms of RNA processing, editing, localization and stability.[763–767]

Some of the first characterized non-coding RNAs were found to regulate transcription by modulating RNA polymerase II activity, directly (such as RNAs from B2 SINE and Alu elements) or indirectly through interaction with transcription factors.[768,769] 7SK, for instance, acts primarily by sequestering and inactivating the transcription elongation factor b (P-TEFb), a heterodimer composed of cyclin-dependent kinase 9 (Cdk9) and cyclin T1, which connects transcription to the cell cycle and chromatin architecture.[770–777] It controls stress-induced transcriptional reprogramming[778] and regulates several aspects of the expression not just of mRNAs, but also of snRNAs, bidirectional and enhancer RNAs,[777,779] and acts as a multi-functional RNA scaffold that regulates neuron homeostasis.[780]

Genomically associated RNAs have been shown to regulate gene expression by other mechanisms. Transcripts spanning the cyclin D1 (*CCND1*) regulatory promoter sequences recruit and allosterically regulate the TLS RNA-binding protein and induce chromatin modification to repress *CCND1* expression.[781] At the *DHFR* locus, a non-coding RNA initiated from an upstream minor promoter forms a stable RNA-DNA triplex within the major promoter to repress DHFR expression.[782] Likewise, among others,[783] the lncRNAs ANRIL (CDKN2B-AS1), Khps1 and CISAL regulate expression of the cyclin-dependent kinase inhibitor *CDKN2B*, the proto-oncogene *SPHK1* and the tumor suppressor *BRAC1*, respectively, via triplex-mediated changes in chromatin structure.[481,784–786]

Non-coding RNAs from intergenic spacers and promoter regions of rDNA genes in humans establish and maintain heterochromatin structure at specific rDNA promoters via the recognition of RNA secondary structures and formation of triplexes with target DNA sequences that recruit DNA methyltransferase DNMT3b and more.[148–151] Indeed, lncRNAs play central roles in the formation and function of heterochromatic domains, including telomeres[ad] and centromeres, in all eukaryotes.

Many lncRNAs associate with enzymes and complexes that impart histone modifications and DNA methylation.[99,796,797] Both small and long non-coding RNAs control the target specificity of and the interplay between repressive Polycomb group (PcG) and activating Trithorax group (TrxG) protein complexes (Chapter 16), chromatin modifiers that maintain silent and active expression states of genes during development[99,178,796,798–800] (Chapter 14).

For example, it has been shown that the mouse chromodomain-containing PRC1 component Cbx7 binds RNA and that its association with the inactive X chromosome depends on interaction with RNA.[801] Imprinting of loci by lncRNAs such as

---

[ad] Maintenance of telomeres, in addition to the telomerase RNA component TERC (Chapter 9) involves transcripts named TERRA (telomeric repeat-containing RNAs) transcribed from subtelomeric regions in a developmentally regulated fashion. TERRA RNAs contain repeat rich sequences that form G-quartet structures and regulate local heterochromatin stability, telomerase activity and telomere length, as well as biological processes such as the induction and maintenance of pluripotency.[787–793] Although first discovered and characterized in mammalian cells, analogous RNAs have similar functions in different organisms, including fungi.[794,795]

Air and Kcnq1ot1 (Chapter 9) and X-chromosome epigenetic silencing by *Xist*-locus derived RNAs involve recruitment of PcG and other chromatin-modifying complexes (Chapter 16). The ~3.8 kb lncRNA ANRIL, some of whose exons are primate-specific,[389] is transcribed from a GWAS region associated with autoimmune and other disorders[231,425,426,429] and recruits components of the Polycomb Repressor Complexes 1 and 2 (PRC1/2) to epigenetically silence INK4B/ARF/INK4A tumor suppressor cluster.[802–804] Exon 8 of ANRIL, which is mainly comprised of repeat elements, mediates ANRIL's association with target loci to modulate their expression through H3K27me3 deposition.[805] The lncRNA Chaer controls hypertrophic heart growth by binding to and inhibiting the function of PRC2.[674]

Expression of the genes in the *Hox* clusters are controlled by enhancer elements present in intergenic regions that bind regulatory proteins, which are thought to activate nearby protein-coding genes *in cis* but that are also co-linearly transcribed into non-coding RNAs during development.[806–812] Hundreds of lncRNAs associate with PRC2 complexes, including functionally validated lncRNAs such as TUG1, Meg3/Gtl2 and HOTAIR[813–815] (Chapter 16).

HOTAIR is a ~2.2 kb spliced RNA transcribed from the *HOXC* locus, antisense to the flanking genes *HOXC11* and *HOXC12*, which was originally shown to direct heterochromatin formation *in trans* across a 40 kb domain of the *HOXD* cluster in human fibroblasts.[89] HOTAIR was later shown also to influence gene expression at other sites around the genome by recruitment of PRC2 and LSD1/CoREST/REST repressive chromatin-modifying complexes.[813,816,817] As with Xist, HOTAIR has different functional domains, with a 5' domain that binds PRC2 and a 3' domain that binds LSD1, and has been proposed to act as a scaffold for protein complexes,[816,818] likely a general function of lncRNAs[800,819] (Chapter 16).

Other intergenic or antisense lncRNAs transcribed from homeotic loci (*Evx1*, *HoxA13* and *HoxB5/B6* and many others – Chapter 16) have been shown to bind to TrxG;[260,341] the spliced lncRNA HOTTIP (~3.8 kb) is transcribed from a region immediately downstream of the human *HOXA13* gene, interacts with TrxG MLL component WDR5 and directs the complex to activate *HOXA13* and additional neighboring *HOXA*

genes by a mechanism that involves chromosomal looping.[260]

In plants, lncRNAs also control many aspects of development and environmental responses,[820] exemplified by the lncRNAs COOLAIR transcribed antisense to the major repressor of flowering, *FLOWERING LOCUS C (FLC)*, and COLDAIR transcribed from the first intron of *FLC*, which mediate cold-induced epigenetic repression ('vernalization') of flowering time.[821] COOLAIR and COLDAIR contain conserved modular secondary structures and act by recruitment of PRC2 and other epigenetic regulators, RNA-DNA R-loop formation, chromatin looping and the formation of phase-separated condensates.[822–834] A distal COOLAIR variant sequesters TrxG into condensates away from the promoter[835] and unspliced COOLAIR forms "clouds" around the locus,[836] similar to Xist (see Chapter 16). COOLAIR is also differentially spliced in *Arabidopsis* variants adapted to different climes.[837,838] Another lncRNA, FLAIL, represses flowering time.[839]

These are prominent examples among recurrent themes for lncRNAs, incorporating their functions as scaffolds, epigenetic guides, chromatin organizers and control devices, allosteric regulators and ribozymes, as well as decoys that sequester regulatory factors,[97,175,800,819,840,841] acting as 'target mimics', 'miRNA sponges' or 'competing endogenous RNAs'.[611,620,842,843]

THE WILD WEST

In the 'Insights of the Decade' section of the special issue of *Science* magazine in 2010, less than 10 years after the publication of the draft human genome sequence, it was noted that

> Many mysteries about the genome's dark matter are still under investigation. Even so, the overall picture is clear: 10 years ago, genes had the spotlight all to themselves. Now they have to share it with a large, and growing, ensemble.[845]

Indeed, it progressively became evident that lncRNAs are ubiquitously involved in differentiation and development processes in eukaryotes (Figure 13.9).

A recent review observed

> "The prior widely held perception that they are predominantly junk [should] also [be] factored in" to the analysis of such experiments and that [although] "there have since

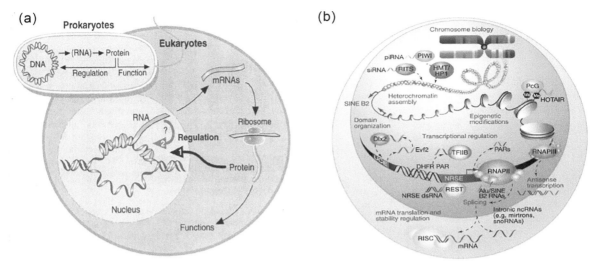

FIGURE 13.9 Depictions of eukaryotic RNA regulation in 1994 and less than 15 years later. (a) At a time when the understanding of genetic information was still largely based on bacterial studies (upper left), proteins were thought to perform all the functions in cells, including regulation (arrow 1, black) in eukaryotic cells, with speculation about a possible role of RNA in gene regulation represented as a question mark (arrow 2, red). (Reproduced from Nowak[844] with permission from the American Association for the Advancement of Science.) (b) After the turn of the century, many additional examples of regulatory RNAs emerged and shown to act in many levels of gene expression in the nucleus and cytoplasm of eukaryotic cells. (Reproduced from Amaral et al.[180] with permission from the American Association for the Advancement of Science.)

been more than a thousand publications on the functions of these lncRNAs, both in cis and trans, many (molecular biologists) are still only aware of the earlier dismissive publications".[846]

Most lncRNAs still remain experimentally untouched or poorly characterized – such as LINC02476 (GenBank CB338058), composed of at least five exons spanning 288 kb, found in 2003 associated with autism, in patient breakpoints that disrupt this non-coding RNA transcript,[847] but only now being studied due to its differential expression in cancer.[848]

The bottom line is that these highly important regulatory RNAs were present all along but, despite the cases documented in the closing decades of the 20th century and many more detected by transcriptomic, enhancer 'traps' and other biochemical and functional screens in the first decade of the 21st century, they were not generally, until recently, taken seriously. They have also been studied in all sorts of ways, with all sorts of specific and general hypotheses, dubbed "The Wild West" by Jeannie Lee[799] and

"The Noncoding RNA Revolution" by Tom Cech and Joan Steitz.[849]

As Stent noted in relation to the unexpected discovery that DNA is the genetic material,[850] the finding of dynamic and differential genome-wide transcription of intergenic, antisense and overlapping lncRNAs, like the related discovery of intervening sequences, was "premature" in the sense that it could not be readily incorporated into the existing conceptual fabric. To put the plethora of regulatory RNAs into full perspective and to integrate them into a contemporary framework for the genetic programming of complex organisms, we must first consider the epigenome and the amount of information required for multicellular development.

FURTHER READING

Amaral P.P. and Mattick J.S. (2008) Noncoding RNA in development. *Mammalian Genome* 19: 454–92.
Andergassen D. and Rinn J.L. (2021) From genotype to phenotype: Genetics of mammalian long non-coding RNAs in vivo. *Nature Reviews Genetics* 23: 229–43.

Briggs J.A., Wolvetang E.J., Mattick J.S., Rinn J.L. and Barry G. (2015) Mechanisms of long non-coding RNAs in mammalian nervous system development, plasticity, disease, and evolution. *Neuron* 88: 861–77.

Clark M.B. and Mattick J.S. (2011) Long noncoding RNAs in cell biology. *Seminars in Cell and Developmental Biology* 22: 366–76.

Deveson I.W., Hardwick S.A., Mercer T.R. and Mattick J.S. (2017) The dimensions, dynamics, and relevance of the mammalian noncoding transcriptome. *Trends in Genetics* 33: 464–78.

Mattick J. (2010) Video Q&A: Non-coding RNAs and eukaryotic evolution - a personal view. *BMC Biology* 8: 67.

Mattick J.S (2009) The genetic signatures of noncoding RNAs. *PLOS Genetics* 5: e1000459.

Mercer T.R., Dinger M.E. and Mattick J.S. (2009) Long noncoding RNAs: insights into function. *Nature Reviews Genetics* 10: 155–9.

Morris K.V. and Mattick J.S. (2014) The rise of regulatory RNA. *Nature Reviews Genetics* 15: 423–37.

Rinn J.L. and Chang H.Y. (2020) Long noncoding RNAs: Molecular modalities to organismal functions. *Annual Review of Biochemistry* 89: 283–308.

Unfried J.P. and Ulitsky I. (2022) Substoichiometric action of long noncoding RNAs. *Nature Cell Biology* 24: 608-15.

Winkle M., El-Daly S.M., Fabbri M. and Calin G.A. (2021) Noncoding RNA therapeutics—challenges and potential solutions. *Nature Reviews Drug Discovery* 20: 629–51.

14 The Epigenome

Early studies in *Drosophila* and other organisms showed that the patterns of gene expression vary in different cell types, which define their identity and fate, and that these patterns can be maintained following DNA replication and subsequently through mitosis. That is, there is a secondary form of genomically encoded heritable information, termed 'epigenetic' information, which is embedded in chromatin modifications and manifested as canalized pathway choices during differentiation and development, first proposed by Conrad Waddington in the 1940s.[1-5]

The developmentally regulated packaging of eukaryotic DNA into compacted heterochromatin[a] and more transcriptionally active euchromatin had been known since the early 20th century, with different regions of the genome thought to be open or closed for business, akin to a library compactus.[8,9]

CHROMATIN STRUCTURE

In 1974, Ada and Donald Olins[10] and Roger Kornberg[11,12] reported that eukaryotic chromatin appears like "linear arrays of spheroid units" or "beads-on-a-string", respectively, and that the DNA is wound like cotton around a spool into 11 nm diameter 'nucleosomes', which contain four pairs of histones.[11] The Olins also credited another investigator, Christopher Woodcock, who had obtained similar images. Woodcock's paper[13] was, however, rejected by the journal *Nature*, a reviewer asserting that to accept the article would require "rewriting our textbooks on cytology and genetics" and that "such a naïve paper … should not be published anywhere".[14]

It was known from the 1960s from the work of Vincent Allfrey, Alfred Mirsky and others that histones can be methylated or acetylated, sometimes in response to external stimuli, and that these modifications affect transcription,[15-17] although the extraordinary range of histone modifications was not apparent until much later.

Pluripotent cells have relatively open chromatin, as do cancer cells, whereas the extent of closed chromatin increases as cells differentiate.[8,18] Nucleosomes in heterochromatin are compacted into higher-order structures, initially described as 30 nm fibers, but the exact nature of these structures remains controversial.[19-22] Chromatin is further compacted during meiosis and mitosis.[23-26] While the mechanisms controlling chromatin condensation and decondensation are not well understood, it is clear that histone modifications and non-histone proteins play important roles.[27,28] Moreover, in all eukaryotes – from yeast to plants and animals – RNAs have been shown to be associated with chromatin, degradation of which by RNase changes the patterns of exposed DNA.[29-31]

The fine-scale organization of the eukaryotic nucleus, chromosomes and chromatin becomes more elaborate with increased developmental complexity, documented by Torbjorn Caspersson, Julie Korenberg, Mary Rycowski, Georgio Bernardi, Wendy Bickmore and others, who also showed that cytological 'banding' patterns, gene density, intron density, protein density, GC content, CpG island and repeat distributions vary widely across chromosomes.[32-40]

The classical banding patterns correlate with the distribution of repeats. In human chromosomes Alu elements are concentrated in the so-called Reverse or R-bands, especially in the T-bands, the most intensely stained and most GC-rich fraction of the R-bands. LINE1 elements are concentrated in the alternating Giemsa or G-bands[33,35,39-41] and sequester genes with specialized functions in the nucleolus and inactive lamina-associated domains (see below), indicating a global role of transposable elements in orchestrating the function, regulation and expression of their host genes.[42]

TOPOLOGICAL DOMAINS

In situ fluorescent hybridization studies by Thomas and Marion Cremer, Bickmore and others from the 1990s showed that chromosomes occupy defined

---

[a] There are two types of heterochromatin: facultative heterochromatin, which is developmentally regulated (such as occurs in X-chromosome inactivation and at many other discrete loci during differentiation and development), and constitutive heterochromatin (such as occurs in centromeric and telomeric regions of chromosomes).[6,7]

DOI: 10.1201/9781003109242-14

175

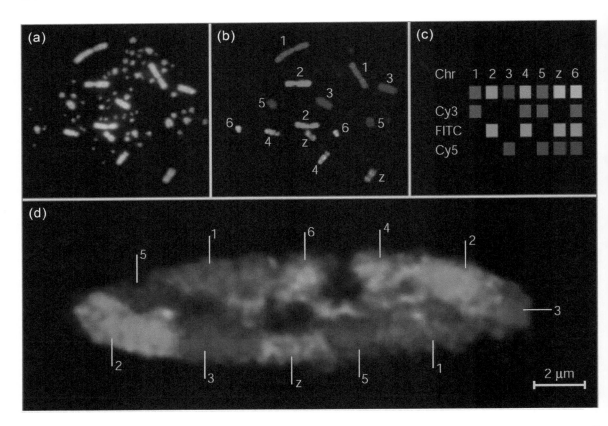

FIGURE 14.1   Chromosome territories (CTs) in the chicken fibroblast nucleus. (Reproduced from Cremer and Cremer[44] with permission of Springer Nature.)

'territories' in the nuclei of animal and plant cells[43–48] (Figure 14.1), confirming the conclusions drawn by the cytogeneticists Carl Rabi and Theodor Boveri a century before.[49–51] These studies, refined and expanded by new techniques, also revealed radial segregation of chromosomal domains: gene-rich and actively transcribed chromosomal regions are located in the center of the nucleus, whereas gene-poor and genetically quiescent heterochromatic regions are sequestered at the periphery, associated with the nuclear membrane.[47,51–56]

Job Dekker and colleagues showed that, in both animals and plants, euchromatic and heterochromatic regions are partitioned into megabase-sized active 'A' and inactive 'B' compartments, respectively,[57,58] which encompass smaller three-dimensional 'topologically associated domains', or TADs, with high-frequency intra-chromatin interactions.[22,57–71]

A striking example is the discovery by Elphège Nora, Dekker, Edith Heard and colleagues that the X-inactivation center (XIC), which they failed for

decades to define using cloned transgenes of up to 500 kb, spans bipartite TADs[b] that occupy ~800 kb of genomic territory: the promoter for the *Xist* gene, which triggers X inactivation, lies in one TAD of ~500 kb, whereas its antisense regulator *Tsix* lies in another TAD of ~300 kb.[76,77] They also proposed that TADs underlie many properties of the long-range transcriptional regulation that occurs in animals and plants,[61,76] a prediction that coalesces with later observations that subnuclear and subcellular compartment organization is at least partly driven by RNA-mediated phase separation[78–82] (Chapter 16). The topological organization of chromatin during development is also reliant on repetitive elements and their interaction with the heterochromatin 1 (HP1) protein family.[41,83]

---

[b] The structure of these domains on the X chromosome and others on autosomes is regulated by the lncRNAs Dxz4 and Firre.[72–75]

TADs have an average size of ~0.5–1 Mb, shown by proximity ligation (cross-linking the DNA *in situ* to identify sequences physically adjacent in three-dimensional space),[60] with higher resolution analyses revealing finer scale internal TAD organization.[70,84] TADs appear to demarcated by boundary regions anchored by the 'insulator' protein CTCF[c] and the 'cohesin' complex (Figure 14.2), which interact and control chromatin loop extrusion,[64,86–91] involving phase-separation,[92] evident as "architectural stripes", where loop anchors secure topological domains and link enhancers (see below) to cognate promoters.[90] A cell lineage-specific subset of CTCF binding sites[d] and TAD boundaries are controlled by DNA methylation,[e] indicating an interplay between epigenetic modifications, chromatin organization and transcript isoforms during development.[94–100]

CTCF is also associated with attachment to the nuclear lamina, a filamentous protein network underlying the nuclear membrane in animal cells, where it demarcates 'laminar-associated domains' (LADs). LADs have low transcriptional activity,[101] consistent with the earlier observations that gene-poor and quiescent genomic regions are located at the nuclear periphery. The composition of the nuclear lamina varies in different tissues, and mutations in laminar proteins result in a range of conditions including muscular dystrophies and neurological disorders.[102] Lamin-like proteins also occur in plants and dynamically tether heterochromatin to the nuclear periphery in response to environmental and developmental signals.[55]

Vertebrate genomes are also partitioned into 'isochores', megabase-sized domains of different G+C content, which are most pronounced in mammals.[103] Isochores may correlate with TADs and LADs, with the G+C distribution apparently playing a role in "moulding" chromatin accessibility, although the relationship is unclear.[104]

The number of LADs, TADs and replication domains (~2,000) in the human genome is similar to the number of chromosome bands observed in prometaphase chromosomes.[105,106] TADs and TAD boundaries also correspond with the bands and inter-bands seen on *Drosophila* polytene chromosomes,[107] as well as with 'chromomeres' – locally coiled chromatin domains observed in mitotic and meiotic prophase chromosomes,[108,109] supported by the observation that TADs are condensed chromatin domains separated by regions of active chromatin.[85]

Some reports suggest that TADS are stable across evolution, cell types and independent of gene expression, and may represent DNA replication modules,[8,63,111–114] whereas others indicate that TADs, and to a lesser extent A and B compartments, vary among cell types and are reorganized during differentiation and development.[48,60,64,71,115] TADs may be equivalent to the chromatin domains formed by enhancer action[84,116] (see below). TADs in human pluripotent stem cells are demarcated, at least in part, by transcriptionally active HERV-H retrotransposons[117] and regulated by the RNAi pathway via AGO1 association with expressed enhancers.[118] Some evidence suggests that megabase-scale TADs are largely cell-type invariant, whereas 'subTADs' reconfigure in a cell type-specific manner.[110] TAD reconfiguration at the *HoxD* locus appears to regulate limb development,[119] and cell-type specialization is encoded by chromatin topologies.[115]

TADs are also reorganized in response to physiological parameters, such as hormone signaling and neuronal activation.[120–122] They are also the functional units of the DNA damage response, required for the one-sided cohesin-mediated loop extrusion of chromatin domains containing the double-strand break-specific histone variant, phosphorylated H2A.X (see below), a process that involves transcription of non-coding RNAs.[123] Mutations affecting TAD boundaries are associated with human developmental disorders and cancers, apparently due to aberrant promoter-enhancer interactions.[124,125]

ENHANCERS

'Enhancers' are upstream, downstream or intronic non-protein-coding genomic regions in animals and (to a lesser extent[126]) plants that control developmental cell-type-specific spatiotemporal expression patterns of protein-coding and non-protein-coding genes in their neighborhood, by altering the organization of chromatin.[127–133] Enhancers can be located hundreds of kilobases away from their target genes and are (local) position and orientation-independent.[126,134–140]

Enhancers were classically recognized and genetically defined by their developmental effects, rather

---

[c] There is also conflicting data, with other studies showing a poor correlation between CTCF binding sites and TAD boundaries.[85]

[d] Many CTCF binding sites are derived from transposable elements.[93]

[e] DNA methylation also regulates alternative polyadenylation via CTCF and the cohesin complex.[94]

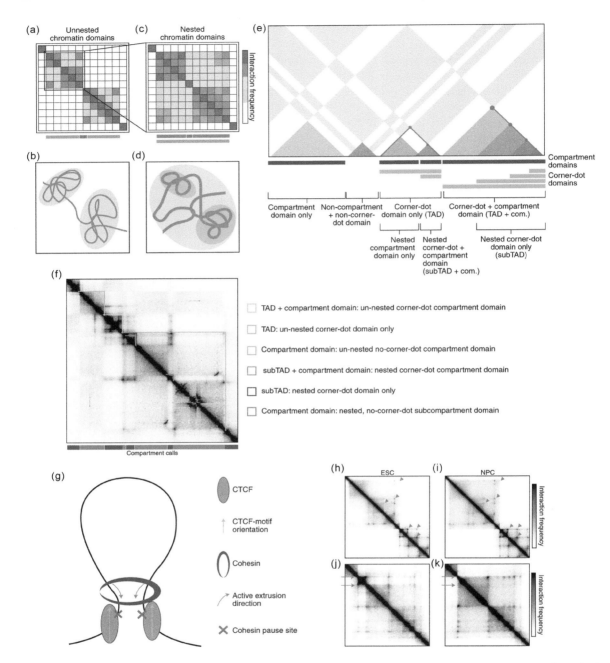

FIGURE 14.2 The structural features of topologically associating domains. (a–d) Heat-map representations (top) and schematized globular interactions (bottom) of TADs (a,b) and nested subTADs (c,d). (e) Cartoon representation of different classes of contact domains parsed by their structural features and degree of nesting. (f) Identification of contact-domain classes from e in cortical neuron Hi-C data,[84] binned at 10-kb resolution. (g) Cohesin translocation extrudes DNA in an ATP-dependent manner into long-range looping interactions that form the topological basis for TAD and subTAD loop domains. (h–k) Contact frequency heat maps of high-resolution Hi-C data from embryonic stem cells (ESC, h,j) and neural progenitor cells (NPC; i,k).[84] (h,i) Green arrows denote the corners of a subset of the nested chromatin domains evident in this genomic region. (j,k) Green arrows annotate a high-insulation-strength, cell-type-invariant TAD boundary. Blue arrows point to a lower-insulation-strength, cell-type-dynamic subTAD boundary. (Reproduced from Beagan and Phillips-Cremins[110] with permission of Springer Nature.)

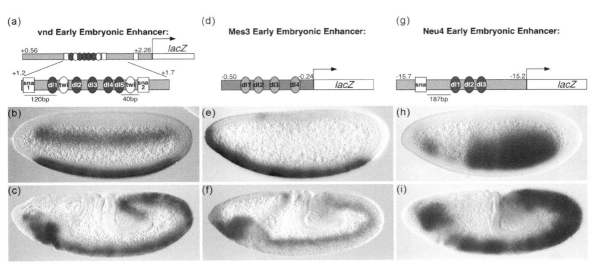

FIGURE 14.3 Restricted expression patterns of embryonic enhancers at two different developmental stages visualized by *lacZ* expression in transgenic *Drosophila* embryos. (Reproduced from Stathopoulos et al.[152] with permission of Elsevier.)

than their biochemical properties or mode of action. Although not described as such, enhancer activity was first observed in the *bithorax* complex of *Drosophila*,[141–143] but it was only in the early 1980s that the term was coined to describe the unexpected ability of SV40 viral DNA sequences to increase the expression of a cloned β-globin gene.[144] Many tissue-specific enhancers, often containing repetitive elements similar to those in viral enhancers,[f] were subsequently identified in mammalian immunoglobulin and globin gene[g] loci, as well as in the Drosophila *bithorax* complex and other genes that show restricted expression patterns during development (Figure 14.3), initially using deletion strategies.[134,135,146–151]

Many enhancers have since been identified by other genetic and bioinformatic approaches,[153–155] the former using insertions of transposons with reporter genes, called 'enhancer trapping', in *Drosophila*,[143,152,156,157] plants[158] and vertebrates.[159] More recently attempts have been made to characterize known enhancers and identify others by genome-wide analysis of the binding positions of presumed signature proteins (the 'transcriptional co-activators'

P300 and Mediator[h]) combined with the presence of correlated histone modifications,[167–171] the presence of nucleosome-depleted regions and/or the expression of 'enhancer RNAs' (eRNAs),[172–179] which yield different prediction sets and blur the distinction between enhancer and (protein-coding) gene promoters[127,139,154,180] (see below and Chapter 16).

The appearance of enhancers has been linked to the emergence of animal multicellularity and phenotypic diversity,[181,182] neuronal expansion in vertebrates[183] and the recent evolution of primates.[184] Positive selection for nucleotide changes in enhancers has contributed, for example, to the uniquely human aspects of thermoregulation (sweat glands in the skin)[185] and digit and limb patterning, including the increase in size and rotation of the thumb toward the palm for enhanced dexterity.[186] Body plan specification is controlled by multiple enhancers to ensure precise patterns of gene expression.[151,187] Clusters of enhancers, such as those at the beta globin locus, but also many others, have been dubbed "super-enhancers", "stretch enhancers" or "enhancer jungles".[127,188–193] Enhancers also play

---

[f] Endogenous retroviruses have been shown to be a source of enhancers.[145]

[g] Globin enhancers were originally and are still often referred to as 'locus control regions'.[146]

[h] P300 is a histone modifying enzyme.[160] Mediator is a highly modular multi-subunit complex that appears to connect distal transcription factors with the transcription initiation machinery.[161–163] Both P300 and Mediator bind RNA, which is required for their chromosomal localization, TAD juxtaposition and local chromatin modification[164–166] (Chapter 16).

a role in the etiology of cancer,[127,192,194] and disruptions of chromatin topological domains cause rewiring of gene-enhancer interactions with pathogenic consequences.[64,122,131,136,195,196]

Enhancers are still incompletely defined, physically and conceptually,[128,153,154] but have been described as "DNA logic gates".[197] Mechanistically, enhancers were originally conceived as clusters of transcription factor binding sites that are brought into contact with target protein-coding gene promoters by long-distance DNA looping, a model first proposed by Mark Ptashne to reconcile enhancer function with transcription factor control of gene expression.[198] The persistence and vagaries of the initial interpretation of how enhancers work[124,128,131,136,169,199] has been referred to by Marc Halfon[i] as a case of "founder fallacy" and "validation creep",[154] by no means the first in molecular biology or science generally.

There is good evidence that enhancer action leads to the juxtaposition of distal chromosomal sequences in three-dimensional space, and to consequent transcriptional activation of genes in their orbit.[200] Enhancer-mediated DNA looping may be equivalent to TADs[131,201] but enhancers can exert their action across TAD boundaries, which may in turn play a role in mediating formation, reorganization and/or juxtaposition of such domains,[122,131,132,139,143,157,202,203] although genome topology and gene expression can be uncoupled.[204] Enhancers also recruit histone-modifying chromatin remodeling proteins, such as the CREB-binding protein (CBP, see below).[166,205]

However, evidence for the direct interaction of transcription factors bound at enhancers with target protein-coding gene promoters is limited, in some cases contradictory,[206] and intimate contact may be more an enduring presumption than an accurate mechanistic description,[154] especially in view of the fact that enhancers are transcribed in the cells in which they are active.[134,135,146–150,172–179,207–209] Indeed, enhancers have many if not all of the characteristics of *bona fide* genes, including promoters.[210,211] Most lncRNAs originate from enhancers[209,212] and enhancer RNA production is considered the most reliable indicator of enhancer action.[172–179] How enhancers select their targets is unknown, but likely

involves RNA-DNA, RNA-RNA and RNA-protein interactions[213–215] (Chapter 16).

Strikingly, the number of mammalian enhancers, estimated to be in the hundreds of thousands,[130,170,172,180,192,216–219] far outweighs the number of protein-coding genes, which indicates that distal sequences that regulate developmental expression patterns occupy a much larger fraction of the genome than those constituting the proximal promoters of protein-coding genes.

NUCLEOSOMES AND HISTONES

Partial digestion of exposed DNA in chromatin with micrococcal nuclease yields a ladder of modal DNA lengths in multiples of ~180bp, reflecting 147bp of DNA supercoiled around the outside of the nucleosome core particle and ~35bp of linker DNA between (in mammals), although the average length of the linker sequence varies between species and cell types.[220]

There are approximately 30 million nucleosomes in a human cell.[221] Canonical nucleosomes are composed of an octamer of four small, highly basic proteins: histones H2A, H2B, H3 and H4; the central H3-H4 tetramer is sandwiched between two H2A-H2B dimers and the N-terminal tails of the histones, which protrude beyond the DNA shell and are the major sites for post-translational modifications[222–224] (Figure 14.4) (see below). Canonical histones are produced during the replicative S-phase of the cell cycle and are among the most highly conserved proteins in evolution.[225] Interestingly, the genes encoding the canonical (but not the variant) histones are some of the few genes that lack introns, possibly as their constitutive production with chromosomal replication does not require efference signals to be transmitted in parallel.

Archaeal histones form a structure similar to the eukaryotic H3-H4 tetramer, but, unlike eukaryotic histones, lack extended N-terminal tails and post-translational modifications.[227] Both possess a copper (Cu2+) binding site at the H3 dimerization interface and have been shown to have copper reductase activity,[228] suggesting that they originated as a mechanism for copper utilization under oxidizing conditions.[229]

Another histone, H1, binds to the outside of the nucleosome at the entry and exit sites of the DNA to stabilize the particle and/or play a role in coiling of nucleosomes into higher-order structures.[230,231] A homolog of histone H1 exists in bacteria, and also

---

[i] Halfon notes, for example, that "a recent paper erroneously states that enhancers 'were first described as nucleosome-depleted regions with a high density of sequence motifs recognized by DNA-binding transcription factors'."[154]

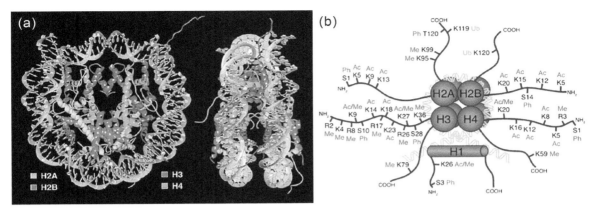

FIGURE 14.4   (a) Nucleosome structure showing histone octamer core, encircling DNA and protruding histone tails. (Reproduced from Luger et al.[223] with permission of Springer Nature.) (b) Some of the many modifications of histone N-terminal tails. (Reproduced from Zhao et al.[226] with permission of Springer Nature.)

appears to have been acquired by eukaryotes at the time of their origin.[232]

Nucleosomes were initially thought to be simply a means of compacting genomes – there is ~2.5 m of DNA in a mammalian cell – and this is likely an important function. Nonetheless, they are not static but dynamic structures, histones being exchanged and differentially modified during differentiation and development.[231,233–235]

The promoters of protein-coding genes and developmental enhancers initially appeared to be 'nucleosome-free regions' based on their sensitivity to DNase digestion and accessibility to transcription factors.[236] However, more sensitive approaches have revealed that nucleosomes do occur in the vicinity of promoters but are "unstable" and subject to higher turnover.[31,237–243]

There are also variant forms of nucleosomes, mostly involving H2A. H2A can be replaced by H2A.Z, which, unlike other histones, is multi-exonic and produced throughout the cell cycle. H2A.Z is present in most tissues, but most highly expressed in embryos, essential for development in insects, vertebrates and plants[244–248] and associated with memory formation.[249]

There exist two H2A.Z genes encoding almost identical proteins (three amino acid differences) in chordates, one of which expresses a primate-specific alternatively spliced isoform in the brain.[250,251] The two H2A.Z subtypes display differential occupancy at the promoters of protein-coding genes and enhancers, and regulate genes involved in early embryological, neural crest and craniofacial development

development,[246,252] as well as the progression of some types of cancers.[253] Just one of the three amino acid differences between the H2A.Z subtypes is sufficient to rescue the developmental abnormalities caused by mutations in an enzyme that catalyzes replacement of the canonical H2A-H2B dimer with the H2A.Z-H2B dimer.[252]

The H2A variant H2A.X is recruited to double-stranded DNA breaks and its phosphorylated form is required for their repair, a process that is also involved in programmed genomic rearrangements during immune cell development.[254,255]

Another H2A variant ('macroH2A') contains an additional and highly conserved large C-terminal domain that has homologs in all kingdoms of life and is covalently linked to its N-terminal histone homology domain, which is also highly conserved but quite different from that in the canonical H2A.[256] The macro domain has ADP-ribosylation activity and possibly RNA-binding activity.[257] MacroH2As are encoded by two multi-exonic genes, one of which is alternatively spliced in the macro domain. They associate with the inactive X chromosome of female mammalian cells and inactive genes and appear to have a role in maintaining heterochromatin.[258,259]

There are also short variants, H2A.B, H2A.L, H2A.P and H2A.Q and splice isoforms thereof, which lack the C-terminal tail of the core H2A. These variants appeared in mammals and are tissue-specific, being expressed in the testis and, in the case of H2A.B, also in the brain.[260] The H2A.B variant binds RNA, replaces H2A.Z in nucleosomes at transcription start sites and intron-exon boundaries in the testis and the brain, and interacts with RNA polymerase II to promote the activation of transcription.[261–263] It is also involved in biparental inheritance controlling embryonic development in mice.[264] H2A.B has a propensity for chromatin decompaction[261,265] and co-localizes with the RNAi proteins Miwi and Dicer in spermatids,[260] indicating a relationship between regulatory RNAs, chromatin organization and splicing pathways.

The H2A.L.2 variant also has an RNA-binding domain and appears to be guided to its sites of incorporation by RNA.[266] In sperm development it dimerizes with the H2B testis-specific variant TH2B as a prelude to nucleosome displacement by other highly basic proteins called protamines,[260] originally discovered by Miescher,[267] which mediate the extreme compaction of the chromosomes.

Histone H3 is replaced in nucleosomes by H3.3 (which differs from H3 by only four amino acids) in telomeres and pericentromeric regions and when chromatin assembly occurs at times other than replication,[268–272] including in meiotic sex chromosome inactivation.[273] Histone H2A-H2B is bound to an essential telomerase RNA domain, which suggests a role for histones in the folding and function of the telomerase RNA component.[274]

H3.3/H2A.Z double variant-containing nucleosomes are enriched in active promoters and enhancers.[237,238] Loss of H3.3 results in fertility and/or defects in gastrulation or neural crest development in flies, fish and frogs.[237,275–277] In mammals, H3.3 also accumulates in neurons, reaching near saturation by adolescence, where it controls neuronal- and glial-specific gene expression patterns, with an essential role in plasticity and cognition.[278] Rare missense mutations in H3.3 have been shown to cause neurologic dysfunction and congenital anomalies.[279] Mutations of lysines in the tail of H3.3 are commonly observed in glioblastomas.[280,281]

In flies and vertebrates, there are two seemingly redundant genes encoding H3.3, H3A[j] and H3B, which vary in their 3'UTRs,[272] whose individual loss in mammals causes infertility and reduced viability.[283,284] Loss of both causes embryonic lethality, due to heterochromatic dysfunction at telomeres and centromeres,[285] the latter of which can be rescued by injecting dsRNA derived from pericentromeric

transcripts, indicating a functional link with the silencing of such regions by an RNAi pathway.[284]

In centromeres, H3 is replaced by another variant, CENP-A, which is essential for kinetochore formation required for chromosome segregation during mitosis and meiosis.[286,287] In plants, epigenetic memory is reset by replacing H3 with the variant H3.10 (which is refractory to lysine 27 methylation) during sperm maturation to globally reprogram paternal gametes.[288]

We could go on. The bottom line is that there is a high degree of complexity in the composition of nucleosomes, dynamic histone exchange and remodeling of chromatin during development.[289] However, little is known about the decisional processes and mechanisms that determine when and where different histones are incorporated into particular nucleosomes, other than that RNA and 'pioneer transcription factors' are involved (Chapter 16).

NUCLEOSOME REMODELING

Pioneer or 'architectural' transcription factors, such as Sox2 and Sox11,[k] which are 'high mobility group' (HMG) proteins, bend DNA structure and initiate the opening of chromatin by eviction of the linker histone H1.[290–293] Other pioneer factors, such as the winged helix/forkhead box (Fox) proteins, bind to DNA within nucleosomes in promoters and enhancers leading to their destabilization, also by histone H1 displacement, and recruit Mediator and cohesin to permit chromatin access for tissue-specific remodeling factors such as FoxA, with different targets in different cells at different stages of development.[294–298]

Histones are escorted to nucleosomes by companion or 'chaperone' proteins,[268,286,299,300] and histone exchange requires a conserved family of ATP-dependent 'chromatin remodeling enzymes' variously known as SWI/SNF, ISWI, NuRD, CHD and INO80,[301] many now known to be regulated by cis- and trans-acting non-coding RNAs (Chapter 16).

HISTONE MODIFICATIONS

Histone modification by methylation and acetylation was first observed and proposed to have a regulatory function in the 1960s, and nucleosomes were known to affect transcription,[302–304] but it was not

---

[j] The coding sequence of H3A appears to have evolved under strong purifying selection in the lobe finned fish and tetrapods, without any change to the amino acid sequence.[282]

[k] Sox2 and Sox11 are involved in the maintenance of pluripotency and neuronal differentiation, respectively.

until 1991 that Michael Grunstein and colleagues provided definitive evidence of gene regulation by histone acetylation.[305] In 1996, David Allis and colleagues isolated a histone acetyl transferase, making use of the fact that histones in the macronucleus of *Tetrahymena* cells are highly acetylated whereas those in the micronuclei are not, which for the first time directly linked a transcriptional regulator to a histone-modifying enzyme.[306,307] A reciprocal histone deacetylase (HDAC) activity was reported a month later by Stuart Schreiber and colleagues.[308]

Shortly thereafter it was shown that the mammalian 'transcriptional co-activators' CBP, P300 and the yeast 'transcriptional adaptor protein' Gcn5 function in multi-subunit complexes to acetylate histones in nucleosomes,[160,309,310] linking what had been vaguely referred to as 'transcription factors' to chromatin modification. These findings changed the perception of the nucleosome from being simply a mechanism for genome compaction to a major player in regulating its expression.

The 1990s and 2000s saw the identification of a bewildering array of histone modifications, mainly by mass spectrometry – many of which still remain to be characterized[311,312] – at last count in over 60 different positions, mainly in the N-terminal tails of the histones, which are intrinsically disordered[313,314] (Chapter 16) and exposed beyond the periphery of the nucleosome.[l] These modifications span mono-, di- and tri-methylation, acetylation, ADP ribosylation, ubiquitylation and/or sumoylation of various lysines in histones H2A, H2A.X, H2B, H3 and H4, mono- and di-methylation, acetylation and deimination of arginines (to citrulline[m]) in H2A, H3 and H4, phosphorylation of serines, threonines, tyrosines and one lysine in H2A, H2A.X, H2B, H3 and H4, isomerization of prolines in H3, and O-palmitoylation of a serine in H4.[27,320,321]

Histone modifications also include propionylation, butyrylation, malonylation, formylation, glutathionylation, tyrosine hydroxylation and lysine crotonylation,[311,312,322,323] the latter at 28 different lysines in H1, H2A, H2B, H3 and H4.[324] Many of these modifications are in low abundance, suggesting particular contextual functions. An example of their impact, however, is that citrullination of histone H1 leads to its displacement from the nucleosome and the decondensation of chromatin in pluripotent cells and during developmental reprogramming.[316,322]

The shorthand nomenclature for modifications is histone > amino acid (single letter code) > position > modification – for example, the methylation of arginine 11 on histone H4 is written as H4R11me, and the acetylation of lysine 5 on histone H2B is written as H2BK5ac, etc.

Other more exotic modifications have been discovered, such as histone ufmylation (the conjugation of UFM1 ubiquitin-like protein to H4 to promote DNA repair),[325,326] the covalent conjugation of the metabolite lactate at 28 sites on core histones (histone "lactylation")[327] and the conjugation of the neurotransmitters serotonin and dopamine to H3 glutamine 5 (H3Q5ser) and trimethylated lysine 4 (H3K4me3Q5ser) in specific regions of the brain[328,329] Cocaine administration, which causes dopamine release from the ventral tegmental area (VTA), induces hyperacetylation[n] of H3 and H4 at genes associated with cocaine addiction in the *nucleus accumbens*, a brain 'reward' region.[331] Moreover, rats undergoing withdrawal from cocaine dopaminylate histone H3 glutamine 5 (H3Q5dop) in the VTA, inhibition of which reverses cocaine-mediated gene expression changes, attenuates dopamine release in the *nucleus accumbens*, and reduces cocaine-seeking behavior.[328] These are potentially profound observations for understanding brain function – neurotransmitters have lasting epigenetic effects.

There are over 100 enzymes known to catalyze histone modifications at particular amino acid positions in mammals (called code 'writers'), and dozens more that remove them ('erasers'),[o] mostly acting on histone H3,[332] with a similar albeit less extensive repertoire in other animals, plants and fungi. Many of these proteins are encoded by homologs of genes first

---

[l] Some modifications also occur in the internal globular domains of the histones.[311]

[m] Citrulline is also an intermediate in the urea cycle. Citrullination catalyzed by peptidylarginine deiminases (PADs) neutralizes arginine's positive charge, can antagonize arginine methylation for local gene regulation and global chromatin decompaction, with implications for cell pluripotency and differentiation.[315,316] The peptidylarginine deiminase PAD4 is essential for the remarkable formation of neutrophil extracellular (chromatin) traps (NETs) that are phagocytosed by macrophages to stimulate innate immune responses during infection.[317–319]

[n] Alcohol consumption also increases histone acetylation in fetal and adult mouse brain.[330]

[o] The discovery of histone modification erasers was unexpected by many, as were the discoveries of DNA demethylases (see below) and RNA modification erasers (Chapter 17). Indeed, most levels of regulation beyond transcription factors were initially met with skepticism, and then largely shoehorned into the transcription factor paradigm.

identified as critical for *Drosophila* development, notably *Polycomb*, *Trithorax* and *Zeste* (Chapter 5). The two multi-subunit Polycomb complexes in mammals, PRC1 and PRC2, act non-redundantly at target genes to maintain transcriptional programs and cellular identity. PRC2 methylates lysine 27 on histone H3 (H3K27me), while PRC1 ubiquitinates histone H2A at lysine 119 (H2AK119ub),[333] both preferentially at unmethylated CpG islands,[334] with a complex interplay between them, including, for example, with the core PRC component EED, which recruits histone deacetylases.[335,336] Trithorax proteins, which activate gene expression, contain the SET domain, which methylates H3K4 and is found in all eukaryotes.[337]

Substantial innovations in the subunit composition of chromatin-modifying complexes have accompanied increased developmental complexity. Histone modification writer, reader and eraser complexes are more elaborate and diverse in mammals than invertebrates. The *Drosophila* PRC1 complex, for example, has just one version of its constituent subunits, whereas mammalian PRC1 can incorporate any one of two RING subunits, three PHC subunits, six PCGF subunits and five CBX subunits,[338–340] the latter of which interact with the neural gene repression factor REST[341] and appear to be involved in the formation of local phase-separated domains[342] (Chapter 16).

Similar increases in subunit complexity and/or the numbers of orthologs or genomic binding sites also occurred in the Mediator complex in metazoans,[162,343] CTCF in bilaterians,[344] the HUSH complex for heterochromatin regulation in vertebrates,[345] and the major expansions of the fast evolving zinc-finger transcription factors (one of the largest gene families in humans), many of which have associated metazoan-specific BTB, tetrapod-specific KRAB or mammal-specific SCAN domains.[346,347]

Different types of histone modifications are recognized by over 70 known 'reader' proteins, many of which contain Tudor, PHD finger, MBT, bromo or chromo domains that occur in a range of chromatin remodeling and histone-modifying factors.[332,348–352] PHD fingers read the tail of histone H3, primarily the methylation state of H3K4 (K4me3/2), and to a lesser extent the methylation state of H3R2 (R2me2)[p] and

the acetylation state of H3K14.[352] Bromo domains[q] primarily recognize acetylated lysine residues,[356] and occur along with acetyltransferase domains in the pioneer factors CBP and P300.[357] Chromo, Tudor and MBT domains are part of an extended family that evolved from a common ancestor and recognize methylated lysines.[358,359]

Underscoring their importance, mutations in histone modification writers, readers and erasers cause developmental abnormalities, intellectual disabilities and cancers.[360–364] For example, 10% of leukemias are caused by translocations and ectopic fusions of the Trithorax homolog KMT2A (lysine-specific methyltransferase 2A), previously called MLL1 – for 'mixed lineage leukemia' 1.[365] Dysregulation of the chromatin-binding PHD finger protein JARID1, which binds H3K4me2/3, also causes leukemias.[366] Haploinsufficiency of histone deacetylase 4 (HDAC4) results in brachydactyly mental retardation syndrome.[367] A number of drugs that inhibit histone deacetylases have been licensed for use against hematopoietic cancers, particularly lymphomas and myelomas.[368]

THE HISTONE CODE

In 2000, David Allis and Brian Strahl proposed the 'histone code hypothesis':

> First, the establishment of … a combinatorial pattern of histone modification, i.e., the histone code, in a given cellular or developmental context … Second, the specific interpretation or the 'reading' of the histone code … (which) function broadly to set up an epigenetic landscape that determines cell fate decision-making during embryogenesis and development.[370]

The last sentence is the key and far-reaching conclusion, which takes gene regulation in eukaryotes well beyond conventional transcription factors and suggests that epigenetic processes comprise the senior level of control of developmental trajectories, notwithstanding the fact that the differentiation state of cells can be changed by ectopic expression of transcription factors (Chapter 15).

It has taken a long time for this view of the regulation of cell fate during development to overcome

---

[p] The 7SK RNA/P-TEFb complex has also been reported to be a 'reader' of the H4R3me2 modification.[353,354]

[q] Bromodomain proteins have been explored as a target for anticancer drugs, with mixed results.[355]

the hegemony of transcription factors, and there has been staunch opposition to it. As Allis later recalled:

> Chromatin studies in this era paled in comparison with the more exciting studies on transacting transcription factors that were all the rage ... Moreover, well defined paradigms of gene regulation had been elegantly worked out in prokaryotic models ... Histone proteins were viewed as only being in the way of where all of this exciting action took place. My career choice to study histone biology was a steep uphill climb, especially given the popular notion that histones did not really matter in gene regulation.[371]

Even after histone modifications were shown to have a role in the regulation of the expression of iconic genes involved in development, their action was widely interpreted in terms of nucleosome control of transcription factor accessibility, rather than considering what might regulate nucleosome position and histone modification state in the first place.

The emphasis on transcription initiation as the main focus of 'gene regulation' and the resistance to the suggestion that epigenetic regulation may determine which genes are available to be transcribed are perhaps best illustrated by a 2013 article by Mark Ptashne, who pioneered the characterization of transcription factor binding to DNA in bacteria and yeast.[198,372–374] Ptashne's article, entitled 'Epigenetics: Core Misconcept',[375] stated:

> Development of an organism from a fertilized egg is driven primarily by the actions of regulatory proteins called transcription factors ... Rather, it is said, chemical modifications to DNA ... and to histones ... drive gene regulation. This obviously cannot be true because the enzymes that impose such modifications lack the essential specificity ... and so these enzymes would have no way, on their own, of specifying which genes to regulate under any given set of conditions.[375]

The latter point is correct, but Ptashne and many others overlooked the possibility that the specificity might be supplied by trans-acting RNAs, despite the fact that he had elsewhere recognized that RNA molecules can act as a transcriptional co-activators.[376] Of course, regulation of chromatin organization and transcription initiation is not mutually exclusive nor separable; the factors involved act in concert to govern the complex patterns of gene expression during development (Chapter 15).

Deciphering the histone code is a huge challenge, not the least because of the difficulty of analyzing the modifications and their effects on gene expression at the nucleosome level, the dependency of the context of the large combination possibilities of chromatin marks, and the heterogeneity of the samples. Nonetheless, the growing popularity of the field not only led to the rapid discovery of the many enzymes and complexes involved[377,378] but also the roles of modifications by a number of pioneering labs,[r] using *in vitro* approaches (e.g., with reconstituted nucleosomes) and modification-specific antibodies for global analysis of the *in vivo* distribution of nucleosomes containing the modification.[27,320,379–381]

The latter revealed non-random patterns of modifications in different tissues and developmental stages, such as in the *Neurod2* gene in the brain (Figure 14.5), hypoacetylation of the inactive X chromosome in female mammals and silent mating type genes in yeast, and hyperacetylation of the upregulated X chromosome in *Drosophila* males or transcribed globin genes in erythrocytes.[381]

High-resolution mapping by sequencing of immunoprecipitated chromatin ('ChIP-seq') has shown that histone modifications are differently imposed in complex patterns at millions of different genomic positions in different tissues or cell types at different stages of differentiation and development.[382–386] There is clearly also 'crosstalk' between histone modifications,[387–389] which may occur in modules, but little is yet understood of the lexicon or syntax.[384]

Active genes are characterized by acetylation of various lysines or arginines, which neutralizes their charge interactions and makes chromatin more accessible.[368,390] Different acetylations are found in different regions of genes and regulatory regions: H2AK9ac, H2BK5ac, H3K9ac, H3K18ac, H3K27ac, H3K36ac and H4K91ac are mainly located in the region surrounding the transcription start site, whereas H2BK12ac, H2BK20ac, H2BK120ac, H3K4ac, H4K5ac, H4K8ac, H4K12ac and H4K16ac are elevated in the promoter and transcribed regions of active genes.[384]

Nonetheless, even the roles of well-studied modifications, including acetylation, of different histones and residues by distinct complexes in different cell

---

r   Including those of Allis, Shelley Berger, Rudi Jaenisch, Thomas Jenuwein, Manolis Kellis, Tony Kouzarides, Bob Kingston, Danny Reinberg, Bing Ren, Bryan Turner, Rick Young, Jerry Workman, Shi Yang and many others.

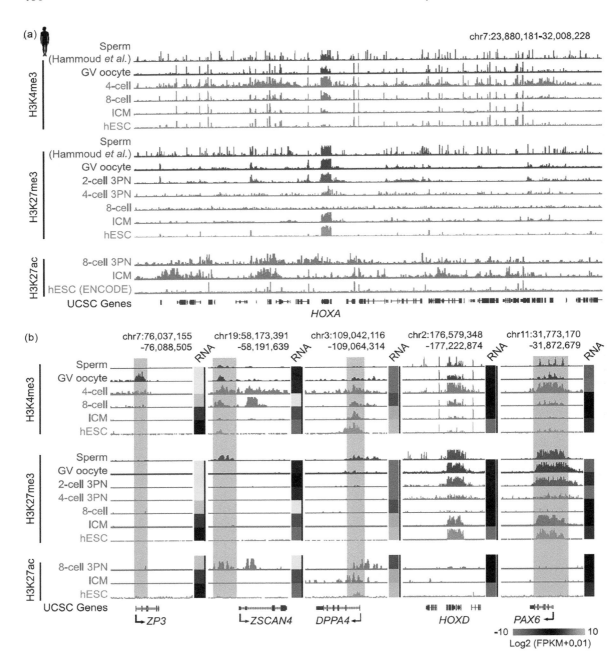

FIGURE 14.5 Dynamic landscape of histone modifications at the mouse *Neuro2d* (*Neuronal Differentiation 2*) locus in different tissues and during development. (Reproduced from ENCODE Project Consortium[369] under Creative Commons CC BY license.)

types and species are far from fully characterized. For example, in human cells the histone acetyltransferase KAT8[s] modifies different H4 residues (H4K5 and H4K8 vs H4K16) depending on its associated proteins, with different regulatory and pleiotropic effects.[394] H3K27ac marks are generally thought to distinguish active enhancers from inactive/poised enhancers that contain H3K4me1 alone,[167,395] although H3K27ac alone is insufficient to permit enhancer activity[396] (see below).

Conversely trimethylation of the same lysine (H3K27me3) marks facultative heterochromatin (regions that are differentially expressed in development and/or differentiation), such as the inactive X chromosome.[397] This modification is imposed by PRC2 through one of two alternative catalytic subunits, EZH1 or EZH2,[t] which are expressed at different stages of development.[398] Mutations in EZH2 cause Weaver Syndrome, which is characterized by skeletal and cognitive abnormalities.[399] Mutually exclusive acetylation and methylation also occur at other lysines including H2BK5, H3K4, H3K9 and H3K36, all of which are acetylated at active promoters.[384]

H3K27me3 has been widely implicated in restraining the expression of lineage-specifying and cell-state defining loci from plants to animals,[288,400–403] and mutation of this residue recapitulates PRC2 transformations.[404] Its role in regulating the timing of the differentiation of progenitor cells has also been linked with epigenetic switches controlled by opposing PRC2 and Kdm6a/b demethylase activities, for example in regulating T cell commitment timing in mammals.[405] H3K27me3 repression of gene expression also appears to be confined within TADs.[402,406]

There have been various attempts to use signature histone marks to identify enhancers, with initial correlations with the binding of the transcriptional

co-activator P300 suggesting that enhancers are characterized by the presence of monomethylated histone H3 lysine 4 (H3K4me1) and the absence of trimethylated H3K4 (H3K4me3).[170] However, subsequent studies showed that H3K4me3 is enriched, whereas H3K4me1 is reduced, in highly active enhancers,[169,407] and that characterized enhancer regions contain a variety of histone modifications in different combinations, not necessarily the presumed canonical H3K4me1 or H3K27ac marks.[171,384,408,409] Bioinformatic predictions of enhancers based on histone modification patterns alone have low validation rates.[155,407,410]

The H3K4me3 modification is not only associated with active enhancers but also with actively transcribed protein-coding genes,[411] or genes 'poised'[u] for transcriptional activation.[415–417] H3K4me3-modified histones exhibit a peak around transcriptional start sites[390] and interact with RNA polymerase subunit TFIID.[349,418–420] Transcription start sites also exhibit a typical flanking bimodal pattern of H3K4me2- and H3K4me3-marked nucleosomes.[420,421]

So-called 'bivalent domains' containing both H3K4 and H3K27 methylation occur around conserved non-coding sequences associated with developmentally important transcription factors, suggesting that chromatin state is important for maintaining embryonic pluripotency.[422,423] Recent data shows that pluripotent states are determined by interactions between chromatin modifications and enhancer expression to reconfigure the target specificity of the pioneer transcription factors Oct4, Sox2 and Nanog.[424] Other modifications, such as H4K16ac occur in active enhancers and protein-coding genes,[171] further obscuring the distinction between them.

H3K14pr and H3K14bu are (also) preferentially enriched at promoters of active genes[323] and H2AK119ub1 guides maternal inheritance and zygotic deposition of H3K27me3 in mouse embryos.[425,426] Histone H4 lysine 16 acetylation (H4K16ac), a hallmark of decondensed, transcriptionally permissive chromatin, directly stimulates the Dot1 histone H3K79 methyltransferase.[389] H3.3 variants are phosphorylated at S31 in gene bodies for high-level activation of rapidly induced genes, shown in macrophages to be coordinated with SETD2 methylation of H3K36

---

[s] KAT8, also known as MOF or MYST1, is classically associated with H4K16 acetylation in transcriptional activation, notably in the MSL complex that executes the roX RNA-directed global upregulation of the expression of the X-chromosome in *Drosophila* males for dosage compensation. Disruption of the orthologous human MSL complex also impairs H4K16 acetylation and results in an X-linked syndrome marked by developmental delay, gait disturbance and facial dysmorphism,[391] as well as tumor maintenance by exacerbating chromosomal instability,[392,393] the latter exemplifying that histone modifications have other roles in chromosome biology beyond the regulation of gene expression.

[t] EZH1 and EZH2 contain the lysine-specific SET (Su(var)3-9, Enhancer of Zeste, Trithorax) domain that uses the cofactor S-adenosyl-L-methionine (SAM) as the methyl donor.

[u] It is also clear that transcriptional 'pausing' and modulation of elongation rates plays an important role in the dynamic control of gene expression, including splicing.[79,218,412–414]

to effect recruitment and ejection of chromatin regulators.[427]

Constitutive heterochromatin in genomic regions such as the centromeres and telomeres contain high levels of H3K20me3 and H3K9me3,[398,417] the latter of which binds the repressive HP1 protein via its chromodomain.[348,428] H3K4me3 and H3K9me3 mark imprinting control regions.[417] Histone sumoylation appears to act as a repressive mark by recruiting HDACs to gene promoters,[429] and H3K36me is present in nucleosomes along the body of transcribed genes, and is necessary for efficient constitutive pre-mRNA splicing by recruiting the chromo domain protein Eaf3 to mediate interaction with the splicing machinery.[430,431]

These are just some examples. The patterns are complex and studies are becoming more sophisticated.[369] Targeted deposition or removal of histone modifications using CRISPR/Cas9-fusions and related approaches, such as single-cell CRISPR screens and chromatin modification profiling by mass spectrometry, are starting to allow the dissection of causal roles for individual modifications.[432–439]

An important discovery was that nucleosomes are preferentially positioned over exons,[440–443] suggesting that histone modifications convey exon-specific information, and that epigenetic control of gene expression extends to the level of individual exons. This offers a mechanistic explanation for the observed coupling of chromatin structure, transcription and splicing,[444] including the physical co-location of alternatively spliced exons with promoters,[445] and a basis for exon selection by histone modifications at different stages of development in different cell types and different conditions,[443,446–453] which appears to be controlled in part by small RNAs.[454–456]

Chromatin-modifying proteins have a profound impact on developmental processes because they lie at the functional center of epigenetic regulatory networks. They do not make (although they do convey) locus-specific regulatory decisions but rather are directed by other information that does. How histone modification writers and erasers select particular nucleosomes at particular genomic positions for particular modifications in different cell types is unknown, but is likely RNA guided (Chapter 16). The histone modifications and nucleosome positioning must be tightly controlled during development, as developmental trajectories are precise (Chapter 15), although histone modifications are also influenced by metabolic and physiological factors.[457–460] Moreover,

histone-modifying proteins are themselves subject to post-translational modifications.[461,462] which suggests yet more layers of developmental control and environmental tuning.

Histone modifications are often inherited through meiosis and mitosis, to transmit information between generations[425,426,463,464] (Chapter 17) and to 'bookmark' loci for reactivation or maintenance of heterochromatin after cell division.[465–467] The available evidence is that the parental core H3-H4 tetramer is split and segregated strand-specifically at the replication fork, that parental histones are recycled to sister chromatids and re-incorporated near their original positions, maintaining their acetylation and methylation marks, possibly asymmetrically to alter cell fate in daughter cells.[468,469]

Replication timing maintains the global epigenetic state in human cells.[470] It is still unclear, however, how the histone modifications are inherited, particularly in view of the report that epigenetic memory is independent of symmetric histone inheritance replication,[471] although histone-modifying enzymes remain associated with DNA during replication.[472,473] Epigenetic marks are also erased and reset with every round of transcription (which involves similar disassembly of or navigation through nucleosomes),[469,473–476] again possibly involving RNA direction.[234,353,477]

The imposition and maintenance of this information clearly involves histone modification writers but this does not explain their locus specificity, which probably operates to exon level. Whatever the mechanism(s), the amount of information involved, and stored in the genome, must be enormous.

DNA METHYLATION

All four bases of DNA are subject to modifications, more than 20 of which have been identified,[478,479] the most common being methylation of cytosines and adenosines. In bacteria,[v] DNA methylation (mainly m[6]A) is used to protect endogenous sequences against restriction endonuclease cleavage (Chapter 6), but also has roles in DNA replication and gene expression.[481–484] Adenosine methylation has been reported in protists, plants and animals,[485–492] although emerging evidence suggests that the source may be microbial or RNA contamination.[493,494]

---

[v] Bacterial DNA can also be modified by phosphorothioation, as part of an alternative restriction-modification defense system.[480]

5-Methylcytosine (m^5C) occurs widely in eukaryotic genomes and is the best studied. In fungi and plants, cytosine methylation is used to silence viruses and transposons, as well as to regulate development[495–497] and environmental responses,[498,499] likely related processes. In maize, the cycling of transposable elements between active and inactive states to regulate local gene expression is determined by the methylation state of the element,[495,500,501] which may also be in part the role of TEs in animals.

Most invertebrate genomes are not heavily methylated and some species such as *Drosophila* and *C. elegans* appear to lack DNA methylation, indicating that it does not play a role in their development, although it is used for genome defense and gene regulation in other invertebrates.[502,503] For as yet unknown reasons, a major evolutionary transition from fractional to global methylation occurred at the origin of the vertebrates,[504] as did the appearance of regional variation in GC content.[103]

DNA methylation as a major player in gene regulation in mammals first came to light in the 1970s with the observation that there is differential methylation of the mammalian X chromosomes.[505,506] Later studies showed that there is widespread erasure of methylation in the mammalian germ line[w] and in early development,[513–516] selective reimposition of methylation at different loci including enhancers in different cell lineages,[497,516–518] and reactivation of genes (and induction of tumors) by a cytosine analog (5-aza-cytidine) that cannot be methylated.[519–521] Embryonic stem cells maintain their pluripotent state in the absence of DNA methylation, but cannot differentiate.[496]

In mammals, methylation is primarily, but not solely, associated with repression of gene expression, notably in inactivated X chromosomes,[x] pericentromeric heterochromatin, imprinted loci and the regulation of transposons (which are related, as most of the targets of methylation are TE-derived) and occurs mostly and symmetrically in cytosines in CpG dinucleotides, except those clustered in so-called 'CpG islands'.[523–528] Deamination of methylcytosine

yields thymidine, which is thought to account for the underrepresentation of CpG dinucleotides in mammalian genomes.[523,529]

The sequence symmetry of CpG enables propagation of the methylation mark through cell division, which combined with its complex interplay with Polycomb repressive and other histone-modifying complexes[530,531] and its differential patterns during development, led to the proposal that CpG methylation comprises a pathway for cellular memory of transcriptional states.[505,506]

CpG islands occur mainly in promoters (including those of enhancers[219]), especially those of broadly expressed housekeeping genes,[529,532,533] methylation of which correlates negatively with gene expression, although repression of these promoters appears to occur primarily by H3K27me3 histone modifications.[496,531,534] Genes with CpG island promoters also have other characteristic epigenetic signatures, including high levels of H4K20me1, H2BK5me1 and H3K79me1/2/3 at their 5′ end.[535]

In contrast, tissue-specific protein-coding genes usually, but not always, lack islands.[529,532,533] In transcriptionally active genes, CpG islands are devoid of methylation and enriched for permissive nucleosome modifications such as H3K4 methylation. On the other hand, DNA methylation is enriched in the body of highly transcribed genes, often associated with H3K36 methylation,[430,535–538] where it influences nucleosome positioning[539] and alternative splicing,[540] phenomena that may be linked. It has been recently shown that hypomethylated CpG dinucleotides preserve an archive of tissue-specific developmental enhancers in adult mouse cells, marking decommissioned sites and enabling recovery of epigenetic memory,[541] a process involving the pioneering factor FoxA and TET2/3 methylcytosine dioxygenases[542] (see below).

Cytosine methylation is carried out by DNA methyltransferases, of which vertebrates have three: two 'establishment' DNA methylases (Dnmt3a and 3b), and one 'maintenance' methylase (Dnmt1) that recognizes hemi-methylated CpG sites following DNA replication. All three are required for embryonic development, with mutations causing syndromic developmental neurological, sensory and immunological defects, and loss of Dnmt1 in neurons at later stages resulting in cognitive defects.[361,543–545] The histone mark H3K36me2 also recruits Dnmt3a to regulate intergenic DNA methylation[537] and H3K23 ubiquitylation couples

---

[w] In zebrafish, the methylome is erased in oocytes but not in sperm,[507] and the methylome pattern is reconstituted in the zygote (apparently) to match the paternal pattern.[508] Thereafter, methylation appears to be constitutive throughout development,[509–511] as it is also in Xenopus.[512]

[x] The spreading of X-inactivation from the 'X-inactivation center' on the X chromosome in females appears to be mediated by methylation of LINE elements distributed along the chromosome.[522]

maintenance DNA methylation with replication.[546] While DNA methylation is thought to be stable, it is cycled at promoters at high frequency, suggesting an updating mechanism.[547,548]

The methyl-CpG-binding protein MeCP2 links DNA methylation to histone methylation,[530] and is essential for brain development and function.[549,550] Loss of MeCP2, which is encoded on the X chromosome, causes a neurological disorder called Rett Syndrome with variable penetrance in females (due to variable patterns of X inactivation) whereas its loss in males usually leads to severe congenital encephalopathies and early death.[550] Dnmt3a and MeCP2 originated at the onset of vertebrates, with methylation of non-CpG sites being exceptionally high in the mammalian brain and regulating highly conserved developmental genes, with a likely role in the evolution of cognition.[551]

Dnmt2 was originally thought to be a DNA methyltransferase but is, in fact, a tRNA methyltransferase, and it seems likely that the modern DNA methyltransferases evolved from an ancient RNA methyltransferase.[552–554]

Methylcytosine is converted to hydroxymethylcytosine (hmC) by TET proteins,[555] which can also further oxidize hmC to generate 5-formylcytosine and 5-carboxylcytosine.[556] There are three TET proteins in mammals with different expression patterns and different targets during development.[557,558] TET proteins hydroxymethylate DNA at enhancers and telomeres,[559] and TET1 and TET2 associate with Nanog to facilitate reprogramming of somatic cells to pluripotency.[560–562] Formation of 5-hMC is required in embryonal stem cells for the maintenance of pluripotency and inner cell mass specification.[557] It is also required in the brain, especially in Purkinje cells, where it is almost 40% as abundant as meC.[563] TET3 is present in neurons and oligodendrocytes but absent in astrocytes.[564] TET3 regulates behavioral adaptation in the neocortex,[565] as well as synaptic transmission and plasticity in the hippocampus,[566] and its loss results in increased anxiety-like behavior and impaired spatial orientation.[564] Fear extinction, an important form of reversal learning, leads to a dramatic genome-wide redistribution of 5-hmC within the infralimbic prefrontal cortex, and learning-induced accumulation of 5-hmC is associated with the establishment of epigenetic states that promote gene expression and rapid behavioral adaptation.[565]

DNA methylation patterns have been extensively analyzed following the discovery by Marianne Frommer and colleagues that bisulfite treatment of DNA converts cytosine, but not meC, to uracil, which sequences as T,[567,568] and more recently by direct DNA sequencing using nanopore technology, which can distinguish modified from unmodified bases.[569,570] For this reason (technical ease of analysis) and its earlier discovery, DNA methylation has been more widely studied than histone modifications, notably in the 'Human Epigenome Project', which revealed differences in methylation patterns in different cell types and interplay between genetic variations and epigenetic state during development and aging,[y] in the brain, in cancer and other diseases such as arthritis[226,545,572–575] (Figure 14.6). Akin to the insights now routinely offered by RNA-seq, new techniques will increasingly reveal the variety and dynamics of epigenetic states and transcription factor occupancy at single-cell resolution during development and in diseases such as cancer.[576,577]

However, as with histone modifications, there is little known about the signaling pathways that direct the locus-specific imposition or removal of cytosine methylation by generic enzymes during development, learning and disease, except that the RNAi pathway is involved (Chapter 16).

THE REGULATION OF DEVELOPMENT

The roles of the (admittedly at the time vague) organization of chromatin in the regulation of gene expression, and the mechanisms that might be involved, were rarely considered when the bacterial model was extrapolated to developmentally complex eukaryotes. Accordingly, since then, the regulation of gene activity by chromatin architecture has been viewed predominantly through the lens of DNA-binding transcription factors.

This interpretative lens led to the loose and confusing description of many proteins that are required to mediate the patterns of gene expression during development, such as those that organize chromosomal domains or modify chromatin, as 'pioneer transcription factors' or 'transcriptional

---

[y] The link between genetic variations and epigenetic state of regulatory elements affecting gene and trait expression is illustrated by the classic example of lactase non-persistence in mammals and the selection for lactase persistence during aging observed in many Europeans and other pastoral cultures, which involves non-coding variations, a specific lncRNA (LOC100507600 or Lactase antisense RNA 1), RNA interference and DNA methylation in intronic enhancers.[435,571]

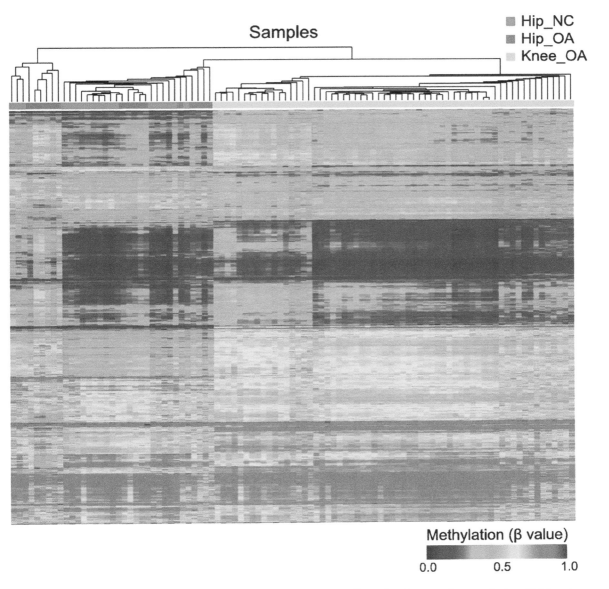

FIGURE 14.6 Aberrant methylation patterns in enhancer loci in cartilage chondrocytes from patients with hip osteo-arthritis (OA) and knee OA, compared to healthy controls (NC). (Reproduced with permission from Lin et al.[575] under Creative Commons CC BY license.)

co-activators',[304,578–581] despite the fact that they have no intrinsic or only vague DNA-binding specificity (Chapter 16). The implied assumption biased the interpretations of experimental observations and retarded the understanding of the control of gene expression during development by placing proteins that are required to instruct genome architecture into the same conceptual and mechanistic basket as those that bind to specific sequences to activate or inhibit transcriptional initiation.

Appreciating that chromatin modifications play a central role in the regulation of gene expression during development has also been confused by the term 'epigenetic inheritance', implying that it is separate from 'genetic' (DNA-based) inheritance, obscuring the fact that the unfolding cascade of epigenetic modifications must be instructed by information that is encoded in the genome.

The key challenges are to consider how much information is required to orchestrate organismal ontogeny (Chapter 15) and to identify the pathways that connect chromatin modifications, enhancers, effector proteins and other layers of genome regulation during ontogeny (Chapter 16). How this information is modulated by the environment and during learning is addressed in Chapter 17.

FURTHER READING

Allis C.D., Caparros M.-L., Jenuwein T., and Reinberg D., eds. (2015) *Epigenetics*, second edition (Cold Spring Harbor, NY: Cold Spring Harbor Laboratory Press).

Cremer T. and Cremer C. (2001) Chromosome territories, nuclear architecture and gene regulation in mammalian cells. *Nature Reviews Genetics* 2: 292–301.

Halfon M.S. (2019) Studying transcriptional enhancers: The founder fallacy, validation creep, and other biases. *Trends in Genetics* 35: 93–103.

Olins D.E. and Olins A.L. (2003) Chromatin history: Our view from the bridge. *Nature Reviews Molecular Cell Biology* 4: 809–14.

Waddington C.H. (1966) *Principles of Development and Differentiation* (Macmillan, London)

15 The Programming of Development

AUTOPOIESIS

The word 'autopoiesis' was coined in 1972 by Humberto Maturana and Francisco Varela from the Greek αὐτο (auto-, meaning 'self') and ποίησις (poiesis, meaning 'creation') to refer to the property, ability and process of self-replication,[1] the defining feature of life, the unit of which is the cell.

Cellular life on Earth is estimated to date back at least 3.7 billion years,[2] not all that long after the crust cooled sufficiently for water to condense – suggesting that either self-replicating entities can evolve quickly in an aqueous environment or, as proposed by the British astronomers Fred Hoyle and Chandra Wickramasinghe, that life evolved elsewhere and arrived on dust particles from space, taking hold when the conditions were right.[3]

As described in Chapter 4, developmentally complex eukaryotes appeared and evolved over the past billion years, first documented in metazoans as trace fossils of sponges around 900 million years ago, then segmented wormlike creatures in the Ediacaran Period around 600 million years ago. Phenotypic diversity exploded 520 million years ago in the Cambrian when recognizable representatives of all modern phyla appear almost 'overnight', with rapid secondary radiations also occurring after subsequent major extinction events. Design variations were explored using a relatively stable proteome, albeit not without some innovations and expansions of particular gene families. While it is widely thought that the greater energy capacity enabled by the photosynthetic production of oxygen and electron transport in mitochondria was an enabling step, or at least a precondition, it is likely that new mechanisms had to evolve to manage cellular growth, organization and differentiation into an integrated three-dimensional assemblage where each cell has a specialized function.

THE OVERARCHING QUESTION

Evidently there must be differences in the decisional systems that control spatially organized cell division and specific commands to exit division and enter differentiation in multicellular organisms, in contrast to simple microbial growth and reproduction allowed by nutrient supply.[a] These differences must also be reflected in the structure of the information held in the genomes of plants and animals, which show some universal features: the use of 'enhancers' to direct gene expression patterns during development; the organization of the genome into chromatin territories, topological domains and nucleosomes; the post-translational modification of histones and many other proteins (Chapters 14 and 16); large numbers of 'repetitive' sequences derived from transposable elements that drive phenotypic innovations in, e.g., vertebrate,[5,6] tetrapod,[7] mammalian[8,9] and primate[b] evolution;[10,11] the partitioning of protein-coding and non-protein-coding genes into exons and introns (Chapter 7); and the increase in the numbers of small and large regulatory RNAs with developmental complexity (Chapters 12 and 13).

Why are there so many histone variants and modifications and how do chromatin marks relate to the expression and function of 'genes' and 'enhancers'? How is site selection achieved in the deployment of these variants and modifications at millions of locations around the genome during development? What is the role of different types of repetitive elements? Why are there large numbers of small RNAs and lncRNAs expressed in highly cell-specific patterns? Are these phenomena intertwined?

These questions are subsets of the overarching question of the nature and scaling of the information and decisional hierarchies that direct multicellular development, especially since it is the non-coding, not the coding, portion of the genome that expands with increasing developmental and cognitive complexity. How much information, then, is required to program human development, how is it encoded and how is it manifested?

Consideration of these questions, and even awareness of their importance to the understanding of

---

[a] Genome engineering has created a bacterial cell with only 473 protein-coding genes, the smallest genome of any free-living organism. Many of these essential genes have unknown functions.[4]

[b] 63% of the primate-specific hypersensitive regions (open chromatin associated with promoters) are occupied by TE remnants.[10]

DOI: 10.1201/9781003109242-15

molecular biology and genome biology, has been given scant attention in the face of the emphasis on 'genes' and proteins, the widespread assumption that RNA functions purely a link between the two, and the focus of most biomedical research on biochemistry, physiology and associated diseases. However, the specification of metabolic and physiological pathways surely comprises a minority of the genomic information in developmentally complex organisms. Most of the human genome – perhaps not unreasonably 99% – must be devoted to the regulation of development – generating ~30–40 trillion cells[12,13] with highly specific arrangements characteristics and connections in an adult capable of reproduction and parenting. In mammals, the developmental symphony is largely conducted *in utero* and largely completed at puberty, with only reproduction, menopause and senescence to follow, and physiological homeostasis in between.

A compounding problem is that most studies of the regulation of gene expression in humans and other mammals have been carried out with cultured cells, often transformed or derived from tumors, and maintained in artificial environments divorced from developmental processes. This is, however, changing with the arrival of increasingly sophisticated systems for *in vitro* embryonic development beyond gastrulation[14–18] involving four-dimensional organoid cultures[19,20] capable of forming (albeit imperfectly) almost any type of complex tissue[c] – such as lung,[22] liver,[23] intestine and colon,[24,25] kidney,[26] retina[27] and brain[28–33] – with well-defined signaling environments and development recapitulation protocols, as well as strategies for 'program unlocking' with cell de-differentiation, differentiation and trans-differentiation. Cerebral organoids derived from human, gorilla and chimpanzee cells have been used to study developmental mechanisms[d] driving brain expansion,[34] although there are still major limitations to these systems and barriers to be overcome.[42]

TISSUE ARCHITECTURE AND CELL IDENTITY

It is commonly asserted that humans have ~200 different cell 'types' – muscle cells, bone cells (osteoblasts and osteoclasts), neurons, astrocytes, keratinocytes, fibroblasts, hepatocytes, etc. (see, e.g., [43]). However, this a gross underestimate of the range of spatial identities of cells that follow different and highly precise trajectories during development, as Waddington envisaged.[44] The differential expression of *Hox* genes gives some insight into the positional specificity of superficially similar cells: there are 118 subtypes of neurons in *C. elegans* (which has only 308 neurons), each of which expresses a different combination of *Hox* genes.[45] Human skin fibroblasts express different *HOX* genes in different parts of the body, reflecting their "positional identities relative to major anatomic axes",[46] which are cell autonomous and epigenetically controlled.[47,48]

Although the transcriptional profile of only a small fraction has been examined, the increasing resolution of single-cell RNA sequencing and chromatin analyses is confirming the large range of cell states and identities at different developmental stages,[e] and in different regions of the brain,[52–56] revealing the differential expression not only of transcription factors, Hox genes and other regulatory proteins,[57,58] but also the highly restricted expression of enhancers,[59] lncRNAs and TE-derived RNAs.[57,60–68]

An example is the expression of the imprinted maternal allele-expressed lncRNA Meg3/Gtl2 during the embryonic development of the inner ear, which is "localized to the spiral ganglion, stria vascularis, Reissner's membrane, and greater epithelial ridge in the cochlear duct", leading to the conclusion that "Meg3/Gtl2 RNA functions as a noncoding regulatory RNA … that plays a role in pattern specification and differentiation of cells during otocyst development, as well as in the maintenance of a number of terminally differentiated cochlear cell types".[69]

Human cortical pyramidal neurons exhibit heterogeneity in their electrophysiological properties.[70] The liver, kidney, lungs, cortex, hippocampus, eyes, tongue, ovaries and testes all have different but specific designs. Each muscle is comprised of

---

[c] With ethical implications and limitations increasingly more prominent than technical constraints.[21]

[d] This study found that ZEB2 is an evolutionary regulator of the key delayed morphological transition with shorter cell cycles underlying human brain organoid expansion.[34] ZEB2 is a highly conserved 'master transcription factor' involved in the epithelial-mesenchymal transition (EMT), which has been under positive selection pressure in the primate lineage[34] and is regulated by non-coding RNAs,[35–37] as are other aspects of EMT[38–41] (Chapter 16).

---

[e] Single-cell RNA sequencing is also allowing the investigation of the specialization and divergence of cell types across a range of evolutionary distances,[49,50] revealing greater complexity of gene expression and regulation in humans relative to mouse.[51]

similar cell types, but each has a different architecture (Figure 15.1), in which the component cells must have highly precise positional placement and identity.

Bones in particular, while comprised of the same types of cells, exhibit extraordinary diversity and precision of structure and architecture. Each vertebra in the spine has a distinct and characteristic shape with complex grooves and projections. The long bones in ribs, arms, hands, legs and feet, the short cuboidal bones in the wrist and ankle, the flat but still precisely curved bones of the face, cranium, shoulder blades, sternum and pelvis, and the exquisitely fine ossicles of the inner ear, all have specific irregular shapes. Moreover, different bones have different internal structures – dense lamellae, honey-combed trabeculae, or marrow – adapted to different weight-bearing, torsional or other functions, with a legion of tendons and muscles to match.

Equivalent bones in other species are different from those humans, but just as idiosyncratically and precisely sculpted (Figure 15.2). The precision of this architecture is likely to require much more, and far more precise, information than simply the differential expression of *Hox* proteins or relatively generic transcription factors (TFs) such as the muscle specific MyoD (see below).

The challenge for mammalian development is to direct the ontogeny, architecture, arrangement, and interconnections of each of the bones, muscles, and organs from a single fertilized cell to a functional adult. Cell state and fate has to be computed and specified by the genome at every stage, tuned by cell-cell interactions and feedback loops (involving intercellular ligand-receptor signaling,[71,72] cell contact area detection[73] and other forms of communication, including bioelectric circuits and mechanical

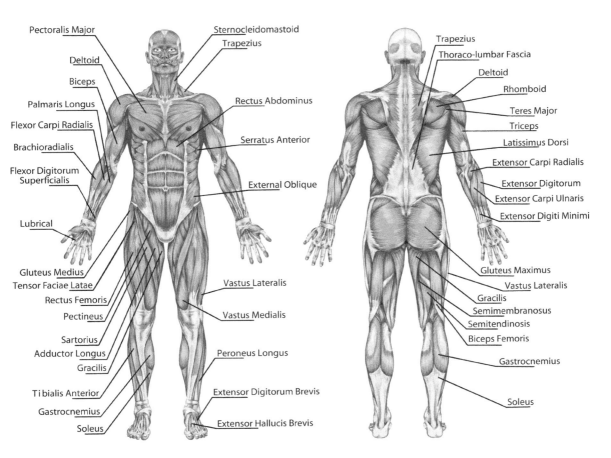

FIGURE 15.1   The exquisite patterning of human muscles (Image: Shutterstock).

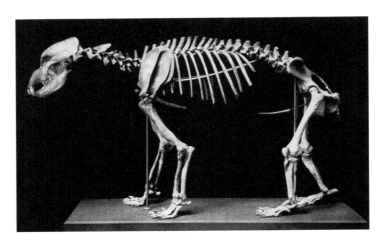

FIGURE 15.2   The skeleton of the prehistoric cave bear, showing the unique architecture of each bone. (Image from the 1906 Annual Report of the Director to the Board of Trustees of the Field Museum of Natural History, Chicago.)

forces[74–76]), without which the endogenous program would be degraded by stochastic errors.[f]

PROGRAMMED ONTOGENY

While embryogenesis has been studied in many animals, in only one has the ontogeny of all cells in the adult been characterized to date.[g] This is the nematode *C. elegans*, where John Sulston and Robert Horvitz determined the progression from the fertilized egg to the 959 and 1,031 cells that comprise the adult male and female hermaphrodite, respectively[79–82] (Figure 15.3). The results show that the ontogeny of *C. elegans* is tightly programmed and (mutations aside) invariant, which is required as the positions and capacities of every cell are critical to the functioning of the organism.

Imagine the complexity of the equivalent graph for a human.

Programmed ontogeny occurs with high precision in all animals, including humans, evidenced not only by the architectural reproducibility of the myriad of muscles, bones and organs but also by the fact that monozygotic twins are essentially phenocopies (Figure 15.4), with the same body and facial shape, dimples and hairlines, etc., notwithstanding a small number of post-zygotic mutations[83] and context-dependent clonal amplifications of such cells as adipocytes or lymphocytes.[h] The ontogeny of a human must be as precise and hard-wired, or nearly so, as that of the worm, even if orders of magnitude more complex and multi-layered over many more cell divisions, which implies orders of magnitude more regulatory information. The same must apply to carp, axolotls, crocodiles, iguanas, chickens, emus, eagles, echidnas, dogs, horses, whales, etc., each with idiosyncratic ontogeny graphs.

The phenotypic reproducibility means that there may be as many as ~10^{13} distinct cell identities (as opposed to cell 'types') in a human,[12,13] and ~10^{13} leaves (Waddington valleys) in the ontogeny graph, compared to 10^3 in *C. elegans*. Every decision to divide or differentiate must be calculated with high precision.[i] Developmental reproducibility also shows that, if there is any 'noise' in gene expression, as often claimed, such noise does not have significant impact on phenotype.

---

[f] A thought experiment is to imagine the difference between a robot programmed to build a motor vehicle that has no feedback or sensory mechanisms, versus one that does and can refine its actions accordingly. This explains why disruption of cell-cell signaling leads to aberrant development, why morphogenetic gradients of diffusible factors are employed and why sequential expression of Hox proteins is used to specify positional identity,[77] but does not mean that positional information conveyed by cell-cell communication is the main mechanism of developmental control, as is often assumed.[78]

[g] Cell divisions in the early stages of embryogenesis (up to gastrulation) have been documented in the ascidian *Phallusia mammillata*.[57,73]

[h] The phenotypic diversity of humans (leaving aside body weight) must therefore be a function of variations in genome sequences, which, on average, differ by 0.1%.[84]

[i] Embryonic ontogeny requires establishment of asymmetry in the zygote, which is achieved in different ways in different lineages[85–87] and asymmetry of cell division generally.[88]

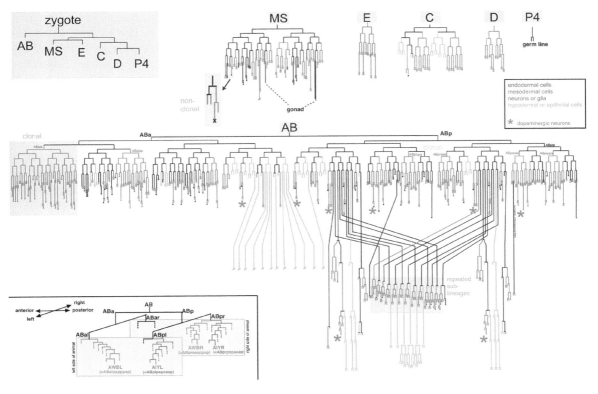

FIGURE 15.3   The invariant ontogeny of *C. elegans*. (Reproduced from Hobert[82] under Creative Commons license.)

FIGURE 15.4   The precision of human ontogeny: monozygotic twins. (Images: Shutterstock.)

Indeed, it is likely that the systems that control development have evolved to suppress noise,[89-91] as happened in the transition from analog to digital computers.[92] Developmental programming must also be extremely robust, to ensure that enough progeny develop to reproductive capacity, especially in birds and mammals, which have limited numbers of offspring. Indeed such robustness appears, in part, to be provided by conservation and resilience of TF networks[93,94] as well as by miRNAs and siRNAs[95-97] and 'shadow' enhancers,[98-102] which provide yet more examples of the non-random exaptation and non-neutral evolution of transposable elements.[60,103,104]

Therefore, most of the genomic information required for development is that which orchestrates cell lineages, cell fates and tissue architectures: the feed-forward programming of decisions at each stage that specify when and in what plane cells divide, the developmental trajectory it is in (for example, mesoderm or ectodermal cell types) and when it ceases to divide and enters terminal differentiation to form a myoblast, fibroblast, osteocyte or a neuron. In addition to cell fate trajectories there must be other information that determines how a cell is positioned in relation to adjacent cells and, in the nervous system, with which other cells synaptic connections are made.

The organizing centers of cell division in animals are the specialized ribonucleoprotein organelles called centrosomes (see below). They are associated with the eukaryotic centromere, a (usually) central region of highly specialized chromatin that are also associated with internal granules called 'centrioles', which attach to kinetochores for spindle formation and chromatid pairing and separation to daughter cells during mitosis and meiosis.[105] Centromeres from phase-separated domains[106] comprised of long tandem arrays derived from retrotransposons,[107] contain specialized histones, express non-coding RNAs, and are epigenetically controlled by complex networks of non-coding RNAs and the RNA interference pathway.[108-123] In animal cells, centromeres are attached to the centrosomes, which are not essential for spindle formation or chromosome segregation but, importantly, contain cell cycle regulators and signaling molecules, many of which are post-translationally modified in 'intrinsically disordered regions' (Chapter 16). Centrosomes are connected to and direct the movements of microtubules and other cytoskeletal structures and proteins, including primary cilia,[72] to control the spatial organization of cells, cell division, cell polarity and cell migration, with developmental and neurological consequences.[124-126]

Centrosomes are essential for the spatially precise execution of cell division. The animal centrosome evolved with the transition to complex multicellularity, as a hybrid organelle with the ability to act as a plasma membrane-associated primary cilium organizer and a juxtanuclear microtubule-organizing center, enabling the connection between and integration of extracellular and intracellular signals with cell-autonomous (developmentally programmed) information.[127]

Plants and animals also have many different types of TFs that act to license transcription during development, unlike bacteria (and unicellular eukaryotes) where TFs mostly regulate genes that are expressed under particular environmental conditions, such as the *lac* operon. Some eukaryotic TFs are undoubtedly required for gene expression responses to physiological and environmental variables, but that is only a minor function in developmentally complex organisms, and it is likely that the mechanisms of transcriptional control during animal and plant development are fundamentally different from those in bacteria.[128,129]

Cell differentiation is readily accomplished by turning on stage-specific (stem cell, mesoderm, genital ridge, neural crest, muscle, etc.) TFs to express the repertoire of proteins required for the function of that cell type – myosins in muscle, keratins in skin, synaptic receptors in neurons, etc. The differentiation state of cells can also be altered, and developmental programs overruled or reversed in culture, by ectopic expression of master or 'pioneer' TFs, such as the conversion of fibroblasts to myoblasts (muscle cells) by MyoD[130] or to stem cells[j] by the reprogramming 'Yamanaka factors', Oct3/4, Sox2, Klf4 and c-Myc.[132,133]

While this is superficially taken to indicate the primacy of transcription factors, MyoD and Yamanaka factors have little or no DNA-binding specificity, but rather change the differentiation state or reactivate the pluripotency program[k] by opening

---

[j] Interestingly, reprogramming of fibroblasts to stem cells is also modulated and (less efficiently) achieved by non-coding RNAs, presumably related to their role in epigenetic remodeling.[131]

[k] Which also occurs at different points during development *in vivo*, such as during the formation of the cranial neural crest.[134]

chromatin and/or recruiting chromatin modifying proteins,[135–139] whereas developmentally precise anatomical patterns and cell-type specification are dictated *in vivo* by chromatin state, enhancers and regulatory RNAs[140–142] (Chapter 16). A compelling example is the patterning of the human face and brain.[143–147]

Plant development also needs to be programmed but, while exhibiting remarkable diversity and versatility, is not and does not need to be as precise as that in animals, which have strict design requirements for their mobility and (consequently) rapid signal processing (i.e., cognitive ability). By contrast, plant development must have flexibility to adapt its formation to a constrained location and situation, such as available light and other environmental challenges, which may account for its more relaxed use of retrotransposons and polyploidy.[148–150]

LINEAGE SPECIFICATION

The systems required for lineage specification began to appear in single-celled eukaryotes and basal multicellular eukaryotes, wherein non-coding RNAs were exapted to control cell fate decisions to generate specialized forms in organisms such as yeasts[151–155] and fungal pathogens,[156,157] the social amoeba *Dictyostelium discoideum*,[158,159] the ciliate *Oxytricha trifallax*[160–162] and protozoan parasites,[163] the latter notable for their paucity of conventional transcription factors.[164]

The life cycle transitions of *Capsaspora owczarzaki*, a unicellular relative of animals with the largest known protein-coding gene repertoire for transcriptional regulation, are associated with changing chromatin states, differential lncRNA expression and transcription factor network interconnections.[165] *Capsaspora*, however, lacks animal promoter types and its cis-regulatory sites are small, proximal and lack signatures of animal enhancers, indicating "that the emergence of animal multicellularity was linked to a major shift in genome cis-regulatory complexity, most notably the appearance of distal enhancer regulation".[165] Indeed enhancers appear at the dawn of animal evolution,[166–168] and enhancers and complex epigenetic regulation of chromatin structure are the defining features of developmentally complex organisms, along with, and likely driven by, the widespread exaptation of transposon-derived sequences.

The foundational transition in animal development[l] is the maternal-zygotic transition (MZT),[175,176] wherein the ovum (by far the largest cell produced in most organisms) supports the growing embryo after fertilization by providing the resources required, including the initial signals derived from localized maternal proteins and transcripts[177,178] and some sperm RNAs.[179,180] At all of the early stages, miRNAs and lncRNAs (in particular TE-derived RNAs) play an important role, from the splitting of the zygote into two-cell and subsequent partitioning into more cells, forming the morula, blastocyst and the differentiation of the ESCs in the inner cell mass into different lineages.[176] Notably, the maternally loaded transcriptome shows more variability across species than that expressed by the zygote during its middle developmental stages, and the maternal transcripts that are degraded during the MZT are less conserved than those that remain after the onset of ZGA, supporting the 'hourglass model'[m] of early development.[176,181–185] The hourglass model is also supported by the temporal expression patterns of miRNAs in *Drosophila*.[185]

HOW MUCH INFORMATION IS REQUIRED?

A world away, the Boolean network models of gene regulation put forward by Stuart Kauffman (Chapter 5), based on the control of gene expression in bacteria, suggested that combinatorial control by 'transcription factors' and other regulatory proteins would suffice to regulate complex developmental processes.[186–192] This was supported by Eric Davidson's analysis of transcription factor expression and gene regulatory networks (GRNs) in sea urchin development,[193–196] as well as by others,[191] for example, in the regulation of *Drosophila* body plan,[197] plant flower development[198] and the yeast cell cycle.[199,200]

---

[l] Another important aspect of developmental programming and morphological innovation is the concept of 'heterochrony', modulation of developmental timing, in which regulation by non-coding RNAs figures prominently.[169–174]

[m] The hourglass model of embryonic evolution predicts an hourglass-like divergence of gene expression during animal embryogenesis – with embryos being more divergent at the earliest and latest stages of development but conserved during a mid-embryonic (phylotypic) period that serves as a source of the basic body plan within a phylum.[181,182]

In bacteria, most transcription factors are repressors[201] and Boolean logic gates such as AND, OR and NAND can be accomplished at promoters by simple combinations of interactions between just two transcription factors,[202,203] such as observed in the *lac* operon, although it is possible to have more than two transcription factors and associated binding sites that act independently on an adjacent promoter.[202] Consequently, a common presumption has been that promoters in complex organisms can be targeted by multiple transcription factors (and it is clear that this is the case[204–207]), and that an individual transcription factor can address many promoters differentially in different cells. How such discrimination is achieved is unclear, but posited to be enabled by site "accessibility",[206,208–210] presumably a function of epigenetic processes controlling chromatin structure.[209] Similarly, particular miRNAs can target the 3'UTRs of many mRNAs and many mRNAs can be targeted by many different miRNAs, although the decisional logic of such multilateral networks is unclear.

The theoretical foundations of GRNs can be traced to Ptashne's studies in the 1960s of the bistable switch that controls the lysogenic and lytic states of lambda phage in *E. coli*.[211–214] The underlying assumption in GRN models of development is that transcription factors play largely invariant roles in the GRNs in which they function.[213]

This assumption can accommodate alternative splice variants for increased complexity but does not accommodate the fact that most eukaryotic transcription factors and other proteins involved in developmental regulation contain extensive intrinsically disordered regions, which are major sites of post-translational modifications[215–217] (Chapter 16). The presence of IDRs in eukaryotic regulatory proteins confers "promiscuity",[215,218,219] i.e., flexibility in binding partners, and implies, as does the cell-type variation in their binding sites,[220] that the recognition of (or access to) cis-regulatory elements by TFs during development is context-dependent, i.e., requires additional information, leading to the conclusion that "developmental determination by GRNs alone is untenable".[217]

The response to the surprising lack of increase in the numbers of genes encoding transcription factors and other regulatory proteins between nematodes and humans[n] has been to assert that the power of 'combinatorial' control is sufficient to enable "a dramatic expansion in regulatory complexity",[221,222] which, although not formalized mathematically nor defined mechanistically, implies a superlinear relationship between the number of regulatory factors and the number of regulatable events. Moreover, in this conception, the developmental programming of more complex organisms is assumed simply to involve an expansion in the numbers of *cis*-acting sequences in DNA (and RNA) recognized by regulatory proteins,[191,222] which, along with alterations in the expression of these proteins and the networks in which they participate, is posited to lie at the heart of the development and phenotypic diversity of animals and plants.[192,196,223–230]

In trying to frame the characteristics of such networks in abstract terms, many theoretical gene regulatory models refer to 'attractors', defined as 'stationary (i.e., stable) network states that represent biologically meaningful properties, such as cell identity',[200,231,232] which in humans presumably number in the trillions.

The potential number of states in random Boolean networks is 2^R where R is the number of regulatory variables, and the number of attractor states scales superlinearly (R^x for x>1).[189] It is mechanistically implausible that any genetic event can be an arbitrarily complex Boolean function of other events. Kauffman recognized this and constrained his modeling to a maximum number of inputs, initially set at two.[232] It was later shown that the number of attractor states in Kauffman networks is a superpolynomial function of system size.[233] Others have suggested that combinatorial control by transcription factors might be equivalent to a class of computing called a Boltzmann machine,[234] although the most plausible substrate for cellular computing is RNA.[235]

In any case, while there is little doubt that expansion of the regulatory superstructure underpins the emergence and divergence of developmentally complex organisms, the implied logic of 'combinatorial control' is that the number of possible cell states increases exponentially or factorially with the number of regulatory proteins. In animals this number is so vast – 1000 regulatory proteins is potentially 2^{1000} Boolean states or 1000! combinations, both of which are astronomically large numbers, far greater than the estimated number of atoms in the observable universe, ($\sim 10^{80}$)[o] – as to be more than capable, even if heavily discounted, of providing

---

[n] Although we now know that there has been a massive expansion in the number of regulatory RNAs with increased developmental complexity.

[o] The 'Eddington Number'.[236]

sufficient power to program the development of a worm or a human, so no need for further concern or consideration regarding the constraints of regulatory proteins on complexity.

While it is clear that many factors can influence a decision to transcribe a gene, so there is some sort of multifactorial control,[231] the use of the term 'combinatorial' is ambiguous. It is by no means clear that regulatory factors operate combinatorically[p] at the point of decision,[q] as opposed to the hierarchical, binary and sequential control of gene expression in the temporal (developmental) sense,[239–242] nor that the amount of required regulatory information scales as an inverse factorial as implicitly assumed by the former. Indeed, the assumption has never been clearly articulated, nor (consequently) been justified or formalized mathematically by reference to decision theory or control system theory, or mechanistically; nor has it been subjected to critical scrutiny, as the assumption has sat comfortably with mainstream preconceptions.

Decisional control systems are usually structured as binary hierarchies, as are developmental ontogenies, which is not inconsistent with sequential Boolean control. Sorting algorithms in binary trees scale as NlogN, where N is the total number of states (branches) in the ontogeny tree. If a functional network maintains the same proportional connectivity (i.e., where N nodes are connected to the same fraction of other nodes in the network), the numbers of connections, presumably transacted by regulatory factors, scales as N^2. Alternatively, the number of decisions in an ontogeny tree scale as a geometric function $[2^{(N+1)} - 1]$ ($\sim 2^N$) of the number of branches, whose exact number may depend on the length of the branches but, in any case, is a very large number.

How does the amount of regulatory information scale with organismal complexity? The empirical evidence suggests that it is the opposite to that which is commonly assumed. Prokaryotic genomes are predominantly composed of protein-coding sequences,

and therefore it is possible to do a first approximation analysis of the relationship between the numbers of genes encoding regulatory factors (ignoring the small number of regulatory RNAs) and the total numbers of genes in cells of different genetic complexity.

Consistent with an NlogN or N^2 relationship,[r] the numbers of regulatory proteins R (as assessed computationally by the presence of DNA and RNA binding domains) in bacterial genomes increases quasi-quadratically with gene number G: $R = O(G^{1.96})$[243–247] (Figure 15.5).

CONSTRAINTS IMPOSED BY THE SUPERLINEAR SCALING OF REGULATORY INFORMATION

The superlinear scaling of regulatory information has many implications for the evolution and development of organisms, and indeed all functionally integrated complex systems.[245,248,249] Since the number of genes involved in gene regulation scales roughly twice as fast as the total number of genes, and there is no hint of any deviation at the top end of the range, there must be a limit (as the number of regulatory genes cannot exceed the total), which is ostensibly (and likely explains) the observed upper limit of bacterial genome sizes (about 10 Mb, ~9000 genes), where over 20% of the genes are regulatory.[s] This limit was also likely reached early in evolution.[243]

There is no reason to think that eukaryotes can or have escaped the greater than linear scaling of regulatory factors with complexity, and the number of transcription factors in eukaryotes also exhibit a power law relationship with the number of protein-coding genes, albeit with a lower exponent $[R = O(N^{1.23})]$,[247] although (and likely because) it is presently impossible to account the number of genes encoding regulatory RNAs in eukaryotic genomes.[250,251]

In any case, to be able to realize the benefits and ecological opportunities of multicellularity, complex eukaryotes must have found solutions to the scaling problem of constructing elaborate decision trees in developmental programming. One is to introduce additional layers of regulation and added flexibility for regulatory evolution, which has clearly

---

[p] Although transcription factor pairings can alter their target specificity.[237,238]

[q] That is, the regulation of the expression of individual genes at particular levels. This conception of combinatorial control implies a 'voting function' at regulatory elements in the gene promoter, and other sites of regulatory action. Indeed, such a voting function breaks down into *Go* (activate), *No-Go* (repress) and *Tie-breaker*, after which any more inputs are redundant, which fits with the observed two-three factor transcription factor control of bacterial gene expression.

[r] It is difficult to distinguish NlogN from N^2 over two orders of magnitude of bacterial genome sizes.

[s] Presumably the point where the added benefit of increased genetic complexity is offset by the disproportionally higher burden of regulatory overhead.

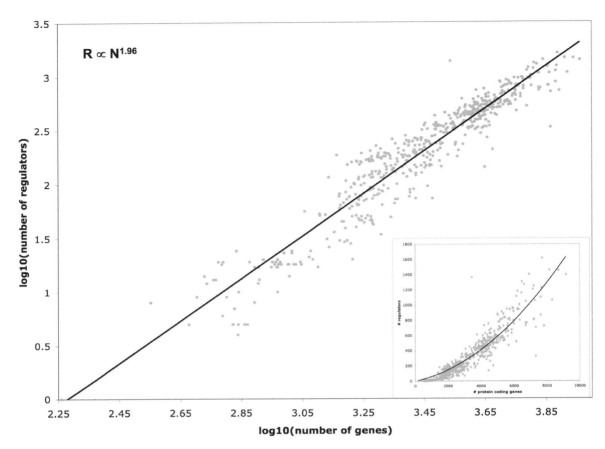

FIGURE 15.5   Quasi-quadratic relationship between genes encoding regulatory proteins and total numbers of protein-coding genes in 683 prokaryotic genomes.[243] The analysis was computationally assessed by comparing all predicted proteins with Pfam annotated regulatory domains (such as DNA-binding domains) with all annotated open reading frames. The insert is the linear relationship. The main plot is a log-log graph of the data. The line of best fit has a slope of 1.96 (R ∝ N^1.96). (Analysis by Larry Croft, reproduced with permission.)

occurred, in respect of chromatin state, epigenetic and other post-translational modifications, enhancers, lncRNAs and miRNAs. The second is modularity, enabling the reuse of subroutines where possible. The third is to separate the regulatory signal from the analog functions directed by it, to use the infrastructure most efficiently. The fourth is compartmentalization, which appears to be far more widespread and sophisticated than realized from simple ultrastructural images of cells, particularly the partitioning of the cell and chromatin into phase-separated domains that have the potential to focus interactions and reduce noise and cross-talk in gene regulatory transactions[90] (see next chapter). This is fertile ground for both theoretical and mechanistic studies of how autopoietic programs are best designed and executed.

If prokaryotes and eukaryotes are subject to the same laws of regulatory scaling, as the developmental complexity of organisms[t] increases, the fraction of the genome devoted to regulation must increase, which would, in principle, explain the genomic expansion and increasing domination of non-protein-coding sequences across orders of magnitude of developmental complexity.

HOW MUCH INFORMATION IS THERE IN THE HUMAN GENOME?

It is worth reflecting on the amount of information that is stored in the genomes of humans and other

---

[t] Metabolic, differentiation, developmental or cognitive complexity.

animals. In computational terms, the haploid human genome contains around 6.6 gigabits of data (after conversion of 3.3 billion AGCT nucleotides to binary characters – 00,01,10,11) or 825 megabytes, less than required for the storage of a few hundred images, and far less than the capacity of a smart phone. Put in these terms, it seems incredible that such a compact suite of data can program the development, physiology, cognitive capacity and reproduction of a human. Indeed, it is likely that genomic information is dense in complex organisms, as it is in bacteria and viruses.

GENOMES AS .ZIP FILES OF TRANSCRIPTOMES

We suggest that not only is the vast majority of the human genome devoted to the regulation of development – it is not junk – but also that the necessary huge expansion in regulatory information could only be achieved by separating regulatory signals from consequent actions (aided by the emergence of the nucleus and the physical separation of transcription and translation in eukaryotes – Chapter 4), using RNA to direct generic effector proteins to specific locations in the genome and in other RNAs, exemplified in simple form by the RNAi and CRISPR systems. Indeed, the ability to target a regulatory action to different sites using RNA guides, either small RNAs or modules within lncRNAs (Chapter 16), constitutes a large increase in regulatory power at minimal cost, at the same time also enabling enormous flexibility in developmental programming and evolutionary adaptation.

Chiara Alberti and Luisa Cochella summarized it well with respect to miRNAs (which can be applied equally to lncRNAs – our addition):

> we present a view of miRNAs (and lncRNAs) in the context of development as a hierarchical and canalized series of gene regulatory networks. In this scheme, only a fraction

of embryonic miRNAs [and lncRNAs] act at the top of this hierarchy, with their loss resulting in broad developmental defects, whereas most other miRNAs [and lncRNAs] are expressed with high cellular specificity and play roles at the periphery of development, affecting the terminal features of specialized cells.[252]

Finally, we suggest that genomes are better viewed not as repositories of protein-coding genes but highly compacted transcriptomes that are unzipped during development. Indeed, as we show in the next chapter, the production of an army of small RNAs and lncRNAs, interacting with an expanded repertoire of effector proteins, lies at the heart of all regulatory and organizational adaptations that allowed eukaryotes to access and explore the dimensions of multicellularity and developmental complexity, and regulate every aspect of their four-dimensional ontogeny. These innovations comprise a major addition to the nature of genetic information, at the same time increasing the robustness and evolvability of these systems.

FURTHER READING

Maturana H.R. and Varela F.J. (1972) *Autopoiesis and Cognition: The Realization of the Living* (D. Reidel Publishing Company, Dordrecht).

Salthe S. (1993) *Development and Evolution: Complexity and Change in Biology* (MIT Press, Cambridge, MA).

Edelman GM (1978) The Mindful Brain: Cortical Organization and the Group-selective Theory of Higher Brain Function (MIT Press).

Fedoroff NV (2012) Transposable elements, epigenetics, and genome evolution. *Science* 338: 758–67.

Erwin DH and Davidson EH (2009) The evolution of hierarchical gene regulatory networks. *Nature Reviews Genetics* 10: 141–8.

Alberti C and Cochella L (2017) A framework for understanding the roles of miRNAs in animal development. *Development* 144: 2548–59.

16 RNA Rules

RNA IS A CORE COMPONENT OF CHROMATIN

DNA and proteins have been the focus of the study of chromosome structure, but RNA is also a major component, essential to the organization of chromatin and the 'nuclear matrix'.[1–18] As long ago as 1989, Sheldon Penman and colleagues demonstrated that transcription is required to maintain nuclear structure, and that chromatin integrity is destroyed by treatment with RNase, noting that "ribonucleoprotein granules were dispersed throughout the euchromatic regions" and suggesting "that RNA is a structural component of the nuclear matrix, which in turn may organize the higher order structure of chromatin"[4] (Chapter 4).

Genome-wide mapping and sequencing studies subsequently showed that there are many chromatin-bound RNAs in animal cells and that the locations of long non-coding RNAs in chromatin are "focal, sequence-specific and numerous",[19] with thousands of "tightly associated" non-coding RNAs tethered adjacent to active genes.[11,20,21] Well-studied non-coding RNAs such as 7SK, U1, B2 and Alu RNAs, Gas5 and SRA, and more recently a large coterie of enhancer-derived and other lncRNAs, have been shown to be involved in the regulation of transcription initiation, elongation, termination and splicing.[22–25] The stress response induces the transcription downstream of protein-coding genes of thousands of lncRNAs that remain chromatin bound.[26] Chromatin-associated RNAs, which include those transcribed from enhancers and repeats, have been shown to have roles in genome organization via enhancer-promoter interactions and the formation of transcription hubs, heterochromatin and nuclear bodies (or 'granules')[11–13,15,16,21,27–30] through their interaction with proteins containing intrinsically disordered regions and the formation of phase-separated domains, as set out below.

REGULATION OF CHROMOSOME STRUCTURE

The scaffolding of euchromatin involves highly abundant ('CoT1') repeat RNAs, predominantly from 5' truncated LINE elements,[31,32] the expression of which varies during development and is regulated by other RNAs.[33,34] Chromatin-associated RNA proximity ligation reveals an RNA-DNA contact map similar to that observed by DNA-DNA ligation in topologically associated domains.[13] LncRNAs have been shown to regulate TAD formation,[35–37] and a recent analysis identified more than 10,000 RNA–chromatin interactions mediated by protein-coding RNAs and non-coding RNAs.[38] The RNAi machinery has also been shown to regulate nuclear topology.[39,40]

Many binding sites for CTCF, a zinc-finger containing protein (see below) that appears to anchor boundary sequences in TADs[41] (Chapter 14), are derived from transposable elements[42] and transcriptionally active HERV-H retrotransposons demarcate TADs in human pluripotent stem cells.[43] Similar to that observed with 'enhancer' RNAs (see below), lncRNAs have been reported to regulate neighboring genes through interaction with the Mediator complex,[44,45] a master coordinator of transcription and cell lineage commitment that also organizes chromosome topology (Chapter 14).

LINE- and centromere-derived repeat RNAs are structural and functional components of centromeric chromatin.[46–49] Heterochromatin formation generally requires the expression of repetitive sequences[50] and the RNAi pathway,[51–55] and RNA binding is required for heterochromatic localization of HP1 and the Suv39h histone methyltransferase.[56–59] Chromatin compaction is also controlled by lncRNAs that target IAP retrotransposons.[60] Telomere formation and maintenance requires specialized non-coding RNAs,[61,62] as does pairing of

DOI: 10.1201/9781003109242-16

homologous chromosomes in meiosis[63,64] and many, if not most, chromatin-associated proteins bind RNA,[65] including those involved in other chromatin-regulated process such as DNA stability and damage repair.[66]

RNA GUIDANCE OF CHROMATIN REMODELING

Chromatin structure is modulated during development by 'pioneer transcription factors' that alter cell fate in plants and animals by targeting nucleosomes and/or common DNA motifs.[67–71] The best known examples of reprogramming proteins are the 'Yamanaka' factors, Oct4 (*Pou5f1* gene), Sox2, Klf4 and c-Myc, which are (collectively) capable of converting differentiated cells to 'induced pluripotent stem cells' (iPSCs),[72–74] a process enhanced by inclusion of the RNA-binding protein, Lin28.[75] Oct4 is also involved, *inter alia*, in the differentiation of pluripotent cells to form the cranial neural crest.[76]

Another key pluripotency and reprogramming factor is Nanog, a homeobox-containing protein.[77,78] Homeoboxes are helix-loop-helix DNA-binding domains that exhibit a preference, but not specificity, for the common motif TAAT,[79,80] in the case of Nanog TAAT(G/T)(G/T).[81] Oct4 is also a homeobox-containing protein that recognizes the loose consensus sequence TTT(G/T)(G/C)(T/A)T(T/A), which occurs at thousands of sites around the genome.[82–84]

The expression of Oct4, Nanog and other pluripotency factors[a] is regulated by non-coding RNAs,[b] including pseudogene-derived lncRNAs,[93–98] one of which recruits the histone-lysine N-methyltransferase SUV39H1 to epigenetically silence Oct4 expression.[97,98] Reciprocally Oct4 and Nanog regulate the expression of lncRNAs that modulate pluripotency.[99] Oct4 and Nanog also have multiple pseudogenes,[100–103] some of which are differentially expressed in pluripotent and tumor cell lines.[102,104]

There are 16 classes of genes encoding homeobox proteins in animals, 11 in plants, with hundreds of orthologs in the human genome, most of which contain additional domains.[80] As noted already (Chapter 5),

Hox proteins are 'master controllers' of gene expression patterns during animal and plant development, and regulate the expression of many genes at different developmental stages. While they recognize similar sequences *in vitro*, Hox proteins display wide functional diversity and identification of their *in vivo* genomic targets has proven elusive, as has the identification of the targets of Oct4 and Sox2.[80,83,105–109] Analysis of genome-wide DNase I hypersensitivity profiles and transcription factor (TF)-binding sites identified 120 and validated eight 'pioneer' TF families that dynamically open chromatin (including Sox2, Oct4 and Hoxa1), and identified 'settler' TFs (including c-Myc), and the nuclear hormone receptor RXR:RAR and NF-κB families, whose genomic binding is dependent on chromatin opening by pioneer TFs.[110]

The targets of Sox2, Oct4, Nanog and other Hox proteins change with developmental stage,[77,111,112] all of which suggests that other factors are involved in determining their locus specificity. In this context, it may not be an outlier observation that the *Drosophila* Hox protein Bicoid (which controls anterior-posterior patterning) binds RNA via its homeodomain,[113,114] nor that highly conserved lncRNAs are produced in vertebrate endoderm lineages from paralogous regions in *HOXA* and *HOXB* clusters.[115]

Sox2 is a member of a subclass of 'high mobility group' (HMG) proteins, the most abundant chromatin-associated proteins after histones. HMG proteins bend DNA structure, initiate chromatin opening and facilitate nucleosome remodeling.[116–118] There are three classes, one of which (HMG-A) is abundant in embryonic cells and binds AT-rich sequences, another (HMG-N) binds nucleosomes, and the third (HMG-B, which includes the Sox proteins) binds the DNA helix minor groove with no sequence specificity.[119,120] Sox2 influences development not only in pluripotent stem cells but also in the lung, ear and eye, and in neural lineages, but how it and other HMG-B proteins achieve their tissue-specific versatility is unclear.[107]

While Sox2 has low affinity for DNA,[116] it binds RNA with high affinity through its HMG domain,[121,122] as do other members of the HMG-B family,[123] "which requires a reassessment of how these proteins establish proper patterns of gene expression across the genome".[121] The HMG-B domain of the mammalian sex-determining protein Sry is homologous to the RNA-binding domain of a viral protein,[124] suggesting that their target selection *in vivo* is guided by trans-acting RNA signals. There are well-documented

---

[a] These factors have distinct roles in cell lineage specification[77] and the regulation of their expression is intertwined.[74,83,85–89] Nanog exerts its action in part via TET1/2 methylcytosine hydroxylases.[90]

[b] Noncoding RNAs also regulate the expression of the nuclear hormone receptor ESR[191] and the CEBPA (CCAAT enhancer-binding protein alpha).[92]

examples of lncRNAs that interact with Sox2 to regulate pluripotency, neurogenesis, neuronal differentiation and brain development,[122,125–128] and a lncRNA has been shown to interact with a chromatin-remodeling complex to induce nucleosome repositioning.[129]

Sox2, Nanog and Oct4 are often found at super enhancers[130] and state-specific differences in enhancer activity correspond with reconfiguration of Sox2, Nanog and Oct4 binding and target gene expression.[111] The lncRNA Evf2 selectively represses genes across megabase distances by coupling recruitment and sequestration of Sox2 into phase-separated domains (see below), affecting enhancer targeting and activity, with genome-wide effects.[122] In human embryonic stem cells Oct4 and Nanog associate with transcripts of the human endogenous retrovirus subfamily H (HERV-H) transposable elements, which are required to maintain stem cell identity and whose terminal repeats function as enhancers.[131,132]

The classic master switch transcription factor, MyoD, which can reprogram fibroblasts into muscle cells and is central to muscle differentiation *in vivo*,[133] is regulated by lncRNAs,[134–136] as are other aspects of muscle gene expression.[137–139] The pioneer transcription factor CBP also binds RNAs, including those transcribed from enhancers, to stimulate histone acetylation and transcription.[140]

Nucleosome repositioning and remodeling is accomplished by the ATP-dependent imitation switch (ISWI), chromodomain helicase DNA-binding (CHD), SWI/SNF (switch/'sucrose non-fermentable') (SWI/SNF) and INO80 complexes.[141,142] These complexes are directed to specific sites in chromatin or antagonized by lncRNAs, including Xist and enhancer RNAs, in processes as diverse as rRNA synthesis, myogenic differentiation and proliferation, endothelial proliferation, migration and angiogenic function, atherosclerosis, cardiomyopathy, liver regeneration and stem cell renewal, immunity and inflammation, and various cancers,[129,136,143–160] leading one group to conclude that "every cell type expresses precise lncRNA signatures to control lineage-specific regulatory programs".[160]

However, the patchy data on the binding of RNAs by the various proteins that control chromatin remodeling during development reflects limited investigations because of the expectation that all 'transcription factors' bind to DNA, rather than be directed by RNA-DNA and other RNA-mediated interactions.

GUIDANCE OF TRANSCRIPTION FACTORS

Loose DNA sequence specificities are a feature of eukaryotic transcription factors generally. While eukaryotic genomes are orders of magnitude larger than those of prokaryotes, their more conventional TFs have shorter DNA recognition sites (6–10bp versus 15–25bp in *E. coli*[161]), often expressed as a 'consensus' sequence, but better represented by multiple sequences, with many TFs recognizing different primary and secondary motifs.[162–166]

Moreover, high-throughput chromatin immunoprecipitation experiments with antibodies against specific TFs show different patterns of binding in different cell types, so additional factors must be involved. Such factors can be either (or both) transacting signals or chromatin accessibility, the latter supported by the observation that TF-binding sites are nucleosome depleted and DNase-sensitive, indicating that epigenomic decisions precede TF factor binding.[167–170]

The largest class of TFs in animals and plants contain 'zinc-finger' (ZF) domains,[c] specifically the C2H2 class, of which there are over 700 encoded in the human genome, and which recognize more sequence motifs than all other transcription factors combined.[171] Human C2H2-ZF proteins contain an average ~10 C2H2 domains (ranging from 1 to 30), classified into three groups: 'triple', 'multiple-adjacent', and 'separated-paired' C2H2 finger proteins, enabling some to bind multiple ligands.

It is thought that most ZFs bind to DNA, although most of the binding sequences are unidentified,[165] but many ZFs also bind to RNA or protein, and some to RNA only.[162,172,173] The classic example is TFIIIA (Chapter 8), which is required for the transcription of 5S rRNA genes and is titrated off DNA by its higher affinity for 5S rRNA, the first demonstration of the regulation of TFs by RNAs.[174,175]

A large fraction of C2H2-ZF TFs have been shown to regulate alternative splicing.[176] A splice variant that introduces three additional amino acids (KTS) between the third and fourth ZFs of the Wilm's tumor protein WT1[d] changes the specificity of the WT1 protein from DNA to spliceosomes,[178] presumably by binding RNA, given that WT1 also contains

---

c So-called because they have a domain shaped like a finger that is structured by a coordinated zinc ion.

d Frequently mutated in pediatric kidney tumors and urinogenitary developmental disorders.[177]

an RNA recognition motif[179] and transcription and splicing are coupled.[180] Disturbance of the ratio of +/-KTS isoforms causes a developmental syndrome, affecting kidney and genital development.[181] Both isoforms bind DNA and RNA *in vitro*,[182–184] shuffle between the nucleus and translating polysomes in the cytoplasm,[185] and their subnuclear location is RNase- but not DNase-sensitive.[182]

A 1994 analysis by Yigong Shi and Jeremy Berg of two representative C2H2-ZF proteins, one of which was Sp1 (which controls the expression of many housekeeping, tissue-specific, cell cycle and signaling pathway response genes[186]), showed that they have a higher affinity for RNA-DNA hybrids than for double-stranded DNA and that this increased affinity was strand-specific, i.e., dependent on which strand is RNA.[187]

The C2H2-ZF transcription factor YY1, which regulates the expression of various genes during embryogenesis, cell differentiation and proliferation,[188] binds chromatin in an RNA-enhanced fashion[189] and appears to play a major role in mediating enhancer-promoter loops.[190] YY1 also interacts with an RNA-binding protein involved in splicing regulation, depletion of which attenuates YY1 chromatin binding and YY1-dependent DNA looping and transcription.[29] The ZF-containing TAD insulator CTCF has also been shown to be a high-affinity RNA-binding protein.[65,191–195]

Later studies confirmed that over 800 human proteins bind RNA-DNA hybrids and over 300 prefer binding RNA-DNA hybrids over dsDNA.[196] These observations raise the possibility, if not the likelihood, that trans-acting RNAs are involved in the exposure and selection of genomic TF-binding sites, explaining the differential locus specificity of TF binding and the reason for a loose consensus sequence,[e] as well as enabling directionality of action by strand selection.

RNA-DNA hybrids[f] (which form 'R-loops' with the displaced DNA strand) occur widely throughout the human genome[198,199] and even encompass 8% of the yeast genome.[200] RNA-DNA hybrids are enriched at unmethylated CpG-rich promoters, transcription start sites and regions enriched for activating histone modifications such as H3K4me1/2/3,

H3K9ac and H3K27ac.[201] RNA-DNA hybrids regulate genome stability and DNA repair,[197,202,203] promoter-proximal chromatin architecture and cellular differentiation,[204] transcriptional activation[205] and "are enriched at loci with ... potential transcriptional regulatory properties ... supporting a model of certain transcription factors binding preferentially to the RNA:DNA conformation".[206] The formation and stability of RNA-DNA hybrids are in turn regulated by RNA methylation and other modifications[197,207] (Chapter 17).

Nucleic acid triplex structures[g] (wherein a single stranded DNA or RNA forms 'Hoogsteen' hydrogen bonds with the purine-rich strand of polypyrimidine-polypurine tracts in the major groove of duplex DNA) also occur *in vivo*, as first detected by the sequence-specific binding of RNA to 'native' dsDNA.[214] Triplex-forming sequences are overrepresented in eukaryotic but not bacterial genomes, notably in regulatory regions and promoters.[215–217] Antibody and sequencing studies have also shown that triplex structures abound in eukaryotic chromosomes[218–221] (Figure 16.1). Triplex hotspots targeted by lncRNAs have been proposed to contribute to chromatin compartmentalization in conjunction with 'architectural' TFs such as CTCF[222] and the positions of lncRNA:DNA triplex-forming sites have been shown to be predictors for TADs.[223] Triplex-forming oligonucleotides have been shown to alter cell division, inhibit tumor growth, stimulate recombination and modulate target gene expression.[224–227]

Many lncRNAs, including those expressed from enhancers, have been shown to interact sequence-specifically with DNA[228] to regulate various processes through R-loop or triplex formation, including chromatin architecture, transcription, radiation response, cell proliferation, cell differentiation and organ development, in some cases (at least) intersecting with epigenetic pathways.[204,217,229–240] Triplexes are also involved in small RNA-mediated transcriptional gene silencing.[241]

---

[e] A subset of which can be specifically addressed by exact match to a transacting RNA, either a small RNA or an RNA sequence within a longer RNA.

[f] Interestingly, the stability of genomic RNA-DNA hybrids *in vivo* is controlled by methylation of the RNA (see below).[197]

[g] There are other alternative and multi-stranded structures in eukaryotic genomes, including Z-DNA (binding domains for which occur in RNA editing enzymes, see Chapter 17), G-quadruplexes, I-motifs and cruciform structures, which, regrettably, despite the availability of specific antibodies, have not been mapped in genome-wide studies of genomic features and their dynamic relationship to cell type.[208–212] Many of these alternative DNA structures are formed by simple sequence repeats, which also abound in the genomes of plants and animals.[213]

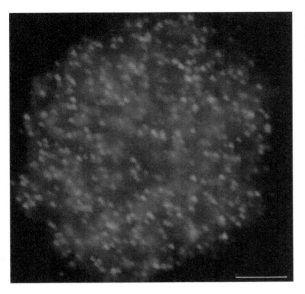

FIGURE 16.1 Triplex-forming DNAs in the interphase nucleus of a human monocytic leukemia cell visualized *in situ* by an anti-triplex monoclonal antibody. The bar represents 5 μm. (Reproduced from Ohno et al.[220] with permission of Springer Nature.)

A good example, from plants, is the lncRNA APOLO, which coordinates the expression of multiple genes in response to cold through sequence complementarity and R-loop formation, decoys Polycomb and binds transcription factors at the promoter of a master regulator of root hair formation.[242–244] Amazingly, APOLO function can be partly mimicked by the sequence-unrelated lncRNA UPAT, which interacts with orthologous proteins in mammals, indicating conservation of regulatory structures and lncRNA functions across kingdoms.[244]

The enhancer lncRNA KHPS1 forms a triplex with enhancer DNA sequences to activate expression of the neighboring *SPHK1* gene, by evicting CTCF, which insulates the enhancer from the SPHK1 promoter. Deletion of the triplex-forming sequence attenuates SPHK1 expression, leading to decreased cell migration and invasion, and the targeting of KHPS1 lncRNA can be switched by swapping the triplex-forming promoter sequence to other genes.[232,233]

Other classes of 'transcription factors' such as Y-box proteins also bind RNA, and known RNA-binding proteins such as hnRNP K (better known as a 'splicing factor') also act as transcription factors.[245,246] The 'paired-box' transcription factor Pax5, which is a 'master regulator' of B-cell development by recruiting chromatin-remodeling, histone-modifying and basal transcription factor complexes to its target genes,[247] is hijacked to the Epstein Barr Virus genome by a viral-encoded non-coding RNA.[248] Another 'transcription factor', the nuclear hormone receptor ESR1 (estrogen receptor α), which is commonly activated in breast cancer, is also an RNA-binding protein.[249]

Dual RNA-DNA or ambiguous RNA/DNA-binding proteins also include p53,[172,250] the 'guardian of the genome', possibly the most intensively studied gene and protein in human molecular biology, which binds a lncRNA ('damage-induced noncoding RNA', DINO).[251] The dual DNA/RNA-binding protein TLS/FUS (Translocated in LipoSarcoma/FUsed in Sarcoma) is allosterically regulated by lncRNA pncRNA-D.[252–254] Even RNA polymerase is regulated by RNAs. In mammals, RNA polymerase II is repressed by short RNA polymerase III transcripts derived from mouse B2 and human Alu repeat (SINE) elements.[255–258] These elements also provide mobile RNA polymerase II promoters.[259]

Importantly, approximately half (~350) of the human C2H2-ZF proteins, many of which are unique to primates, contain a KRAB transcriptional repression domain, which binds TEs,[h] and evolved by recurrent TE capture that partners them with emergent TE-mediated regulatory networks, influencing genomic imprinting, placental growth and brain development.[171,260–270] Most eukaryotic TFs also contain intrinsically disordered domains that overlap their DNA-binding domain and direct their target specificity,[271–275] likely by interaction with guide RNAs (see below).

Presumably, different types of DNA/RNA-binding proteins recognize different types of nucleic acid structures and transact a different type of signal in different contexts within the decisional systems that control cell division and differentiation during development, as well as in physiological responses. The fact that most eukaryotic 'transcription factors' have confusing and enigmatic functions attests to the likelihood that they have been interpreted in the wrong conceptual framework, with RNA the missing link.[276]

---

[h] KZFPs (KRAB domain-containing zinc finger proteins) control the pleiotropic activation of TE-derived transcriptional cis-regulator sequences, some of which are primate-specific, during early embryogenesis, in part through histone H3K9me3-dependent heterochromatin formation and DNA methylation.[260,261] Primate-specific KZFPs also regulate gene expression in neurons.[262]

GUIDANCE OF DNA METHYLATION

Transcriptional gene silencing in fungi and plants by RNA-directed DNA methylation was well established in the 1980s and 1990s (Chapter 12). These studies eventually showed that the enzymes that methylate DNA are directed to their sites of action by small RNAs interacting with the RNAi protein AGO4.[277,278] Small RNAs (miRNAs, siRNAs and piRNAs) also induce site-specific DNA methylation in animals,[52,53,279,280] which again involves Argonaute proteins,[281–285] suggesting that what had originated as an RNA-based mechanism for defense against viruses has been co-opted as a means of genome regulation.[i]

In 2004, Linda Jeffery and Sara Nakielny showed that the *de novo* DNA methylases Dnmt3a and Dnmt3b, but not the maintenance methylase Dnmt1, bind siRNAs with high affinity.[287] Later others reported that Dnmt1 (which restores methylation at hemi-methylated CpG sites after DNA replication) binds lncRNAs to alter DNA methylation patters at cognate loci.[288–291]

Demethylation also appears to be an active process guided by RNAs.[292–295] RNA-directed DNA demethylation has also been reported to involve R-loop formation,[239,294] and recruitment of the TET2 dioxygenase/demethylase (which unlike other TET enzymes does not contain a DNA-binding domain, but does bind RNA[296,297]) by RNAs transcribed from endogenous retroviruses.[298]

Some methyl-CpG-binding proteins, including MeCP2, bind siRNAs and other RNAs, mediated through a domain distinct from the methyl-CpG-binding domain with, interestingly, RNA and methyl-CpG binding being mutually exclusive,[287,299] although there seems to be variations of the regulatory mechanisms, including RNA-mediated recruitment to phase-separated heterochromatin compartments (see below). LncRNAs have also been shown to link DNA methylation with histone modification through triplex formation with target sequences.[300]

GUIDANCE OF HISTONE MODIFICATIONS

A range of histone variants and over 100 different histone modifications are differentially incorporated into nucleosomes located at millions of different positions in different cell types and different stages of development and differentiation (Chapter 14). However, like DNA methylation enzymes, histone-modifying enzymes also have no intrinsic DNA-binding capacity or specificity, which is often assumed to be provided by sequence-specific DNA-binding proteins or transcription factors that interact with them. On the other hand, like DNA methylation enzymes, many histone modification writers and readers contain domains that bind RNA and/or contain RNA-binding modules. These include RNA recognition motifs,[301] chromodomains,[296,302,303] bromodomains,[296] Tudor domains,[304] PRC2 subunits EZH2, EED, Suz12 and Jarid2,[305–310] the H3K20 trimethylase Suv4–20h[60] and other histone-modifying complexess.[311]

RNA binding to histones was first reported in the mid-1960s[312,313] (Chapter 4). Around the turn of the century, a number of groups showed that PcG (Polycomb group) proteins from *C. elegans* and vertebrates also bind RNA, and that this binding is essential for their chromatin localization and repression of homeotic genes.[314–316] In 2005, Renato Paro and colleagues showed that the switch from the silenced to the activated state of a Polycomb response element in the *Drosophila bithorax* locus (Chapter 5) requires non-coding transcription.[317]

In 2007, John Rinn and colleagues showed that lncRNAs transcribed from human homeotic gene loci, like those in *Drosophila*, are expressed along developmental axes and demarcate active and silent chromosomal domains that have different H3K27me3 profiles and RNA polymerase accessibility, the exemplar of which, HOTAIR (Chapter 13), interacts with PRC2 and is required for PRC2 occupancy and histone H3K27 trimethylation at the *HOXD* locus.[318] Other studies showed, for example, that retinoic acid-induced expression of lncRNAs follows the collinear activation state and correlates with loss of Polycomb repression at the *HOXA* locus.[319]

In 2008, the groups of Chandrasekhar Kanduri and Peter Fraser showed that lncRNAs differentially expressed from parentally imprinted loci, specifically the 105 kb Air RNA and the 91 kb Kcnq1ot1 RNA (formerly KvLQT1-AS, Chapter 9), and later others,

---

[i] Transcriptional gene silencing can be induced by siRNAs in the absence of DNA methylation,[286] indicating that small RNAs participate in other pathways that control chromatin state and architecture.

also bind PRC2 to repress the relevant alleles.[320–322] In the same year, we showed that lncRNAs expressed from the antisense strand of homeotic gene loci during embryonic stem cell differentiation are associated with both the chromatin-activating Trithorax MLL1 complex and activated chromatin containing H3K4me3 marks.[323] Subsequently other lncRNAs (including enhancer RNAs and small RNAs derived from them) were also shown to associate with Trithorax complexes, including RNAs involved in maintenance of stem cell fates and lineage specification (such as Evx1-as and HOTTIP),[14,45,324–334] and even grain yield in rice.[335]

In 2009, Rinn and colleagues surveyed over 3,300 lncRNAs and showed that ~20% (but only ~2% of mRNAs) interact with PRC2, and that others are bound by other chromatin-modifying complexes.[336] Moreover, knocking down a selection of these RNAs caused derepression of genes normally silenced by PRC2.[336] Over 9,000 RNAs bind PRC2 in embryonic stem cells,[306] with many individual cases, including the lncRNAs H19, MEG3, ANRIL[j] and HOTAIR, subsequently characterized in some detail.[303,309,337–342]

RNA has also been shown to be required for PRC2 chromatin occupancy, PRC2 function and cell state definition.[343] Short RNAs transcribed from Polycomb-repressed loci resemble PRC2-binding sites in *Xist*, and interact with PRC2 through its subunit Suz12.[307] PRC2 binds G-quadruplex structures in RNA,[344] which inhibit PRC2 activity and are antagonized by allosteric activation of PRC2 by H3K27me3 and regulators of histone methyltransferases,[345,346] indicating complex decisional transactions.

PRC1 function also appears to be controlled by RNA[303] and PRC1 resides in membrane-less phase-separated nuclear organelles[347] that are likely to be RNA nucleated (see below).

PRC2[k] binds many RNAs 'promiscuously',[348] a description[l] that does not mean 'non-specifically'.[339,349] The association of Polycomb and Trithorax complexes with many RNAs, likely through orthologous domains (see below), is consistent with their function as guide molecules for RNA-directed site-specific histone and DNA modifications. It is also consistent with the fact that Polycomb and Trithorax proteins are involved in many differentiation and developmental decisions, from cell cycle regulation to embryogenesis and body plan specification,[350–355] so their binding to many different guide RNAs would be expected. Again, these include RNA-DNA, RNA-RNA and RNA-protein interactions, recruitment or eviction of histone modifiers and other chromatin-modifying proteins, alteration of DNA topology, allosteric inhibition, and reorganization of phase-separated domains.[342,343,345,346,356–364]

LncRNAs also control switching between Polycomb and Trithorax response elements.[329,356] Other histone modifications are also regulated by lncRNAs,[m] including during memory formation.[366] An intriguing observation, not inconsistent with RNA involvement, is that histone-modifying enzymes, rather than the parental histones, may remain associated with DNA through replication to re-establish the epigenetic information on the newly assembled chromatin.[367]

XIST AS THE EXEMPLAR

While initially thought to be a special case, the best characterized and most illustrative example of the complex interplay between lncRNAs, chromatin structure and gene expression is Xist.[368,369] Xist has eight exons and is 17 kb in length.[370] It has a highly modular structure, including a number of types of conserved 'repeat' sequences, and interacts with over 80 different proteins including cohesin, Polycomb and other chromatin remodelers,[370–374] at low copy number.[375,376]

Xist-mediated silencing of the inactive X chromosome in mammals requires its repeat sequences[377–382] and involves Polycomb recruitment, deacetylation of H3K27ac and H3K27 methylation on the silenced chromosome.[305,368,382–387] Spreading involves the partitioning of chromatin topology[388–392] by the formation of phase-separated domains[376,393,394] and the interaction with RNA-binding proteins with repetitive elements (in particular LINE-1 elements) in the X chromosome to recruit silencing mechanisms targeted to repeats,[379,394–399] as first proposed by Mary

---

[j] ANRIL also binds PRC1.

[k] For many reasons, PRC2 has been the most intensively studied of all of the histone-modifying complexes with respect to the role of lncRNAs in epigenetic regulation of gene expression.[339]

[l] The presence of intrinsically disordered domains in most proteins involved in development has also been described as conferring promiscuity, a functional trait that allows flexible interactions in regulatory networks (see below).

[m] LncRNAs also control the methylation of a number of non-histone proteins involved in cell signaling, gene expression and RNA processing.[365]

Lyon,[400,401] likely via triplex formation.[402] Xist also acts as a suppressor of hematological cancers[403] and is essential to maintain X-inactivation of immune genes, dysregulated in females suffering systemic lupus erythematosus or COVID-19 infection.[385]

Xist expression and action is controlled and effected by other lncRNAs[369] that are expressed antisense to *Xist* (Tsix, which blocks RepA RNA binding to PRC2[305,404,405]) or from adjacent loci on the active or inactive X chromosomes[406–408] (Figure 16.2). The Jpx lncRNA, whose gene resides ~10 kb upstream of *Xist*, activates *Xist* by regulating CTCF anchor site selection to alter the topography of chromosome loops.[37,409,410] A primate-specific TE-derived lncRNA, XACT, coats active X chromosomes in pluripotent cells and is connected into the pluripotency regulatory network in humans by primate-specific retroviral enhancer.[411] Another lncRNA expressed from the X-inactivation center, Tsx, functions in germ and stem cell development as well as in learning and behavior.[412] The lncRNA Firre, which also contains a number of repeats, one of which interacts with the nuclear matrix factor hnRNPU, anchors the inactive X chromosome near the nucleolus and is required for the maintenance of its repressive H3K27me3 marks.[413] Firre is also required for the topological organization of other chromosomal regions,[414,415] and is involved in other developmental processes including adipogenesis and hematopoiesis.[416]

X-inactivation also involves the RNAi enzyme Dicer[368,417,418] and methylation of *Xist* transcripts,[419]

implicating small RNAs, the RNA interference pathway and RNA modifications in a complex set of decisional pathways that control chromatin architecture from yeast to humans.[420,421] Chromosomal dosage compensation in *Drosophila*, which involves global activation of the single X chromosome in males is controlled by the lncRNAs roX1 and roX2,[422] via a conserved predicted stem-loop structure required for histone H4K16 acetylation of the X chromosome[423] and selective X-chromosome subnuclear compartmentalization.[424]

ENHANCER RNAs AND CHROMATIN STRUCTURE

As discussed in Chapter 14, enhancers play a key role in specifying cell identity and are a signature feature of the regulation of gene expression during development.

Enhancers were initially identified by their activity, rather than their physical manifestation, but have been interpreted in terms of the initial speculations about the latter, postulated and promulgated by Mark Ptashne (and widely accepted) to comprise cluster of binding sites for TFs that act at a distance by 'looping' to make contact with the promoters of target genes, some of which can be located hundreds of kilobases distant.[425–431]

However, early studies had shown that lncRNAs are transcribed from enhancer regions in well-studied loci,[432–434] with supporting evidence accumulating

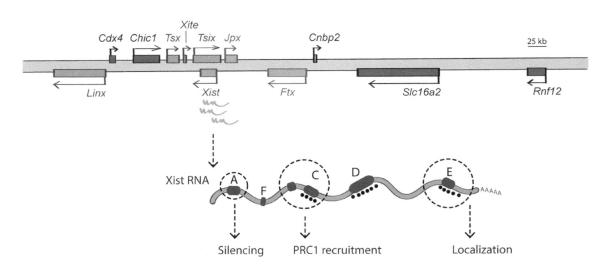

FIGURE 16.2 The organization of the *Xist* locus. The *Xist*, *Tsix*, *Jpx*, *Xite*, *Tsx* and *Ftx* genes specify lncRNAs. (Reproduced from Loda and Heard[399] under Creative Commons attribution license.)

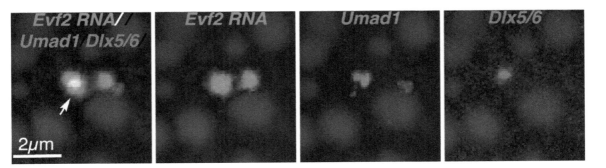

FIGURE 16.3   Cloud formed by the enhancer lncRNA Evf2 and its localization to activated (*Umad1*, 1.6 Mb distant) and repressed (*Akr1b8*, 27 Mb distant) target protein-coding genes. (Reproduced with permission from Cajigas et al[473] with permission of Elsevier.)

with improving transcriptomic and chromatin analysis technologies. Although sometimes referred to as (protein-coding) 'gene deserts',[435] enhancers exhibit the characteristics of *bona fide* genes, including nucleosome-depleted promoter regions that bind transcription factors and the transcription of adjacent sequences.[436–441] Indeed, the epigenetic architecture of, and the features of transcription initiation at, the promoters of conventional protein-coding genes and enhancers are almost indistinguishable.[436,441–443]

Enhancers and 'super-enhancers' are transcribed to produce non-coding RNAs specifically in the cells in which they are active[317,435,437,441–447] and their expression is considered the best molecular indicator of enhancer activity in developmental processes[437,443,448–453] and cancers.[91,454–457]

Enhancers recruit RNA polymerase[458] and produce short unstable bidirectional transcripts ('eRNAs') from their promoters,[437,444,459–462] as do protein-coding genes,[430,463–465] and it is uncertain whether these transcripts play a role in enhancer action or simply mark active promoters and/or reflect promiscuous RNA polymerase initiation at accessible chromatin.[441,461,464,466–468] On the other hand, enhancers also express multi-exonic lncRNAs, the half-life of which is exosome regulated,[466] and many if not most lncRNAs likely derive from enhancers.[441,444,446,469–479] Enhancers with tissue-specific activity are enriched in introns, suggesting that "the genomic location of active enhancers is key for the tissue-specific control of gene expression".[480]

There is good evidence that enhancers regulate chromosome looping and local chromatin reorganization to alter cell fate,[435,441,451,481,482] which is consistent with but does not demonstrate direct contact between enhancer TF-binding sites and the promoters

of target genes. It is also consistent with enhancer RNAs organizing looping with target genes, which has been demonstrated in at least two cases,[482,483] and/or the formation of topologically associated, possibly phase-separated, chromatin domains[16] as local hubs of transcription regulation (Figure 16.3). Recent studies suggest that there is no direct contact between TFs bound at the enhancer and the promoter of genes regulated by enhancer action,[484] and that maintenance of enhancer-promoter interactions and activation of transcription are separable events.[485]

At the heart of the debates about the mechanism of enhancer action has been the question of whether the RNAs transcribed from active enhancers are simply a passive by-product of TF occupancy, or whether it is the 'act of enhancer transcription' or the enhancer RNAs themselves that mediate enhancer action.[442,447,486,487] The evidence for these possibilities, which are not mutually exclusive, has been widely canvassed and variously interpreted, with the former initially favored because it fitted the TF paradigm of protein-coding gene regulation and did not require acceptance of large numbers of regulatory RNAs.[441,476,487–489] In line with these preconceptions, some studies have reported that the transcribed enhancer RNA sequences can be partly/substantially (it is difficult to be sure[n]) deleted, truncated or replaced with no obvious effect (see, e.g., [490,491]).

Other studies showed that enhancer RNAs are required for enhancer activity.[92,473,479,482,492–498] For example, deletion in mice of the multi-exonic lncRNA Maenli, which is expressed from an enhancer that controls limb development and is deleted in a human

---

[n] Due to incomplete characterization of the enhancer RNA/transcription unit.

developmental disorder, recapitulates the human phenotype.[479] Deletion of internal exons of the enhancer-derived lncRNA ThymoD blocks T-cell development and causes developmental malignancies.[482] Truncation of the lncRNA Evf2, which is transcribed from the highly conserved Dlx5/6 enhancer that spatially organizes the expression of a 27 Mb region on chr6 during mouse forebrain development, abrogates the action of the enhancer.[473,499] The latter study also showed that the 5' and 3' ends of the Evf2 enhancer RNA had different functions,[o] and that Evf2 formed an "RNA cloud" encompassing its target genes.[473] It has also been shown that methylation and splicing of enhancer RNAs are required for enhancer function and chromatin organization.[490,500–502]

siRNA-mediated knockdown of enhancer RNA also abrogates or reduces enhancer action, demonstrating the involvement of the RNA.[477,503–507] RNA has been shown to be required for the formation of enhancer-target promoter contacts by the transcription factor YY1[190] (which itself regulates the expression of many lncRNAs[508]) and ectopic expression of enhancer RNAs upregulates expression of the genes normally targeted by the enhancer.[472,509]

Careful analysis shows that the variable phenotypic consequences of enhancer lncRNA knockdown, truncation or ablation depend on the details.[496,510,511] For example, a short deletion of the promoter and first two exons of the 17 kb lncRNA Hand2os1, which is expressed from an enhancer essential for heart morphogenesis, did not produce discernable heart phenotypes, but deletion of exons 4 and 5 caused severe contraction defects in adult heart that worsened with age, and deletion of the entire *Hand2os1* sequence led to dysregulated cardiac gene expression, septum lesion, heart hypoplasia and perinatal death.[510]

It has been shown that enhancer RNAs produced in response to immune signaling bind the bromodomains of BRD4 (and other epigenetic reader bromodomain-containing proteins) to augment BRD4 enhancer recruitment and transcriptional cofactor activity.[512] BRD4 also cooperates with oncogenic fusions of MLL1 to induce transcriptional activation of enhancer RNAs, one of which has been shown to bind histone H4K31ac to promote histone recognition and oncogene transcription.[513] Other mechanisms

may involve the interplay between different types of regulatory RNAs and chromatin-associated proteins, as suggested by interactions between NEAT1 and BRD4/WDR5[p] complexes, and enhancer RNAs with cohesin, with effects on specific target genes.[474] Finally, there is also evidence of concerted action of *cis*-acting enhancer RNAs with other transcripts that have trans-acting roles.[514–516]

While there may not yet be universal acceptance, the evidence is accumulating that enhancer RNAs are integral to enhancer function,[517] and that enhancer RNAs are simply a (large) class of lncRNAs that regulate chromatin architecture and the expression of protein-coding and (other) lncRNAs, albeit through physical mechanisms that are not yet well understood, but involve recognition of effector proteins and sequence-specific RNA-DNA contacts via R-loops or triplexes,[17,221,236,518] and formation of topologically associated domains to form developmental stage-specific transcriptional hubs.[431,519,520] It is also evident that transcription itself modulates chromosome topology and phase transition-driven nuclear body assembly.[521–524]

There are ~400,000 enhancers (and ~400,000 differentially accessible chromatin elements,[525,526] which likely correspond to promoters) in the human genome.[438,441,444,448,456,460,527–533] This is similar to the number of lncRNAs expressed from the human genome (Chapter 13). Indeed, apart from the fact that they do not encode proteins,[q] enhancers might be properly viewed as genes, which, together with the multitude of other genes expressing functional non-coding RNAs, resolves the G-value enigma.

mRNAs may also have enhancer function,[12,490] which would not be surprising given the interwoven nature of the expression of genetic information during the complex ontogenies that underpin animal and, to a lesser extent, plant development.

RNA SCAFFOLDING OF PHASE-SEPARATED DOMAINS

It has been known for many years that there are many ribonucleoprotein (RNP) complexes that exist in defined territories in the nucleus and cytoplasm of eukaryotic cells, prominent examples of which

---

[o] Evf2, which has a human homolog, also interacts with Sox2 to alter its target specificity, regulates transcription of the homeodomain transcription factors Dlx5 and Dlx6 as well as cohesin binding, and influences chromatin remodeling in the formation of GABA-dependent neuronal circuitry.[122,145,473,499]

[p] WDR5 is a a core subunit of the human MLL1-4 histone H3K4 methyltransferase complexes.
[q] New mechanisms that control the expression and function of RNAs expressed from promoters of protein-coding genes and enhancers are still being identified, affecting splicing, elongation, termination and processing/half-life.[466,534–536]

include nucleoli, spliceosomes, paraspeckles and stress granules,[537-539] none of which are membrane-bound. One of the most important advances of recent years, first canvassed by Harry Walter and Donald Brooks in 1995, and later demonstrated by Clifford Brangwynne, Anthony Hyman and colleagues, is that these and other focal or 'punctate' organelles are phase-separated condensates or 'coacervates'[540,541] that compartmentalize biochemical and regulatory hubs,[542] although not without some controversy and uncertainty.[543]

Phase-separated condensates, which are heterogeneous in constitution and properties, are commonly referred to as 'phase-separated domains' (PSDs). They are also called 'liquid crystal domains', 'liquid droplets', 'biomolecular condensates', 'nuclear clouds' or 'nuclear bodies',[544] and exist in an aqueous state distinct from the surrounding environment, the biological manifestation of liquid or soft matter physics.[545]

In vitro PSDs form spontaneously by association of oppositely charged molecules such as negatively charged RNA interacting with positively charged proteins.[544] *In vivo* they are formed by interactions between RNAs,[r] RNA-binding proteins and proteins with intrinsically disordered domains (IDRs),[27,539,542,544,546-553] explaining the latter's previously mysterious function.[539]

IDRs lack rigid tertiary structure and are characterized by a high proportion of small, polar and positively charged amino acids (arginine, histidine and lysine), often in the form of RGG/RG, histidine-rich domains or other repeats.[554-557] IDRs are promiscuous, i.e., they interact with and are tunable by many partners.[550,556-561] Intrinsically disordered RGG/RG domains mediate specificity in RNA binding,[562,563] and IDRs flank the DNA-binding domains of transcription factors,[273] direct TF binding and, *inter alia*, the temporal regulation of transcription complexes that specify neuronal subtypes.[564]

IDRs and PSDs occur in bacteria and archaea,[565-568] but there have been sharp increases in the fraction of the proteome containing IDRs between prokaryotes, simple eukaryotes and multicellular organisms, and the number of proteins containing IDRs correlates with the number of cell types, suggesting co-evolution of IDR-mediated transactions with developmental complexity.[559,560]

IDRs are usually located at the N- or C-terminal region of the protein. IDRs are present in and essential for the function of nearly all of the proteins involved in animal and plant development,[s] including RNA polymerase, most transcription factors, Hox proteins, histones, histone-modifying proteins, other chromatin-binding proteins, the Mediator complex, RNA-binding proteins, splicing factors, membrane receptors, cytoskeletal proteins and nuclear hormone receptors.[271,272,274,539,551,560,563,570-577]

Surprisingly, the majority of proteins subject to alternative splicing contain IDRs.[560] Moreover, IDRs are overrepresented in alternatively spliced exons subject to tissue- and lineage-specific regulation,[578-581] especially in exons that are alternatively spliced in mammals but constitutively spliced in other vertebrates,[582] which changes the subcellular localization of the isoform and coordinates phase transitions within the cell.[583,584]

IDRs also occur in proteins that are flexibly involved in signal transduction and transport,[539] such as the Ras-GTPase-activating proteins (SH3 domain)-binding proteins G3BP (which forms phase-separated domains),[585] clathrin-mediated endocytosis[586] and synapsins, which are required for the maintenance of synaptic vesicle clusters in neurons by IDR-mediated phase separation.[587]

IDRs are major sites of post-translational modifications and many biological processes, including the regulation of the cell cycle and circadian clocks,[539] have been shown to be dependent on post-translational modification of IDRs.[539,588-590] Post-translational modifications modulate RNA binding[591] and alter the propensity to nucleate PSDs,[592] adding layers of complexity to their interactions and regulation.[539,557,593,594]

The known post-translational modifications not only include the 100 or so found in histones, but also 95 in the IDR of the axonal microtubule–associated protein Tau, which is involved in Alzheimer's and other neurodegenerative diseases.[595] Tandem RNA-binding sites in the RNA-binding protein, TIA-1, facilitate PSD stress granule formation[596] and reduction of this protein protects against Tau-mediated neurodegeneration.[597]

Other proteins involved in neurological functions and disorders, such as TDP-43, ataxin, c9orf72 and FMRP (fragile X mental retardation protein), also contain IDRs that are involved in phase separation,

---

[r] 'RNA regulates the formation, identity, and localization of phase-separated granules'.[542]

[s] A much higher percentage than found in the rest of the proteome.[272,569]

controlled in part by post-translational modifications.[598,599] For example, the IDR of TDP-43 binds RNAs,[600] and the loss of its RNA-binding ability by mutations or post-translational acetylation leads to its sequestration into PSDs.[601] The IDRs of ataxin mediate formation of neuronal mRNP assemblies, and are essential for long-term memory formation as well as c9orf72-induced neurodegeneration.[602] Neuronal-specific micro-exons overlapping IDRs in the translation initiation factor eIF4G regulate the coalescence of phase-separated granules to repress translation, are misregulated in autism, and their deletion in mice leads to altered hippocampal synaptic plasticity and deficits in social behavior, learning, and memory.[603]

Aberrant promiscuity of IDR-containing proteins (IDPs) and perturbations of PSD formation may underlie the dosage sensitivity of oncogenes and other proteins[604,605] as well as neurodegenerative disorders such as Alzheimer's Disease, Parkinson's Disease, Frontotemporal Dementia, Muscular Dystrophy and Amyotrophic Lateral Sclerosis, where repeat expansions affecting RNA and/or their encoded proteins result in pathological aggregates.[598,606,607]

Mutations in RBPs that cause human monogenic diseases are observed more commonly in IDRs than globular domains,[586,608] indicating that, despite their relatively simple composition, IDRs have strong sequence constraints. It is also clear that RNA nucleates the formation, and is the structural scaffold, of PSDs.[27,542,547,549,551,553,609–614]

PSDs encompass a range of nuclear compartments:[27,615] DNA replication initiation sites;[616] telomeres;[617] centrosomes[618] and meiotic chromosomal pairing foci;[619,620] germ granules;[592,621] nucleoli, Cajal bodies and 'histone locus bodies';[540,622–624] 'extranucleolar droplets';[523] spliceosomes ('nuclear speckles'[t]);[626] specialized spliceosomes (via lncRNAs Gomafu[u] and Malat1);[627–629] paraspeckles[v] (Neat1);[630–634] heterochromatin;[609,635] Polycomb bodies;[347] primate-specific nuclear stress bodies;[636,637] nuclear glucocorticoid receptor foci;[638] SARS-Cov2 viral assembly domains;[639] and others, including in plants.[640] They also include cytoplasmic organelles[537,641,642] such as P-granules,[643,644] G-bodies,[645] stress granules,[646] polar bodies (whose formation is dependent on a lncRNA),[647,648] localized

mRNP translational assemblies[649,650] and synaptic compartments.[651]

It has been proposed that lncRNAs play a central role in organizing the three-dimensional genome,[521] including the formation of spatial compartments and transcriptional condensates[610,614,652–655] (Figure 16.4) and hence the four-dimensional patterns of gene expression during differentiation and development.[w] It has been shown that phase separation drives chromatin looping[657] and is required for the action of enhancers and super-enhancers;[551,610,658–660] that transcription factors activate genes through the phase-separation capacity of their activation domains by forming PSDs with RNA polymerase II;[661,662] that Mediator and RNA polymerase II associate in transcription-dependent condensates;[658,661–663] that phase separation of RNA-binding protein promotes polymerase binding and transcription;[664] and that PSDs scaffolded by lncRNAs, including repeat-derived RNAs, mediate heterochromatin formation,[32,609,614,665–669] euchromatin formation,[670] nucleolar structure,[671–674] splicing[675] and DNA damage repair.[676–678]

For example, it has been shown that the cytoplasmic lncRNA NORAD, which is induced by DNA damage and required for genome stability, prevents aberrant mitosis by sequestering Pumilio proteins (which bind many RNAs to regulate stem cell fate, development and neurological functions[680]) into PSDs via multiple repeats.[681–684] Lack of NORAD accelerates aging in mice.[685] Similarly sequestration of the double-strand beak enzyme RAG1 into nucleoli modulates V(D)J recombination activity.[686] Many natural antisense lncRNAs with embedded mammalian interspersed repeats are overrepresented at loci linked to neurodegeneration and/or encoding IDPs.[687]

The PARP1 superfamily,[x] one of the most abundant proteins in the eukaryotic nucleus,[688–690] which catalyzes the polymerization of ADP-ribose units and attachment of poly (ADP-ribose) polymers to arginines in target proteins,[y] including histones, for DNA repair, stabilization of replication forks

---

[t] p53 target gene association with nuclear speckles is driven by p53.[625]

[u] Japanese for 'spotted pattern'.

[v] Involving RNA-protein and RNA-RNA interactions.[613,630,631]

[w] PSDs may also serve to reduce noise in biological signal processing and control.[656]

[x] There are 18 members of the PARP superfamily encoded in the human genome. PARP2 is also involved in chromatin modification, PARP3 is a core component of centrosomes, and PARP4 is associated with vault particles (Chapter 8).[688]

[y] Interestingly, a similar enzyme in bacteriophage has been reported to add entire RNA chains to a host ribosomal protein to modulate the phage replication cycle.[691]

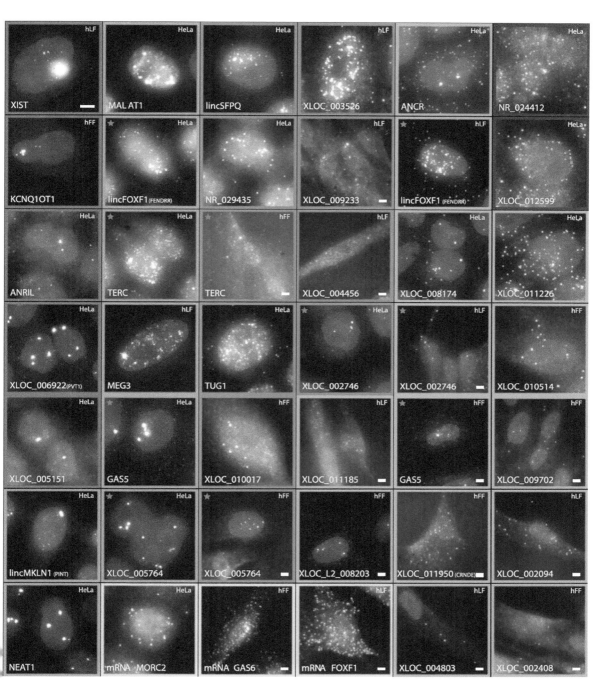

FIGURE 16.4 Subcellular and subnuclear localization of RNAs in punctate domains. (Reproduced with permission from Cabili et al.[679] with the permission of the authors under Creative Commons attribution license.)

and the modification of chromatin, also binds lncRNAs,[692–695] regulates RNA metabolism[696,697] and modulates the phase-separation properties of RNA-binding proteins.[698,699]

Many lncRNAs appear to be localized to defined nuclear and cytoplasmic foci that resemble liquid droplets,[336,414,473,679,700–702] and a genome-wide study identified hundreds of non-coding RNAs forming nuclear compartments near their transcriptional loci, in dozens of cases guiding cooperating proteins into these 3D compartments and regulating the expression of genes contained within them [228,614] (Figure 16.5).

X-chromosome dosage compensation in *Drosophila* requires the formation of a phase-separated coacervate by the lncRNAs roX1 and roX2 interacting with the IDR of a specific partner protein (MSL2, 'male sex lethal 2'). Moreover, replacing the IDR of the mammalian ortholog of MSL2 with that from *Drosophila* along with expression of roX2 is sufficient to nucleate ectopic dosage compensation in mammalian cells, showing that the roX–MSL2 IDR interaction is the primary determinant for compartmentalization of the X chromosome, and a likely exemplar of lncRNA-IDR interactions in general.[424]

As further evidence that the eukaryotic nucleus (and indeed the eukaryotic cytoplasm) is finely organized, and that many more phase-separated domains remain to be discovered, a recent report has shown that two related RNA modification enzymes that normally reside (in one case) in the nucleolus and (in the other) in an unknown cytoplasmic domain proximal to mitochondria, both relocate upon nerve cell depolarization to different small unknown punctate nuclear domains.[703]

The complexity is extraordinary and literature on this topic is burgeoning, but it is now clear that PSDs comprise a major and until recently unappreciated fine-scale and dynamic spatial regulation of subcellular and chromatin organization, "the active chromatin hub", first proposed by Wouter de Laat and Frank Grosveld in 1993 based on the study of globin enhancers,[704] which "unifies the roles of active promoters and enhancers".[520] It has also been proposed, with experimental support, that "ribonucleoprotein complexes can act as block copolymers to form RNA-scaffolding biomolecular condensates with optimal sizes and structures in cells".[634]

AN ADDITION TO THE ANCIENT RNA WORLD HYPOTHESIS

The ability of RNA to nucleate phase-separated domains adds a third dimension to its role in the origin of life.[705] While it has been widely accepted that RNA was likely the primordial informational and catalytic molecule of life, its advent would also have enabled the formation of a pre-cellular phase-separated privileged environment wherein organic reactions could be concentrated and evolve. Indeed, compartmentalized RNA catalysis has been demonstrated in membrane-free coacervate protocells.[706,707]

The development of RNA-nucleated coacervates likely involved its interaction with positively charged (particularly arginine-rich) disordered proteins.[708]

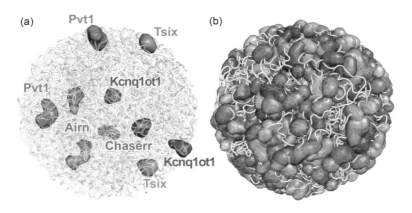

FIGURE 16.5 RNA promotes the formation of spatial compartments in the nucleus. (a) A 3D space filling nuclear structure model of selected lncRNAs. (b) A 3D space filling nuclear structure model of 543 lncRNAs that display at least 50-fold enrichment in the nucleus. Each sphere corresponds to a 1 Mb region or larger where each lncRNA is enriched. (Reproduced from Quinodoz et al.[614] with permission of Elsevier.)

Intrinsically disordered proteins are encoded by the most ancient codons and appear to be the first polypeptides, likely to have functioned initially as chaperones, with catalysis transferred first from RNA to ribonucleoprotein complexes and then to proteins,[554,555,557] a process that may have been interactive.[709]

STRUCTURE-FUNCTION RELATIONSHIPS IN LNCRNAS

The length of lncRNAs varies enormously although in many cases their true length and structure are unknown due to their cell-type specificity and low representation in RNA sequencing datasets. However, high depth sequencing has shown that most are multi-exonic,[710] and some are over 100 kb in length (post splicing), so-called macroRNAs, which have a mean length of 92 kb and are predominantly localized in the nucleus.[711]

Why are lncRNAs so long? The likely answer is that they contain a set of modular domains for binding proteins and guiding them to target sequences in DNA or (other) RNAs[228,712–714] (Figure 16.6).

First, although rapidly evolving (under relaxed structure-function constrains and positive selection for adaptive radiation), lncRNAs exhibit common motifs and motif combinations across vertebrates,[715] and at least 18% of the human genome is conserved at the level of predicted RNA structure.[716] For example, it has been shown that conserved pseudoknots in lncRNA MEG3 are essential for stimulation of the p53 pathway.[717]

Second, similar and potentially paralogous predicted RNA structures occur at many places throughout the genome.[718–720]

Third, lncRNAs are enriched for repeat sequences, which have highly non-random distributions in them.[131,681,721] A notable feature of *Xist*, for example, is that its most highly conserved sequences are the repeat elements, whereas its unique sequences have evolved rapidly,[722] and many of its biological functions, including PRC2 binding, are mediated through its modular repeat elements.[371,379,380,387,397,723]

Many of the lncRNAs referred to earlier have also been shown to be modular, with common features being rapid sequence evolution and structural divergence while retaining related functions, sometimes across large evolutionary distances, and the use of TE-derived sequences as protein-binding domains.[410,721,724,725]

Indeed, TE-derived sequences and tandem repeats participate in many RNA-protein interactions,[34,675,726,727] which leads to the reasonable conclusion that repeat sequences act as RNA-, DNA-, and protein-binding domains that are the essential components of lncRNA function,[721] and that TEs are key building blocks of lncRNAs[397,728] (as well as fulfilling many other modular functions in gene control and gene expression; Chapter 10). Transposition is an efficient means of mobilizing functional cassettes[270] and allowing evolution to explore phenotypic space by modulation of the epigenetic control of developmental trajectories. As pointed out by Neil Brockdorf, "tandem repeat amplification has been exploited to allow orthodox RBPs [RNA binding proteins] to confer new functions for Xist-mediated chromosome inactivation … with potential generality of tandem repeat expansion in the evolution of functional long non-coding RNAs".[397]

Fourth, many lncRNAs bind chromatin-modifying proteins, transcription factors, nuclear matrix proteins and RNA-binding proteins, in those cases that are well studied, like Xist, roX and HOTAIR, to exert functional consequences.[309,311,342,371,727] It has also been reported that an mRNA can act as a scaffold to assemble adaptor protein assemblies to regulate intracellular transport.[729]

Fifth, chemical probing has shown that lncRNAs, including Xist, physically have a modular structure,[371,374,730,731] and the chemical data matches that predicted by evolutionary conservation of secondary structure, validating both.[374]

Finally, the extensive alternative splicing of lncRNAs strongly imputes a modular structure[710,732–735] and alternative splicing has, unsurprisingly, been shown to alter the function of lncRNAs.[735–738]

If the complex ontogeny of a human requires a large number of guide RNAs then it is not surprising that many have similar protein-binding modules, with variation in repertoire and genomic target specificity, which may only require short stretches of nucleotide complementarity, given the high strength of RNA-RNA and RNA-DNA interactions.[739] These modules may also include enzymes that have adjunct roles in epigenetic transactions: for example, the developmentally regulated lncRNA *H19* binds to and inhibits S-adenosylhomocysteine hydrolase, a feedback inhibitor of DNA methyltransferases.[740] The alternative splicing of lncRNA exons (which itself must be epigenetically controlled[741]) permits

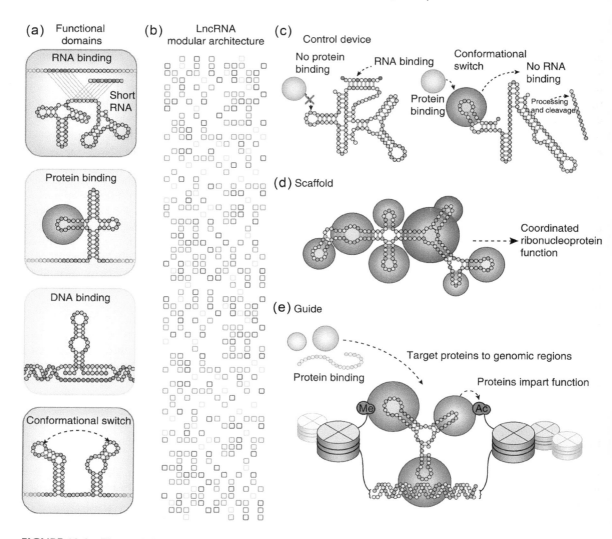

FIGURE 16.6   The modular domain structure and interactions of lncRNAs. (Reproduced from Mercer and Mattick.[714])

selection of specific protein-binding modules and target sites for feed-forward control of protein and regulatory RNA gene expression at many different loci and ultimately cell fate (hold, divide or differentiate) decisions at every stage of developmental ontogeny. There is no cogent model for such fine control by proteins alone.

The challenge now is to determine the repertoire of RNA structures using RNA folding programs, evolutionary conservation, physical and chemical analyses, and machine learning.[374,716,718–720,742–747] The challenge is to determine which RNA structures bind which proteins or which DNA or RNA targets, with a range of techniques becoming available to map RNA localization and interactions,[13,371,518,614,748–755] so

that lncRNA biology can be parsed and understood, and thereby construct an expanded Rfam ('RNA family') database,[756] like the Pfam protein domain database[757] that has proved so useful in identifying protein function.

A NEW VIEW OF THE GENOME OF COMPLEX ORGANISMS

It is increasingly evident that lncRNAs, enhancers, topologically associated chromatin domains, transposon-derived sequences and other repeats, chromatin-remodeling and epigenetic information are merging into the same conceptual and mechanistic space.

Imagine the versatility and temporal precision that could be achieved if the locus specificity or local accessibility of proteins that control gene expression during development is guided by stage-specific modular RNAs. The strength of RNA is its potential to address targets through sequence-specific duplex or triplex base-pairing while at the same time recruiting and directing effector proteins to specific genomic locations.

We propose that regulatory RNAs, including the 'repeat' sequences within them, are the evolutionarily and developmentally flexible platforms expressed from the genome in an unfolding symphony to direct and execute the extraordinarily complex decisions required for the precise ontogeny of trillions of differentiated cells in a human, and similarly in other mammals, vertebrates, invertebrates and plants. That is, the epigenetic marks that control development, physiological adaptations and brain function are positioned and controlled by RNAs (among their many other functions in cell biology and gene regulation), and that the proportion of the genome devoted to specifying regulatory and architectural RNAs increases with developmental and cognitive complexity.[758,759]

Small RNAs (miRNAs, sgRNAs, etc.) are simple sequence-specific guides for a single type of effector, such as RISC or CRISPR/Cas. LncRNAs not only have target sequence specificity but are also scaffolds for a range of proteins, notably chromatin-modifying complexes, with both targets and cargoes being four-dimensionally regulated by alternative splicing of lncRNA exons in a feed-forward cascade that directs the next cell fate decision during development. This is a highly efficient system that, like RNAi and CRISPR but in a far more sophisticated and modular manner, directs generic protein effectors to their sites of action.

The misleading historical perspective on the relationship between RNAs and proteins was best expressed by Ewa Grzybowska and colleagues, who concluded: "The current perception of RNA-protein interactions is strongly biased toward a protein-centric approach, in which proteins regulate the expression and activity of RNA, not the other way around."[563]

FURTHER READING

Bah A. and Forman-Kay J.D. (2016) Modulation of intrinsically disordered protein function by post-translational modifications. *Journal of Biological Chemistry* 291: 6696–705.

Cosby R.L., et al. (2021) Recurrent evolution of vertebrate transcription factors by transposase capture. *Science* 371: eabc6405.

Davidovich C., Zheng L., Goodrich K.J. and Cech T.R. (2013) Promiscuous RNA binding by Polycomb repressive complex 2. *Nature Structural & Molecular Biology* 20: 1250–7.

Guttman M. and Rinn J.L. (2012) Modular regulatory principles of large non-coding RNAs. *Nature* 482: 339–46.

Hahn S. (2018) Phase separation, protein disorder, and enhancer function. *Cell* 175: 1723–5.

Kapusta A., et al. (2013) Transposable elements are major contributors to the origin, diversification, and regulation of vertebrate long noncoding RNAs. *PLOS Genetics* 9: e1003470.

Kim T.-K., Hemberg M. and Gray J.M. (2015) Enhancer RNAs: A class of long noncoding RNAs synthesized at enhancers. *Cold Spring Harbor Perspectives in Biology* 7: a018622.

Mercer T.R. and Mattick J.S. (2013) Structure and function of long noncoding RNAs in epigenetic regulation. *Nature Structural & Molecular Biology* 20: 300–7.

Niklas K.J., Dunker A.K. and Yruela I. (2018) The evolutionary origins of cell type diversification and the role of intrinsically disordered proteins. *Journal of Experimental Botany* 69: 1437–46.

Polymenidou M. (2018) The RNA face of phase separation. *Science* 360: 859–60.

Quinodoz S.A., et al. (2021) RNA promotes the formation of spatial compartments in the nucleus. *Cell* 184: 5775–90.

Sabari B.R., et al. (2018) Coactivator condensation at super-enhancers links phase separation and gene control. *Science* 361: eaar3958.

Staby L., et al. (2017) Eukaryotic transcription factors: Paradigms of protein intrinsic disorder. *Biochemical Journal* 474: 2509–32.

Trizzino M., et al. (2017) Transposable elements are the primary source of novelty in primate gene regulation. *Genome Research* 27: 1623–33.

Uversky V.N. (2016) Dancing protein clouds: The strange biology and chaotic physics of intrinsically disordered proteins. *Journal of Biological Chemistry* 291: 6681–8.

17 Plasticity

Gene-environment interactions occur in all organisms, ranging from bacterial transcriptional responses to nutrient availability to epigenetic changes in eukaryotes in response to environmental circumstances, such as that observed in the increased production of erythrocytes and increased hemoglobin expression at high altitudes[1–3] or the advent of type 2 diabetes upon prolonged obesity,[4] although genetic factors are also involved.[2,5]

The major focus of the studies of molecular basis of long-term gene-environment interactions in humans and other eukaryotes has been changes in the patterns of DNA methylation[6–8] and, more recently, histone modifications.[9,10] However, it is clear that RNA is also modified, over a far wider chemical range than DNA. If DNA methylation and histone modifications are RNA-directed (Chapter 16), as are some RNA modifications (Chapter 8), it is logical that RNA is the major conduit for epigenome-environment interactions[11] and, by extension, that the expansion of RNA modifications and RNA editing in complex organisms underpins phenotypic plasticity, learning, and cognition.[12] It also appears increasingly likely that RNA is the vehicle for transgenerational soft-wired inheritance of experience.[13,14]

RNA MODIFICATIONS AND THE UNKNOWN EPITRANSCRIPTOME

There are over 25 cell-specific versions of the 5' cap structure of RNAs and non-canonical initiating nucleotides with functional consequences in RNA processing, export, stability and translation.[15–18] There are also over 140 known chemical modifications of internal ribonucleotides[19–22] (Figure 17.1). These modifications occur on all four standard bases as well as on the ribose, and were detected initially in the highly abundant rRNAs and tRNAs, and later in snoRNAs and snRNAs. For decades RNA modifications were thought to be irreversible decorations important for structural stability and/or catalytic function, including in prebiotic evolution,[23] but this view changed with the discoveries that rRNA modifications are context-specific,[24] that modifications

occur in thousands of mRNAs, lncRNAs, enhancer RNAs, vault RNAs, miRNAs and other non-coding RNAs,[25–37] and, especially, that RNA modifications are reversible,[38–40] leading to the birth of the term 'epitranscriptome' to describe the collective of regulated RNA modifications.[41]

There are technical challenges in identifying RNA modifications in sequencing datasets.[42] Most of what is presently known stems from the study of m6A, and to a lesser extent m1A, m6Am (N6,2'-O-dimethyladenosine), m5C and pseudouridylation modifications, for which there are specific antibodies or reagents available[a] for immunocapture to identify the positions of modified bases.[25,26,28–33,43–46]

m6A modifications in mRNAs occur typically near the stop codon, but also in 5'UTRs, coding sequences, introns and 3'UTRs.[25,26,32,47] m6A modifications have been found to regulate miRNA processing[34,48] and mRNA polyadenylation, processing, splicing, stability, translation and export[48–59] by destabilizing RNA duplexes and altering RNA-protein interactions.[60–62] m6A modification of mRNAs, promoter-associated RNAs, enhancer RNAs and repeat RNAs have been shown to regulate chromatin state, phase separation, the stability of R-loops (and their role in the repair of double-strand breaks) and transcription.[63–69]

m6A modifications have been shown to regulate yeast meiosis,[70] the activity of endogenous retroviruses,[71] heterochromatin formation and TE function in embryonic stem cells and early embryos,[72–74] mammalian stem cell renewal, differentiation and development,[44,52,54,75–78] spermatogenesis, oogenesis and fertility,[50,56,57,79,80] embryonic development,[81,82] adipogenesis,[39] circadian rhythms,[49,83,84] stress responses,[85–87] hematopoietic differentiation,[88] immune responses[89] and inflammation,[90,91] IgH recombination,[92] neurogenesis,[93–95] neural differentiation[96,97] and neural circuitry,[98,99] spatiotemporal control of mRNA translation in neurons,[100,101] cerebellar development,[102,103] learning and memory,[104–109] cancer stem cell differentiation,[110] neuronal functions

---

[a] In the case of m5C, RNA bisulfite sequencing.[27]

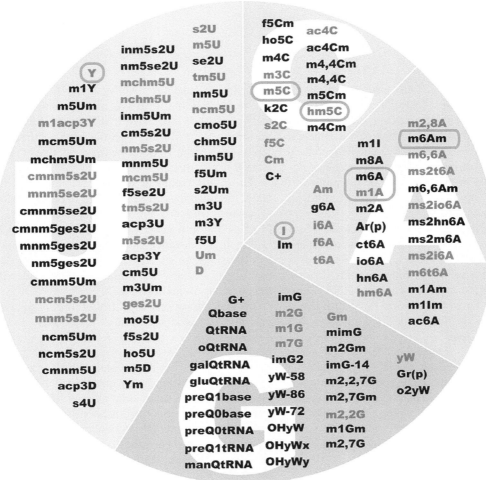

FIGURE 17.1 The set of known RNA modifications classified by their reference nucleotide, highlighting those that have been associated to a human disease (red), as well as those for which a transcriptome-wide detection method has been established (green). (Reproduced from Jonkhout et al.[19])

and sex determination in flies[111–113] and even rice and potato yield,[114] in some cases involving interplay with histone modifications.[90]

Although most current techniques favor highly expressed RNAs, m6A modifications are increasingly being detected in non-coding regions of pre-mRNAs and lncRNAs in tissues such as placenta, kidney, liver and brain, with evidence of modulation of lncRNA functions and properties, including splicing regulation and possible effects of sequence polymorphisms.[115] Specific examples include m6A

modification of MALAT1 as a structural switch affecting its protein binding and phase separation properties,[116,117] enhancement of *Xist* repressive activity,[118] and modulation of the stress-induced repressor of cyclin D1 pncRNA-D to induce cell cycle arrest.[119]

Other RNA modifications that have been studied have been shown to similarly affect a wide range of processes including chromatin organization, mRNA stability, tRNA and miRNA processing, and to be involved in neurological and other disorders.[30,36,46,77,120–126] Many of these modifications have been documented to affect the function of regulatory RNAs, such as SRA,[127] 7SK[128,129] and vault RNAs.[28]

As with histone modification enzymes, there are a range of RNA modification writers, readers and erasers, the loss or perturbation of which, including in rRNAs and tRNAs, results in a range of diseases, including cancer, intellectual disability and developmental disorders.[19,20,130–132]

The repertoire, substrate range (often from tRNA to mRNA) and deployment of RNA modification enzymes have been expanded by successive gene duplications that have occurred at the base of the eukaryote, metazoan, vertebrate and primate lineages, with 90 cataloged in the human genome.[133]

In mammals there are multiple m6A writers (the METTL protein family), readers and erasers[134,135] with far-reaching functions. For example, METTL3 regulates heterochromatin in embryonic stem cells[72] and promotes homologous recombination-mediated repair of double-strand breaks by modulating DNA-RNA hybrid accumulation.[67] METTL16, which is essential for mouse embryonic development,[82] regulates the expression of an enzyme that produces the methyl donor, S-adenosyl methionine.[136] The m6A reader Ythdc1 regulates the scaffolding function of LINE1 RNA in mouse ESCs and early embryos.[73] The m6A reader Prrc2a controls oligodendroglial specification and myelination.[97] The ALKBH5 m6A eraser controls translation[40] and the splicing and stability of long 3′UTR mRNAs in male germ cells.[56]

There are eight mammalian m5C writers, Nsuns1–7 and Dnmt2. Nsun1, 2, 5 and Dnmt2 are present in all eukaryotes, whereas the other Nsuns are specific to multicellular organisms and are differentially expressed during development, particularly in the brain: Nsuns1–4 participate in embryonic development, cell proliferation and differentiation; disabling mutations in Nsun2[b] and Nsun7 cause intellectual disability and male sterility, respectively; Nsun5 is essential for normal growth and cerebellar development; Nsun6 associates with the Golgi apparatus and catalyzes the formation of m5C72 in specific tRNAs. It also methylates mRNAs, particularly in their 3′UTRs, but is apparently dispensable for normal development.[36,46,75,121,137–141] Nsun1 binds RNA polymerase II (RNAPII) and Nsun3 and Dnmt2 bind hnRNPK, which interacts with lineage-specific transcription factors, and with CDK9/P-TEFb to recruit RNAPII to active chromatin hubs.[124] The subcellular localizations of enzymes that methylate guanosine are altered by neuronal stimulation[142] (Figure 17.2).

The field is in its infancy: most RNA sequencing protocols involve conversion to DNA, with concomitant loss of modification information, although mismatch patterns and blockage of reverse transcription can provide an indication.[143] A solution is at hand with the advent of direct RNA sequencing using nanopore technology,[c] and the (altered) signals from some base modifications are now being identified.[42,146–150] One can only speculate on the processes controlled by the many other RNA modifications that are yet to be analyzed.

In 2005, Katalin Karikó and colleagues discovered that RNAs containing modified nucleotides [m5C, m6A, m5U, s2U or 1-methylpseudouridine[d] (m1Ψ, a naturally occurring component of eukaryotic 18S rRNA)] do not activate the mammalian Toll-like receptor 7 that detects single-stranded RNAs, innate immune receptors,[152] potentiating the development of mRNA-based vaccines.[151] Proteins produced from mRNA vaccines stimulate both innate and adaptive immune responses, and the platform is much more flexible and scalable than attenuated viruses.[153] mRNA vaccines can be produced within a day or two of a viral sequence being available and formed the frontline against SARS-CoV-2, and will likely be a platform for the delivery of other vaccines, autoimmune rectification, targeted cancer therapies and even the treatment of heart failure in the future,[154–158] using lipid nanoparticles for delivery.[159] There is also increasing appreciation of the potential for non-coding RNA therapeutics, given the central role of RNA in most biological processes, and growing evidence for the efficacy of non-coding RNA interventions.[160–163]

THE EXPANSION OF RNA EDITING IN COGNITIVE EVOLUTION

An important subset of RNA modifications is base deamination, referred to as RNA 'editing'.[e] There are two classes: adenosine deamination to inosine (A>I), which registers as guanosine in translation

---

[b] Loss of the Nsun2 ortholog in *Drosophila* causes short-term memory deficits.

[c] Excitingly, nanopore sequencing has recently been adapted to read protein sequences at single amino acid resolution.[144,145]

[d] Replacement of uridine with m1Ψ in synthetic mRNA improves its immunogenicity, translational capacity and stability.[151]

[e] The term RNA editing was coined by Rob Benne and colleagues in 1986 to describe the small RNA-guided site-specific insertion or deletion of uridines in mRNAs in mitochondria of Trypanosomes and related protists.[164–166]

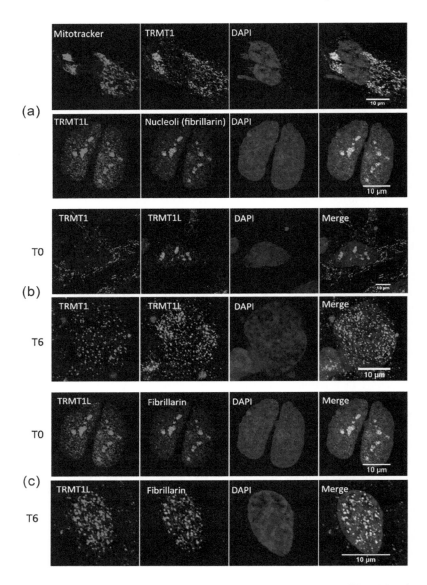

FIGURE 17.2 Dynamic changes in subcellular location of two enzymes (TRMT1 and TRMT1L) that catalyze N2,N2-dimethylguanosine (m2,2G) RNA modification. (a) TMRT1 localizes to compartments associated with mitochondria and TRMT1L to nucleoli in resting human neuroblastoma cells. (b) Depolarization of the cells causes the mitochondrial TRMT to relocate to punctate domains in the nucleus, and the nucleoli to fragment (c). (Adapted from Jonkhout et al.[142])

and RNA sequencing, but has differences that may be important *in vivo*; and cytosine deamination to form uracil (C>U), or methylcytosine to thymine (meC>T). Alterations in both A>I and C>U editing feature prominently in human cancers.[167–171]

A>I EDITING

A>I editing was discovered in the late 1980s by Brenda Bass, Hal Weintraub, David Kimelman and Marc Kirschner who showed that synthetic and natural RNA duplexes formed between sense and antisense RNAs expressed from the *bFGF* locus in *Xenopus* (toad) eggs, and double-stranded viral RNAs, are substrates for an enzyme that deaminates adenosines.[172–174] A>I editing has since been extensively characterized by Bass, Kazuko Nishikura, Mary O'Connell, Robert Reenan, Charles Samuel, Peter Seeberg, Marie Öhman, Gerhardt Wagner and colleagues.

A>I editing is performed by animal-specific enzymes called ADARs, which have evolved from enzymes that deaminate adenosines in tRNAs.[168,175–180] Invertebrates have one or two ADARs, whereas vertebrates have three (ADAR1–3). The expression of the ADARs varies across development and tissues in mammals. ADAR1 is widely expressed throughout the body and is the most highly expressed ADAR outside the central nervous system.[181]

Editing appears to occur both co- and post-transcriptionally.[182,183] The basic substrate is imperfect double-stranded RNA, especially A:C mismatches, which includes dsRNA regions in pre-mRNAs and lncRNAs, often involving intron-exon base pairing.[184] The substrate specificities of the different ADAR orthologs is not well understood,[175] although evidence suggests that ADAR1 imposes symmetrical editing at positions in dsRNAs 30–35 bp away from structural disruptions[185] and that ADAR2 substrate recognition involves a GCU(A/C)A pentaloop conserved in mammals and birds.[186]

A>I RNA editing is widespread in RNAs encoding proteins involved in neurotransmission, including presynaptic release machineries and voltage- and ligand-gated ion channels,[168,176,187,188] where it alters codons or splicing patterns[f] and therefore protein structure-function relationships[172,187,191–197] to modulate the electrophysiological properties of the synapse and neuronal connections in response to activity,[168,191,192,196,198] and to adapt to environmental conditions.[199]

The substrate range also includes RNAs encoding proteins involved in brain patterning, neural cell identity, maturation and function, as well as in DNA repair, implying a role for RNA editing not only in neural transmission and network plasticity but also in brain development and memory consolidation.[200] The RNAs encoding ADARs are themselves also edited,[201–203] and RNA editing is regulated by m^6A modifications,[204] indicating feedback loops and interplay between RNA editing and modification systems.

A>I RNA editing occurs commonly in introns, where it influences nuclear retention and splicing, including that of ADAR2 itself.[176,183,205–207] RNA editing also alters miRNA processing, expression and target specificity.[208–211]

The regulatory pathways that control A>I RNA editing are not understood. Presumably, editing alters the structure and information content of coding and regulatory RNAs in response to environmental signals and experience, a possibility supported by the observations that ADAR2 has inositol hexakisphosphate (InsP6) complexed in its active site[212] and that InsP6 regulates AMPA/glutamate receptors,[213] indicating that ADAR activity and/or target selection is linked to cell signaling pathways.

Vertebrate ADAR1 and ADAR2 are widely expressed, most highly in brain, where their editing profiles overlap;[181,214] both are mainly localized in the nucleus, although a longer isoform of ADAR1 shuttles between the nucleus and the cytoplasm,[179] where it modulates the innate immune response (see below). ADAR1 and ADAR2 form homo- or heterodimers *in vivo* and dimerization is required for catalysis.[176–178]

In addition to the deaminase domain and RNA-binding domains[g] present in other ADARs, vertebrate ADAR1 also contains one (in the constitutively expressed shorter isoform) or two domains (in the inducible longer isoform) that recognize an alternate left-handed helical DNA or RNA structure, termed Z,[216–220] which occurs naturally through the genome.[h] ADAR1 binds to Z-DNA and Z-RNA[i] in or near repetitive elements, especially Alu elements.[223]

The longer isoform of ADAR1 can be induced by interferon[j] and edits non-coding regions of endogenous dsRNAs at Alu repeats, which are 'flipped' into Z-conformation to distinguish self and suppress inappropriate activation by the MDA5 helicase of the innate immune response that occurs in the presence of unmodified viral dsRNAs.[225–228] It is also induced during learning and its knockdown leads to a reduction in Z-DNA at sites where ADAR1 is recruited and an inability to modify previously

---

[f] ADAR2 also regulates the stability of mRNAs through editing of Alu elements in their 3'UTRs[189] with cross-talk from lncRNAs.[190]

[g] There are three dsRNA-binding domains in ADAR1, and two each in ADAR2 and ADAR3.[215]

[h] There are also other DNA structures that differ from the canonical B-form right-handed double helix described by Watson and Crick, such as G quadruplexes and A-form helices, which also occur in double-stranded RNAs and in DNA-RNA hybrids. These were discovered in the decade from 1979 to 1989,[221] and presumably have functional significance, but their distribution in genomes, and how they might vary during differentiation and development, is as yet only poorly characterized, a blind spot in the ENCODE projects.

[i] Z-DNA is commonly found in introns, and RNA editing commonly involves dsRNAs formed between exons and introns in pre-mRNAs, adding to the selective pressure on their sequences.[222]

[j] The longer isoform may be constitutively expressed and even be the dominant isoform in some tissues.[224]

acquired memory.[229] Translation of the longer iso-form of ADAR1 is potentiated by m⁶A modification of a conserved site in the ADAR1 transcript,[228] with cross-talk between modification systems during host responses to viral infections.[230] Loss of ADAR1 in mice results in embryonic lethality, due to a failure of hematopoiesis.[231,232] Mutations in human ADAR1 are one of the defined genetic causes of Aicardi-Goutières syndrome, an autoinflammatory disorder characterized by spontaneous interferon production and neurological problems.[233,234] Curiously, however, the loss of the editing capacity of ADAR1 has little developmental effect if the innate immune system is prevented from sensing the unedited dsRNAs.[214,235]

Vertebrate ADAR2 is required for the editing of neuroreceptor mRNAs, especially that encoding the AMPA receptor subunit GluA2 (Gria2),[236] which has the functional consequence of rendering it Ca⁺⁺ impermeable.[192] ADAR2 activity is also regulated in part by snoRNAs and nucleolar sequestration.[237,238] Mutations in human ADAR2 cause microcephaly and neurodevelopmental disorders.[239,240] Its defi-ciency in mice causes seizures and early lethality, which can be rescued by hard-wiring the single nucleotide change in the *Gria2* gene.[241] This obser-vation begs the question of why evolution has not imposed this change in the first place, but rather con-served the surrounding intronic sequences that are required for editing.[242]

While GluA2 receptors in adults are almost universally edited,[k] such editing may be a mecha-nism for pruning synaptic trees during the postna-tal maturation of neuronal circuitry in response to experience,[244–249] so that only the relevant survive.[l] ADAR2 expression increases as neurons mature and is spatially regulated within neurons.[198,251] Mice that lack ADAR2 but have the genomically encoded com-pensatory codon change in GluA2 exhibit changes in behavior, hearing ability and the expression profiles of RNAs in the brain, including those encoding pro-teins involved in synaptic trafficking.[252] There is also an N-terminal extension of ADAR2 that is expressed most highly in the cerebellum (through the alterna-tive splicing of an upstream exon), which harbors a sequence motif closely related to the single-stranded RNA-binding domain of ADAR3, which is also

expressed most highly in the cerebellum.[253] Another splice variant of ADAR2 has an insertion of an Alu cassette within the deaminase domain, which alters its catalytic activity.[254]

The single ADAR in *Drosophila* is homologous to vertebrate ADAR2 but also has similarities to ADAR1, in that it mediates suppression of both innate immune responses and brain functions.[255,256] Each neuronal population in *Drosophila* has a different editing signa-ture[188] and *Drosophila* lacking ADAR are morpholog-ically normal but exhibit extreme behavioral deficits including temperature-sensitive paralysis, locomo-tor defects, tremors and neurodegeneration.[250,256,257] Bees exhibit widespread A>I editing during foraging and brood caring task performance.[258] *C. elegans* has two ADARs, most similar to mammalian ADAR2 and ADAR3, which edit germline and neuronal tran-scripts (including 3'UTRs), the loss of which results in chemotaxis defects and reduced lifespan.[259–261]

ADAR3 is vertebrate- and brain-specific, con-tains both single- and double-stranded RNA-binding domains, and is thought to be catalytically inactive,[m] although its deaminase domain appears normal, pos-sibly because it does not dimerize and may act as an inhibitor of ADAR1 and ADAR2.[262–264] Loss of ADAR3 in mice causes no obvious developmental deficiencies, but does affect learning and memory.[265]

Interestingly, cephalopods such as squid, cuttle-fish and octopus, highly intelligent invertebrates, use A>I RNA editing far more extensively than mam-mals to modulate the sequence of mRNAs specify-ing proteins involved in nerve-cell development and signal transmission.[251,266–269]

Editing of non-coding sequences, on the other hand, has expanded enormously in primates, espe-cially humans.[199] While for many years it was thought that RNA editing is primarily directed at altering protein sequences, analysis of large-scale cDNA sequencing datasets revealed that it occurs in thousands of transcripts, largely in non-coding sequences,[270–274] altering RNA structure,[275] and pre-sumably regulatory circuits and networks, thereby influencing RNA-directed epigenetic memory. These analyses also showed that there is a massive expan-sion (a 35-fold increase) of A>I editing of human RNAs compared to mouse, mainly in the brain and mostly in Alu sequences.[270–274]

---

k GluA2 receptors are under-edited in malignant brain tumors.[243]
l Evidence from knockout of the *Drosophila* ortholog of ADAR2 also suggests that "edited isoforms of CNS proteins are required for optimum synaptic response capabilities in the brain during the behaviorally complex adult life stage".[250]

m There is another ADAR-like gene in mammals, TENR, which is expressed in the male germline and lacks a key catalytic resi-due in the deaminase domain.[168,175,177–179]

Alu elements invaded the genome in three waves during primate evolution and occupy 10.5% of the human genome (~1.2 million largely sequence-unique copies).[277,278] The editing of Alu sequences is higher in human transcripts compared to nonhuman primates, and new editable human-specific Alu insertions, subsequent to the human-chimpanzee split, are enriched in genes related to neuronal functions and neurological diseases[276] (Figure 17.3). Virtually all adenosines within double-stranded regions of Alu transcripts undergo A-to-I editing, although most sites exhibit editing at only low levels (<1%), and it has been estimated that there are over 100 million Alu RNA editing sites distributed across most human genes.[279]

Alu elements (and the related B2 SINE elements in mice) have been linked to a wide variety of functions, including new exons, splice junctions, promoters, nuclear localization, differentiation signals and stress responses, both within longer transcripts and as separate small RNAs[280–288] (Chapters 10, 13 and 16). They also have self-cleaving property[289] and their increased processing has been linked to neurological disorders.[288,290] However, their intense editing suggests a wider role as modular cassettes that permit the superimposition of plasticity on an otherwise hard-wired RNA regulatory system, which has been positively selected for physiological adaptation and cognitive advance. Alu elements are derived from a dimeric fusion of 7SL RNA, which is part of the signal recognition particle involved in targeting proteins for export (Chapter 8), and the conservation of its core structure[277,291,292] may provide a clue as to its function and success.

C>U Editing

C>U editing is performed by a family of proteins called APOBECs,[293–296] so named because the first to be discovered altered the sequence of Apolipoprotein B ('ApoB editing complex 1'),[n] a key constituent of circulating lipid transport vesicles (produced in the liver), which introduces a stop codon to produce a shorter version in the intestine.[297–299]

APOBECs arose at the beginning of the vertebrate radiation, although there is evidence of precursors in invertebrates,[300] and can edit both RNA

and single-stranded DNA (C>T), which muddies the functional waters. APOBECs can also edit meC (but not the TET-oxidized bases 5-hydroxymethylcytosine, 5-formylcytosine and 5-carboxylcytosine) to form T,[301] an editing event that cannot be distinguished in bisulfate-based methylation assays, and there may be cross-talk between methylation and editing systems *in vivo*.[302]

There are five families of APOBECs: AID and APOBEC2 occur in all vertebrates, APOBEC4 and APOBEC5 in tetrapods, APOBEC1[o] in the amniotes (birds and mammals) and reptiles, and APOBEC3 (some members of which have two deaminase domains) in placental mammals.[293–296,304,305] Expansions of the latter in different lineages correlates with the extent of germline colonization by retroviruses, most notably in primates, which have seven APOBEC3 paralogs (A, B, C, D/E, F, G, H)[295,296,306] (Figure 17.4), which exhibit some of the strongest signatures of positive selection in the human genome.[307,308] Other mammals show intermediate extents of APOBEC3 duplication.[306,309,310] The APOBECs show tissue-specific expression, in the immune system, muscle, liver and brain,[296,311] with APOBEC3G expression primarily in neurons[312] and APOBEC4 in testis.[305]

The ancestral gene, AID, arose in the jawless fishes as a central feature of adaptive immunity, and is involved in the somatic rearrangement and hypermutation of immunoglobulin domains in B cells and T cells, processes that are also heavily regulated by lncRNAs (Chapters 13 and 16).

The other APOBECs are generally thought to provide a defense against exogenous retroviruses and the mobilization of retrotransposons, such as endogenous retroviruses and SINE and LINE retroelements,[306,313–315] although why the APOBEC3 family expanded under strong selection in primates is unclear. Perhaps the clue is the widespread cooption of transposable elements in neuronal differentiation and function.[316]

Recent studies have shown that L1 retroelements are mobilized in neurons in culture to induce somatic mosaicism, a process that is controlled in part by methyl-CpG-binding protein 2 (MeCP2) (Chapter 14).[317–319] L1 elements are also differentially and dynamically methylated and histone modified during stem cell reprogramming and

---

[n] In 1987, by Lawrence Chan and colleagues, and by Lyn Powell, James Scott and colleagues, who reported that apolipoprotein B (apoB) mRNA contained a tissue-specific C>U base-modification that is not genomically encoded.[297,298]

[o] APOBEC1 deficiency has transgenerational epigenetic effects on testicular germ cell tumor susceptibility and embryonic viability.[303]

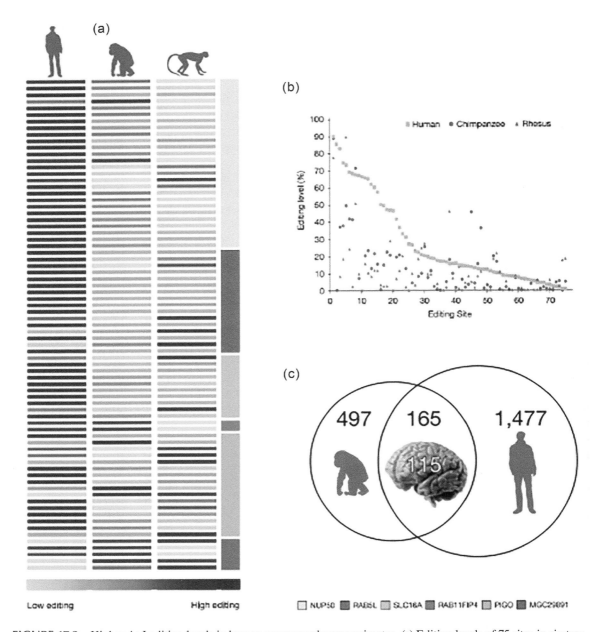

FIGURE 17.3 Higher A>I editing levels in human versus nonhuman primates. (a) Editing levels of 75 sites in six transcripts originating from cerebellum tissue of four humans, two chimpanzees, and two rhesus monkeys. (b) Editing level per site for humans, chimpanzees and rhesus monkeys. (c) Number of common human and chimpanzee genes showing new (independent) Alu element insertions, 115 of which (out of 165) occur in genes with neurological function and/or associated with neurological disease. (Reproduced from Paz-Yaacov et al.[276] with permission of the authors.)

neurodifferentiation.[320,321] L1 and Alu retroelements are mobilized in the human brain,[322] which suggests the possibility that the APOBECs, especially the APOBEC3 family, like KRAB zinc finger proteins,[323] might have evolved to domesticate

retroelements, and manage their activity[p] in response to environmental cues, not simply suppress them.

---

[p] While avoiding neurodevelopmental or neuroinflammatory disorders due to inappropriate L1 expression of neurotoxic retroviral sequences and proteins.[324–327]

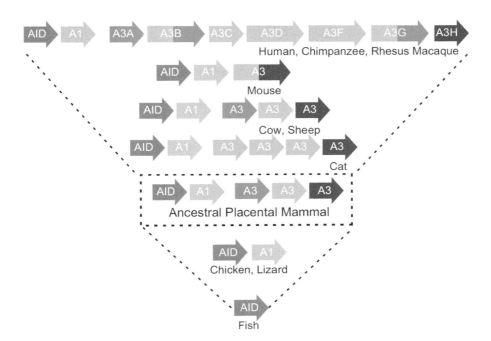

FIGURE 17.4   The expansion of ABOBEC genes in vertebrate, mammalian and primate lineages (colors denoting subfamilies). (Reproduced from Harris and Dudley[313] with permission of Elsevier.)

Moreover, DNA demethylation in human neural progenitor cells leads to transcriptional activation and chromatin remodeling of hominoid-specific L1 elements,[q] while older L1s and other classes of transposable elements remain silent; these activated L1s act as alternative promoters for many protein-coding genes involved in neuronal functions, "revealing a hominoid-specific L1-based transcriptional network controlled by DNA methylation that influences neuronal protein-coding genes".[329]

THE BRAIN

Despite a century since Santiago Ramón y Cajal's meticulous depictions of the architecture and complexity of the central nervous system,[330] and many subsequent developments in neurophysiology, we are still nowhere near understanding the molecular basis of high-level brain function. While the general architecture of the brain is hard-wired,[331] its fine connections are selected and evolve in response to experience, as proposed by Gerry Edelman.[332,333]

The brain is plastic[334] and has most complex molecular transactions. Except for the testis (itself

another world), the coding and non-coding transcriptome is most varied and complex in the brain, as is the extent of RNA splicing, trafficking, modification and editing. The cell-type specificity of lncRNA expression is also most pronounced in brain (Chapter 13) and the genomic sequence variants affecting neuropsychiatric functions, neurodegenerative diseases and some neurodevelopmental disorders[335] primarily lie in non-coding regions (Chapter 11). Many neurodegenerative diseases appear to be linked to dysregulation of lnc/enhancer RNAs and/or be a consequence of aberrant RNA-protein interactions and formation of inclusion bodies associated with expansions of simple repeat sequences.[336–339] As with non-coding genomic regions that show accelerated evolution or are specific to primates or humans[340,341] (Chapters 10 and 13), single cell studies confirm the highly regulated expression of non-coding RNAs in specific brain areas relevant to human evolution and neurological diseases.[342–345]

There is considerable evidence for the involvement of regulatory RNAs in brain evolution, development and function "by virtue of their abundant sequence innovation in mammals and plausible mechanistic connections to the adaptive processes that occurred recently in the primate and human lineages",[346,347] including the widespread use of lncRNAs to nucleate

---

[q] L1 expression is differentially regulated in pluripotent stem cells of humans and other great apes.[328]

specialized domains in the neuronal nucleus;[348] the expression of primate-restricted KRAB zinc finger proteins in specific regions of the developing and adult brain, which target cognate TE regulatory elements to control neuronal differentiation;[323,326,349,350] the widespread use of 3'UTRs as regulatory RNAs[351,352] to, e.g., maintain axonal integrity;[353] and the massive expansion of RNA editing during vertebrate evolution, especially in primates.

Small RNAs, lncRNAs, TEs and networks thereof are involved in neuronal differentiation, synaptic plasticity,[r] long-term potentiation, learning and behavior,[346,354–366] examples being the role of *Malat1* in synapse formation;[367] the response of Gomafu to neuronal activity and its modulation of schizophrenia-associated alternative splicing and response to methamphetamine;[368,369] the response of Neat1 to neuronal activity and its role in mediating histone methylation, age-related memory impairment and behavioral responses to stress;[370–372] the role of the lncRNA Gas5 in cocaine action and addiction;[373] the role of the TE-derived neuronal lncRNA BC1[s] in anxiety, exploratory behavior[377] and memory;[378] the regulation of impulsive and aggressive behaviors by lncRNA MAALIN;[379] the downregulation of the primate-specific lncRNA LINC00473 in the prefrontal cortex of depressed females but not males, accompanied by female-specific changes in synaptic function;[380] blockage by the loss of function of lncRNA Meg3 of the glycine-induced increase of the GluA1 subunit of AMPA receptors on the plasma membrane, a major hallmark of LTP;[381] the role of lncRNAs Tsx in hippocampal short-term memory formation[382] and LoNA in long-term memory formation;[383] the regulation of social hierarchy in mice by lncRNA AtLAS;[384] the regulation of locust aggregation by lncRNA PAHAL;[385] and lncRNA dynamics in the behavioral transition from nurses to foragers in *Drosophila*.[386] Moreover, epigenetic processes (which are likely RNA-directed, Chapter 16) are required for synaptic processes, cognition, learning and memory – if epigenetic processes are disrupted, learning is also disrupted.[7,10,13,387–390]

Consequently, it appears that the brain adapted processes and regulatory mechanisms that are relatively hard-wired in development, rendering them

soft-wired to enable the formation of synaptic networks that are tuned by environmental cues and cell communication to process, store and recall information.

There are several other intriguing aspects of brain molecular biology.

First, most if not all components of the innate and adaptive immune systems have paralogs and orthologs expressed in the brain, most of which also occur in invertebrates (before the appearance of the adaptive immune system in vertebrates), suggesting that the adaptive immune system is a specialized offshoot of cell recognition pathways that first evolved for communication in the nervous system.[t]

For example, the immunoglobulin (Ig) fold is present in most neuronal adhesion and receptor molecules, including the N-CAMs, myelin-associated glycoproteins, nectins, telencephalin, contactin and neuroglian, as well as in other proteins that are found in the immune system but also occur in brain, including the Toll-like receptors, CD4, Thy-1, the major histocompatibility complex and the complement family.[395–401]

Toll-like receptors, thought mainly to activate innate immune responses, also regulate development, neural morphogenesis and neural connectivity,[402,403] in part by recognizing neurotrophins to control neuronal survival and death and acting as adhesion molecules to instruct axon and dendrite targeting and synaptic partner matching.[404]

Cytokines, some of which are expressed in invertebrate glial cells,[405] are also present in the brain, where they affect neurotransmitter production, leading to changes in motor activity, anxiety, arousal and alarm. They also regulate sleep and a variety of neuroendocrine functions, as well as neuronal development.[406,407]

The RAG1 and RAG2 proteins that have their origins as transposases and mediate VDJ recombination in B-cell and T-cell receptors are also expressed in the nervous system[408–411] (Figure 17.5). V(D)J recombination, like programmed genomic rearrangements and DNA repair in other organisms, is RNA directed.[412–414] Lack of RAG-1 impairs memory formation[415] and lack of RAG-2[u] impairs

---

[r] Which "may drive species-specific changes in cognition".[354]
[s] BC1[374] regulates dopamine receptors[375] and associates with the fragile X syndrome protein FMRP to regulate the translation of specific mRNAs at synapses.[376]

[t] The blood brain barrier may be a mechanism to prevent the two systems from interfering with each other,[391,392] as they may in pathological situations.[393,394]
[u] RAG2 activity is regulated by a PHD domain that binds histone H3K4me3 modifications, indicating an interplay between epigenetic information and DNA recombination.[416]

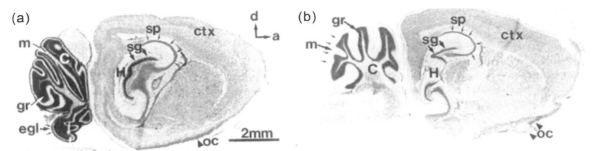

FIGURE 17.5 *Rag1* expression in mouse brain. *In situ* hybridization of parasagittal sections of (a) 10 day postnatal and (b) adult brain hybridized with RAG-1 antisense riboprobes, counterstained with cresyl violet to reveal the location of cell bodies. The strongest antisense hybridization was observed in the cerebellum and hippocampus. (Reproduced from Chun et al.[408] with permission of Elsevier.)

retinal development, axonal growth and navigation.[411] Somatic gene recombination also occurs in human neurons[v] with similarities to V(D)J recombination but with a different mechanism that involves an RNA intermediate and reverse transcription.[418]

Second, while still largely a black box, neurons regulate activity differentially at thousands of synapses, which involves transport along microtubules in both neurons and associated oligodendrocytes[w] of RNA granules[421,422] containing mRNAs and non-coding RNAs (including miRNAs, antisense pseudogene transcripts, and Alu- and other TE-containing RNAs) via various RNA-binding proteins and motor proteins.[354,364,423–427] There is synapse-specific local translation and, in all likelihood, context-dependent processing, editing and modifications of RNAs in response to activity.[354,428–430] A number of lncRNAs have been shown to be regulated by neuronal activity and to regulate synaptic protein localization and translation, as well as synapse density, morphology, dendritic tree complexity, activity, plasticity and stability.[376,378,384,431–434] The neuronally expressed gene *Arc*, which is essential for synaptic plasticity and memory formation,[435–437] encodes a repurposed retrotransposon-derived protein that mediates intercellular RNA transfer.[438]

Synaptic protein synthesis associated with memory formation is also regulated by the RNA interference pathway,[439] including by Mili-bound (26–28nt) piRNAs, thought to be mainly involved in repressing TEs in the formation of germ cells but also highly

expressed in the brain.[356,440–442] The piRNA pathway is also required for adult neurogenesis in mice[358] and is involved in the regulation of transposon mobilization in *Drosophila* brain.[443] The loss of Mili results in hypomethylation of LINE1 promoters and behavioral deficits such as hyperactivity and reduced anxiety.[444]

Third, neuronal and protein transport between the cell body and synapses is bidirectional,[445] which has been interpreted as a "sushi-train" mechanism to patrol synapses,[446,447] but retrograde transport back to the nucleus may also return information, as a mechanism for consolidating long-term memory[200] (see below). Loss of the RNA-binding and transport proteins Staufen and Pumilio leads to the inability to form long-term memory.[448] Staufen also binds Alu sequences.[449]

Fourth, there is widespread transcription at neuronal activity-regulated enhancers.[450] Neuronal enhancers are hotspots for DNA single-strand break repair[451] and such hotspots occur at sites involved in neuronal identity, synapse function and neural cell adhesion, which are enriched in RNA-binding proteins.[452,453] Brain activity and fear conditioning causes DNA double-strand breaks in neurons[454,455] and activity-induced DNA breaks govern the expression of neuronal genes.[456] An lncRNA is required for DNA damage response in neurons, the loss of which causes Purkinje cell degeneration and impairs motor function.[457] The DNA repair-associated protein gadd45γ is required for the consolidation of associative fear memory.[458] DNA repair is focused on transcribed genes and declines with age – possibly associated with reduced learning activity – with deficiencies in DNA repair linked to both developmental and age-associated neurodegenerative diseases.[459]

---

[v] Widespread mosaic somatic gene recombination in neurons was first detected in the Alzheimer's disease-related gene *APP*, which encodes amyloid precursor protein.[417]

[w] There is also evidence of RNA transport between glial cells and neurons.[419,420]

Moreover, there are many unusual 'DNA repair' enzymes and DNA polymerases in the brain,[x] some with reverse transcriptase activity.[200] While these phenomena are usually interpreted in terms of protecting the genomic integrity of post-mitotic neurons, an alternative (and not mutually exclusive) possibility is that RNA-directed changes to DNA ("re-writing to disc") is involved in long-term memory formation.[200] The human DNA polymerase Polθ, which occurs in the brain, was recently shown to reverse transcribe RNA and promote RNA-templated DNA repair.[461] Polθ also promotes repeat expansions in Huntington's Disease and other neurodegenerative disorders.[462]

RNA-DIRECTED TRANSGENERATIONAL EPIGENETIC INHERITANCE

The foundational work on RNA interference in *C. elegans* showed that small RNA-mediated gene silencing can be inherited for many generations (Chapter 12),[14,463–465] most stably when provoked by maternal piRNAs.[466] piRNAs are required to initiate inheritance but not to maintain it, although maintenance is dependent on the nuclear RNAi pathway.[466–469] Similar transgenerational inheritance is observed plants where the inheritance is confined to dsRNAs targeting methylation of gene promoters rather than the transcribed sequence.[470] The existence of many imprinted alleles shows that epigenetic information can also be transmitted through meiosis in mammals to control gene expression in the next generation.

Although their evolutionary significance has not been widely discussed, the observations in *C. elegans* refute the long-standing assumption that the soma cannot communicate with the germline, since the inherited dsRNA triggers can be delivered by injection into somatic cells or ingestion of engineered bacteria producing dsRNAs, opening a new frontier for understanding the nature of both hard- and soft-wired inheritance.

Unusual non-Mendelian patterns of inheritance were, in fact, first observed in peas by Bateson in 1915,[471] and occasionally thereafter in other plants, but were not studied in a systematic way until Robert Brink, who coined the term 'paramutation' in the 1950s to describe the atypical inheritance of traits displayed by particular alleles in maize (where it is best studied),[472–474] tomato and other species[475–477] (Chapter 5).

Paramutation is RNA-directed transgenerational inheritance.[478] It may be summarized as the transfer of epigenetic information from one allele of a gene to another to induce metastable silencing, which is heritable for generations but incompletely penetrant and reversible.[474] Importantly, paramutation is a somatic phenomenon, which is transmitted to the germline.[474] Mechanistically, paramutation is trans-acted by small RNAs that direct RNA-processing and/or chromatin-modifying proteins to RNA transcripts or DNA in a sequence-specific manner, with self-reinforcing feedback loops that involve differential methylation and the biogenesis of particular classes of sRNAs, including piRNAs, mediated by Argonaute proteins.[477,479,480] Moreover, mutations in *Mop1* ('Mediator of paramutation1'), a component of the RNA-directed DNA methylation pathway responsible for methylation of TEs adjacent to transcriptionally active genes, alter the distribution and frequency of meiotic recombination in maize,[481] indicating a link between transposons and plasticity of inheritance.

While initially thought to be confined to plants, paramutation also occurs in animals[477,482,483] (Figure 17.6). In *C. elegans*, where animal RNA interference was discovered, the phenotypic consequences of ectopically introduced small dsRNAs can persist for many generations, as with plant paramutation, without any change to the underlying sequence.[466,484] Paramutation has also been described in *Drosophila* and mouse, the latter affecting a wide spectrum of characteristics such as pigmentation, cardiac hypertrophy, embryo development and axonal growth (the latter via a lncRNA), again mediated through the RNA interference pathway and small interfering RNAs.[477,482,483,485–487] The inheritance of paramutations, at least in mouse, requires the RNA methyltransferase Dnmt2.[488]

Paramutation is associated with, and its strength is often dependent upon the length of, simple (usually dinucleotide or trinucleotide) tandem sequence repeats (STRs)[y] in the locus,[477] an interesting obser-

---

[x] These enzymes include DNA polymerase ζ, which performs RNA-directed DNA modification.[460] Some are themselves subject to RNA editing, suggesting a molecular mechanism for context-dependent strength of memory formation.[200]

[y] STRs are also referred to as 'variable number tandem repeats' (VNTRs) or 'microsatellites', variation in which was exploited for DNA fingerprinting by Alex Jeffries in the late 1980s.[489]

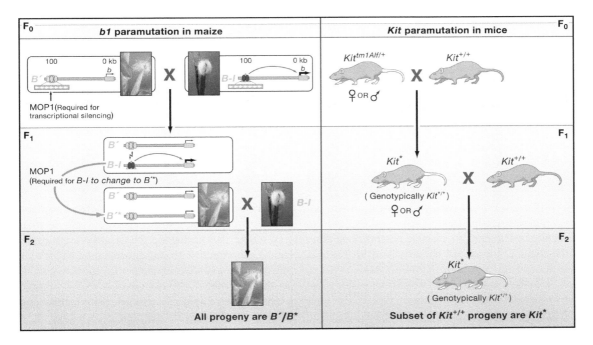

FIGURE 17.6 Paramutation at the *b1* locus in maize (left; arrow at top left indicates the position of seven tandem repeats of a 853 bp sequence unique to this location within the maize genome) and (right) at the *kit* locus in mouse (paramutable and paramutagenic *Kit^tm1Alf* alleles), which confers white tail tips. (Reproduced from Chandler[475] with permission of Elsevier.)

vation in view of the vast number of STR sequences that occur in animal and plant genomes.

The human genome contains over one million highly polymorphic and plastic STRs, whose mitotic (somatic) and meiotic (germline) expansion/contraction rates can be orders of magnitude higher than single nucleotide mutations, and impart a continuum of effects.[490–493] Variation in STRs is associated with neurological and psychiatric disorders[z] such as autism, and cancer, as well as the modulation of physiological and neurological traits, including circadian rhythms, sociosexual interactions, intelligence, hormone sensitivity, cognition, personality, addiction, neuronal differentiation, brain development and behavioral evolution.[492–499] An STR has also been shown to regulate the transcription of the hTERT (human telomerase reverse transcriptase) gene in a cell-context–dependent manner, the absence of which results in telomere shortening, cellular senescence and impaired tumor growth.[500]

STRs are enriched in promoters and enhancers, associated with differential DNA methylation, and account for 10%–15% of the genetic variation observed in complex traits, making them substantial contributors to the missing heritability in genome-wide association studies that only poll haplotype blocks.[493,496,501–503] This is exemplified by the paramutation-like behavior imposed by untransmitted paternal alleles of the human insulin gene that have expanded numbers of repeats associated with a reduced incidence of type 1 diabetes.[504] Interestingly, a positively selected brain-specific lncRNA contains a tandem repeat that varies among individuals and affects its stability.[505] A large and highly polymorphic human-specific 30bp tandem repeat located within an intron of a gene encoding a calcium channel has been shown to act as an enhancer and to be associated with bipolar disorder and schizophrenia.[495]

The mutation rate in STRs is controlled, in part, by epigenetic processes, and the length of STRs, including in humans, can be modulated by environmental parameters, notably (as the main one studied) stress.[476,477,506,507] As summarized by Jay Hollick:

---

[z] Reasonably proposed to be 'the pathological ends of phenotypic bell curves in which healthy individuals occupy the middle territory'.[493]

"The ability of heritable epigenetic regulatory information on one homologue to be copied to the other represents a potential adaptation of the diploid condition to rapidly disseminate a memory of environmental responses to future generations".[477]

The extent of STR modulation of gene expression may be underestimated and underappreciated. It is usually only reported in model organisms where structured pedigrees and genotypes can be constructed and monitored. A genome-wide analysis in tomato identified "thousands of candidate regions for paramutation-like behaviour... the methylation patterns for a subset of (which) segregate with non-Mendelian ratios, consistent with secondary paramutation-like interactions to variable extents depending on the locus".[508] It would be surprising if this was not a general phenomenon, with enormous implications for understanding plant and animal biology and evolvability (Chapter 18).

While most studies of paramutation have involved focused analyses of physical traits, there have been a number of reports that transgenerational epigenetic inheritance (in animals as diverse as worms, planarians and mammals) extends to metabolic, endocrine, immunological and cognitive experience, including trauma and stress, as well as learned behaviors.[13,509–518] There is both male and female transmission.[14,519–522] Various studies have implicated lncRNAs, miRNAs, CTCF recruitment to an *Fto*[aa] enhancer, peroxisome proliferator-activated receptor (PPAR) pathways, cysteine synthases and tRNA fragments, and modifications thereof, as being involved.[478,516,519,525–534]

One of these RNAs, the vault RNA VTRNA2-1, was identified in genome-wide screens as being "as a top environmentally responsive epiallele",[535] which has been associated with effects on oocytes of pre-conceptual alcohol consumption[522] and elsewhere with cancer etiology and outcomes.[536,537] The progeny of mice that survived a sublethal systemic infection with *Candida albicans* or an endotoxin dose exhibited cellular, developmental, transcriptional and epigenetic changes in the myeloid progenitor cell compartment, with enhanced responsiveness to endotoxin challenge and improved protection against systemic heterologous bacterial infections.[538] The sperm DNA of parental male mice infected with *C. albicans* showed DNA methylation differences linked to immune gene loci.[538] piRNAs derived from ancient processed viral pseudogenes transmit transgenerational sequence-specific immune memory in rodents and primates.[539] There are also innate fears and phobias that clearly have appreciable genetic components.[540,541]

It has been shown that specific tRNA fragments (Chapter 12) in mice are transferred from the epididymis to maturing sperm through small vesicles called epididymosomes. These tRNA fragments are then conveyed by the sperm to the fertilized egg where they influence the expression of specific genes associated with the retroelement MERV during embryonic development. In mouse, changes to the paternal diet modulate the tRNA fragment composition of epididymosomes, with a consequent alteration of specific metabolic pathways in the offspring.[526,527]

The mechanisms by which experience is conveyed intergenerationally are ill-defined, and the field will remain controversial until this is understood. Not only is it difficult to separate genetic and epigenetic effects from cultural and environmental influences,[479,542–545] it is also difficult to conceive a mechanism to transmit complex traits, even considering the possibility of RNA signaling between the soma and the germline. Relevant perhaps is that, like the brain, the testis is an immunologically privileged tissue, with a tight barrier between blood vessels and the Sertoli cells in the seminiferous tubule, which isolates the later stages of sperm development.[546,547]

Whatever the details, it is nonetheless already clear that there are two forms of inheritance – gene alleles and epialleles – hard-wired DNA sequence information and RNA-directed epigenetic information that is responsive to and influenced by environmental factors, directly contradicting the fixed inheritance of variation assumed by the Modern Synthesis.

FURTHER READING

Conine C.C. and Rando O.J. (2021) Soma-to-germline RNA communication. *Nature Reviews Genetics* 23: 73–88.

Hannan A.J. (2018) Tandem repeats mediating genetic plasticity in health and disease. *Nature Reviews Genetics* 19: 286–98.

[aa] *Fto* encodes an N6-methyladenosine demethylase that is associated with obesity[523] and is required for adipogenesis.[39] It also regulates neurogenesis, neural circuitry, memory formation and locomotor responses.[94,98,105,524] Recruitment of CTCF to an *Fto* enhancer is involved in transgenerational inheritance of obesity.[516]

Hollick J.B. (2017) Paramutation and related phenomena in diverse species. *Nature Reviews Genetics* 18: 5–23

Kim S. and Kaang B.-K. (2017) Epigenetic regulation and chromatin remodeling in learning and memory. *Experimental & Molecular Medicine* 49: e281.

Levenson J.M. and Sweatt J.D. (2005) Epigenetic mechanisms in memory formation. *Nature Reviews Neuroscience* 6: 108–18.

Nishikura K. (2015) A-to-I editing of coding and non-coding RNAs by ADARs. *Nature Reviews Molecular Cell Biology* 17: 83–96.

18 Beyond the Jungle of Dogmas

"In science, 'perverse' cracking of pots is sometimes necessary to design new experiments, and to imagine new ideas that could help one to forge ahead, beyond the actual jungle of dogmas and preconceived opinions."

(Klaus Scherrer[1])

THE MISUNDERSTANDING OF MOLECULAR BIOLOGY

It seems that the nature of genetic information in complex organisms has been misunderstood since the inception of molecular biology, because of the assumption that most genetic information is transacted by proteins. This assumption holds largely true for prokaryotes and to a lesser extent for eukaryotic microorganisms, which mainly must organize a cell to obtain nutrients and reproduce, albeit itself no mean feat. However, developmentally complex organisms, especially motile animals, have had to evolve much more sophisticated mechanisms to orchestrate cell differentiation and their assembly into highly organized ensembles.[2,3]

The foundational assumption that genes (only) encode proteins led to many subsidiary assumptions, primarily that the vast tracts of non-coding and repetitive sequences in the genomes of complex organisms are graveyards of evolutionary junk colonized by molecular parasites. This rationalization was not disturbed by contrary genetic and molecular evidence assembled by Barbara McClintock, Roy Britten and Eric Davidson, Ed Lewis and others whose intuitions were ignored. It persisted in handwaving about the power of combinatorial control of gene expression by transcription factors. It persisted in founder fallacies and validation creep, a notable one being that developmental 'enhancers' bring transcription factors bound at their promoters into contact with the promoters of protein-coding genes whose expression they control, rather than the now increasingly clear alternative that they produce regulatory RNAs that organize local transcription and splicing hubs. Indeed, understanding enhancers is key to understanding the programming of differentiation and development.

Contrary to the view that the genomes of humans and other complex organisms are full of non-functional evolutionary detritus, they are in fact replete with information and dynamic activity.[a] The human genome makes trillions of cell fate decisions during development with high specificity and near-perfect reproducibility. This precision is effected by epigenetic mechanisms directed by regulatory RNAs, whose versatility may be largely achieved by programmed (cell state-specific) alternative splicing to alter target sites and modular recruitment of different types of effector proteins, incorporated into decisional hierarchies networked by sequences coopted from and distributed by transposable elements.

The separation of signal from consequent action, exemplified by RNAi and CRISPR, and writ large by enhancers and other types of lncRNAs, is a highly, and likely the most, efficient and versatile means of gene regulation. The advent of these advanced RNA-based regulatory systems permitted the emergence of developmentally complex organisms, following which evolution experimented with more sophisticated designs to colonize new niches, leading to the extraordinary biodiversity that we see today, embellished by sexual (mate) selection as proposed by Darwin.[4] Most differences between species and individuals are embedded in variations in their regulatory architecture, the extent of which expands with developmental complexity, like increasingly elaborate building plans using a relatively generic set of component parts, albeit with occasional important innovations, such as the immunoglobulin domain, Arc RNA transfer and RNA editing proteins.

Meanwhile, behind the scenes, while all organisms benefit from information processing, animals were climbing the next mountain, cognition, by superimposing plasticity on hardwired genomic

---

[a] This is not to say there every sequence is functional. There will, of course, be recent duplications and transpositions that have not yet been subject to evolutionary selection.

DOI: 10.1201/9781003109242-18

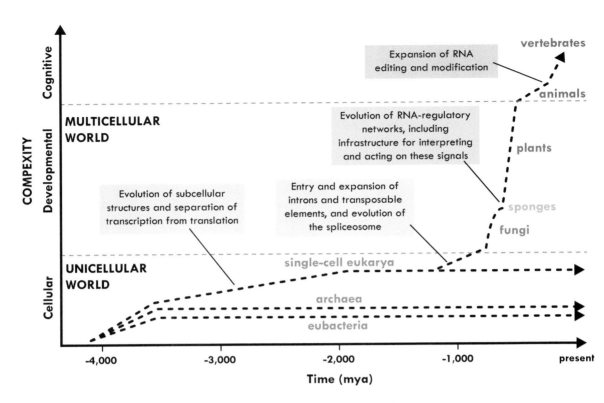

FIGURE 18.1   The evolution of complexity. (Image modified from Mattick[10].)

information (Figure 18.1), with selection strength dependent on mobility and boosted by dexterity.[5,6] It may be no accident, contrary to Steven Jay Gould's proposition of contingent history,[7–9] that the most cognitively advanced vertebrates and invertebrates are the primates and cephalopods.

It is increasingly clear that, rather than a simple intermediate between gene and protein, RNA is the computational engine of the cell, development, cognition and evolution.[11] The challenge is not only to understand the principles of how RNA interacts with effector proteins in the formation and function of dedicated cellular domains to orchestrate cell division or differentiate decisions, but to decipher the code itself. This will be a bit like climbing inside a computer to try to work out how it can produce and project three-dimensional images, except that the human genome does so while building itself and its internal computer the brain, with far more precision and complexity, from an amazingly compact information suite that is roughly equivalent in size to that which can be held on a 1 Gb thumb drive.

THE EVOLUTION OF EVOLVABILITY

The other long-standing assumptions have been that mutations are random, and that experience cannot be communicated to modulate the phenotype of subsequent generations, asserted in the formative years of evolutionary biology and molecular genetics based on preconception rather than evidence. Both assumptions are clearly incorrect, with non-random mutation and epigenetic inheritance now well documented in both plants and animals,[12–18] and evidence of a relationship.[19]

This raises the question of whether there is interplay between genetic and epigenetic inheritance to accelerate evolutionary processes. It is obvious that evolution cannot have proceeded by random search alone – the number of variables is too great. This problem was recognized generically two decades ago by Rodney Downey and Michael Fellows, who pointed out that in large complex systems random searches become computationally intractable ('NP-hard') because of the exponential increase in the possibilities.[20,21] This problem must also apply to evolution if it operates, as described by Dennett, as a grinding algorithm of generate and test,[22] and becomes increasingly acute in organisms, notably birds and mammals, that have long generation times and small numbers of progeny.[b]

Downey and Fellows's proposed solution to the problem is to define the most productive subspace and optimal tactics to decrease the complexity of the search and increase the chances of productive outcomes, termed 'Parameterized Complexity'.[20,21] Logically, in the case of evolution, any (initially random) event that enhanced the evolvability of the lineage concerned must have been subject to second-order selection on the basis of its strategic advantage. By extension, any lineage that stumbled into such strategic advantage would come to dominate the evolutionary landscape and, by definition, be part of the toolkit of most, if not all, extant lineages in the biosphere.

The evidence to support this logic is fragmentary, and the topic of the evolution of evolvability has been subject to considerable speculation and debate.[26] Evolutionary computer science has shown that random mutation, recombination and selection are not universally effective in improving complex systems and that for adaptation to occur, these systems must acquire evolvability.[27,28] Moreover, the modeling suggests that simple genotype–phenotype mapping is suboptimal, whereas the use of indirect developmental representations allow the reuse of code (modularity), and scaling up of the complexity of artificially evolved phenotypes, for example, in robotics, artificial life and morphogenetic engineering.[29] Indeed one important enabler of biological evolution is modularity,[28,30–32] itself an evolved characteristic,[33] classically and graphically exemplified by simple homeotic mutations that convert insects from having two wings to four.[34]

Different organisms have tuned their innate mutation frequencies to optimize the trade-off between survival and evolvability,[14,35,36] but this is only a blunderbuss approach. However, mutation frequency and transposon distribution vary across the genome and over time,[15,37] as judged by indices of neutral evolution, although these indices are unreliable and the extant distribution is difficult to disentangle from selection and differential repair and recombination.[16,38–42] Nonetheless, such variation occurs in the extensive non-coding regions of the genomes of complex organisms, evidence of selection at one level or another.

Rapid evolution has been documented in many species. It is, for example, observed in the human lineage,[43] correlating with bursts of Alu element invasion, reflecting the huge positive selection value of cognitive advancement in intra- and inter-species competition.[6,43]

The interplay between genetic and epigenetic inheritance changes the dynamics of natural selection, as argued by Eva Jablonka and others.[44–50] It also changes the interplay between genes and environment, and the dichotomy of 'Nature versus Nurture' assessments, with profound evolutionary and social implications. Epigenome-associated mutation bias reduces the occurrence of deleterious mutations in essential genes in *Arabidopsis*[18] and recent evidence indicates that mutation sites in sperm are non-random, associated with adaptation.[51] There is also evidence that epigenetic changes can increase drug resistance in cancer cells until a genetic solution is found.[52]

If epigenetic information can provide transgenerational phenotypic advantage, it is not a long stretch to suggest that evolution may have found ways to convert this information into hardwired genetic changes to speed adaptive change. In fact, it would be foolish to reject the possibility out of hand. The infrastructure – RNA-mediated epigenetic inheritance and RNA-templated DNA repair – is in place. Has evolution learned how to learn?[5]

Genomes contain biological software encompassing codes for components, self-assembly, differentiation and reproduction, supplemented by information in parental cells and epigenetic memories. Not only has the data evolved, but also the data structures, implementation systems and search algorithms. We have some way to go to understand the complexity and beauty of genetic programming, but the best places to start are to accept that RNA plays a major role in the evolution and mechanics of developmental control and cognitive processes, and to keep an open and receptive mind, especially when we are once again surprised.

---

[b] The numbers of variation options in such organisms per generation must be miniscule without compensatory mechanisms, such as the enigmatic double round of piRNA expression in sperm development, posited to allow controlled transposon mobilization and subsequent siRNA-mediated transcriptional proofing (which most sperm fail), to generate viable options for evolutionary selection in small populations with long generation times.[23,24] There are high primary rates of retrotransposition in mammals.[25]

A new scientific truth does not triumph by convincing its opponents and making them see the light, but rather because its opponents eventually die, and a new generation grows up that is familiar with it.

(Max Planck[53])

FURTHER READING

Blount Z.D., Lenski R.E. and Losos J.B. (2018) Contingency and determinism in evolution: Replaying life's tape. *Science* 362: eaam5979.

Dennett D. (1995) *Darwin's Dangerous Idea: Evolution and the Meanings of Life* (Simon Schuster, New York).

Erwin D.H. (2015) Novelty and innovation in the history of life. *Current Biology* 25: R930–940.

Jablonka E. (2017) The evolutionary implications of epigenetic inheritance. *The Royal Society Interface Focus* 7: 20160135.

Jablonka E. and Lamb M. (1994) *Epigenetic Inheritance and Evolution—The Lamarckian Dimension* (Oxford University Press, Oxford).

Mattick J.S. (2007) A new paradigm for developmental biology. *Journal of Experimental Biology* 210: 1526–47.

Mattick J.S. (2009) Deconstructing the dogma: A new view of the evolution and genetic programming of complex organisms. Annals of the New York Academy of Science 1178: 29–46.

Mattick J.S. (2009) Has evolution learnt how to learn? *EMBO Reports* 10: 665.

Mattick J.S. (2011) The central role of RNA in human development and cognition. *FEBS Letters* 585: 1600–16.

Payne J.L. and Wagner A. (2019) The causes of evolvability and their evolution. *Nature Reviews Genetics* 20: 24–38.

Skinner M.K. (2015) Environmental epigenetics and a unified theory of the molecular aspects of evolution: A neo-Lamarckian concept that facilitates neo-Darwinian evolution. *Genome Biology and Evolution* 7: 1296–302.

References

PREFACE

1. Weaver W. (1970) Molecular biology: Origin of the term. *Science* 170: 581–2.

CHAPTER 2

1. Lavoisier A.L. (1789) *Traité Élémentaire de Chimie* (Chez Cuchet, Combloux).
2. Proust J.L. (1794) Recherches sur le Bleu de Prusse. *Journal de Physique, de Chimie, et d'Histoire Naturelle* 2: 334–41.
3. Rocke A.J. (2005) In search of El Dorado: John Dalton and the origins of the atomic theory. *Social Research* 72: 125–58.
4. Avogadro A. (1811) Essai d'une manière de déterminer les masses relatives des molécules élémentaires des corps, et les proportions selon lesquelles elles entrent dans ces combinaisons. *Journal de Physique, de Chimie et d'Histoire naturelle* 73: 58–76.
5. Hinshelwood C.N. and Pauling L. (1956) Amedeo Avogadro. *Science* 124: 708–13.
6. Marshall J.L. and Marshall V.R. Rediscovery of the elements: The periodic table. *The Hexagon* 98: 23–9.
7. Nes W.D. (2011) Biosynthesis of cholesterol and other sterols. *Chemical Reviews* 111: 6423–51.
8. Braconnot H. (1815) Sur la nature des corps gras. *Annales de Chimie* 2 (XCIII): 225–77.
9. Chevreul M.E. (1823) *Recherches Chimiques 1* (Wentworth Press, Sydney, 2018).
10. Gobley T.N. (1847) Examen comparatif du jaune d'oeuf et de la matière cérébrale. *Journal de Pharmacie et de Chimie* t11: 409–12.
11. Bertrand G. Projet de reforme de la nomenclature de chimie biologique. *Bulletin de la Société de Chimie Biologique* 5: 96–109.
12. Barnett J.A. (2003) Beginnings of microbiology and biochemistry: The contribution of yeast research. *Microbiology* 149: 557–67.
13. Payen A. and Persoz J.-F. (1833) Mémoire sur la diastase, les principaux produits de ses réactions et leurs applications aux arts industriels. *Annales de Chimie et de Physique* 53: 73–92.
14. Dumas J.-B. (1839) Rapport sur un mémoire de M. Payen, relatif à la composition de la matière ligneuse *Comptes Rendus* 8: 51–3.
15. Payen A. (1839) Mémoire sur les applications théoriques et pratiques des propriétés du tissu élémentaire des végétaux. *Comptes Rendus* 8: 59–61.
16. Young F.G. (1957) Claude Bernard and the discovery of glycogen; a century of retrospect. *British Medical Journal* 1: 1431–7.
17. Wöhler F. (1828) Ueber künstliche Bildung des Harnstoffs. *Annalen der Physik und Chemie* 88: 253–6.
18. Toby S. (2000) Acid test finally wiped out vitalism, and yet. *Nature* 408: 767.
19. Mulder G.J. (1839) Ueber die Zusammensetzung einiger thierischen Substanzen. *Journal für Praktische Chemie* 16: 129–52.
20. Teich M. and Needham D. (1992) *A Documentary History of Biochemistry 1770–1940* (Leicester University Press, Leicester).
21. Vickery H.B. (1950) The origin of the word protein. *Yale Journal of Biology and Medicine* 22: 387–93.
22. Hartley H. (1951) Origin of the word 'protein'. *Nature* 168: 244.
23. Holmes F.L. (1979) Early theories of protein metabolism. *Annals of the New York Academy of Sciences* 325: 171–88.
24. Rosenfeld L. (2003) Justus Liebig and animal chemistry. *Clinical Chemistry* 49: 1696–707.
25. Glas E. (1999) The evolution of a scientific concept. *Journal for General Philosophy of Science* 30: 37–58.
26. Welch G.R. and Clegg J.S. (2010) From protoplasmic theory to cellular systems biology: A 150-year reflection. *American Journal of Physiology: Cell Physiology* 298: C1280–90.
27. Huxley T.H. and Snell A.L.F. (1909) *Autobiography and Selected Essays* (Houghton Miflin Company, Boston).
28. Darwin F. (1887) *The Life and Letters of Charles Darwin, Including an Autobiographical* (John Murray, London).
29. Holmes D.S., Mayfield J.E. and Bonner J. (1974) Sequence composition of rat ascites chromosomal ribonucleic acid. *Biochemistry* 13: 849–55.
30. Globus N. and Blandford R.D. (2020) The chiral puzzle of life. *The Astrophysical Journal* 895: L11.
31. Fischer E. (1894) Einfluβ der Konfiguration auf die Wirkung der Enzyme. *Berichte der Deutschen Chemischen Gesellschaft* 27: 2985–93.
32. Kohler R. (1971) The background to Eduard Buchner's discovery of cell-free fermentation. *Journal of the History of Biology* 4: 35–61.
33. Vickery H.B. and Schmidt C.L.A. (1931) The history of the discovery of the amino acids. *Chemical Reviews* 9: 169–318.
34. Vickery H.B. and Osborne T.B. (1928) A review of hypotheses of the structure of proteins. *Physiological Reviews* 8: 393–446.
35. Fruton J.S. (1985) Contrasts in scientific style. Emil Fischer and Franz Hofmeister: Their research groups and their theory of protein structure. *Proceedings of the American Philosophical Society* 129: 313–70.
36. Sumner J.B. (1926) The isolation and crystallization of the enzyme urease. *Journal of Biological Chemistry* 69: 435–41.
37. Andras P. and Andras C. (2005) The origins of life – the 'protein interaction world' hypothesis: Protein interactions were the first form of self-reproducing life and nucleic acids evolved later as memory molecules. *Medical Hypotheses* 64: 678–88.
38. Oparin A.I. (1968) *Genesis and Evolutionary Development of Life* (Academic Press, Cambridge, MA).
39. Lanham U.N. (1952) Oparin's hypothesis and the evolution of nucleoproteins. *American Naturalist* 86: 213–8.

40. Tirard S. (2017) J. B. S. Haldane and the origin of life. *Journal of Genetics* 96: 735–9.

41. Powner M.W., Gerland B. and Sutherland J.D. (2009) Synthesis of activated pyrimidine ribonucleotides in prebiotically plausible conditions. *Nature* 459: 239–42.

42. Patel B.H., Percivalle C., Ritson D.J., Duffy C.D. and Sutherland J.D. (2015) Common origins of RNA, protein and lipid precursors in a cyanosulfidic protometabolism. *Nature Chemistry* 7: 301–7.

43. Ferus M., et al. (2017) Formation of nucleobases in a Miller–Urey reducing atmosphere. *Proceedings of the National Academy of Sciences USA* 114: 4306–11.

44. Becker S., et al. (2019) Unified prebiotically plausible synthesis of pyrimidine and purine RNA ribonucleotides. *Science* 366: 76–82.

45. Miller S.L. (1953) A production of amino acids under possible primitive earth conditions. *Science* 117: 528–9.

46. Miller S.L. and Urey H.C. (1959) Organic compound synthesis on the primitive Earth. *Science* 130: 245–51.

47. Weaver W. (1970) Molecular biology: Origin of the term. *Science* 170: 581–2.

48. Turing A.M. (1952) The chemical basis of morphogenesis. *Philosophical Transactions of the Royal Society B: Biological Sciences* 237: 37–72.

49. Kornberg H. (2000) Krebs and his trinity of cycles. *Nature Reviews Molecular Cell Biology* 1: 225–8.

50. Wakil S.J. (1962) Lipid metabolism. *Annual Review of Biochemistry* 31: 369–406.

51. Bender D.A. (2012) *Amino Acid Metabolism*, 3rd edition (John Wiley & Sons, Hoboken, NJ).

52. Nelson D.A. and Cox M. (2016) *Lehninger Principles of Biochemistry*, 7th edition: International Edition (Macmillan, London).

53. Yasuo T., Kusuhara Y., Yasumatsu K. and Ninomiya Y. (2008) Multiple receptor systems for glutamate detection in the taste organ. *Biological & Pharmaceutical Bulletin* 31: 1833–7.

54. Olby R. (1994) *The Path to the Double Helix: The Discovery of DNA* (Dover Publications, Mineola, NY).

55. Hunter G.K. (2000) *Vital Forces: The Discovery of the Molecular Basis of Life* (Academic Press, Cambridge, MA).

56. Dahm R. (2008) Discovering DNA: Friedrich Miescher and the early years of nucleic acid research. *Human Genetics* 122: 565–81.

57. Dahm R. (2010) From discovering to understanding. Friedrich Miescher's attempts to uncover the function of DNA. *EMBO Reports* 11: 153–60.

58. Veigl S.J., Harman O. and Lamm E. (2020) Friedrich Miescher's discovery in the historiography of Genetics: From contamination to confusion, from nuclein to DNA. *Journal of the History of Biology* 53: 451–84.

59. Lamm E., Harman O. and Veigl S.J. (2020) Before Watson and Crick in 1953 came Friedrich Miescher in 1869. *Genetics* 215: 291–6.

60. Flemming W. (1878) Zur Kenntniss der Zelle und ihrer Theilungs-Erscheinungen. *Schriften des Naturwissenschaftlichen Vereins für Schleswig-Holstein* 3: 23–7.

61. Flemming W. (1882) *Zellsubstanz, Kern und Zelltheilung (Cell substance, nucleus and cell division)* (F. C. W. Vogel, Leipzig).

62. Paweletz N. (2001) Walther flemming: Pioneer of mitosis research. *Nature Reviews Molecular Cell Biology* 2: 72–5.

63. Flemming W. (1875) Studien uber die Entwicklungsgeschichte der Najaden. *Sitzungsberichte der Kaiserlichen Akademie der Wissenschaften Wien* 71: 81–147.

64. Van Beneden E. (1876) Contribution a l'histoire de la vesiculaire germinative et du premier noyau embryonnaire. *Bulletins de l'Académie royale des sciences, des lettres et des beaux-arts de Belgique (2me série)* 42: 35–97.

65. Boveri T. (1888) Zellen-Studien II. Die Befruchtung und Teilung des Eies von *Ascaris megalocephala*. *Jenaische Zeitschrift für Naturwissenschaft* 22: 685–882.

66. Boveri T. (1901) Zellen-studien IV: Über die Natur der Centrosomen. *Jenaische Zeitschrift für Naturwissenschaft* 35: 1–220.

67. Waldeyer W. (1888) Ueber Karyokinese und ihre Beziehungen zu den Befruchtungsvorgängen. *Archiv für mikroskopische Anatomie* 32: 1.

68. Kossel A. (1967) The chemical composition of the cell nucleus, in *Nobel Lectures Physiology or Medicine 1901–1921* (Elsevier Publishing Company, Amsterdam). https://www.nobelprize.org/prizes/medicine/1910/kossel/lecture/

69. Jones M.E. (1953) Albrecht Kossel, a biographical sketch. *Yale Journal of Biology and Medicine* 26: 80–97.

70. Roberts H.F. (1929) *Plant Hybridization before Mendel* (Princeton University Press, Princeton, NJ).

71. Simunek M., Hoßfeld U., Thümmler F. and Breidbach O. (2011) *The Mendelian Dioskuri: Correspondence of Armin with Erich von Tschermak-Seysenegg, 1898–1951* (Institute of Contemporary History of the Academy of Sciences Prague & Pavel Mervart, Červený Kostelec, Czech Republic). https://www-pavelmervart-cz.translate.goog/kontakt/?_x_tr_sl=cs&_x_tr_tl=en&_x_tr_hl=en&_x_tr_pto=sc&_x_tr_sch=http

72. Simunek M., Hoßfeld U., Thümmler F. and Sekerák J. (2011) *The Letters on G. J. Mendel – Correspondence of William Bateson, Hugo Iltis, and Erich von Tschermak-Seysenegg with Alois and Ferdinand Schindler, 1902–1935* (Institute of Contemporary History of the Academy of Sciences Prague & Pavel Mervart).

73. Friedrich-Schiller-Universitaet Jena (2011) Early history of genetics revised: New light shed on 'rediscovery' of Mendel's laws of heredity. *ScienceDaily* 3 May 2011.

74. Mendel G. (1865) Versuche über Plflanzen-hybriden (Experiments in Plant Hybridization, Bateson Translation). *Verhandlungen des naturforschenden Vereines in Brünn* IV: 3–47.

75. Bateson W. (1909) *Mendel's Principles of Heredity* (Cambridge University Press, Cambridge).

76. Satzinger H. (2008) Theodor and Marcella Boveri: Chromosomes and cytoplasm in heredity and development. *Nature Reviews Genetics* 9: 231–8.

77. Rabi C. (1885) On cell division. *Morphological Yearbook* 10: 214–330.

78. Boveri T. (1909) Die Blastomerenkerne von *Ascaris megalocephala* und die Theorie der Chromosomenindividualität. *Archiv für Zellforschung* 3: 181–268.

79. Hansford S. and Huntsman D.G. (2014) Boveri at 100: Theodor Boveri and genetic predisposition to cancer. *Journal of Pathology* 234: 142–5.

80. Sutton W.S. (1903) The chromosomes in heredity. *Biological Bulletin* 4: 231–51.

81. Halliburton W.D. (1895) Nucleo-proteids. *Journal of Physiology* 18: 306–18.
82. Johannsen W. (1909) *Elemente der exakten Erblichkeitslehre* (Gustav Fischer, Stuttgart).
83. Johannsen W. (1911) The genotype conception of heredity. *American Naturalist* 45: 129–59.
84. Brush S.G. (1978) Nettie M. Stevens and the discovery of sex determination by chromosomes. *Isis* 69: 163–72.
85. Wilson E.B. (1905) The chromosomes in relation to the determination of sex in insects. *Science* 22: 500–2.
86. Bridges C.B. (1914) Direct proof through non-disjunction that the sex-linked genes of *Drosophila* are borne by the X-chromosome. *Science* 40: 107–9.
87. Bridges C.B. (1916) Non-disjunction as proof of the chromosome theory of heredity. *Genetics* 1: 1–52.
88. Rubin G.M. and Lewis E.B. (2000) A brief history of *Drosophila's* contributions to genome research. *Science* 287: 2216–8.
89. Sturtevant A.H. (1913) The linear arrangement of six sex-linked factors in *Drosophila*, as shown by their mode of association. *Journal of Experimental Zoology* 14: 43–59.
90. Morgan T.H., Bridges C.B., Muller H.J. and Sturtevant A.H. (1915) *The Mechanism of Mendelian Heredity* (Henry Holt and Company, New York).
91. Morgan T.H. (1922) C.roonian lecture:—On the mechanism of heredity. *Proceedings of the Royal Society B: Biological Sciences* 94: 162–97.
92. Muller .H.J. (1930) Types of visible variations induced by X-rays in *Drosophila*. *Journal of Genetics* 22: 299–334.
93. Edwards A.W.F. (2012) Reginald Crundall Punnett: First Arthur Balfour Professor of Genetics, Cambridge, 1912. *Genetics* 192: 3–13.
94. Strauss B.S. (2016) Biochemical genetics and molecular biology: The contributions of George Beadle and Edward Tatum. *Genetics* 203: 13–20.
95. Strauss B.S. (2017) A physicist's quest in biology: Max Delbrück and "Complementarity". *Genetics* 206: 641–50.
96. Timoféeff-Ressovsky N.W. (1927) Studies on the phenotypic manifestation of hereditary factors. I. On the phenotypic manifestation of the genovariation *radius incompletus* in *Drosophila funebris*. *Genetics* 12: 128–98.
97. Timoféeff-Ressovsky N.V., Zimmer K.G. and Delbrück M. (1935) Ueber die Natur der Genmutation und der Genstruktur. *Nachrichten von der Gesellschaft der Wissenschaften zu Göttingen, Mathematisch-Physikalische Klasse, Fachgruppe VI* 1: 189–245.
98. Ratner V.A. (2001) Nikolay Vladimirovich Timoféeff-Ressovsky (1900–1981): Twin of the century of genetics. *Genetics* 158: 933–9.
99. Elgin S.C.R. and Reuter G. (2013) Position-effect variegation, heterochromatin formation, and gene silencing in *Drosophila*. *Cold Spring Harbor Perspectives in Biology* 5: a017780.
100. Muller H.J., Prokofyeva A. and Raffel D. (1935) Minute intergenic rearrangement as a cause of apparent gene mutation. *Nature* 135: 253–5.
101. Goldschmidt R. (1937) Spontaneous chromatin rearrangements in *Drosophila*. *Nature* 140: 767.
102. Dietrich M.R. (2003) Richard goldschmidt: Hopeful monsters and other 'heresies'. *Nature Reviews Genetics* 4: 68–74.
103. Barrell B.G., Air G.M. and Hutchison 3rd, C.A. (1976) Overlapping genes in bacteriophage phiX174. *Nature* 264: 34–41.
104. Henikoff S., Keene M.A., Fechtel K. and Fristrom J.W. (1986) Gene within a gene: Nested *Drosophila* genes encode unrelated proteins on opposite DNA strands. *Cell* 44: 33–42.
105. Gott J.M. et al. (1988) Genes within genes: Independent expression of phage T4 intron open reading frames and the genes in which they reside. *Genes & Development* 2: 1791–9.
106. Kapranov P. et al. (2005) Examples of the complex architecture of the human transcriptome revealed by RACE and high-density tiling arrays. *Genome Research* 15: 987–97.
107. Carninci P. et al. (2005) The transcriptional landscape of the mammalian genome. *Science* 309: 1559–63.
108. Denoeud F. et al. (2007) Prominent use of distal 5′ transcription start sites and discovery of a large number of additional exons in ENCODE regions. *Genome Research* 17: 746–59.
109. Kapranov P., et al. (2007) RNA maps reveal new RNA classes and a possible function for pervasive transcription. *Science* 316: 1484–8.
110. Kumar A. (2009) An overview of nested genes in eukaryotic genomes. *Eukaryotic Cell* 8: 1321–9.
111. Goldschmidt R. (1935) Gen und Außeneigenschaft (Untersuchungen anDrosophila) I. *Zeitschrift für induktive Abstammungs- und Vererbungslehre* 69: 38–69.
112. Lamm E. (2008) Hopeful heretic–Richard Goldschmidt's genetic metaphors. *History and Philosophy of the Life Sciences* 30: 387–405.
113. Ayala F.J. and Fitch W.M. (1997) Genetics and the origin of species: An introduction. *Proceedings of the National Academy of Sciences USA* 94: 7691–7.
114. Dobzhansky T. (1973) Nothing in biology makes sense except in the light of evolution. *The American Biology Teacher* 35: 125–9.
115. Huxley J. (1942) *Evolution: The Modern Synthesis* (George Allen & Unwin Ltd., Crows Nest).
116. Provine W.B. (1971) *The Origins of Theoretical Population Genetics* (University of Chicago Press, Chicago, IL).
117. Visscher P.M. and Goddard M.E. (2019) From R.A. Fisher's 1918 paper to GWAS a century later. *Genetics* 211: 1125–30.
118. Darlington C.D. and Ford E.B. (1956) Natural populations and the breakdown of classical genetics. *Proceedings of the Royal Society B: Biological Sciences* 145: 350–64.
119. Harman O.S. (2005) Cyril Dean Darlington: The man who 'invented' the chromosome. *Nature Reviews Genetics* 6: 79–85.
120. Haldane J.B.S. (1924) A mathematical theory of natural and artificial selection. Part II the influence of partial self-fertilisation, inbreeding, assortative mating, and selective fertilisation on the composition of mendelian populations, and on natural selection. *Biological Reviews* 1: 158–63.
121. v'ant Hof A.E., et al. (2016) The industrial melanism mutation in British peppered moths is a transposable element. *Nature* 534: 102–5.
122. Fisher R.A. (1918) The correlation between relatives on the supposition of Mendelian inheritance. *Transactions of the Royal Society of Edinburgh* 53: 399–433.
123. Fisher R.A. (1930) *The Genetical Theory of Natural Selection* (Clarendon Press, Oxford).
124. Edwards A.W. (2000) The genetical theory of natural selection. *Genetics* 154: 1419–26.
125. Darwin C. (1859) *On the Origin of Species by Means of Natural Selection* (John Murray, London).

126. Fitzgerald D.M. and Rosenberg S.M. (2019) What is mutation? A chapter in the series: How microbes "jeopardize" the modern synthesis. *PLOS Genetics* 15: e1007995.

127. Weismann A. (1889) *Essays Upon Heredity and Kindred Biological Problems.* Authorized translation (Clarendon Press, Oxford).

128. Nilsson E.E., Maamar M.B. and Skinner M.K. (2020) Environmentally induced epigenetic transgenerational inheritance and the Weismann Barrier: The dawn of Neo-Lamarckian theory. *Journal of Developmental Biology* 8: 28.

129. Waddington C.H. (1942) Canalization of development and the inheritance of acquired characters. *Nature* 150: 563–5.

130. Waddington C.H. (1953) Genetic assimilation of an acquired character. *Evolution* 7: 118–26.

131. Waddington C.H. (1956) Genetic assimilation of the bithorax phenotype. *Evolution* 10: 1–13.

132. Cairns J. (2008) The foundations of molecular biology: A 50th anniversary. *Current Biology* 18: R234–6.

133. Delbruck M. (1940) Radiation and the hereditary mechanism. *The American Naturalist* 74: 350–62.

134. Lederberg J. and Lederberg E.M. (1952) Replica plating and indirect selection of bacterial mutants. *Journal of Bacteriology* 63: 399–406.

135. Galton F. (1875) The history of twins, as a criterion of the relative powers of nature and nurture. *Fraser's Magazine* 12: 566–76.

136. Durant J.R. (1979) Scientific naturalism and social reform in the thought of Alfred Russel Wallace. *British Journal for the History of Science* 12: 31–58.

137. Provine W.B. and Russell E.S. (1986) Geneticists and race. *American Zoologist* 26: 857–87.

138. Caspari E.W. and Marshak R.E. (1965) The rise and fall of Lysenko. *Science* 149: 275–8.

139. Soyfer V.N. (2001) The consequences of political dictatorship for Russian science. *Nature Reviews Genetics* 2: 723–9.

140. Reichard P. (2002) Oswald T. Avery and the Nobel Prize in medicine. *Journal of Biological Chemistry* 277: 13355–62.

141. Levene P.A. (1909) Uber die Hefenucleinsaure. *Biochemische Zeitschrift* 17: 120–31.

142. Levene P.A. (1919) The structure of yeast nucleic acid. *Journal of Biological Chemistry* 40: 415–24.

143. Levene P.A. and Jacobs W.A. (1912) On the structure of thymus nucleic acid. *Journal of Biological Chemistry* 12: 411–20.

144. Stent G.S. (1970) DNA. *Daedalus* 99: 909–37.

145. Hargittai I. (2009) The tetranucleotide hypothesis: A centennial. *Structural Chemistry* 20: 753–6.

146. Levene P.A. and Jacobs W.A. (1909) Über die Hefe-Nucleinsäure. *Berichte der deutschen chemischen Gesellschaft* 42: 2474–8.

147. Frixione E. and Ruiz-Zamarripa L. (2019) The "scientific catastrophe" in nucleic acids research that boosted molecular biology. *Journal of Biological Chemistry* 294: 2249–55.

148. Signer R., Caspersson T. and Hammarsten E. (1938) Molecular shape and size of thymonucleic acid. *Nature* 141: 122.

149. Kerr S.E., Seraidarian K. and Wargon M. (1949) Studies on ribonucleic acid: III. On the composition of the ribonucleic acid of beef pancreas, with notes on the action of ribonuclease. *Journal of Biological Chemistry* 181: 773–80.

150. Levene P.A. and Bass L.W. (1931) *Nucleic Acids* (Chemical Catalog Company, New York).

151. Davidson J.N. and Waymouth C. (1943) Ribonucleic acids in animal tissues. *Nature* 152: 47–8.

152. Allen F.W. (1941) The biochemistry of the nucleic acids, purines, and pyrimidines. *Annual Review of Biochemistry* 10: 221–44.

153. Chargaff E. (1950) Chemical specificity of nucleic acids and mechanism of their enzymatic degradation. *Experientia* 6: 201–9.

154. Feulgen R. and Rossenbeck H. (1924) Mikroskopisch-chemischer Nachweis einer Nucleinsäure von Typus der Thymonucleinsäure und die darauf beruhende elektive Färbung von Zellkernen in Mikroskopischer Praäparaten. *Zeitschrift für physiologische Chemie* 135: 203–48.

155. Kasten F.H. (2003) Robert Feulgen and his histochemical reaction for DNA. *Biotechnic & Histochemistry* 78: 45–9.

156. Brachet J. (1933) Recherches sur la synthese de l'acide thymonucleique pendant le developpement de l'oeuf d'oursin. *Archives de Biologie* 44: 519–76.

157. Brachet J. (1941) La localisation des acides pentosenucléiques dans les tissus animaux et les oeufs d'amphibiens en voie de développement. *Archives de Biologie* 53: 207–57.

158. Brachet J. (1941) La detection histochimique et la microdosage des acides pentosenucléiques. *Enzymologia* 10: 87–96.

159. Brachet J. (1950) The localization and the role of ribonucleic acid in the cell. *Annals of the New York Academy of Sciences* 50: 861–9.

160. Brachet J.L.A. (1960) *The Biological Role of Ribonucleic Acids.* Sixth Weizmann Memorial Lecture Series (Elsevier, Amsterdam).

161. Belozersky A.N. and Spirin A.S. (1958) Correlation between the compositions of deoxyribonucleic and ribonucleic acids. *Nature* 182: 111–2.

162. Sevag M.G., Smolens J. and Lackman D.B. (1940) The nucleic acid content and distribution in *Streptococcus pyogenes*. *Journal of Biological Chemistry* 134: 523–9.

163. Westhof E. and Fritsch V. (2000) RNA folding: Beyond Watson–crick pairs. *Structure* 8: R55–65.

164. Leontis N.B., Lescoute A. and Westhof E. (2006) The building blocks and motifs of RNA architecture. *Current Opinion in Structural Biology* 16: 279–87.

165. Brown D.M. and Todd A.R. (1955) Nucleic acids. *Annual Review of Biochemistry* 24: 311–38.

166. Todd A. (1958) Synthesis in the study of nucleotides; basic work on phosphorylation opens the way to an attack on nucleic acids and nucleotide coenzymes. *Science* 127: 787–92.

167. Bawden F.C. and Pirie N.W. (1937) The isolation and some properties of liquid crystalline substances from solanaceous plants infected with three strains of tobacco mosaic virus. *Proceedings of the Royal Society B: Biological Sciences* 123: 274–320.

168. Stanley W.M. (1935) Isolation of a crystalline protein possessing the properties of tobacco-mosaic virus. *Science* 81: 644–5.

169. Gierer A. and Schramm G. (1956) Infectivity of ribonucleic acid from tobacco mosaic virus. *Nature* 177: 702–3.

170. Markham R. (1953) Virus nucleic acids. *Advances in Virus Research* 1: 315–32.

171. Garrod A.E. (1902) The incidence of alkaptonuria a study in chemical individuality. *Lancet* 2: 1616–20.

172. Garrod A. (1908) On inborn errors of metabolism. The croonian lectures. *Lancet* 172: 1–7.
173. Lederberg J. and McCray A.T. (2001) 'Ome Sweet 'Omics—a genealogical treasury of words. *The Scientist* 15: 8.
174. Beadle G.W. and Tatum E.L. (1941) Genetic control of biochemical reactions in *Neurospora*. *Proceedings of the National Academy of Sciences USA* 27: 499–506.
175. Wright S. (1917) Color inheritance in mammals: II. The mouse—better adapted to experimental work than any other mammal—seven sets of mendelian allelomorphs identified—factorial hypothesis framed by Cuenot on basis of his work with mice. *Journal of Heredity* 8: 373–8.
176. Goldschmidt R. (1951) L. Cuénot: 1866–1951. *Science* 113: 309–10.
177. Buican D. (1982) Mendelism in France and the work of Lucien Cuénot. *Scientia* 76: 129–37.
178. Hickman M. and Cairns J. (2003) The centenary of the one-gene one-enzyme hypothesis. *Genetics* 163: 839–41.
179. Morgan H.D., Sutherland H.G.E., Martin D.I.K. and Whitelaw E. (1999) Epigenetic inheritance at the agouti locus in the mouse. *Nature Genetics* 23: 314–8.
180. Whitelaw E. and Martin D.I. (2001) Retrotransposons as epigenetic mediators of phenotypic variation in mammals. *Nature Genetics* 27: 361–5.
181. Horowitz N.H. (1991) Fifty years ago: The *Neurospora* revolution. *Genetics* 127: 631–5.
182. Davis R.H. and Perkins D.D. (2002) Neurospora: A model of model microbes. *Nature Reviews Genetics* 3: 397–403.
183. Whitaker R.J. and Barton H.A. (2018) *Women in Microbiology* (American Society of Microbiology, Washington DC).
184. Lederberg J. and Tatum E.L. (1946) Gene recombination in *Escherichia coli*. *Nature* 158: 558.
185. Caspersson T. and Schultz J. (1938) Nucleic acid metabolism of the chromosomes in relation to gene reproduction. *Nature* 142: 294–95.
186. Twort F.W. (1915) An investigation on the nature of ultra-microscopic viruses. *Lancet* 186: 1241–3.
187. Schrödinger E. (1944) *What Is Life? The Physical Aspect of the Living Cell* (Cambridge Univ. Press, Cambridge).
188. Cobb M. (2016) *Life's Greatest Secret: The Race to Crack the Genetic Code* (Profile Books, London).
189. Griffith F. (1928) The significance of Pneumococcal types. *Journal of Hygiene* 27: 113–59.
190. Avery O.T., MacLeod C.M. and McCarty M. (1944) Studies on the chemical nature of the substance inducing transformation of Pneumococcal types: Induction of transformation by a desoxyribonucleic acid fraction isolated from Pneumococcus type III. *Journal of Experimental Medicine* 79: 137–58.
191. Boivin A., Vendrely R. and Lehoult Y. (1945) L'acide thymonucléique hautement polymerisé, principe capable de conditionner la specificité sérologique et l'équipement enzymatique des Bactéries. Conséquences pour la biochimie de l'hérédité. *Comptes Rendus de l'Académie des Sciences* 221: 646–8.
192. Stent G.S. (1972) Prematurity and uniqueness in scientific discovery. *Scientific American* 227: 84–93.
193. Cobb M. (2014) Oswald Avery, DNA, and the transformation of biology. *Current Biology* 24: R55–60.
194. Various (1983) Nature conference: Thirty years of DNA. *Nature* 302: 651–4.
195. Hershey A.D. and Chase M. (1952) Independent functions of viral protein and nucleic acid in growth of bacteriophage. *Journal of General Physiology* 36: 39–56.
196. Johnson T.B. and Coghill R.D. (1925) Researches on pyrimidines. C111. The discovery of 5-methyl-cytosine in tuberculinic acid, the nucleic acid of the tubercle bacillus. *Journal of the American Chemical Society* 47: 2838–44.
197. Hotchkiss R.D. (1948) The quantitative separation of purines, pyrimidines, and nucleosides by paper chromatography. *Journal of Biological Chemistry* 175: 315–32.
198. Wyatt G.R. (1950) Occurrence of 5-methylcytosine in nucleic acids. *Nature* 166: 237–8.
199. Rudner R., Karkas J.D. and Chargaff E. (1968) Separation of *B. subtilis* DNA into complementary strands. 3. Direct analysis. *Proceedings of the National Academy of Sciences USA* 60: 921–2.
200. Rogerson A.C. (1991) There appear to be conserved constraints on the distribution of nucleotide sequences in cellular genomes. *Journal of Molecular Evolution* 32: 24–30.
201. Fickett J.W., Torney D.C. and Wolf D.R. (1992) Base compositional structure of genomes. *Genomics* 13: 1056–64.
202. Prabhu V.V. (1993) Symmetry observations in long nucleotide sequences. *Nucleic Acids Research* 21: 2797–800.
203. Qi D. and Cuticchia A.J. (2001) Compositional symmetries in complete genomes. *Bioinformatics* 17: 557–9.
204. Albrecht-Buehler G. (2006) Asymptotically increasing compliance of genomes with Chargaff's second parity rules through inversions and inverted transpositions. *Proceedings of the National Academy of Sciences USA* 103: 17828–33.
205. Shporer S., Chor B., Rosset S. and Horn D. (2016) Inversion symmetry of DNA k-mer counts: Validity and deviations. *BMC Genomics* 17: 696.
206. Cristadoro G., Degli Esposti M. and Altmann E.G. (2018) The common origin of symmetry and structure in genetic sequences. *Scientific Reports* 8: 15817.
207. Vischer E. and Chargaff E. (1948) The composition of the pentose nucleic acids of yeast and pancreas. *Journal of Biological Chemistry* 176: 715–34.
208. Chargaff E., Vischer E., Doniger R., Green C. and Misani F. (1949) The composition of the desoxypentose nucleic acids of thymus and spleen. *Journal of Biological Chemistry* 177: 405–16.
209. Vischer E., Zamenhof S. and Chargaff E. (1949) Microbial nucleic acids: The desoxypentose nucleic acids of avian tubercle bacilli and yeast. *Journal of Biological Chemistry* 177: 429–38.
210. Maddox B. (2002) *Rosalind Franklin: The Dark Lady of DNA* (HarperCollins, New York).
211. Williams G. (2019) *Unravelling the Double Helix: The Lost Heroes of DNA* (Weidenfeld & Nicolson, London).
212. Astbury W.T. and Bell F.O. (1938) X-ray study of thymonucleic acid. *Nature* 141: 747–8.
213. Astbury W.T. (1947) X-ray studies of nucleic acids. *Symposia of the Society for Experimental Biology* 1: 66–76.
214. Watson J.D. and Crick F.H. (1953) Molecular structure of nucleic acids; a structure for deoxyribose nucleic acid. *Nature* 171: 737–8.
215. Franklin R.E. and Gosling R.G. (1953) Molecular configuration in sodium thymonucleate. *Nature* 171: 740–1.
216. Fuller W. (2003) Who said 'helix'? *Nature* 424: 876–8.
217. Cobb M. (2017) 60 years ago, Francis Crick changed the logic of biology. *PLOS Biology* 15: e2003243.

218. Olby R. (2003) Quiet debut for the double helix. *Nature* 421: 402–5.

219. Meselson M. and Stahl F.W. (1958) The replication of DNA in *Escherichia coli*. *Proceedings of the National Academy of Sciences USA* 44: 671–82.

220. Bessman M.J., Kornberg A., Lehman I.R. and Simms E.S. (1956) Enzymic synthesis of deoxyribonucleic acid. *Biochimica et Biophysica Acta* 21: 197–8.

221. Kornberg A., Kornberg S.R. and Simms E.S. (1956) Metaphosphate synthesis by an enzyme from Escherichia coli. *Biochimica et Biophysica Acta* 20: 215–27.

222. Lehman I.R. (2003) Discovery of DNA polymerase. *Journal of Biological Chemistry* 278: 34733–8.

223. Peacocke A. (2005) Historical article: Titration studies and the structure of DNA. *Trends in Biochemical Sciences* 30: 160–2.

224. Creeth J.M. (1948) *Some Physico-chemical Studies on Nucleic Acids and Related Substances*. Ph.D. Dissertation, University of London. http://ethos.bl.uk/OrderDetails. do?did=1&uin=uk.bl.ethos.729338.

225. Harding S.E., Channell G. and Phillips-Jones M.K. (2018) The discovery of hydrogen bonds in DNA and a re-evaluation of the 1948 Creeth two-chain model for its structure. *Biochemical Society Transactions* 46: 1171–82.

226. Shing Ho P. and Carter M. (2011) DNA structure: Alphabet soup for the cellular soul, in *DNA Replication - Current Advances* (InTechOpen).

227. Rich A. (2009) The era of RNA awakening: Structural biology of RNA in the early years. *Quarterly Reviews of Biophysics* 42: 117–37.

228. Nikolova E.N. et al. (2013) A historical account of Hoogsteen base-pairs in duplex DNA. *Biopolymers* 99: 955–68.

229. Herbert A. and Rich A. (1999) Left-handed Z-DNA: Structure and function. *Genetica* 106: 37–47.

230. Spiegel J., Adhikari S. and Balasubramanian S. (2020) The structure and function of DNA G-quadruplexes. *Trends in Chemistry* 2: 123–36.

231. Zeraati M. et al. (2018) I-motif DNA structures are formed in the nuclei of human cells. *Nature Chemistry* 10: 631–7.

232. Calladine C.R., Drew H.R., Luisi B.F. and Travers A.A. (2004) *Understanding DNA*. 3rd Edition. The Molecule and How it Works (Academic Press, Cambridge, MA).

233. Doty P., Marmur J., Eigner J. and Schildkraut C. (1960) Strand separation and specific recombination in deoxyribonucleic acids: Physical chemical studies. *Proceedings of the National Academy of Sciences USA* 46: 461–76.

234. Marmur J. and Doty P. (1961) Thermal renaturation of deoxyribonucleic acids. *Journal of Molecular Biology* 3: 585–94.

CHAPTER 3

1. Brachet J. and Chantrenne H. (1951) Protein synthesis in nucleated and non-nucleated halves of *Acetabularia mediterranea* studied with carbon-14 dioxide. *Nature* 168: 950.

2. Hammerling J. (1953) Nucleo-cytoplasmic relationships in the development of *Acetabularia*. *International Review of Cytology* 2: 475–98.

3. Caspersson T. and Schultz J. (1939) Pentose nucleotides in the cytoplasm of growing tissues. *Nature* 143: 602–3.

4. Brachet J. (1941) La localisation des acides pentosenucléiques dans les tissus animaux et les oeufs d'amphibiens en voie de développement. *Archives de Biologie* 53: 207–57.

5. Brachet J. (1941) La detection histochimique et la microdosage des acides pentosenucléiques. *Enzymologia* 10: 87–96.

6. Caspersson T. (1941) Studien über den Eiweißumsatz der Zelle. *Naturwissenschaften* 29: 33–43.

7. Claude A. (1938) Concentration and purification of chicken tumor I agent. *Science* 87: 467–8.

8. Claude A. (1943) The constitution of protoplasm. *Science* 97: 451–6.

9. Claude A. (1946) Fractionation of mammalian liver cells by differential centrifugation: I. Problems, methods, and preparation of extract. *Journal of Experimental Medicine* 84: 51–9.

10. Brachet J. (1950) The localization and the role of ribonucleic acid in the cell. *Annals of the New York Academy of Sciences* 50: 861–9.

11. Palade G.E. and Siekevitz P. (1956) Liver microsomes - An integrated morphological and biochemical study. *Journal of Biophysical and Biochemical Cytology* 2: 171–200.

12. Roberts R.B. (1958) Introduction, in R.B. Roberts (ed.) *Microsomal Particles and Protein Synthesis* (Pergamon Press, Oxford).

13. Brimacombe R., Stoffler G. and Wittmann H.G. (1978) Ribosome structure. *Annual Review of Biochemistry* 47: 217–49.

14. Marko A.M. and Butler G.C. (1951) The isolation of sodium desoxyribonucleate with sodium dodecyl sulfate. *Journal of Biological Chemistry* 190: 165–76.

15. Colter J.S. and Brown R.A. (1956) Preparation of nucleic acids from Ehrlich ascites tumor cells. *Science* 124: 1077–8.

16. Georgiev G.P., Samarina O.P., Lerman M.I., Smirnov M.N. and Severtzov A.N. (1963) Biosynthesis of messenger and ribosomal ribonucleic acids in the nucleolochromosomal apparatus of animal cells. *Nature* 200: 1291–4.

17. Holley R.W., Apgar J. and Merrill S.H. (1961) Evidence for the liberation of a nuclease from human fingers. *Journal of Biological Chemistry* 236: PC42–3.

18. Parks M.M., et al. (2018) Variant ribosomal RNA alleles are conserved and exhibit tissue-specific expression. *Science Advances* 4: eaao0665.

19. Guimaraes J.C. and Zavolan M. (2016) Patterns of ribosomal protein expression specify normal and malignant human cells. *Genome Biology* 17: article 236.

20. Panda A., et al. (2020) Tissue- and development-stage-specific mRNA and heterogeneous CNV signatures of human ribosomal proteins in normal and cancer samples. *Nucleic Acids Research* 48: 7079–98.

21. Siekevitz P. and Zamecnik P.C. (1981) Ribosomes and protein synthesis. *Journal of Cell Biology* 91: 53s–65s.

22. Jacob F. (1988) *The Statue Within: An Autobiography* (Basic Books, New York).

23. Carroll S.B. (2013) *Brave Genius: A Scientist, a Philosopher, and Their Daring Adventures from the French Resistance to the Nobel Prize* (Crown Publishers, New York).

24. Cobb M. (2015) Who discovered messenger RNA? *Current Biology* 25: R526–32.

25. Caldwell P.C. and Hinshelwood C. (1950) Some considerations on autosynthesis in bacteria. *Journal of the Chemical Society* 1950: 3156–9.

26. Pauling L., Itano H.A., Singer S.J. and Wells I.C. (1949) Sickle cell anemia, a molecular disease. *Science* 110: 543–8.

27. Hunt J.A. and Ingram V.M. (1958) Allelomorphism and the chemical differences of the human haemoglobins A, S and C. *Nature* 181: 1062–3.

28. Ingram V.M. (1958) Abnormal human haemoglobins. *Proceedings of the Royal Society of Medicine* 51: 645–6.

29. Hardison R.C., et al. (2002) HbVar: A relational database of human hemoglobin variants and thalassemia mutations at the globin gene server. *Human Mutation* 19: 225–33.

30. Giardine B., et al. (2011) Systematic documentation and analysis of human genetic variation in hemoglobinopathies using the microattribution approach. *Nature Genetics* 43: 295–301.

31. Prescott D.M. and Mazia D. (1954) The permeability of nucleated and enucleated fragments of *Amoeba proteus* to D_2O. *Experimental Cell Research* 6: 117–26.

32. Boivin A. and Vendrely R. (1947) Sur le rôle possible des deux acides nucléiques dans la cellule vivante. *Experientia* 3: 32–4.

33. Dounce A.L. (1952) Duplicating mechanism for peptide chain and nucleic acid synthesis. *Enzymologia* 15: 251–8.

34. Dounce A.L. (1953) Nucleic acid template hypotheses. *Nature* 172: 541.

35. Work T.S. and Campbell P.N. (1953) Nucleic acid template hypotheses. *Nature* 172: 541–2.

36. Rich A. and Watson J.D. (1954) Some relations between DNA and RNA. *Proceedings of the National Academy of Sciences USA* 40: 759–64.

37. Brachet J.L.A. (1960) *The Biological Role of Ribonucleic Acids*. Sixth Weizmann Memorial Lecture Series (Elsevier, Amsterdam).

38. Watson J.D. (1963) Involvement of RNA in synthesis of proteins. *Science* 140: 17–26.

39. Brenner S. (1957) On the impossibility of all overlapping triplet codes in information transfer from nucleic acid to proteins. *Proceedings of the National Academy of Sciences USA* 43: 687–94.

40. Crick F.H. (1958) On protein synthesis. *Symposia of the Society for Experimental Biology* 12: 138–63.

41. Hoagland M.B., Stephenson M.L., Scott J.F., Hecht L.I. and Zamecnik P.C. (1958) Soluble ribonucleic acid intermediate in protein synthesis. *Journal of Biological Chemistry* 231: 241–57.

42. Twort F.W. (1915) An investigation on the nature of ultra-microscopic viruses. *Lancet* 186: 1241–3.

43. D'Herelle F. (2007) On an invisible microbe antagonistic toward dysenteric bacilli. *Research in Microbiology* 158: 553–4.

44. Cairns J. and Watson J.D. (2007) *Phage and the Origins of Molecular Biology* (Cold Spring Harbor Laboratory Press, Cold Spring Harbor, NY).

45. Pardee A.B. (1954) Nucleic acid precursors and protein synthesis. *Proceedings of the National Academy of Sciences USA* 40: 263–70.

46. Jacob F. and Monod J. (1961) Genetic regulatory mechanisms in the synthesis of proteins. *Journal of Molecular Biology* 3: 318–56.

47. Brenner S., Jacob F. and Meselson M. (1961) An unstable intermediate carrying information from genes to ribosomes for protein synthesis. *Nature* 190: 576–81.

48. Hershey A.D., Dixon J. and Chase M. (1953) Nucleic acid economy in bacteria infected with bacteriophage T2. I. Purine and pyrimidine composition. *Journal of General Physiology* 36: 777–89.

49. Volkin E. and Astrachan L. (1956) Phosphorus incorporation in *Escherichia coli* ribonucleic acid after infection with bacteriophage T2. *Virology* 2: 149–61.

50. Pardee A.B., Jacob F. and Monod J. (1959) The genetic control and cytoplasmic expression of "Inducibility" in the synthesis of β-galactosidase by *E. coli*. *Journal of Molecular Biology* 1: 165–78.

51. Pardee A.B. (1985) Roots: Molecular basis of gene expression: Origins from the Pajama experiment. *BioEssays* 2: 86–9.

52. Nomura M., Hall B.D. and Spiegelman S. (1960) Characterization of RNA synthesized in *Escherichia coli* after bacteriophage T2 infection. *Journal of Molecular Biology* 2: 306–26.

53. Spiegelman S., Hall B.D. and Storck R. (1961) The occurrence of natural DNA-RNA complexes in *E. coli* infected with T2. *Proceedings of the National Academy of Sciences USA* 47: 1135–41.

54. Hall B.D. and Spiegelman S. (1961) Sequence complementarity of T2-DNA and T2-specific RNA. *Proceedings of the National Academy of Sciences USA* 47: 137–46.

55. Gros F., et al. (1961) Unstable ribonucleic acid revealed by pulse labelling of *Escherichia coli*. *Nature* 190: 581–5.

56. Weiss S.B. and Gladstone L. (1959) A mammalian system for the incorporation of cytidine triphosphate into ribonucleic acid. *Journal of the American Chemical Society* 81: 4118–9.

57. Stevens A. (1960) Incorporation of the adenine ribonucleotide into RNA by cell fractions from *E. coli* B. *Biochemical and Biophysical Research Communications* 3: 92–6.

58. Allfrey V.G. and Mirsky A.E. (1962) Evidence for complete DNA-dependence of RNA synthesis in isolated thymus nuclei. *Proceedings of the National Academy of Sciences USA* 48: 1590–6.

59. Olby R. (1994) *The Path to the Double Helix: The Discovery of DNA* (Dover Publications, Mineola, NY).

60. Kay L. (2000) *Who Wrote the Book of Life? A History of the Genetic Code* (Stanford University Press, Redwood City, CA).

61. Strauss B.S. (2019) Martynas Yčas: The "Archivist" of the RNA tie club. *Genetics* 211: 789–95.

62. Cobb M. (2016) *Life's Greatest Secret: The Race to Crack the Genetic Code* (Profile Books, London).

63. Shannon C.E. (1940) *An algebra for theoretical genetics*. PhD thesis, Massachusetts Institute of Technology.

64. Shannon C.E. (1948) A mathematical theory of communication. *Bell System Technical Journal* 27: 379–423.

65. Shannon C.E. and Weaver W. (1949) *A Mathematical Theory of Communication* (University of Illinois Press, Champaign, IL).

66. Wiener N. (1948) *Cybernetics: Or Control and Communication in the Animal and the Machine* (2nd revised ed. 1961). (MIT Press, Cambridge, MA).

67. Mayr E. (1961) Cause and effect in biology. *Science* 134: 1501–6.

68. Sanger F. and Tuppy H. (1951) The amino-acid sequence in the phenylalanyl chain of insulin. I. The identification of lower peptides from partial hydrolysates. *Biochemical Journal* 49: 463–81.

69. Sanger F. and Tuppy H. (1951) The amino-acid sequence in the phenylalanyl chain of insulin. 2. The investigation of peptides from enzymic hydrolysates. *Biochemical Journal* 49: 481–90.

70. Sanger F. and Thompson E.O.P. (1953) The amino-acid sequence in the glycyl chain of insulin. 1. The identification of lower peptides from partial hydrolysates. *Biochemical Journal* 53: 353–66.

71. Sanger F. and Thompson E.O.P. (1953) The amino-acid sequence in the glycyl chain of insulin. 2. The investigation of peptides from enzymic hydrolysates. *Biochemical Journal* 53: 366–74.

72. Benzer S. (1957) The elementary units of heredity, in W. D. McElroy and B. Glass (eds.) *A Symposium on The Chemical Basis of Heredity* (Johns Hopkins University Press, Baltimore, MD).

73. Benzer S. (1959) On the topology of the genetic fine structure. *Proceedings of the National Academy of Sciences USA* 45: 1607–20.

74. Benzer S. (1960) Genetic fine structure. *Harvey Lectures* 56: 1–21.

75. Benzer S. (1961) On the topography of the genetic fine structure. *Proceedings of the National Academy of Sciences USA* 47: 403–15.

76. Benzer S. (1962) The fine structure of the gene. *Scientific American* 206: 70–84.

77. Hayes W. (1968) *The Genetics of Bacteria and Their Viruses* (2nd Edition). (Blackwell Scientific Publications, Hoboken, NJ).

78. Harris W.A. (2008) Seymour Benzer 1921–2007 The man who took us from genes to behaviour. *PLOS Biology* 6: e41.

79. Benzer S. (1957) The elementary units of heredity, in W. D. McElroy and B. Glass (eds.) *The Chemical Basis of Heredity* (Johns Hopkins Press, Baltimore, MD).

80. Yanofsky C. (2007) Establishing the triplet nature of the genetic code. *Cell* 128: 815–8.

81. Crick F.H., Barnett L., Brenner S. and Watts-Tobin R.J. (1961) General nature of the genetic code for proteins. *Nature* 192: 1227–32.

82. Tsugita A. and Fraenkel-Conrat H. (1960) The amino acid composition and C-terminal sequence of a chemically evoked mutant of Tmv. *Proceedings of the National Academy of Sciences USA* 46: 636–42.

83. Crick F.H.C. (1966) Codon—anticodon pairing: The wobble hypothesis. *Journal of Molecular Biology* 19: 548–55.

84. Crick F.H., Griffith J.S. and Orgel L.E. (1957) Codes without commas. *Proceedings of the National Academy of Sciences USA* 43: 416–21.

85. Nirenberg M.W. and Matthaei J.H. (1961) The dependence of cell-free protein synthesis in *E. coli* upon naturally occurring or synthetic polyribonucleotides. *Proceedings of the National Academy of Sciences USA* 47: 1588–602.

86. Matthaei J.H., Jones O.W., Martin R.G. and Nirenberg M.W. (1962) Characteristics and composition of RNA coding units. *Proceedings of the National Academy of Sciences USA* 48: 666–77.

87. Leder P. and Nirenberg M. (1964) RNA codewords and protein synthesis II. Nucleotide sequence of a valine RNA codeword. *Proceedings of the National Academy of Sciences USA* 52: 420–7.

88. Nirenberg M., et al. (1965) RNA codewords and protein synthesis VII. On the general nature of the RNA code. *Proceedings of the National Academy of Sciences USA* 53: 1161–8.

89. Nirenberg M., et al. (1966) The RNA code and protein synthesis. *Cold Spring Harbor Symposia on Quantitative Biology* 31: 11–24.

90. Crick F.H.C. (1966) Genetic code - yesterday today and tomorrow. *Cold Spring Harbor Symposia on Quantitative Biology* 31: 3–9.

91. Marshall J. (2014) The genetic code. *Proceedings of the National Academy of Sciences USA* 111: 5760.

92. Lwoff A.M. (1977) Jacques Lucien Monod, 9 February 1910–31 May 1976. *Biographical Memoirs of Fellows of the Royal Society* 23: 385–412.

93. Stedman E. and Stedman E. (1950) Cell specificity of histones. *Nature* 166: 780–1.

94. McClintock B. (1956) Controlling elements and the gene. *Cold Spring Harbor Symposia on Quantitative Biology* 21: 197–216.

95. Comfort N.C. (2003) *The Tangled Field: Barbara McClintock's Search for the Patterns of Genetic Control* (Harvard University Press, Cambridge, MA).

96. Rich A. and Davies D.R. (1956) A new two-stranded helical structure: Polyadenylic acid and polyuridylic acid. *Journal of the American Chemical Society* 78: 3548–9.

97. Rich A. (1960) A hybrid helix containing both deoxyribose and ribose polynucleotides and its relation to the transfer of information between the nucleic acids. *Proceedings of the National Academy of Sciences USA* 46: 1044–53.

98. Rich A. (2004) The excitement of discovery. *Annual Review of Biochemistry* 73: 1–37.

99. Fedoroff N., Wellauer P.K. and Wall R. (1977) Intermolecular duplexes in heterogeneous nuclear RNA from HeLa cells. *Cell* 10: 597–610.

100. Calvet J.P. and Pederson T. (1981) Base-pairing interactions between small nuclear RNAs and nuclear RNA precursors as revealed by psoralen cross-linking *in vivo*. *Cell* 26: 363–70.

101. Lu Z. and Chang H.Y. (2018) The RNA base-pairing problem and base-pairing solutions. *Cold Spring Harbor Perspectives in Biology* 10: a034926.

102. Felsenfeld G., Davies D.R. and Rich A. (1957) Formation of a three-stranded polynucleotide molecule. *Journal of the American Chemical Society* 79: 2023–4.

103. Hoogsteen K. (1963) The crystal and molecular structure of a hydrogen-bonded complex between 1-methylthymine and 9-methyladenine. *Acta Crystallographica* 16: 907–16.

104. Rich A. and RajBhandary U.L. (1976) Transfer RNA: Molecular structure, sequence, and properties. *Annual Review of Biochemistry* 45: 805–60.

105. Sanger F., Brownlee G.G. and Barrell B.G. (1965) A two-dimensional fractionation procedure for radioactive nucleotides. *Journal of Molecular Biology* 13: 373–98.

106. Holley R.W., et al. (1965) Structure of a ribonucleic acid. *Science* 147: 1462–5.

107. Hori H. (2014) Methylated nucleosides in tRNA and tRNA methyltransferases. *Frontiers in Genetics* 5: 144.

108. Kim S.H., et al. (1974) Three-dimensional tertiary structure of yeast phenylalanine transfer RNA. *Science* 185: 435–40.

109. Robertus J.D., et al. (1974) Structure of yeast phenylalanine tRNA at 3 Å resolution. *Nature* 250: 546–51.

110. Kim S.H., et al. (1974) The general structure of transfer RNA molecules. *Proceedings of the National Academy of Sciences USA* 71: 4970–4.

111. Barciszewska M.Z., Perrigue P.M. and Barciszewski J. (2016) tRNA – the golden standard in molecular biology. *Molecular BioSystems* 12: 12–7.

112. Suzuki T. (2021) The expanding world of tRNA modifications and their disease relevance. *Nature Reviews Molecular Cell Biology* 22: 375–92.

113. Stout C.D., et al. (1976) Atomic coordinates and molecular conformation of yeast phenylalanyl tRNA. An independent investigation. *Nucleic Acids Research* 3: 1111–23.

114. Holbrook S.R., Sussman J.L., Warrant R.W. and Kim S.-H. (1978) Crystal structure of yeast phenylalanine transfer RNA II: Structural features and functional implications. *Journal of Molecular Biology* 123: 631–60.

115. Westhof E. and Fritsch V. (2000) RNA folding: Beyond Watson–Crick pairs. *Structure* 8: R55–65.

116. Leontis N.B., Lescoute A. and Westhof E. (2006) The building blocks and motifs of RNA architecture. *Current Opinion in Structural Biology* 16: 279–87.

117. Pardee A.B. and Prestidge L.S. (1959) On the nature of the repressor of beta-galactosidase synthesis in *Escherichia coli*. *Biochimica and Biophysica Acta* 36: 545–7.

118. Szilard L. (1960) The control of the formation of specific proteins in bacteria and in animal cells. *Proceedings of the National Academy of Sciences USA* 46: 277–92.

119. Rich A. (1961) The transfer of information between the nucleic acids, in D. Rudnick (ed.) *Molecular and Cellular Synthesis* (Ronald Press, New York).

120. Paigen K. (1962) On regulation of DNA transcription. *Journal of Theoretical Biology* 3: 268–82.

121. Sypherd P.S. and Strauss N. (1963) The role of RNA in repression of enzyme synthesis. *Proceedings of the National Academy of Sciences USA* 50: 1059–66.

122. Leppek K., et al. (2020) Gene- and species-specific Hox mRNA translation by ribosome expansion segments. *Molecular Cell* 80: 980–95.

123. Monod J. and Jacob F. (1961) General conclusions: Teleonomic mechanisms in cellular metabolism, growth, and differentiation. *Cold Spring Harbor Symposia on Quantitative Biology* 26: 389–401.

124. Monod J., Changeux J.P. and Jacob F. (1963) Allosteric proteins and cellular control systems. *Journal of Molecular Biology* 6: 306–29.

125. Monod J., Wyman J. and Changeux J.-P. (1965) On the nature of allosteric transitions: A plausible model. *Journal of Molecular Biology* 12: 88–118.

126. Stent G.S. (1964) The operon: On its third anniversary. Modulation of transfer RNA species can provide a workable model of an operator-less operon. *Science* 144: 816–20.

127. Gilbert W. and Muller-Hill B. (1966) Isolation of the *lac* repressor. *Proceedings of the National Academy of Sciences USA* 56: 1891–8.

128. Ptashne M. (1967) Isolation of the lambda phage repressor. *Proceedings of the National Academy of Sciences USA* 57: 306–13.

129. Ptashne M. (1967) Specific binding of the lambda phage repressor to lambda DNA. *Nature* 214: 232–4.

130. Gilbert W. and Muller-Hill B. (1967) The lac operator is DNA. *Proceedings of the National Academy of Sciences USA* 58: 2415–21.

131. Englesberg E., Irr J., Power J. and Lee N. (1965) Positive control of enzyme synthesis by gene C in l-arabinose system. *Journal of Bacteriology* 90: 946–57.

132. Hahn S. (2014) Ellis Englesberg and the discovery of positive control in gene regulation. *Genetics* 198: 455–60.

133. Dunn J.J. and Bautz E.K. (1969) DNA-dependent RNA polymerase from *E. coli*: Studies on the role of sigma in chain initiation. *Biochemical and Biophysical Research Communications* 36: 925–30.

134. Burgess R.R., Travers A.A., Dunn J.J. and Bautz E.K. (1969) Factor stimulating transcription by RNA polymerase. *Nature* 221: 43–6.

135. Reich P.R., Forget B.G. and Weissman S.M. (1966) RNA of low molecular weight in KB cells infected with adenovirus type 2. *Journal of Molecular Biology* 17: 428–39.

136. Hindley J. (1967) Fractionation of 32P-labelled ribonucleic acids on polyacrylamide gels and their characterization by fingerprinting. *Journal of Molecular Biology* 30: 125–36.

137. Wassarman K.M. and Storz G. (2000) 6S RNA regulates *E. coli* RNA polymerase activity. *Cell* 101: 613–23.

138. Ikemura T. and Dahlberg J.E. (1973) Small ribonucleic-acids of *Escherichia coli* I. Characterization by polyacrylamide-gel electrophoresis and fingerprint analysis. *Journal of Biological Chemistry* 248: 5024–32.

139. Ikemura T. and Dahlberg J.E. (1973) Small ribonucleic-acids of *Escherichia coli* II. Noncoordinate accumulation during stringent control. *Journal of Biological Chemistry* 248: 5033–41.

140. Møller T., Franch T., Udesen C., Gerdes K. and Valentin-Hansen P. (2002) Spot 42 RNA mediates discoordinate expression of the *E. coli galactose* operon. *Genes & Development* 16: 1696–706.

141. Beisel C.L. and Storz G. (2011) The base-pairing RNA Spot 42 participates in a multioutput feedforward loop to help enact catabolite repression in *Escherichia coli*. *Molecular Cell* 41: 286–97.

142. Wang X., Ji S.C., Jeon H.J., Lee Y. and Lim H.M. (2015) Two-level inhibition of *galK* expression by Spot 42: Degradation of mRNA mK2 and enhanced transcription termination before the *galK* gene. *Proceedings of the National Academy of Sciences USA* 112: 7581–6.

143. Stretton A.O.W. (2002) The first sequence: Fred Sanger and insulin. *Genetics* 162: 527–32.

144. Edman P. (1950) Method for determination of amino acid seqence in peptides. *Acta Chemica Scandinavica* 4: 283–93.

145. Edman P. and Begg G. (1967) A protein sequenator. *European Journal of Biochemistry* 1: 80–91.

146. Bragg L.W. (1913) The structure of some crystals as indicated by their diffraction of X-rays. *Proceedings of the Royal Society A: Mathematical, Physical and Engineering Sciences* 89: 248–77.

147. Thomas J.M. (2012) The birth of X-ray crystallography. *Nature* 491: 186–7.

148. Kendrew J.C., et al. (1958) A three-dimensional model of the myoglobin molecule obtained by x-ray analysis. *Nature* 181: 662–6.

149. Perutz M.F., et al. (1960) Structure of hæmoglobin: A three-dimensional Fourier synthesis at 5.5-Å resolution, obtained by X-ray analysis. *Nature* 185: 416–22.

150. Kendrew J.C., et al. (1960) Structure of myoglobin: A three-dimensional Fourier synthesis at 2 Å resolution. *Nature* 185: 422–7.

151. Kirk R. (2014) The first of its kind. *Nature* 511: 13.

152. de Chadarevian S. (2018) John Kendrew and myoglobin: Protein structure determination in the 1950s. *Protein Science* 27: 1136–43.

153. Wüthrich K. (2001) The way to NMR structures of proteins. *Nature Structural Biology* 8: 923–5.

154. Clore M.G. (2011) Adventures in biomolecular NMR, in D.M. Grant and R.K. Harris (eds.) *Encyclopedia of Magnetic Resonance* (John Wiley & Sons, New York).

155. Nogales E. (2018) Profile of Joachim Frank, Richard Henderson, and Jacques Dubochet, 2017 Nobel laureates in chemistry. *Proceedings of the National Academy of Sciences USA* 115: 441–4.

156. Maier T., Leibundgut M., Boehringer D. and Ban N. (2010) Structure and function of eukaryotic fatty acid synthases. *Quarterly Reviews of Biophysics* 43: 373–422.

157. Murata K. and Wolf M. (2018) Cryo-electron microscopy for structural analysis of dynamic biological macromolecules. *Biochimica et Biophysica Acta* 1862: 324–34.

158. Benjin X. and Ling L. (2020) Developments, applications, and prospects of cryo-electron microscopy. *Protein Science* 29: 872–82.

159. Ho C.-M., et al. (2020) Bottom-up structural proteomics: CryoEM of protein complexes enriched from the cellular milieu. *Nature Methods* 17: 79–85.

160. Cobb M. (2017) 60 years ago, Francis Crick changed the logic of biology. *PLOS Biology* 15: e2003243.

161. Watson J.D., Gann A. and Witkowski J. (2012) *Double Helix, Annotated and Ilustrated* (Simon & Schuster, New York).

162. Watson J.D. (1965) *Molecular Biology of the Gene* (Benjamin, New York).

163. Crick F. (1970) Central dogma of molecular biology. *Nature* 227: 561–3.

164. Burness A.T.H., Vizoso A.D. and Clothier F.W. (1963) Encephalomyocarditis virus and its ribonucleic acid: Sedimentation characteristics: Sedimentation coefficients of encephalomyocarditis virus and its ribonucleic acid. *Nature* 197: 1177–8.

165. Montagnier L. and Sanders F.K. (1963) Encephalomyocarditis virus and its ribonucleic acid: Sedimentation characteristics: Sedimentation properties of infective ribonucleic acid extracted from Encephalomyocarditis virus. *Nature* 197: 1178–81.

166. Montagnier L. and Sanders F.K. (1963) Replicative form of Encephalomyocarditis virus ribonucleic acid. *Nature* 199: 664–7.

167. Baltimore D., Becker Y. and Darnell J.E. (1964) Virus-specific double-stranded RNA in poliovirus-infected cells. *Science* 143: 1034–6.

168. Weissmann C., Borst P., Burdon R.H., Billeter M.A. and Ochoa S. (1964) Replication of viral RNA. III. Double-stranded replicative form of MS2 phage RNA. *Proceedings of the National Academy of Sciences USA* 51: 682–90.

169. Stern R. and Friedman R.M. (1970) Double-stranded RNA synthesized in animal cells in the presence of actinomycin D. *Nature* 226: 612–6.

170. Montagnier L. (1968) Presence of a double chain ribonucleic acid in animal cells. *Comptes Rendus de l'Académie des Sciences* 267: 1417–20.

171. Kolakofsky D. (2015) A short biased history of RNA viruses. *RNA* 21: 667–9.

172. Astier-Manifacier S. and Cornuet P. (1971) RNA-dependent RNA polymerase in Chinese cabbage. *Biochimica et Biophysica Acta* 232: 484–93.

173. Mattick J.S. and Mehler M.F. (2008) RNA editing, DNA recoding and the evolution of human cognition. *Trends in Neuroscience* 31: 227–33.

174. Boeke J.D., Garfinkel D.J., Styles C.A. and Fink G.R. (1985) Ty elements transpose through an RNA intermediate. *Cell* 40: 491–500.

175. Temin H.M. and Mizutani S. (1970) RNA-dependent DNA polymerase in virions of Rous sarcoma virus. *Nature* 226: 1211–3.

176. Baltimore D. (1970) Viral RNA-dependent DNA polymerase: RNA-dependent DNA polymerase in virions of RNA tumour viruses. *Nature* 226: 1209–11.

177. Coffin J.M. and Fan H. (2016) The discovery of reverse transcriptase. *Annual Review of Virology* 3: 29–51.

178. Wood W.B. (1988) *The Nematode Caenorhabditis Elegans* (Cold Spring Harbor Laboratory, Cold Spring Harbor, NY).

179. Stent G.S. (1968) That was the molecular biology that was. *Science* 160: 390–5.

180. Brenner S. (1997) Centaur biology. *Current Biology* 7: R454.

181. Friedmann H.C. (2004) From *Butyribacterium* to *E. coli*: An essay on unity in biochemistry. *Perspectives in Biology and Medicine* 47: 47–66.

CHAPTER 4

1. Bianconi E., et al. (2013) An estimation of the number of cells in the human body. *Annals of Human Biology* 40: 463–71.

2. Sender R., Fuchs S. and Milo R. (2016) Revised estimates for the number of human and bacteria cells in the body. *PLOS Biology* 14: e1002533.

3. Williams R.W. and Herrup K. (1988) The control of neuron number. *Annual Review of Neuroscience* 11: 423–53.

4. Andersen B.B., Korbo L. and Pakkenberg B. (1992) A quantitative study of the human cerebellum with unbiased stereological techniques. *Journal of Comparative Neurology* 326: 549–60.

5. Azevedo F.A.C., et al. (2009) Equal numbers of neuronal and nonneuronal cells make the human brain an isometrically scaled-up primate brain. *Journal of Comparative Neurology* 513: 532–41.

6. Levin P.A. and Angert E.R. (2015) Small but mighty: Cell size and bacteria. *Cold Spring Harbor Perspectives in Biology* 7: a019216.

7. Land M., et al. (2015) Insights from 20 years of bacterial genome sequencing. *Functional & Integrative Genomics* 15: 141–61.

8. Cairns J. (1963) The bacterial chromosome and its manner of replication as seen by autoradiography. *Journal of Molecular Biology* 6: 208–13.

9. Lederberg J. (1952) Cell genetics and hereditary symbiosis. *Physiological Reviews* 32: 403–30.

10. Yanofsky C. (1981) Attenuation in the control of expression of bacterial operons. *Nature* 289: 751–8.

11. Dacks J.B. and Field M.C. (2007) Evolution of the eukary-otic membrane-trafficking system: Origin, tempo and mode. *Journal of Cell Science* 120: 2977–85.

12. Parton R.G. and del Pozo M.A. (2013) Caveolae as plasma membrane sensors, protectors and organizers. *Nature Reviews Molecular Cell Biology* 14: 98–112.

13. Naslavsky N. and Caplan S. (2018) The enigmatic endo-some – sorting the ins and outs of endocytic trafficking. *Journal of Cell Science* 131: jcs216499.

14. Schrader M. and Fahimi H.D. (2008) The peroxisome: Still a mysterious organelle. *Histochemistry and Cell Biology* 129: 421–40.

15. de Duve C. (2005) The lysosome turns fifty. *Nature Cell Biology* 7: 847–9.

16. Xu H. and Ren D. (2015) Lysosomal physiology. *Annual Review of Physiology* 77: 57–80.

17. Perera R.M. and Zoncu R. (2016) The lysosome as a regu-latory hub. *Annual Review of Cell and Developmental Biology* 32: 223–53.

18. Schwarz D.S. and Blower M.D. (2016) The endoplasmic reticulum: Structure, function and response to cellular sig-naling. *Cellular and Molecular Life Sciences* 73: 79–94.

19. Gallo A., Vannier C. and Galli T. (2016) Endoplasmic reticulum-plasma membrane associations: Structures and functions. *Annual Review of Cell and Developmental Biology* 32: 279–301.

20. Mellman I. and Simons K. (1992) The Golgi complex: In vitro veritas? *Cell* 68: 829–40.

21. Guo Y., Sirkis D.W. and Schekman R. (2014) Protein sort-ing at the trans-Golgi network. *Annual Review of Cell and Developmental Biology* 30: 169–206.

22. Golgi C. (1898) Intorno alla struttura delle cellule ner-vose. *Bollettino della Società Medico-Chirurgica di Pavia* 13: 316.

23. Spinelli J.B. and Haigis M.C. (2018) The multifaceted contributions of mitochondria to cellular metabolism. *Nature Cell Biology* 20: 745–54.

24. Jensen P.E. and Leister D. (2014) Chloroplast evolution, structure and functions. *F1000Prime Reports* 6: 40.

25. Vance J.E. (1990) Phospholipid synthesis in a mem-brane fraction associated with mitochondria. *Journal of Biological Chemistry* 265: 7248–56.

26. Kornmann B., et al. (2009) An ER-mitochondria tethering complex revealed by a synthetic biology screen. *Science* 325: 477–81.

27. Valm A.M., et al. (2017) Applying systems-level spectral imaging and analysis to reveal the organelle interactome. *Nature* 546: 162–7.

28. Wu Y., et al. (2017) Contacts between the endoplasmic reticulum and other membranes in neurons. *Proceedings of the National Academy of Sciences USA* 114: E4859–67.

29. Shai N., et al. (2018) Systematic mapping of contact sites reveals tethers and a function for the peroxisome-mito-chondria contact. *Nature Communications* 9: 1761.

30. Islinger M., Godinho L., Costello J. and Schrader M. (2015) The different facets of organelle interplay—an overview of organelle interactions. *Frontiers in Cell and Developmental Biology* 3: 56.

31. Pederson T. (2011) The nucleolus. *Cold Spring Harbor Perspectives in Biology* 3: a000638.

32. Surovtsev I.V. and Jacobs-Wagner C. (2018) Subcellular organization: A critical feature of bacterial cell replica-tion. *Cell* 172: 1271–93.

33. Greening C. and Lithgow T. (2020) Formation and func-tion of bacterial organelles. *Nature Reviews Microbiology* 18: 677–89.

34. Flechsler J., Heimerl T., Huber H., Rachel R. and Berg I.A. (2021) Functional compartmentalization and meta-bolic separation in a prokaryotic cell. *Proceedings of the National Academy of Sciences USA* 118: e2022114118.

35. Guo Q., et al. (2018) *In situ* structure of neuronal C9orf72 poly-GA aggregates reveals proteasome recruitment. *Cell* 172: 696–705.

36. Deniston C.K., et al. (2020) Structure of LRRK2 in Parkinson's disease and model for microtubule interac-tion. *Nature* 588: 344–9.

37. Hoffman D.P., et al. (2020) Correlative three-dimensional super-resolution and block-face electron microscopy of whole vitreously frozen cells. *Science* 367: eaaz5357.

38. Tegunov D., Xue L., Dienemann C., Cramer P. and Mahamid J. (2021) Multi-particle cryo-EM refinement with M visualizes ribosome-antibiotic complex at 3.5 Å in cells. *Nature Methods* 18: 186–93.

39. Heinrich L., et al. (2021) Whole-cell organelle segmenta-tion in volume electron microscopy. *Nature* 599: 141–6.

40. Schrödinger E. (1944) *What Is Life? The Physical Aspect of the Living Cell* (Cambridge Univ. Press, Cambridge).

41. Ball P. (2018) Schrödinger's cat among biology's pigeons: 75 years of What Is Life? *Nature* 560: 548–50.

42. Russell M.J. and Hall A.J. (1997) The emergence of life from iron monosulphide bubbles at a submarine hydro-thermal redox and pH front. *Journal of the Geological Society* 154: 377–402.

43. Barge L.M., Flores E., Baum M.M., VanderVelde D.G. and Russell M.J. (2019) Redox and pH gradients drive amino acid synthesis in iron oxyhydroxide mineral systems. *Proceedings of the National Academy of Sciences USA* 116: 4828–33.

44. Lane N. (2015) *The Vital Question: Why is Life the Way It Is?* (Profile Books, London).

45. Cartwright J.H.E. and Russell M.J. (2019) The origin of life: The submarine alkaline vent theory at 30. *The Royal Society Interface Focus* 9: 20190104.

46. Dodd M.S., et al. (2017) Evidence for early life in Earth's oldest hydrothermal vent precipitates. *Nature* 543: 60–4.

47. Pearce B.K.D., Pudritz R.E., Semenov D.A. and Henning T.K. (2017) Origin of the RNA world: The fate of nucleo-bases in warm little ponds. *Proceedings of the National Academy of Sciences USA* 114: 11327–32.

48. Kim H.-J., et al. (2011) Synthesis of carbohydrates in mineral-guided prebiotic cycles. *Journal of the American Chemical Society* 133: 9457–68.

49. Mulkidjanian A.Y., Bychkov A.Y., Dibrova D.V., Galperin M.Y. and Koonin E.V. (2012) Origin of first cells at ter-restrial, anoxic geothermal fields. *Proceedings of the National Academy of Sciences USA* 109: E821–30.

50. Van Kranendonk M.J., Deamer D.W. and Djokic T. (2017) Life springs. *Scientific American* 317: 28–35.

51. Djokic T., Van Kranendonk M.J., Campbell K.A., Walter M.R. and Ward C.R. (2017) Earliest signs of life on land preserved in ca. 3.5 Ga hot spring deposits. *Nature Communications* 8: 15263.

52. Forsythe J.G., et al. (2017) Surveying the sequence diver-sity of model prebiotic peptides by mass spectrometry. *Proceedings of the National Academy of Sciences USA* 114: E7652–9.

53. Yi R., et al. (2020) A continuous reaction network that produces RNA precursors. *Proceedings of the National Academy of Sciences USA* 117: 13267–74.
54. Damer B. and Deamer D.W. (2020) The hot spring hypothesis for an origin of life. *Astrobiology* 20: 429–52.
55. Van Kranendonk M.J., et al. (2021) Elements for the origin of life on land: A deep-time perspective from the Pilbara Craton of Western Australia. *Astrobiology* 21: 39–59.
56. Galtier N., Tourasse N. and Gouy M. (1999) A nonhyperthermophilic common ancestor to extant life forms. *Science* 283: 220–1.
57. Patel A., et al. (2017) ATP as a biological hydrotrope. *Science* 356: 753–6.
58. Woese C.R. (1967) *The Genetic Code the Molecular Basis for Genetic Expression* (Harper, New York).
59. Woese C.R. and Fox G.E. (1977) Phylogenetic structure of the prokaryotic domain: The primary kingdoms. *Proceedings of the National Academy of Sciences USA* 74: 5088–90.
60. Woese C.R., Kandler O. and Wheelis M.L. (1990) Towards a natural system of organisms: Proposal for the domains Archaea, Bacteria, and Eucarya. *Proceedings of the National Academy of Sciences USA* 87: 4576–9.
61. Stanier R.Y. and Van Niel C.B. (1962) The concept of a bacterium. *Archiv für Mikrobiologie* 42: 17–35.
62. Sapp J. (2005) The prokaryote-eukaryote dichotomy: Meanings and mythology. *Microbiology and Molecular Biology Reviews* 69: 292–305.
63. Eme L., Sharpe S.C., Brown M.W. and Roger A.J. (2014) On the age of eukaryotes: Evaluating evidence from fossils and molecular clocks. *Cold Spring Harbor Perspectives in Biology* 6: a016139.
64. Margulis L. (1996) Archaeal-eubacterial mergers in the origin of Eukarya: Phylogenetic classification of life. *Proceedings of the National Academy of Sciences USA* 93: 1071–6.
65. Olsen G.J. and Woese C.R. (1997) Archaeal genomics: An overview. *Cell* 89: 991–4.
66. Koonin E.V. (2010) The origin and early evolution of eukaryotes in the light of phylogenomics. *Genome Biology* 11: 209.
67. Archibald J.M. (2015) Endosymbiosis and eukaryotic cell evolution. *Current Biology* 25: R911–21.
68. Eme L., Spang A., Lombard J., Stairs C.W. and Ettema T.J.G. (2017) Archaea and the origin of eukaryotes. *Nature Reviews Microbiology* 15: 711–23.
69. Zaremba-Niedzwiedzka K., et al. (2017) Asgard archaea illuminate the origin of eukaryotic cellular complexity. *Nature* 541: 353–8.
70. Mattiroli F., et al. (2017) Structure of histone-based chromatin in Archaea. *Science* 357: 609–12.
71. Brunk C.F. and Martin W.F. (2019) Archaeal histone contributions to the origin of eukaryotes. *Trends in Microbiology* 27: 703–14.
72. Imachi H., et al. (2020) Isolation of an archaeon at the prokaryote–eukaryote interface. *Nature* 577: 519–25.
73. Williams T.A., Cox C.J., Foster P.G., Szöllősi G.J. and Embley T.M. (2020) Phylogenomics provides robust support for a two-domains tree of life. *Nature Ecology & Evolution* 4: 138–47.
74. Stevens K.M., et al. (2020) Histone variants in archaea and the evolution of combinatorial chromatin complexity. *Proceedings of the National Academy of Sciences USA* 117: 33384–95.
75. Colson P., et al. (2018) Ancestrality and mosaicism of giant viruses supporting the definition of the fourth TRUC of microbes. *Frontiers in Microbiology* 9: 2668.
76. Watson T. (2019) The trickster microbes that are shaking up the tree of life. *Nature* 569: 322–4.
77. Martin W. and Müller M. (1998) The hydrogen hypothesis for the first eukaryote. *Nature* 392: 37–41.
78. Doolittle W.F. (1998) A paradigm gets shifty. *Nature* 392: 15–6.
79. Cavalier-Smith T. (2002) Chloroplast evolution: Secondary symbiogenesis and multiple losses. *Current Biology* 12: R62–4.
80. McFadden G.I. and van Dooren G.G. (2004) Evolution: Red algal genome affirms a common origin of all plastids. *Current Biology* 14: R514–R6.
81. Sagan L. (1967) On the origin of mitosing cells. *Journal of Theoretical Biology* 14: 255–74.
82. Schwartz R.M. and Dayhoff M.O. (1978) Origins of prokaryotes, eukaryotes, mitochondria, and chloroplasts. *Science* 199: 395–403.
83. Newton K.J. (1988) Plant mitochondrial genomes: Organization, expression and variation. *Annual Review of Plant Physiology and Plant Molecular Biology* 39: 503–32.
84. Boore J.L. (1999) Animal mitochondrial genomes. *Nucleic Acids Research* 27: 1767–80.
85. Gray M.W. (2012) Mitochondrial evolution. *Cold Spring Harbor Perspectives in Biology* 4: a011403.
86. Daniell H., Lin C.S., Yu M. and Chang W.J. (2016) Chloroplast genomes: Diversity, evolution, and applications in genetic engineering. *Genome Biology* 17: 134.
87. Björkholm P., Harish A., Hagström E., Ernst A.M. and Andersson S.G.E. (2015) Mitochondrial genomes are retained by selective constraints on protein targeting. *Proceedings of the National Academy of Sciences USA* 112: 10154–61.
88. Allen J.F. (2015) Why chloroplasts and mitochondria retain their own genomes and genetic systems: Colocation for redox regulation of gene expression. *Proceedings of the National Academy of Sciences USA* 112: 10231–8.
89. Cavalier-Smith T. (1987) The origin of eukaryotic and archaebacterial cells. *Annals of the New York Academy of Science* 503: 17–54.
90. Cavalier-Smith T. (2009) Predation and eukaryote cell origins: A coevolutionary perspective. *International Journal of Biochemistry and Cell Biology* 41: 307–22.
91. McCauley D.E., Sundby A.K., Bailey M.F. and Welch M.E. (2007) Inheritance of chloroplast DNA is not strictly maternal in *Silene vulgaris* (Caryophyllaceae): Evidence from experimental crosses and natural populations. *American Journal of Botany* 94: 1333–7.
92. Smith J.M. (1978) *The Evolution of Sex* (Cambridge University Press, Cambridge).
93. Morran L.T., Schmidt O.G., Gelarden I.A., Parrish 2nd R.C., and Lively C.M. (2011) Running with the Red Queen: Host-parasite coevolution selects for biparental sex. *Science* 333: 216–8.
94. Markov A.V. (2014) Horizontal gene transfer as a possible evolutionary predecessor of sexual reproduction. *Paleontological Journal* 48: 219–33.
95. Speijer D., Lukes J. and Elias M. (2015) Sex is a ubiquitous, ancient, and inherent attribute of eukaryotic life. *Proceedings of the National Academy of Sciences USA* 112: 8827–34.

96. Ishikawa F. and Naito T. (1999) Why do we have linear chromosomes? A matter of Adam and Eve. *Mutation Research* 434: 99–107.

97. Burgers P.M.J. and Kunkel T.A. (2017) Eukaryotic DNA replication fork. *Annual Review of Biochemistry* 86: 417–38.

98. Holliday R.A. (1964) A mechanism for gene conversion in fungi. *Genetics Research* 5: 282–304.

99. Holliday R. (1977) Recombination and meiosis. *Philosophical Transactions of the Royal Society B: Biological Sciences* 277: 359–70.

100. Heyer W.-D., Ehmsen K.T. and Liu J. (2010) Regulation of homologous recombination in eukaryotes. *Annual Review of Genetics* 44: 113–39.

101. Cavalier-Smith T. (2002) Origins of the machinery of recombination and sex. *Heredity* 88: 125–41.

102. Zinder N.D. and Lederberg J. (1952) Genetic exchange in *Salmonella*. *Journal of Bacteriology* 64: 679–99.

103. Parkinson J.S. (2016) Classic spotlight: The discovery of bacterial transduction. *Journal of Bacteriology* 198: 2899–900.

104. Weiss R.A. (2006) The discovery of endogenous retroviruses. *Retrovirology* 3: 67.

105. Forterre P. (2006) The origin of viruses and their possible roles in major evolutionary transitions. *Virus Research* 117: 5–16.

106. Forterre P. (2006) Three RNA cells for ribosomal lineages and three DNA viruses to replicate their genomes: A hypothesis for the origin of cellular domain. *Proceedings of the National Academy of Sciences USA* 103: 3669–74.

107. Koonin E.V., Senkevich T.G. and Dolja V.V. (2006) The ancient virus world and evolution of cells. *Biology Direct* 1: 29.

108. Koonin E.V. and Krupovic M. (2018) The depths of virus exaptation. *Current Opinion in Virology* 31: 1–8.

109. Krupovic M., Dolja V.V. and Koonin E.V. (2019) Origin of viruses: Primordial replicators recruiting capsids from hosts. *Nature Reviews Microbiology* 17: 449–58.

110. Quammen D. (2018) *The Tangled Tree: A Radical New History of Life* (Simon Schuster, New York).

111. Parfrey L.W., Lahr D.J.G., Knoll A.H. and Katz L.A. (2011) Estimating the timing of early eukaryotic diversification with multigene molecular clocks. *Proceedings of the National Academy of Sciences USA* 108: 13624–9.

112. Strother P.K., Battison L., Brasier M.D. and Wellman C.H. (2011) Earth's earliest non-marine eukaryotes. *Nature* 473: 505–9.

113. Knauth L.P. and Kennedy M.J. (2009) The late Precambrian greening of the Earth. *Nature* 460: 728–32.

114. Doyle J.A. (2012) Molecular and fossil evidence on the origin of Angiosperms. *Annual Review of Earth and Planetary Sciences* 40: 301–26.

115. Consortium (2013) The *Amborella* genome and the evolution of flowering plants. *Science* 342: 1241089.

116. Leebens-Mack J.H. et al. (2019) One thousand plant transcriptomes and the phylogenomics of green plants. *Nature* 574: 679–85.

117. Osborne C.P. and Beerling D.J. (2006) Nature's green revolution: The remarkable evolutionary rise of C4 plants. *Philosophical Transactions of the Royal Society B: Biological Sciences* 361: 173–94.

118. Turner E.C. (2021) Possible poriferan body fossils in early Neoproterozoic microbial reefs. *Nature* 596: 87–91.

119. Glaessner M.F. (1959) The oldest fossil faunas of South Australia. *Geologische Rundschau* 47: 522–31.

120. Shen B., Dong L., Xiao S. and Kowalewski M. (2008) The Avalon explosion: Evolution of Ediacara morphospace. *Science* 319: 81–4.

121. Wainright P.O., Hinkle G., Sogin M.L. and Stickel S.K. (1993) Monophyletic origins of the metazoa: An evolutionary link with fungi. *Science* 260: 340–2.

122. Xiao S. and Laflamme M. (2009) On the eve of animal radiation: Phylogeny, ecology and evolution of the Ediacara biota. *Trends in Ecology & Evolution* 24: 31–40.

123. Chang Y., et al. (2015) Phylogenomic analyses indicate that early fungi evolved digesting cell walls of algal ancestors of land plants. *Genome Biology and Evolution* 7: 1590–601.

124. Torruella G., et al. (2015) Phylogenomics reveals convergent evolution of lifestyles in close relatives of animals and fungi. *Current Biology* 25: 2404–10.

125. Glaessner M.F. (1961) Pre-Cambrian animals. *Scientific American* 204: 72–8.

126. McMenamin M.A. (1996) Ediacaran biota from Sonora, Mexico. *Proceedings of the National Academy of Sciences USA* 93: 4990–3.

127. Chen Z., Zhou C., Yuan X. and Xiao S. (2019) Death march of a segmented and trilobate bilaterian elucidates early animal evolution. *Nature* 573: 412–5.

128. Watson T. (2020) These bizarre ancient species are rewriting animal evolution. *Nature* 586: 662–5.

129. Evans S.D., Hughes I.V., Gehling J.G. and Droser M.L. (2020) Discovery of the oldest bilaterian from the Ediacaran of South Australia. *Proceedings of the National Academy of Sciences USA* 117: 7845–50.

130. Conway-Morris S. (1993) Ediacaran-like fossils in Cambrian Burgess Shale-type faunas of North America. *Palaeontology* 36: 593–695.

131. Jensen S., Gehling J.G. and Droser M.L. (1998) Ediacara-type fossils in Cambrian sediments. *Nature* 393: 567–9.

132. Erwin D.H., et al. (2011) The Cambrian conundrum: Early divergence and later ecological success in the early history of animals. *Science* 334: 1091–7.

133. Gehling J.G. and Droser M.L. (2018) Ediacaran scavenging as a prelude to predation. *Emerging Topics in Life Sciences* 2: 213–22.

134. Morris S.C. (1989) Burgess Shale faunas and the Cambrian Explosion. *Science* 246: 339–46.

135. Jablonski D. and Bottjer D.J. (1988) The ecology of evolutionary innovation - the fossil record, in *11th Annual Spring Systematics Symp - Evolutionary Innovations: Patterns and Processes*, pp. 253–88.

136. Caron J.-B. and Rudkin D. (2009) *A Burgess Shale Primer - History, Geology, and Research Highlights* (The Burgess Shale Consortium, Toronto).

137. Peterson K.J., Dietrich M.R. and McPeek M.A. (2009) MicroRNAs and metazoan macroevolution: Insights into canalization, complexity, and the Cambrian explosion. *BioEssays* 31: 736–47.

138. Lee M.S.Y., Soubrier J. and Edgecombe G.D. (2013) Rates of phenotypic and genomic evolution during the Cambrian explosion. *Current Biology* 23: 1889–95.

139. Caron J.-B., Gaines R.R., Aria C., Mángano M.G. and Streng M. (2014) A new phyllopod bed-like assemblage from the Burgess Shale of the Canadian Rockies. *Nature Communications* 5: 3210.

140. Peterson K.J., McPeek M.A. and Evans D.A.D. (2005) Tempo and mode of early animal evolution: Inferences from rocks, Hox, and molecular clocks. *Paleobiology* 31: 36–55.

141. Benton M.J. (2005) *Vertebrate Palaeontology* (Blackwell, Oxford).

142. Krug A.Z. and Jablonski D. (2012) Long-term origination rates are reset only at mass extinctions. *Geology* 40: 731–4.

143. Kaiho K. and Oshima N. (2017) Site of asteroid impact changed the history of life on earth: The low probability of mass extinction. *Scientific Reports* 7: 14855.

144. Chiarenza A.A., et al. (2020) Asteroid impact, not volcanism, caused the end-Cretaceous dinosaur extinction. *Proceedings of the National Academy of Sciences USA* 117: 17084–93.

145. Bozdag G.O., Libby E., Pineau R., Reinhard C.T. and Ratcliff W.C. (2021) Oxygen suppression of macroscopic multicellularity. *Nature Communications* 12: 2838.

146. Lyons T.W., Reinhard C.T. and Planavsky N.J. (2014) The rise of oxygen in Earth's early ocean and atmosphere. *Nature* 506: 307–15.

147. Sperling E.A., et al. (2015) Statistical analysis of iron geochemical data suggests limited late Proterozoic oxygenation. *Nature* 523: 451–4.

148. Fox D. (2016) What sparked the Cambrian explosion? *Nature* 530: 268–70.

149. Raymond J. and Segrè D. (2006) The effect of oxygen on biochemical networks and the evolution of complex life. *Science* 311: 1764–7.

150. Jabłońska J. and Tawfik D.S. (2021) The evolution of oxygen-utilizing enzymes suggests early biosphere oxygenation. *Nature Ecology & Evolution* 5: 442–8.

151. Butterfield N.J. (2009) Oxygen, animals and oceanic ventilation: An alternative view. *Geobiology* 7: 1–7.

152. Zhang S., et al. (2016) Sufficient oxygen for animal respiration 1,400 million years ago. *Proceedings of the National Academy of Sciences USA* 113: 1731.

153. Brown S.W. (1966) Heterochromatin. *Science* 151: 417–25.

154. Passarge E. (1979) Emil Heitz and the concept of heterochromatin: Longitudinal chromosome differentiation was recognized fifty years ago. *American Journal of Human Genetics* 31: 106–15.

155. Zhimulev I.F., et al. (2004) Polytene chromosomes: 70 years of genetic research, in K. W. Jeon (ed.) *International Review of Cytology* (Academic Press, Cambridge, MA).

156. Eagen K.P., Hartl T.A. and Kornberg R.D. (2015) Stable chromosome condensation revealed by chromosome conformation capture. *Cell* 163: 934–46.

157. Heitz E. (1928) Das Heterochromatin der Moose. *Jahrbücher für Wissenschaftliche Botanik* 69: 762–818.

158. Heitz E. (1933) Die somatische Heteropyknose bei *Drosophila melanogaster* und ihre genetische Bedeutung. *Cell and Tissue Research* 20: 237–87.

159. Hsu T.C. (1962) Differential rate in RNA synthesis between euchromatin and heterochromatin. *Experimental Cell Research* 27: 332–4.

160. Poulson D.F. and Metz C.W. (1938) Studies on the structure of nucleolus-forming regions and related structures in the giant salivary gland chromosomes of Diptera. *Journal of Morphology* 63: 363–95.

161. Beermann W. (1952) Chromomerenkonstanz und spezifische Modifikation der Chromosomenstruktur in der Entwicklung und Organdifferenzierung von *Chironomus tentans*. *Chromosoma* 5: 139–98.

162. Sass H. (1980) Features of in vitro puffing and RNA synthesis in polytene chromosomes of *Chironomus*. *Chromosoma* 78: 33–78.

163. Pavan C. and Breuer M.E. (1952) Polytene chromosomes: In different tissues of rhynchosciara. *Journal of Heredity* 43: 151–8.

164. Breuer M.E. and Pavan C. (1955) Behavior of polytene chromosomes of *Rhynchosciara angelae* at different stages of larval development. *Chromosoma* 7: 371–86.

165. Ritossa F. (1962) New puffing pattern induced by temperature shock and DNP in drosophila. *Experientia* 18: 571–3.

166. Stocker A.J. and Pavan C. (1974) The influence of ecdysterone on gene amplification, DNA synthesis, and puff formation in the salivary gland chromosomes of Rhynchosciara hollaenderi. *Chromosoma* 45: 295–319.

167. Beermann W. (1963) Cytological aspects of information transfer in cellular differentiation. *American Zoologist* 3: 23–32.

168. Clever U. (1965) Puffing changes in incubated and in ecdysone treated *Chironomus tentans* salivary glands. *Chromosoma* 17: 309–22.

169. Clever U. (1965) Chromosomal changes associated with differentiation. *Brookhaven Symposia in Biology* 18: 242–53.

170. van Breugel F.M.A. (1966) Puff induction in larval salivary gland chromosomes of Drosophila *hydei sturtevant*. *Genetica* 37: 17–28.

171. Kroeger H. and Lezzi M. (1966) Regulation of gene action in insect development. *Annual Review of Entomology* 11: 1–22.

172. Jamrich M., Greenleaf A.L. and Bautz E.K. (1977) Localization of RNA polymerase in polytene chromosomes of *Drosophila melanogaster*. *Proceedings of the National Academy of Sciences USA* 74: 2079–83.

173. Graessmann A., Graessmann M. and Larat F.J.S. (1973) Involvement of RNA in the process of puff induction in polytene chromosomes, in B.A. Hamkalo and J. Papaconstantinou (eds.) *Molecular Cytogenetics* (Springer, Boston, MA).

174. Lakhotia S.C. and Mukherjee T. (1982) Absence of novel translation products in relation to induced activity of the 93D puff in *Drosophila melanogaster*. *Chromosoma* 85: 369–74.

175. Lakhotia S.C. and Sharma A. (1996) The 93D (hsr-omega) locus of *Drosophila*: Non-coding gene with house-keeping functions. *Genetica* 97: 339–48.

176. Malik H.S. and Henikoff S. (2009) Major evolutionary transitions in centromere complexity. *Cell* 138: 1067–82.

177. Talbert P.B. and Henikoff S. (2020) What makes a centromere? *Experimental Cell Research* 389: 111895.

178. Wunderlich V. (2002) JMM: Past and present. Chromosomes and cancer: Theodor Boveri's predictions 100 years later. *Journal of Molecular Medicine* 80: 545–8.

179. Scheer U. (2014) Historical roots of centrosome research: Discovery of Boveri's microscope slides in Würzburg. *Philosophical Transactions of the Royal Society B: Biological Sciences* 369: 20130469.

180. McKinley K.L. and Cheeseman I.M. (2016) The molecular basis for centromere identity and function. *Nature Reviews Molecular Cell Biology* 17: 16–29.

181. Ohno S., Kaplan W.D. and Kinosita R. (1959) Formation of the sex chromatin by a single X-chromosome in liver cells of Rattus norvegicus. *Experimental Cell Research* 18: 415–8.

182. Barr M.L. and Bertram E.G. (1949) A morphological distinction between neurones of the male and female, and the behaviour of the nucleolar satellite during accelerated nucleoprotein synthesis. *Nature* 163: 676–7.

183. Lyon M.F. (1961) Gene action in the X-chromosome of the mouse (*Mus musculus*). *Nature* 190: 372–3.

184. Lyon M.F. (1962) Sex chromatin and gene action in the mammalian X-chromosome. *American Journal of Human Genetics* 14: 135–48.

185. Centerwall W.R. and Benirschke K. (1973) Male Tortoiseshell and Calico (T-C) Cats: Animal models of sex chromosome mosaics, aneuploids, polyploids, and chimerics. *Journal of Heredity* 64: 272–8.

186. Basta M. and Pandya A.M. (2020) Genetics, X-linked inheritance, in *StatPearls* (StatPearls Publishing, Treasure Island, FL).

187. Dobyns W.B. et al. (2004) Inheritance of most X-linked traits is not dominant or recessive, just X-linked. *American Journal of Medical Genetics* 129A: 136–43.

188. Lucchesi J.C. (1973) Dosage compensation in Drosophila. *Annual Review of Genetics* 7: 225–37.

189. Brown S.W. and Nur U. (1964) Heterochromatic chromosomes in the Coccids. *Science* 145: 130–6.

190. Swift H. (1965) Molecular morphology of the chromosome. *In Vitro* 1: 26–49.

191. Gall J.G. (1956) On the submicroscopic structure of chromosomes. *Brookhaven Symposia in Biology* 8: 17–32.

192. Singh G., Pratt G., Yeo G.W. and Moore M.J. (2015) The clothes make the mRNA: Past and present trends in mRNP fashion. *Annual Review of Biochemistry* 84: 325–54.

193. Flemming W. (1882) *Zellsubstanz, Kern und Zelltheilung (Cell Substance, Nucleus and Cell Division)* (F. C. W. Vogel, Leipzig).

194. Morgan G.T. (2002) Lampbrush chromosomes and associated bodies: New insights into principles of nuclear structure and function. *Chromosome Research* 10: 177–200.

195. Dinger M.E., Baillie G.J. and Musgrave D.R. (2000) Growth phase-dependent expression and degradation of histones in the thermophilic archaeon Thermococcus zilligii. *Molecular Microbiology* 36: 876–85.

196. Attar N., et al. (2020) The histone H3-H4 tetramer is a copper reductase enzyme. *Science* 369: 59–64.

197. Rudolph J. and Luger K. (2020) The secret life of histones. *Science* 369: 33.

198. Oudet P., Gross-Bellard M. and Chambon P. (1975) Electron microscopic and biochemical evidence that chromatin structure is a repeating unit. *Cell* 4: 281–300.

199. Kossel A. (1894) Uber einen peptonartigen bestandteil des zellkerns. *Zeitschrift fur physiologische Chemie* 8: 611–5.

200. Olins A.L. and Olins D.E. (1974) Spheroid chromatin units (ν bodies). *Science* 183: 330–2.

201. Olins D.E. and Olins A.L. (1978) Nucleosomes: The structural quantum in chromosomes: Virtually all the DNA of eukaryotic cells is organized into a repeating array of nucleohistone particles called nucleosomes. These chromatin subunits are close-packed into higher-order fibers and are modified during chromosome expression. *American Scientist* 66: 704–11.

202. Olins D.E. and Olins A.L. (2003) Chromatin history: Our view from the bridge. *Nature Reviews Molecular Cell Biology* 4: 809–14.

203. Kornberg R.D. (1974) Chromatin structure: A repeating unit of histones and DNA. *Science* 184: 868–71.

204. Kornberg R.D. and Thomas J.O. (1974) Chromatin structure; oligomers of the histones. *Science* 184: 865–8.

205. Stein G.S., Spelsberg T.C. and Kleinsmith L.J. (1974) Nonhistone chromosomal proteins and gene regulation. *Science* 183: 817–24.

206. Stedman E. and Stedman E. (1950) Cell specificity of histones. *Nature* 166: 780–1.

207. Huang R.C. and Bonner J. (1962) Histone, a suppressor of chromosomal RNA synthesis. *Proceedings of the National Academy of Sciences USA* 48: 1216–22.

208. Allfrey V.G., Littau V.C. and Mirsky A.E. (1963) On the role of of histones in regulation ribonucleic acid synthesis in the cell nucleus. *Proceedings of the National Academy of Sciences USA* 49: 414–21.

209. Frenster J.H., Allfrey V.G. and Mirsky A.E. (1963) Repressed and active chromatin isolated from interphase lymphocytes. *Proceedings of the National Academy of Sciences USA* 50: 1026–32.

210. Littau V.C., Burdick C.J., Allfrey V.G. and Mirsky S.A. (1965) The role of histones in the maintenance of chromatin structure. *Proceedings of the National Academy of Sciences USA* 54: 1204–12.

211. Pardon J.F., Wilkins M.H. and Richards B.M. (1967) Super-helical model for nucleohistone. *Nature* 215: 508–9.

212. Pardon J.F. and Wilkins M.H. (1972) A super-coil model for nucleohistone. *Journal of Molecular Biology* 68: 115–24.

213. Allis C.D. (2015) "Modifying" my career toward chromatin biology. *Journal of Biological Chemistry* 290: 15904–8.

214. Allfrey V.G. and Mirsky A.E. (1964) Structural modifications of histones and their possible role in the regulation of RNA synthesis. *Science* 144: 559.

215. Allfrey V.G., Faulkner R. and Mirsky A.E. (1964) Acetylation and methylation of histones and their possible role in the regulation of RNA synthesis. *Proceedings of the National Academy of Sciences USA* 51: 786–94.

216. Pogo B.G., Allfrey V.G. and Mirsky A.E. (1966) RNA synthesis and histone acetylation during the course of gene activation in lymphocytes. *Proceedings of the National Academy of Sciences USA* 55: 805–12.

217. Allfrey V.G., Pogo B.G., Littau V.C., Gershey E.L. and Mirsky A.E. (1968) Histone acetylation in insect chromosomes. *Science* 159: 314–6.

218. Riggs A.D. (1975) X inactivation, differentiation, and DNA methylation. *Cytogenetics and Cell Genetics* 14: 9–25.

219. Holliday R. and Pugh J.E. (1975) DNA modification mechanisms and gene activity during development. *Science* 187: 226–32.

220. Brownell J.E., et al. (1996) Tetrahymena histone acetyltransferase A: A homolog to yeast Gcn5p linking histone acetylation to gene activation. *Cell* 84: 843–51.

221. Jones P.A. and Taylor S.M. (1980) Cellular differentiation, cytidine analogs and DNA methylation. *Cell* 20: 85–93.

222. Bird A., Taggart M., Frommer M., Miller O.J. and Macleod D. (1985) A fraction of the mouse genome that is derived from islands of nonmethylated, CpG-rich DNA. *Cell* 40: 91–9.

223. Bird A.P. (1986) CpG-rich islands and the function of DNA methylation. *Nature* 321: 209–13.

224. Brachet J. (1941) La detection histochimique et la micro-dosage des acides pentosenucléiques. *Enzymologia* 10: 87–96.

225. Brachet J. (1950) The localization and the role of ribonucleic acid in the cell. *Annals of the New York Academy of Sciences* 50: 861–9.

226. Mirsky A.E. and Ris H. (1947) The chemical composition of isolated chromosomes. *Journal of General Physiology* 31: 7–18.

227. Feinendegen L.E. and Bond V.P. (1963) Observations on nuclear RNA during mitosis in human cancer cells in culture (HeLa-S3), studied with tritiated cytidine. *Experimental Cell Research* 30: 393–404.

228. Edström J.-E. (1964) Chromosomal RNA and other nuclear RNA fractions, in M. Locke (ed.) *The Role of Chromosomes in Development* (Academic Press, Cambridge, MA).

229. Frenster J.H. (1965) A model of specific de-repression within interphase chromatin. *Nature* 206: 1269–70.

230. Frenster J.H. (1965) Nuclear polyanions as de-repressors of synthesis of ribonucleic acid. *Nature* 206: 680–3.

231. Huang R.C. and Bonner J. (1965) Histone-bound RNA, a component of native nucleohistone. *Proceedings of the National Academy of Sciences USA* 54: 960–7.

232. Prestayko A.W. and Busch H. (1968) Low molecular weight RNA of the chromatin fraction from Novikoff hepatoma and rat liver nuclei. *Biochimica et Biophysica Acta* 169: 327–37.

233. Huang R.C. and Huang P.C. (1969) Effect of protein-bound RNA associated with chick embryo chromatin on template specificity of the chromatin. *Journal of Molecular Biology* 39: 365–78.

234. Dahmus M.E. and McConnell D.J. (1969) Chromosomal ribonucleic acid of rat ascites cells. *Biochemistry* 8: 1524–34.

235. Bekhor I., Bonner J. and Dahmus G.K. (1969) Hybridization of chromosomal RNA to native DNA. *Proceedings of the National Academy of Sciences USA* 62: 271–7.

236. Holmes D.S., Mayfield J.E., Sander G. and Bonner J. (1972) Chromosomal RNA: Its properties. *Science* 177: 72–4.

237. Bonner J. and Widholm J. (1967) Molecular complementarity between nuclear DNA and organ-specific chromosomal RNA. *Proceedings of the National Academy of Sciences USA* 57: 1379–85.

238. Bonner J., et al. (1968) The biology of isolated chromatin: Chromosomes, biologically active in the test tube, provide a powerful tool for the study of gene action. *Science* 159: 47–56.

239. Mayfield J.E. and Bonner J. (1971) Tissue differences in rat chromosomal RNA. *Proceedings of the National Academy of Sciences USA* 68: 2652–5.

240. Weinberg R.A. and Penman S. (1968) Small molecular weight monodisperse nuclear RNA. *Journal of Molecular Biology* 38: 289–304.

241. Jacobson R.A. and Bonner J. (1968) The occurrence of dihydro ribothymidine in chromosomal RNA. *Biochemical and Biophysical Research Communications* 33: 716–20.

242. Sivolap Y.M. and Bonner J. (1971) Association of chromosomal RNA with repetitive DNA. *Proceedings of the National Academy of Sciences USA* 68: 387–9.

243. Holmes D.S., Mayfield J.E. and Bonner J. (1974) Sequence composition of rat ascites chromosomal ribonucleic acid. *Biochemistry* 13: 849–55.

244. Benjamin W., Levander O.A., Gellhorn A. and DeBellis R.H. (1966) An RNA-histone complex in mammalian cells: The isolation and characterization of a new RNA species. *Proceedings of the National Academy of Sciences USA* 55: 858–65.

245. Sypherd P.S. and Strauss N. (1963) The role of RNA in repression of enzyme synthesis. *Proceedings of the National Academy of Sciences USA* 50: 1059–66.

246. Goldstein L. (1976) Role for small nuclear RNAs in "programming" chromosomal information? *Nature* 261: 519–21.

247. Kanehisa T., Fujitani H., Sano M. and Tanaka T. (1971) Studies on low molecular weight RNA of chromatin. Effects of template activity of chick liver chromatin. *Biochimica et Biophysica Acta* 240: 46–55.

248. Kanehisa T., Tanaka T. and Kano Y. (1972) Low molecular RNA associated with chromatin: Purification and characterization of RNA that stimulates RNA synthesis. *Biochimica et Biophysica Acta* 277: 584–9.

249. Kanehisa T., Oki Y. and Ikuta K. (1974) Partial specificity of low-molecular weight RNA that stimulates RNA synthesis in various tissues. *Archives of Biochemistry and Biophysics* 165: 146–52.

250. Bekhor I. (1973) Physical studies on the effect of chromosomal RNA on reconstituted nucleohistones. *Archives of Biochemistry and Biophysics* 155: 39–46.

251. Paul J. and Duerksen J.D. (1975) Chromatin-associated RNA content of heterochromatin and euchromatin. *Molecular and Cellular Biochemistry* 9: 9–16.

252. Herman R., Weymouth L. and Penman S. (1978) Heterogeneous nuclear RNA-protein fibers in chromatin-depleted nuclei. *Journal of Cell Biology* 78: 663–74.

253. Pederson T. and Bhorjee J.S. (1979) Evidence for a role of RNA in eukaryotic chromosome structure. Metabolically stable, small nuclear RNA species are covalently linked to chromosomal DNA in HeLa cells. *Journal of Molecular Biology* 128: 451–80.

254. Monahan J.J. and Hall R.H. (1973) Fractionation of chromatin components. *Canadian Journal of Biochemistry* 51: 709–20.

255. Marzluff Jr. W.F., White E.L., Benjamin R. and Huang R.C. (1975) Low molecular weight RNA species from chromatin. *Biochemistry* 14: 3715–24.

256. Monahan J.J. and Hall R.H. (1974) Fractionation of L-cell chromatin into DNA, RNA, and protein fractions on Cs₂SO₄ equilibruim density gradients. *Analytical Biochemistry* 62: 217–39.

257. Pierpont M.E. and Yunis J.J. (1977) Localization of chromosomal RNA in human G-banded metaphase chromosomes. *Experimental Cell Research* 106: 303–8.

258. Bachellerie J.P., Puvion E. and Zalta J.P. (1975) Ultrastructural organization and biochemical characterization of chromatin - RNA - protein complexes isolated from mammalian cell nuclei. *European Journal of Biochemistry* 58: 327–37.

259. Bonner J. (1971) Problematic chromosomal RNA. *Nature* 231: 543–4.

260. von Heyden H.W. and Zachau H.G. (1971) Characterization of RNA in fractions of calf thymus chromatin. *Biochimica et Biophysica Acta* 232: 651–60.

261. Tolstoshev P. and Wells J.R. (1974) Nature and origins of chromatin-associated ribonucleic acid of avian reticulocytes. *Biochemistry* 13: 103–11.

262. Artman M. and Roth J.S. (1971) Chromosomal RNA: An artifact of preparation? *Journal of Molecular Biology* 60: 291–301.

263. Alfageme C.R. and Infante A.A. (1975) Nuclear RNA in sea urchin embryos. II. Absence of "cRNA". *Experimental Cell Research* 96: 263–70.

264. Pederson T. (2009) The discovery of eukaryotic genome design and its forgotten corollary–the postulate of gene regulation by nuclear RNA. *FASEB Journal* 23: 2019–21.

265. Getz M.J. and Saunders G.F. (1973) Origins of human leukocyte chromatin-associated RNA. *Biochimica et Biophysica Acta* 312: 555–73.

266. Bynum J.W. and Volkin E. (1980) Chromatin-associated RNA: Differential extraction and characterization. *Biochimica et Biophysica Acta* 607: 304–18.

267. Pederson T. (1977) Isolation and characterization of chromatin from the cellular slime mold, *Dictyostelium discoideum*. *Biochemistry* 16: 2771–7.

268. Elgin S.C. and Bonner J. (1970) Limited heterogeneity of the major nonhistone chromosomal proteins. *Biochemistry* 9: 4440–7.

269. Bakke A.C. and Bonner J. (1979) Purification and the histones of *Dictyostelium discoideum* chromatin. *Biochemistry* 18: 4556–62.

270. Zbarskii I.B. and Debov S.S. (1948) On the proteins of the cell nucleus. *Doklady Akademii Nauk SSSR* 63: 795–8.

271. Pederson T. (2000) Half a century of "the nuclear matrix". *Molecular Biology of the Cell* 11: 799–805.

272. Soyer M.-O. and Haapala O.K. (1974) Structural changes of dinoflagellate chromosomes by pronase and ribonuclease. *Chromosoma* 47: 179–92.

273. Penman S., et al. (1982) Cytoplasmic and nuclear architecture in cells and tissue: Form, functions, and mode of assembly. *Cold Spring Harbor Symposia on Quantitative Biology* 46: 1013–28.

274. Nickerson J.A., Krochmalnic G., Wan K.M. and Penman S. (1989) Chromatin architecture and nuclear RNA. *Proceedings of the National Academy of Sciences USA* 86: 177–81.

275. Scherrer K. (1989) A unified matrix hypothesis of DNA-directed morphogenesis, protodynamism and growth control. *Bioscience Reports* 9: 157–88.

276. Scherrer K. (2003) Historical review: The discovery of 'giant' RNA and RNA processing: 40 years of enigma. *Trends in Biochemical Sciences* 28: 566–71.

277. Scherrer K. and Darnell J.E. (1962) Sedimentation characteristics of rapidly labelled RNA from HeLa cells. *Biochemical and Biophysical Research Communications* 7: 486–90.

278. Georgiev G.P. and Mantieva V.L. (1962) The isolation of DNA-like RNA and ribosomal RNA from the nucleolochromosomal apparatus of mammalian cells. *Biochimica et Biophysica Acta* 61: 153–4.

279. Scherrer K., Darnell J.E. and Latham H. (1963) Demonstration of an unstable RNA and of a precursor to ribosomal RNA in Hela cells. *Proceedings of the National Academy of Sciences USA* 49: 240–8.

280. Georgiev G.P., Samarina O.P., Lerman M.I., Smirnov M.N. and Severtzov A.N. (1963) Biosynthesis of messenger and ribosomal ribonucleic acids in the nucleolochromosomal apparatus of animal cells. *Nature* 200: 1291–4.

281. Scherrer K. and Marcaud L. (1968) Messenger RNA in avian erythroblasts at transcriptional and translational levels and problem of regulation in animal cells. *Journal of Cellular Physiology* 72: 181–212.

282. Soeiro R., Birnboim H.C. and Darnell J.E. (1966) Rapidly labeled HeLa cell nuclear RNA. II. Base composition and cellular localization of a heterogeneous RNA fraction. *Journal of Molecular Biology* 19: 362–72.

283. Warner J.R., Soeiro R., Birnboim H.C., Girard M. and Darnell J.E. (1966) Rapidly labeled HeLa cell nuclear RNA. I. Identification by zone sedimentation of a heterogeneous fraction separate from ribosomal precursor RNA. *Journal of Molecular Biology* 19: 349–61.

284. Attardi G., Parnas H., Hwang M.I. and Attardi B. (1966) Giant-size rapidly labeled nuclear ribonucleic acid and cytoplasmic messenger ribonucleic acid in immature duck erythrocytes. *Journal of Molecular Biology* 20: 145–82.

285. Penman S., Smith I. and Holtzman E. (1966) Ribosomal RNA synthesis and processing in a particulate site in the HeLa cell nucleus. *Science* 154: 786–9.

286. Penman S. (1966) RNA metabolism in the HeLa cell nucleus. *Journal of Molecular Biology* 17: 117–30.

287. Weinberg R.A., Loening U., Willems M. and Penman S. (1967) Acrylamide gel electrophoresis of HeLa cell nucleolar RNA. *Proceedings of the National Academy of Sciences USA* 58: 1088–95.

288. Scherrer K., Marcaud L., Zajdela F., London I.M. and Gros F. (1966) Patterns of RNA metabolism in a differentiated cell: A rapidly labeled, unstable 60S RNA with messenger properties in duck erythroblasts. *Proceedings of the National Academy of Sciences USA* 56: 1571–8.

289. Houssais J.F. and Attardi G. (1966) High Molecular weight nonribosomal-type nuclear RNA and cytoplasmic messenger RNA in Hela cells. *Proceedings of the National Academy of Sciences USA* 56: 616–23.

290. Bryson V. and Vogel H.J. (1965) Evolving genes and proteins. *Science* 157: 68–71.

291. Harris H. (2013) History: Non-coding RNA foreseen 48 years ago. *Nature* 497: 188.

292. Penman S., Vesco C. and Penman M. (1968) Localization and kinetics of formation of nuclear heterodisperse RNA, cytoplasmic heterodisperse RNA and polyribosome-associated messenger RNA in HeLa cells. *Journal of Molecular Biology* 34: 49–60.

293. Penman S., Rosbash M. and Penman M. (1970) Messenger and heterogeneous nuclear RNA in HeLa cells: Differential inhibition by cordycepin. *Proceedings of the National Academy of Sciences USA* 67: 1878–85.

294. Lengyel J. and Penman S. (1975) hnRNA size and processing as related to different DNA content in two dipterans: Drosophila and Aedes. *Cell* 5: 281–90.

295. Allfrey V.G. and Mirsky A.E. (1962) Evidence for complete DNA-dependence of RNA synthesis in isolated thymus nuclei. *Proceedings of the National Academy of Sciences USA* 48: 1590–6.

296. Shearer R.W. and McCarthy B.J. (1967) Evidence for ribonucleic acid molecules restricted to the cell nucleus. *Biochemistry* 6: 283–9.

297. Bantle J.A. and Hahn W.E. (1976) Complexity and characterization of polyadenylated RNA in the mouse brain. *Cell* 8: 139–50.

298. Levy B., Johnson C.B. and McCarthy B.J. (1976) Diversity of sequences in total and polyadenylated nuclear RNA from *Drosophila* cells. *Nucleic Acids Research* 3: 1777–89.

299. Lewis E.B. (1963) Genes and developmental pathways. *American Zoologist* 3: 33–56.

300. Jacob F. and Monod J. (1961) Genetic regulatory mechanisms in the synthesis of proteins. *Journal of Molecular Biology* 3: 318–56.

301. Mayr E. (1961) Cause and effect in biology. *Science* 134: 1501–6.

302. Kauffman S. (1974) The large scale structure and dynamics of gene control circuits: An ensemble approach. *Journal of Theoretical Biology* 44: 167–90.

303. Kauffman S. (1969) Homeostasis and differentiation in random genetic control networks. *Nature* 224: 177–8.

304. Georgiev G.P. (1969) On the structural organization of operon and the regulation of RNA synthesis in animal cells. *Journal of Theoretical Biology* 25: 473–90.

305. Britten R.J. and Kohne D.E. (1968) Repeated sequences in DNA. Hundreds of thousands of copies of DNA sequences have been incorporated into the genomes of higher organisms. *Science* 161: 529–40.

306. Darnell J.E. (2002) The surprises of mammalian molecular cell biology. *Nature Medicine* 8: 1068–71.

307. Sarkar N. (1997) Polyadenylation of mRNA in prokaryotes. *Annual Review of Biochemistry* 66: 173–97.

308. Singer R.H. and Penman S. (1972) Stability of HeLa cell mRNA in actinomycin. *Nature* 240: 100–2.

309. Carninci P., et al. (1996) High-efficiency full-length cDNA cloning by biotinylated CAP trapper. *Genomics* 37: 327–36.

310. Milcarek C., Price R. and Penman S. (1974) The metabolism of a poly(A) minus mRNA fraction in HeLa cells. *Cell* 3: 1–10.

311. Katinakis P.K., Slater A. and Burdon R.H. (1980) Nonpolyadenylated mRNAs from eukaryotes. *FEBS Letters* 116: 1–7.

312. Salditt-Georgieff M., Harpold M.M., Wilson M.C. and Darnell Jr. J.E. (1981) Large heterogeneous nuclear ribonucleic acid has three times as many 5' caps as polyadenylic acid segments, and most caps do not enter polyribosomes. *Molecular Cell Biology* 1: 179–87.

313. Cheng J., et al. (2005) Transcriptional maps of 10 human chromosomes at 5-nucleotide resolution. *Science* 308: 1149–54.

314. Lee S.Y., Mendecki J. and Brawerman G. (1971) A polynucleotide segment rich in adenylic acid in the rapidly-labeled polyribosomal RNA component of mouse sarcoma 180 ascites cells. *Proceedings of the National Academy of Sciences USA* 68: 1331–5.

315. Darnell J.E., Wall R. and Tushinski R.J. (1971) An adenylic acid-rich sequence in messenger RNA of HeLa cells and its possible relationship to reiterated sites in DNA. *Proceedings of the National Academy of Sciences USA* 68: 1321–25.

316. Edmonds M., Vaughan Jr. M.H., and Nakazato H. (1971) Polyadenylic acid sequences in the heterogeneous nuclear RNA and rapidly-labeled polyribosomal RNA of HeLa cells: Possible evidence for a precursor relationship. *Proceedings of the National Academy of Sciences USA* 68: 1336–40.

317. Perry R.P., Bard E., Hames B.D., Kelly D.E. and Schibler U. (1976) The relationship between hnRNA and mRNA. *Progress in Nucleic Acid Research and Molecular Biology* 19: 275–92.

318. Salditt-Georgieff M., et al. (1976) Methyl labeling of HeLa cell hnRNA: A comparison with mRNA. *Cell* 7: 227–37.

319. Bachenheimer S. and Darnell J.E. (1975) Adenovirus-2 mRNA is transcribed as part of a high-molecular-weight precursor RNA. *Proceedings of the National Academy of Sciences USA* 72: 4445–9.

320. Weber J., Jelinek W. and Darnell J.E. (1977) The definition of a large viral transcription unit late in Ad2 infection of HeLa cells: Mapping of nascent RNA molecular labeled in isolated nuclei. *Cell* 10: 611–6.

321. Nevins J.R. and Darnell J.E. (1978) Groups of adenovirus type 2 mRNAs derived from a large primary transcript: Probable nuclear origin and possible common 3' ends. *Journal of Virology* 25: 811–23.

322. Ramanathan A., Robb G.B. and Chan S.-H. (2016) mRNA capping: Biological functions and applications. *Nucleic Acids Research* 44: 7511–26.

323. Wang J., et al. (2019) Quantifying the RNA cap epitranscriptome reveals novel caps in cellular and viral RNA. *Nucleic Acids Research* 47: e130.

324. Miura K.-I., Watanabe K. and Sugiura M. (1974) 5'-Terminal nucleotide sequences of the double-stranded RNA of silkworm cytoplasmic polyhedrosis virus. *Journal of Molecular Biology* 86: 31–48.

325. Miura K.-I., Watanabe K., Sugiura M. and Shatkin A.J. (1974) The 5'-terminal nucleotide sequences of the double-stranded RNA of human reovirus. *Proceedings of the National Academy of Sciences USA* 71: 3979–83.

326. Desrosiers R., Friderici K. and Rottman F. (1974) Identification of methylated nucleosides in messenger RNA from Novikoff hepatoma cells. *Proceedings of the National Academy of Sciences USA* 71: 3971–5.

327. Furuichi Y., et al. (1975) Methylated, blocked 5 termini in HeLa cell mRNA. *Proceedings of the National Academy of Sciences USA* 72: 1904–8.

328. Furuichi Y. and Miura K.-I. (1975) A blocked structure at the 5' terminus of mRNA from cytoplasmic polyhedrosis virus. *Nature* 253: 374–5.

329. Furuichi Y. (2015) Discovery of m7G-cap in eukaryotic mRNAs. *Proceedings of the Japan Academy, Series B* 91: 394–409.

330. Konarska M.M., Padgett R.A. and Sharp P.A. (1984) Recognition of cap structure in splicing in vitro of mRNA precursors. *Cell* 38: 731–6.

331. Fechter P. and Brownlee G.G. (2005) Recognition of mRNA cap structures by viral and cellular proteins. *Journal of General Virology* 86: 1239–49.

332. Adams J.M. and Cory S. (1975) Modified nucleosides and bizarre 5'-termini in mouse myeloma mRNA. *Nature* 255: 28–33.

333. Shatkin A. (1976) Capping of eucaryotic mRNAs. *Cell* 9: 645–53.

334. Banerjee A.K. (1980) 5'-terminal cap structure in eucaryotic messenger ribonucleic acids. *Microbiological Reviews* 44: 175–205.

335. Perry R. and Kelley D. (1974) Existence of methylated messenger RNA in mouse L cells. *Cell* 1: 37–42.

CHAPTER 5

1. Atkin N.B., Mattinson G., Becak W. and Ohno S. (1965) The comparative DNA content of 19 species of placental mammals, reptiles and birds. *Chromosoma* 17: 1–10.
2. Cui J., Schlub T.E. and Holmes E.C. (2014) An allometric relationship between the genome length and virion volume of viruses. *Journal of Virology* 88: 6403–10.
3. Campillo-Balderas J.A., Lazcano A. and Becerra A. (2015) Viral genome size distribution does not correlate with the antiquity of the host lineages. *Frontiers in Ecology and Evolution* 3: 143.
4. Mohanta T.K. and Bae H. (2015) The diversity of fungal genome. *Biological Procedures Online* 17: 8.
5. Corradi N., Pombert J.-F., Farinelli L., Didier E.S. and Keeling P.J. (2010) The complete sequence of the smallest known nuclear genome from the microsporidian *Encephalitozoon intestinalis*. *Nature Communications* 1: 77.
6. Pedersen R.A. (1971) DNA content, ribosomal gene multiplicity, and cell size in fish. *Journal of Experimental Zoology* 177: 65–78.
7. Pellicer J., Fay M.F. and Leitch I.J. (2010) The largest eukaryotic genome of them all? *Botanical Journal of the Linnean Society* 164: 10–5.
8. Elliott T.A. and Gregory T.R. (2015) What's in a genome? The C-value enigma and the evolution of eukaryotic genome content. *Philosophical Transactions of the Royal Society B: Biological Sciences* 370: 20140331.
9. Ahn A.H. and Kunkel L.M. (1993) The structural and functional diversity of dystrophin. *Nature Genetics* 3: 283–91.
10. Pozzoli U., et al. (2003) Comparative analysis of vertebrate dystrophin loci indicate intron gigantism as a common feature. *Genome Research* 13: 764–72.
11. Nakabayashi K. and Scherer S.W. (2001) The human contactin-associated protein-like 2 gene (CNTNAP2) spans over 2 Mb of DNA at chromosome 7q35. *Genomics* 73: 108–12.
12. Vogel F. (1964) A preliminary estimate of the number of human genes. *Nature* 201: 847.
13. Watson J.D. (1965) *Molecular Biology of the Gene* (Benjamin, Amsterdam).
14. Kauffman S.A. (1969) Metabolic stability and epigenesis in randomly constructed genetic nets. *Journal of Theoretical Biology* 22: 437–67.
15. Britten R.J. and Davidson E.H. (1969) Gene regulation for higher cells: A theory. *Science* 165: 349–57.
16. Waring M. and Britten R.J. (1966) Nucleotide sequence repetition: A rapidly reassociating fraction of mouse DNA. *Science* 154: 791–4.
17. Britten R.J. and Kohne D.E. (1968) Repeated sequences in DNA. Hundreds of thousands of copies of DNA sequences have been incorporated into the genomes of higher organisms. *Science* 161: 529–40.
18. Scheller R.H., Costantini F.D., Kozlowski M.R., Britten R.J. and Davidson E.H. (1978) Specific representation of cloned repetitive DNA sequences in sea urchin RNAs. *Cell* 15: 189–203.
19. Jurka J., Kapitonov V.V., Kohany O. and Jurka M.V. (2007) Repetitive sequences in complex genomes: Structure and evolution. *Annual Review of Genomics and Human Genetics* 8: 241–59.
20. Cuny G., Soriano P., Macaya G. and Bernardi G. (1981) The major components of the mouse and human genomes. 1. Preparation, basic properties and compositional heterogeneity. *European Journal of Biochemistry* 115: 227–33.
21. Bernardi G., et al. (1985) The mosaic genome of warm-blooded vertebrates. *Science* 228: 953–8.
22. Wells J.N. and Feschotte C. (2020) A field guide to eukaryotic transposable elements. *Annual Review of Genetics* 54: 539–61.
23. McClintock B. (1934) The relationship of a particular chromosomal element to the development of the nucleoli in *Zea mays*. *Zeitschrift für Zellforschung und mikroskopische Anatomie* 21: 294–328.
24. McStay B. (2016) Nucleolar organizer regions: Genomic 'dark matter' requiring illumination. *Genes & Development* 30: 1598–610.
25. McClintock B. (1951) Chromosome organization and genic expression. *Cold Spring Harbor Symposia on Quantitative Biology* 16: 13–47.
26. McClintock B. (1956) Controlling elements and the gene. *Cold Spring Harbor Symposia on Quantitative Biology* 21: 197–216.
27. McClintock B. (1961) Some parallels between gene control systems in maize and in bacteria. *American Naturalist* 95: 265–77.
28. McClintock B. (1938) The production of homozygous deficient tissues with mutant characteristics by means of the aberrant mitotic behavior of ring-shaped chromosomes. *Genetics* 23: 315–76.
29. McClintock B. (1941) The stability of broken ends of chromosomes in *Zea mays*. *Genetics* 26: 234–82.
30. Creighton H.B. and McClintock B. (1931) A correlation of cytological and genetical crossing-over in *Zea mays*. *Proceedings of the National Academy of Sciences USA* 17: 492–7.
31. Fedoroff N.V. (2012) McClintock's challenge in the 21st century. *Proceedings of the National Academy of Sciences USA* 109: 20200–3.
32. Chomet P. and Martienssen R. (2017) Barbara McClintock's final years as nobelist and mentor: A memoir. *Cell* 170: 1049–54.
33. McClintock B. (1950) The origin and behavior of mutable loci in maize. *Proceedings of the National Academy of Sciences USA* 36: 344–55.
34. Ravindran S. (2012) Barbara McClintock and the discovery of jumping genes. *Proceedings of the National Academy of Sciences USA* 109: 20198–9.
35. Witkin E.M. (2002) Chances and choices: Cold Spring Harbor 1944–1955. *Annual Review of Microbiology* 56: 1–15.
36. Brink R.A. and Nilan R.A. (1952) The relation between light variegated and medium variegated pericarp in maize. *Genetics* 37: 519–44.
37. Bennett E.A., et al. (2008) Active Alu retrotransposons in the human genome. *Genome Research* 18: 1875–83.
38. Houck C.M., Rinehart F.P. and Schmid C.W. (1979) A ubiquitous family of repeated DNA sequences in the human genome. *Journal of Molecular Biology* 132: 289–306.

39. Fedoroff N.V. (1989) About maize transposable elements and development. *Cell* 56: 181–91.

40. Bai L. and Brutnell T.P. (2011) The activator/dissociation transposable elements comprise a two-component gene regulatory switch that controls endogenous gene expression in maize. *Genetics* 187: 749–59.

41. Fedoroff N.V. (1999) The Suppressor-mutator element and the evolutionary riddle of transposons. *Genes to Cells* 4: 11–9.

42. Jones R.N. (2005) McClintock's controlling elements: The full story. *Cytogenetic and Genome Research* 109: 90–103.

43. Fedoroff N., Wessler S. and Shure M. (1983) Isolation of the transposable maize controlling elements Ac and Ds. *Cell* 35: 235–42.

44. Green M.M. (1977) Genetic instability in *Drosophila melanogaster*: De novo induction of putative insertion mutations. *Proceedings of the National Academy of Sciences USA* 74: 3490–3.

45. Green M.M. (1986) Genetic instability in *Drosophila melanogaster*: The genetics of an MR element that makes complete P insertion mutations. *Proceedings of the National Academy of Sciences USA* 83: 1036–40.

46. Mérel V., Boulesteix M., Fablet M. and Vieira C. (2020) Transposable elements in *Drosophila*. *Mobile DNA* 11: 23.

47. Shapiro J.A. (1969) Mutations caused by the insertion of genetic material into the galactose operon of *Escherichia coli*. *Journal of Molecular Biology* 40: 93–105.

48. Kleckner N. (1981) Transposable elements in prokaryotes. *Annual Review of Genetics* 15: 341–404.

49. Engels W.R. and Preston C.R. (1981) Identifying P factors in *Drosophila* by means of chromosome breakage hotspots. *Cell* 26: 421–8.

50. Ryder E. and Russell S. (2003) Transposable elements as tools for genomics and genetics in *Drosophila*. *Briefings in Functional Genomics and Proteomics* 2: 57–71.

51. Comfort N.C. (2003) *The Tangled Field: Barbara McClintock's Search for the Patterns of Genetic Control* (Harvard University Press, Cambridge, MA).

52. McClintock B. (1965) Components of action of the regulators Spm and Ac. *Carnegie Institution of Washington Year Book* 64: 527–6.

53. McClintock B. (1966) Regulation of pattern of gene expression by controlling elements in maize. *Carnegie Institution of Washington Yearbook* 65: 568–77.

54. Mcclintock B. (1958) The suppressor-mutator system of control of gene action in maize. *Carnegie Institution of Washington Year Book* 57: 415–29.

55. McClintock B. (1953) Induction of instability at selected loci in maize. *Genetics* 38: 579–99.

56. Martienssen R., Barkan A., Taylor W.C. and Freeling M. (1990) Somatically heritable switches in the DNA modification of Mu transposable elements monitored with a suppressible mutant in maize. *Genes & Development* 4: 331–43.

57. Volpe T.A. et al. (2002) Regulation of heterochromatic silencing and histone H3 lysine-9 methylation by RNAi. *Science* 297: 1833–7.

58. Lippman Z. et al. (2004) Role of transposable elements in heterochromatin and epigenetic control. *Nature* 430: 471–6.

59. Lippman Z. and Martienssen R. (2004) The role of RNA interference in heterochromatic silencing. *Nature* 431: 364–70.

60. Slotkin R.K. and Martienssen R. (2007) Transposable elements and the epigenetic regulation of the genome. *Nature Reviews Genetics* 8: 272–85.

61. McClintock B. (1984) The significance of responses of the genome to challenge. *Science* 226: 792–801.

62. Bateson W. and Pellew C. (1915) On the genetics of "Rogues" among culinary peas (Pisum sativum). *Journal of Genetics* 5: 13–36.

63. Bateson W. and Pellew C. (1920) The genetics of "rogues" among culinary peas (i). *Proceedings of The Royal Society B: Biological Sciences* 91: 186–95.

64. Brink R.A. (1956) A genetic change associated with the R locus in maize which is directed and potentially reversible. *Genetics* 41: 872–89.

65. Brink R.A. (1958) Paramutation at the R locus in maize. *Cold Spring Harbor Symposia on Quantitative Biology* 23: 379–91.

66. Chandler V.L. (2007) Paramutation: From maize to mice. *Cell* 128: 641–5.

67. Pilu R. (2015) Paramutation phenomena in plants. *Seminars in Cell & Developmental Biology* 44: 2–10.

68. Hollick J.B. (2017) Paramutation and related phenomena in diverse species. *Nature Reviews Genetics* 18: 5–23.

69. Slack J.M.W. (2002) Conrad Hal Waddington: The last Renaissance biologist? *Nature Reviews Genetics* 3: 889–95.

70. Waddington C.H. (1956) Genetic assimilation of the bithorax phenotype. *Evolution* 10: 1–13.

71. Waddington C.H. (1953) Genetic assimilation of an acquired character. *Evolution* 7: 118–26.

72. Bard J.B.L. (2017) C.H. Waddington's differences with the creators of the modern evolutionary synthesis: A tale of two genes. *History and Philosophy of the Life Sciences* 39: 18.

73. Banta J.A. and Richards C.L. (2018) Quantitative epigenetics and evolution. *Heredity* 121: 210–24.

74. Nishikawa K. and Kinjo A.R. (2018) Mechanism of evolution by genetic assimilation: Equivalence and independence of genetic mutation and epigenetic modulation in phenotypic expression. *Biophysical Reviews* 10: 667–76.

75. Fanti L., Piacentini L., Cappucci U., Casale A.M. and Pimpinelli S. (2017) Canalization by selection of *de novo* induced mutations. *Genetics* 206: 1995–2006.

76. Vigne P., et al. (2021) A single-nucleotide change underlies the genetic assimilation of a plastic trait. *Science Advances* 7: eabd9941.

77. Beddington R.S.P., Rashbass P. and Wilson V. (1992) *Brachyury* - a gene affecting mouse gastrulation and early organogenesis. *Development* 116: 157–65.

78. Johnson D.R. (1974) Hairpin-tail: A case of post-reductional gene action in the mouse egg? *Genetics* 76: 795–805.

79. Lyon M.F. and Glenister P.H. (1977) Factors affecting the observed number of young resulting from adjacent-2 disjunction in mice carrying a translocation. *Genetical Research* 29: 83–92.

80. Surani M.A.H., Barton S.C. and Norris M.L. (1984) Development of reconstituted mouse eggs suggests imprinting of the genome during gametogenesis. *Nature* 308: 548–50.

81. McGrath J. and Solter D. (1984) Completion of mouse embryogenesis requires both the maternal and paternal genomes. *Cell* 37: 179–83.

82. Barton S.C., Surani M.A.H. and Norris M.L. (1984) Role of paternal and maternal genomes in mouse development. *Nature* 311: 374–6.

83. Ferguson-Smith A.C. and Bourc'his D. (2018) The discovery and importance of genomic imprinting. *eLife* 7: e42368.

84. Reik W. and Walter J. (2001) Evolution of imprinting mechanisms: The battle of the sexes begins in the zygote. *Nature Genetics* 27: 255–6.

85. Reik W. and Lewis A. (2005) Co-evolution of X-chromosome inactivation and imprinting in mammals. *Nature Reviews Genetics* 6: 403–10.

86. Renfree M.B., Suzuki S. and Kaneko-Ishino T. (2013) The origin and evolution of genomic imprinting and viviparity in mammals. *Philosophical Transactions of the Royal Society B: Biological Sciences* 368: 20120151.

87. Patten M.M., et al. (2014) The evolution of genomic imprinting: Theories, predictions and empirical tests. *Heredity* 113: 119–28.

88. Morison I.M., Ramsay J.P. and Spencer H.G. (2005) A census of mammalian imprinting. *Trends in Genetics* 21: 457–65.

89. Koerner M.V., Pauler F.M., Huang R. and Barlow D.P. (2009) The function of non-coding RNAs in genomic imprinting. *Development* 136: 1771–83.

90. Bogutz A.B., et al. (2019) Evolution of imprinting via lineage-specific insertion of retroviral promoters. *Nature Communications* 10: 5674.

91. Greally J.M. (2002) Short interspersed transposable elements (SINEs) are excluded from imprinted regions in the human genome. *Proceedings of the National Academy of Sciences USA* 99: 327–32.

92. Renfree M.B., Papenfuss A.T., Shaw G. and Pask A.J. (2009) Eggs, embryos and the evolution of imprinting: Insights from the platypus genome. *Reproduction, Fertility and Development* 21: 935–42.

93. Vogels A. and Fryns J.P. (2002) The Prader-Willi syndrome and the Angelman syndrome. *Genetic Counseling* 13: 385–96.

94. Eggermann T., et al. (2015) Imprinting disorders: A group of congenital disorders with overlapping patterns of molecular changes affecting imprinted loci. *Clinical Epigenetics* 7: 123.

95. Butler M.G. (2020) Imprinting disorders in humans: A review. *Current Opinion in Pediatrics* 32: 719–29.

96. Cockett N.E., et al. (1996) Polar overdominance at the ovine *callipyge* locus. *Science* 273: 236–8.

97. Charlier C., et al. (2001) The callipyge mutation enhances the expression of coregulated imprinted genes in cis without affecting their imprinting status. *Nature Genetics* 27: 367–9.

98. Georges M., Charlier C. and Cockett N. (2003) The callipyge locus: Evidence for the trans interaction of reciprocally imprinted genes. *Trends in Genetics* 19: 248–52.

99. Royo H. and Cavaille J. (2008) Non-coding RNAs in imprinted gene clusters. *Biology of the Cell* 100: 149–66.

100. Meyer P. and Saedler H. (1996) Homology-dependent gene silencing in plants. *Annual Review of Plant Physiology and Plant Molecular Biology* 47: 23–48.

101. Bingham P.M. (1997) Cosuppression comes to the animals. *Cell* 90: 385–7.

102. Garrick D., Fiering S., Martin D.I.K. and Whitelaw E. (1998) Repeat-induced gene silencing in mammals. *Nature Genetics* 18: 56–9.

103. Selker E.U. (2002) Repeat-Induced gene silencing in fungi. *Advances in Genetics* 46: 439–50.

104. Fulci V. and Macino G. (2007) Quelling: Post-transcriptional gene silencing guided by small RNAs in *Neurospora crassa*. *Current Opinion in Microbiology* 10: 199–203.

105. Matzke M.A. and Mosher R.A. (2014) RNA-directed DNA methylation: An epigenetic pathway of increasing complexity. *Nature Reviews Genetics* 15: 394–408.

106. Ashe H.L., Monks J., Wijgerde M., Fraser P. and Proudfoot N.J. (1997) Intergenic transcription and transinduction of the human beta-globin locus. *Genes & Development* 11: 2494–509.

107. Bridges C.B. and Morgan T.H. (1923) *The Third-Chromosome Group of Mutant Characters of Drosophila Melanogaster* (Carnegie Institution of Washington, Washington, DC).

108. Lewis E.B. (1963) Genes and developmental pathways. *American Zoologist* 3: 33–56.

109. Lewis E.B. (2007) *Genes, Development and Cancer, The Life and Work of Edward B. Lewis* (Springer, Dordrecht).

110. Bateson W. (1894) *Materials for the Study of Variation Treated with Especial Regard to Discontinuity in the Origin of Species* (Macmillan, London).

111. Lewis E.B. (1994) Homeosis: The first 100 years. *Trends in Genetics* 10: 341–3.

112. Crow J.F. and Bender W. (2004) Edward B. Lewis, 1918–2004. *Genetics* 168: 1773–83.

113. Lewis E.B. (1978) A gene complex controlling segmentation in *Drosophila*. *Nature* 276: 565–70.

114. Lewis E.B. (1985) Regulation of the genes of the bithorax complex in *Drosophila*. *Cold Spring Harbor Symposia on Quantitative Biology* 50: 155–64.

115. Peifer M., Karch F. and Bender W. (1987) The bithorax complex: Control of segmental identity. *Genes & Development* 1: 891–8.

116. Duncan I. (1987) The bithorax complex. *Annual Review of Genetics* 21: 285–319.

117. Lewis E.B. (1998) The bithorax complex: The first fifty years. *International Journal of Developmental Biology* 42: 403–15.

118. Maeda R.K. and Karch F. (2006) The ABC of the *BX-C*: The bithorax complex explained. *Development* 133: 1413–22.

119. Maeda R.K. and Karch F. (2009) The *bithorax* complex of *Drosophila*: An exceptional *Hox* cluster. *Current Topics in Developmental Biology* 88: 1–33.

120. Akbari O.S., Bousum A., Bae E. and Drewell R.A. (2006) Unraveling cis-regulatory mechanisms at the *abdominal-A* and *Abdominal-B* genes in the *Drosophila bithorax* complex. *Developmental Biology* 293: 294–304.

121. Akam M. (1989) *Hox* and *HOM*: Homologous gene clusters in insects and vertebrates. *Cell* 57: 347–9.

122. Lewis E.B. (1955) Some aspects of position pseudoallelism. *American Naturalist* 89: 73–89.

123. Ingham P. (1985) *Abdominal* gene organization. *Nature* 313: 98–9.

124. Lewis E.B. (1951) Pseudoallelism and gene evolution. *Cold Spring Harbor Symposia on Quantitative Biology* 16: 159–74.

125. Lewis E.B. (1954) The theory and application of a new method of detecting chromosomal rearrangements in *Drosophila melanogaster*. *American Naturalist* 88: 225–39.

126. Zhou J., Ashe H., Burks C. and Levine M. (1999) Characterization of the transvection mediating region of the *abdominal-B* locus in *Drosophila*. *Development* 126: 3057–65.

127. Duncan IW (2002) Transvection effects in Drosophila. *Annual Review of Genetics* 36: 521–56.

128. Galouzis C.C. and Prud'homme B. (2021) Transvection regulates the sex-biased expression of a fly X-linked gene. *Science* 371: 396–400.

129. Hopmann R., Duncan D. and Duncan I. (1995) Transvection in the iab-5,6,7 region of the bithorax complex of *Drosophila*: Homology independent interactions in trans. *Genetics* 139: 815–33.

130. Muller M., Hagstrom K., Gyurkovics H., Pirrotta V. and Schedl P. (1999) The mcp element from the *Drosophila melanogaster* bithorax complex mediates long-distance regulatory interactions. *Genetics* 153: 1333–56.

131. Gemkow M.J., Verveer P.J. and Arndt-Jovin D.J. (1998) Homologous association of the Bithorax-Complex during embryogenesis: Consequences for transvection in *Drosophila melanogaster*. *Development* 125: 4541–52.

132. Duncan I. and Montgomery G. (2002) E. B. Lewis and the bithorax complex: Part II. From cis-trans test to the genetic control of development. *Genetics* 161: 1–10.

133. Lewis E.B. (1964) Genetic control and regulation of developmental pathways, in M. Locke (ed.) *The Role of Chromosomes in Development* (Academic Press, Cambridge, MA).

134. Lawrence P.A. (1992) *The Making of a Fly: The Genetics of Animal Design* (Blackwell Scientific, Hoboken, NJ).

135. Akam M.E., Martinez-Arias A., Weinzierl R. and Wilde C.D. (1985) Function and expression of ultrabithorax in the *Drosophila* embryo. *Cold Spring Harbor Symposia on Quantitative Biology* 50: 195–200.

136. Hogness D.S., et al. (1985) Regulation and products of the Ubx domain of the bithorax complex. *Cold Spring Harbor Symposia on Quantitative Biology* 50: 181–94.

137. Lipshitz H.D., Peattie D.A. and Hogness D.S. (1987) Novel transcripts from the Ultrabithorax domain of the bithorax complex. *Genes & Development* 1: 307–22.

138. Irish V.F., Martinez-Arias A. and Akam M. (1989) Spatial regulation of the *Antennapedia* and *Ultrabithorax* homeotic genes during *Drosophila* early development. *EMBO Journal* 8: 1527–37.

139. Tillib S., et al. (1999) Trithorax- and Polycomb-group response elements within an Ultrabithorax transcription maintenance unit consist of closely situated but separable sequences. *Molecular and Cellular Biology* 19: 5189–202.

140. Sanchez-Herrero E. and Akam M. (1989) Spatially ordered transcription of regulatory DNA in the *bithorax* complex of *Drosophila*. *Development* 107: 321–9.

141. Bae E., Calhoun V.C., Levine M., Lewis E.B. and Drewell R.A. (2002) Characterization of the intergenic RNA profile at *abdominal-A* and *Abdominal-B* in the *Drosophila* bithorax complex. *Proceedings of the National Academy of Sciences USA* 99: 16847–52.

142. Graveley B.R. et al. (2011) The developmental transcriptome of *Drosophila melanogaster*. *Nature* 471: 473–9.

143. Micol J.L. and García-Bellido A. (1988) Genetic analysis of "transvection effects" involving contrabithorax mutations in *Drosophila melanogaster*. *Proceedings of the National Academy of Sciences USA* 85: 1146–50.

144. Micol J.L., Castelli-Gair J.E. and Garcia-Bellido A. (1990) Genetic analysis of transvection effects involving cis-regulatory elements of the *Drosophila* Ultrabithorax gene. *Genetics* 126: 365–73.

145. Mattick J.S. and Gagen M.J. (2001) The evolution of controlled multitasked gene networks: The role of introns and other noncoding RNAs in the development of complex organisms. *Molecular Biology and Evolution* 18: 1611–30.

146. Peifer M. and Bender W. (1986) The anterobithorax and bithorax mutations of the bithorax complex. *EMBO Journal* 5: 2293–303.

147. Geyer P.K., Green M.M. and Corces V.G. (1990) Tissue-specific transcriptional enhancers may act in trans on the gene located in the homologous chromosome: The molecular basis of transvection in *Drosophila*. *EMBO Journal* 9: 2247–56.

148. McCall K., O'Connor M.B. and Bender W. (1994) Enhancer traps in the *Drosophila bithorax* complex mark parasegmental domains. *Genetics* 138: 387–99.

149. Judd B.H. (1995) Mutations of zeste that mediate transvection are recessive enhancers of position-effect variegation in *Drosophila melanogaster*. *Genetics* 141: 245–53.

150. Lim B., Heist T., Levine M. and Fukaya T. (2018) Visualization of transvection in living *Drosophila* embryos. *Molecular Cell* 70: 287–96.

151. Judd B.H. (1988) Transvection: Allelic cross talk. *Cell* 53: 841–3.

152. Wu C.T. and Morris J.R. (1999) Transvection and other homology effects. *Current Opinion in Genetics and Development* 9: 237–46.

153. Lewis P.H. (1947) New mutants report. *Drosophila Information Service* 21: 69.

154. Gorfinkiel N., et al. (2004) The *Drosophila* Polycomb group gene *Sex combs extra* encodes the ortholog of mammalian Ring1 proteins. *Mechanisms of Development* 121: 449–62.

155. Peterson A.J., et al. (2004) Requirement for sex comb on midleg protein interactions in *Drosophila* polycomb group repression. *Genetics* 167: 1225–39.

156. Sinclair D.A., et al. (1998) The *Additional sex combs* gene of *Drosophila* encodes a chromatin protein that binds to shared and unique Polycomb group sites on polytene chromosomes. *Development* 125: 1207–16.

157. Ingham P.W. (1981) Trithorax: A new homoeotic mutation of *Drosophila melanogaster* II. The role *of trx* (+) after embryogenesis. *Wilhelm Roux's Archives of Developmental Biology* 190: 365–9.

158. Ingham P.W. (1983) Differential expression of *bithorax* complex genes in the absence of the *extra sex combs* and *trithorax* genes. *Nature* 306: 591–3.

159. Ingham P.W. (1984) A gene that regulates the *bithorax* complex differentially in larval and adult cells of *Drosophila*. *Cell* 37: 815–23.

160. Ingham P.W. (1985) Genetic control of the spatial pattern of selector gene expression in *Drosophila*. *Cold Spring Harbor Symposia on Quantitative Biology* 50: 201–8.

161. Struhl G. and Akam M. (1985) Altered distributions of Ultrabithorax transcripts in extra sex combs mutant embryos of Drosophila. *EMBO Journal* 4: 3259–64.

162. Kennison J.A. and Tamkun J.W. (1988) Dosage-dependent modifiers of polycomb and *antennapedia* mutations in *Drosophila*. *Proceedings of the National Academy of Sciences USA* 85: 8136–40.

163. Shearn A. (1989) The *ash-1, ash-2* and *trithorax* genes of *Drosophila melanogaster* are functionally related. *Genetics* 121: 517–25.

164. Kennison J.A. (1995) The Polycomb and trithorax group proteins of *Drosophila*: Trans-regulators of homeotic gene function. *Annual Review of Genetics* 29: 289–303.

165. Ingham P.W. (1985) A clonal analysis of the requirement for the *trithorax* gene in the diversification of segments in *Drosophila. Journal of Embryology and Experimental Morphology* 89: 349–65.

166. Schuettengruber B., Bourbon H.-M., Di Croce L. and Cavalli G. (2017) Genome regulation by Polycomb and Trithorax: 70 years and counting. *Cell* 171: 34–57.

167. Schuettengruber B., Chourrout D., Vervoort M., Leblanc B. and Cavalli G. (2007) Genome regulation by polycomb and trithorax proteins. *Cell* 128: 735–45.

168. Erwin D.H. (2015) Retrospective. Eric Davidson (1937–2015). *Science* 350: 517.

169. Hood L. and Rothenberg E.V. (2015) Developmental biologist Eric H. Davidson, 1937–2015. *Proceedings of the National Academy of Sciences USA* 112: 13423–5.

170. Crippa M., Davidson E.H. and Mirsky A.E. (1967) Persistence in early amphibian embryos of informational RNAs from lampbrush chromosome stage of oogenesis. *Proceedings of the National Academy of Sciences USA* 57: 885–92.

171. Davidson E.H. (1968) *Gene Activity in Early Development* (Academic Press, Cambridge, MA).

172. Davidson E.H. and Erwin D.H. (2006) Gene regulatory networks and the evolution of animal body plans. *Science* 311: 796–800.

173. Denis H. (1966) Gene expression in amphibian development II. Release of the genetic information in growing embryos. *Journal of Molecular Biology* 22: 285–304.

174. Mirsky A.E. and Ris H. (1951) The desoxyribonucleic acid content of animal cells and Its evolutionary significance. *Journal of General Physiology* 34: 451–62.

175. Taft R.J. and Mattick J.S. (2003) Increasing biological complexity is positively correlated with the relative genome-wide expansion of non-protein-coding DNA sequences. *Genome Biology Preprint Depository*. http://genomebiology.com/2003/5/1/P1.

176. Taft R.J., Pheasant M. and Mattick J.S. (2007) The relationship between non-protein-coding DNA and eukaryotic complexity. *BioEssays* 29: 288–99.

177. Liu G., Mattick J.S. and Taft R.J. (2013) A meta-analysis of the genomic and transcriptomic composition of complex life. *Cell Cycle* 12: 2061–72.

178. Davidson E.H., Klein W.H. and Britten R.J. (1977) Sequence organization in animal DNA and a speculation on hnRNA as a coordinate regulatory transcript. *Developmental Biology* 55: 69–84.

179. Davidson E.H. and Britten R.J. (1979) Regulation of gene expression: Possible role of repetitive sequences. *Science* 204: 1052–9.

180. Davidson E.H. and Britten R.J. (1973) Organization, transcription, and regulation in the animal genome. *Quarterly Review of Biology* 48: 565–613.

181. Davidson E.H. (1971) Note on the control of gene expression during development. *Journal of Theoretical Biology* 32: 123–30.

182. Britten R.J. and Davidson E.H. (1971) Repetitive and non-repetitive DNA sequences and a speculation on origins of evolutionary novelty. *Quarterly Review of Biology* 46: 111–38.

183. Fedoroff N., Wellauer P.K. and Wall R. (1977) Intermolecular duplexes in heterogeneous nuclear RNA from HeLa cells. *Cell* 10: 597–610.

184. Pederson T. (2009) The discovery of eukaryotic genome design and its forgotten corollary–the postulate of gene regulation by nuclear RNA. *FASEB Journal* 23: 2019–21.

185. Davidson E.H. (2006) *The Regulatory Genome: Gene Regulatory Networks In Development And Evolution* (Academic Press, Cambridge, MA).

186. Erwin D.H. and Davidson E.H. (2009) The evolution of hierarchical gene regulatory networks. *Nature Reviews Genetics* 10: 141–8.

187. Peter I.S. and Davidson E.H. (2011) Evolution of gene regulatory networks controlling body plan development. *Cell* 144: 970–85.

188. King M.C. and Wilson A.C. (1975) Evolution at two levels in humans and chimpanzees. *Science* 188: 107–16.

189. Jacob F. (1977) Evolution and tinkering. *Science* 196: 1161–6.

190. Arthur W. (2002) The emerging conceptual framework of evolutionary developmental biology. *Nature* 415: 757–64.

191. Carroll S.B. (2008) Evo-devo and an expanding evolutionary synthesis: A genetic theory of morphological evolution. *Cell* 134: 25–36.

192. Loehlin D.W., Ames J.R., Vaccaro K. and Carroll S.B. (2019) A major role for noncoding regulatory mutations in the evolution of enzyme activity. *Proceedings of the National Academy of Sciences USA* 116: 12383–9.

193. Waddington C.H. (1969) Gene regulation in higher cells. *Science* 166: 639–40.

194. Kauffman S. (1969) Homeostasis and differentiation in random genetic control networks. *Nature* 224: 177–8.

195. Kauffman S. (1971) Gene regulation networks: A theory for their global structure and behaviors, in G. Schatten (ed.) *Current Topics in Developmental Biology* (Academic Press, Cambridge, MA).

196. Kauffman S. (1974) The large scale structure and dynamics of gene control circuits: An ensemble approach. *Journal of Theoretical Biology* 44: 167–90.

197. Levine M. and Davidson E.H. (2005) Gene regulatory networks for development. *Proceedings of the National Academy of Sciences USA* 102: 4936–42.

198. Peter I.S. and Davidson E.H. (2011) A gene regulatory network controlling the embryonic specification of endoderm. *Nature* 474: 635–9.

199. Peter I.S., Faure E. and Davidson E.H. (2012) Predictive computation of genomic logic processing functions in embryonic development. *Proceedings of the National Academy of Sciences USA* 109: 16434–42.

200. Albert R. and Othmer H.G. (2003) The topology of the regulatory interactions predicts the expression pattern of the segment polarity genes in *Drosophila melanogaster. Journal of Theoretical Biology* 223: 1–18.

201. Davidich M.I. and Bornholdt S. (2008) Boolean network model predicts cell cycle sequence of fission yeast. *PLOS ONE* 3: e1672.

202. W., R. (1974) Nuclear and messenger RNA. *Nature* 247: 506–7. https://doi.org/10.1038/247506a0

203. Crick F. (1971) General model for the chromosomes of higher organisms. *Nature* 234: 25–7.

204. Kolodny G.M. (1973) Regulation of gene expression in eukaryotic cells. *Journal of Cell Biology* 59: 175a.

205. Kolodny G.M. (1975) The regulation of gene expression in eukaryotic cells. *Medical Hypotheses* 1: 15–22.

206. Wise G.E. and Goldstein L. (1973) Electron microscope localization of nuclear RNA's that shuttle between cytoplasm and nucleus and nuclear RNA's that do not. *Journal of Cell Biology* 56: 129–38.

207. Goldstein L., Wise G.E. and Beeson M. (1973) Proof that certain RNAs shuttle non-randomly between cytoplasm and nucleus. *Experimental Cell Research* 76: 281–8.

208. Goldstein L. (1976) Role for small nuclear RNAs in "programming" chromosomal information? *Nature* 261: 519–21.

209. Goldstein L. and Ko C. (1975) The characteristics of shuttling RNAs confirmed. *Experimental Cell Research* 96: 297–302.

210. Mandel P. and Métais P. (1948) Les acides nucleiques du plasma sanguin chez l'homme. *Comptes Rendus de l'Académie des Sciences* 142: 241–3.

211. Kolodny G.M., Culp L.A. and Rosenthal L.J. (1972) Secretion of RNA by normal and transformed cells. *Experimental Cell Research* 73: 65–72.

212. Kolodny G.M. (1971) Evidence for transfer of macromolecular RNA between mammalian cells in culture. *Experimental Cell Research* 65: 313–24.

213. Judd B.H. and Young M.W. (1974) An examination of the one cistron: One chromomere concept. *Cold Spring Harbor Symposia on Quantitative Biology* 38: 573–9.

214. Daneholt B. (1972) Giant RNA transcript in a Balbiani ring. *Nature New Biology* 240: 229–32.

215. Heywood S.M., Kennedy D.S. and Bester A.J. (1975) Studies concerning the mechanism by which translational-control RNA regulates protein synthesis in embryonic muscle. *European Journal of Biochemistry* 58: 587–93.

216. Bester A.J., Kennedy D.S. and Heywood S.M. (1975) Two classes of translational control RNA: Their role in the regulation of protein synthesis. *Proceedings of the National Academy of Sciences USA* 72: 1523–7.

217. Brawerman G. (1976) A model for the control of transcription during development. *Cancer Research* 36: 4278–81.

218. Dickson E. and Robertson H.D. (1976) Potential regulatory roles for RNA in cellular development. *Cancer Research* 36: 3387–93.

219. Robertson H.D. and Dickson E. (1975) RNA processing and the control of gene expression. *Brookhaven Symposia in Biology* 26: 240–66.

220. Rothenberg E.V. (2016) Eric Davidson: Steps to a gene regulatory network for development. *Developmental Biology* 412: S7–19.

221. Jacob F., Perrin D., Sanchez C. and Monod J. (2005) The operon: A group of genes with expression coordinated by an operator [Facsimile of the article in Comptes Rendus de l'Académie des Sciences de Paris 250 (1960) 1727–1729]. *Comptes Rendus Biologies* 328: 514–20.

222. Jacob F. and Monod J. (1961) Genetic regulatory mechanisms in the synthesis of proteins. *Journal of Molecular Biology* 3: 318–56.

223. Monod J., Changeux J.P. and Jacob F. (1963) Allosteric proteins and cellular control systems. *Journal of Molecular Biology* 6: 306–29.

224. Ptashne M. (1967) Specific binding of the lambda phage repressor to lambda DNA. *Nature* 214: 232–4.

225. Maniatis T., Ptashne M. and Maurer R. (1974) Control elements in the DNA of bacteriophage lambda. *Cold Spring Harbor Symposia on Quantitative Biology* 38: 857–68.

226. Ptashne M., et al. (1980) How the lambda repressor and cro work. *Cell* 19: 1–11.

227. Deichmann U. (2016) Interview with Eric Davidson. *Developmental Biology* 412: S20–9.

228. Peter I.S. and Davidson E.H. (2009) Genomic control of patterning. *International Journal of Developmental Biology* 53: 707–16.

229. Peter I.S. and Davidson E.H. (2009) Modularity and design principles in the sea urchin embryo gene regulatory network. *FEBS Letters* 583: 3948–58.

230. Peter I.S. and Davidson E.H. (2010) The endoderm gene regulatory network in sea urchin embryos up to mid-blastula stage. *Developmental Biology* 340: 188–99.

231. Peter I.S. and Davidson E.H. (2016) Implications of developmental gene regulatory networks inside and outside developmental biology. *Current Topics in Developmental Biology* 117: 237–51.

232. Peter I.S. and Davidson E.H. (2017) Assessing regulatory information in developmental gene regulatory networks. *Proceedings of the National Academy of Sciences USA* 114: 5862–9.

233. Faure E., Peter I.S. and Davidson E.H. (2013) A new software package for predictive gene regulatory network modeling and redesign. *Journal of Computational Biology* 20: 419–23.

234. Calzone F.J., Lee J.J., Le N., Britten R.J. and Davidson E.H. (1988) A long, nontranslatable poly(A) RNA stored in the egg of the sea urchin *Strongylocentrotus purpuratus*. *Genes & Development* 2: 305–18.

235. Britten R.J., Baron W.F., Stout D.B. and Davidson E.H. (1988) Sources and evolution of human Alu repeated sequences. *Proceedings of the National Academy of Sciences USA* 85: 4770–4.

236. Britten R. (2006) Transposable elements have contributed to thousands of human proteins. *Proceedings of the National Academy of Sciences USA* 103: 1798–803.

237. Britten R.J. (2010) Transposable element insertions have strongly affected human evolution. *Proceedings of the National Academy of Sciences USA* 107: 19945–8.

CHAPTER 6

1. Yi D. (2008) Cancer, viruses, and mass migration: Paul Berg's venture into eukaryotic biology and the advent of recombinant DNA research and technology, 1967–1980. *Journal of the History of Biology* 41: 589–636.

2. Luria S.E. and Human M.L. (1952) A nonhereditary, host-induced variation of bacterial viruses. *Journal of Bacteriology* 64: 557–69.

3. Bertani G. and Weigle J.J. (1953) Host controlled variation in bacterial viruses. *Journal of Bacteriology* 65: 113–21.

4. Weigle J.J. and Bertani G. (1953) Variations of bacteriophage conditioned by host bacteria. *Annales de l'Institut Pasteur* 84: 175–9.

5. Arber W. and Dussoix D. (1962) Host specificity of DNA produced by *Escherichia coli*. I. Host controlled modification of bacteriophage lambda. *Journal of Molecular Biology* 5: 18–36.

6. Dussoix D. and Arber W. (1962) Host specificity of DNA produced by *Escherichia coli*. II. Control over acceptance of DNA from infecting phage lambda. *Journal of Molecular Biology* 5: 37–49.

7. Arber W., Hattman S. and Dussoix D. (1963) On the host-controlled modification of bacteriophage lambda. *Virology* 21: 30–5.

8. Dussoix D. and Arber W. (1965) Host specificity of DNA produced by *Escherichia coli*. IV. Host specificity of infectious DNA from bacteriophage lambda. *Journal of Molecular Biology* 11: 238–46.

9. Arber W. (1965) Host-controlled modification of bacteriophage. *Annual Review of Microbiology* 19: 365–78.

10. Kuhnlein U., Linn S. and Arber W. (1969) Host specificity of DNA produced by *Escherichia coli*. XI. In vitro modification of phage fd replicative form. *Proceedings of the National Academy of Sciences USA* 63: 556–62.

11. Arber W. and Linn S. (1969) DNA modification and restriction. *Annual Review of Biochemistry* 38: 467–500.

12. Lederberg S. and Meselson M. (1964) Degradation of non-replicating bacteriophage DNA in non-accepting cells. *Journal of Molecular Biology* 8: 623–8.

13. Meselson M. and Yuan R. (1968) DNA restriction enzyme from *E. coli*. *Nature* 217: 1110–4.

14. Loenen W.A.M., Dryden D.T.F., Raleigh E.A., Wilson G.G. and Murray N.E. (2014) Highlights of the DNA cutters: A short history of the restriction enzymes. *Nucleic Acids Research* 42: 3–19.

15. Kennaway C.K. et al. (2012) Structure and operation of the DNA-translocating type I DNA restriction enzymes. *Genes & Development* 26: 92–104.

16. Loenen W.A.M., Dryden D.T.F., Raleigh E.A. and Wilson G.G. (2014) Type I restriction enzymes and their relatives. *Nucleic Acids Research* 42: 20–44.

17. Smith H.O. and Wilcox K.W. (1970) A restriction enzyme from *Hemophilus influenzae*. I. Purification and general properties. *Journal of Molecular Biology* 51: 379–91.

18. Kelly Jr. T.J., and Smith H.O. (1970) A restriction enzyme from *Hemophilus influenzae*. II. Base sequence of the recognition site. *Journal of Molecular Biology* 51: 393–409.

19. Danna K. and Nathans D. (1971) Specific cleavage of simian virus 40 DNA by restriction endonuclease of *Hemophilus influenzae*. *Proceedings of the National Academy of Sciences USA* 68: 2913–7.

20. Weiss B. and Richardson C.C. (1967) Enzymatic breakage and joining of deoxyribonucleic acid, I. Repair of single-strand breaks in DNA by an enzyme system from *Escherichia coli* infected with T4 bacteriophage. *Proceedings of the National Academy of Sciences USA* 57: 1021–8.

21. Jackson D.A., Symons R.H. and Berg P. (1972) Biochemical method for inserting new genetic information into DNA of Simian Virus 40: Circular SV40 DNA molecules containing lambda phage genes and the galactose operon of *Escherichia coli*. *Proceedings of the National Academy of Sciences USA* 69: 2904–9.

22. Cohen S.N. (2013) DNA cloning: A personal view after 40 years. *Proceedings of the National Academy of Sciences USA* 110: 15521–9.

23. Boyer H.W. (1971) DNA restriction and modification mechanisms in bacteria. *Annual Review of Microbiology* 25: 153–76.

24. Cohen S.N., Chang A.C.Y., Boyer H.W. and Helling R.B. (1973) Construction of biologically functional bacterial plasmids *in vitro*. *Proceedings of the National Academy of Sciences USA* 70: 3240–4.

25. Chang A.C.Y. and Cohen S.N. (1974) Genome construction between bacterial species *in vitro*: Replication and expression of *Staphylococcus* plasmid genes in *Escherichia coli*. *Proceedings of the National Academy of Sciences USA* 71: 1030–4.

26. Morrow J.F., et al. (1974) Replication and transcription of eukaryotic DNA in *Esherichia coli*. *Proceedings of the National Academy of Sciences USA* 71: 1743–7.

27. Berg P. and Singer M.F. (1995) The recombinant DNA controversy: Twenty years later. *Proceedings of the National Academy of Sciences USA* 92: 9011–3.

28. Berg P. (2008) Asilomar 1975: DNA modification secured. *Nature* 455: 290–1.

29. Bolivar F., et al. (1977) Construction and characterization of new cloning vehicles. II. A multipurpose cloning system. *Gene* 2: 95–113.

30. Norrander J., Kempe T. and Messing J. (1983) Construction of improved M13 vectors using oligodeoxynucleotide-directed mutagenesis. *Gene* 26: 101–6.

31. Vieira J. and Messing J. (1982) The pUC plasmids, an M13mp7-derived system for insertion mutagenesis and sequencing with synthetic universal primers. *Gene* 19: 259–68.

32. Yanisch-Perron C., Vieira J. and Messing J. (1985) Improved M13 phage cloning vectors and host strains: Nucleotide sequences of the M13mp18 and pUC19 vectors. *Gene* 33: 103–19.

33. Dean D. (1981) A plasmid cloning vector for the direct selection of strains carrying recombinant plasmids. *Gene* 15: 99–102.

34. Polisky B., Bishop R.J. and Gelfand D.H. (1976) A plasmid cloning vehicle allowing regulated expression of eukaryotic DNA in bacteria. *Proceedings of the National Academy of Sciences USA* 73: 3900–4.

35. Young C.L., Britton Z.T. and Robinson A.S. (2012) Recombinant protein expression and purification: A comprehensive review of affinity tags and microbial applications. *Biotechnology Journal* 7: 620–34.

36. Maniatis T., Fritsch E.F. and Sambrook J. (1982) *Molecular Cloning: A Laboratory Manual* (Cold Spring Harbor Laboratory, Cold Spring Harbor, NY).

37. Various (1988) *Current Protocols in Molecular Biology* (John Wiley & Sons, Hoboken, NJ).

38. Gill K.P. and Denham M. (2020) Optimized transgene delivery using third-generation lentiviruses. *Current Protocols in Molecular Biology* 133: e125.

39. Nodelman I.M., Patel A., Levendosky R.F. and Bowman G.D. (2020) Reconstitution and purification of nucleosomes with recombinant histones and purified DNA. *Current Protocols in Molecular Biology* 133: e130.

40. Reimer K.A. and Neugebauer K.M. (2020) Preparation of mammalian nascent RNA for long read sequencing. *Current Protocols in Molecular Biology* 133: e128.

41. Vasquez C.A., Cowan Q.T. and Komor A.C. (2020) Base editing in human cells to produce single-nucleotide-variant clonal cell lines. *Current Protocols in Molecular Biology* 133: e129.

42. Baltimore D. (1970) Viral RNA-dependent DNA polymerase: RNA-dependent DNA polymerase in virions of RNA tumour viruses. *Nature* 226: 1209–11.

43. Temin H.M. and Mizutani S. (1970) RNA-dependent DNA polymerase in virions of Rous sarcoma virus. *Nature* 226: 1211–3.
44. Michelson A.M. and Todd A.R. (1955) Nucleotides part XXXII. Synthesis of a dithymidine dinucleotide containing a 3′: 5′-internucleotidic linkage. *Journal of the Chemical Society (Resumed)*: 2632–8. https://doi.org/10.1039/JR9550002632
45. Hall R.H., Todd A. and Webb R.F. (1957) Nucleotides. Part XLI. Mixed anhydrides as intermediates in the synthesis of dinucleoside phosphates. *Journal of the Chemical Society* 1957: 3291–6.
46. Todd A. (1958) Synthesis in the study of nucleotides; basic work on phosphorylation opens the way to an attack on nucleic acids and nucleotide coenzymes. *Science* 127: 787–92.
47. Gilham P.T. and Khorana H.G. (1958) Studies on polynucleotides. I. A new and general method for the chemical synthesis of the C5″-C3″ internucleotidic linkage. Syntheses of deoxyribodinucleotides. *Journal of the American Chemical Society* 80: 6212–22.
48. Letsinger R.L. and Ogilvie K.K. (1969) Nucleotide chemistry. XIII. Synthesis of oligothymidylates via phosphotriester intermediates. *Journal of the American Chemical Society* 91: 3350–5.
49. Letsinger R.L, Finnan J.L., Heavner G.A. and Lunsford W.B. (1975) Nucleotide chemistry. XX. Phosphite coupling procedure for generating internucleotide links. *Journal of the American Chemical Society* 97: 3278–9.
50. Reese C.B. (1978) The chemical synthesis of oligo- and poly-nucleotides by the phosphotriester approach. *Tetrahedron* 34: 3143–79.
51. Reese C.B. (2005) Oligo- and poly-nucleotides: 50 years of chemical synthesis. *Organic & Biomolecular Chemistry* 3: 3851–68.
52. Matteucci M.D. and Caruthers M.H. (1981) Synthesis of deoxyoligonucleotides on a polymer support. *Journal of the American Chemical Society* 103: 3185–91.
53. Beaucage S.L. and Caruthers M.H. (1981) Deoxynucleoside phosphoramidites—A new class of key intermediates for deoxypolynucleotide synthesis. *Tetrahedron Letters* 22: 1859–62.
54. Goodchild J. (1990) Conjugates of oligonucleotides and modified oligonucleotides: A review of their synthesis and properties. *Bioconjugate Chemistry* 1: 165–87.
55. Beaucage S.L. and Iyer R.P. (1992) Advances in the synthesis of oligonucleotides by the phosphoramidite approach. *Tetrahedron* 48: 2223–311.
56. Roy S. and Caruthers M. (2013) Synthesis of DNA/RNA and their analogs via phosphoramidite and H-phosphonate chemistries. *Molecules* 18: 14268–84.
57. Roy B., Depaix A., Périgaud C. and Peyrottes S. (2016) Recent trends in nucleotide synthesis. *Chemical Reviews* 116: 7854–97.
58. Smith C.I.E. and Zain R. (2019) Therapeutic oligonucleotides: State of the art. *Annual Review of Pharmacology and Toxicology* 59: 605–30.
59. Bonde M.T. et al. (2015) Direct mutagenesis of thousands of genomic targets using microarray-derived oligonucleotides. *ACS Synthetic Biology* 4: 17–22.
60. Palluk S. et al. (2018) De novo DNA synthesis using polymerase-nucleotide conjugates. *Nature Biotechnology* 36: 645–50.
61. Eisenstein M. (2020) Enzymatic DNA synthesis enters new phase. *Nature Biotechnology* 38: 1113–5.
62. Kosuri S. and Church G.M. (2014) Large-scale de novo DNA synthesis: Technologies and applications. *Nature Methods* 11: 499–507.
63. Southern E.M. (1975) Detection of specific sequences among DNA fragments separated by gel electrophoresis. *Journal of Molecular Biology* 98: 503–17.
64. Alwine J.C., Kemp D.J. and Stark G.R. (1977) Method for detection of specific RNAs in agarose gels by transfer to diazobenzyloxymethyl-paper and hybridization with DNA probes. *Proceedings of the National Academy of Sciences USA* 74: 5350–4.
65. Towbin H., Staehelin T. and Gordon J. (1979) Electrophoretic transfer of proteins from polyacrylamide gels to nitrocellulose sheets: Procedure and some applications. *Proceedings of the National Academy of Sciences USA* 76: 4350–4.
66. Burnette W.N. (1981) "Western Blotting": Electrophoretic transfer of proteins from sodium dodecyl sulfate-polyacrylamide gels to unmodified nitrocellulose and radiographic detection with antibody and radioiodinated protein A. *Analytical Biochemistry* 112: 195–203.
67. Moritz C.P. (2020) 40 years Western blotting: A scientific birthday toast. *Journal of Proteomics* 212: 103575.
68. Bowen B., Steinberg J., Laemmli U.K. and Weintraub H. (1980) The detection of DNA-binding proteins by protein blotting. *Nucleic Acids Research* 8: 1–20.
69. Qian Z.W. and Wilusz J. (1993) Cloning of a cDNA encoding an RNA binding protein by screening expression libraries using a Northwestern strategy. *Analytical Biochemistry* 212: 547–54.
70. Grunstein M. and Hogness D.S. (1975) Colony hybridization: A method for the isolation of cloned DNAs that contain a specific gene. *Proceedings of the National Academy of Sciences USA* 72: 3961–5.
71. Brownlee G.G., Sanger F. and Barrell B.G. (1967) Nucleotide sequence of 5S-ribosomal RNA from *Escherichia coli*. *Nature* 215: 735–6.
72. Brownlee G.G., Sanger F. and Barrell B.G. (1968) The sequence of 5s ribosomal ribonucleic acid. *Journal of Molecular Biology* 34: 379–412.
73. Wu R. and Kaiser A.D. (1968) Structure and base sequence in the cohesive ends of bacteriophage lambda DNA. *Journal of Molecular Biology* 35: 523–37.
74. Wu R. and Taylor E. (1971) Nucleotide sequence analysis of DNA. II. Complete nucleotide sequence of the cohesive ends of bacteriophage lambda DNA. *Journal of Molecular Biology* 57: 491–511.
75. Wu R. (1972) Nucleotide sequence analysis of DNA. *Nature New Biology* 236: 198–200.
76. Min Jou W., Haegeman G., Ysebaert M. and Fiers W. (1972) Nucleotide sequence of the gene coding for the bacteriophage MS2 coat protein. *Nature* 237: 82–8.
77. Fiers W. et al. (1976) Complete nucleotide sequence of bacteriophage MS2 RNA: Primary and secondary structure of the replicase gene. *Nature* 260: 500–7.
78. Maxam A.M. and Gilbert W. (1977) A new method for sequencing DNA. *Proceedings of the National Academy of Sciences USA* 74: 560–4.
79. Sanger F. et al. (1977) Nucleotide sequence of bacteriophage phi X174 DNA. *Nature* 265: 687–95.

80. Sanger F., Nicklen S. and Coulson A.R. (1977) DNA sequencing with chain-terminating inhibitors. *Proceedings of the National Academy of Sciences USA* 74: 5463–7.

81. Smith L.M. et al. (1986) Fluorescence detection in automated DNA sequence analysis. *Nature* 321: 674–9.

82. Prober J.M. et al. (1987) A system for rapid DNA sequencing with fluorescent chain-terminating dideoxynucleotides. *Science* 238: 336–41.

83. Watson J.D. and Cook-Deegan R.M. (1991) Origins of the human genome project. *FASEB Journal* 5: 8–11.

84. Bentley D.R. et al. (2008) Accurate whole human genome sequencing using reversible terminator chemistry. *Nature* 456: 53–9.

85. Heather J.M. and Chain B. (2016) The sequence of sequencers: The history of sequencing DNA. *Genomics* 107: 1–8.

86. Foret F., Klepárník K. and Minárik M. (2019) Nucleic acids: Chromatographic and electrophoretic methods, in P. Worsfold, A. Townshend and C. Poole (eds.) *Encyclopedia of Analytical Science* (Third Edition) (Academic Press, Cambridge, MA).

87. Wilson J.T., Forget B.G., Wilson L.B. and Weissman S.M. (1977) Human globin messenger RNA: Importance of cloning for structural analysis. *Science* 196: 200–2.

88. Begley C.G. and Ellis L.M. (2012) Raise standards for preclinical cancer research. *Nature* 483: 531–3.

89. Freedman L.P., Cockburn I.M. and Simcoe T.S. (2015) The economics of reproducibility in preclinical research. *PLOS Biology* 13: e1002165.

90. Begley C.G. and Ioannidis John P.A. (2015) Reproducibility in science. *Circulation Research* 116: 116–26.

91. Ioannidis J.P.A. (2017) Acknowledging and overcoming nonreproducibility in basic and preclinical research. *Journal of the American Medical Association* 317: 1019–20.

92. Morange M. (1998) *A History of Molecular Biology* (Harvard University Press, Cambridge, MA).

93. Finn R.D. et al. (2008) The Pfam protein families database. *Nucleic Acids Research* 36: D281–8.

94. El-Gebali S. et al. (2019) The Pfam protein families database in 2019. *Nucleic Acids Research* 47: D427–32.

95. Rubin G.M. and Lewis E.B. (2000) A brief history of *Drosophila's* contributions to genome research. *Science* 287: 2216–8.

96. Bender W., Spierer P., Hogness D.S. and Chambon P. (1983) Chromosomal walking and jumping to isolate DNA from the *Ace* and *rosy* loci and the *bithorax* complex in *Drosophila melanogaster*. *Journal of Molecular Biology* 168: 17–33.

97. Bender W. et al. (1983) Molecular genetics of the bithorax complex in *Drosophila melanogaster*. *Science* 221: 23–9.

98. Garber R.L., Kuroiwa A. and Gehring W.J. (1983) Genomic and cDNA clones of the homeotic locus *Antennapedia* in *Drosophila*. *EMBO Journal* 2: 2027–36.

99. McGinnis W., Levine M.S., Hafen E., Kuroiwa A. and Gehring W.J. (1984) A conserved DNA sequence in homoeotic genes of the *Drosophila Antennapedia* and *Bithorax* complexes. *Nature* 308: 428–33.

100. Scott M.P. and Weiner A.J. (1984) Structural relationships among genes that control development - sequence homology between the *Antennapedia*, *Ultrabithorax*, and *Fushi Tarazu* loci of *Drosophila*. *Proceedings of the National Academy of Sciences USA* 81: 4115–9.

101. Akam M.E., Martinez-Arias A., Weinzierl R. and Wilde C.D. (1985) Function and expression of ultrabithorax in the *Drosophila* embryo. *Cold Spring Harbor Symposia on Quantitative Biology* 50: 195–200.

102. Irish V.F., Martinez-Arias A. and Akam M. (1989) Spatial regulation of the *Antennapedia* and *Ultrabithorax* homeotic genes during *Drosophila* early development. *EMBO Journal* 8: 1527–37.

103. Sanchez-Herrero E. and Akam M. (1989) Spatially ordered transcription of regulatory DNA in the *bithorax* complex of *Drosophila*. *Development* 107: 321–9.

104. Müller M.M., Carrasco A.E. and DeRobertis E.M. (1984) A homeo-box-containing gene expressed during oogenesis in xenopus. *Cell* 39: 157–62.

105. Gehring W.J. et al. (1994) Homeodomain-DNA recognition. *Cell* 78: 211–23.

106. Gehring W.J. and Ikeo K. (1999) *Pax 6*: Mastering eye morphogenesis and eye evolution. *Trends in Genetics* 15: 371–7.

107. Duboule D. and Dollé P. (1989) The structural and functional organization of the murine *HOX* gene family resembles that of *Drosophila* homeotic genes. *EMBO Journal* 8: 1497–505.

108. Graham A., Papalopulu N. and Krumlauf R. (1989) The murine and *Drosophila* homeobox gene complexes have common features of organization and expression. *Cell* 57: 367–78.

109. Akam M. (1989) *Hox* and *HOM*: Homologous gene clusters in insects and vertebrates. *Cell* 57: 347–9.

110. Tournier-Lasserve E., Odenwald W.F., Garbern J., Trojanowski J. and Lazzarini R.A. (1989) Remarkable intron and exon sequence conservation in human and mouse homeobox *Hox1.3* genes. *Molecular and Cellular Biology* 9: 2273–8.

111. Potter S.S. and Branford W.W. (1998) Evolutionary conservation and tissue-specific processing of *Hoxa 11* antisense transcripts. *Mammalian Genome* 9: 799–806.

112. Hahn H. et al. (1996) Mutations of the human homolog of *Drosophila patched* in the nevoid basal cell carcinoma syndrome. *Cell* 85: 841–51.

113. Johnson R.L. et al. (1996) Human homolog of *patched*, a candidate gene for the basal cell nevus syndrome. *Science* 272: 1668–71.

114. Davies P.C.W. and Lineweaver C.H. (2011) Cancer tumors as Metazoa 1.0: Tapping genes of ancient ancestors. *Physical Biology* 8: 015001.

115. Trigos A.S., Pearson R.B., Papenfuss A.T. and Goode D.L. (2019) Somatic mutations in early metazoan genes disrupt regulatory links between unicellular and multicellular genes in cancer. *eLife* 8: e40947.

116. Rous P. (1911) A sarcoma of the fowl transmissible by an agent separable from the tumor cells. *Journal of Experimental Medicine* 13: 397–411.

117. Ellerman V. and Bang O. (1908) Experimentelle leukämie bei Hühnern. *Zentralblatt für Bakteriologie, Mikrobiologie und Hygiene* 46: 595–609.

118. Fujinami A. and Inamoto K. (1914) Untersuchungen über das Vorkommen und die klinische Bedeutung der Sarkome beim Hausgeflügel. *Zeitschrift für Krebsforschung* 14: 94–119.

119. Martin G.S. (2004) The road to Src. *Oncogene* 23: 7910–7.

120. Irby R.B. and Yeatman T.J. (2000) Role of Src expression and activation in human cancer. *Oncogene* 19: 5636–42.

121. Croce C.M. (2008) Oncogenes and cancer. *New England Journal of Medicine* 358: 502–11.
122. Peiris M.N., Li F. and Donoghue D.J. (2019) BCR: A promiscuous fusion partner in hematopoietic disorders. *Oncotarget* 10: 2738–54.
123. Knudson A.G. (1971) Mutation and cancer: Statistical study of retinoblastoma. *Proceedings of the National Academy of Sciences USA* 68: 820–3.
124. Du W. and Pogoriler J. (2006) Retinoblastoma family genes. *Oncogene* 25: 5190–200.
125. Berry J.L. et al. (2019) The RB1 story: Characterization and cloning of the first tumor suppressor gene. *Genes* 10: 879.
126. Levine A.J. and Oren M. (2009) The first 30 years of p53: Growing ever more complex. *Nature Reviews Cancer* 9: 749–58.
127. Lane D. and Levine A. (2010) p53 research: The past thirty years and the next thirty years. *Cold Spring Harbor Perspectives in Biology* 2: a000893.
128. Sherr C.J. (2004) Principles of tumor suppression. *Cell* 116: 235–46.
129. Friedrich G. and Soriano P. (1991) Promoter traps in embryonic stem cells: A genetic screen to identify and mutate developmental genes in mice. *Genes & Development* 5: 1513–23.
130. Irion S. et al. (2007) Identification and targeting of the *ROSA26* locus in human embryonic stem cells. *Nature Biotechnology* 25: 1477–82.
131. Li X. et al. (2014) *Rosa26*-targeted swine models for stable gene over-expression and Cre-mediated lineage tracing. *Cell Research* 24: 501–4.
132. Yang D. et al. (2016) Identification and characterization of rabbit *ROSA26* for gene knock-in and stable reporter gene expression. *Scientific Reports* 6: 25161.
133. Zambrowicz B.P. et al. (1997) Disruption of overlapping transcripts in the ROSA βgeo 26 gene trap strain leads to widespread expression of β-galactosidase in mouse embryos and hematopoietic cells. *Proceedings of the National Academy of Sciences USA* 94: 3789–94.
134. Dolgin E. (2017) The most popular genes in the human genome. *Nature* 551: 427–31.
135. Levitt N.C. and Hickson I.D. (2002) Caretaker tumour suppressor genes that defend genome integrity. *Trends in Molecular Medicine* 8: 179–86.
136. Papamichos-Chronakis M. and Peterson C.L. (2013) Chromatin and the genome integrity network. *Nature Reviews Genetics* 14: 62–75.
137. Oki E., Oda S., Maehara Y. and Sugimachi K. (1999) Mutated gene-specific phenotypes of dinucleotide repeat instability in human colorectal carcinoma cell lines deficient in DNA mismatch repair. *Oncogene* 18: 2143–7.
138. Schumacher T.N., Scheper W. and Kvistborg P. (2019) Cancer neoantigens. *Annual Review of Immunology* 37: 173–200.
139. Bhullar K.S. et al. (2018) Kinase-targeted cancer therapies: Progress, challenges and future directions. *Molecular Cancer* 17: 48.
140. Samstein R.M. et al. (2019) Tumor mutational load predicts survival after immunotherapy across multiple cancer types. *Nature Genetics* 51: 202–6.
141. Blass E. and Ott P.A. (2021) Advances in the development of personalized neoantigen-based therapeutic cancer vaccines. *Nature Reviews Clinical Oncology* 18: 215–29.
142. Sterner R.C. and Sterner R.M. (2021) CAR-T cell therapy: Current limitations and potential strategies. *Blood Cancer Journal* 11: 69.
143. Nemazee D. (2000) Receptor selection in B and T lymphocytes. *Annual Review of Immunology* 18: 19–51.
144. Cumano A. et al. (2019) New molecular insights into immune cell development. *Annual Review of Immunology* 37: 497–519.
145. Cohn M. et al. (2007) Reflections on the clonal-selection theory. *Nature Reviews Immunology* 7: 823–30.
146. Oppenheim J.J. (2001) Cytokines: Past, present, and future. *International Journal of Hematology* 74: 3–8.
147. Köhler G. and Milstein C. (1975) Continuous cultures of fused cells secreting antibody of predefined specificity. *Nature* 256: 495–7.
148. Winter G. and Harris W.J. (1993) Humanized antibodies. *Immunology Today* 14: 243–6.
149. Tsien R.Y. (1998) The green fluorescent protein. *Annual Review of Biochemistry* 67: 509–44.
150. Ow D.W. et al. (1986) Transient and stable expression of the firefly luciferase gene in plant cells and transgenic plants. *Science* 234: 856–9.
151. Roda A., Pasini P., Mirasoli M., Michelini E. and Guardigli M. (2004) Biotechnological applications of bioluminescence and chemiluminescence. *Trends in Biotechnology* 22: 295–303.
152. Sherkow J.S. and Greely H.T. (2015) The history of patenting genetic material. *Annual Review of Genetics* 49: 161–82.
153. Rehman W., Arfons L.M. and Lazarus H.M. (2011) The rise, fall and subsequent triumph of thalidomide: Lessons learned in drug development. *Therapeutic Advances in Hematology* 2: 291–308.
154. Mullis K. et al. (1986) Specific enzymatic amplification of DNA in vitro: The polymerase chain reaction. *Cold Spring Harbor Symposia on Quantitative Biology* 51: 263–73.
155. Saiki R.K. et al. (1988) Primer-directed enzymatic amplification of DNA with a thermostable DNA polymerase. *Science* 239: 487–91.
156. Brock T.D. (1995) The road to Yellowstone—and beyond. *Annual Review of Microbiology* 49: 1–29.
157. NHGRI (2020) *Polymerase chain reaction (PCR) fact sheet.* https://www.genome.gov/about-genomics/fact-sheets/Polymerase-Chain-Reaction-Fact-Sheet.
158. Adams M.D. et al. (1992) Sequence identification of 2,375 human brain genes. *Nature* 355: 632–4.
159. Roberts L. (1991) Genome patent fight erupts. *Science* 254: 184–6.
160. Rounsley S.D. et al. (1996) The construction of *Arabidopsis* expressed sequence tag assemblies. A new resource to facilitate gene identification. *Plant Physiology* 112: 1177–83.
161. Adams M.D. et al. (1994) A model for high-throughput automated DNA sequencing and analysis core facilities. *Nature* 368: 474–5.
162. Venter J.C., Smith H.O. and Hood L. (1996) A new strategy for genome sequencing. *Nature* 381: 364–6.
163. Venter J.C. et al. (1998) Shotgun sequencing of the human genome. *Science* 280: 1540–2.
164. Weber J.L. and Myers E.W. (1997) Human whole-genome shotgun sequencing. *Genome Research* 7: 401–9.
165. Fritsch E.F., Lawn R.M. and Maniatis T. (1979) Characterisation of deletions which affect the expression of fetal globin genes in man. *Nature* 279: 598–603.

166. Bell G.I., Karam J.H. and Rutter W.J. (1981) Polymorphic DNA region adjacent to the 5′ end of the human insulin gene. *Proceedings of the National Academy of Sciences USA* 78: 5759–63.
167. Jagadeeswaran P., Tuan D., Forget B.G. and Weissman S.M. (1982) A gene deletion ending at the midpoint of a repetitive DNA sequence in one form of hereditary persistence of fetal haemoglobin. *Nature* 296: 469–70.
168. Rees A., Stocks J., Shoulders C.C., Galton D.J. and Baralle F.E. (1983) DNA polymorphism adjacent to human apoprotein A-1 gene: Relation to hypertriglyceridaemia. *Lancet* 321: 444–6.
169. Clark M.B. et al. (2015) Quantitative gene profiling of long noncoding RNAs with targeted RNA sequencing. *Nature Methods* 12: 339–42.
170. Gloss B.S. and Dinger M.E. (2016) The specificity of long noncoding RNA expression. *Biochimica et Biophysica Acta* 1859: 16–22.
171. Deveson I.W., Hardwick S.A., Mercer T.R. and Mattick J.S. (2017) The dimensions, dynamics, and relevance of the mammalian noncoding transcriptome. *Trends in Genetics* 33: 464–78.
172. Müller J.B. et al. (2020) The proteome landscape of the kingdoms of life. *Nature* 582: 592–6.
173. Gerstberger S., Hafner M. and Tuschl T. (2014) A census of human RNA-binding proteins. *Nature Reviews Genetics* 15: 829–45.
174. Hentze M.W., Castello A., Schwarzl T. and Preiss T. (2018) A brave new world of RNA-binding proteins. *Nature Reviews Molecular Cell Biology* 19: 327–41.

CHAPTER 7

1. Swift H. (1950) The constancy of desoxyribose nucleic acid in plant nuclei. *Proceedings of the National Academy of Sciences USA* 36: 643–54.
2. Britten R.J. and Davidson E.H. (1969) Gene regulation for higher cells: A theory. *Science* 165: 349–57.
3. Sparrow A.H., Price H.J. and Underbrink A.G. (1972) A survey of DNA content per cell and per chromosome of prokaryotic and eukaryotic organisms: Some evolutionary considerations. *Brookhaven Symposia in Biology* 23: 451–94.
4. Tjio J.H. and Levan A. (1956) The chromosome number of man. *Hereditas* 42: 1–6.
5. Gartler S.M. (2006) The chromosome number in humans: A brief history. *Nature Reviews Genetics* 7: 655–60.
6. Mirsky A.E. and Ris H. (1951) The desoxyribonucleic acid content of animal cells and its evolutionary significance. *Journal of General Physiology* 34: 451–62.
7. Gregory T.R. (2001) Coincidence, coevolution, or causation? DNA content, cellsize, and the C-value enigma. *Biological Reviews* 76: 65–101.
8. Palazzo A.F. and Gregory T.R. (2014) The case for junk DNA. *PLOS Genetics* 10: e1004351.
9. Thomas Jr. C.A. (1971) The genetic organization of chromosomes. *Annual Review of Genetics* 5: 237–56.
10. Gregory T.R. and Hebert P.D.N. (1999) The modulation of DNA content: Proximate causes and ultimate consequences. *Genome Research* 9: 317–24.
11. Cavalier-Smith T. (1978) Nuclear volume control by nucleoskeletal DNA, selection for cell volume and cell growth rate, and the solution of the DNA C-value paradox. *Journal of Cell Science* 34: 247–78.
12. Cavalier-Smith T. (1982) Skeletal DNA and the evolution of genome size. *Annual Review of Biophysics and Bioengineering* 11: 273–302.
13. Beaton M.J. and Cavalier-Smith T. (1999) Eukaryotic noncoding DNA is functional: Evidence from the differential scaling of cryptomonad genomes. *Proceedings of the Royal Society B: Biological Sciences* 266: 2053–9.
14. Hardie D.C. and Hebert P.D.N. (2003) The nucleotypic effects of cellular DNA content in cartilaginous and ray-finned fishes. *Genome* 46: 683–706.
15. Marshall W.F. et al. (2012) What determines cell size? *BMC Biology* 10: 101.
16. Fedoroff N.V. (2012) Transposable elements, epigenetics, and genome evolution. *Science* 338: 758–67.
17. Moore G.P. (1984) The C-value paradox. *BioScience* 34: 425–9.
18. Byers T.J. (1986) Molecular biology of DNA in *Acanthamoeba*, *Amoeba*, *Entamoeba*, and *Naegleria*. *International Review of Cytology* 99: 311–41.
19. Otto S.P. and Whitton J. (2000) Polyploid incidence and evolution. *Annual Review of Genetics* 34: 401–37.
20. Adams K.L. and Wendel J.F. (2005) Polyploidy and genome evolution in plants. *Current Opinion in Plant Biology* 8: 135–41.
21. Taft R.J., Pheasant M. and Mattick J.S. (2007) The relationship between non-protein-coding DNA and eukaryotic complexity. *BioEssays* 29: 288–99.
22. Metcalfe C.J., Filée J., Germon I., Joss J. and Casane D. (2012) Evolution of the Australian lungfish (*Neoceratodus forsteri*) genome: A major role for CR1 and L2 LINE elements. *Molecular Biology and Evolution* 29: 3529–39.
23. Sun C. et al. (2012) LTR retrotransposons contribute to genomic gigantism in plethodontid salamanders. *Genome Biology and Evolution* 4: 168–83.
24. Wu C. and Lu J. (2019) Diversification of transposable elements in Arthropods and its impact on genome evolution. *Genes* 10: 338.
25. Petersen M. et al. (2019) Diversity and evolution of the transposable element repertoire in arthropods with particular reference to insects. *BMC Evolutionary Biology* 19: 11.
26. Wong W.Y. et al. (2019) Expansion of a single transposable element family is associated with genome-size increase and radiation in the genus Hydra. *Proceedings of the National Academy of Sciences USA* 116: 22915–7.
27. Liu G., Mattick J.S. and Taft R.J. (2013) A meta-analysis of the genomic and transcriptomic composition of complex life. *Cell Cycle* 12: 2061–72.
28. Bird A.P. (1995) Gene number, noise reduction and biological complexity. *Trends in Genetics* 11: 94–100.
29. Ohno S. (1970) *Evolution by Gene Duplication* (Springer-Verlag, Berlin).
30. Zhang J. (2003) Evolution by gene duplication: An update. *Trends in Ecology & Evolution* 18: 292–8.
31. Wolfe K.H. (2015) Origin of the yeast whole-genome duplication. *PLOS Biology* 13: e1002221.
32. McLysaght A., Hokamp K. and Wolfe K.H. (2002) Extensive genomic duplication during early chordate evolution. *Nature Genetics* 31: 200–4.

33. Sharp A.J. et al. (2005) Segmental duplications and copy-number variation in the human genome. *American Journal of Human Genetics* 77: 78–88.

34. Ozkan H., Levy A.A. and Feldman M. (2001) Allopolyploidy-induced rapid genome evolution in the wheat (Aegilops-Triticum) group. *Plant Cell* 13: 1735–47.

35. Roose M.L. and Gottlieb L.D. (1976) Genetic and biochemical consequences of polyploidy in Tragopogon. *Evolution* 30: 818–30.

36. Levin D.A. (1983) Polyploidy and novelty in flowering plants. *American Naturalist* 122: 1–25.

37. Lamichhaney S. et al. (2018) Rapid hybrid speciation in Darwin's finches. *Science* 359: 224–8.

38. Jacques P.-É., Jeyakani J. and Bourque G. (2013) The majority of primate-specific regulatory sequences are derived from transposable elements. *PLOS Genetics* 9: e1003504.

39. Trizzino M. et al. (2017) Transposable elements are the primary source of novelty in primate gene regulation. *Genome Research* 27: 1623–33.

40. Etchegaray E., Naville M., Volff J.-N. and Haftek-Terreau Z. (2021) Transposable element-derived sequences in vertebrate development. *Mobile DNA* 12: 1.

41. Cosby R.L. et al. (2021) Recurrent evolution of vertebrate transcription factors by transposase capture. *Science* 371: eabc6405.

42. Hermant C. and Torres-Padilla M.-E. (2021) TFs for TEs: The transcription factor repertoire of mammalian transposable elements. *Genes & Development* 35: 22–39.

43. Aziz R.K., Breitbart M. and Edwards R.A. (2010) Transposases are the most abundant, most ubiquitous genes in nature. *Nucleic Acids Research* 38: 4207–17.

44. Serrato-Capuchina A. and Matute D.R. (2018) The role of transposable elements in speciation. *Genes* 9: 254.

45. Wells J.N. and Feschotte C. (2020) A field guide to eukaryotic transposable elements. *Annual Review of Genetics* 54: 539–61.

46. Britten R.J., Baron W.F., Stout D.B. and Davidson E.H. (1988) Sources and evolution of human Alu repeated sequences. *Proceedings of the National Academy of Sciences USA* 85: 4770–4.

47. Feschotte C. (2008) Transposable elements and the evolution of regulatory networks. *Nature Reviews Genetics* 9: 397–405.

48. Patoori S., Barnada S. and Trizzino M. (2021) Young transposable elements rewired gene regulatory networks in human and chimpanzee hippocampal intermediate progenitors. *bioRxiv* epub: 2021.11.24.469877.

49. Ohno S. (1972) So much "junk" DNA in our genome. *Brookhaven Symposia in Biology* 23: 366–70.

50. King J.L. and Jukes T.H. (1969) Non-darwinian evolution. *Science* 164: 788–98.

51. Muller H.J. (1950) Our load of mutations. *American Journal of Human Genetics* 2: 111–76.

52. Muller H.J. (1964) The relation of recombination to mutational advance. *Mutation Research* 1: 2–9.

53. Ohno S. (1972) An argument for the genetic simplicity of man and other mammals. *Journal of Human Evolution* 1: 651–62.

54. Haldane J.B.S. (1957) The cost of natural selection. *Journal of Genetics* 55: 511.

55. Nei M. (1969) Gene duplication and nucleotide substitution in evolution. *Nature* 221: 40–2.

56. Nei M. (2005) Selectionism and neutralism in molecular evolution. *Molecular Biology and Evolution* 22: 2318–42.

57. Eddy S.R. (2012) The C-value paradox, junk DNA and ENCODE. *Current Biology* 22: R898–9.

58. Graur D. (2017) An upper limit on the functional fraction of the human genome. *Genome Biology and Evolution* 9: 1880–5.

59. Charlesworth B., Charlesworth D., Coyne J.A. and Langley C.H. (2016) Hubby and Lewontin on protein variation in natural populations: When molecular genetics came to the rescue of population genetics. *Genetics* 203: 1497–503.

60. Pigliucci M. and Schlichting C.D. (1997) On the limits of quantitative genetics for the study of phenotypic evolution. *Acta Biotheoretica* 45: 143–60.

61. Paul J. (1972) General theory of chromosome structure and gene activation in Eukaryotes. *Nature* 238: 444–6.

62. Southern E. (1974) *Eukaryotic DNA* (University Park Press, Baltimore).

63. Freeling M., Xu J., Woodhouse M. and Lisch D. (2015) A solution to the C-Value paradox and the function of junk DNA: The genome balance hypothesis. *Molecular Plant* 8: 899–910.

64. Veitia R.A. and Bottani S. (2009) Whole genome duplications and a 'function' for junk DNA? Facts and hypotheses. *PLOS ONE* 4: e8201.

65. Brenner S. (1998) Refuge of spandrels. *Current Biology* 8: R669.

66. Brosius J. (2003) How significant is 98.5% 'junk' in mammalian genomes? *Bioinformatics* 19(Suppl 2): ii35.

67. Brosius J. (2003) The contribution of RNAs and retroposition to evolutionary novelties. *Genetica* 118: 99–116.

68. Brosius J. (2005) Waste not, want not - transcript excess in multicellular eukaryotes. *Trends in Genetics* 21: 287–8.

69. Strasser B.J. (2010) Collecting, comparing, and computing sequences: The making of Margaret O. Dayhoff's Atlas of Protein Sequence and Structure, 1954–1965. *Journal of the History of Biology* 43: 623–60.

70. Cobb M. (2017) 60 years ago, Francis Crick changed the logic of biology. *PLOS Biology* 15: e2003243.

71. Zuckerkandl E. and Pauling L. (1962) Molecular evolution, in M. Kasha and B. Pullman (eds.) *Horizons in Biochemistry* (Academic Press, Cambridge, MA).

72. Margoliash E. (1963) Primary structure and evolution of cytochrome c. *Proceedings of the National Academy of Sciences USA* 50: 672–9.

73. Dickerson R.E. (1971) The structure of cytochrome c and the rates of molecular evolution. *Journal of Molecular Evolution* 1: 26–45.

74. Kumar S. (2005) Molecular clocks: Four decades of evolution. *Nature Reviews Genetics* 6: 654–62.

75. Ho S.Y.W. and Duchêne S. (2014) Molecular-clock methods for estimating evolutionary rates and timescales. *Molecular Ecology* 23: 5947–65.

76. Alvarez-Ponce D. (2021) Richard Dickerson, molecular clocks, and rates of protein evolution. *Journal of Molecular Evolution* 89: 122–6.

77. Kimura M. (1968) Evolutionary rate at the molecular level. *Nature* 217: 624–6.

78. Kimura M. (1979) The neutral theory of molecular evolution. *Scientific American* 241: 98–126.

79. Kimura M. (1983) *The Neutral Theory of Molecular Evolution* (Cambridge University Press, Cambridge).

80. Kimura M. (1986) DNA and the neutral theory. *Philosophical Transactions of the Royal Society B: Biological Sciences* 312: 343–54.

81. Kimura M. (1989) The neutral theory of molecular evolution and the world view of the neutralists. *Genome* 31: 24–31.

82. O'Donald P. (1969) "Haldane's Dilemma" and the rate of natural selection. *Nature* 221: 815–6.

83. Ohta T. (1973) Slightly deleterious mutant substitutions in evolution. *Nature* 246: 96–8.

84. Ohta T. (2002) Near-neutrality in evolution of genes and gene regulation. *Proceedings of the National Academy of Sciences USA* 99: 16134–7.

85. Lynch M. (2002) Intron evolution as a population-genetic process. *Proceedings of the National Academy of Sciences USA* 99: 6118–23.

86. Lynch M. and Conery J.S. (2003) The origins of genome complexity. *Science* 302: 1401–4.

87. Lynch M. (2006) The origins of eukaryotic gene structure. *Molecular Biology and Evolution* 23: 450–68.

88. Lynch M. et al. (2016) Genetic drift, selection and the evolution of the mutation rate. *Nature Reviews Genetics* 17: 704–14.

89. Koonin E.V. (2016) Splendor and misery of adaptation, or the importance of neutral null for understanding evolution. *BMC Biology* 14: 114.

90. Sung W., Ackerman M.S., Miller S.F., Doak T.G. and Lynch M. (2012) Drift-barrier hypothesis and mutation-rate evolution. *Proceedings of the National Academy of Sciences USA* 109: 18488–92.

91. Xu C. and Zhang J. (2018) Alternative polyadenylation of mammalian transcripts is generally deleterious, not adaptive. *Cell Systems* 6: 734–42.

92. Liu Z. and Zhang J. (2018) Human C-to-U coding RNA editing is largely nonadaptive. *Molecular Biology and Evolution* 35: 963–9.

93. Liu Z. and Zhang J. (2018) Most m⁶A RNA modifications in protein-coding regions are evolutionarily unconserved and likely nonfunctional. *Molecular Biology and Evolution* 35: 666–75.

94. Xu C., Park J.-K. and Zhang J. (2019) Evidence that alternative transcriptional initiation is largely nonadaptive. *PLOS Biology* 17: e3000197.

95. Xu C. and Zhang J. (2020) A different perspective on alternative cleavage and polyadenylation. *Nature Reviews Genetics* 21: 63.

96. Barton N.H., Etheridge A.M. and Véber A. (2017) The infinitesimal model: Definition, derivation, and implications. *Theoretical Population Biology* 118: 50–73.

97. Hurst L.D. (2009) Genetics and the understanding of selection. *Nature Reviews Genetics* 10: 83–93.

98. Lewontin R.C. (1974) *The Genetic Basis of Evolutionary Change* (Columbia University Press, New York).

99. Dover G. (2000) How genomic and developmental dynamics affect evolutionary processes. *BioEssays* 22: 1153–9.

100. Dover G. (2000) *Dear Mr Darwin: Letters on the Evolution of Life and Human Nature* (Weidenfeld & Nicolson, London).

101. Mattick J.S., Taft R.J. and Faulkner G.J. (2010) A global view of genomic information--moving beyond the gene and the master regulator. *Trends in Genetics* 26: 21–8.

102. Yusuf L. et al. (2020) Noncoding regions underpin avian bill shape diversification at macroevolutionary scales. *Genome Research* 30: 553–65.

103. Gilad Y., Oshlack A., Smyth G.K., Speed T.P. and White K.P. (2006) Expression profiling in primates reveals a rapid evolution of human transcription factors. *Nature* 440: 242–5.

104. Barton R.A. and Venditti C. (2014) Rapid evolution of the cerebellum in humans and other great apes. *Current Biology* 24: 2440–4.

105. Lou D.I. et al. (2014) Rapid evolution of BRCA1 and BRCA2 in humans and other primates. *BMC Evolutionary Biology* 14: 155.

106. Raia P. et al. (2018) Unexpectedly rapid evolution of mandibular shape in hominins. *Scientific Reports* 8: 7340.

107. Cechova M. et al. (2020) Dynamic evolution of great ape Y chromosomes. *Proceedings of the National Academy of Sciences USA* 117: 26273–80.

108. Chintalapati M. and Moorjani P. (2020) Evolution of the mutation rate across primates. *Current Opinion in Genetics and Development* 62: 58–64.

109. Bowling D.L. et al. (2020) Rapid evolution of the primate larynx? *PLOS Biology* 18: e3000764.

110. Mayr E. (1970) *Populations, Species, and Evolution. An Abridgment of Animal Species and Evolution* (Harvard University Press, Cambridge, MA).

111. Jacob F. (1977) Evolution and tinkering. *Science* 196: 1161–6.

112. Niklas K.J. (2014) The evolutionary-developmental origins of multicellularity. *American Journal of Botany* 101: 6–25.

113. Ohta T. and Gillespie J.H. (1996) Development of neutral and nearly neutral theories. *Theoretical Population Biology* 49: 128–42.

114. Kreitman M. (1983) Nucleotide polymorphism at the alcohol dehydrogenase locus of *Drosophila melanogaster*. *Nature* 304: 412–7.

115. Chamary J.V. and Hurst L.D. (2005) Evidence for selection on synonymous mutations affecting stability of mRNA secondary structure in mammals. *Genome Biology* 6: R75.

116. Chamary J.V, Parmley J.L. and Hurst L.D. (2006) Hearing silence: Non-neutral evolution at synonymous sites in mammals. *Nature Reviews Genetics* 7: 98–108.

117. Lawrie D.S., Messer P.W., Hershberg R. and Petrov D.A. (2013) Strong purifying selection at synonymous sites in *D. melanogaster*. *PLOS Genetics* 9: e1003527.

118. Dhindsa R.S., Copeland B.R., Mustoe A.M. and Goldstein D.B. (2020) Natural selection shapes codon usage in the human genome. *American Journal of Human Genetics* 107: 83–95.

119. LaBella A.L., Opulente D.A., Steenwyk J.L., Hittinger C.T. and Rokas A. (2019) Variation and selection on codon usage bias across an entire subphylum. *PLOS Genetics* 15: e1008304.

120. Bollenbach T., Vetsigian K. and Kishony R. (2007) Evolution and multilevel optimization of the genetic code. *Genome Research* 17: 401–4.

121. Itzkovitz S. and Alon U. (2007) The genetic code is nearly optimal for allowing additional information within protein-coding sequences. *Genome Research* 17: 405–12.

122. Itzkovitz S., Hodis E. and Segal E. (2010) Overlapping codes within protein-coding sequences. *Genome Research* 20: 1582–9.

123. Loehlin D.W., Ames J.R., Vaccaro K. and Carroll S.B. (2019) A major role for noncoding regulatory mutations in the evolution of enzyme activity. *Proceedings of the National Academy of Sciences USA* 116: 12383–9.

124. Siddiq M.A. and Thornton J.W. (2019) Fitness effects but no temperature-mediated balancing selection at the polymorphic *Adh* gene of *Drosophila melanogaster*. *Proceedings of the National Academy of Sciences USA* 116: 21634–40.

125. Callier V. (2018) in *Quantamagazine*, November 8, 2018.

126. Garud N.R., Messer P.W. and Petrov D.A. (2021) Detection of hard and soft selective sweeps from *Drosophila melanogaster* population genomic data. *PLOS Genetics* 17: e1009373.

127. Kern A.D. and Hahn M.W. (2018) The neutral theory in light of natural selection. *Molecular Biology and Evolution* 35: 1366–71.

128. Jensen J.D. et al. (2019) The importance of the Neutral Theory in 1968 and 50 years on: A response to Kern and Hahn 2018. *Evolution* 73: 111–4.

129. Pheasant M. and Mattick J.S. (2007) Raising the estimate of functional human sequences. *Genome Research* 17: 1245–53.

130. Yona A.H., Alm E.J. and Gore J. (2018) Random sequences rapidly evolve into *de novo* promoters. *Nature Communications* 9: 1530.

131. Frith M.C. et al. (2006) Evolutionary turnover of mammalian transcription start sites. *Genome Research* 16: 713–22.

132. Ponting C.P. and Hardison R.C. (2011) What fraction of the human genome is functional? *Genome Research* 21: 1769–76.

133. Kutter C. et al. (2012) Rapid turnover of long noncoding RNAs and the evolution of gene expression. *PLOS Genetics* 8: e1002841.

134. Quinn J.J. et al. (2016) Rapid evolutionary turnover underlies conserved lncRNA-genome interactions. *Genes & Development* 30: 191–207.

135. Glinsky G. and Barakat T.S. (2019) The evolution of Great Apes has shaped the functional enhancers' landscape in human embryonic stem cells. *Stem Cell Research* 37: 101456.

136. Fisher S., Grice E.A., Vinton R.M., Bessling S.L. and McCallion A.S. (2006) Conservation of RET regulatory function from human to zebrafish without sequence similarity. *Science* 312: 276–9.

137. Carroll S.B. (2008) Evo-devo and an expanding evolutionary synthesis: A genetic theory of morphological evolution. *Cell* 134: 25–36.

138. Carroll S.B., Prud'homme B. and Gompel N (2008) Regulating evolution. *Scientific American* 298: 60–7.

139. Jeong S. et al. (2008) The evolution of gene regulation underlies a morphological difference between two *Drosophila* sister species. *Cell* 132: 783–93.

140. Shubin N., Tabin C. and Carroll S. (2009) Deep homology and the origins of evolutionary novelty. *Nature* 457: 818–23.

141. Shibata Y. et al. (2012) Extensive evolutionary changes in regulatory element activity during human origins are associated with altered gene expression and positive selection. *PLOS Genetics* 8: e1002789.

142. King M.C. and Wilson A.C. (1975) Evolution at two levels in humans and chimpanzees. *Science* 188: 107–16.

143. Wray G.A. (2007) The evolutionary significance of cis-regulatory mutations. *Nature Reviews Genetics* 8: 206–16.

144. Wolfe K.H., Sharp P.M. and Li W.-H. (1989) Mutation rates differ among regions of the mammalian genome. *Nature* 337: 283–5.

145. Ellegren H., Smith N.G. and Webster M.T. (2003) Mutation rate variation in the mammalian genome. *Current Opinion in Genetics and Development* 13: 562–8.

146. Supek F. and Lehner B. (2015) Differential DNA mismatch repair underlies mutation rate variation across the human genome. *Nature* 521: 81–4.

147. Perera D. et al. (2016) Differential DNA repair underlies mutation hotspots at active promoters in cancer genomes. *Nature* 532: 259–63.

148. Oldmeadow C., Mengersen K., Mattick J.S. and Keith J.M. (2010) Multiple evolutionary rate classes in animal genome evolution. *Molecular Biology and Evolution* 27: 942–53.

149. Barriere A., Gordon K.L. and Ruvinsky I. (2011) Distinct functional constraints partition sequence conservation in a cis-regulatory element. *PLOS Genetics* 7: e1002095.

150. Pouyet F., Aeschbacher S., Thiéry A. and Excoffier L. (2018) Background selection and biased gene conversion affect more than 95% of the human genome and bias demographic inferences. *eLife* 7: e36317.

151. Monroe J.G. et al. (2022) Mutation bias reflects natural selection in *Arabidopsis thaliana*. *Nature* 602: 101-5.

152. Zhang J. (2022) Important genomic regions mutate less often than do other regions. *Nature*. News & Views: https://doi.org/10.1038/d41586-022-00017-6.

153. Vierstra J. et al. (2020) Global reference mapping of human transcription factor footprints. *Nature* 583: 729–36.

154. Zuckerkandl E. (1976) Gene control in eukaryotes and the c-value paradox "excess" DNA as an impediment to transcription of coding sequences. *Journal of Molecular Evolution* 9: 73–104.

155. Doolittle W.F. and Sapienza C. (1980) Selfish genes, the phenotype paradigm and genome evolution. *Nature* 284: 601–3.

156. Comings D.E. (1972) The structure and function of chromatin, in *Advances in Human Genetics* (Springer).

157. Gilbert W. (1978) Why genes in pieces? *Nature* 271: 501.

158. Loomis Jr. W.F. (1973) Vestigial DNA? *Developmental Biology* 30: 3–4.

159. Hutchinson J., Narayan R.K.J. and Rees H. (1980) Constraints upon the composition of supplementary DNA. *Chromosoma* 78: 137–45.

160. Jain H.K. (1980) Incidental DNA. *Nature* 288: 647–8.

161. Vanin E.F. (1985) Processed pseudogenes: Characteristics and evolution. *Annual Review of Genetics* 19: 253–72.

162. Jacq C., Miller J.R. and Brownlee G.G. (1977) A pseudogene structure in 5S DNA of *Xenopus laevis*. *Cell* 12: 109–20.

163. Ohno S. (1985) Dispensable genes. *Trends in Genetics* 1: 160–4.

164. Cheetham S.W., Faulkner G.J. and Dinger M.E. (2020) Overcoming challenges and dogmas to understand the functions of pseudogenes. *Nature Reviews Genetics* 21: 191–201.

165. McCarrey J.R. and Riggs A.D. (1986) Determinator-inhibitor pairs as a mechanism for threshold setting in development: A possible function for pseudogenes. *Proceedings of the National Academy of Sciences USA* 83: 679–83.

166. Nepveu A. and Marcu K.B. (1986) Intragenic pausing and anti-sense transcription within the murine *c-myc* locus. *EMBO Journal* 5: 2859–65.

167. Zhou B.-S., Beidler D.R. and Cheng Y.-C. (1992) Identification of antisense RNA transcripts from a human DNA topoisomerase I pseudogene. *Cancer Research* 52: 4280–5.

168. Korneev S.A., Park J.H. and O'Shea M. (1999) Neuronal expression of neural nitric oxide synthase (nNOS) protein is suppressed by an antisense RNA transcribed from an NOS pseudogene. *Journal of Neuroscience* 19: 7711–20.

169. Korneev S.A. et al. (2008) Novel noncoding antisense RNA transcribed from human anti-NOS2A locus is differentially regulated during neuronal differentiation of embryonic stem cells. *RNA* 14: 2030–7.

170. Koop B.F., Goodman M., Xu P., Chan K. and Slightom J.L. (1986) Primate η-globin DNA sequences and man's place among the great apes. *Nature* 319: 234–8.

171. Giannopoulou E. et al. (2012) A single nucleotide polymorphism in the *HbbP1* gene in the human β-globin locus is associated with a mild β-thalassemia disease phenotype. *Hemoglobin* 36: 433–45.

172. Moleirinho A. et al. (2013) Evolutionary constraints in the β-globin cluster: The signature of purifying selection at the δ-globin (HBD) locus and its role in developmental gene regulation. *Genome Biology and Evolution* 5: 559–71.

173. Ma Y. et al. (2021) Genome-wide analysis of pseudogenes reveals HBBP1's human-specific essentiality in erythropoiesis and implication in β-thalassemia. *Developmental Cell* 56: 478–93.

174. O'Brien S.J. (1973) On estimating functional gene number in Eukaryotes. *Nature New Biology* 242: 52–4.

175. Gerdes S.Y. et al. (2003) Experimental determination and system level analysis of essential genes in *Escherichia coli* MG1655. *Journal of Bacteriology* 185: 5673–84.

176. Giaever G. et al. (2002) Functional profiling of the *Saccharomyces cerevisiae* genome. *Nature* 418: 387–91.

177. Brown S.W. (1966) Heterochromatin. *Science* 151: 417–25.

178. Peacock W.J. et al. (1978) Fine structure and evolution of DNA in heterochromatin. *Cold Spring Harbor Symposia on Quantitative Biology* 42: 1121–35.

179. Williams G.C. (1966) *Adaptation and Natural Selection* (Princeton University Press, Princeton, NJ).

180. Hamilton W.D. (1970) Selfish and spiteful behaviour in an evolutionary model. *Nature* 228: 1218–20.

181. Dawkins R. (1976) *The Selfish Gene* (Oxford University Press, Oxford).

182. Orgel L.E. and Crick F.H. (1980) Selfish DNA: The ultimate parasite. *Nature* 284: 604–7.

183. Orgel L.E., Crick F.H. and Sapienza C. (1980) Selfish DNA. *Nature* 288: 645–6.

184. Fedoroff N.V. (2012) McClintock's challenge in the 21st century. *Proceedings of the National Academy of Sciences USA* 109: 20200–3.

185. Stoye J.P., Fenner S., Greenoak G.E., Moran C. and Coffin J.M. (1988) Role of endogenous retroviruses as mutagens: The hairless mutation of mice. *Cell* 54: 383–91.

186. Kleckner N. (1981) Transposable elements in prokaryotes. *Annual Review of Genetics* 15: 341–404.

187. Shapiro J.A. (1992) Natural genetic engineering in evolution. *Genetica* 86: 99–111.

188. Witzany G. (2009) *Natural Genetic Engineering and Natural Genome Editing*, Annals of the New York Academy of Sciences, Vol. 1178 (Wiley-Blackwell, Hoboken, NJ).

189. Charlesworth B., Sniegowski P. and Stephan W. (1994) The evolutionary dynamics of repetitive DNA in eukaryotes. *Nature* 371: 215–20.

190. Iranzo J. and Koonin E.V. (2018) How genetic parasites persist despite the purge of natural selection. *Europhysics Letters* 122: 58001.

191. Palazzo A.F. and Koonin E.V. (2020) Functional long non-coding RNAs evolve from junk transcripts. *Cell* 183: 1151–61.

192. Comfort N.C. (2001) From controlling elements to transposons: Barbara McClintock and the Nobel Prize. *Trends in Biochemical Sciences* 26: 454–7.

193. Comfort N.C. (2003) *The Tangled Field: Barbara McClintock's Search for the Patterns of Genetic Control* (Harvard University Press, Cambridge, MA).

194. Brandt J. et al. (2005) Transposable elements as a source of genetic innovation: Expression and evolution of a family of retrotransposon-derived neogenes in mammals. *Gene* 345: 101–11.

195. Gall J.G. (1981) Chromosome structure and the C-value paradox. *Journal of Cell Biology* 91: S3–S14.

196. Hurst G.D.D. and Werren J.H. (2001) The role of selfish genetic elements in eukaryotic evolution. *Nature Reviews Genetics* 2: 597–606.

197. Williamson B. (1977) DNA insertions and gene structure. *Nature* 270: 295–7.

198. Darnell Jr. J.E., (2013) Reflections on the history of pre-mRNA processing and highlights of current knowledge: A unified picture. *RNA* 19: 443–60.

199. Bachenheimer S. and Darnell J.E. (1975) Adenovirus-2 mRNA is transcribed as part of a high-molecular-weight precursor RNA. *Proceedings of the National Academy of Sciences USA* 72: 4445–9.

200. Berget S.M., Moore C. and Sharp P.A. (1977) Spliced segments at the 5′ terminus of adenovirus 2 late mRNA. *Proceedings of the National Academy of Sciences USA* 74: 3171–5.

201. Chow L.T., Gelinas R.E., Broker T.R. and Roberts R.J. (1977) An amazing sequence arrangement at the 5′ ends of adenovirus 2 messenger RNA. *Cell* 12: 1–8.

202. Berk A.J. (2016) Discovery of RNA splicing and genes in pieces. *Proceedings of the National Academy of Sciences USA* 113: 801–5.

203. Westphal H. and Lai S.-P. (1978) Displacement loops in adenovirus DNA-RNA hybrids. *Cold Spring Harbor Symposia on Quantitative Biology* 42: 555–8.

204. Jeffreys A.J. and Flavell R.A. (1977) Rabbit beta-globin gene contains a large insert in coding sequence. *Cell* 12: 1097–108.

205. Kinniburgh A.J., Mertz J.E. and Ross J. (1978) The precursor of mouse β-globin messenger RNA contains two intervening RNA sequences. *Cell* 14: 681–93.

206. Breathnach R., Mandel J.L. and Chambon P. (1977) Ovalbumin gene is split in chicken DNA. *Nature* 270: 314–9.
207. Doel M.T., Houghton M., Cook E.A. and Carey N.H. (1977) The presence of ovalbumin mRNA coding sequences in multiple restriction fragments of chicken DNA. *Nucleic Acids Research* 4: 3701–13.
208. Baldacci P. et al. (1979) Isolation of the lysozyme gene of chicken. *Nucleic Acids Research* 6: 2667–81.
209. Tonegawa S., Maxam A.M., Tizard R., Bernard O. and Gilbert W. (1978) Sequence of a mouse germ-line gene for a variable region of an immunoglobulin light chain. *Proceedings of the National Academy of Sciences USA* 75: 1485–9.
210. Glover D.M. and Hogness D.S. (1977) Novel arrangement of 18s and 28s sequences in a repeating unit of *Drosophila-melanogaster* rDNA. *Cell* 10: 167–76.
211. Goldberg S., Weber J. and Darnell J.E. (1977) The definition of a large viral transcription unit late in Ad2 infection of HeLa cells: Mapping by effects of ultraviolet irradiation. *Cell* 10: 617–21.
212. Goldberg S., Schwartz H. and Darnell Jr. J.E., (1977) Evidence from UV transcription mapping in HeLa cells that heterogeneous nuclear RNA is the messenger RNA precursor. *Proceedings of the National Academy of Sciences USA* 74: 4520–3.
213. Konkel D.A., Tilghman S.M. and Leder P. (1978) The sequence of the chromosomal mouse β-globin major gene: Homologies in capping, splicing and poly(A) sites. *Cell* 15: 1125–32.
214. Crick F. (1979) Split genes and RNA splicing. *Science* 204: 264–71.
215. Sharp P.A. (2005) The discovery of split genes and RNA splicing. *Trends in Biochemical Sciences* 30: 279–81.
216. Cavalier-Smith T. (1985) Selfish DNA and the origin of introns. *Nature* 315: 283–4.
217. Mattick J.S. (1994) Introns: Evolution and function. *Current Opinion in Genetics and Development* 4: 823–31.
218. Gallegos J.E. and Rose A.B. (2015) The enduring mystery of intron-mediated enhancement. *Plant Science* 237: 8–15.
219. Darnell J.E. (2002) The surprises of mammalian molecular cell biology. *Nature Medicine* 8: 1068–71.
220. Ohno S. (1980) Gene duplication, junk DNA, intervening sequences and the universal signal for their removal. *Revista Brasileira De Genetica* 3: 99–114.
221. Milcarek C., Price R. and Penman S. (1974) The metabolism of a poly(A) minus mRNA fraction in HeLa cells. *Cell* 3: 1–10.
222. Salditt-Georgieff M., Harpold M.M., Wilson M.C. and Darnell Jr. J.E., (1981) Large heterogeneous nuclear ribonucleic acid has three times as many 5' caps as polyadenylic acid segments, and most caps do not enter polyribosomes. *Molecular Cell Biology* 1: 179–87.
223. Cheng J. et al. (2005) Transcriptional maps of 10 human chromosomes at 5-nucleotide resolution. *Science* 308: 1149–54.
224. Padgett R.A., Grabowski P.J., Konarska M.M., Seiler S. and Sharp P.A. (1986) Splicing of messenger RNA precursors. *Annual Review of Biochemistry* 55: 1119–50.
225. Sharp P.A. et al. (1987) Splicing of messenger RNA precursors. *Cold Spring Harbor Symposium on Quantitative Biology* 52: 277–85.
226. Xing Y., Johnson C.V., Dobner P.R. and Lawrence J.B. (1993) Higher level organization of individual gene transcription and RNA splicing. *Science* 259: 1326–30.
227. St Laurent G. et al. (2012) Intronic RNAs constitute the major fraction of the non-coding RNA in mammalian cells. *BMC Genomics* 13: 504.
228. Heilig R., Mursaskowsky R., Kloepfer C. and Mandel J.L. (1982) The ovalbumin gene family: Complete sequence and structure of the Y gene. *Nucleic Acids Research* 10: 4363–82.
229. Leff S.E., Rosenfeld M.G. and Evans R.M. (1986) Complex transcriptional units: Diversity in gene expression by alternative RNA processing. *Annual Review of Biochemistry* 55: 1091–117.
230. Croft L. et al. (2000) ISIS, the intron information system, reveals the high frequency of alternative splicing in the human genome. *Nature Genetics* 24: 340–1.
231. Lareau L.F., Green R.E., Bhatnagar R.S. and Brenner S.E. (2004) The evolving roles of alternative splicing. *Current Opinion in Structural Biology* 14: 273–82.
232. Blencowe B.J. (2006) Alternative splicing: New insights from global analyses. *Cell* 126: 37–47.
233. Darnell Jr. J.E., (1978) Implications of RNA-RNA splicing in evolution of eukaryotic cells. *Science* 202: 1257–60.
234. Doolittle W.F. (1978) Genes in pieces - were they ever together. *Nature* 272: 581–2.
235. Blake C.C.F. (1978) Do genes-in-pieces imply proteins-in-pieces? *Nature* 273: 267.
236. Blake C.C. (1979) Exons encode protein functional units. *Nature* 277: 598.
237. Blake C.C. (1981) Exons and the structure, function and evolution of haemoglobin. *Nature* 291: 616.
238. Holland S.K. and Blake C.C. (1987) Proteins, exons and molecular evolution. *Biosystems* 20: 181–206.
239. Stoltzfus A., Spencer D.F., Zuker M., Logsdon Jr. J.M., and Doolittle W.F. (1994) Testing the exon theory of genes: The evidence from protein structure. *Science* 265: 202–7.
240. Deveson I.W. et al. (2018) Universal alternative splicing of noncoding exons. *Cell Systems* 6: 245–55.
241. Palmer J.D. and Logsdon Jr. J.M., (1991) The recent origins of introns. *Current Opinion in Genetics and Development* 1: 470–7.
242. Cavalier-Smith T. (1985) Eukaryotic gene numbers, non-coding DNA and genome size, in T. Cavalier-Smith, (ed.) *The Evolution of Genome Size* (John Wiley & Sons, Chichester).
243. Gilbert W., Marchionni M. and McKnight G. (1986) On the antiquity of introns. *Cell* 46: 151–4.
244. Brenner S., Dove W., Herskowitz I. and Thomas R. (1990) Genes and development: Molecular and logical themes. *Genetics* 126: 479–86.
245. Ferat J.L. and Michel F.(1993) Group II self-splicing introns in bacteria. *Nature* 364: 358–61.
246. Lambowitz A.M. and Belfort M. (1993) Introns as mobile genetic elements. *Annual Review of Biochemistry* 62: 587–622.
247. Cavalier-Smith T. (1991) Intron phylogeny: A new hypothesis. *Trends in Genetics* 7: 145–8.
248. Logsdon J.M.J. (1998) The recent origins of spliceosomal introns revisited. *Current Opinion in Genetics and Development* 8: 637–48.

249. Ng S.Y. et al. (1985) Evolution of the functional human beta-actin gene and its multi-pseudogene family: Conservation of noncoding regions and chromosomal dispersion of pseudogenes. *Molecular and Cellular Biology* 5: 2720–32.

250. Lloyd C. and Gunning P. (1993) Noncoding regions of the gamma-actin gene influence the impact of the gene on myoblast morphology. *Journal of Cell Biology* 121: 73–82.

251. Dutton J.R. et al. (2008) An evolutionarily conserved intronic region controls the spatiotemporal expression of the transcription factor Sox10. *BMC Developmental Biology* 8: 105.

252. Hsu A.P. et al. (2013) GATA2 haploinsufficiency caused by mutations in a conserved intronic element leads to MonoMAC syndrome. *Blood* 121: 3830–7.

253. Ridley M. (1999) *Genome: The Autobiography of a Species* in 23 Chapters (HarperCollins, New York).

254. Sharp P.A. (1985) On the origin of RNA splicing and introns. *Cell* 42: 397–400.

255. Bridgeman B. (1995) A review of the role of efference copy in sensory and oculomotor control systems. *Annals of Biomedical Engineering* 23: 409–22.

256. Mattick J.S. and Gagen M.J. (2001) The evolution of controlled multitasked gene networks: The role of introns and other noncoding RNAs in the development of complex organisms. *Molecular Biology and Evolution* 18: 1611–30.

257. Mattick J.S. (2001) Non-coding RNAs: The architects of eukaryotic complexity. *EMBO Reports* 2: 986–91.

258. Mattick J.S. (2003) Challenging the dogma: The hidden layer of non-protein-coding RNAs in complex organisms. *BioEssays* 25: 930–9.

259. Mattick J.S. (2004) RNA regulation: A new genetics? *Nature Reviews Genetics* 5: 316–23.

260. Mattick J.S. (2004) The hidden genetic program of complex organisms. *Scientific American* 291: 60–7.

261. Mattick J.S. (2007) A new paradigm for developmental biology. *Journal of Experimental Biology* 210: 1526–47.

262. Amaral P.P., Dinger M.E., Mercer T.R. and Mattick J.S. (2008) The eukaryotic genome as an RNA machine. *Science* 319: 1787–9.

263. Amaral P.P. and Mattick J.S. (2008) Noncoding RNA in development. *Mammalian Genome* 19: 454–92.

264. Mattick J.S. (2009) Deconstructing the dogma: A new view of the evolution and genetic programming of complex organisms. *Annals of the New York Academy of Science* 1178: 29–46.

265. Mattick J.S. (2011) The central role of RNA in human development and cognition. *FEBS Letters* 585: 1600–16.

266. Morris K.V. and Mattick J.S. (2014) The rise of regulatory RNA. *Nature Reviews Genetics* 15: 423–37.

267. Borsari B. et al. (2021) Enhancers with tissue-specific activity are enriched in intronic regions. *Genome Research* 31: 1325–36.

268. Lin S.L., Kim H. and Ying S.Y. (2008) Intron-mediated RNA interference and microRNA (miRNA). *Frontiers in Bioscience* 13: 2216–30.

CHAPTER 8

1. Chomczynski P. and Sacchi N. (1987) Single-step method of RNA isolation by acid guanidinium thiocyanate-phenol-chloroform extraction. *Analytical Biochemistry* 162: 156–9.

2. Moriyama Y., Hodnett J.L., Prestayko A.W. and Busch H. (1969) Studies on the nuclear 4 to 7S RNA of the Novikoff hepatoma. *Journal of Molecular Biology* 39: 335–49.

3. Bishop J.M. et al. (1970) The low molecular weight RNAs of Rous sarcoma virus. II. The 7S RNA. *Virology* 42: 927–37.

4. Prestayko A.W. and Busch H. (1968) Low molecular weight RNA of the chromatin fraction from Novikoff hepatoma and rat liver nuclei. *Biochimica et Biophysica Acta* 169: 327–37.

5. Nakamura T., Prestayko A.W. and Busch H. (1968) Studies on nucleolar 4 to 6S ribonucleic acid of Novikoff hepatoma cells. *Journal of Biological Chemistry* 243: 1368–75.

6. Weinberg R.A. and Penman S. (1968) Small molecular weight monodisperse nuclear RNA. *Journal of Molecular Biology* 38: 289–304.

7. Dingman C.W. and Peacock A.C. (1968) Analytical studies on nuclear ribonucleic acid using polyacrylamide gel electrophoresis. *Biochemistry* 7: 659–68.

8. Prestayko A.W., Tonato M. and Busch H. (1970) Low molecular weight RNA associated with 28S nucleolar RNA. *Journal of Molecular Biology* 47: 505–15.

9. Pederson T. and Bhorjee J.S. (1979) Evidence for a role of RNA in eukaryotic chromosome structure. Metabolically stable, small nuclear RNA species are covalently linked to chromosomal DNA in HeLa cells. *Journal of Molecular Biology* 128: 451–80.

10. Weinberg R.A. (1973) Nuclear RNA metabolism. *Annual Review of Biochemistry* 42: 329–54.

11. Zieve G., Benecke B.J. and Penman S. (1977) Synthesis of two classes of small RNA species *in vivo* and *in vitro*. *Biochemistry* 16: 4520–5.

12. Reichel R. and Benecke B.J. (1980) Reinitiation of synthesis of small cytoplasmic RNA species K and L in isolated HeLa cell nuclei *in vitro*. *Nucleic Acids Research* 8: 225–34.

13. Elder J.T., Pan J., Duncan C.H. and Weissman S.M. (1981) Transcriptional analysis of interspersed repetitive polymerase III transcription units in human DNA. *Nucleic Acids Research* 9: 1171–89.

14. Haynes S.R. and Jelinek W.R. (1981) Low molecular weight RNAs transcribed in vitro by RNA polymerase III from Alu-type dispersed repeats in Chinese hamster DNA are also found *in vivo*. *Proceedings of the National Academy of Sciences USA* 78: 6130–4.

15. Lerner M.R. and Steitz J.A. (1979) Antibodies to small nuclear RNAs complexed with proteins are produced by patients with systemic lupus erythematosus. *Proceedings of the National Academy of Sciences USA* 76: 5495–9.

16. Lerner M.R., Boyle J.A., Mount S.M., Wolin S.L. and Steitz J.A. (1980) Are snRNPs involved in splicing? *Nature* 283: 220–4.

17. Rogers J. and Wall R. (1980) A mechanism for RNA splicing. *Proceedings of the National Academy of Sciences USA* 77: 1877–9.

18. Shine J. and Dalgarno L. (1975) Determinant of cistron specificity in bacterial ribosomes. *Nature* 254: 34–8.

19. Steitz J.A. and Jakes K. (1975) How ribosomes select initiator regions in mRNA: Base pair formation between the 3′ terminus of 16S rRNA and the mRNA during initiation of protein synthesis in *Escherichia coli*. *Proceedings of the National Academy of Sciences USA* 72: 4734–8.

20. Tan E.M. and Kunkel H.G. (1966) Characteristics of a soluble nuclear antigen precipitating with sera of patients with Systemic Lupus Erythematosus. *Journal of Immunology* 96: 464–71.

21. Lerner M.R., Boyle J.A., Hardin J.A. and Steitz J.A. (1981) Two novel classes of small ribonucleoproteins detected by antibodies associated with lupus erythematosus. *Science* 211: 400–2.

22. Hardin J.A. et al. (1982) Antibodies from patients with connective tissue diseases bind specific subsets of cellular RNA-protein particles. *Journal of Clinical investigation* 70: 141–7.

23. Reeves W.H., Narain S. and Satoh M. (2003) Henry Kunkel, Stephanie Smith, clinical immunology, and split genes. *Lupus* 12: 213–7.

24. Thore S., Mayer C., Sauter C., Weeks S. and Suck D. (2003) Crystal structures of the *Pyrococcus abyssi* Sm core and its complex with RNA: Common features of RNA binding in archaea and eukarya. *Journal of Biological Chemistry* 278: 1239–47.

25. Padgett R.A. (2001) *mRNA Splicing: Role of snRNAs* (John Wiley & Sons, Ltd, Hoboken, NJ).

26. Yang V.W., Lerner M.R., Steitz J.A. and Flint S.J. (1981) A small nuclear ribonucleoprotein is required for splicing of adenoviral early RNA sequences. *Proceedings of the National Academy of Sciences USA* 78: 1371–5.

27. Padgett R.A., Mount S.M., Steitz J.A. and Sharp P.A. (1983) Splicing of messenger RNA precursors is inhibited by antisera to small nuclear ribonucleoprotein. *Cell* 35: 101–7.

28. Calvet J.P. and Pederson T. (1981) Base-pairing interactions between small nuclear RNAs and nuclear RNA precursors as revealed by psoralen cross-linking *in vivo*. *Cell* 26: 363–70.

29. Calvet J.P., Meyer L.M. and Pederson T. (1982) Small nuclear RNA U2 is base-paired to heterogeneous nuclear RNA. *Science* 217: 456–8.

30. Padgett R.A., Grabowski P.J., Konarska M.M., Seiler S. and Sharp P.A. (1986) Splicing of messenger RNA precursors. *Annual Review of Biochemistry* 55: 1119–50.

31. Fica S.M. and Nagai K. (2017) Cryo-electron microscopy snapshots of the spliceosome: Structural insights into a dynamic ribonucleoprotein machine. *Nature Structural & Molecular Biology* 24: 791–9.

32. Wilkinson M.E., Charenton C. and Nagai K. (2020) RNA splicing by the spliceosome. *Annual Review of Biochemistry* 89: 359–88.

33. Bai R. et al. (2021) Structure of the activated human minor spliceosome. *Science* 371: eabg0879.

34. Kwek K.Y. et al. (2002) U1 snRNA associates with TFIIH and regulates transcriptional initiation. *Nature Structural Biology* 9: 800–5.

35. Jobert L. et al. (2009) Human U1 snRNA forms a new chromatin-associated snRNP with TAF15. *EMBO Reports* 10: 494–500.

36. Mondal T., Rasmussen M., Pandey G.K., Isaksson A. and Kanduri C. (2010) Characterization of the RNA content of chromatin. *Genome Research* 20: 899–907.

37. Chinen M., Morita M., Fukumura K. and Tani T. (2010) Involvement of the spliceosomal U4 small nuclear RNA in heterochromatic gene silencing at fission yeast centromeres. *Journal of Biological Chemistry* 285: 5630–8.

38. Kaida D. et al. (2010) U1 snRNP protects pre-mRNAs from premature cleavage and polyadenylation. *Nature* 468: 664–8.

39. Berg M.G. et al. (2012) U1 snRNP determines mRNA length and regulates isoform expression. *Cell* 150: 53–64.

40. Yin Y. et al. (2020) U1 snRNP regulates chromatin retention of noncoding RNAs. *Nature* 580: 147–50.

41. Caizzi L. et al. (2021) Efficient RNA polymerase II pause release requires U2 snRNP function. *Molecular Cell* 81: 1920–34.

42. Hall S.L. and Padgett R.A. (1994) Conserved sequences in a class of rare eukaryotic nuclear introns with non-consensus splice sites. *Journal of Molecular Biology* 239: 357–65.

43. Tarn W.-Y. and Steitz J.A. (1996) A novel spliceosome containing U11, U12, and U5 snRNPs excises a minor class (AT–AC) intron *in vitro*. *Cell* 84: 801–11.

44. Burge C.B., Padgett R.A. and Sharp P.A. (1998) Evolutionary fates and origins of U12-type introns. *Molecular Cell* 2: 773–85.

45. Turunen J.J., Niemelä E.H., Verma B. and Frilander M.J. (2013) The significant other: Splicing by the minor spliceosome. *Wiley Interdisciplinary Reviews RNA* 4: 61–76.

46. Jutzi D., Akinyi M.V., Mechtersheimer J., Frilander M.J. and Ruepp M.-D. (2018) The emerging role of minor intron splicing in neurological disorders. *Cell Stress* 2: 40–54.

47. Baumgartner M. et al. (2018) Minor spliceosome inactivation causes microcephaly, owing to cell cycle defects and death of self-amplifying radial glial cells. *Development* 145: dev166322.

48. Li L. et al. (2020) Defective minor spliceosomes induce SMA-associated phenotypes through sensitive intron-containing neural genes in Drosophila. *Nature Communications* 11: 5608.

49. Pene J.J., Knight Jr. E., and Darnell Jr. J.E. (1968) Characterization of a new low molecular weight RNA in HeLa cell ribosomes. *Journal of Molecular Biology* 33: 609–23.

50. Bachellerie J.P., Michot B. and Raynal F. (1983) Recognition signals for mouse pre-rRNA processing. A potential role for U3 nucleolar RNA. *Molecular Biology Reports* 9: 79–86.

51. Kiss T. (2002) Small nucleolar RNAs: An abundant group of noncoding RNAs with diverse cellular functions. *Cell* 109: 145–8.

52. Rimer J.M. et al. (2018) Long-range function of secreted small nucleolar RNAs that direct 2′-O-methylation. *Journal of Biological Chemistry* 293: 13284–96.

53. Ochs R.L., Lischwe M.A., Spohn W.H. and Busch H. (1985) Fibrillarin: A new protein of the nucleolus identified by autoimmune sera. *Biology of the Cell* 54: 123–33.

54. Tyc K. and Steitz J.A. (1989) U3, U8 and U13 comprise a new class of mammalian snRNPs localized in the cell nucleolus. *EMBO Journal* 8: 3113–9.

55. Fatica A., Galardi S., Altieri F. and Bozzoni I. (2000) Fibrillarin binds directly and specifically to U16 box C/D snoRNA. *RNA* 6: 88–95.

56. Kass S., Tyc K., Steitz J.A. and Sollner-Webb B. (1990) The U3 small nucleolar ribonucleoprotein functions in the first step of preribosomal RNA processing. *Cell* 60: 897–908.

57. Kiss-Laszlo Z., Henry Y., Bachellerie J.P., Caizergues-Ferrer M. and Kiss T. (1996) Site-specific ribose methylation of preribosomal RNA: A novel function for small nucleolar RNAs. *Cell* 85: 1077–88.

58. Tollervey D. and Kiss T. (1997) Function and synthesis of small nucleolar RNAs. *Current Opinion in Cell Biology* 9: 337–42.

59. Fournier M.J. and Maxwell E.S. (1993) The nucleolar snRNAs: Catching up with the spliceosomal snRNAs. *Trends in Biochemical Sciences* 18: 131–5.

60. Bachellerie J.P., Cavaille J. and Huttenhofer A. (2002) The expanding snoRNA world. *Biochimie* 84: 775–90.

61. Lafontaine D.L.J. and Tollervey D. (1998) Birth of the snoRNPs: The evolution of the modification-guide snoRNAs. *Trends in Biochemical Sciences* 23: 383–8.

62. Omer A.D. *et al.* (2000) Homologs of small nucleolar RNAs in Archaea. *Science* 288: 517–22.

63. Terns M.P. and Terns R.M. (2002) Small nucleolar RNAs: Versatile trans-acting molecules of ancient evolutionary origin. *Gene Expression* 10: 17–39.

64. Abel Y. and Rederstorff M. (2019) SnoRNAs and the emerging class of sdRNAs: Multifaceted players in oncogenesis. *Biochimie* 164: 17–21.

65. Venema J. and Tollervey D. (1999) Ribosome synthesis in *Saccharomyces cerevisiae*. *Annual Review of Genetics* 33: 261–311.

66. Falaleeva M. et al. (2016) Dual function of C/D box small nucleolar RNAs in rRNA modification and alternative pre-mRNA splicing. *Proceedings of the National Academy of Sciences USA* 113: E1625–34.

67. Sharma S. et al. (2017) Specialized box C/D snoRNPs act as antisense guides to target RNA base acetylation. *PLOS Genetics* 13: e1006804.

68. Gall J.G., Bellini M., Wu Z. and Murphy C. (1999) Assembly of the nuclear transcription and processing machinery: Cajal bodies (coiled bodies) and transcriptosomes. *Molecular Biology of the Cell* 10: 4385–402.

69. Jády B.E. and Kiss T. (2001) A small nucleolar guide RNA functions both in 2′-O-ribose methylation and pseudouridylation of the U5 spliceosomal RNA. *EMBO Journal* 20: 541–51.

70. Mowry K.L. and Steitz J.A. (1987) Identification of the human U7 snRNP as one of several factors involved in the 3′ end maturation of histone premessenger RNA's. *Science* 238: 1682–7.

71. Gall J.G. and Callan H.G. (1989) The sphere organelle contains small nuclear ribonucleoproteins. *Proceedings of the National Academy of Sciences USA* 86: 6635–9.

72. Bratkovič T., Božič J. and Rogelj B. (2019) Functional diversity of small nucleolar RNAs. *Nucleic Acids Research* 48: 1627–51.

73. Vitali P. and Kiss T. (2019) Cooperative 2′-O-methylation of the wobble cytidine of human elongator tRNAMet(CAT) by a nucleolar and a Cajal body-specific box C/D RNP. *Genes & Development* 33: 741–6.

74. Nostramo R.T. and Hopper A.K. (2019) Beyond rRNA and snRNA: tRNA as a 2′-O-methylation target for nucleolar and Cajal body box C/D RNPs. *Genes & Development* 33: 739–40.

75. Mitchell J.R., Cheng J. and Collins K. (1999) A box H/ACA small nucleolar RNA-like domain at the human telomerase RNA 3′ end. *Molecular and Cellular Biology* 19: 567–76.

76. Jady B.E., Bertrand E. and Kiss T. (2004) Human telomerase RNA and box H/ACA scaRNAs share a common Cajal body-specific localization signal. *Journal of Cell Biology* 164: 647–52.

77. Ghanim G.E. et al. (2021) Structure of human telomerase holoenzyme with bound telomeric DNA. *Nature* 593: 449–53.

78. Jády B.E., Ketele A. and Kiss T. (2012) Human intron-encoded Alu RNAs are processed and packaged into Wdr79-associated nucleoplasmic box H/ACA RNPs. *Genes & Development* 26: 1897–910.

79. Schubert T. et al. (2012) Df31 protein and snoRNAs maintain accessible higher-order structures of chromatin. *Molecular Cell* 48: 434–44.

80. Jorjani H. et al. (2016) An updated human snoRNAome. *Nucleic Acids Research* 44: 5068–82.

81. Kufel J. and Grzechnik P. (2019) Small nucleolar RNAs tell a different tale. *Trends in Genetics* 35: 104–17.

82. Maxwell E.S. and Fournier M.J. (1995) The small nucleolar RNAs. *Annual Review of Biochemistry* 64: 897–934.

83. Tycowski K.T., Shu M.D. and Steitz J.A. (1996) A mammalian gene with introns instead of exons generating stable RNA products. *Nature* 379: 464–6.

84. Bortolin M.L. and Kiss T. (1998) Human U19 intron-encoded snoRNA is processed from a long primary transcript that possesses little potential for protein coding. *RNA* 4: 445–54.

85. Pelczar P. and Filipowicz W. (1998) The host gene for intronic U17 small nucleolar RNAs in mammals has no protein-coding potential and is a member of the 5′-terminal oligopyrimidine gene family. *Molecular and Cellular Biology* 18: 4509–18.

86. Rebane A., Tamme R., Laan M., Pata I. and Metspalu A. (1998) A novel snoRNA (U73) is encoded within the introns of the human and mouse ribosomal protein S3a genes. *Gene* 210: 255–63.

87. Tanaka R. et al. (2000) Intronic U50 small-nucleolar-RNA (snoRNA) host gene of no protein-coding potential is mapped at the chromosome breakpoint t(3;6)(q27;q15) of human B-cell lymphoma. *Genes to Cells* 5: 277–87.

88. Cavaille J., Seitz H., Paulsen M., Ferguson-Smith A.C. and Bachellerie J.P. (2002) Identification of tandemly-repeated C/D snoRNA genes at the imprinted human 14q32 domain reminiscent of those at the Prader-Willi/Angelman syndrome region. *Human Molecular Genetics* 11: 1527–38.

89. Zhang X.-O. et al. (2014) Species-specific alternative splicing leads to unique expression of sno-lncRNAs. *BMC Genomics* 15: 287.

90. Smith C.M. and Steitz J.A. (1998) Classification of gas5 as a multi-small-nucleolar-RNA (snoRNA) host gene and a member of the 5′-terminal oligopyrimidine gene family reveals common features of snoRNA host genes. *Molecular and Cellular Biology* 18: 6897–909.

91. Cavaille J. et al. (2000) Identification of brain-specific and imprinted small nucleolar RNA genes exhibiting an unusual genomic organization. *Proceedings of the National Academy of Sciences USA* 97: 14311–6.

92. Cavaille J., Vitali P., Basyuk E., Huttenhofer A. and Bachellerie J.P. (2001) A novel brain-specific box C/D small nucleolar RNA processed from tandemly repeated introns of a noncoding RNA gene in rats. *Journal of Biological Chemistry* 276: 26374–83.

93. Lee S., Walker C.L. and Wevrick R. (2003) Prader-Willi syndrome transcripts are expressed in phenotypically significant regions of the developing mouse brain. *Gene Expression Patterns* 3: 599–609.

94. Kawaji H. et al. (2008) Hidden layers of human small RNAs. *BMC Genomics* 9: 157.

95. Taft R.J. et al. (2009) Small RNAs derived from snoR-NAs. *RNA* 15: 1233–40.

96. Ender C. et al. (2008) A human snoRNA with microRNA-like functions. *Molecular Cell* 32: 519–28.

97. Yin Q.-F. et al. (2012) Long noncoding RNAs with snoRNA ends. *Molecular Cell* 48: 219–30.

98. Xing Y.-H. *et al.* (2017) SLERT regulates DDX21 rings associated with Pol I transcription. *Cell* 169: 664–78.

99. Wu M. et al. (2021) lncRNA SLERT controls phase separation of FC/DFCs to facilitate Pol I transcription. *Science* 373: 547–55.

100. Talross G.J.S., Deryusheva S. and Gall J.G. (2021) Stable lariats bearing a snoRNA (slb-snoRNA) in eukaryotic cells: A level of regulation for guide RNAs. *Proceedings of the National Academy of Sciences USA* 118: e2114156118.

101. Runte M. et al. (2001) The IC-SNURF-SNRPN transcript serves as a host for multiple small nucleolar RNA species and as an antisense RNA for UBE3A. *Human Molecular Genetics* 10: 2687–700.

102. Gallagher R.C., Pils B., Albalwi M. and Francke U. (2002) Evidence for the role of PWCR1/HBII-85 C/D box small nucleolar RNAs in Prader-Willi syndrome. *American Journal of Human Genetics* 71: 669–78.

103. Kishore S. and Stamm S. (2006) The snoRNA HBII-52 regulates alternative splicing of the serotonin receptor 2C. *Science* 311: 230–2.

104. Zieve G. and Penman S. (1976) Small RNA species of the HeLa cell: Metabolism and subcellular localization. *Cell* 8: 19–31.

105. Erikson E., Erikson R.L., Henry B. and Pace N.R. (1973) Comparison of oligonucleotides produced by RNase T1 digestion of 7S RNA from avian and murine oncornaviruses and from uninfected cells. *Virology* 53: 40–6.

106. Walter P. and Blobel G. (1982) Signal recognition particle contains a 7S RNA essential for protein translocation across the endoplasmic reticulum. *Nature* 299: 691–8.

107. Matlin K.S. (2013) History of the signal hypothesis, in *eLS* (John Wiley & Sons). 10.1002/9780470015902.a0025089.

108. Li W.Y., Reddy R., Henning D., Epstein P. and Busch H. (1982) Nucleotide sequence of 7S RNA. Homology to Alu DNA and La 4.5S RNA. *Journal of Biological Chemistry* 257: 5136–42.

109. Batey R.T., Rambo R.P., Lucast L., Rha B. and Doudna J.A. (2000) Crystal structure of the ribonucleoprotein core of the signal recognition particle. *Science* 287: 1232–9.

110. Walter P. and Blobel G. (1983) Disassembly and reconstitution of signal recognition particle. *Cell* 34: 525–33.

111. Larsen N. and Zwieb C. (1991) SRP-RNA sequence alignment and secondary structure. *Nucleic Acids Research* 19: 209–15.

112. Luirink J. and Dobberstein B. (1994) Mammalian and *Escherichia coli* signal recognition particles. *Molecular Microbiology* 11: 9–13.

113. Onafuwa-Nuga A.A., Telesnitsky A. and King S.R. (2006) 7SL RNA, but not the 54-kd signal recognition particle protein, is an abundant component of both infectious HIV-1 and minimal virus-like particles. *RNA* 12: 542–6.

114. Tian C., Wang T., Zhang W. and Yu X.-F. (2007) Virion packaging determinants and reverse transcription of SRP RNA in HIV-1 particles. *Nucleic Acids Research* 35: 7288–302.

115. Wang T. et al. (2007) 7SL RNA mediates virion packaging of the antiviral cytidine deaminase APOBEC3G. *Journal of Virology* 81: 13112–24.

116. Wang T., Tian C., Zhang W., Sarkis P.T.N. and Yu X.-F. (2008) Interaction with 7SL RNA but not with HIV-1 genomic RNA or P Bodies Is required for APOBEC3F virion packaging. *Journal of Molecular Biology* 375: 1098–112.

117. Itano M.S., Arnion H., Wolin S.L. and Simon S.M. (2018) Recruitment of 7SL RNA to assembling HIV-1 virus-like particles. *Traffic* 19: 36–43.

118. Abdelmohsen K. et al. (2014) 7SL RNA represses p53 translation by competing with HuR. *Nucleic Acids Research* 42: 10099–111.

119. Weiner A.M. (1980) An abundant cytoplasmic 7S RNA is complementary to the dominant interspersed middle repetitive DNA-sequence family in the human genome. *Cell* 22: 209–18.

120. Ullu E. and Tschudi C. (1984) Alu sequences are processed 7SL RNA genes. *Nature* 312: 171–2.

121. Quentin Y. (1992) Origin of the Alu family: A family of Alu-like monomers gave birth to the left and the right arms of the Alu elements. *Nucleic Acids Research* 20: 3397–401.

122. Wassarman D.A. and Steitz J.A. (1991) Structural analyses of the 7SK ribonucleoprotein (RNP), the most abundant human small RNP of unknown function. *Molecular Cell Biology* 11: 3432–45.

123. Ringuette M., Liu W.C., Jay E., Yu K.K. and Krause M.O. (1980) Stimulation of transcription of chromatin by specific small nuclear RNAs. *Gene* 8: 211–24.

124. Sohn U., Szyszko J., Coombs D. and Krause M. (1983) 7S-K nuclear RNA from simian virus 40-transformed cells has sequence homology to the viral early promoter. *Proceedings of the National Academy of Sciences USA* 80: 7090–4.

125. Blencowe B.J. (2002) Transcription: Surprising role for an elusive small nuclear RNA. *Current Biology* 12: R147–9.

126. Yang Z., Zhu Q., Luo K. and Zhou Q. (2001) The 7SK small nuclear RNA inhibits the CDK9/cyclin T1 kinase to control transcription. *Nature* 414: 317–22.

127. Nguyen V.T., Kiss T., Michels A.A. and Bensaude O. (2001) 7SK small nuclear RNA binds to and inhibits the activity of CDK9/cyclin T complexes. *Nature* 414: 322–5.

128. Kohoutek J. (2009) P-TEFb- the final frontier. *Cell Division* 4: 19.

129. Lenasi T. and Barboric M. (2010) P-TEFb stimulates transcription elongation and pre-mRNA splicing through multilateral mechanisms. *RNA Biology* 7: 145–50.

130. Peterlin B.M., Brogie J.E. and Price D.H. (2012) 7SK snRNA: A noncoding RNA that plays a major role in regulating eukaryotic transcription. *Wiley Interdisciplinary Reviews RNA* 3: 92–103.

131. Castelo-Branco G. et al. (2013) The non-coding snRNA 7SK controls transcriptional termination, poising, and bidirectionality in embryonic stem cells. *Genome Biology* 14: R98.

132. Quaresma A.J.C., Bugai A. and Barboric M. (2016) Cracking the control of RNA polymerase II elongation by 7SK snRNP and P-TEFb. *Nucleic Acids Research* 44: 7527–39.

133. Flynn R.A. et al. (2016) 7SK-BAF axis controls pervasive transcription at enhancers. *Nature Structural & Molecular Biology* 23: 231–8.

134. Egloff S., Studniarek C. and Kiss T. (2018) 7SK small nuclear RNA, a multifunctional transcriptional regulatory RNA with gene-specific features. *Transcription* 9: 95–101.

135. Gruber A.R. et al. (2008) Invertebrate 7SK snRNAs. *Journal of Molecular Evolution* 66: 107–15.

136. Yazbeck A.M., Tout K.R. and Stadler P.F. (2018) Detailed secondary structure models of invertebrate 7SK RNAs. *RNA Biology* 15: 158–64.

137. Miller J., McLachlan A.D. and Klug A. (1985) Repetitive zinc-binding domains in the protein transcription factor IIIA from Xenopus oocytes. *EMBO Journal* 4: 1609–14.

138. Klug A. (2010) The discovery of Zinc Fingers and their applications in gene regulation and genome manipulation. *Annual Review of Biochemistry* 79: 213–31.

139. Hu S., Wang X. and Shan G. (2016) Insertion of an Alu element in a lncRNA leads to primate-specific modulation of alternative splicing. *Nature Structural & Molecular Biology* 23: 1011–9.

140. Pelham H.R. and Brown D.D. (1980) A specific transcription factor that can bind either the 5S RNA gene or 5S RNA. *Proceedings of the National Academy of Sciences USA* 77: 4170–4.

141. Honda B.M. and Roeder R.G. (1980) Association of a 5S gene transcription factor with 5S RNA and altered levels of the factor during cell differentiation. *Cell* 22: 119–26.

142. Hendrick J.P., Wolin S.L., Rinke J., Lerner M.R. and Steitz J.A. (1981) Ro small cytoplasmic ribonucleoproteins are a subclass of La ribonucleoproteins: Further characterization of the Ro and La small ribonucleoproteins from uninfected mammalian cells. *Molecular Cell Biology* 1: 1138–49.

143. Maraia R.J., Sasaki-Tozawa N., Driscoll C.T., Green E.D. and Darlington G.J. (1994) The human Y4 small cytoplasmic RNA gene is controlled by upstream elements and resides on chromosome 7 with all other hY scRNA genes. *Nucleic Acids Research* 22: 3045–52.

144. Stein A.J., Fuchs G., Fu C., Wolin S.L. and Reinisch K.M. (2005) Structural insights into RNA quality control: The Ro autoantigen binds misfolded RNAS via its central cavity. *Cell* 121: 529–39.

145. Perreault J., Perreault J.P. and Boire G. (2007) Ro-associated Y RNAs in metazoans: Evolution and diversification. *Molecular Biology and Evolution* 24: 1678–89.

146. Krude T. (2010) Non-coding RNAs: New players in the field of eukaryotic DNA replication, in H.P. Nasheuer (ed.) *Genome Stability and Human Diseases. Subcellular Biochemistry*, vol. 50, pp. 105–18 (Springer, Dordrecht).

147. Chen X. et al. (2003) The Ro autoantigen binds misfolded U2 small nuclear RNAs and assists mammalian cell survival after UV Irradiation. *Current Biology* 13: 2206–11.

148. Fuchs G., Stein A.J., Fu C., Reinisch K.M. and Wolin S.L. (2006) Structural and biochemical basis for misfolded RNA recognition by the Ro autoantigen. *Nature Structural & Molecular Biology* 13: 1002–9.

149. Sim S. et al. (2009) The subcellular distribution of an RNA quality control protein, the Ro autoantigen, is regulated by noncoding Y RNA binding. *Molecular Biology of the Cell* 20: 1555–64.

150. Chen X. et al. (2007) An ortholog of the Ro autoantigen functions in 23S rRNA maturation in *D. radiodurans*. *Genes & Development* 21: 1328–39.

151. Wurtmann E.J. and Wolin S.L. (2010) A role for a bacterial ortholog of the Ro autoantigen in starvation-induced rRNA degradation. *Proceedings of the National Academy of Sciences USA* 107: 4022–7.

152. Chen X. et al. (2013) An RNA degradation machine sculpted by Ro autoantigen and noncoding RNA. *Cell* 153: 166–77.

153. Sim S., Wolin S.L., Storz G. and Papenfort K. (2018) Bacterial Y RNAs: Gates, tethers, and tRNA mimics. *Microbiology Spectrum* 6. doi: 10.1128/microbiolspec. RWR-0023-2018.

154. Kedersha N.L. and Rome L.H. (1986) Isolation and characterization of a novel ribonucleoprotein particle: Large structures contain a single species of small RNA. *Journal of Cell Biology* 103: 699–709.

155. Kedersha N.L., Miquel M.C., Bittner D. and Rome L.H. (1990) Vaults. II. Ribonucleoprotein structures are highly conserved among higher and lower eukaryotes. *Journal of Cell Biology* 110: 895–901.

156. van Zon A., Mossink M.H., Scheper R.J., Sonneveld P. and Wiemer E.A. (2003) The vault complex. *Cellular and Molecular Life Sciences* 60: 1828–37.

157. Horos R. et al. (2019) The small non-coding Vault RNA1-1 acts as a riboregulator of autophagy. *Cell* 176: 1054–67.

158. Mizushima N. (2007) Autophagy: Process and function. *Genes & Development* 21: 2861–73.

159. Amort M. et al. (2015) Expression of the vault RNA protects cells from undergoing apoptosis. *Nature Communications* 6: 7030.

160. Bracher L. et al. (2020) Human vtRNA1-1 levels modulate signaling pathways and regulate apoptosis in human cancer cells. *Biomolecules* 10: 614.

161. Wakatsuki S. et al. (2021) Small noncoding vault RNA modulates synapse formation by amplifying MAPK signaling. *Journal of Cell Biology* 220: e201911078.

162. Nallagatla S.R., Toroney R. and Bevilacqua P.C. (2011) Regulation of innate immunity through RNA structure and the protein kinase PKR. *Current Opinion in Structural Biology* 21: 119–27.

163. Reich P.R., Forget B.G. and Weissman S.M. (1966) RNA of low molecular weight in KB cells infected with adenovirus type 2. *Journal of Molecular Biology* 17: 428–39.

164. Ohe K. and Weissman S.M. (1971) The nucleotide sequence of a low molecular weight ribonucleic acid from cells infected with adenovirus 2. *Journal of Biological Chemistry* 246: 6991–7009.

165. Mathews M.B. (1975) Genes for VA-RNA in adenovirus 2. *Cell* 6: 223–9.

166. Akusjarvi G., Mathews M.B., Andersson P., Vennstrom B. and Pettersson U. (1980) Structure of genes for virus-associated RNAI and RNAII of adenovirus type 2. *Proceedings of the National Academy of Sciences USA* 77: 2424–8.

167. Thimmappaya B., Weinberger C., Schneider R.J. and Shenk T. (1982) Adenovirus VAI RNA is required for efficient translation of viral mRNAs at late times after infection. *Cell* 31: 543–51.

168. Mathews M.B. and Shenk T. (1991) Adenovirus virus-associated RNA and translation control. *Journal of Virology* 65: 5657–62.

169. Hood I.V. et al. (2019) Crystal structure of an adenovirus virus-associated RNA. *Nature Communications* 10: 2871.

170. Krayev A.S. et al. (1982) Ubiquitous transposon-like repeats B1 and B2 of the mouse genome: B2 sequencing. *Nucleic Acids Research* 10: 7461–75.

171. Kramerov D.A., Grigoryan A.A., Ryskov A.P. and Georgiev G.P. (1979) Long double-stranded sequences (dsRNA-B) of nuclear pre-mRNA consist of a few highly abundant classes of sequences: Evidence from DNA cloning experiments. *Nucleic Acids Research* 6: 697–713.

172. Fornace Jr. A.J., and Mitchell J.B. (1986) Induction of B2 RNA polymerase III transcription by heat shock: Enrichment for heat shock induced sequences in rodent cells by hybridization subtraction. *Nucleic Acids Research* 14: 5793–811.

173. Sutcliffe J.G., Milner R.J., Gottesfeld J.M. and Lerner R.A. (1984) Identifier sequences are transcribed specifically in brain. *Nature* 308: 237–41.

174. DeChiara T.M. and Brosius J. (1987) Neural BC1 RNA: cDNA clones reveal nonrepetitive sequence content. *Proceedings of the National Academy of Sciences USA* 84: 2624–8.

175. Tiedge H., Fremeau Jr. R.T., Weinstock P.H., Arancio O. and Brosius J. (1991) Dendritic location of neural BC1 RNA. *Proceedings of the National Academy of Sciences USA* 88: 2093–7.

176. Tiedge H., Chen W. and Brosius J. (1993) Primary structure, neural-specific expression, and dendritic location of human BC200 RNA. *Journal of Neuroscience* 13: 2382–90.

177. Volff J.N. and Brosius J. (2007) Modern genomes with retro-look: Retrotransposed elements, retroposition and the origin of new genes. *Gene and Protein Evolution* 3: 175–90.

178. Lewejohann L. et al. (2004) Role of a neuronal small non-messenger RNA: Behavioural alterations in BC1 RNA-deleted mice. *Behavioural Brain Research* 154: 273–89.

179. Brutlag D., Schekman R. and Kornberg A. (1971) A possible role for RNA polymerase in the initiation of M13 DNA synthesis. *Proceedings of the National Academy of Sciences USA* 68: 2826–9.

180. Sugino A., Hirose S. and Okazaki R. (1972) RNA-linked nascent DNA fragments in Escherichia coli. *Proceedings of the National Academy of Sciences USA* 69: 1863–7.

181. Zakian V.A. (2012) Telomeres: The beginnings and ends of eukaryotic chromosomes. *Experimental Cell Research* 318: 1456–60.

182. Yu G.L., Bradley J.D., Attardi L.D. and Blackburn E.H. (1990) In vivo alteration of telomere sequences and senescence caused by mutated *Tetrahymena* telomerase RNAs. *Nature* 344: 126–32.

183. Greider C.W. and Blackburn E.H. (1985) Identification of a specific telomere terminal transferase activity in Tetrahymena extracts. *Cell* 43: 405–13.

184. Nakamura T.M. and Cech T.R. (1998) Reversing time: Origin of telomerase. *Cell* 92: 587–90.

185. Curcio M.J. and Belfort M. (2007) The beginning of the end: Links between ancient retroelements and modern telomerases. *Proceedings of the National Academy of Sciences USA* 104: 9107–8.

186. Belfort M., Curcio M.J. and Lue N.F. (2011) Telomerase and retrotransposons: Reverse transcriptases that shaped genomes. *Proceedings of the National Academy of Sciences USA* 108: 20304–10.

187. McEachern M.J., Krauskopf A. and Blackburn E.H. (2000) Telomeres and their control. *Annual Review of Genetics* 34: 331–58.

188. Qi X. et al. (2012) RNA/DNA hybrid binding affinity determines telomerase template-translocation efficiency. *EMBO Journal* 31: 150–61.

189. Logeswaran D., Li Y., Podlevsky J.D. and Chen J.J.L. (2021) Monophyletic origin and divergent evolution of animal telomerase RNA. *Molecular Biology and Evolution* 38: 215–28.

190. Washietl S., Hofacker I.L. and Stadler P.F. (2005) Fast and reliable prediction of noncoding RNAs. *Proceedings of the National Academy of Sciences USA* 102: 2454–9.

191. Miao Z. et al. (2015) RNA-Puzzles Round II: Assessment of RNA structure prediction programs applied to three large RNA structures. *RNA* 21: 1066–84.

192. Backofen R. et al. (2017) RNA-bioinformatics: Tools, services and databases for the analysis of RNA-based regulation. *Journal of Biotechnology* 261: 76–84.

193. Smith M.A., Seemann S.E., Quek X.C. and Mattick J.S. (2017) DotAligner: Identification and clustering of RNA structure motifs. *Genome Biology* 18: 244.

194. Griffiths-Jones S. et al. (2005) Rfam: Annotating noncoding RNAs in complete genomes. *Nucleic Acids Research* 33: D121–4.

195. Kalvari I. et al. (2018) Rfam 13.0: Shifting to a genome-centric resource for non-coding RNA families. *Nucleic Acids Research* 46: D335–42.

196. The RNAcentral Consortium (2018) RNAcentral: A hub of information for non-coding RNA sequences. *Nucleic Acids Research* 47: D221–9.

197. Crick F.H.C. (1966) Genetic code - yesterday today and tomorrow. *Cold Spring Harbor Symposia on Quantitative Biology* 31: 3–9.

198. Falk R. and Lazcano A. (2012) The forgotten dispute: A.I. Oparin and H.J. Muller on the origin of life. *Hist Philos Life Sci* 34: 373–90.

199. Lazcano A. (2012) The biochemical roots of the RNA world: From zymonucleic acid to ribozymes. *History and Philosophy of the Life Sciences* 34: 407–23.

200. Rich A. (2004) The excitement of discovery. *Annual Review of Biochemistry* 73: 1–37.

201. Orgel L.E. (1968) Evolution of the genetic apparatus. *Journal of Molecular Biology* 38: 381–93.

202. Woese C.R. (1970) The problem of evolving a genetic code. *Bioscience* 20: 471–85.

203. Orgel L.E. (2004) Prebiotic chemistry and the origin of the RNA world. *Critical Reviews in Biochemistry and Molecular Biology* 39: 99–123.

204. Kruger K. et al. (1982) Self-splicing RNA: Autoexcision and autocyclization of the ribosomal RNA intervening sequence of *Tetrahymena*. *Cell* 31: 147–57.

205. Michel F. and Dujon B. (1983) Conservation of RNA secondary structures in two intron families including mitochondrial-, chloroplast- and nuclear-encoded members. *EMBO Journal* 2: 33–8.

206. Michel F. and Ferat J.-L. (1995) Structure and activities of group II introns. *Annual Review of Biochemistry* 64: 435–61.

207. Lambowitz A.M. and Zimmerly S. (2004) Mobile group II introns. *Annual Review of Genetics* 38: 1–35.

208. Bonen L. and Vogel J. (2001) The ins and outs of group II introns. *Trends in Genetics* 17: 322–31.

209. Sharp P.A. (1985) On the origin of RNA splicing and introns. *Cell* 42: 397–400.

210. Cavalier-Smith T. (1991) Intron phylogeny: A new hypothesis. *Trends in Genetics* 7: 145–8.

211. Fica S.M., Mefford M.A., Piccirilli J.A. and Staley J.P. (2014) Evidence for a group II intron–like catalytic triplex in the spliceosome. *Nature Structural & Molecular Biology* 21: 464–71.

212. Galej W.P. et al. (2016) Cryo-EM structure of the spliceosome immediately after branching. *Nature* 537: 197–201.

213. Smathers C.M. and Robart A.R. (2019) The mechanism of splicing as told by group II introns: Ancestors of the spliceosome. *Biochimica et Biophysica Acta* 1862: 194390.

214. Ferat J.L. and Michel F. (1993) Group II self-splicing introns in bacteria. *Nature* 364: 358–61.

215. Irimia M. and Roy S.W. (2014) Origin of spliceosomal introns and alternative splicing. *Cold Spring Harbor Perspectives in Biology* 6: a016071.

216. Guerrier-Takada C., Gardiner K., Marsh T., Pace N. and Altman S. (1983) The RNA moiety of ribonuclease P is the catalytic subunit of the enzyme. *Cell* 35: 849–57.

217. Walker S.C. and Engelke D.R. (2006) Ribonuclease P: The evolution of an ancient RNA enzyme. *Critical Reviews in Biochemistry and Molecular Biology* 41: 77–102.

218. Reddy R. et al. (1981) Characterization and subcellular localization of 7–8S RNAs of Novikoff hepatoma. *Journal of Biological Chemistry* 256: 8452–7.

219. Lan P. et al. (2020) Structural insight into precursor ribosomal RNA processing by ribonuclease MRP. *Science* 369: 656–63.

220. Chang D.D. and Clayton D.A. (1989) Mouse RNAase MRP RNA is encoded by a nuclear gene and contains a decamer sequence complementary to a conserved region of mitochondrial RNA substrate. *Cell* 56: 131–9.

221. Martin A.N. and Li Y. (2007) RNase MRP RNA and human genetic diseases. *Cell Research* 17: 219–26.

222. Takagi Y., Warashina M., Stec W.J., Yoshinari K. and Taira K. (2001) Recent advances in the elucidation of the mechanisms of action of ribozymes. *Nucleic Acids Research* 29: 1815–34.

223. Gilbert W. (1986) The RNA World. *Nature* 319: 618.

224. Tjhung K.F., Shokhirev M.N., Horning D.P. and Joyce G.F. (2020) An RNA polymerase ribozyme that synthesizes its own ancestor. *Proceedings of the National Academy of Sciences USA* 117: 2906–13.

225. Becker S. et al. (2019) Unified prebiotically plausible synthesis of pyrimidine and purine RNA ribonucleotides. *Science* 366: 76–82.

226. Lazcano A. (2015) The RNA World and the origin of life: A short history of a tidy evolutionary narrative. *BIO Web of Conferences* 4: 00013.

227. Fedor M.J. and Williamson J.R. (2005) The catalytic diversity of RNAs. *Nature Reviews Molecular Cell Biology* 6: 399–412.

228. Salehi-Ashtiani K., Lupták A., Litovchick A. and Szostak J.W. (2006) A genomewide search for ribozymes reveals an HDV-like sequence in the human *CPEB3* gene. *Science* 313: 1788–92.

229. Webb C.-H.T., Riccitelli N.J., Ruminski D.J. and Lupták A. (2009) Widespread occurrence of self-cleaving ribozymes. *Science* 326: 953.

230. de la Peña M. and García-Robles I. (2010) Intronic hammerhead ribozymes are ultraconserved in the human genome. *EMBO Reports* 11: 711–6.

231. De la Peña M., García-Robles I. and Cervera A. (2017) The hammerhead ribozyme: A long history for a short RNA. *Molecules* 22: 78.

232. Perreault J. et al. (2011) Identification of hammerhead ribozymes in all domains of life reveals novel structural variations. *PLOS Computational Biology* 7: e1002031.

233. Hernandez A.J. et al. (2020) B2 and ALU retrotransposons are self-cleaving ribozymes whose activity is enhanced by EZH2. *Proceedings of the National Academy of Sciences USA* 117: 415–25.

234. Chen Y. et al. (2021) Hovlinc is a recently evolved class of ribozyme found in human lncRNA. *Nature Chemical Biology* 17: 601–7.

235. Scheitl C.P.M., Ghaem Maghami M., Lenz A.-K. and Höbartner C. (2020) Site-specific RNA methylation by a methyltransferase ribozyme. *Nature* 587: 663–7.

236. Wolk S.K. et al. (2020) Modified nucleotides may have enhanced early RNA catalysis. *Proceedings of the National Academy of Sciences USA* 117: 8236–42.

237. Noller H.F., Hoffarth V. and Zimniak L. (1992) Unusual resistance of peptidyl transferase to protein extraction procedures. *Science* 256: 1416–9.

238. Steitz T.A. and Moore P.B. (2003) RNA, the first macromolecular catalyst: The ribosome is a ribozyme. *Trends in Biochemical Sciences* 28: 411–8.

239. Lambowitz A.M. and Belfort M. (1993) Introns as mobile genetic elements. *Annual Review of Biochemistry* 62: 587–622.

240. Wallis M.G. and Schroeder R. (1997) The binding of antibiotics to RNA. *Progress in Biophysics and Molecular Biology* 67: 141–54.

241. Gallego J. and Varani G. (2001) Targeting RNA with small-molecule drugs: therapeutic promise and chemical challenges. *Accounts of Chemical Research* 34: 836–43.

242. Schneider-Poetsch T., Chhipi-Shrestha J.K. and Yoshida M. (2021) Splicing modulators: On the way from nature to clinic. *The Journal of Antibiotics* 74: 603–16.

243. Endo Y., Mitsui K., Motizuki M. and Tsurugi K. (1987) The mechanism of action of ricin and related toxic lectins on eukaryotic ribosomes. The site and the characteristics of the modification in 28S ribosomal RNA caused by the toxins. *Journal of Biological Chemistry* 262: 5908–12.

244. Westhof E. and Fritsch V. (2000) RNA folding: Beyond Watson–Crick pairs. *Structure* 8: R55–65.

245. Westhof E., Masquida B. and Jossinet F. (2011) Predicting and modeling RNA architecture. *Cold Spring Harbor Perspectives in Biology* 3: a003632.

246. Westhof E. and Fritsch V. (2011) The endless subtleties of RNA-protein complexes. *Structure* 19: 902–3.

247. Crossley M.P., Bocek M. and Cimprich K.A. (2019) R-loops as cellular regulators and genomic threats. *Molecular Cell* 73: 398–411.

248. Cetin N.S. et al. (2019) Isolation and genome-wide characterization of cellular DNA:RNA triplex structures. *Nucleic Acids Research* 47: 2306–21.

249. Tuerk C. and Gold L. (1990) Systematic evolution of ligands by exponential enrichment: RNA ligands to bacteriophage T4 DNA polymerase. *Science* 249: 505–10.

250. Ellington A.D. and Szostak J.W. (1990) In vitro selection of RNA molecules that bind specific ligands. *Nature* 346: 818–22.

251. Muotri A.R., da Veiga Pereira L., dos Reis Vasques L. and Menck C.F.M. (1999) Ribozymes and the anti-gene therapy: How a catalytic RNA can be used to inhibit gene function. *Gene* 237: 303–10.

252. Gawande B.N. et al. (2017) Selection of DNA aptamers with two modified bases. *Proceedings of the National Academy of Sciences USA* 114: 2898–903.

253. Gold L., Singer B., He Y.Y. and Brody E. (1997) SELEX and the evolution of genomes. *Current Opinion in Genetics and Development* 7: 848–51.

254. Diener T.O. (1971) Potato spindle tuber "virus". IV. A replicating, low molecular weight RNA. *Virology* 45: 411–28.

255. Sanger H.L., Klotz G., Riesner D., Gross H.J. and Kleinschmidt A.K. (1976) Viroids are single-stranded covalently closed circular RNA molecules existing as highly base-paired rod-like structures. *Proceedings of the National Academy of Sciences USA* 73: 3852–6.

256. Diener T.O. (2016) Viroids: "living fossils" of primordial RNAs? *Biology Direct* 11: 15.

257. Mills D.R., Peterson R.L. and Spiegelman S. (1967) An extracellular Darwinian experiment with a self-duplicating nucleic acid molecule. *Proceedings of the National Academy of Sciences USA* 58: 217–24.

258. Kaper J.M. and Waterworth H.E. (1977) Cucumber mosaic virus associated RNA 5: Causal agent for tomato necrosis. *Science* 196: 429–31.

259. Hillman B.I., Carrington J.C. and Morris T.J. (1987) A defective interfering RNA that contains a mosaic of a plant virus genome. *Cell* 51: 427–33.

260. Simon A.E., Roossinck M.J. and Havelda Z. (2004) Plant virus satellite and defective interfering RNAs: New paradigms for a new century. *Annual Review of Phytopathology* 42: 415–37.

261. Joyce G.F. (1991) The rise and fall of the RNA world. *New Biologist* 3: 399–407.

262. Penman S. (1991) If genes just make proteins and our proteins are the same, then why are we so different? *Journal of Cellular Biochemistry* 47: 95–8.

CHAPTER 9

1. Marx J.L. (1984) New ways to "mutate" genes. *Science* 225: 819.

2. Zamecnik P.C. and Stephenson M.L. (1978) Inhibition of Rous sarcoma virus replication and cell transformation by a specific oligodeoxynucleotide. *Proceedings of the National Academy of Sciences USA* 75: 280–4.

3. Zamecnik P.C., Goodchild J., Taguchi Y. and Sarin P.S. (1986) Inhibition of replication and expression of human T-cell lymphotropic virus type III in cultured cells by exogenous synthetic oligonucleotides complementary to viral RNA. *Proceedings of the National Academy of Sciences USA* 83: 4143–6.

4. Hindley J. (1967) Fractionation of 32P-labelled ribonucleic acids on polyacrylamide gels and their characterization by fingerprinting. *Journal of Molecular Biology* 30: 125–36.

5. Wassarman K.M. and Storz G. (2000) 6S RNA regulates *E. coli* RNA polymerase activity. *Cell* 101: 613–23.

6. Trotochaud A.E. and Wassarman K.M. (2005) A highly conserved 6S RNA structure is required for regulation of transcription. *Nature Structural & Molecular Biology* 12: 313–9.

7. Steuten B. et al. (2014) Regulation of transcription by 6S RNAs. *RNA Biology* 11: 508–21.

8. Schleich T. and Goldstein J. (1964) Gel filtration properties of ccd-prepared *E. coli* B sRNA. *Proceedings of the National Academy of Sciences USA* 52: 744–9.

9. Goldstein J. and Harewood K. (1969) Another species of ribonucleic acid in *Escherichia coli*. *Journal of Molecular Biology* 39: 383–7.

10. Spiegelman W.G. et al. (1972) Bidirectional transcription and the regulation of phage lambda repressor synthesis. *Proceedings of the National Academy of Sciences USA* 69: 3156–60.

11. Shaw D.C. et al. (1978) Gene K, a new overlapping gene in bacteriophage G4. *Nature* 272: 510–5.

12. Sanger F. et al. (1977) Nucleotide sequence of bacteriophage phi X174 DNA. *Nature* 265: 687–95.

13. Fiers W. et al. (1978) Complete nucleotide sequence of SV40 DNA. *Nature* 273: 113–20.

14. Contreras R., Rogiers R., Van de Voorde A. and Fiers W. (1977) Overlapping of the *VP2-VP3* gene and the *VP1* gene in the SV40 genome. *Cell* 12: 529–38.

15. Oh B.-K. and Apirion D. (1991) 10Sa RNA, a small stable RNA of *Escherichia coli*, is functional. *Molecular and General Genetics* 229: 52–6.

16. Komine Y., Kitabatake M., Yokogawa T., Nishikawa K. and Inokuchi H. (1994) A tRNA-like structure is present in 10Sa RNA, a small stable RNA from Escherichia coli. *Proceedings of the National Academy of Sciences USA* 91: 9223–7.

17. Muto A. et al. (1996) Structure and function of 10Sa RNA: Trans-translation system. *Biochimie* 78: 985–91.

18. Moriano-Gutierrez S. et al. (2020) The noncoding small RNA SsrA is released by *Vibrio fischeri* and modulates critical host responses. *PLOS Biology* 18: e3000934.

19. Walker S.C. and Engelke D.R. (2006) Ribonuclease P: evolution of an ancient RNA enzyme. *Critical Reviews in Biochemistry and Molecular Biology* 41: 77–102.

20. Ray B.K. and Apirion D. (1979) Characterization of 10S RNA: A new stable RNA molecule from *Escherichia coli*. *Molecular and General Genetics* 174: 25–32.

21. Jain S.K., Gurevitz M. and Apirion D. (1982) A small RNA that complements mutants in the RNA processing enzyme ribonuclease P. *Journal of Molecular Biology* 162: 515–33.

22. Gralla J., Steitz J.A. and Crothers D.M. (1974) Direct physical evidence for secondary structure in an isolated fragment of R17 bacteriophage mRNA. *Nature* 248: 204–8.

23. Lemaire G., Gold L. and Yarus M. (1978) Autogenous translational repression of bacteriophage T4 gene *32* expression in *vitro*. *Journal of Molecular Biology* 126: 73–90.

24. Fallon A.M., Jinks C.S., Strycharz G.D. and Nomura M. (1979) Regulation of ribosomal protein synthesis in *Escherichia coli* by selective mRNA inactivation. *Proceedings of the National Academy of Sciences USA* 76: 3411–5.

25. Olins P.O. and Nomura M. (1981) Translational regulation by ribosomal protein S8 in *Escherichia coli*: Structural homology between rRNA binding site and feedback target on mRNA. *Nucleic Acids Research* 9: 1757–64.

26. Meyer B.J. (2019) rRNA mimicry in RNA regulation of gene expression, in G. Storz and K. Papenfort (ed.) *Regulating with RNA in Bacteria and Archaea* (ASM Press, Washington, DC).

27. Bester A.J., Kennedy D.S. and Heywood S.M. (1975) Two classes of translational control RNA: Their role in the regulation of protein synthesis. *Proceedings of the National Academy of Sciences USA* 72: 1523–7.

28. Heywood S.M., Kennedy D.S. and Bester A.J. (1975) Studies concerning the mechanism by which translational-control RNA regulates protein synthesis in embryonic muscle. *European Journal of Biochemistry* 58: 587–93.

29. McCarthy T.L., Siegel E., Mroczkowski B. and Heywood S.M. (1983) Characterization of translational-control ribonucleic acid isolated from embryonic chick muscle. *Biochemistry* 22: 935–41.

30. Heywood S.M. (1986) tcRNA as a naturally occurring antisense RNA in eukaryotes. *Nucleic Acids Research* 14: 6771–2.

31. Tomizawa J. and Itoh T. (1981) Plasmid ColE1 incompatibility determined by interaction of RNA I with primer transcript. *Proceedings of the National Academy of Sciences USA* 78: 6096–100.

32. Stougaard P., Molin S. and Nordstrom K. (1981) RNAs involved in copy-number control and incompatibility of plasmid R1. *Proceedings of the National Academy of Sciences USA* 78: 6008–12.

33. Rosen J., Ryder T., Ohtsubo H. and Ohtsubo E. (1981) Role of RNA transcripts in replication incompatibility and copy number control in antibiotic resistance plasmid derivatives. *Nature* 290: 794–7.

34. Simons R.W. and Kleckner N. (1983) Translational control of IS10 transposition. *Cell* 34: 683–91.

35. Mizuno T., Chou M.Y. and Inouye M. (1984) A unique mechanism regulating gene expression: Translational inhibition by a complementary RNA transcript (micRNA). *Proceedings of the National Academy of Sciences USA* 81: 1966–70.

36. Guo P.X., Erickson S. and Anderson D. (1987) A small viral RNA is required for in vitro packaging of bacteriophage phi 29 DNA. *Science* 236: 690–4.

37. Simons R.W. and Kleckner N. (1988) Biological regulation by antisense RNA in prokaryotes. *Annual Review of Genetics* 22: 567–600.

38. Inouye M. and Delihast N. (1988) Small RNAs in the prokaryotes: A growing list of diverse roles. *Cell* 53: 5–7.

39. Papenfort K. and Vogel J. (2010) Regulatory RNA in bacterial pathogens. *Cell Host & Microbe* 8: 116–27.

40. Novick R.P. et al. (1993) Synthesis of staphylococcal virulence factors is controlled by a regulatory RNA molecule. *EMBO journal* 12: 3967–75.

41. Chevalier C. et al. (2010) *Staphylococcus aureus* RNAIII binds to two distant regions of coa mRNA to arrest translation and promote mRNA degradation. *PLOS Pathogens* 6: e1000809.

42. Wassarman K.M., Zhang A. and Storz G. (1999) Small RNAs in *Escherichia coli*. *Trends in Microbiology* 7: 37–45.

43. Waters L.S. and Storz G. (2009) Regulatory RNAs in bacteria. *Cell* 136: 615–28.

44. Livny J. and Waldor M.K. (2007) Identification of small RNAs in diverse bacterial species. *Current Opinion in Microbiology* 10: 96–101.

45. Sharma C.M. and Vogel J. (2009) Experimental approaches for the discovery and characterization of regulatory small RNA. *Current Opinion in Microbiology* 12: 536–46.

46. Li L. et al. (2012) BSRD: A repository for bacterial small regulatory RNA. *Nucleic Acids Research* 41: D233–8.

47. Tree J.J., Granneman S., McAteer S.P., Tollervey D. and Gally D.L. (2014) Identification of bacteriophage-encoded anti-sRNAs in pathogenic *Escherichia coli*. *Molecular Cell* 55: 199–213.

48. Melamed S. et al. (2016) Global mapping of small RNA-target interactions in bacteria. *Molecular Cell* 63: 884–97.

49. Waters S.A. et al. (2017) Small RNA interactome of pathogenic *E. coli* revealed through crosslinking of RNase E. *EMBO Journal* 36: 374–87.

50. Hör J. and Vogel J. (2017) Global snapshots of bacterial RNA networks. *EMBO Journal* 36: 245–7.

51. Sledjeski D.D., Gupta A. and Gottesman S. (1996) The small RNA, DsrA, is essential for the low temperature expression of RpoS during exponential growth in Escherichia coli. *EMBO Journal* 15: 3993–4000.

52. Sledjeski D. and Gottesman S. (1995) A small RNA acts as an antisilencer of the H-NS-silenced *rcsA* gene of *Escherichia coli*. *Proceedings of the National Academy of Sciences USA* 92: 2003–7.

53. Majdalani N., Cunning C., Sledjeski D., Elliott T. and Gottesman S. (1998) DsrA RNA regulates translation of RpoS message by an anti-antisense mechanism, independent of its action as an antisilencer of transcription. *Proceedings of the National Academy of Sciences USA* 95: 12462–7.

54. Altuvia S., WeinsteinFischer D., Zhang A.X., Postow L. and Storz G. (1997) A small, stable RNA induced by oxidative stress: Role as a pleiotropic regulator and antimutator. *Cell* 90: 43–53.

55. Zhang A. et al. (1998) The OxyS regulatory RNA represses rpoS translation and binds the Hfq (HF-I) protein. *EMBO Journal* 17: 6061–8.

56. Franze de Fernandez M.T., Eoyang L. and August J.T. (1968) Factor fraction required for the synthesis of bacteriophage Qβ-RNA. *Nature* 219: 588–90.

57. Møller T. et al. (2002) Hfq: A bacterial Sm-like protein that mediates RNA-RNA interaction. *Molecular Cell* 9: 23–30.

58. Gottesman S. (2004) The small RNA regulators of *Escherichia coli*: Roles and mechanisms. *Annual Review of Microbiology* 58: 303–28.

59. Majdalani N., Vanderpool C.K. and Gottesman S. (2005) Bacterial small RNA regulators. *Critical Reviews in Biochemistry and Molecular Biology* 40: 93–113.

60. Wassarman K.M., Repoila F., Rosenow C., Storz G. and Gottesman S. (2001) Identification of novel small RNAs using comparative genomics and microarrays. *Genes & Development* 15: 1637–51.

61. Zhang A., Wassarman K.M., Ortega J., Steven A.C. and Storz G. (2002) The Sm-like Hfq protein increases OxyS RNA interaction with target mRNAs. *Molecular Cell* 9: 11–22.

62. Geissmann T.A. and Touati D. (2004) Hfq, a new chaperoning role: Binding to messenger RNA determines access for small RNA regulator. *EMBO Journal* 23: 396–405.

63. Lenz D.H. *et al.* (2004) The small RNA chaperone Hfq and multiple small RNAs control quorum sensing in *Vibrio harveyi* and *Vibrio cholerae*. *Cell* 118: 69–82.

64. Chao Y. and Vogel J. (2010) The role of Hfq in bacterial pathogens. *Current Opinion in Microbiology* 13: 24–33.

65. Fuchs M. et al. (2021) An RNA-centric global view of *Clostridioides difficile* reveals broad activity of Hfq in a clinically important gram-positive bacterium. *Proceedings of the National Academy of Sciences USA* 118: e2103579118.

66. Lucchetti-Miganeh C., Burrowes E., Baysse C. and Ermel G. (2008) The post-transcriptional regulator CsrA plays a central role in the adaptation of bacterial pathogens to different stages of infection in animal hosts. *Microbiology* 154: 16–29.

67. Sy B.M., Lan R. and Tree J.J. (2020) Early termination of the Shiga toxin transcript generates a regulatory small RNA. *Proceedings of the National Academy of Sciences USA* 117: 25055–65.

68. Zhao E.M. et al. (2021) RNA-responsive elements for eukaryotic translational control. *Nature Biotechnology* epub ahead of print: https://doi.org/10.1038/s41587-021-01068-2.

69. Huang F., Yang Z. and Yarus M. (1998) RNA enzymes with two small-molecule substrates. *Chemistry & Biology* 5: 669–78.

70. Winkler W.C., Nahvi A., Roth A., Collins J.A. and Breaker R.R. (2004) Control of gene expression by a natural metabolite-responsive ribozyme. *Nature* 428: 281–6.

71. Winkler W., Nahvi A. and Breaker R.R. (2002) Thiamine derivatives bind messenger RNAs directly to regulate bacterial gene expression. *Nature* 419: 952–6.

72. Mironov A.S. et al. (2002) Sensing small molecules by nascent RNA: A mechanism to control transcription in bacteria. *Cell* 111: 747–56.

73. Winkler W.C., Nahvi A., Sudarsan N., Barrick J.E. and Breaker R.R. (2003) An mRNA structure that controls gene expression by binding S-adenosylmethionine. *Nature Structural Biology* 10: 701–7.

74. Winkler W.C. (2005) Riboswitches and the role of non-coding RNAs in bacterial metabolic control. *Current Opinion in Chemical Biology* 9: 594–602.

75. Nudler E. and Mironov A.S. (2004) The riboswitch control of bacterial metabolism. *Trends in Biochemical Sciences* 29: 11–7.

76. Chowdhury S., Maris C., Allain F.H. and Narberhaus F. (2006) Molecular basis for temperature sensing by an RNA thermometer. *EMBO Journal* 25: 2487–97.

77. Loh E. et al. (2009) A trans-acting riboswitch controls expression of the virulence regulator *PrfA* in *Listeria monocytogenes*. *Cell* 139: 770–9.

78. Stormo G.D. (2003) New tricks for an old dogma: Riboswitches as cis-only regulatory systems. *Molecular Cell* 11: 1419–20.

79. Winkler W.C. and Breaker R.R. (2005) Regulation of bacterial gene expression by riboswitches. *Annual Review of Microbiology* 59: 487–517.

80. Nudler E. (2006) Flipping riboswitches. *Cell* 126: 19–22.

81. Serganov A. and Patel D.J. (2007) Ribozymes, riboswitches and beyond: Regulation of gene expression without proteins. *Nature Reviews Genetics* 8: 776–90.

82. Brewer K.I. et al. (2021) Comprehensive discovery of novel structured noncoding RNAs in 26 bacterial genomes. *RNA Biology* 18: 2417–32.

83. Chauvier A. et al. (2021) Monitoring RNA dynamics in native transcriptional complexes. *Proceedings of the National Academy of Sciences USA* 118: e2106564118.

84. Montange R.K. and Batey R.T. (2006) Structure of the S-adenosylmethionine riboswitch regulatory mRNA element. *Nature* 441: 1172–5.

85. Tang D.-J. et al. (2020) A SAM-I riboswitch with the ability to sense and respond to uncharged initiator tRNA. *Nature Communications* 11: 2794.

86. Bocobza S. et al. (2007) Riboswitch-dependent gene regulation and its evolution in the plant kingdom. *Genes & Development* 21: 2874–9.

87. Donovan P.D. et al. (2018) TPP riboswitch-dependent regulation of an ancient thiamin transporter in *Candida*. *PLOS Genetics* 14: e1007429.

88. McRose D. et al. (2014) Alternatives to vitamin B1 uptake revealed with discovery of riboswitches in multiple marine eukaryotic lineages. *The ISME Journal* 8: 2517–29.

89. Cheah M.T., Wachter A., Sudarsan N. and Breaker R.R. (2007) Control of alternative RNA splicing and gene expression by eukaryotic riboswitches. *Nature* 447: 497–500.

90. Pham H.L. *et al.* (2017) Engineering a riboswitch-based genetic platform for the self-directed evolution of acid-tolerant phenotypes. *Nature Communications* 8: 411.

91. Dhamodharan V., Nomura Y., Dwidar M. and Yokobayashi Y. (2018) Optochemical control of gene expression by photo-caged guanine and riboswitches. *Chemical Communications* 54: 6181–3.

92. Kim D.S., Gusti V., Pillai S.G. and Gaur R.K. (2005) An artificial riboswitch for controlling pre-mRNA splicing. *RNA* 11: 1667–77.

93. Vitreschak A.G., Rodionov D.A., Mironov A.A. and Gelfand M.S. (2004) Riboswitches: The oldest mechanism for the regulation of gene expression? *Trends in Genetics* 20: 44–50.

94. Breaker R.R. (2012) Riboswitches and the RNA world. *Cold Spring Harbor Perspectives in Biology* 4: a003566.

95. Plesner P., Goodchild J., Kalckar H.M. and Zamecnik P.C. (1987) Oligonucleotides with rapid turnover of the phosphate groups occur endogenously in eukaryotic cells. *Proceedings of the National Academy of Sciences USA* 84: 1936–9.

96. Paterson B.M., Roberts B.E. and Kuff E.L. (1977) Structural gene identification and mapping by DNA-mRNA hybrid-arrested cell-free translation. *Proceedings of the National Academy of Sciences USA* 74: 4370–4.

97. Both G.W., Mattick J.S. and Bellamy A.R. (1983) Serotype-specific glycoprotein of simian 11 rotavirus: Coding assignment and gene sequence. *Proceedings of the National Academy of Sciences USA* 80: 3091–5.

98. Pestka S., Daugherty B.L., Jung V., Hotta K. and Pestka R.K. (1984) Anti-mRNA: Specific inhibition of translation of single mRNA molecules. *Proceedings of the National Academy of Sciences USA* 81: 7525–8.

99. Rosenberg U.B., Preiss A., Seifert E., Jackle H. and Knipple D.C. (1985) Production of phenocopies by Kruppel anti-sense RNA injection into *Drosophila* embryos. *Nature* 313: 703–6.

100. Melton D.A. (1985) Injected anti-sense RNAs specifically block messenger RNA translation in vivo. *Proceedings of the National Academy of Sciences USA* 82: 144–8.

101. McGarry T.J. and Lindquist S. (1986) Inhibition of heat shock protein synthesis by heat-inducible antisense RNA. *Proceedings of the National Academy of Sciences USA* 83: 399–403.

102. Kim S.K. and Wold B.J. (1985) Stable reduction of thymidine kinase activity in cells expressing high levels of anti-sense RNA. *Cell* 42: 129–38.

103. Izant J.G. and Weintraub H. (1984) Inhibition of thymidine kinase gene expression by anti-sense RNA: A molecular approach to genetic analysis. *Cell* 36: 1007–15.

104. Izant J.G. and Weintraub H. (1985) Constitutive and conditional suppression of exogenous and endogenous genes by anti-sense RNA. *Science* 229: 345–52.

105. Crooke S.T. (2011) The Isis manifesto. *Bioentrepreneur.* https://doi.org/10.1038/bioe.2011.7.

106. Crooke S.T., Liang X.-H., Baker B.F. and Crooke R.M. (2021) Antisense technology: A review. *Journal of Biological Chemistry* 296: 100416.

107. Kurreck J. (2003) Antisense technologies. *European Journal of Biochemistry* 270: 1628–44.

108. Crooke S.T. (2004) Progress in antisense technology. *Annual Review of Medicine* 55: 61–95.

109. Crooke S.T. (2017) Molecular mechanisms of antisense oligonucleotides. *Nucleic Acid Therapeutics* 27: 70–7.

110. Bennett C.F. (2019) Therapeutic antisense oligonucleotides are coming of age. *Annual Review of Medicine* 70: 307–21.

111. Shen W. et al. (2019) Chemical modification of PS-ASO therapeutics reduces cellular protein-binding and improves the therapeutic index. *Nature Biotechnology* 37: 640–50.

112. Travers A. (1984) Regulation by anti-sense RNA. *Nature* 311: 410.

113. Rosen C.A., Sodroski J.G. and Haseltine W.A. (1985) The location of cis-acting regulatory sequences in the human T cell lymphotropic virus type III (HTLV-III/LAV) long terminal repeat. *Cell* 41: 813–23.

114. Kao S.Y., Calman A.F., Luciw P.A. and Peterlin B.M. (1987) Anti-termination of transcription within the long terminal repeat of HIV-1 by tat gene product. *Nature* 330: 489–93.

115. Feng S. and Holland E.C. (1988) HIV-1 tat trans-activation requires the loop sequence within tar. *Nature* 334: 165–7.

116. Tycowski K.T. et al. (2015) Viral noncoding RNAs: More surprises. *Genes & Development* 29: 567–84.

117. Greenaway P.J. and Wilkinson G.W.G. (1987) Nucleotide sequence of the most abundantly transcribed early gene of human cytomegalovirus strain AD169. *Virus Research* 7: 17–31.

118. Reeves M.B., Davies A.A., McSharry B.P., Wilkinson G.W. and Sinclair J.H. (2007) Complex I binding by a virally encoded RNA regulates mitochondria-induced cell death. *Science* 316: 1345–8.

119. Diatchenko L. et al. (1996) Suppression subtractive hybridization: A method for generating differentially regulated or tissue-specific cDNA probes and libraries. *Proceedings of the National Academy of Sciences USA* 93: 6025–30.

120. Henikoff S., Keene M.A., Fechtel K. and Fristrom J.W. (1986) Gene within a gene: Nested *Drosophila* genes encode unrelated proteins on opposite DNA strands. *Cell* 44: 33–42.

121. Williams T. and Fried M. (1986) A mouse locus at which transcription from both DNA strands produces mRNAs complementary at their 3' ends. *Nature* 322: 275–9.

122. Spencer C.A., Gietz R.D. and Hodgetts R.B. (1986) Overlapping transcription units in the dopa decarboxylase region of *Drosophila. Nature* 322: 279–81.

123. Spencer C.A., Gietz R.D. and Hodgetts R.B. (1986) Analysis of the transcription unit adjacent to the 3′-end of the dopa decarboxylase gene in *Drosophila melanogaster. Developmental Biology* 114: 260–4.

124. Carninci P. et al. (2005) The transcriptional landscape of the mammalian genome. *Science* 309: 1559–63.

125. Frith M.C., Pheasant M. and Mattick J.S. (2005) The amazing complexity of the human transcriptome. *European Journal of Human Genetics* 13: 894–7.

126. Mattick J.S. and Makunin I.V. (2006) Non-coding RNA. *Human Molecular Genetics* 15: R17–29.

127. Nepveu A. and Marcu K.B. (1986) Intragenic pausing and anti-sense transcription within the murine *c-myc* locus. *EMBO Journal* 5: 2859–65.

128. Kindy M.S., McCormack J.E., Buckler A.J., Levine R.A. and Sonenshein G.E. (1987) Independent regulation of transcription of the two strands of the *c-myc* gene. *Molecular and Cellular Biology* 7: 2857–62.

129. Khochbin S. and Lawrence J.J. (1989) An antisense RNA involved in p53 mRNA maturation in murine erythroleukemia cells induced to differentiate. *EMBO Journal* 8: 4107–14.

130. Skeiky Y.A. and Iatrou K. (1990) Silkmoth chorion antisense RNA. Structural characterization, developmental regulation and evolutionary conservation. *Journal of Molecular Biology* 213: 53–66.

131. Thrash-Bingham C.A. and Tartof K.D. (1999) aHIF: A natural antisense transcript overexpressed in human renal cancer and during hypoxia. *Journal of the National Cancer Institute* 91: 143–51.

132. Silverman T.A., Noguchi M. and Safer B. (1992) Role of sequences within the first intron in the regulation of expression of eukaryotic initiation factor 2 alpha. *Journal of Biological Chemistry* 267: 9738–42.

133. Farrell C.M. and Lukens L.N. (1995) Naturally occurring antisense transcripts are present in chick embryo chondrocytes simultaneously with the down-regulation of the alpha 1 (I) collagen gene. *Journal of Biological Chemistry* 270: 3400–8.

134. Hildebrandt M. and Nellen W. (1992) Differential antisense transcription from the Dictyostelium EB4 gene locus: Implications on antisense-mediated regulation of mRNA stability. *Cell* 69: 197–204.

135. Lagrutta A.A., McCarthy J.G., Scherczinger C.A. and Heywood S.M. (1989) Identification and developmental expression of a novel embryonic myosin heavy-chain gene in chicken. *DNA* 8: 39–50.

136. Rogers J.C. (1988) RNA complementary to α-amylase mRNA in barley. *Plant Molecular Biology* 11: 125–38.

137. Schulz R.A. and Butler B.A. (1989) Overlapping genes of *Drosophila melanogaster*: Organization of the *z600-gonadal-Eip28/29* gene cluster. *Genes & Development* 3: 232–42.

138. Miyajima N. et al. (1989) Two erbA homologs encoding proteins with different T3 binding capacities are transcribed from opposite DNA strands of the same genetic locus. *Cell* 57: 31–9.

139. Eveleth D.D. and Marsh J.L. (1987) Overlapping transcription units in Drosophila: Sequence and structure of the *Cs* gene. *Molecular and General Genetics* 209: 290–8.

140. Potts J.D., Vincent E.B., Runyan R.B. and Weeks D.L. (1992) Sense and antisense TGF beta 3 mRNA levels correlate with cardiac valve induction. *Developmental Dynamics* 193: 340–5.

141. Werner A. (2005) Natural antisense transcripts. *RNA Biology* 2: 53–62.

142. Engstrom P.G. et al. (2006) Complex loci in human and mouse genomes. *PLOS Genetics* 2: e47.

143. Ashe H.L., Monks J., Wijgerde M., Fraser P. and Proudfoot N.J. (1997) Intergenic transcription and transinduction of the human beta-globin locus. *Genes & Development* 11: 2494–509.

144. Kong S., Bohl D., Li C. and Tuan D. (1997) Transcription of the HS2 enhancer toward a cis-linked gene is independent of the orientation, position, and distance of the enhancer relative to the gene. *Molecular and Cellular Biology* 17: 3955–65.

145. Adelman J.P., Bond C.T., Douglass J. and Herbert E. (1987) Two mammalian genes transcribed from opposite strands of the same DNA locus. *Science* 235: 1514–7.

146. Lewis E.B. (1978) A gene complex controlling segmentation in *Drosophila*. *Nature* 276: 565–70.

147. Akam M.E., Martinez-Arias A., Weinzierl R. and Wilde C.D. (1985) Function and expression of ultrabithorax in the *Drosophila* embryo. *Cold Spring Harbor Symposia on Quantitative Biology* 50: 195–200.

148. Hogness D.S. et al. (1985) Regulation and products of the Ubx domain of the bithorax complex. *Cold Spring Harbor Symposia on Quantitative Biology* 50: 181–94.

149. Lipshitz H.D., Peattie D.A. and Hogness D.S. (1987) Novel transcripts from the Ultrabithorax domain of the bithorax complex. *Genes & Development* 1: 307–22.

150. Garaulet D. and Lai E. (2015) Hox miRNA regulation within the *Drosophila Bithorax* Complex: Patterning behavior. *Mechanisms of Development* 138: 151–9.

151. Sanchez-Herrero E. and Akam M. (1989) Spatially ordered transcription of regulatory DNA in the *bithorax* complex of *Drosophila*. *Development* 107: 321–9.

152. Cumberledge S., Zaratzian A. and Sakonju S. (1990) Characterization of two RNAs transcribed from the cis-regulatory region of the *abd-A* domain within the *Drosophila bithorax* complex. *Proceedings of the National Academy of Sciences USA* 87: 3259–63.

153. Kornienko A.E., Guenzl P.M., Barlow D.P. and Pauler F.M. (2013) Gene regulation by the act of long non-coding RNA transcription. *BMC Biology* 11: 59.

154. Li W., Notani D. and Rosenfeld M.G. (2016) Enhancers as non-coding RNA transcription units: Recent insights and future perspectives. *Nature Reviews Genetics* 17: 207–23.

155. Tillib S. et al. (1999) Trithorax- and Polycomb-group response elements within an Ultrabithorax transcription maintenance unit consist of closely situated but separable sequences. *Molecular and Cellular Biology* 19: 5189–202.

156. Zhou J., Ashe H., Burks C. and Levine M. (1999) Characterization of the transvection mediating region of the *abdominal-B* locus in Drosophila. *Development* 126: 3057–65.

157. Maeda R.K. and Karch F. (2006) The ABC of the *BX-C*: The *bithorax* complex explained. *Development* 133: 1413–22.

158. Garbe J.C. and Pardue M.L. (1986) Heat shock locus 93D of *Drosophila melanogaster*: A spliced RNA most strongly conserved in the intron sequence. *Proceedings of the National Academy of Sciences USA* 83: 1812–6.

159. Lakhotia S.C. and Mukherjee T. (1982) Absence of novel translation products in relation to induced activity of the 93D puff in *Drosophila melanogaster*. *Chromosoma* 85: 369–74.

160. Garbe J.C., Bendena W.G., Alfano M. and Pardue M.L. (1986) A *Drosophila* heat shock locus with a rapidly diverging sequence but a conserved structure. *Journal of Biological Chemistry* 261: 16889–94.

161. Fini M.E., Bendena W.G. and Pardue M.L. (1989) Unusual behavior of the cytoplasmic transcript of hsr omega: An abundant, stress-inducible RNA that is translated but yields no detectable protein product. *Journal of Cell Biology* 108: 2045–57.

162. Bendena W.G., Ayme-Southgate A., Garbe J.C. and Pardue M.L. (1991) Expression of heat-shock locus hsr-omega in nonstressed cells during development in *Drosophila melanogaster*. *Developmental Biology* 144: 65–77.

163. Lakhotia S.C. and Sharma A. (1996) The 93D (hsr-omega) locus of *Drosophila*: Non-coding gene with house-keeping functions. *Genetica* 97: 339–48.

164. Prasanth K.V., Rajendra T.K., Lal A.K. and Lakhotia S.C. (2000) Omega speckles - a novel class of nuclear speckles containing hnRNPs associated with noncoding hsr-omega RNA in *Drosophila*. *Journal of Cell Science* 113: 3485–97.

165. Kemp D.J., Harris A.W. and Adams J.M. (1980) Transcripts of the immunoglobulin C mu gene vary in structure and splicing during lymphoid development. *Proceedings of the National Academy of Sciences USA* 77: 7400–4.

166. Lennon G.G. and Perry R.P. (1985) Cμ-containing transcripts initiate heterogeneously within the IgH enhancer region and contain a novel 5′-nontranslatable exon. *Nature* 318: 475–8.

167. Yancopoulos G.D. and Alt F.W. (1985) Developmentally controlled and tissue-specific expression of unrearranged VH gene segments. *Cell* 40: 271–81.

168. Reaban M.E. and Griffin J.A. (1990) Induction of RNA-stabilized DMA conformers by transcription of an immunoglobulin switch region. *Nature* 348: 342–4.

169. Bolland D.J. et al. (2004) Antisense intergenic transcription in V(D)J recombination. *Nature Immunology* 5: 630–7.

170. Abarrategui I. and Krangel M.S. (2007) Noncoding transcription controls downstream promoters to regulate T-cell receptor alpha recombination. *EMBO Journal* 26: 4380–90.

171. Yewdell W.T. and Chaudhuri J. (2017) A transcriptional serenAID: The role of noncoding RNAs in class switch recombination. *International Immunology* 29: 183–96.

172. Rothschild G. et al. (2020) Noncoding RNA transcription alters chromosomal topology to promote isotype-specific class switch recombination. *Science Immunology* 5: eaay5864.

173. Calzone F.J., Lee J.J., Le N, Britten R.J. and Davidson E.H. (1988) A long, nontranslatable poly(A) RNA stored in the egg of the sea urchin *Strongylocentrotus purpuratus*. *Genes & Development* 2: 305–18.

174. Cory S., Graham M., Webb E., Corcoran L. and Adams J.M. (1985) Variant (6;15) translocations in murine plasmacytomas involve a chromosome 15 locus at least 72 kb from the *c-myc* oncogene. *EMBO Journal* 4: 675–81.

175. Graham M. and Adams J.M. (1986) Chromosome 8 breakpoint far 3′ of the *c-myc* oncogene in a Burkitt's lymphoma 2;8 variant translocation is equivalent to the murine *pvt-1* locus. *EMBO Journal* 5: 2845–51.

176. Shtivelman E., Henglein B., Groitl P., Lipp M. and Bishop J.M. (1989) Identification of a human transcription unit affected by the variant chromosomal translocations 2;8 and 8;22 of Burkitt lymphoma. *Proceedings of the National Academy of Sciences USA* 86: 3257–60.

177. Shtivelman E. and Bishop J.M. (1989) The *PVT* gene frequently amplifies with *MYC* in tumor cells. *Molecular and Cellular Biology* 9: 1148–54.

178. Shtivelman E. and Bishop J.M. (1990) Effects of translocations on transcription from *PVT*. *Molecular and Cellular Biology* 10: 1835–9.

179. Huppi K., Pitt J., Wahlberg B. and Caplen N. (2012) The 8q24 gene desert: An oasis of non-coding transcriptional activity. *Frontiers in Genetics* 3: 69.

180. Tseng Y.-Y. et al. (2014) *PVT1* dependence in cancer with *MYC* copy-number increase. *Nature* 512: 82–6.

181. Tolomeo D., Agostini A., Visci G., Traversa D. and Tiziana Storlazzi C. (2021) PVT1: A long non-coding RNA recurrently involved in neoplasia-associated fusion transcripts. *Gene* 779: 145497.

182. Tam W., Ben-Yehuda D. and Hayward W.S. (1997) *bic*, a novel gene activated by proviral insertions in avian leukosis virus-induced lymphomas, is likely to function through its noncoding RNA. *Molecular and Cellular Biology* 17: 1490–502.

183. Tam W. (2001) Identification and characterization of human BIC, a gene on chromosome 21 that encodes a noncoding RNA. *Gene* 274: 157–67.

184. Eis P.S. et al. (2005) Accumulation of miR-155 and BIC RNA in human B cell lymphomas. *Proceedings of the National Academy of Sciences USA* 102: 3627–32.

185. Tam W. and Dahlberg J.E. (2006) miR-155/BIC as an oncogenic microRNA. *Genes Chromosomes Cancer* 45: 211–2.

186. Takei Y., Ishikawa S., Tokino T., Muto T. and Nakamura Y. (1998) Isolation of a novel TP53 target gene from a colon cancer cell line carrying a highly regulated wild-type TP53 expression system. *Genes, Chromosomes and Cancer* 23: 1–9.

187. Diaz-Lagares A. et al. (2016) Epigenetic inactivation of the p53-induced long noncoding RNA TP53 target 1 in human cancer. *Proceedings of the National Academy of Sciences USA* 113: E7535–44.

188. Bussemakers M.J. et al. (1999) DD3: A new prostate-specific gene, highly overexpressed in prostate cancer. *Cancer Research* 59: 5975–9.

189. Lee G.L., Dobi A. and Srivastava S. (2011) Diagnostic performance of the PCA3 urine test. *Nature Reviews Urology* 8: 123–4.

190. Beck-Engeser G.B. et al. (2008) Pvt1-encoded microRNAs in oncogenesis. *Retrovirology* 5: 4.

191. Velleca M.A., Wallace M.C. and Merlie J.P. (1994) A novel synapse-associated noncoding RNA. *Molecular and Cellular Biology* 14: 7095–104.

192. Rodriguez A., Griffiths-Jones S., Ashurst J.L. and Bradley A. (2004) Identification of mammalian microRNA host genes and transcription units. *Genome Research* 14: 1902–10.

193. Brannan C.I., Dees E.C., Ingram R.S. and Tilghman S.M. (1990) The product of the *H19* gene may function as an RNA. *Molecular and Cellular Biology* 10: 28–36.

194. Leibovitch M.P. et al. (1991) The human ASM (Adult Skeletal Muscle) gene expression and chromosomal assignment to 11p15. *Biochemical and Biophysical Research Communications* 180: 1241–50.

195. Gascoigne D.K. et al. (2012) Pinstripe: A suite of programs for integrating transcriptomic and proteomic datasets identifies novel proteins and improves differentiation of protein-coding and non-coding genes. *Bioinformatics* 28: 3042–50.

196. Bartolomei M.S., Zemel S. and Tilghman S.M. (1991) Parental imprinting of the mouse H19 gene. *Nature* 351: 153–5.

197. Hao Y., Crenshaw T., Moulton T., Newcomb E. and Tycko B. (1993) Tumour-suppressor activity of H19 RNA. *Nature* 365: 764–7.

198. Leighton P.A., Ingram R.S., Eggenschwiler J., Efstratiadis A. and Tilghman S.M. (1995) Disruption of imprinting caused by deletion of the H19 gene region in mice. *Nature* 375: 34–9.

199. Wrana J.L. (1994) H19, a tumour suppressing RNA? *BioEssays* 16: 89–90.

200. Forné T. et al. (1997) Loss of the maternal *H19* gene induces changes in *Igf2* methylation in both *cis* and *trans*. *Proceedings of the National Academy of Sciences USA* 94: 10243–8.

201. Hur S.K. et al. (2016) Humanized *H19/Igf2* locus reveals diverged imprinting mechanism between mouse and human and reflects Silver–Russell syndrome phenotypes. *Proceedings of the National Academy of Sciences USA* 113: 10938–43.

202. Brunkow M.E. and Tilghman S.M. (1991) Ectopic expression of the *H19* gene in mice causes prenatal lethality. *Genes & Development* 5: 1092–101.

203. Brown C.J. et al. (1991) A gene from the region of the human X inactivation centre is expressed exclusively from the inactive X chromosome. *Nature* 349: 38–44.

204. Brown C.J. et al. (1992) The human *XIST* gene: Analysis of a 17kb inactive X-specific RNA that contains conserved repeats and is highly localized within the nucleus. *Cell* 71: 527–42.

205. Clemson C.M., McNeil J.A., Willard H.F. and Lawrence J.B. (1996) XIST RNA paints the inactive X chromosome at interphase: Evidence for a novel RNA involved in nuclear/chromosome structure. *Journal of Cell Biology* 132: 259–75.

206. Hall L.L. and Lawrence J.B. (2003) The cell biology of a novel chromosomal RNA: Chromosome painting by XIST/Xist RNA initiates a remodeling cascade. *Seminars in Cell & Developmental Biology* 14: 369–78.

207. Chaumeil J., Le Baccon P., Wutz A. and Heard E. (2006) A novel role for Xist RNA in the formation of a repressive nuclear compartment into which genes are recruited when silenced. *Genes & Development* 20: 2223–37.

208. Wutz A. (2011) Gene silencing in X-chromosome inactivation: Advances in understanding facultative heterochromatin formation. *Nature Reviews Genetics* 12: 542–53.

209. Galupa R. and Heard E. (2018) X-chromosome inactivation: A crossroads between chromosome architecture and gene regulation. *Annual Review of Genetics* 52: 535–66.

210. Heard E. and Disteche C.M. (2006) Dosage compensation in mammals: Fine-tuning the expression of the X chromosome. *Genes & Development* 20: 1848–67.

211. Fan G. and Tran J. (2011) X chromosome inactivation in human and mouse pluripotent stem cells. *Human Genetics* 130: 217–22.

212. Duret L., Chureau C., Samain S., Weissenbach J. and Avner P. (2006) The *Xist* RNA gene evolved in eutherians by pseudogenization of a protein-coding gene. *Science* 312: 1653–5.

213. Elisaphenko E.A. et al. (2008) A dual origin of the *Xist* gene from a protein-coding gene and a set of transposable elements. *PLOS ONE* 3: e2521.

214. Meller V.H., Wu K.H., Roman G., Kuroda M.I. and Davis R.L. (1997) roX1 RNA paints the X chromosome of male Drosophila and is regulated by the dosage compensation system. *Cell* 88: 445–57.

215. Kelley R.L. et al. (1999) Epigenetic spreading of the *Drosophila* dosage compensation complex from *roX* RNA genes into flanking chromatin. *Cell* 98: 513–22.

216. Meller V.H. (2003) Initiation of dosage compensation in Drosophila embryos depends on expression of the roX RNAs. *Mechanisms of Development* 120: 759–67.

217. Kelley R.L. and Kuroda M.I. (2003) The *Drosophila* roX1 RNA gene can overcome silent chromatin by recruiting the male-specific lethal dosage compensation complex. *Genetics* 164: 565–74.

218. Park Y., Oh H., Meller V.H. and Kuroda M.I. (2005) Variable splicing of non-coding roX2 RNAs influences targeting of MSL dosage compensation complexes in *Drosophila. RNA Biology* 2: 157–64.

219. Ilik I.A. et al. (2013) Tandem stem-loops in roX RNAs act together to mediate X chromosome dosage compensation in *Drosophila. Molecular Cell* 51: 156–73.

220. Akhtar A., Zink D. and Becker P.B. (2000) Chromodomains are protein-RNA interaction modules. *Nature* 407: 405–9.

221. Akhtar A. (2003) Dosage compensation: An intertwined world of RNA and chromatin remodelling. *Current Opinion in Genetics and Development* 13: 161–9.

222. Valsecchi C.I.K. et al. (2020) RNA nucleation by MSL2 induces selective X chromosome compartmentalization. *Nature* 589: 137–42.

223. Erdmann V.A., Szymanski M., Hochberg A., de Groot N. and Barciszewski J. (1999) Collection of mRNA-like non-coding RNAs. *Nucleic Acids Research* 27: 192–5.

224. Erdmann V.A., Szymanski M., Hochberg A., Groot N. and Barciszewski J. (2000) Non-coding, mRNA-like RNAs database Y2K. *Nucleic Acids Research* 28: 197–200.

225. Watanabe Y. and Yamamoto M. (1994) *S. pombe mei2+* encodes an RNA-binding protein essential for premeiotic DNA synthesis and meiosis I, which cooperates with a novel RNA species meiRNA. *Cell* 78: 487–98.

226. Ding D.Q. et al. (2012) Meiosis-specific noncoding RNA mediates robust pairing of homologous chromosomes in meiosis. *Science* 336: 732–6.

227. Hiraoka Y. (2020) Phase separation drives pairing of homologous chromosomes. *Current Genetics* 66: 881–7.

228. Kumimoto H., Yoshida H. and Okamoto K. (1995) RNA polymerase II transcribes *Dictyostelium* untranslatable gene, *dutA*, specifically in the developmental phase. *Biochemical and Biophysical Research Communications* 216: 273–8.

229. Kumimoto H., Yoshida H. and Okamoto K. (1996) Expression of *Dictyostelium* early gene, *dutA*, is independent of cAMP pulses but dependent on protein kinase A. *FEMS Microbiology Letters* 140: 121–4.

230. Yoshida H., Kumimoto H. and Okamoto K. (1994) dutA RNA functions as an untranslatable RNA in the development of Dictyostelium discoideum. *Nucleic Acids Research* 22: 41–6.

231. Su X.Z. et al. (1995) The large diverse gene family *var* encodes proteins involved in cytoadherence and antigenic variation of *Plasmodium falciparum*-infected erythrocytes. *Cell* 82: 89–100.

232. Crespi M.D. et al. (1994) enod40, a gene expressed during nodule organogenesis, codes for a non-translatable RNA involved in plant growth. *EMBO Journal* 13: 5099–112.

233. Campalans A., Kondorosi A. and Crespi M. (2004) Enod40, a short open reading frame-containing mRNA, induces cytoplasmic localization of a nuclear RNA binding protein in *Medicago truncatula. Plant Cell* 16: 1047–59.

234. Gultyaev A.P. and Roussis A. (2007) Identification of conserved secondary structures and expansion segments in enod40 RNAs reveals new enod40 homologues in plants. *Nucleic Acids Research* 35: 3144–52.

235. Zhou B.-S., Beidler D.R. and Cheng Y.-C. (1992) Identification of antisense RNA transcripts from a human DNA topoisomerase I pseudogene. *Cancer Research* 52: 4280–5.

236. Korneev S.A., Park J.H. and O'Shea M. (1999) Neuronal expression of neural nitric oxide synthase (nNOS) protein is suppressed by an antisense RNA transcribed from an NOS pseudogene. *Journal of Neuroscience* 19: 7711–20.

237. Korneev S.A. et al. (2008) Novel noncoding antisense RNA transcribed from human anti-NOS2A locus is differentially regulated during neuronal differentiation of embryonic stem cells. *RNA* 14: 2030–7.

238. Kloc M., Spohr G. and Etkin L.D. (1993) Translocation of repetitive RNA sequences with the germ plasm in *Xenopus* oocytes. *Science* 262: 1712–4.

239. Kloc M. and Etkin L.D. (1994) Delocalization of Vg1 mRNA from the vegetal cortex in *Xenopus* oocytes after destruction of Xlsirt RNA. *Science* 265: 1101–3.

240. Swalla B.J. and Jeffery W.R. (1995) A maternal RNA localized in the yellow crescent is segregated to the larval muscle cells during ascidian development. *Developmental Biology* 170: 353–64.

241. Swalla B.J. and Jeffery W.R. (1996) PCNA mRNA has a 3'UTR antisense to yellow crescent RNA and is localized in ascidian eggs and embryos. *Developmental Biology* 178: 23–34.

242. Coccia E.M. et al. (1992) Regulation and expression of a growth arrest-specific gene (*gas5*) during growth, differentiation, and development. *Molecular and Cellular Biology* 12: 3514–21.

243. Smith C.M. and Steitz J.A. (1998) Classification of *gas5* as a multi-small-nucleolar-RNA (snoRNA) host gene and a member of the 5'-terminal oligopyrimidine gene family reveals common features of snoRNA host genes. *Molecular and Cellular Biology* 18: 6897–909.

244. Raho G., Barone V., Rossi D., Philipson L. and Sorrentino V. (2000) The *gas5* gene shows four alternative splicing patterns without coding for a protein. *Gene* 256: 13–7.

245. Sang L. et al. (2021) Mitochondrial long non-coding RNA GAS5 tunes TCA metabolism in response to nutrient stress. *Nature Metabolism* 3: 90–106.

246. Lanz R.B. et al. (1999) A steroid receptor coactivator, SRA, functions as an RNA and is present in an SRC-1 complex. *Cell* 97: 17–27.

247. Lanz R.B., Razani B., Goldberg A.D. and O'Malley B.W. (2002) Distinct RNA motifs are important for coactivation of steroid hormone receptors by steroid receptor RNA activator (SRA). *Proceedings of the National Academy of Sciences USA* 99: 16081–6.

248. Ghosh S.K., Patton J.R. and Spanjaard R.A. (2012) A small RNA derived from RNA coactivator SRA blocks steroid receptor signaling via inhibition of Pus1p-mediated pseudouridylation of SRA: Evidence of a novel RNA binding domain in the N-terminus of steroid receptors. *Biochemistry* 51: 8163–72.

249. Novikova I.V., Hennelly S.P. and Sanbonmatsu K.Y. (2012) Structural architecture of the human long non-coding RNA, steroid receptor RNA activator. *Nucleic Acids Research* 40: 5034–51.

250. Sengupta D.J. et al. (1996) A three-hybrid system to detect RNA-protein interactions in vivo. *Proceedings of the National Academy of Sciences USA* 93: 8496–501.

251. Sengupta D.J., Wickens M. and Fields S. (1999) Identification of RNAs that bind to a specific protein using the yeast three-hybrid system. *RNA* 5: 596–601.

252. Saha S., Ansari A.Z., Jarrell K.A. and Ptashne M. (2003) RNA sequences that work as transcriptional activating regions. *Nucleic Acids Research* 31: 1565–70.

253. Hsieh-Li H.M. et al. (1995) Hoxa 11 structure, extensive antisense transcription, and function in male and female fertility. *Development* 121: 1373–85.

254. Potter S.S. and Branford W.W. (1998) Evolutionary conservation and tissue-specific processing of *Hoxa 11* antisense transcripts. *Mammalian Genome* 9: 799–806.

255. Bedford M., Arman E., Orr-Urtreger A. and Lonai P. (1995) Analysis of the *Hoxd-3* gene: Structure and localization of its sense and natural antisense transcripts. *DNA and Cell Biology* 14: 295–304.

256. Liu J.K., Ghattas I., Liu S., Chen S. and Rubenstein J.L. (1997) *Dlx* genes encode DNA-binding proteins that are expressed in an overlapping and sequential pattern during basal ganglia differentiation. *Developmental Dynamics* 210: 498–512.

257. McGuinness T. et al. (1996) Sequence, organization, and transcription of the *Dlx-1* and *Dlx-2* locus. *Genomics* 35: 473–85.

258. Lee J.T., Davidow L.S. and Warshawsky D. (1999) *Tsix*, a gene antisense to *Xist* at the X-inactivation centre. *Nature Genetics* 21: 400–4.

259. Lee J.T., Lu N. and Han Y. (1999) Genetic analysis of the mouse X inactivation center defines an 80-kb multifunction domain. *Proceedings of the National Academy of Sciences USA* 96: 3836–41.

260. Moore T. et al. (1997) Multiple imprinted sense and antisense transcripts, differential methylation and tandem repeats in a putative imprinting control region upstream of mouse *Igf2*. *Proceedings of the National Academy of Sciences USA* 94: 12509–14.

261. Ainscough J.F., Koide T., Tada M., Barton S. and Surani M.A. (1997) Imprinting Cof *Igf2* and *H19* from a 130kb YAC transgene. *Development* 124: 3621–32.

262. Rougeulle C., Cardoso C., Fontes M., Colleaux L. and Lalande M. (1998) An imprinted antisense RNA overlaps *UBE3A* and a second maternally expressed transcript. *Nature Genetics* 19: 15–6.

263. Smith R.J., Dean W., Konfortova G. and Kelsey G. (2003) Identification of novel imprinted genes in a genome-wide screen for maternal methylation. *Genome Research* 13: 558–69.

264. Schoenfelder S., Smits G., Fraser P., Reik W. and Paro R. (2007) Non-coding transcripts in the *H19* imprinting control region mediate gene silencing in transgenic *Drosophila*. *EMBO Reports* 8: 1068–73.

265. Wutz A. et al. (1997) Imprinted expression of the *Igf2r* gene depends on an intronic CpG island. *Nature* 389: 745–9.

266. Lyle R. et al. (2000) The imprinted antisense RNA at the *Igf2r* locus overlaps but does not imprint *Mas1*. *Nature Genetics* 25: 19–21.

267. Stöger R. et al. (1993) Maternal-specific methylation of the imprinted mouse *Igf2r* locus identifies the expressed locus as carrying the imprinting signal. *Cell* 73: 61–71.

268. Peters J. et al. (1999) A cluster of oppositely imprinted transcripts at the *Gnas* locus in the distal imprinting region of mouse chromosome 2. *Proceedings of the National Academy of Sciences USA* 96: 3830–5.

269. Kim J., Bergmann A., Wehri E., Lu X. and Stubbs L. (2001) Imprinting and evolution of two Kruppel-type zinc-finger genes, *ZIM3* and *ZNF264*, located in the *PEG3/USP29* imprinted domain. *Genomics* 77: 91–8.

270. Tierling S. et al. (2006) High-resolution map and imprinting analysis of the *Gtl2-Dnchc1* domain on mouse chromosome 12. *Genomics* 87: 225–35.

271. Nesterova T.B., Barton S.C., Surani M.A. and Brockdorff N. (2001) Loss of *Xist* imprinting in diploid parthenogenetic preimplantation embryos. *Developmental Biology* 235: 343–50.

272. Miyoshi N. et al. (2000) Identification of an imprinted gene, *Meg3/Gtl2* and its human homologue *MEG3*, first mapped on mouse distal chromosome 12 and human chromosome 14q. *Genes to Cells* 5: 211–20.

273. Schuster-Gossler K., Simon-Chazottes D., Guenet J.L., Zachgo J. and Gossler A. (1996) Gtl2lacZ, an insertional mutation on mouse chromosome 12 with parental origin-dependent phenotype. *Mammalian Genome* 7: 20–4.

274. Schuster-Gossler K., Bilinski P., Sado T., Ferguson-Smith A. and Gossler A. (1998) The mouse *Gtl2* gene is differentially expressed during embryonic development, encodes multiple alternatively spliced transcripts, and may act as an RNA. *Developmental Dynamics* 212: 214–28.

275. Smilinich N.J. et al. (1999) A maternally methylated CpG island in KvLQT1 is associated with an antisense paternal transcript and loss of imprinting in Beckwith-Wiedemann syndrome. *Proceedings of the National Academy of Sciences USA* 96: 8064–9.

276. Thakur N. et al. (2004) An antisense RNA regulates the bidirectional silencing property of the *Kcnq1* imprinting control region. *Molecular and Cellular Biology* 24: 7855–62.

277. Mancini-Dinardo D., Steele S.J., Levorse J.M., Ingram R.S. and Tilghman S.M. (2006) Elongation of the Kcnq1ot1 transcript is required for genomic imprinting of neighboring genes. *Genes & Development* 20: 1268–82.

278. Pandey R.R. et al. (2008) Kcnq1ot1 antisense noncoding RNA mediates lineage-specific transcriptional silencing through chromatin-level regulation. *Molecular Cell* 32: 232–46.

279. Kanduri C. (2011) Kcnq1ot1: A chromatin regulatory RNA. *Seminars in Cell & Developmental Biology* 22: 343–50.

280. Reik W. and Constancia M. (1997) Genomic imprinting. Making sense or antisense? *Nature* 389: 669–71.

281. Sleutels F., Zwart R. and Barlow D.P. (2002) The non-coding Air RNA is required for silencing autosomal imprinted genes. *Nature* 415: 810–3.

282. Nagano T. et al. (2008) The Air noncoding RNA epigenetically silences transcription by targeting G9a to chromatin. *Science* 322: 1717–20.

283. Mohammad F., Mondal T., Guseva N., Pandey G.K. and Kanduri C. (2010) Kcnq1ot1 noncoding RNA mediates transcriptional gene silencing by interacting with Dnmt1. *Development* 137: 2493–9.

284. Deveson I.W. et al. (2018) Universal alternative splicing of noncoding exons. *Cell Systems* 6: 245–55.

285. Mazumder B., Seshadri V. and Fox P.L. (2003) Translational control by the 3′-UTR: The ends specify the means. *Trends in Biochemical Sciences* 28: 91–8.

286. Siepel A. et al. (2005) Evolutionarily conserved elements in vertebrate, insect, worm, and yeast genomes. *Genome Research* 15: 1034–50.

287. Chen C.-Y., Chen S.-T., Juan H.-F. and Huang H.-C. (2012) Lengthening of 3′UTR increases with morphological complexity in animal evolution. *Bioinformatics* 28: 3178–81.

288. Mayr C. (2016) Evolution and biological roles of alternative 3′UTRs. *Trends in Cell Biology* 26: 227–37.

289. Mayr C. (2017) Regulation by 3′-untranslated regions. *Annual Review of Genetics* 51: 171–94.

290. Rastinejad F., Conboy M.J., Rando T.A. and Blau H.M. (1993) Tumor suppression by RNA from the 3′ untranslated region of alpha-tropomyosin. *Cell* 75: 1107–17.

291. Rastinejad F. and Blau H.M. (1993) Genetic complementation reveals a novel regulatory role for 3′ untranslated regions in growth and differentiation. *Cell* 72: 903–17.

292. Fan H., Villegas C., Huang A. and Wright J.A. (1996) Suppression of malignancy by the 3′ untranslated regions of ribonucleotide reductase R1 and R2 messenger RNAs. *Cancer Research* 56: 4366–9.

293. Jupe E.R., Liu X.T., Kiehlbauch J.L., McClung J.K. and Dell'Orco R.T. (1996) Prohibitin in breast cancer cell lines: Loss of antiproliferative activity is linked to 3′ untranslated region mutations. *Cell Growth and Differentiation* 7: 871–8.

294. Jenny A. et al. (2006) A translation-independent role of oskar RNA in early *Drosophila* oogenesis. *Development* 133: 2827–33.

295. Mercer T.R. et al. (2011) Expression of distinct RNAs from 3′ untranslated regions. *Nucleic Acids Research* 39: 2393–403.

296. Miura P., Shenker S., Andreu-Agullo C., Westholm J.O. and Lai E.C. (2013) Widespread and extensive lengthening of 3′ UTRs in the mammalian brain. *Genome Research* 23: 812–25.

297. Kocabas A., Duarte T., Kumar S. and Hynes M.A. (2015) Widespread differential expression of coding region and 3′UTR sequences in neurons and other tissues. *Neuron* 88: 1149–56.

298. Vilborg A., Passarelli M.C., Yario T.A., Tycowski K.T. and Steitz J.A. (2015) Widespread inducible transcription downstream of human genes. *Molecular Cell* 59: 449–61.

299. Malka Y. et al. (2017) Post-transcriptional 3′-UTR cleavage of mRNA transcripts generates thousands of stable uncapped autonomous RNA fragments. *Nature Communications* 8: 2029.

300. Sun H.-X., Li Y., Niu Q.-W. and Chua N.-H. (2017) Dehydration stress extends mRNA 3′ untranslated regions with noncoding RNA functions in *Arabidopsis*. *Genome Research* 27: 1427–36.

301. Jeyapalan Z. et al. (2011) Expression of CD44 3′-untranslated region regulates endogenous microRNA functions in tumorigenesis and angiogenesis. *Nucleic Acids Research* 39: 3026–41.

302. Vilborg A. and Steitz J.A. (2017) Readthrough transcription: How are DoGs made and what do they do? *RNA Biology* 14: 632–6.

303. Andreassi C. et al. (2021) Cytoplasmic cleavage of IMPA1 3′ UTR is necessary for maintaining axon integrity. *Cell Reports* 34: 108778.

304. Lee R.C., Feinbaum R.L. and Ambros V. (1993) The *C. elegans* heterochronic gene *lin-4* encodes small RNAs with antisense complementarity to *lin-14*. *Cell* 75: 843–54.

305. Wightman B., Ha I. and Ruvkun G. (1993) Posttranscriptional regulation of the heterochronic gene *lin-14* by *lin-4* mediates temporal pattern formation in *C. elegans*. *Cell* 75: 855–62.

306. Ruvkun G., Wightman B. and Ha I. (2004) The 20 years it took to recognize the importance of tiny RNAs. *Cell* 116: S93–6.

307. Ambros V. (2008) The evolution of our thinking about microRNAs. *Nature Medicine* 14: 1036–40.

308. Olsen P.H. and Ambros V. (1999) The *lin-4* regulatory RNA controls developmental timing in *Caenorhabditis elegans* by blocking LIN-14 protein synthesis after the initiation of translation. *Developmental Biology* 216: 671–80.

309. Lee R., Feinbaum R. and Ambros V. (2004) A short history of a short RNA. *Cell* 116: S89–92.

310. Holley R.W. et al. (1965) Structure of a ribonucleic acid. *Science* 147: 1462–5.

311. Wickens M. and Takayama K. (1994) RNA. Deviants - or emissaries. *Nature* 367: 17–8.

312. Nowak R. (1994) Mining treasures from 'junk DNA'. *Science* 263: 608–10.

CHAPTER 10

1. Smith C., Econome J., Schutt A., Klco S. and Cantor C. (1987) A physical map of the *Escherichia coli* K12 genome. *Science* 236: 1448–53.

2. Römling U., Grothues D., Bautsch W. and Tümmler B. (1989) A physical genome map of *Pseudomonas aeruginosa* PAO. *EMBO Journal* 8: 4081–9.

3. Link A.J. and Olson M.V. (1991) Physical map of the *Saccharomyces cerevisiae* genome at 110-kilobase resolution. *Genetics* 127: 681–98.

4. Yanofsky M.F. et al. (1990) The protein encoded by the *Arabidopsis* homeotic gene *agamous* resembles transcription factors. *Nature* 346: 35–9.

5. Wensink P.C., Finnegan D.J., Donelson J.E. and Hogness D.S. (1974) A system for mapping DNA sequences in the chromosomes of *Drosophila melanogaster*. *Cell* 3: 315–25.

6. Grunstein M. and Hogness D.S. (1975) Colony hybridization: A method for the isolation of cloned DNAs that contain a specific gene. *Proceedings of the National Academy of Sciences USA* 72: 3961–5.

7. Bender W., Spierer P., Hogness D.S. and Chambon P. (1983) Chromosomal walking and jumping to isolate DNA from the *Ace* and *rosy* loci and the *bithorax* complex in *Drosophila melanogaster*. *Journal of Molecular Biology* 168: 17–33.

8. Bender W. et al. (1983) Molecular genetics of the bithorax complex in *Drosophila melanogaster*. *Science* 221: 23–9.

9. Rubin G.M. and Lewis E.B. (2000) A brief history of *Drosophila*'s contributions to genome research. *Science* 287: 2216–8.

10. Nüsslein-Volhard C. and Wieschaus E. (1980) Mutations affecting segment number and polarity in *Drosophila*. *Nature* 287: 795–801.

11. Bingham P.M., Levis R. and Rubin G.M. (1981) Cloning of DNA sequences from the *white* locus of *D. melanogaster* by a novel and general method. *Cell* 25: 693–704.

12. Cooley L., Kelley R. and Spradling A. (1988) Insertional mutagenesis of the *Drosophila* genome with single P elements. *Science* 239: 1121–8.

13. Robertson E., Bradley A., Kuehn M. and Evans M. (1986) Germ-line transmission of genes introduced into cultured pluripotent cells by retroviral vector. *Nature* 323: 445–8.

14. Folger K.R., Wong E.A., Wahl G. and Capecchi M.R. (1982) Patterns of integration of DNA microinjected into cultured mammalian cells: Evidence for homologous recombination between injected plasmid DNA molecules. *Molecular and Cellular Biology* 2: 1372–87.

15. Mansour S.L., Thomas K.R. and Capecchi M.R. (1988) Disruption of the proto-oncogene int-2 in mouse embryo-derived stem cells: A general strategy for targeting mutations to non-selectable genes. *Nature* 336: 348–52.

16. Kuehn M.R., Bradley A., Robertson E.J. and Evans M.J. (1987) A potential animal model for Lesch–Nyhan syndrome through introduction of HPRT mutations into mice. *Nature* 326: 295–8.

17. Koller B.H. et al. (1989) Germ-line transmission of a planned alteration made in a hypoxanthine phosphoribosyltransferase gene by homologous recombination in embryonic stem cells. *Proceedings of the National Academy of Sciences USA* 86: 8927–31.

18. Smithies O., Gregg R.G., Boggs S.S., Koralewski M.A. and Kucherlapati R.S. (1985) Insertion of DNA sequences into the human chromosomal β-globin locus by homologous recombination. *Nature* 317: 230–4.

19. Thomas K.R. and Capecchi M.R. (1990) Targeted disruption of the murine *int-1* proto-oncogene resulting in severe abnormalities in midbrain and cerebellar development. *Nature* 346: 847–50.

20. Thompson S., Clarke A.R., Pow A.M., Hooper M.L. and Melton D.W. (1989) Germ line transmission and expression of a corrected HPRT gene produced by gene targeting in embryonic stem cells. *Cell* 56: 313–21.

21. Schwartzberg P.L., Goff S.P. and Robertson E.J. (1989) Germ-line transmission of a *c-abl* mutation produced by targeted gene disruption in ES cells. *Science* 246: 799–803.

22. O'Kane C.J. and Gehring W.J. (1987) Detection in situ of genomic regulatory elements in *Drosophila*. *Proceedings of the National Academy of Sciences USA* 84: 9123–7.

23. Brand A.H. and Perrimon N. (1993) Targeted gene expression as a means of altering cell fates and generating dominant phenotypes. *Development* 118: 401–15.

24. Anderson S. et al. (1981) Sequence and organization of the human mitochondrial genome. *Nature* 290: 457–65.

25. Shinozaki K. et al. (1986) The complete nucleotide sequence of the tobacco chloroplast genome: Its gene organization and expression. *EMBO Journal* 5: 2043–9.

26. Smith L.M. et al. (1986) Fluorescence detection in automated DNA sequence analysis. *Nature* 321: 674–9.

27. Fleischmann R.D. et al. (1995) Whole-genome random sequencing and assembly of *Haemophilus influenzae* Rd. *Science* 269: 496–512.

28. Adams M.D. et al. (1992) Sequence identification of 2,375 human brain genes. *Nature* 355: 632–4.

29. Blattner F.R. et al. (1997) The complete genome sequence of *Escherichia coli* K-12. *Science* 277: 1453–62.

30. Andersson S.G.E. et al. (1998) The genome sequence of *Rickettsia prowazekii* and the origin of mitochondria. *Nature* 396: 133–40.

31. Nakabachi A. et al. (2006) The 160-kilobase genome of the bacterial endosymbiont *Carsonella*. *Science* 314: 267.

32. Bennett G.M. and Moran N.A. (2013) Small, smaller, smallest: The origins and evolution of ancient dual symbioses in a phloem-feeding insect. *Genome Biology and Evolution* 5: 1675–88.

33. Han K. et al. (2013) Extraordinary expansion of a *Sorangium cellulosum* genome from an alkaline milieu. *Scientific Reports* 3: 2101.

34. Land M. et al. (2015) Insights from 20 years of bacterial genome sequencing. *Functional & Integrative Genomics* 15: 141–61.

35. Hugenholtz P., Goebel B.M. and Pace N.R. (1998) Impact of culture-independent studies on the emerging phylogenetic view of bacterial diversity. *Journal of Bacteriology* 180: 4765–74.

36. Tyson G.W. et al. (2004) Community structure and metabolism through reconstruction of microbial genomes from the environment. *Nature* 428: 37–43.

37. Lukjancenko O., Wassenaar T.M. and Ussery D.W. (2010) Comparison of 61 sequenced *Escherichia coli* genomes. *Microbial Ecology* 60: 708–20.

38. Rouli L., Merhej V., Fournier P.E. and Raoult D. (2015) The bacterial pangenome as a new tool for analysing pathogenic bacteria. *New Microbes and New Infections* 7: 72–85.

39. Coelho L.P. et al. (2022) Towards the biogeography of prokaryotic genes. *Nature* 601: 252–6.

40. Pellicer J., Fay M.F. and Leitch I.J. (2010) The largest eukaryotic genome of them all? *Botanical Journal of the Linnean Society* 164: 10–5.

41. Goffeau A. et al. (1996) Life with 6000 genes. *Science* 274: 546–67.

42. Mewes H.W. et al. (1997) Overview of the yeast genome. *Nature* 387: 7–8.

43. Zhang Z., Hesselberth J.R. and Fields S. (2007) Genome-wide identification of spliced introns using a tiling microarray. *Genome Research* 17: 503–9.

44. Parenteau J. et al. (2008) Deletion of many yeast introns reveals a minority of genes that require splicing for function. *Molecular Biology of the Cell* 19: 1932–41.

45. Hayashi S., Mori S., Suzuki T., Suzuki T. and Yoshihisa T. (2019) Impact of intron removal from tRNA genes on *Saccharomyces cerevisiae*. *Nucleic Acids Research* 47: 5936–49.

46. Parenteau J. et al. (2019) Introns are mediators of cell response to starvation. *Nature* 565: 612–7.

47. Morgan J.T., Fink G.R. and Bartel D.P. (2019) Excised linear introns regulate growth in yeast. *Nature* 565: 606–11.

48. Wood V. et al. (2002) The genome sequence of *Schizosaccharomyces pombe*. *Nature* 415: 871–80.

49. Botstein D., Chervitz S.A. and Cherry J.M. (1997) Yeast as a model organism. *Science* 277: 1259–60.

50. Puddu F. et al. (2019) Genome architecture and stability in the *Saccharomyces cerevisiae* knockout collection. *Nature* 573: 416–20.

51. Bing J., Han P.-J., Liu W.-Q., Wang Q.-M. and Bai F.-Y. (2014) Evidence for a Far East Asian origin of lager beer yeast. *Current Biology* 24: R380–1.

52. Gallone B. et al. (2018) Origins, evolution, domestication and diversity of *Saccharomyces* beer yeasts. *Current Opinion in Biotechnology* 49: 148–55.

53. Galagan J.E. et al. (2003) The genome sequence of the filamentous fungus *Neurospora crassa*. *Nature* 422: 859–68.

54. Gardner M.J. et al. (2002) Genome sequence of the human malaria parasite *Plasmodium falciparum*. *Nature* 419: 498–511.

55. Holt R.A. et al. (2002) The genome sequence of the malaria mosquito *Anopheles gambiae*. *Science* 298: 129–49.

56. The *C. elegans* Sequencing Consortium (1998) Genome sequence of the nematode *C. elegans*: A platform for investigating biology. *Science* 282: 2012–8.

57. Sulston J.E. and Horvitz H.R. (1977) Post-embryonic cell lineages of the nematode, *Caenorhabditis elegans*. *Developmental Biology* 56: 110–56.

58. Check E. (2002) Worm cast in starring role for Nobel prize. *Nature* 419: 548.

59. Feng Z. et al. (2006) A *C. elegans* model of nicotine-dependent behavior: Regulation by TRP-family channels. *Cell* 127: 621–33.

60. Rauthan M. et al. (2017) MicroRNA regulation of nAChR expression and nicotine-dependent behavior in *C. elegans*. *Cell Reports* 21: 1434–41.

61. Tissenbaum H.A. (2015) Using *C. elegans* for aging research. *Invertebrate Reproduction & Development* 59: 59–63.

62. Adams M.D. et al. (2000) The genome sequence of *Drosophila melanogaster*. *Science* 287: 2185–95.

63. Reiter L.T., Potocki L., Chien S., Gribskov M. and Bier E. (2001) A systematic analysis of human disease-associated gene sequences in *Drosophila melanogaster*. *Genome Research* 11: 1114–25.

64. Miklos G.L.G. and Maleszka R. (2000) Deus ex genomix. *Nature Neuroscience* 3: 424–5.

65. Neuman S., Kovalio M., Yaffe D. and Nudel U. (2005) The *Drosophila* homologue of the *dystrophin* gene – Introns containing promoters are the major contributors to the large size of the gene. *FEBS Letters* 579: 5365–71.

66. Pozzoli U. et al. (2003) Comparative analysis of vertebrate dystrophin loci indicate intron gigantism as a common feature. *Genome Research* 13: 764–72.

67. Lander E.S. et al. (2001) Initial sequencing and analysis of the human genome. *Nature* 409: 860–921.

68. The Arabidopsis Genome I (2000) Analysis of the genome sequence of the flowering plant *Arabidopsis thaliana*. *Nature* 408: 796–815.

69. Yu J. et al. (2002) A draft sequence of the rice genome (*Oryza sativa* L. ssp. *indica*). *Science* 296: 79–92.

70. Goff S.A. et al. (2002) A draft sequence of the rice genome (*Oryza sativa* L. ssp. *japonica*). *Science* 296: 92–100.

71. Consortium (2001) Initial sequencing and analysis of the human genome. *Nature* 409: 860–921.

72. Venter J.C. et al. (2001) The sequence of the human genome. *Science* 291: 1304–51.

73. Internation Human Genome Sequencing Consortium (2004) Finishing the euchromatic sequence of the human genome. *Nature* 431: 931–45.

74. Waterston R.H. et al. (2002) Initial sequencing and comparative analysis of the mouse genome. *Nature* 420: 520–62.

75. Gibbs R.A. et al. (2004) Genome sequence of the Brown Norway rat yields insights into mammalian evolution. *Nature* 428: 493–521.

76. Lindblad-Toh K. et al. (2005) Genome sequence, comparative analysis and haplotype structure of the domestic dog. *Nature* 438: 803–19.

77. Elsik C.G., Tellam R.L. and Worley K.C. (2009) The genome sequence of taurine cattle: A window to ruminant biology and evolution. *Science* 324: 522–8.

78. Chimpanzee Sequencing and Analysis Consortium (2005) Initial sequence of the chimpanzee genome and comparison with the human genome. *Nature* 437: 69–87.

79. Hillier L.W. et al. (2004) Sequence and comparative analysis of the chicken genome provide unique perspectives on vertebrate evolution. *Nature* 432: 695–716.

80. Aparicio S. et al. (2002) Whole-genome shotgun assembly and analysis of the genome of *Fugu rubripes*. *Science* 297: 1301–10.

81. Dehal P. et al. (2002) The draft genome of *Ciona intestinalis*: Insights into chordate and vertebrate origins. *Science* 298: 2157–67.

82. Brudno M. et al. (2004) Automated whole-genome multiple alignment of rat, mouse, and human. *Genome Research* 14: 685–92.

83. Demuth J.P., Bie T.D., Stajich J.E., Cristianini N. and Hahn M.W. (2006) The evolution of mammalian gene families. *PLOS ONE* 1: e85.

84. Lutfalla G. et al. (2003) Comparative genomic analysis reveals independent expansion of a lineage-specific gene family in vertebrates: The class II cytokine receptors and their ligands in mammals and fish. *BMC Genomics* 4: 29.

85. Niimura Y. and Nei M. (2007) Extensive gains and losses of olfactory receptor genes in mammalian evolution. *PLOS ONE* 2: e708.

86. Hahn M.W. and Wray G.A. (2002) The g-value paradox. *Evolution & Development* 4: 73–5.

87. Bird A.P. et al. (1995) Transcriptional noise and the evolution of gene number. *Philosophical Transactions of the Royal Society B: Biological Sciences* 349: 249–53.

88. Hodgkin J. (2001) What does a worm want with 20,000 genes? *Genome Biology* 2: comment2008.1.

89. Stein L.D. et al. (2003) The genome sequence of *Caenorhabditis briggsae*: A platform for comparative genomics. *PLOS Biology* 1: e45.

90. Hillier L.W. et al. (2005) Genomics in *C. elegans*: So many genes, such a little worm. *Genome Research* 15: 1651–60.

91. Srivastava M. et al. (2010) The *Amphimedon queenslandica* genome and the evolution of animal complexity. *Nature* 466: 720–6.

92. Chen X.-G. et al. (2015) Genome sequence of the Asian Tiger mosquito, *Aedes albopictus*, reveals insights into its biology, genetics, and evolution. *Proceedings of the National Academy of Sciences USA* 112: E5907–15.

93. Colbourne J.K. et al. (2011) The ecoresponsive genome of *Daphnia pulex*. *Science* 331: 555–61.

94. Bianconi E. et al. (2013) An estimation of the number of cells in the human body. *Annals of Human Biology* 40: 463–71.

95. Azevedo F.A.C. et al. (2009) Equal numbers of neuronal and nonneuronal cells make the human brain an isometrically scaled-up primate brain. *Journal of Comparative Neurology* 513: 532–41.

96. Goodstadt L. and Ponting C.P. (2006) Phylogenetic reconstruction of orthology, paralogy, and conserved synteny for dog and human. *PLOS Computational Biology* 2: e133.

97. Clamp M. et al. (2007) Distinguishing protein-coding and noncoding genes in the human genome. *Proceedings of the National Academy of Sciences USA* 104: 19428–33.

98. Church D.M. et al. (2009) Lineage-specific biology revealed by a finished genome assembly of the mouse. *PLOS Biology* 7: e1000112.

99. Zimin A.V. et al. (2009) A whole-genome assembly of the domestic cow, *Bos taurus*. *Genome Biology* 10: R42.

100. Pertea M. and Salzberg S.L. (2010) Between a chicken and a grape: Estimating the number of human genes. *Genome Biology* 11: 206.

101. Montague M.J. et al. (2014) Comparative analysis of the domestic cat genome reveals genetic signatures underlying feline biology and domestication. *Proceedings of the National Academy of Sciences USA* 111: 17230–5.

102. Willyard C. (2018) New human gene tally reignites debate. *Nature* 558: 354–5.

103. Taft R.J., Pheasant M. and Mattick J.S. (2007) The relationship between non-protein-coding DNA and eukaryotic complexity. *BioEssays* 29: 288–99.

104. Liu G., Mattick J.S. and Taft R.J. (2013) A meta-analysis of the genomic and transcriptomic composition of complex life. *Cell Cycle* 12: 2061–72.

105. Rogers J.C. (1988) RNA complementary to α-amylase mRNA in barley. *Plant Molecular Biology* 11: 125–38.

106. Vinogradov A.E. and Anatskaya O.V. (2007) Organismal complexity, cell differentiation and gene expression: Human over mouse. *Nucleic Acids Research* 35: 6350–6.

107. Bell G. and Mooers A.O. (2008) Size and complexity among multicellular organisms. *Biological Journal of the Linnean Society* 60: 345–63.

108. Schad E., Tompa P. and Hegyi H. (2011) The relationship between proteome size, structural disorder and organism complexity. *Genome Biology* 12: R120.

109. Schmucker D. et al. (2000) Drosophila Dscam is an axon guidance receptor exhibiting extraordinary molecular diversity. *Cell* 101: 671–84.

110. Hattori D. et al. (2007) Dscam diversity is essential for neuronal wiring and self-recognition. *Nature* 449: 223–7.

111. Deveson I.W. et al. (2018) Universal alternative splicing of noncoding exons. *Cell Systems* 6: 245–55.

112. Tress M.L., Abascal F. and Valencia A. (2017) Alternative splicing may not be the key to proteome complexity. *Trends in Biochemical Sciences* 42: 98–110.

113. Levine M. and Tjian R. (2003) Transcription regulation and animal diversity. *Nature* 424: 147–51.

114. Solé R.V., Fernández P. and Kauffman S.A. (2003) Adaptive walks in a gene network model of morphogenesis: Insights into the Cambrian explosion. *International Journal of Developmental Biology* 47: 685–93.

115. Krebs A.R. (2021) Studying transcription factor function in the genome at molecular resolution. *Trends in Genetics* 37: 798–806.

116. Clark A.G. et al. (2007) Evolution of genes and genomes on the *Drosophila* phylogeny. *Nature* 450: 203–18.

117. Kim B.Y. et al. (2021) Highly contiguous assemblies of 101 drosophilid genomes. *eLife* 10: e66405.

118. Bejerano G. et al. (2004) Ultraconserved elements in the human genome. *Science* 304: 1321–5.

119. Stephen S., Pheasant M., Makunin I.V. and Mattick J.S. (2008) Large-scale appearance of ultraconserved elements in tetrapod genomes and slowdown of the molecular clock. *Molecular Biology and Evolution* 25: 402–8.

120. Katzman S. et al. (2007) Human genome ultraconserved elements are ultraselected. *Science* 317: 915.

121. Bejerano G. et al. (2006) A distal enhancer and an ultraconserved exon are derived from a novel retroposon. *Nature* 441: 87–90.

122. Sandelin A. et al. (2004) Arrays of ultraconserved noncoding regions span the loci of key developmental genes in vertebrate genomes. *BMC Genomics* 5: 99.

123. Poulin F. et al. (2005) In vivo characterization of a vertebrate ultraconserved enhancer. *Genomics* 85: 774–81.

124. Pennacchio L.A. et al. (2006) In vivo enhancer analysis of human conserved non-coding sequences. *Nature* 444: 499–502.

125. Feng J. et al. (2006) The Evf-2 noncoding RNA is transcribed from the Dlx-5/6 ultraconserved region and functions as a Dlx-2 transcriptional coactivator. *Genes & Development* 20: 1470–84.

126. Visel A. et al. (2008) Ultraconservation identifies a small subset of extremely constrained developmental enhancers. *Nature Genetics* 40: 158–60.

127. Calin G.A. et al. (2007) Ultraconserved regions encoding ncRNAs are altered in human leukemias and carcinomas. *Cancer Cell* 12: 215–29.

128. Ferdin J. et al. (2013) HINCUTs in cancer: Hypoxia-induced noncoding ultraconserved transcripts. *Cell Death & Differentiation* 20: 1675–87.

129. Liz J. et al. (2014) Regulation of pri-miRNA processing by a long noncoding RNA transcribed from an ultraconserved region. *Molecular Cell* 55: 138–47.

130. Fiorenzano A. et al. (2018) An ultraconserved element containing lncRNA preserves transcriptional dynamics and maintains ESC self-renewal. *Stem Cell Reports* 10: 1102–14.

131. Zhang C., Peng Y., Wang Y., Xu H. and Zhou X. (2020) Transcribed ultraconserved noncoding RNA uc.153 is a new player in neuropathic pain. *Pain* 161: 1744–54.

132. Derti A., Roth F.P., Church G.M. and Wu C.T. (2006) Mammalian ultraconserved elements are strongly depleted among segmental duplications and copy number variants. *Nature Genetics* 38: 1216–20.

133. Wang X. and Goldstein D.B. (2020) Enhancer domains predict gene pathogenicity and inform gene discovery in complex disease. *American Journal of Human Genetics* 106: 215–33.

134. Ahituv N. et al. (2007) Deletion of ultraconserved elements yields viable mice. *PLOS Biology* 5: e234.

135. Dickel D.E. et al. (2018) Ultraconserved enhancers are required for normal development. *Cell* 172: 491–9.

136. Snetkova V. et al. (2021) Ultraconserved enhancer function does not require perfect sequence conservation. *Nature Genetics* 53: 521–8.

137. Pittman M. and Pollard K.S. (2021) Ultraconservation of enhancers is not ultranecessary. *Nature Genetics* 53: 429–30.

138. Pollard K.S. et al. (2006) An RNA gene expressed during cortical development evolved rapidly in humans. *Nature* 443: 167–72.

139. Prabhakar S., Noonan J.P., Paabo S. and Rubin E.M. (2006) Accelerated evolution of conserved noncoding sequences in humans. *Science* 314: 786.

140. Rakic P. (2009) Evolution of the neocortex: A perspective from developmental biology. *Nature Reviews Neuroscience* 10: 724–35.

141. Molnár Z. and Pollen A. (2014) How unique is the human neocortex? *Development* 141: 11–6.

142. Prabhakar S. et al. (2008) Human-specific gain of function in a developmental enhancer. *Science* 321: 1346–50.

143. Doan R.N. et al. (2016) Mutations in human accelerated regions disrupt cognition and social behavior. *Cell* 167: 341–54.

144. Suzuki S., Miyabe E. and Inagaki S. (2018) Novel brain-expressed noncoding RNA, HSTR1, identified at a human-specific variable number tandem repeat locus with a human accelerated region. *Biochemical and Biophysical Research Communications* 503: 1478–83.

145. Song J.H.T., Lowe C.B. and Kingsley D.M. (2018) Characterization of a human-specific tandem repeat associated with Bipolar Disorder and Schizophrenia. *American Journal of Human Genetics* 103: 421–30.

146. Lipovich L., Vanisri R.R., Kong S.L., Lin C.Y. and Liu E.T. (2006) Primate-specific endogenous cis-antisense transcription in the human 5q31 protocadherin gene cluster. *Journal of Molecular Evolution* 62: 73–88.

147. Zhang Z., Pang A.W. and Gerstein M. (2007) Comparative analysis of genome tiling array data reveals many novel primate-specific functional RNAs in human. *BMC Evolutionary Biology* 7: S14.

148. Derrien T. et al. (2012) The GENCODE v7 catalog of human long noncoding RNAs: Analysis of their gene structure, evolution, and expression. *Genome Research* 22: 1775–89.

149. Wang J. et al. (2014) Primate-specific endogenous retrovirus-driven transcription defines naive-like stem cells. *Nature* 516: 405–9.

150. Durruthy-Durruthy J. et al. (2016) The primate-specific noncoding RNA HPAT5 regulates pluripotency during human preimplantation development and nuclear reprogramming. *Nature Genetics* 48: 44–52.

151. Field A.R. et al. (2019) Structurally conserved primate lncRNAs are transiently expressed during human cortical differentiation and influence cell-type-specific genes. *Stem Cell Reports* 12: 245–57.

152. Hennessy E.J. et al. (2019) The long noncoding RNA CHROME regulates cholesterol homeostasis in primates. *Nature Metabolism* 1: 98–110.

153. Casanova M. et al. (2019) A primate-specific retroviral enhancer wires the XACT lncRNA into the core pluripotency network in humans. *Nature Communications* 10: 5652.

154. Issler O. et al. (2020) Sex-specific role for the long noncoding RNA LINC00473 in depression. *Neuron* 106: 912–26.

155. Wilson K.D. et al. (2020) Endogenous retrovirus-derived lncRNA BANCR promotes cardiomyocyte migration in humans and non-human primates. *Developmental Cell* 54: 694–709.

156. Liu S. et al. (2020) Identifying common genome-wide risk genes for major psychiatric traits. *Human Genetics* 139: 185–98.

157. Torrents D., Suyama M., Zdobnov E. and Bork P. (2003) A genome-wide survey of human pseudogenes. *Genome Research* 13: 2559–67.

158. Vanin E.F. (1985) Processed pseudogenes: Characteristics and evolution. *Annual Review of Genetics* 19: 253–72.

159. Harrison P.M. et al. (2002) Molecular fossils in the human genome: Identification and analysis of the pseudogenes in chromosomes 21 and 22. *Genome Research* 12: 272–80.

160. Zhang Z., Harrison P.M., Liu Y. and Gerstein M. (2003) Millions of years of evolution preserved: A comprehensive catalog of the processed pseudogenes in the human genome. *Genome Research* 13: 2541–58.

161. Carelli F.N. et al. (2016) The life history of retrocopies illuminates the evolution of new mammalian genes. *Genome Research* 26: 301–14.

162. Podlaha O. and Zhang J. (2004) Nonneutral evolution of the transcribed pseudogene Makorin1-p1 in mice. *Molecular Biology and Evolution* 21: 2202–9.

163. Moleirinho A. et al. (2013) Evolutionary constraints in the β-globin cluster: The signature of purifying selection at the δ-globin (HBD) locus and its role in developmental gene regulation. *Genome Biology and Evolution* 5: 559–71.

164. Li W., Yang W. and Wang X.-J. (2013) Pseudogenes: Pseudo or real functional elements? *Journal of Genetics and Genomics* 40: 171–7.

165. Ma Y. et al. (2021) Genome-wide analysis of pseudogenes reveals HBBP1's human-specific essentiality in erythropoiesis and implication in β-thalassemia. *Developmental Cell* 56: 478–93.

166. Korneev S.A., Park J.H. and O'Shea M. (1999) Neuronal expression of neural nitric oxide synthase (nNOS) protein is suppressed by an antisense RNA transcribed from an NOS pseudogene. *Journal of Neuroscience* 19: 7711–20.

167. Hirotsune S. et al. (2003) An expressed pseudogene regulates the messenger-RNA stability of its homologous coding gene. *Nature* 423: 91–6.

168. Yano Y. et al. (2004) A new role for expressed pseudogenes as ncRNA: Regulation of mRNA stability of its homologous coding gene. *Journal of Molecular Medicine* 82: 414–22.

169. Devor E.J. (2006) Primate microRNAs miR-220 and miR-492 lie within processed pseudogenes. *Journal of Heredity* 97: 186–90.

170. Tam O.H. et al. (2008) Pseudogene-derived small interfering RNAs regulate gene expression in mouse oocytes. *Nature* 453: 534–8.

171. Hawkins P.G. and Morris K.V. (2010) Transcriptional regulation of Oct4 by a long non-coding RNA antisense to Oct4-pseudogene 5. *Transcription* 1: 165–75.

172. Poliseno L. et al. (2010) A coding-independent function of gene and pseudogene mRNAs regulates tumour biology. *Nature* 465: 1033–8.

173. Poliseno L. (2012) Pseudogenes: Newly discovered players in human cancer. *Science Signaling* 5: 5.

174. Johnsson P. et al. (2013) A pseudogene long-noncoding-RNA network regulates PTEN transcription and translation in human cells. *Nature Structural & Molecular Biology* 20: 440–6.

175. Cheetham S.W., Faulkner G.J. and Dinger M.E. (2020) Overcoming challenges and dogmas to understand the functions of pseudogenes. *Nature Reviews Genetics* 21: 191–201.

176. Huang C.R.L., Burns K.H. and Boeke J.D. (2012) Active transposition in genomes. *Annual Review of Genetics* 46: 651–75.

177. Wells J.N. and Feschotte C. (2020) A field guide to eukaryotic transposable elements. *Annual Review of Genetics* 54: 539–61.

178. Boeke J.D. and Stoye J.P. (1997) Retrotransposons, Endogenous retroviruses, and the evolution of retroelements, in J.M. Coffin, S.H. Hughes and H.E. Varmus, (eds.) *Retroviruses* (Cold Spring Harbor Laboratory Press, Cold Spring Harbor, NY).

179. Sela N. et al. (2007) Comparative analysis of transposed element insertion within human and mouse genomes reveals Alu's unique role in shaping the human transcriptome. *Genome Biology* 8: R127.

180. Liu G.E., Alkan C., Jiang L., Zhao S. and Eichler E.E. (2009) Comparative analysis of Alu repeats in primate genomes. *Genome Research* 19: 876–85.

181. Fedoroff N.V. (2012) Transposable elements, epigenetics, and genome evolution. *Science* 338: 758–67.

182. Weiner A.M. (2002) SINEs and LINEs: The art of biting the hand that feeds you. *Current Opinion in Cell Biology* 14: 343–50.

183. Beck C.R. et al. (2010) LINE-1 retrotransposition activity in human genomes. *Cell* 141: 1159–70.

184. Burns K.H. and Boeke J.D. (2012) Human transposon tectonics. *Cell* 149: 740–52.

185. Levanon E.Y. et al. (2004) Systematic identification of abundant A-to-I editing sites in the human transcriptome. *Nature Biotechnology* 22: 1001–5.

186. Kim D.D. et al. (2004) Widespread RNA editing of embedded alu elements in the human transcriptome. *Genome Research* 14: 1719–25.

187. Blow M., Futreal P.A., Wooster R. and Stratton M.R. (2004) A survey of RNA editing in human brain. *Genome Research* 14: 2379–87.

188. Athanasiadis A., Rich A. and Maas S. (2004) Widespread A-to-I RNA editing of Alu-containing mRNAs in the human transcriptome. *PLOS Biology* 2: e391.

189. Paz-Yaacov N. et al. (2010) Adenosine-to-inosine RNA editing shapes transcriptome diversity in primates. *Proceedings of the National Academy of Sciences USA* 107: 12174–9.

190. Bazak L. et al. (2014) A-to-I RNA editing occurs at over a hundred million genomic sites, located in a majority of human genes. *Genome Research* 24: 365–76.

191. Zhang X.-O., Gingeras T.R. and Weng Z. (2019) Genome-wide analysis of polymerase III–transcribed Alu elements suggests cell-type–specific enhancer function. *Genome Research* 29: 1402–14.

192. Tomilin N.V. (2008) Regulation of mammalian gene expression by retroelements and non-coding tandem repeats. *BioEssays* 30: 338–48.

193. Boissinot S. and Sookdeo A. (2016) The evolution of LINE-1 in vertebrates. *Genome Biology and Evolution* 8: 3485–507.

194. Tsirigos A. and Rigoutsos I. (2009) Alu and B1 repeats have been selectively retained in the upstream and intronic regions of genes of specific functional classes. *PLOS Computational Biology* 5: e1000610.

195. Shapiro J.A. and von Sternberg R. (2005) Why repetitive DNA is essential to genome function. *Biological Reviews of the Cambridge Philosophical Society* 80: 227–50.

196. Brunet T.D.P. and Doolittle W.F. (2015) Multilevel selection theory and the evolutionary functions of transposable elements. *Genome Biology and Evolution* 7: 2445–57.

197. Kleckner N. (1981) Transposable elements in prokaryotes. *Annual Review of Genetics* 15: 341–404.

198. Shapiro J.A. (1992) Natural genetic engineering in evolution. *Genetica* 86: 99–111.

199. Witzany G. (2009) *Natural Genetic Engineering and Natural Genome Editing*, Annals of the New York Academy of Sciences, vol. 1178 (Wiley-Blackwell, Hoboken, NJ).

200. Roller M. et al. (2021) LINE retrotransposons characterize mammalian tissue-specific and evolutionarily dynamic regulatory regions. *Genome Biology* 22: 62.

201. Cosby R.L. et al. (2021) Recurrent evolution of vertebrate transcription factors by transposase capture. *Science* 371: eabc6405.

202. Brosius J. (2003) How significant is 98.5% 'junk' in mammalian genomes? *Bioinformatics* 19(Suppl 2): ii35.

203. Brosius J. (2003) The contribution of RNAs and retroposition to evolutionary novelties. *Genetica* 118: 99–116.

204. Jurka J. (2004) Evolutionary impact of human Alu repetitive elements. *Current Opinion in Genetics and Development* 14: 603–8.

205. Brandt J. et al. (2005) Transposable elements as a source of genetic innovation: Expression and evolution of a family of retrotransposon-derived neogenes in mammals. *Gene* 345: 101–11.

206. Volff J.N. (2006) Turning junk into gold: Domestication of transposable elements and the creation of new genes in eukaryotes. *BioEssays* 28: 913–22.

207. Volff J.N. and Brosius J. (2007) Modern genomes with retro-look: Retrotransposed elements, retroposition and the origin of new genes. *Gene and Protein Evolution* 3: 175–90.

208. Feschotte C. and Pritham E.J. (2007) DNA transposons and the evolution of eukaryotic genomes. *Annual Review of Genetics* 41: 331–68.

209. Jurka J., Kapitonov V.V., Kohany O. and Jurka M.V. (2007) Repetitive sequences in complex genomes: Structure and evolution. *Annual Review of Genomics and Human Genetics* 8: 241–59.

210. Oliver K.R. and Greene W.K. (2009) Transposable elements: Powerful facilitators of evolution. *BioEssays* 31: 703–14.

211. Zeh D.W., Zeh J.A. and Ishida Y. (2009) Transposable elements and an epigenetic basis for punctuated equilibria. *BioEssays* 31: 715–26.

212. Britten R.J. (2010) Transposable element insertions have strongly affected human evolution. *Proceedings of the National Academy of Sciences USA* 107: 19945–8.

213. Vogel G. (2011) Retrotransposons. Do jumping genes spawn diversity? *Science* 332: 300–1.

214. Johnson R. and Guigo R. (2014) The RIDL hypothesis: Transposable elements as functional domains of long noncoding RNAs. *RNA* 20: 959–76.

215. Dubin M.J., Mittelsten Scheid O. and Becker C. (2018) Transposons: A blessing curse. *Current Opinion in Plant Biology* 42: 23–9.

216. Cosby R.L., Chang N.-C. and Feschotte C. (2019) Host-transposon interactions: Conflict, cooperation, and cooption. *Genes & Development* 33: 1098–116.

217. Peaston A.E. et al. (2004) Retrotransposons regulate host genes in mouse oocytes and preimplantation embryos. *Developmental Cell* 7: 597–606.

218. Feschotte C. (2008) Transposable elements and the evolution of regulatory networks. *Nature Reviews Genetics* 9: 397–405.

219. Faulkner G.J. (2011) Retrotransposons: Mobile and mutagenic from conception to death. *FEBS Letters* 585: 1589–94.

220. Göke J. et al. (2015) Dynamic transcription of distinct classes of endogenous retroviral elements marks specific populations of early human embryonic cells. *Cell Stem Cell* 16: 135–41.

221. Göke J. and Ng H.H. (2016) CTRL+INSERT: Retrotransposons and their contribution to regulation and innovation of the transcriptome. *EMBO Reports* 17: 1131–44.

222. Lowe C.B., Bejerano G. and Haussler D. (2007) Thousands of human mobile element fragments undergo strong purifying selection near developmental genes. *Proceedings of the National Academy of Sciences USA* 104: 8005–10.

223. Simons C., Pheasant M., Makunin I.V. and Mattick J.S. (2006) Transposon-free regions in mammalian genomes. *Genome Research* 16: 164–72.

224. Jordan I.K., Rogozin I.B., Glazko G.V. and Koonin E.V. (2003) Origin of a substantial fraction of human regulatory sequences from transposable elements. *Trends in Genetics* 19: 68–72.

225. Faulkner G.J. et al. (2009) The regulated retrotransposon transcriptome of mammalian cells. *Nature Genetics* 41: 563–71.

226. Xu A.G. et al. (2010) Intergenic and repeat transcription in human, chimpanzee and macaque brains measured by RNA-Seq. *PLOS Computational Biology* 6: e1000843.

227. Skryabin B.V. et al. (1998) The BC200 RNA gene and its neural expression are conserved in Anthropoidea (Primates). *Journal of Molecular Evolution* 47: 677–85.

228. Silva J.C., Shabalina S.A., Harris D.G., Spouge J.L. and Kondrashovi A.S. (2003) Conserved fragments of transposable elements in intergenic regions: Evidence for widespread recruitment of MIR- and L2-derived sequences within the mouse and human genomes. *Genetics Research* 82: 1–18.

229. Xie X., Kamal M. and Lander E.S. (2006) A family of conserved noncoding elements derived from an ancient transposable element. *Proceedings of the National Academy of Sciences USA* 103: 11659–64.

230. Nishihara H., Smit A.F. and Okada N. (2006) Functional noncoding sequences derived from SINEs in the mammalian genome. *Genome Research* 16: 864–74.

231. Brosius J. (1999) RNAs from all categories generate retrosequences that may be exapted as novel genes or regulatory elements. *Gene* 238: 115–34.

232. Kuryshev V.Y., Skryabin B.V., Kremerskothen J., Jurka J. and Brosius J. (2001) Birth of a gene: Locus of neuronal BC200 snmRNA in three prosimians and human BC200 pseudogenes as archives of change in the Anthropoidea lineage. *Journal of Molecular Biology* 309: 1049–66.

233. Britten R. (2006) Transposable elements have contributed to thousands of human proteins. *Proceedings of the National Academy of Sciences USA* 103: 1798–803.

234. Cordaux R., Udit S., Batzer M.A. and Feschotte C. (2006) Birth of a chimeric primate gene by capture of the transposase gene from a mobile element. *Proceedings of the National Academy of Sciences USA* 103: 8101–6.

235. Krull M., Brosius J. and Schmitz J. (2005) Alu-SINE exonization: En route to protein-coding function. *Molecular Biology and Evolution* 22: 1702–11.

236. Hezroni H. et al. (2015) Principles of long noncoding RNA evolution derived from direct comparison of transcriptomes in 17 species. *Cell Reports* 11: 1110–22.

237. Gerdes P., Richardson S.R., Mager D.L. and Faulkner G.J. (2016) Transposable elements in the mammalian embryo: Pioneers surviving through stealth and service. *Genome Biology* 17: 100.

238. Yadav V. et al. (2018) RNAi is a critical determinant of centromere evolution in closely related fungi. *Proceedings of the National Academy of Sciences USA* 115: 3108–13.

239. Hartley G. and O'Neill R.J. (2019) Centromere repeats: Hidden gems of the genome. *Genes* 10: 223.

240. van de Lagemaat L.N., Landry J.-R., Mager D.L. and Medstrand P. (2003) Transposable elements in mammals promote regulatory variation and diversification of genes with specialized functions. *Trends in Genetics* 19: 530–6.

241. Bourque G. et al. (2008) Evolution of the mammalian transcription factor binding repertoire via transposable elements. *Genome Research* 18: 1752–62.

242. Hermant C. and Torres-Padilla M.-E. (2021) TFs for TEs: The transcription factor repertoire of mammalian transposable elements. *Genes & Development* 35: 22–39.

243. Matlik K., Redik K. and Speek M. (2006) L1 antisense promoter drives tissue-specific transcription of human genes. *Journal of Biomedicine and Biotechnology* 2006: 71753.

244. Lunyak V.V. et al. (2007) Developmentally regulated activation of a SINE B2 repeat as a domain boundary in organogenesis. *Science* 317: 248–51.

245. Kelley D. and Rinn J. (2012) Transposable elements reveal a stem cell-specific class of long noncoding RNAs. *Genome Biology* 13: R107.

246. Trizzino M. et al. (2017) Transposable elements are the primary source of novelty in primate gene regulation. *Genome Research* 27: 1623–33.

247. Ferrigno O. et al. (2001) Transposable B2 SINE elements can provide mobile RNA polymerase II promoters. *Nature Genetics* 28: 77–81.

248. Nigumann P., Redik K., Matlik K. and Speek M. (2002) Many human genes are transcribed from the antisense promoter of L1 retrotransposon. *Genomics* 79: 628–34.

249. Speek M. (2001) Antisense promoter of human L1 retrotransposon drives transcription of adjacent cellular genes. *Molecular and Cellular Biology* 21: 1973–85.

250. Ludwig A., Rozhdestvensky T.S., Kuryshev V.Y., Schmitz J. and Brosius J. (2005) An unusual primate locus that attracted two independent Alu insertions and facilitates their transcription. *Journal of Molecular Biology* 350: 200–14.

251. Romanish M.T., Lock W.M., de Lagemaat L.N., Dunn C.A. and Mager D.L. (2007) Repeated recruitment of LTR retrotransposons as promoters by the anti-apoptotic locus NAIP during mammalian evolution. *PLOS Genetics* 3: e10.

252. Whitelaw E. and Martin D.I. (2001) Retrotransposons as epigenetic mediators of phenotypic variation in mammals. *Nature Genetics* 27: 361–5.

253. Schramke V. and Allshire R. (2003) Hairpin RNAs and retrotransposon LTRs effect RNAi and chromatin-based gene silencing. *Science* 301: 1069–74.

254. Lippman Z. et al. (2004) Role of transposable elements in heterochromatin and epigenetic control. *Nature* 430: 471–6.

255. Slotkin R.K. and Martienssen R. (2007) Transposable elements and the epigenetic regulation of the genome. *Nature Reviews Genetics* 8: 272–85.

256. Chueh A.C., Northrop E.L., Brettingham-Moore K.H., Choo K.H. and Wong L.H. (2009) LINE retrotransposon RNA is an essential structural and functional epigenetic component of a core neocentromeric chromatin. *PLOS Genetics* 5: e1000354.

257. Gehring M., Bubb K.L. and Henikoff S. (2009) Extensive demethylation of repetitive elements during seed development underlies gene imprinting. *Science* 324: 1447–51.

258. McCue A.D., Nuthikattu S., Reeder S.H. and Slotkin R.K. (2012) Gene expression and stress response mediated by the epigenetic regulation of a transposable element small RNA. *PLOS Genetics* 8: e1002474.

259. Brind'Amour J. et al. (2018) LTR retrotransposons transcribed in oocytes drive species-specific and heritable changes in DNA methylation. *Nature Communications* 9: 3331.

260. Bogutz A.B. et al. (2019) Evolution of imprinting via lineage-specific insertion of retroviral promoters. *Nature Communications* 10: 5674.

261. Kelley D.R., Hendrickson D.G., Tenen D. and Rinn J.L. (2014) Transposable elements modulate human RNA abundance and splicing via specific RNA-protein interactions. *Genome Biology* 15: 537.

262. Landry J.R., Medstrand P. and Mager D.L. (2001) Repetitive elements in the 5′ untranslated region of a human zinc-finger gene modulate transcription and translation efficiency. *Genomics* 76: 110–6.

263. Lubelsky Y. and Ulitsky I. (2018) Sequences enriched in Alu repeats drive nuclear localization of long RNAs in human cells. *Nature* 555: 107–11.

264. Carlevaro-Fita J. et al. (2019) Ancient exapted transposable elements promote nuclear enrichment of human long noncoding RNAs. *Genome Research* 29: 208–22.

265. Lewejohann L. et al. (2004) Role of a neuronal small non-messenger RNA: Behavioural alterations in BC1 RNA-deleted mice. *Behavioural Brain Research* 154: 273–89.

266. Volpe T.A. et al. (2002) Regulation of heterochromatic silencing and histone H3 lysine-9 methylation by RNAi. *Science* 297: 1833–7.

267. Ferrari R. et al. (2020) TFIIIC binding to Alu elements controls gene expression via chromatin looping and histone acetylation. *Molecular Cell* 77: 475–87.

268. Diehl A.G., Ouyang N. and Boyle A.P. (2020) Transposable elements contribute to cell and species-specific chromatin looping and gene regulation in mammalian genomes. *Nature Communications* 11: 1796.

269. Choudhary M.N.K. et al. (2020) Co-opted transposons help perpetuate conserved higher-order chromosomal structures. *Genome Biology* 21: 16.

270. Brown S.W. (1966) Heterochromatin. *Science* 151: 417–25.

271. Pontis J. et al. (2019) Hominoid-specific transposable elements and KZFPs facilitate human embryonic genome activation and control transcription in naive human ESCs. *Cell Stem Cell* 24: 724–35.

272. Wolf G. et al. (2020) KRAB-zinc finger protein gene expansion in response to active retrotransposons in the murine lineage. *eLife* 9: e56337.

273. Bulut-Karslioglu A. et al. (2012) A transcription factor-based mechanism for mouse heterochromatin formation. *Nature Structural & Molecular Biology* 19: 1023–30.

274. Turelli P. et al. (2020) Primate-restricted KRAB zinc finger proteins and target retrotransposons control gene expression in human neurons. *Science Advances* 6: eaba3200.

275. Babarinde Isaac A. et al. (2021) Transposable element sequence fragments incorporated into coding and noncoding transcripts modulate the transcriptome of human pluripotent stem cells. *Nucleic Acids Research* 49: 9132–53.

276. Lu X. et al. (2014) The retrovirus HERVH is a long noncoding RNA required for human embryonic stem cell identity. *Nature Structural & Molecular Biology* 21: 423–5.

277. Fort A. et al. (2014) Deep transcriptome profiling of mammalian stem cells supports a regulatory role for retrotransposons in pluripotency maintenance. *Nature Genetics* 46: 558–66.

278. Jacques P.-É., Jeyakani J. and Bourque G. (2013) The majority of primate-specific regulatory sequences are derived from transposable elements. *PLOS Genetics* 9: e1003504.

279. Macfarlan T.S. et al. (2012) Embryonic stem cell potency fluctuates with endogenous retrovirus activity. *Nature* 487: 57–63.

280. Modzelewski A.J. et al. (2021) A mouse-specific retrotransposon drives a conserved Cdk2ap1 isoform essential for development. *Cell* 184: 5541–58.

281. Singh K., Carey M., Saragosti S. and Botchan M. (1985) Expression of enhanced levels of small RNA polymerase III transcripts encoded by the B2 repeats in simian virus 40-transformed mouse cells. *Nature* 314: 553–6.

282. Fornace Jr. A.J., and Mitchell J.B. (1986) Induction of B2 RNA polymerase III transcription by heat shock: Enrichment for heat shock induced sequences in rodent cells by hybridization subtraction. *Nucleic Acids Research* 14: 5793–811.

283. Jang K.L. and Latchman D.S. (1989) HSV infection induces increased transcription of Alu repeated sequences by RNA polymerase III. *FEBS Letters* 258: 255–8.

284. Jang K.L. and Latchman D.S. (1992) The herpes simplex virus immediate-early protein ICP27 stimulates the transcription of cellular Alu repeated sequences by increasing the activity of transcription factor TFIIIC. *Biochemical Journal* 284: 667–73.

285. Panning B. and Smiley J.R. (1993) Activation of RNA polymerase III transcription of human Alu repetitive elements by adenovirus type 5: Requirement for the E1b 58-kilodalton protein and the products of E4 open reading frames 3 and 6. *Molecular and Cellular Biology* 13: 3231–44.

286. Russanova V.R., Driscoll C.T. and Howard B.H. (1995) Adenovirus type 2 preferentially stimulates polymerase III transcription of Alu elements by relieving repression: A potential role for chromatin. *Molecular and Cellular Biology* 15: 4282–90.

287. Liu W.M., Chu W.M., Choudary P.V. and Schmid C.W. (1995) Cell stress and translational inhibitors transiently increase the abundance of mammalian SINE transcripts. *Nucleic Acids Research* 23: 1758–65.

288. Li T.-H., Spearow J., Rubin C.M. and Schmid C.W. (1999) Physiological stresses increase mouse short interspersed element (SINE) RNA expression in vivo. *Gene* 239: 367–72.

289. Kimura R.H., Choudary P.V., Stone K.K. and Schmid C.W. (2001) Stress induction of Bm1 RNA in silkworm larvae: SINEs, an unusual class of stress genes. *Cell Stress & Chaperones* 6: 263–72.

290. Rudin C.M. and Thompson C.B. (2001) Transcriptional activation of short interspersed elements by DNA-damaging agents. *Genes, Chromosomes and Cancer* 30: 64–71.

291. Kigami D., Minami N., Takayama H. and Imai H. (2003) MuERV-L is one of the earliest transcribed genes in mouse one-cell embryos. *Biology of Reproduction* 68: 651–4.

292. Marasca F. et al. (2022) LINE1 are spliced in non-canonical transcript variants to regulate T cell quiescence and exhaustion. *Nature Genetics* 50: 180–93.

293. Moore R.S. et al. (2021) The role of the Cer1 transposon in horizontal transfer of transgenerational memory. *Cell* 184: 4697–712.

294. Wong W.Y. et al. (2019) Expansion of a single transposable element family is associated with genome-size increase and radiation in the genus Hydra. *Proceedings of the National Academy of Sciences USA* 116: 22915–7.

295. Etchegaray E., Naville M., Volff J.-N. and Haftek-Terreau Z. (2021) Transposable element-derived sequences in vertebrate development. *Mobile DNA* 12: 1.

296. Wang K. et al. (2021) African lungfish genome sheds light on the vertebrate water-to-land transition. *Cell* 184: 1362–76.

297. Xia B. et al. (2021) The genetic basis of tail-loss evolution in humans and apes. *bioRxiv*: 2021.09.14.460388.

298. Patoori S., Barnada S. and Trizzino M. (2021) Young transposable elements rewired gene regulatory networks in human and chimpanzee hippocampal intermediate progenitors. *bioRxiv* epub: 2021.11.24.469877.

299. Gray M.M., Sutter N.B., Ostrander E.A. and Wayne R.K. (2010) The IGF1 small dog haplotype is derived from Middle Eastern grey wolves. *BMC Biology* 8: 16.

300. Wang W. and Kirkness E.F. (2005) Short interspersed elements (SINEs) are a major source of canine genomic diversity. *Genome Research* 15: 1798–808.

301. Halo J.V. et al. (2021) Long-read assembly of a Great Dane genome highlights the contribution of GC-rich sequence and mobile elements to canine genomes. *Proceedings of the National Academy of Sciences USA* 118: e2016274118.

302. David V.A. et al. (2014) Endogenous retrovirus insertion in the KIT oncogene determines white and white spotting in domestic cats. *G3 (Genes, Genomes, Genetics)* 4: 1881–91.

303. Kratochwil C.F. et al. (2022) An intronic transposon insertion associates with a trans-species color polymorphism in Midas cichlid fishes. *Nature Communications* 13: 296.

304. Hiom K., Melek M. and Gellert M. (1998) DNA transposition by the RAG1 and RAG2 proteins: A possible source of oncogenic translocations. *Cell* 94: 463–70.

305. Kapitonov V.V. and Jurka J. (2005) RAG1 core and V(D)J recombination signal sequences were derived from transib transposons. *PLOS Biology* 3: e181.

306. Sznarkowska A., Mikac S. and Pilch M. (2020) MHC class I regulation: The origin perspective. *Cancers* 12: 1155.

307. Chuong E.B., Elde N.C. and Feschotte C. (2016) Regulatory evolution of innate immunity through co-option of endogenous retroviruses. *Science* 351: 1083–7.

308. Haldane J.B.S. (1924) A mathematical theory of natural and artificial selection. Part I. (The Causes of Evolution. London: Longmans, Green). *Transactions of the Cambridge Philosophical Society* 23: 19–41.

309. Sarkar S. (1992) The founders of theoretical evolutionary genetics, in S. Sarkar (ed.) *The Founders of Evolutionary Genetics* (Springer, Dordrecht).

310. v'ant Hof A.E. et al. (2016) The industrial melanism mutation in British peppered moths is a transposable element. *Nature* 534: 102–5.

311. Nadeau N.J. et al. (2016) The gene cortex controls mimicry and crypsis in butterflies and moths. *Nature* 534: 106–10.

312. v'ant Hof A.E., Edmonds N., Dalikova M., Marec F. and Saccheri I.J. (2011) Industrial melanism in British peppered moths has a singular and recent mutational origin. *Science* 332: 958–60.

313. Xiao H., Jiang N., Schaffner E., Stockinger E.J. and van der Knaap E. (2008) A retrotransposon-mediated gene duplication underlies morphological variation of tomato fruit. *Science* 319: 1527–30.

314. Studer A., Zhao Q., Ross-Ibarra J. and Doebley J. (2011) Identification of a functional transposon insertion in the maize domestication gene *tb1*. *Nature Genetics* 43: 1160–3.

315. Huang C. et al. (2018) ZmCCT9 enhances maize adaptation to higher latitudes. *Proceedings of the National Academy of Sciences USA* 115: E334–41.

316. Kobayashi S., Goto-Yamamoto N. and Hirochika H. (2004) Retrotransposon-induced mutations in grape skin color. *Science* 304: 982.

317. Zhang L. et al. (2019) A high-quality apple genome assembly reveals the association of a retrotransposon and red fruit colour. *Nature Communications* 10: 1494.

318. Butelli E. et al. (2012) Retrotransposons control fruit-specific, cold-dependent accumulation of anthocyanins in blood oranges. *Plant Cell* 24: 1242–55.

319. Baduel P. et al. (2021) Genetic and environmental modulation of transposition shapes the evolutionary potential of *Arabidopsis thaliana*. *Genome Biology* 22: 138.

320. Vitte C., Estep M.C., Leebens-Mack J. and Bennetzen J.L. (2013) Young, intact and nested retrotransposons are abundant in the onion and asparagus genomes. *Annals of Botany* 112: 881–9.

321. Wang Q. and Dooner H.K. (2006) Remarkable variation in maize genome structure inferred from haplotype diversity at the *bz* locus. *Proceedings of the National Academy of Sciences USA* 103: 17644–9.

322. Jiao Y. et al. (2017) Improved maize reference genome with single-molecule technologies. *Nature* 546: 524–7.

323. Stitzer M.C., Anderson S.N., Springer N.M. and Ross-Ibarra J. (2021) The genomic ecosystem of transposable elements in maize. *PLOS Genetics* 17: e1009768.

324. Liedtke H.C., Gower D.J., Wilkinson M. and Gomez-Mestre I. (2018) Macroevolutionary shift in the size of amphibian genomes and the role of life history and climate. *Nature Ecology & Evolution* 2: 1792–9.

325. Lertzman-Lepofsky G., Mooers A.Ø. and Greenberg D.A. (2019) Ecological constraints associated with genome size across salamander lineages. *Proceedings of the Royal Society B: Biological Sciences* 286: 20191780.

326. Canapa A. et al. (2020) Shedding light upon the complex net of genome size, genome composition and environment in chordates. *European Zoological Journal* 87: 192–202.

327. Calarco J.P. et al. (2012) Reprogramming of DNA methylation in pollen guides epigenetic inheritance via small RNA. *Cell* 151: 194–205.

328. Chandler V.L. and Walbot V. (1986) DNA modification of a maize transposable element correlates with loss of activity. *Proceedings of the National Academy of Sciences USA* 83: 1767–71.

329. Chomet P.S., Wessler S. and Dellaporta S.L. (1987) Inactivation of the maize transposable element Activator (Ac) is associated with its DNA modification. *EMBO Journal* 6: 295–302.

330. Fedoroff N.V. (1989) About maize transposable elements and development. *Cell* 56: 181–91.

331. Bertozzi T.M., Takahashi N., Hanin G., Kazachenka A. and Ferguson-Smith A.C. (2021) A spontaneous genetically induced epiallele at a retrotransposon shapes host genome function. *eLife* 10: e65233.

332. Mikkelsen T.S. et al. (2007) Genome of the marsupial *Monodelphis domestica* reveals innovation in non-coding sequences. *Nature* 447: 167–77.

333. Gemmell N.J. et al. (2020) The tuatara genome reveals ancient features of amniote evolution. *Nature* 584: 403–9.

334. Meyer A. et al. (2021) Giant lungfish genome elucidates the conquest of land by vertebrates. *Nature* 590: 284–9.

335. Jeffreys A.J. (1987) Highly variable minisatellites and DNA fingerprints. *Biochemical Society Transactions* 15: 309–17.

336. Willems T. et al. (2014) The landscape of human STR variation. *Genome Research* 24: 1894–904.

337. Kashi Y. and King D.G. (2006) Simple sequence repeats as advantageous mutators in evolution. *Trends in Genetics* 22: 253–9.

338. Vinces M.D., Legendre M., Caldara M., Hagihara M. and Verstrepen K.J. (2009) Unstable tandem repeats in promoters confer transcriptional evolvability. *Science* 324: 1213–6.

339. Gymrek M. et al. (2016) Abundant contribution of short tandem repeats to gene expression variation in humans. *Nature Genetics* 48: 22–9.

340. Bagshaw A.T.M. (2017) Functional mechanisms of microsatellite DNA in eukaryotic genomes. *Genome Biology and Evolution* 9: 2428–43.

341. Quilez J. et al. (2016) Polymorphic tandem repeats within gene promoters act as modifiers of gene expression and DNA methylation in humans. *Nucleic Acids Research* 44: 3750–62.

342. Kim Y.B. et al. (2014) Divergence of *Drosophila melanogaster* repeatomes in response to a sharp microclimate contrast in Evolution Canyon, Israel. *Proceedings of the National Academy of Sciences USA* 111: 10630–5.

343. Budworth H. and McMurray C.T. (2013) A brief history of triplet repeat diseases. *Methods in Molecular Biology* 1010: 3–17.

344. Trost B. et al. (2020) Genome-wide detection of tandem DNA repeats that are expanded in autism. *Nature* 586: 80–6.

345. Tempel S. (2012) Using and understanding RepeatMasker, in W. J. Miller and P. Capy (eds.) *Mobile Genetic Elements: Protocols and Genomic Applications* (Humana Press, Totowa, NJ).

346. http://www.repeatmasker.org.

347. Braslavsky I., Hebert B., Kartalov E. and Quake S.R. (2003) Sequence information can be obtained from single DNA molecules. *Proceedings of the National Academy of Sciences USA* 100: 3960–4.

348. Deamer D., Akeson M. and Branton D. (2016) Three decades of nanopore sequencing. *Nature Biotechnology* 34: 518–24.

349. Soylev A., Le T.M., Amini H., Alkan C. and Hormozdiari F. (2019) Discovery of tandem and interspersed segmental duplications using high-throughput sequencing. *Bioinformatics* 35: 3923–30.

350. Disdero E. and Filée J. (2017) LoRTE: Detecting transposon-induced genomic variants using low coverage PacBio long read sequences. *Mobile DNA* 8: 5.

351. Giani A.M., Gallo G.R., Gianfranceschi L. and Formenti G. (2019) Long walk to genomics: History and current approaches to genome sequencing and assembly. *Computational and Structural Biotechnology Journal* 18: 9–19.

352. Zhou W. et al. (2020) Identification and characterization of occult human-specific LINE-1 insertions using long-read sequencing technology. *Nucleic Acids Research* 48: 1146–63.

353. Jain M. et al. (2018) Nanopore sequencing and assembly of a human genome with ultra-long reads. *Nature Biotechnology* 36: 338–45.

354. Miga K.H. et al. (2020) Telomere-to-telomere assembly of a complete human X chromosome. *Nature* 585: 79–84.

355. Nurk S. et al. (2020) HiCanu: Accurate assembly of segmental duplications, satellites, and allelic variants from high-fidelity long reads. *Genome Research* 30: 1291–305.

356. Bentley D.R. et al. (2008) Accurate whole human genome sequencing using reversible terminator chemistry. *Nature* 456: 53–9.

357. Heather J.M. and Chain B. (2016) The sequence of sequencers: The history of sequencing DNA. *Genomics* 107: 1–8.

358. Shulgina Y. and Eddy S.R. (2021) A computational screen for alternative genetic codes in over 250,000 genomes. *eLife* 10: e71402.

359. Nayfach S. et al. (2021) A genomic catalog of Earth's microbiomes. *Nature Biotechnology* 39: 499–509.

360. Pham V.H.T. and Kim J. (2012) Cultivation of unculturable soil bacteria. *Trends in Biotechnology* 30: 475–84.

361. Browne H.P. et al. (2016) Culturing of 'unculturable' human microbiota reveals novel taxa and extensive sporulation. *Nature* 533: 543–6.

362. Brüssow H. (2009) The not so universal tree of life or the place of viruses in the living world. *Philosophical Transactions of the Royal Society B: Biological Sciences* 364: 2263–74.

363. Koonin E.V. (2009) On the origin of cells and viruses. *Annals of the New York Academy of Sciences* 1178: 47–64.

364. Nasir A., Kim K.M. and Caetano-Anolles G. (2012) Giant viruses coexisted with the cellular ancestors and represent a distinct supergroup along with superkingdoms Archaea, Bacteria and Eukarya. *BMC Evolutionary Biology* 12: 156.

365. Claverie J.-M. (2006) Viruses take center stage in cellular evolution. *Genome Biology* 7: 110.

366. Gregory A.C. et al. (2019) Marine DNA viral macro- and microdiversity from pole to pole. *Cell* 177: 1109–23.

367. Turnbaugh P.J. et al. (2007) The human microbiome project. *Nature* 449: 804–10.

368. Ochoa-Repáraz J. et al. (2009) Role of gut commensal microflora in the development of experimental autoimmune encephalomyelitis. *Journal of Immunology* 183: 6041–50.

369. De Vadder F. et al. (2014) Microbiota-generated metabolites promote metabolic benefits via gut-brain neural circuits. *Cell* 156: 84–96.

370. Singh V. et al. (2016) Microbiota dysbiosis controls the neuroinflammatory response after stroke. *Journal of Neuroscience* 36: 7428–40.

371. Olson C.A. et al. (2018) The gut microbiota mediates the anti-seizure effects of the ketogenic diet. *Cell* 173: 1728–41.

372. Valdes A.M., Walter J., Segal E. and Spector T.D. (2018) Role of the gut microbiota in nutrition and health. *British Medical Journal* 361: j2179.

373. Ding R-X, et al. (2019) Revisit gut microbiota and its impact on human health and disease. *Journal of Food and Drug Analysis* 27: 623–31.

374. Chen P.B. et al. (2020) Directed remodeling of the mouse gut microbiome inhibits the development of atherosclerosis. *Nature Biotechnology* 38: 1288–97.

375. Hsiao E.Y. et al. (2013) Microbiota modulate behavioral and physiological abnormalities associated with neurodevelopmental disorders. *Cell* 155: 1451–63.

376. Sampson T.R. et al. (2016) Gut microbiota regulate motor deficits and neuroinflammation in a model of Parkinson's Disease. *Cell* 167: 1469–80.

377. Kim S. et al. (2017) Maternal gut bacteria promote neurodevelopmental abnormalities in mouse offspring. *Nature* 549: 528–32.

378. Gerhardt S. and Mohajeri M.H. (2018) Changes of colonic bacterial composition in Parkinson's Disease and other neurodegenerative diseases. *Nutrients* 10: 708.

379. Clarke G. et al. (2013) The microbiome-gut-brain axis during early life regulates the hippocampal serotonergic system in a sex-dependent manner. *Molecular Psychiatry* 18: 666–73.

380. Yano J.M. et al. (2015) Indigenous bacteria from the gut microbiota regulate host serotonin biosynthesis. *Cell* 161: 264–76.

381. Desbonnet L., Clarke G., Shanahan F., Dinan T.G. and Cryan J.F. (2014) Microbiota is essential for social development in the mouse. *Molecular Psychiatry* 19: 146–8.

382. Buffington S.A. et al. (2016) Microbial reconstitution reverses maternal diet-induced social and synaptic deficits in offspring. *Cell* 165: 1762–75.

383. Valles-Colomer M. et al. (2019) The neuroactive potential of the human gut microbiota in quality of life and depression. *Nature Microbiology* 4: 623–32.

384. O'Donnell M.P., Fox B.W., Chao P.-H., Schroeder F.C. and Sengupta P. (2020) A neurotransmitter produced by gut bacteria modulates host sensory behaviour. *Nature* 583: 415–20.

385. Schretter C.E. et al. (2018) A gut microbial factor modulates locomotor behaviour in *Drosophila*. *Nature* 563: 402–6.

386. Sudo N. et al. (2004) Postnatal microbial colonization programs the hypothalamic-pituitary-adrenal system for stress response in mice. *Journal of Physiology* 558: 263–75.

387. Turnbaugh P.J. et al. (2009) A core gut microbiome in obese and lean twins. *Nature* 457: 480–4.

388. Barr J.J. et al. (2013) Bacteriophage adhering to mucus provide a non–host-derived immunity. *Proceedings of the National Academy of Sciences USA* 110: 10771–6.

389. Proctor L.M. et al. (2019) The integrative human microbiome project. *Nature* 569: 641–8.

390. Burki F., Roger A.J., Brown M.W. and Simpson A.G.B. (2020) The new tree of Eukaryotes. *Trends in Ecology & Evolution* 35: 43–55.

391. Matasci N. et al. (2014) Data access for the 1,000 Plants (1KP) project. *GigaScience* 3: 17.

392. Chen F. et al. (2018) The sequenced angiosperm genomes and genome databases. *Frontiers in Plant Science* 9: 418.

393. Leebens-Mack J.H. et al. (2019) One thousand plant transcriptomes and the phylogenomics of green plants. *Nature* 574: 679–85.

394. Koepfli K.-P., Benedict Paten, Genome 10K Community of Scientists and O'Brien S.J. (2015) The Genome 10K Project: A way forward. *Annual Review of Animal Biosciences* 3: 57–111.

395. Rhie A. et al. (2021) Towards complete and error-free genome assemblies of all vertebrate species. *Nature* 592: 737–46.

396. Lewin H.A. et al. (2018) Earth BioGenome project: Sequencing life for the future of life. *Proceedings of the National Academy of Sciences USA* 115: 4325–33.

397. Zhang J., Cong Q., Shen J., Opler P.A. and Grishin N.V. (2019) Genomics of a complete butterfly continent. *bioRxiv*: 829887.

398. Ellis E.A., Storer C.G. and Kawahara A.Y. (2021) De novo genome assemblies of butterflies. *GigaScience* 10: 1–8.

399. Genereux D.P. et al. (2020) A comparative genomics multitool for scientific discovery and conservation. *Nature* 587: 240–5.

400. Feng S. et al. (2020) Dense sampling of bird diversity increases power of comparative genomics. *Nature* 587: 252–7.

401. Tettelin H. et al. (2005) Genome analysis of multiple pathogenic isolates of *Streptococcus agalactiae*: Implications for the microbial "pan-genome". *Proceedings of the National Academy of Sciences USA* 102: 13950–5.

402. Morgante M., De Paoli E. and Radovic S. (2007) Transposable elements and the plant pan-genomes. *Current Opinion in Plant Biology* 10: 149–55.

403. Sherman R.M. and Salzberg S.L. (2020) Pan-genomics in the human genome era. *Nature Reviews Genetics* 21: 243–54.

404. Kent W.J., Baertsch R., Hinrichs A., Miller W. and Haussler D. (2003) Evolution's cauldron: Duplication, deletion, and rearrangement in the mouse and human genomes. *Proceedings of the National Academy of Sciences USA* 100: 11484–9.

405. Margulies E.H. et al. (2005) Comparative sequencing provides insights about the structure and conservation of marsupial and monotreme genomes. *Proceedings of the National Academy of Sciences USA* 102: 3354–9.

406. Hughes L.C. et al. (2018) Comprehensive phylogeny of ray-finned fishes (Actinopterygii) based on transcriptomic and genomic data. *Proceedings of the National Academy of Sciences USA* 115: 6249–54.

407. Armstrong J. et al. (2020) Progressive Cactus is a multiple-genome aligner for the thousand-genome era. *Nature* 587: 246–51.

408. Green R.E. et al. (2010) A draft sequence of the Neandertal genome. *Science* 328: 710–22.

409. Fu Q. et al. (2014) Genome sequence of a 45,000-year-old modern human from western Siberia. *Nature* 514: 445–9.

410. Prüfer K. et al. (2014) The complete genome sequence of a Neanderthal from the Altai Mountains. *Nature* 505: 43–9.

411. Pääbo S. (2014) The human condition—a molecular approach. *Cell* 157: 216–26.

412. Sankararaman S., Mallick S., Patterson N. and Reich D. (2016) The combined landscape of Denisovan and Neanderthal ancestry in present-day humans. *Current Biology* 26: 1241–7.

413. Schaefer N.K., Shapiro B. and Green R.E. (2021) An ancestral recombination graph of human, Neanderthal, and Denisovan genomes. *Science Advances* 7: eabc0776.

414. Bergström A. et al. (2020) Insights into human genetic variation and population history from 929 diverse genomes. *Science* 367: eaay5012.

415. Serre D. and Pääbo S. (2004) Evidence for gradients of human genetic diversity within and among continents. *Genome Research* 14: 1679–85.

416. Schuster S.C. et al. (2010) Complete Khoisan and Bantu genomes from southern Africa. *Nature* 463: 943–7.

417. Skoglund P. and Mathieson I. (2018) Ancient genomics of modern humans: The first decade. *Annual Review of Genomics and Human Genetics* 19: 381–404.

418. Chan E.K.F. et al. (2019) Human origins in a southern African palaeo-wetland and first migrations. *Nature* 575: 185–9.

419. Choudhury A. et al. (2020) High-depth African genomes inform human migration and health. *Nature* 586: 741–8.

420. Taliun D. et al. (2021) Sequencing of 53,831 diverse genomes from the NHLBI TOPMed Program. *Nature* 590: 290–9.

421. Anava S. et al. (2020) Illuminating genetic mysteries of the dead sea scrolls. *Cell* 181: 1218–31.

422. van der Valk T. et al. (2021) Million-year-old DNA sheds light on the genomic history of mammoths. *Nature* 591: 265–9.

423. Barlow A. et al. (2021) Middle Pleistocene genome calibrates a revised evolutionary history of extinct cave bears. *Current Biology* 31: 1771–9.

CHAPTER 11

1. Sinsheimer R.L. (2006) To reveal the genomes. *American Journal of Human Genetics* 79: 194–6.

2. Dulbecco R. (1986) A turning point in cancer research: Sequencing the human genome. *Science* 231: 1055–6.

3. Sinsheimer R.L. (1989) The santa cruz workshop—May 1985. *Genomics* 5: 954–6.

4. DoE. (1990). *Human Genome Quarterly* 1 (3): https://web.ornl.gov/sci/techresources/Human_Genome/publicat/hgn/pdfs/Vol1No3.pdf.

5. DeLisi C. (2008) Meetings that changed the world: Santa Fe 1986: Human genome baby-steps. *Nature* 455: 876–7.

6. Tinoco I. et al. (1987) *Report on the Human Genome Initiative, Office of Health and Environmental Research, prepared for Dr. Alvin W. Trivelpiece, Director, Office of Energy Research*, https://web.ornl.gov/sci/techresources/Human_Genome/project/herac2.shtml.

7. Roberts L. (2001) Controversial from the start. *Science* 291: 1182–8.

8. Berg P. (2006) Origins of the human genome project: Why sequence the human genome when 96% of it is junk? *American Journal of Human Genetics* 79: 603–5.

9. Rechsteiner M.C. (1991) The human genome project: Misguided science policy. *Trends in Biochemical Sciences* 16: 455–61.

10. Brenner S. (2007) The human genome: The nature of the enterprise. *Ciba Foundation Symposium 149 - Human Genetic Information: Science, Law and Ethics* 149: 6–12.

11. Sinsheimer R.L. (1986) Human genome sequencing. *Science* 233: 1246.

12. Wheeler D.A. et al. (2008) The complete genome of an individual by massively parallel DNA sequencing. *Nature* 452: 872–6.

13. Watson J.D. (1990) The human genome project: Past, present, and future. *Science* 248: 44–9.

14. Vogel F. (1964) A preliminary estimate of the number of human genes. *Nature* 201: 847.

15. King J.L. and Jukes T.H. (1969) Non-Darwinian evolution. *Science* 164: 788–98.

16. Ohno S. (1972) An argument for the genetic simplicity of man and other mammals. *Journal of Human Evolution* 1: 651–62.

17. Fields C., Adams M.D., White O. and Venter J.C. (1994) How many genes in the human genome? *Nature Genetics* 7: 345–6.

18. Goodfellow P. (1995) A big book of the human genome. *Nature* 377: 285–6.

19. Schuler G.D. (1997) Pieces of the puzzle: Expressed sequence tags and the catalog of human genes. *Journal of Molecular Medicine* 75: 694–8.

20. Maxson Jones K., Ankeny R.A. and Cook-Deegan R. (2018) The Bermuda Triangle: The pragmatics, policies, and principles for data sharing in the history of the Human Genome Project. *Journal of the History of Biology* 51: 693–805. see also https://web.ornl.gov/sci/techresources/Human_Genome/research/bermuda.shtml.

21. Morgan M.J. (2011) A brief (if insular) history of the human genome project. *PLOS Biology* 9: e1000601.

22. Cook-Deegan R. and McGuire A.L. (2017) Moving beyond Bermuda: Sharing data to build a medical information commons. *Genome Research* 27: 897–901.

23. Waterston R.H., Lander E.S. and Sulston J.E. (2002) On the sequencing of the human genome. *Proceedings of the National Academy of Sciences USA* 99: 3712–6.

24. Consortium (2001) Initial sequencing and analysis of the human genome. *Nature* 409: 860–921.

25. Venter J.C. et al. (2001) The sequence of the human genome. *Science* 291: 1304–51.

26. Internation Human Genome Sequencing Consortium (2004) Finishing the euchromatic sequence of the human genome. *Nature* 431: 931–45.

27. Piovesan A., Caracausi M., Antonaros F., Pelleri M.C. and Vitale L. (2016) GeneBase 1.1: A tool to summarize data from NCBI gene datasets and its application to an update of human gene statistics. *Database* 2016: baw153.

28. Pertea M. and Salzberg S.L. (2010) Between a chicken and a grape: Estimating the number of human genes. *Genome Biology* 11: 206.

29. Willyard C. (2018) New human gene tally reignites debate. *Nature* 558: 354–5.

30. Gates A.J., Gysi D.M., Kellis M. and Barabási A.-L. (2021) A wealth of discovery built on the Human Genome Project — by the numbers. *Nature* 590: 212–5.

31. Aparicio S. et al. (2002) Whole-genome shotgun assembly and analysis of the genome of *Fugu rubripes*. *Science* 297: 1301–10.

32. Hillier L.W. et al. (2004) Sequence and comparative analysis of the chicken genome provide unique perspectives on vertebrate evolution. *Nature* 432: 695–716.

33. Duboule D. and Wilkins A.S. (1998) The evolution of 'bricolage'. *Trends in Genetics* 14: 54–9.

34. Waterston R.H. et al. (2002) Initial sequencing and comparative analysis of the mouse genome. *Nature* 420: 520–62.

35. Smith N.G.C., Webster M.T. and Ellegren H. (2002) Deterministic mutation rate variation in the human genome. *Genome Research* 12: 1350–6.

36. Smith N.G., Brandstrom M. and Ellegren H. (2004) Evidence for turnover of functional noncoding DNA in mammalian genome evolution. *Genomics* 84: 806–13.

37. Arndt P.F., Hwa T. and Petrov D.A. (2005) Substantial regional variation in substitution rates in the human genome: Importance of GC content, gene density, and telomere-specific effects. *Journal of Molecular Evolution* 60: 748–63.

38. Duret L. (2009) Mutation patterns in the human genome: More variable than expected. *PLOS Biology* 7: e1000028.

39. Graur D. et al. (2013) On the immortality of television sets: "function" in the human genome according to the evolution-free gospel of encode. *Genome Biology and Evolution* 5: 578–90.

40. Mattick J.S. and Dinger M.E. (2013) The extent of functionality in the human genome. *HUGO Journal* 7: 2.

41. Pheasant M. and Mattick J.S. (2007) Raising the estimate of functional human sequences. *Genome Research* 17: 1245–53.

42. Smit A.F.A. (1999) Interspersed repeats and other mementos of transposable elements in mammalian genomes. *Current Opinion in Genetics and Development* 9: 657–63.

43. Silva J.C., Shabalina S.A., Harris D.G., Spouge J.L. and Kondrashovi A.S. (2003) Conserved fragments of transposable elements in intergenic regions: Evidence for widespread recruitment of MIR- and L2-derived sequences within the mouse and human genomes. *Genetics Research* 82: 1–18.

44. de Koning A.P.J., Gu W., Castoe T.A., Batzer M.A. and Pollock D.D. (2011) Repetitive elements may comprise over two-thirds of the human genome. *PLOS Genetics* 7: e1002384.

45. Hubley R. et al. (2016) The Dfam database of repetitive DNA families. *Nucleic Acids Research* 44: D81–9.

46. Pang K.C. et al. (2007) RNAdb 2.0–an expanded database of mammalian non-coding RNAs. *Nucleic Acids Research* 35: D178–82.

47. Taylor M.S. et al. (2006) Heterotachy in mammalian promoter evolution. *PLOS Genetics* 2: e30.

48. Vierstra J. et al. (2020) Global reference mapping of human transcription factor footprints. *Nature* 583: 729–36.

49. Ross C.J. et al. (2021) Uncovering deeply conserved motif combinations in rapidly evolving noncoding sequences. *Genome Biology* 22: 29.

50. Wong E.S. et al. (2020) Deep conservation of the enhancer regulatory code in animals. *Science* 370: eaax8137.

51. Oldmeadow C., Mengersen K., Mattick J.S. and Keith J.M. (2010) Multiple evolutionary rate classes in animal genome evolution. *Molecular Biology and Evolution* 27: 942–53.

52. Smith M.A., Gesell T., Stadler P.F. and Mattick J.S. (2013) Widespread purifying selection on RNA structure in mammals. *Nucleic Acids Research* 41: 8220–36.

53. Gazal S. et al. (2018) Functional architecture of low-frequency variants highlights strength of negative selection across coding and non-coding annotations. *Nature Genetics* 50: 1600–7.

54. Feng S. et al. (2020) Dense sampling of bird diversity increases power of comparative genomics. *Nature* 587: 252–7.

55. The ENCODE Project Consortium (2007) Identification and analysis of functional elements in 1% of the human genome by the ENCODE pilot project. *Nature* 447: 799–816.

56. Giresi P.G., Kim J., McDaniell R.M., Iyer V.R. and Lieb J.D. (2007) FAIRE (Formaldehyde-Assisted Isolation of Regulatory Elements) isolates active regulatory elements from human chromatin. *Genome Research* 17: 877–85.

57. Check E. (2007) Genome project turns up evolutionary surprises. *Nature* 447: 760–1.

58. Thurman R.E. et al. (2012) The accessible chromatin landscape of the human genome. *Nature* 489: 75–82.

59. Djebali S. et al. (2012) Landscape of transcription in human cells. *Nature* 489: 101–8.

60. Dunham I. et al. (2012) An integrated encyclopedia of DNA elements in the human genome. *Nature* 489: 57–74.

61. Editorial (2014) ENCODE debate revived online. *Nature* 509: 137.

62. Germain P.-L., Ratti E. and Boem F. (2014) Junk or functional DNA? ENCODE and the function controversy. *Biology & Philosophy* 29: 807–31.

63. Eddy S.R. (2012) The C-value paradox, junk DNA and ENCODE. *Current Biology* 22: R898–9.

64. Doolittle W.F. (2013) Is junk DNA bunk? A critique of ENCODE. *Proceedings of the National Academy of Sciences USA* 110: 5294–300.

65. Davidson D.J. and Porteous D.J. (1998) The genetics of cystic fibrosis lung disease. *Thorax* 53: 389–97.

66. Poolman E.M. and Galvani A.P. (2007) Evaluating candidate agents of selective pressure for cystic fibrosis. *Journal of the Royal Society Interface* 4: 91–8.

67. Ferreira A. et al. (2011) Sickle hemoglobin confers tolerance to *Plasmodium* infection. *Cell* 145: 398–409.

68. Mattick J.S. (2009) The genetic signatures of noncoding RNAs. *PLOS Genetics* 5: e1000459.

69. Solomon E. and Bodmer W.F. (1979) Evolution of sickle variant gene. *Lancet* 313: 923.

70. Botstein D., White R.L., Skolnick M. and Davis R.W. (1980) Construction of a genetic linkage map in man using restriction fragment length polymorphisms. *American Journal of Human Genetics* 32: 314–31.

71. Simons M.J. (2001) Intron sequence analysis method for detection of adjacent and remote locus alleles as haplotypes. US patent US20010005626.

72. Lander E.S. and Botstein D. (1986) Strategies for studying heterogeneous genetic traits in humans by using a linkage map of restriction fragment length polymorphisms. *Proceedings of the National Academy of Sciences USA* 83: 7353–7.

73. Davies K.E., Young B.D., Elles R.G., Hill M.E. and Williamson R. (1981) Cloning of a representative genomic library of the human X chromosome after sorting by flow cytometry. *Nature* 293: 374–6.

74. Krumlauf R., Jeanpierre M. and Young B.D. (1982) Construction and characterization of genomic libraries from specific human chromosomes. *Proceedings of the National Academy of Sciences USA* 79: 2971–5.

75. Nabholz M., Miggiano V. and Bodmer W. (1969) Genetic analysis with human—mouse somatic cell hybrids. *Nature* 223: 358–63.

76. Goss S.J. and Harris H. (1975) New method for mapping genes in human chromosomes. *Nature* 255: 680–4.

77. Cox D., Burmeister M., Price E., Kim S. and Myers R. (1990) Radiation hybrid mapping: A somatic cell genetic method for constructing high-resolution maps of mammalian chromosomes. *Science* 250: 245–50.

78. Frazer K.A. et al. (1992) A radiation hybrid map of the region on human chromosome 22 containing the neurofibromatosis type 2 locus. *Genomics* 14: 574–84.

79. Zengerling S. et al. (1987) Mapping of DNA markers linked to the cystic fibrosis locus on the long arm of chromosome 7. *American Journal of Human Genetics* 40: 228–36.

80. Frazer K.A. et al. (2007) A second generation human haplotype map of over 3.1 million SNPs. *Nature* 449: 851–61.

81. Arnheim N., Calabrese P. and Tiemann-Boege I. (2007) Mammalian meiotic recombination hot spots. *Annual Review of Genetics* 41: 369–99.

82. Wall J.D. and Pritchard J.K. (2003) Haplotype blocks and linkage disequilibrium in the human genome. *Nature Reviews Genetics* 4: 587–97.

83. Buetow K.H. et al. (1994) Integrated human genome–wide maps constructed using the CEPH reference panel. *Nature Genetics* 6: 391–3.

84. Gabriel S.B. et al. (2002) The structure of haplotype blocks in the human genome. *Science* 296: 2225–9.

85. Cardon L.R. and Abecasis G.R. (2003) Using haplotype blocks to map human complex trait loci. *Trends in Genetics* 19: 135–40.

86. Lander E. and Botstein D. (1987) Homozygosity mapping: A way to map human recessive traits with the DNA of inbred children. *Science* 236: 1567–70.

87. Monaco A.P. et al. (1986) Isolation of candidate cDNAs for portions of the Duchenne muscular dystrophy gene. *Nature* 323: 646–50.

88. Koenig M. et al. (1987) Complete cloning of the duchenne muscular dystrophy (DMD) cDNA and preliminary genomic organization of the DMD gene in normal and affected individuals. *Cell* 50: 509–17.

89. Hoffman E.P. (2020) The discovery of dystrophin, the protein product of the Duchenne muscular dystrophy gene. *FEBS Journal* 287: 3879–87.

90. Ahn A.H. and Kunkel L.M. (1993) The structural and functional diversity of dystrophin. *Nature Genetics* 3: 283–91.

91. Pozzoli U. et al. (2003) Comparative analysis of vertebrate dystrophin loci indicate intron gigantism as a common feature. *Genome Research* 13: 764–72.

92. Kerem B. et al. (1989) Identification of the cystic fibrosis gene: Genetic analysis. *Science* 245: 1073–80.

93. Tsui L.-C. and Dorfman R. (2013) The cystic fibrosis gene: A molecular genetic perspective. *Cold Spring Harbor Perspectives in Medicine* 3: a009472.

94. van Deutekom J.C. et al. (2001) Antisense-induced exon skipping restores dystrophin expression in DMD patient derived muscle cells. *Human Molecular Genetics* 10: 1547–54.

95. Mann C.J. et al. (2001) Antisense-induced exon skipping and synthesis of dystrophin in the mdx mouse. *Proceedings of the National Academy of Sciences USA* 98: 42–7.

96. De Angelis F.G. et al. (2002) Chimeric snRNA molecules carrying antisense sequences against the splice junctions of exon 51 of the dystrophin pre-mRNA induce exon skipping and restoration of a dystrophin synthesis in Delta 48–50 DMD cells. *Proceedings of the National Academy of Sciences USA* 99: 9456–61.

97. Cirak S. et al. (2011) Exon skipping and dystrophin restoration in patients with Duchenne muscular dystrophy after systemic phosphorodiamidate morpholino oligomer treatment: An open-label, phase 2, dose-escalation study. *Lancet* 378: 595–605.

98. Lee T.W.R., Matthews D.A. and Blair G.E. (2005) Novel molecular approaches to cystic fibrosis gene therapy. *Biochemical Journal* 387: 1–15.

99. Cooney A.L., McCray Jr. P.B. and Sinn P.L. (2018) Cystic fibrosis gene therapy: Looking back, looking forward. *Genes* 9: 538.

100. Verkerk A.J.M.H. et al. (1991) Identification of a gene (FMR-1) containing a CGG repeat coincident with a breakpoint cluster region exhibiting length variation in fragile X syndrome. *Cell* 65: 905–14.

101. Oberle I. et al. (1991) Instability of a 550-base pair DNA segment and abnormal methylation in fragile X syndrome. *Science* 252: 1097–102.

102. Hagerman R.J. et al. (2017) Fragile X syndrome. *Nature Reviews Disease Primers* 3: 17065.

103. Spada A.R.L., Wilson E.M., Lubahn D.B., Harding A.E. and Fischbeck K.H. (1991) Androgen receptor gene mutations in X-linked spinal and bulbar muscular atrophy. *Nature* 352: 77–9.

104. Brook J.D. et al. (1992) Molecular basis of myotonic dystrophy: Expansion of a trinucleotide (CTG) repeat at the 3′ end of a transcript encoding a protein kinase family member. *Cell* 68: 799–808.

105. Liquori C.L. et al. (2001) Myotonic dystrophy type 2 caused by a CCTG expansion in intron 1 of *ZNF9*. *Science* 293: 864–7.

106. MacDonald M.E. et al. (1993) A novel gene containing a trinucleotide repeat that is expanded and unstable on Huntington's disease chromosomes. *Cell* 72: 971–83.

107. Orr H.T. et al. (1993) Expansion of an unstable trinucleotide CAG repeat in spinocerebellar ataxia type 1. *Nature Genetics* 4: 221–6.

108. Koide R. et al. (1994) Unstable expansion of CAG repeat in hereditary dentatorubral–pallidoluysian atrophy (DRPLA). *Nature Genetics* 6: 9–13.

109. Nagafuchi S. et al. (1994) Dentatorubral and pallidoluysian atrophy expansion of an unstable CAG trinucleotide on chromosome 12p. *Nature Genetics* 6: 14–8.

110. Kawaguchi Y. et al. (1994) CAG expansions in a novel gene for Machado-Joseph disease at chromosome 14q32.1. *Nature Genetics* 8: 221–8.

111. Cummings C.J. and Zoghbi H.Y. (2000) Fourteen and counting: Unraveling trinucleotide repeat diseases. *Human Molecular Genetics* 9: 909–16.

112. Bassani S. et al. (2013) The neurobiology of X-linked intellectual disability. *Neuroscientist* 19: 541–52.

113. Paulson H. (2018) Repeat expansion diseases, in D.H. Geschwind, H.L. Paulson and C. Klein (eds.) *Handbook of Clinical Neurology* (Elsevier, Amsterdam).

114. Robinson A. and Linden M.G. (2002) *Clinical Genetics Handbook* (Blackwell Scientific Publications, Boston, MA).

115. van den Berg M.M.J., van Maarle M.C., van Wely M. and Goddijn M. (2012) Genetics of early miscarriage. *Biochimica et Biophysica Acta* 1822: 1951–9.

116. Colley E. et al. (2019) Potential genetic causes of miscarriage in euploid pregnancies: A systematic review. *Human Reproduction Update* 25: 452–72.

117. McCandless S.E., Brunger J.W. and Cassidy S.B. (2004) The burden of genetic disease on inpatient care in a children's hospital. *American Journal of Human Genetics* 74: 121–7.

118. Gjorgioski S. et al. (2020) Genetics and pediatric hospital admissions, 1985 to 2017. *Genetics in Medicine* 22: 1777–85.

119. Rimoin D., Pyeritz R. and Korf B. (2022) Emery and Rimoin's Principles and Practice of Medical Genetics, 7th edition (Elsevier, Amsterdam).

120. Xue Y. et al. (2012) Deleterious- and disease-allele prevalence in healthy individuals: Insights from current predictions, mutation databases, and population-scale resequencing. *American Journal of Human Genetics* 91: 1022–32.

121. Walter K. et al. (2015) The UK10K project identifies rare variants in health and disease. *Nature* 526: 82–90.

122. Lek M. et al. (2016) Analysis of protein-coding genetic variation in 60,706 humans. *Nature* 536: 285–91.

123. Fragoza R. et al. (2019) Extensive disruption of protein interactions by genetic variants across the allele frequency spectrum in human populations. *Nature Communications* 10: 4141.

124. Veltman J.A. and Brunner H.G. (2012) De novo mutations in human genetic disease. *Nature Reviews Genetics* 13: 565–75.

125. Mir Y.R. and Kuchay R.A.H. (2019) Advances in identification of genes involved in autosomal recessive intellectual disability: A brief review. *Journal of Medical Genetics* 56: 567–73.

126. Mattick J.S., Dinger M., Schonrock N. and Cowley M. (2018) Whole genome sequencing provides better diagnostic yield and future value than whole exome sequencing. *Medical Journal of Australia* 209: 197–9.

127. Zhang D. et al. (2020) Incomplete annotation has a disproportionate impact on our understanding of Mendelian and complex neurogenetic disorders. *Science Advances* 6: eaay8299.

128. Smedley D. et al. (2021) 100,000 genomes pilot on rare-disease diagnosis in health care — preliminary report. *New England Journal of Medicine* 385: 1868–80.

129. Murdock D.R. et al. (2021) Transcriptome-directed analysis for Mendelian disease diagnosis overcomes limitations of conventional genomic testing. *Journal of Clinical Investigation* 131: e141500.

130. Woolf L.I. and Adams J. (2020) The early history of PKU. *International Journal of Neonatal Screening* 6: 59.

131. Lo Y.M.D. et al. (1997) Presence of fetal DNA in maternal plasma and serum. *Lancet* 350: 485–7.

132. Lo Y.M.D. et al. (1998) Quantitative analysis of fetal DNA in maternal plasma and serum: Implications for noninvasive prenatal diagnosis. *American Journal of Human Genetics* 62: 768–75.

133. Norwitz E.R. and Levy B. (2013) Noninvasive prenatal testing: The future is now. *Reviews in Obstetrics & Gynecology* 6: 48–62.

134. Mattick J.S. (2006) The future of molecular pathology, in S. Lakhani and S. Fox (eds.) *Molecular Pathology in Cancer Research* (Springer, New York).

135. Cavalli-Sforza L.L. and Edwards A.W.F. (1967) Phylogenetic analysis: Models and estimation procedures. *Evolution* 21: 550–70.

136. Lewontin R.C. (1972) The apportionment of human diversity, in T. Dobzhansky, M.K. Hecht and W.C. Steere (eds.) *Evolutionary Biology* (Springer, New York).

137. Rosenberg N.A. et al. (2002) Genetic structure of human populations. *Science* 298: 2381–5.

138. Sabeti P.C. et al. (2007) Genome-wide detection and characterization of positive selection in human populations. *Nature* 449: 913–8.

139. Karczewski K.J. et al. (2020) The mutational constraint spectrum quantified from variation in 141,456 humans. *Nature* 581: 434–43.

140. Abecasis G.R. et al. (2012) An integrated map of genetic variation from 1,092 human genomes. *Nature* 491: 56–65.

141. Auton A. et al. (2015) A global reference for human genetic variation. *Nature* 526: 68–74.

142. Choudhury A. et al. (2020) High-depth African genomes inform human migration and health. *Nature* 586: 741–8.

143. Polderman T.J.C. et al. (2015) Meta-analysis of the heritability of human traits based on fifty years of twin studies. *Nature Genetics* 47: 702–9.

144. Sahu M. and Prasuna J.G. (2016) Twin studies: A unique epidemiological tool. *Indian Journal of Community Medicine* 41: 177–82.

145. Botstein D. and Risch N. (2003) Discovering genotypes underlying human phenotypes: Past successes for mendelian disease, future approaches for complex disease. *Nature Genetics* 33: 228–37.

146. Hirschhorn J.N. and Daly M.J. (2005) Genome-wide association studies for common diseases and complex traits. *Nature Reviews Genetics* 6: 95–108.

147. Schena M., Shalon D., Davis R.W. and Brown P.O. (1995) Quantitative monitoring of gene expression patterns with a complementary DNA microarray. *Science* 270: 467–70.

148. Gibson G. (2010) Hints of hidden heritability in GWAS. *Nature Genetics* 42: 558–60.

149. Clarke G.M. et al. (2011) Basic statistical analysis in genetic case-control studies. *Nature Protocols* 6: 121–33.

150. Ozaki K. et al. (2002) Functional SNPs in the lymphotoxin-α gene that are associated with susceptibility to myocardial infarction. *Nature Genetics* 32: 650–4.

151. Klein R.J. et al. (2005) Complement factor H polymorphism in Age-Related Macular Degeneration. *Science* 308: 385–9.

152. Burton P.R. et al. (2007) Genome-wide association study of 14,000 cases of seven common diseases and 3,000 shared controls. *Nature* 447: 661–78.

153. Uda M. et al. (2008) Genome-wide association study shows *BCL11A* associated with persistent fetal hemoglobin and amelioration of the phenotype of β-thalassemia. *Proceedings of the National Academy of Sciences USA* 105: 1620–5.

154. Sankaran V.G. et al. (2008) Human fetal hemoglobin expression is regulated by the developmental stage-specific repressor *BCL11A*. *Science* 322: 1839–42.

155. Sankaran V.G. et al. (2011) A functional element necessary for fetal hemoglobin silencing. *New England Journal of Medicine* 365: 807–14.

156. Visscher P.M. et al. (2017) 10 years of GWAS discovery: Biology, function, and translation. *American Journal of Human Genetics* 101: 5–22.

157. Zwir I. et al. (2020) Uncovering the complex genetics of human temperament. *Molecular Psychiatry* 25: 2275–94.

158. Grove J. et al. (2019) Identification of common genetic risk variants for autism spectrum disorder. *Nature Genetics* 51: 431–44.

159. Narita A. et al. (2020) Clustering by phenotype and genome-wide association study in autism. *Translational Psychiatry* 10: 290.

160. Roussos P. et al. (2014) A role for noncoding variation in schizophrenia. *Cell Reports* 9: 1417–29.

161. Zeng B. et al. (2022) Multi-ancestry eQTL meta-analysis of human brain identifies candidate causal variants for brain-related traits. *Nature Genetics* 54: 161–9.

162. Mullins N. et al. (2021) Genome-wide association study of more than 40,000 bipolar disorder cases provides new insights into the underlying biology. *Nature Genetics* 53: 817–29.

163. Demontis D. et al. (2019) Discovery of the first genome-wide significant risk loci for attention deficit/hyperactivity disorder. *Nature Genetics* 51: 63–75.

164. Powell V., Martin J., Thapar A., Rice F. and Anney R.J.L. (2021) Investigating regions of shared genetic variation in attention deficit/hyperactivity disorder and major depressive disorder: A GWAS meta-analysis. *Scientific Reports* 11: 7353.

165. Howard D.M. et al. (2019) Genome-wide meta-analysis of depression identifies 102 independent variants and highlights the importance of the prefrontal brain regions. *Nature Neuroscience* 22: 343–52.

166. Forstner A.J. et al. (2021) Genome-wide association study of panic disorder reveals genetic overlap with neuroticism and depression. *Molecular Psychiatry* 26: 4179–90.

167. Levey D.F. *et al.* (2021) Bi-ancestral depression GWAS in the Million Veteran Program and meta-analysis in >1.2 million individuals highlight new therapeutic directions. *Nature Neuroscience* 24: 954–63.

168. Skuladottir A.T. et al. (2021) A genome-wide meta-analysis uncovers six sequence variants conferring risk of vertigo. *Communications Biology* 4: 1148.

169. Jansen I.E. et al. (2019) Genome-wide meta-analysis identifies new loci and functional pathways influencing Alzheimer's disease risk. *Nature Genetics* 51: 404–13.

170. Andrews S.J., Fulton-Howard B. and Goate A. (2020) Interpretation of risk loci from genome-wide association studies of Alzheimer's disease. *Lancet Neurology* 19: 326–35.

171. Grenn F.P. et al. (2020) The Parkinson's Disease genome-wide association study locus browser. *Movement Disorders* 35: 2056–67.

172. Sud A., Kinnersley B. and Houlston R.S. (2017) Genome-wide association studies of cancer: Current insights and future perspectives. *Nature Reviews Cancer* 17: 692–704.

173. Li Z. and Brown M.A. (2017) Progress of genome-wide association studies of ankylosing spondylitis. *Clinical & Translational Immunology* 6: e163.

174. Paternoster L. et al. (2015) Multi-ancestry genome-wide association study of 21,000 cases and 95,000 controls identifies new risk loci for atopic dermatitis. *Nature Genetics* 47: 1449–56.

175. Dong Z. et al. (2020) Integrated genomics analysis highlights important SNPs and genes implicated in moderate-to-severe asthma based on GWAS and eQTL datasets. *BMC Pulmonary Medicine* 20: 270.

176. Momozawa Y. et al. (2018) IBD risk loci are enriched in multigenic regulatory modules encompassing putative causative genes. *Nature Communications* 9: 2427.

177. Evangelou E. et al. (2018) Genetic analysis of over 1 million people identifies 535 new loci associated with blood pressure traits. *Nature Genetics* 50: 1412–25.

178. Yengo L. et al. (2018) Meta-analysis of genome-wide association studies for height and body mass index in ~700000 individuals of European ancestry. *Human Molecular Genetics* 27: 3641–9.

179. Zhu X., Bai W. and Zheng H. (2021) Twelve years of GWAS discoveries for osteoporosis and related traits: Advances, challenges and applications. *Bone Research* 9: 23.

180. Walters R.K. et al. (2018) Transancestral GWAS of alcohol dependence reveals common genetic underpinnings with psychiatric disorders. *Nature Neuroscience* 21: 1656–69.

181. Kranzler H.R. et al. (2019) Genome-wide association study of alcohol consumption and use disorder in 274,424 individuals from multiple populations. *Nature Communications* 10: 1499.

182. Liu M. et al. (2019) Association studies of up to 1.2 million individuals yield new insights into the genetic etiology of tobacco and alcohol use. *Nature Genetics* 51: 237–44.

183. Cabana-Domínguez J., Shivalikanjli A., Fernàndez-Castillo N. and Cormand B. (2019) Genome-wide association meta-analysis of cocaine dependence: Shared genetics with comorbid conditions. *Progress in Neuro-Psychopharmacology and Biological Psychiatry* 94: 109667.

184. Marees A.T. et al. (2020) Post-GWAS analysis of six substance use traits improves the identification and functional interpretation of genetic risk loci. *Drug and Alcohol Dependence* 206: 107703.

185. Cornelis M.C. et al. (2016) Genome-wide association study of caffeine metabolites provides new insights to caffeine metabolism and dietary caffeine-consumption behavior. *Human Molecular Genetics* 25: 5472–82.

186. Cuellar-Partida G. et al. (2021) Genome-wide association study identifies 48 common genetic variants associated with handedness. *Nature Human Behaviour* 5: 59–70.

187. Jansen P.R. et al. (2019) Genome-wide analysis of insomnia in 1,331,010 individuals identifies new risk loci and functional pathways. *Nature Genetics* 51: 394–403.

188. Deelen J. et al. (2019) A meta-analysis of genome-wide association studies identifies multiple longevity genes. *Nature Communications* 10: 3669.

189. Davies G. et al. (2018) Study of 300,486 individuals identifies 148 independent genetic loci influencing general cognitive function. *Nature Communications* 9: 2098.

190. Davies G. et al. (2011) Genome-wide association studies establish that human intelligence is highly heritable and polygenic. *Molecular Psychiatry* 16: 996–1005.

191. Benyamin B. et al. (2014) Childhood intelligence is heritable, highly polygenic and associated with FNBP1L. *Molecular Psychiatry* 19: 253–8.

192. Davies G. et al. (2015) Genetic contributions to variation in general cognitive function: A meta-analysis of genome-wide association studies in the CHARGE consortium (N=53 949). *Molecular Psychiatry* 20: 183–92.

193. Hill W.D. et al. (2019) A combined analysis of genetically correlated traits identifies 187 loci and a role for neurogenesis and myelination in intelligence. *Molecular Psychiatry* 24: 169–81.

194. Lee J.J. et al. (2018) Gene discovery and polygenic prediction from a genome-wide association study of educational attainment in 1.1 million individuals. *Nature Genetics* 50: 1112–21.

195. Boyle E.A., Li Y.I. and Pritchard J.K. (2017) An expanded view of complex traits: From polygenic to omnigenic. *Cell* 169: 1177–86.

196. Young A.I. (2019) Solving the missing heritability problem. *PLOS Genetics* 15: e1008222.

197. Sohail M. et al. (2019) Polygenic adaptation on height is overestimated due to uncorrected stratification in genome-wide association studies. *eLife* 8: e39702.

198. NCD Risk Factor Collaboration (2016) A century of trends in adult human height. *eLife* 5: e13410.

199. Allen H.L. et al. (2010) Hundreds of variants clustered in genomic loci and biological pathways affect human height. *Nature* 467: 832–8.

200. Visscher P.M., McEvoy B. and Yang J. (2011) From Galton to GWAS: Quantitative genetics of human height. *Genetics Research* 92: 371–9.

201. Wainschtein P. et al. (2019) Recovery of trait heritability from whole genome sequence data. *bioRxiv:* 588020.

202. Sackton T.B. and Hartl D.L. (2016) Genotypic context and epistasis in individuals and populations. *Cell* 166: 279–87.

203. Robinson M.R., Wray N.R. and Visscher P.M. (2014) Explaining additional genetic variation in complex traits. *Trends in Genetics* 30: 124–32.

204. Sinnott-Armstrong N., Naqvi S., Rivas M. and Pritchard J.K. (2021) GWAS of three molecular traits highlights core genes and pathways alongside a highly polygenic background. *eLife* 10: e58615.

205. Yang J. et al. (2015) Genetic variance estimation with imputed variants finds negligible missing heritability for human height and body mass index. *Nature Genetics* 47: 1114–20.

206. Zeevi D. et al. (2019) Analysis of the genetic basis of height in large Jewish nuclear families. *PLOS Genetics* 15: e1008082.

207. Trost B. et al. (2020) Genome-wide detection of tandem DNA repeats that are expanded in autism. *Nature* 586: 80–6.

208. Mitra I. et al. (2021) Patterns of de novo tandem repeat mutations and their role in autism. *Nature* 589: 246–50.

209. Hannan A.J. (2021) Repeat DNA expands our understanding of autism spectrum disorder. *Nature* 589: 200–2.

210. Wilfert A.B. et al. (2021) Recent ultra-rare inherited variants implicate new autism candidate risk genes. *Nature Genetics* 53: 1125–34.

211. Beroukhim R. et al. (2010) The landscape of somatic copy-number alteration across human cancers. *Nature* 463: 899–905.

212. Freedman M.L. et al. (2011) Principles for the post-GWAS functional characterization of cancer risk loci. *Nature Genetics* 43: 513–8.

213. Maurano M.T. et al. (2012) Systematic localization of common disease-associated variation in regulatory DNA. *Science* 337: 1190–5.

214. Cheetham S.W., Gruhl F., Mattick J.S. and Dinger M.E. (2013) Long noncoding RNAs and the genetics of cancer. *British Journal of Cancer* 108: 2419–25.

215. Zhang F. and Lupski J.R. (2015) Non-coding genetic variants in human disease. *Human Molecular Genetics* 24: R102–10.

216. Schierding W., Cutfield W. and O'Sullivan J. (2014) The missing story behind Genome Wide Association Studies: Single nucleotide polymorphisms in gene deserts have a story to tell. *Frontiers in Genetics* 5: 39.

217. Gallagher M.D. and Chen-Plotkin A.S. (2018) The post-GWAS era: From association to function. *American Journal of Human Genetics* 102: 717–30.

218. Fagny M., Platig J., Kuijjer M.L., Lin X. and Quackenbush J. (2020) Nongenic cancer-risk SNPs affect oncogenes, tumour-suppressor genes, and immune function. *British Journal of Cancer* 122: 569–77.

219. Padmanabhan S. and Dominiczak A.F. (2021) Genomics of hypertension: The road to precision medicine. *Nature Reviews Cardiology* 18: 235–50.

220. Watanabe K. et al. (2019) A global overview of pleiotropy and genetic architecture in complex traits. *Nature Genetics* 51: 1339–48.

221. Vernot B. et al. (2012) Personal and population genomics of human regulatory variation. *Genome Research* 22: 1689–97.

222. Trynka G. et al. (2013) Chromatin marks identify critical cell types for fine mapping complex trait variants. *Nature Genetics* 45: 124–30.

223. Bartonicek N. et al. (2017) Intergenic disease-associated regions are abundant in novel transcripts. *Genome Biology* 18: 241.

224. Hardwick S.A. et al. (2019) Targeted, high-resolution RNA sequencing of non-coding genomic regions associated with neuropsychiatric functions. *Frontiers in Genetics* 10: 309.

225. de Goede O.M. et al. (2021) Population-scale tissue transcriptomics maps long non-coding RNAs to complex disease. *Cell* 184: 2633–48.

226. Thomas J., Perron H. and Feschotte C. (2018) Variation in proviral content among human genomes mediated by LTR recombination. *Mobile DNA* 9: 36.

227. Hannan A.J. (2018) Tandem repeats mediating genetic plasticity in health and disease. *Nature Reviews Genetics* 19: 286–98.

228. Jendrzejewski J. et al. (2012) The polymorphism rs944289 predisposes to papillary thyroid carcinoma through a large intergenic noncoding RNA gene of tumor suppressor type. *Proceedings of the National Academy of Sciences USA* 109: 8646–51.

229. Hnisz D. et al. (2013) Super-enhancers in the control of cell identity and disease. *Cell* 155: 934–47.

230. Lai F. et al. (2013) Activating RNAs associate with Mediator to enhance chromatin architecture and transcription. *Nature* 494: 497–501.

231. Smemo S. et al. (2014) Obesity-associated variants within FTO form long-range functional connections with IRX3. *Nature* 507: 371–5.

232. Kim T. et al. (2014) Long-range interaction and correlation between MYC enhancer and oncogenic long noncoding RNA CARLo-5. *Proceedings of the National Academy of Sciences USA* 111: 4173–8.

233. St. Laurent G., Vyatkin Y. and Kapranov P. (2014) Dark matter RNA illuminates the puzzle of genome-wide association studies. *BMC Medicine* 12: 97.

234. Tan J.Y. et al. (2017) Cis-acting complex-trait-associated lincRNA expression correlates with modulation of chromosomal architecture. *Cell Reports* 18: 2280–8.

235. Dong X. et al. (2018) Enhancers active in dopamine neurons are a primary link between genetic variation and neuropsychiatric disease. *Nature Neuroscience* 21: 1482–92.

236. Small K.S. et al. (2018) Regulatory variants at KLF14 influence type 2 diabetes risk via a female-specific effect on adipocyte size and body composition. *Nature Genetics* 50: 572–80.

237. Rahman S. and Mansour M.R. (2019) The role of non-coding mutations in blood cancers. *Disease Models & Mechanisms* 12: dmm041988.

238. Ou M., Li X., Zhao S., Cui S. and Tu J. (2020) Long noncoding RNA CDKN2B-AS1 contributes to atherosclerotic plaque formation by forming RNA-DNA triplex in the CDKN2B promoter. *EBioMedicine* 55: 102694.

239. Ovcharenko I. et al. (2005) Evolution and functional classification of vertebrate gene deserts. *Genome Research* 15: 137–45.

240. Ghoussaini M. et al. (2008) Multiple loci with different cancer specificities within the 8q24 gene desert. *Journal of the National Cancer Institute* 100: 962–6.

241. Huppi K., Pitt J., Wahlberg B. and Caplen N. (2012) The 8q24 gene desert: An oasis of non-coding transcriptional activity. *Frontiers in Genetics* 3: 69.

242. Hon C.-C. et al. (2017) An atlas of human long noncoding RNAs with accurate 5′ ends. *Nature* 543: 199–204.

243. Gandal M.J. et al. (2018) Transcriptome-wide isoform-level dysregulation in ASD, schizophrenia, and bipolar disorder. *Science* 362: eaat8127.

244. Pasmant E., Sabbagh A., Vidaud M. and Bièche I. (2010) ANRIL, a long, noncoding RNA, is an unexpected major hotspot in GWAS. *FASEB Journal* 25: 444–8.

245. Castellanos-Rubio A. et al. (2016) A long noncoding RNA associated with susceptibility to celiac disease. *Science* 352: 91–5.

246. Ballantyne R.L. et al. (2016) Genome-wide interrogation reveals hundreds of long intergenic noncoding RNAs that associate with cardiometabolic traits. *Human Molecular Genetics* 25: 3125–41.

247. Yuan H. et al. (2016) A novel genetic variant in long noncoding RNA gene NEXN-AS1 is associated with risk of lung cancer. *Scientific Reports* 6: 34234.

248. Guo H. et al. (2016) Modulation of long noncoding RNAs by risk SNPs underlying genetic predispositions to prostate cancer. *Nature Genetics* 48: 1142–50.

249. Padua D. et al. (2016) A long noncoding RNA signature for ulcerative colitis identifies IFNG-AS1 as an enhancer of inflammation. *American Journal of Physiology-Gastrointestinal and Liver Physiology* 311: G446–G57.

250. Zhang D.-D. et al. (2017) Long noncoding RNA LINC00305 promotes inflammation by activating the AHRR-NF-κB pathway in human monocytes. *Scientific Reports* 7: 46204.

251. Holdt L.M. and Teupser D. (2018) Long noncoding RNA ANRIL: Lnc-ing genetic variation at the chromosome 9p21 locus to molecular mechanisms of atherosclerosis. *Frontiers in Cardiovascular Medicine* 5: 145.

252. Zhou J. et al. (2019) Whole-genome deep-learning analysis identifies contribution of noncoding mutations to autism risk. *Nature Genetics* 51: 973–80.

253. Ke W. et al. (2021) Genes in human obesity loci are causal obesity genes in *C. elegans*. *PLOS Genetics* 17: e1009736.

254. Saul M.C., Philip V.M., Reinholdt L.G. and Chesler E.J. (2019) High-diversity mouse populations for complex traits. *Trends in Genetics* 35: 501–14.

255. Kundaje A. et al. (2015) Integrative analysis of 111 reference human epigenomes. *Nature* 518: 317–30.

256. Krijger P.H.L. and de Laat W. (2016) Regulation of disease-associated gene expression in the 3D genome. *Nature Reviews Molecular Cell Biology* 17: 771–82.

257. Stunnenberg H.G. et al. (2016) The International Human Epigenome Consortium: A blueprint for scientific collaboration and discovery. *Cell* 167: 1145–9.

258. Zeggini E., Gloyn A.L., Barton A.C. and Wain L.V. (2019) Translational genomics and precision medicine: Moving from the lab to the clinic. *Science* 365: 1409–13.

259. Consortium (2020) The GTEx Consortium atlas of genetic regulatory effects across human tissues. *Science* 369: 1318–30.

260. Boix C.A., James B.T., Park Y.P., Meuleman W. and Kellis M. (2021) Regulatory genomic circuitry of human disease loci by integrative epigenomics. *Nature* 590: 300–7.

261. Yan J. et al. (2021) Systematic analysis of binding of transcription factors to noncoding variants. *Nature* 591: 147–51.

262. Julienne H. et al. (2021) Multitrait GWAS to connect disease variants and biological mechanisms. *PLOS Genetics* 17: e1009713.

263. Tripp S. and Grueber M. (2011) Economic impact of the human genome project. https://www.battelle.org/docs/default-source/misc/battelle-2011-misc-economic-impact-human-genome-project.pdf.

264. Dudchenko O. et al. (2017) De novo assembly of the *Aedes aegypti* genome using Hi-C yields chromosome-length scaffolds. *Science* 356: 92–5.

265. Zhang J. et al. (2011) International Cancer Genome Consortium Data Portal—a one-stop shop for cancer genomics data. *Database* 2011: bar026.

266. Campbell P.J. et al. (2020) Pan-cancer analysis of whole genomes. *Nature* 578: 82–93.

267. Turnbull C. et al. (2018) The 100 000 Genomes Project: Bringing whole genome sequencing to the NHS. *BMJ* 361: k1687.

268. U.K. Department of Health and Social Care (2018) https://www.gov.uk/government/news/matt-hancock-announces-ambition-to-map-5-million-genomes.

269. Denny J.C. et al. (2019) The "All of Us" research program. *New England Journal of Medicine* 381: 668–76.

270. Taliun D. et al. (2021) Sequencing of 53,831 diverse genomes from the NHLBI TOPMed Program. *Nature* 590: 290–9.

271. Momozawa Y. and Mizukami K. (2021) Unique roles of rare variants in the genetics of complex diseases in humans. *Journal of Human Genetics* 66: 11–23.

272. Koch E.M. and Sunyaev S.R. (2021) Maintenance of complex trait variation: Classic theory and modern data. *Frontiers in Genetics* 12: 2198.

273. Khera A.V. et al. (2018) Genome-wide polygenic scores for common diseases identify individuals with risk equivalent to monogenic mutations. *Nature Genetics* 50: 1219–24.

274. Weale M.E. et al. (2021) Validation of an integrated risk tool, including polygenic risk score, for atherosclerotic cardiovascular disease in multiple ethnicities and ancestries. *American Journal of Cardiology* 148: 157–64.

275. Riveros-Mckay F. et al. (2021) Integrated polygenic tool substantially enhances coronary artery disease prediction. *Circulation: Genomic and Precision Medicine* 14: e003304.

276. Wald N.J. and Old R. (2019) The illusion of polygenic disease risk prediction. *Genetics in Medicine* 21: 1705–7.

277. David V. et al. (2021) An analysis of pharmacogenomic-guided pathways and their effect on medication changes and hospital admissions: A systematic review and meta-analysis. *Frontiers in Genetics* 12: 1308.

278. Berger M.F. and Mardis E.R. (2018) The emerging clinical relevance of genomics in cancer medicine. *Nature Reviews Clinical oncology* 15: 353–65.

279. Ding L. et al. (2018) Perspective on oncogenic processes at the end of the beginning of cancer genomics. *Cell* 173: 305–20.

280. Bailey M.H. et al. (2018) Comprehensive characterization of cancer driver genes and mutations. *Cell* 173: 371–85.

281. Liu T., Yuan X. and Xu D. (2016) Cancer-specific telomerase reverse transcriptase (TERT) promoter mutations: Biological and clinical implications. *Genes* 7: 38.

282. Yuan X., Larsson C. and Xu D. (2019) Mechanisms underlying the activation of TERT transcription and telomerase activity in human cancer: Old actors and new players. *Oncogene* 38: 6172–83.

283. Kumar S. et al. (2020) Passenger mutations in more than 2,500 cancer genomes: Overall molecular functional impact and consequences. *Cell* 180: 915–27.

284. Rheinbay E. et al. (2020) Analyses of non-coding somatic drivers in 2,658 cancer whole genomes. *Nature* 578: 102–11.

285. Tsimberidou A.M., Fountzilas E., Nikanjam M. and Kurzrock R. (2020) Review of precision cancer medicine: Evolution of the treatment paradigm. *Cancer Treatment Reviews* 86: 102019.

286. Malone E.R., Oliva M., Sabatini P.J.B., Stockley T.L. and Siu L.L. (2020) Molecular profiling for precision cancer therapies. *Genome Medicine* 12: 8.

287. Waldman A.D., Fritz J.M. and Lenardo M.J. (2020) A guide to cancer immunotherapy: From T cell basic science to clinical practice. *Nature Reviews Immunology* 20: 651–68.

288. Esfahani K. et al. (2020) A review of cancer immunotherapy: From the past, to the present, to the future. *Current Oncology* 27: S87–97.

289. Blass E. and Ott P.A. (2021) Advances in the development of personalized neoantigen-based therapeutic cancer vaccines. *Nature Reviews Clinical Oncology* 18: 215–29.

290. Johnston J.J. et al. (2012) Secondary variants in individuals undergoing exome sequencing: Screening of 572 individuals identifies high-penetrance mutations in cancer-susceptibility genes. *American Journal of Human Genetics* 91: 97–108.

291. Grzymski J.J. et al. (2020) Population genetic screening efficiently identifies carriers of autosomal dominant diseases. *Nature Medicine* 26: 1235–9.

292. Huang K.-L. et al. (2018) Pathogenic germline variants in 10,389 adult cancers. *Cell* 173: 355–70.

293. Moscow J.A., Fojo T. and Schilsky R.L. (2018) The evidence framework for precision cancer medicine. *Nature Reviews Clinical Oncology* 15: 183–92.

294. Miga K.H. et al. (2020) Telomere-to-telomere assembly of a complete human X chromosome. *Nature* 585: 79–84.

295. Logsdon G.A. et al. (2021) The structure, function and evolution of a complete human chromosome 8. *Nature* 593: 101–7.

296. Bergström A. et al. (2020) Insights into human genetic variation and population history from 929 diverse genomes. *Science* 367: eaay5012.

CHAPTER 12

1. Ruvkun G., Wightman B. and Ha I. (2004) The 20 years it took to recognize the importance of tiny RNAs. *Cell* 116: S93–6.

2. Zamore P.D. and Haley B. (2005) Ribo-gnome: The big world of small RNAs. *Science* 309: 1519–24.

3. Rich A. and Davies D.R. (1956) A new two-stranded helical structure: Polyadenylic acid and polyuridylic acid. *Journal of the American Chemical Society* 78: 3548–9.

4. Isaacs A. and Lindenmann J. (1957) Virus interference. 1. The Interferon. *Proceedings of the Royal Society B: Biological Sciences* 147: 258–67.

5. Pestka S. (2007) The Interferons: 50 years after their discovery, there is much more to learn. *Journal of Biological Chemistry* 282: 20047–51.

6. Schneider W.M., Chevillotte M.D. and Rice C.M. (2014) Interferon-stimulated genes: A complex web of host defenses. *Annual Review of Immunology* 32: 513–45.

7. Schroder K., Hertzog P.J., Ravasi T. and Hume D.A. (2004) Interferon-γ: An overview of signals, mechanisms and functions. *Journal of Leukocyte Biology* 75: 163–89.

8. Lewis U.J., Rickes E.L., Williams D.E., McClelland L. and Brink N.G. (1960) Studies on the antiviral agent Helenine. Purification and evidence for ribonucleoprotein nature. *Journal of the American Chemical Society* 82: 5178–82.

9. Tytell A.A., Lampson G.P., Field A.K. and Hilleman M.R. (1967) Inducers of interferon and host resistance. 3. Double-stranded RNA from reovirus type 3 virions (reo 3-RNA). *Proceedings of the National Academy of Sciences USA* 58: 1719–22.

10. Lampson G.P., Tytell A.A., Field A.K., Nemes M.M. and Hilleman M.R. (1967) Inducers of interferon and host resistance. I. Double-stranded RNA from extracts of *Penicillium funiculosum. Proceedings of the National Academy of Sciences USA* 58: 782–9.

11. Field A.K., Lampson G.P., Tytell A.A., Nemes M.M. and Hilleman M.R. (1967) Inducers of interferon and host resistance, IV. Double-stranded replicative form RNA (MS2-Ff-RNA) from *E. coli* infected with MS2 coliphage. *Proceedings of the National Academy of Sciences USA* 58: 2102–8.

12. Hilleman M.R. (1970) Double-stranded RNAs (Poly I:C) in the prevention of viral infections. *Archives of Internal Medicine* 126: 109–24.

13. Ehrenfeld E. and Hunt T. (1971) Double-stranded poliovirus RNA inhibits initiation of protein synthesis by reticulocyte lysates. *Proceedings of the National Academy of Sciences USA* 68: 1075–8.

14. Kerr I.M., Brown R.E. and Ball L.A. (1974) Increased sensitivity of cell-free protein synthesis to double-stranded RNA after interferon treatment. *Nature* 250: 57–9.

15. Buck K.W. (1988) From interferon induction to fungal viruses. *European Journal of Epidemiology* 4: 395–9.

16. Banks G.T. et al. (1968) Viruses in fungi and interferon stimulation. *Nature* 218: 542–5.

17. Wingard S.A. (1928) Hosts and symptoms of ring spot, a virus disease of plants. *Journal of Agricultural Research* 37: 127–54.

18. Hoskins M. (1935) A protective action of neurotropic against viscerotropic Yellow Fever Virus in *Macacus Rhesus. American Journal of Tropical Medicine and Hygiene* s1–15: 675–80.

19. Kerr I.M., Brown R.E. and Hovanessian A.G. (1977) Nature of inhibitor of cell-free protein synthesis formed in response to interferon and double-stranded RNA. *Nature* 268: 540–2.

20. Hovanessian A.G. (1989) The double stranded RNA-activated protein kinase induced by interferon: dsRNA-PK. *Journal of Interferon Research* 9: 641–7.

21. Coleman J., Hirashima A., Inokuchi Y., Green P.J. and Inouye M. (1985) A novel immune system against bacteriophage infection using complementary RNA (micRNA). *Nature* 315: 601–3.

22. Giunta S. and Groppa G. (1985) Interferon and micRNA in cellular defence. *Nature* 318: 237.

23. Takeda K., Kaisho T. and Akira S. (2003) Toll-like receptors. *Annual Review of Immunology* 21: 335–76.

24. Karikó K., Buckstein M., Ni H. and Weissman D. (2005) Suppression of RNA recognition by Toll-like receptors: The impact of nucleoside modification and the evolutionary origin of RNA. *Immunity* 23: 165–75.

25. Kawasaki T. and Kawai T. (2014) Toll-like receptor signaling pathways. *Frontiers in Immunology* 5: 461.

26. Fitzgerald K.A. and Kagan J.C. (2020) Toll-like receptors and the control of immunity. *Cell* 180: 1044–66.

27. Ward A., Hong W., Favaloro V. and Luo L. (2015) Toll receptors instruct axon and dendrite targeting and participate in synaptic partner matching in a *Drosophila* olfactory circuit. *Neuron* 85: 1013–28.

28. Foldi I. et al. (2017) Three-tier regulation of cell number plasticity by neurotrophins and Tolls in *Drosophila. Journal of Cell Biology* 216: 1421–38.

29. Anthoney N., Foldi I. and Hidalgo A. (2018) Toll and Toll-like receptor signalling in development. *Development* 145: dev156018.

30. Chen C.-Y., Shih Y.-C., Hung Y.-F. and Hsueh Y.-P. (2019) Beyond defense: Regulation of neuronal morphogenesis and brain functions via Toll-like receptors. *Journal of Biomedical Science* 26: 90.

31. Karikó K. et al. (2008) Incorporation of pseudouridine Into mRNA yields superior nonimmunogenic vector with increased translational capacity and biological stability. *Molecular Therapy* 16: 1833–40.

32. Fire A.Z. (2007) Gene silencing by double-stranded RNA (Nobel lecture). *Angewandte Chemie-International Edition* 46: 6967–84.

33. Fire A., Albertson D., Harrison S.W. and Moerman D.G. (1991) Production of antisense RNA leads to effective and specific inhibition of gene expression in *C. elegans* muscle. *Development* 113: 503–14.

34. Guo S. and Kemphues K.J. (1995) *par-1,* a gene required for establishing polarity in *C. elegans* embryos, encodes a putative Ser/Thr kinase that is asymmetrically distributed. *Cell* 81: 611–20.

35. Cogoni C. and Macino G. (2000) Post-transcriptional gene silencing across kingdoms. *Current Opinion in Genetics and Development* 10: 638–43.

36. Lindbo J.A., Silva-Rosales L., Proebsting W.M. and Dougherty W.G. (1993) Induction of a highly specific antiviral state in transgenic plants: Implications for regulation of gene expression and virus resistance. *Plant Cell* 5: 1749–59.

37. Covey S.N., AlKaff N.S., Langara A. and Turner D.S. (1997) Plants combat infection by gene silencing. *Nature* 385: 781–2.

38. Ratcliff F., Harrison B.D. and Baulcombe D.C. (1997) A similarity between viral defense and gene silencing in plants. *Science* 276: 1558–60.

39. Ecker J.R. and Davis R.W. (1986) Inhibition of gene expression in plant cells by expression of antisense RNA. *Proceedings of the National Academy of Sciences USA* 83: 5372–6.

40. van der Krol A.R. et al. (1988) An anti-sense chalcone synthase gene in transgenic plants inhibits flower pigmentation. *Nature* 333: 866–9.

41. Jorgensen R.A., Cluster P.D., English J., Que Q. and Napoli C.A. (1996) Chalcone synthase cosuppression phenotypes in petunia flowers: Comparison of sense vs. antisense constructs and single-copy vs. complex T-DNA sequences. *Plant Molecular Biology* 31: 957–73.

42. Scheid O.M. (2019) Illuminating (white and) purple patches. *Plant Cell* 31: 1208–9.

43. Van Blokland R., Vandergeest N., Mol J.N.M. and Kooter J.M. (1994) Transgene-mediated suppression of chalcone synthase expression in petunia-hybrida results from an increase in RNA turnover. *Plant Journal* 6: 861–77.

44. Napoli C., Lemieux C. and Jorgensen R. (1990) Introduction of a chimeric chalcone synthase gene into petunia results in reversible co-suppression of homologous genes *in trans*. *Plant Cell* 2: 279–89.

45. van der Krol A.R., Mur L.A., Beld M., Mol J.N. and Stuitje A.R. (1990) Flavonoid genes in petunia: Addition of a limited number of gene copies may lead to a suppression of gene expression. *Plant Cell* 2: 291–9.

46. Romano N. and Macino G. (1992) Quelling: Transient inactivation of gene expression in *Neurospora crassa* by transformation with homologous sequences. *Molecular Microbiology* 6: 3343–53.

47. Jorgensen R.A., Atkinson R.G., Forster R.L.S. and Lucas W.J. (1998) An RNA-based information superhighway in plants. *Science* 279: 1486–7.

48. Matzke M.A. and Matzke A. (1995) How and why do plants inactivate homologous (trans)genes? *Plant Physiology* 107: 679–85.

49. Matzke M.A., Aufsatz W., Kanno T., Mette M.F. and Matzke A.J. (2002) Homology-dependent gene silencing and host defense in plants. *Advances in Genetics* 46: 235–75.

50. Rocheleau C.E. et al. (1997) Wnt signaling and an APC-related gene specify endoderm in early *C. elegans* embryos. *Cell* 90: 707–16.

51. Pal-Bhadra M., Bhadra U. and Birchler J.A. (1997) Cosuppression in *Drosophila*: Gene silencing of Alcohol dehydrogenase by white-Adh transgenes is Polycomb dependent. *Cell* 90: 479–90.

52. Pal-Bhadra M., Bhadra U. and Birchler J.A. (1999) Cosuppression of nonhomologous transgenes in *Drosophila* involves mutually related endogenous sequences. *Cell* 99: 35–46.

53. Abel P.P. et al. (1986) Delay of disease development in transgenic plants that express the tobacco mosaic virus coat protein gene. *Science* 232: 738–43.

54. Rothstein S.J., Dimaio J., Strand M. and Rice D. (1987) Stable and heritable inhibition of the expression of nopaline synthase in tobacco expressing antisense RNA. *Proceedings of the National Academy of Sciences USA* 84: 8439–43.

55. Smith C.J.S. et al. (1988) Antisense RNA inhibition of polygalacturonase gene-expression in transgenic tomatoes. *Nature* 334: 724–6.

56. Herrera-Estrella L., Depicker A., Van Montagu M. and Schell J. (1983) Expression of chimaeric genes transferred into plant cells using a Ti-plasmid-derived vector. *Nature* 303: 209–13.

57. Smith H.A., Swaney S.L., Parks T.D., Wernsman E.A. and Dougherty W.G. (1994) Transgenic plant virus resistance mediated by untranslatable sense RNAs: Expression, regulation, and fate of nonessential RNAs. *Plant Cell* 6: 1441–53.

58. Smith C.J. et al. (1990) Expression of a truncated tomato polygalacturonase gene inhibits expression of the endogenous gene in transgenic plants. *Molecular and General Genetics* 224: 477–81.

59. Grierson D., Fray R.G., Hamilton A.J., Smith C.J.S. and Watson C.F. (1991) Does co-suppression of sense genes in transgenic plants involve antisense RNA. *Trends in Biotechnology* 9: 122–3.

60. Jorgensen R.A., Que Q. and Stam M. (1999) Do unintended antisense transcripts contribute to sense cosuppression in plants? *Trends in Genetics* 15: 11–2.

61. Wassenegger M., Heimes S., Riedel L. and Sänger H.L. (1994) RNA-directed de novo methylation of genomic sequences in plants. *Cell* 76: 567–76.

62. Baulcombe D.C. (1996) RNA as a target and an initiator of post-transcriptional gene silencing in transgenic plants. *Plant Molecular Biology* 32: 79–88.

63. Meyer P. and Saedler H. (1996) Homology-dependent gene silencing in plants. *Annual Review of Plant Physiology and Plant Molecular Biology* 47: 23–48.

64. Stam M., Mol J.N.M. and Kooter J.M. (1997) The silence of genes in transgenic plants. *Annals of Botany* 79: 3–12.

65. Jones L. et al. (1999) RNA-DNA interactions and DNA methylation in post-transcriptional gene silencing. *Plant Cell* 11: 2291–302.

66. Goyon C. and Faugeron G. (1989) Targeted transformation of *Ascobolus immersus* and de novo methylation of the resulting duplicated DNA sequences. *Molecular and Cellular Biology* 9: 2818–27.

67. Assaad F.F., Tucker K.L. and Signer E.R. (1993) Epigenetic repeat-induced gene silencing (RIGS) in *Arabidopsis*. *Plant Molecular Biology* 22: 1067–85.

68. Stam M., Viterbo A., Mol J.N.M. and Kooter J.M. (1998) Position-dependent methylation and transcriptional silencing of transgenes in inverted t-dna repeats: Implications for posttranscriptional silencing of homologous host genes in plants. *Molecular and Cellular Biology* 18: 6165–77.

69. Mette M.F., Aufsatz W., van Der Winden J., Matzke M.A. and Matzke A.J. (2000) Transcriptional silencing and promoter methylation triggered by double-stranded RNA. *EMBO Journal* 19: 5194–201.

70. de Carvalho F. et al. (1992) Suppression of beta-1,3-glucanase transgene expression in homozygous plants. *EMBO Journal* 11: 2595–602.

71. Sijen T. et al. (2001) Transcriptional and posttranscriptional gene silencing are mechanistically related. *Current Biology* 11: 436–40.

72. Pal-Bhadra M., Bhadra U. and Birchler J.A. (2002) RNAi related mechanisms affect both transcriptional and posttranscriptional transgene silencing in *Drosophila*. *Molecular Cell* 9: 315–27.

73. Voinnet O. and Baulcombe D.C. (1997) Systemic signalling in gene silencing. *Nature* 389: 553.

74. Cogoni C. et al. (1996) Transgene silencing of the al-1 gene in vegetative cells of *Neurospora* is mediated by a cytoplasmic effector and does not depend on DNA-DNA interactions or DNA methylation. *EMBO Journal* 15: 3153–63.

75. Palauqui J.C., Elmayan T., Pollien J.M. and Vaucheret H. (1997) Systemic acquired silencing: Transgene-specific post-transcriptional silencing is transmitted by grafting from silenced stocks to non-silenced scions. *EMBO Journal* 16: 4738–45.

76. Voinnet O., Vain P., Angell S. and Baulcombe D.C. (1998) Systemic spread of sequence-specific transgene RNA degradation in plants is initiated by localized introduction of ectopic promoterless DNA. *Cell* 95: 177–87.

77. Angell S.M. and Baulcombe D.C. (1997) Consistent gene silencing in transgenic plants expressing a replicating potato virus X RNA. *EMBO Journal* 16: 3675–84.

78. Metzlaff M., O'Dell M., Cluster P.D. and Flavell R.B. (1997) RNA-mediated RNA degradation and chalcone synthase A silencing in petunia. *Cell* 88: 845–54.

79. Hamilton A.J. and Baulcombe D.C. (1999) A species of small antisense RNA in posttranscriptional gene silencing in plants. *Science* 286: 950–2.

80. Barry C.S., Llop-Tous M.I. and Grierson D. (2000) The regulation of 1-aminocyclopropane-1-carboxylic acid synthase gene expression during the transition from system-1 to system-2 ethylene synthesis in tomato. *Plant Physiology* 123: 979–86.

81. Fire A. et al. (1998) Potent and specific genetic interference by double-stranded RNA in *Caenorhabditis elegans*. *Nature* 391: 806–11.

82. Waterhouse P.M., Graham M.W. and Wang M.B. (1998) Virus resistance and gene silencing in plants can be induced by simultaneous expression of sense and antisense RNA. *Proceedings of the National Academy of Sciences USA* 95: 13959–64.

83. Ngo H., Tschudi C., Gull K. and Ullu E. (1998) Double-stranded RNA induces mRNA degradation in *Trypanosoma brucei*. *Proceedings of the National Academy of Sciences USA* 95: 14687–92.

84. Kennerdell J.R. and Carthew R.W. (1998) Use of dsRNA-mediated genetic interference to demonstrate that *frizzled* and *frizzled 2* act in the wingless pathway. *Cell* 95: 1017–26.

85. Tuschl T., Zamore P.D., Lehmann R., Bartel D.P. and Sharp P.A. (1999) Targeted mRNA degradation by double-stranded RNA in vitro. *Genes & Development* 13: 3191–7.

86. Kennerdell J.R. and Carthew R.W. (2000) Heritable gene silencing in *Drosophila* using double-stranded RNA. *Nature Biotechnology* 18: 896–8.

87. Yang D., Lu H. and Erickson J.W. (2000) Evidence that processed small dsRNAs may mediate sequence-specific mRNA degradation during RNAi in *Drosophila* embryos. *Current Biology* 10: 1191–200.

88. Wianny F. and Zernicka-Goetz M. (2000) Specific interference with gene function by double-stranded RNA in early mouse development. *Nature Cell Biology* 2: 70–5.

89. Elbashir S.M. et al. (2001) Duplexes of 21-nucleotide RNAs mediate RNA interference in cultured mammalian cells. *Nature* 411: 494–8.

90. Caplen N.J., Parrish S., Imani F., Fire A. and Morgan R.A. (2001) Specific inhibition of gene expression by small double-stranded RNAs in invertebrate and vertebrate systems. *Proceedings of the National Academy of Sciences USA* 98: 9742–7.

91. Grishok A., Tabara H. and Mello C.C. (2000) Genetic requirements for inheritance of RNAi in *C. elegans*. *Science* 287: 2494–7.

92. Hammond S.M., Caudy A.A. and Hannon G.J. (2001) Post-transcriptional gene silencing by double-stranded RNA. *Nature Reviews Genetics* 2: 1110–9.

93. Cogoni C. and Macino G. (1997) Isolation of *quelling-defective (qde)* mutants impaired in posttranscriptional transgene-induced gene silencing in *Neurospora crassa*. *Proceedings of the National Academy of Sciences USA* 94: 10233–8.

94. Cogoni C. and Macino G. (1999) Gene silencing in *Neurospora crassa* requires a protein homologous to RNA-dependent RNA polymerase. *Nature* 399: 166–9.

95. Astier-Manifacier S. and Cornuet P. (1971) RNA-dependent RNA polymerase in Chinese cabbage. *Biochimica et Biophysica Acta* 232: 484–93.

96. Schiebel W. et al. (1998) Isolation of an RNA-directed RNA polymerase-specific cDNA clone from tomato. *Plant Cell* 10: 2087–101.

97. Schiebel W., Haas B., Marinkovic S., Klanner A. and Sanger H.L. (1993) RNA-directed RNA polymerase from tomato leaves. I. Purification and physical properties. *Journal of Biological Chemistry* 268: 11851–7.

98. Brennecke J. et al. (2007) Discrete small RNA-generating loci as master regulators of transposon activity in *Drosophila*. *Cell* 128: 1089–103.

99. Gunawardane L.S. et al. (2007) A slicer-mediated mechanism for repeat-associated siRNA 5′ end formation in *Drosophila*. *Science* 315: 1587–90.

100. Czech B. and Hannon G.J. (2016) One loop to rule them all: The ping-pong cycle and piRNA-guided silencing. *Trends in Biochemical Sciences* 41: 324–37.

101. Dalmay T., Hamilton A., Rudd S., Angell S. and Baulcombe D.C. (2000) An RNA-dependent RNA polymerase gene in Arabidopsis is required for posttranscriptional gene silencing mediated by a transgene but not by a virus. *Cell* 101: 543–53.

102. Fagard M., Boutet S., Morel J.B., Bellini C. and Vaucheret H. (2000) AGO1, QDE-2, and RDE-1 are related proteins required for post-transcriptional gene silencing in plants, quelling in fungi, and RNA interference in animals. *Proceedings of the National Academy of Sciences USA* 97: 11650–4.

103. Shi H. et al. (2000) Genetic interference in *Trypanosoma brucei* by heritable and inducible double-stranded RNA. *RNA* 6: 1069–76.

104. Sijen T. et al. (2001) On the role of RNA amplification in dsRNA-triggered gene silencing. *Cell* 107: 465–76.

105. Zamore P.D. (2002) Ancient pathways programmed by small RNAs. *Science* 296: 1265–9.

106. Poirier E.Z. et al. (2021) An isoform of Dicer protects mammalian stem cells against multiple RNA viruses. *Science* 373: 231–6.

107. Hammond S.M., Bernstein E., Beach D. and Hannon G.J. (2000) An RNA-directed nuclease mediates post-transcriptional gene silencing in *Drosophila* cells. *Nature* 404: 293–6.

108. Zamore P.D., Tuschl T., Sharp P.A. and Bartel D.P. (2000) RNAi: Double-stranded RNA directs the ATP-dependent cleavage of mRNA at 21 to 23 nucleotide intervals. *Cell* 101: 25–33.

109. Bernstein E., Caudy A.A., Hammond S.M. and Hannon G.J. (2001) Role for a bidentate ribonuclease in the initiation step of RNA interference. *Nature* 409: 363–6.

110. Dougherty W.G. and Parks T.D. (1995) Transgenes and gene suppression: Telling us something new? *Current Opinion in Cell Biology* 7: 399–405.

111. Bohmert K. et al. (1998) AGO1 defines a novel locus of Arabidopsis controlling leaf development. *EMBO Journal* 17: 170–80.

112. Farazi T.A., Juranek S.A. and Tuschl T. (2008) The growing catalog of small RNAs and their association with distinct Argonaute/Piwi family members. *Development* 135: 1201–14.

113. Elbashir S.M., Lendeckel W. and Tuschl T. (2001) RNA interference is mediated by 21 and 22 nt RNAs. *Genes & Development* 15: 188–200.

114. Hall T.M. (2005) Structure and function of argonaute proteins. *Structure* 13: 1403–8.

115. Peters L. and Meister G. (2007) Argonaute proteins: Mediators of RNA silencing. *Molecular Cell* 26: 611–23.

116. Meins Jr. F., Si-Ammour A. and Blevins T. (2005) RNA silencing systems and their relevance to plant development. *Annual Review of Cell and Developmental Biology* 21: 297–318.

117. Fei Q., Xia R. and Meyers B.C. (2013) Phased, secondary, small interfering RNAs in posttranscriptional regulatory networks. *Plant Cell* 25: 2400–15.

118. Lee C.H. and Carroll B.J. (2018) Evolution and diversification of small RNA pathways in flowering plants. *Plant and Cell Physiology* 59: 2169–87.

119. Meister G. (2013) Argonaute proteins: Functional insights and emerging roles. *Nature Reviews Genetics* 14: 447–59.

120. Sashital D.G. (2017) Prokaryotic Argonaute uses an all-in-one mechanism to provide host defense. *Molecular Cell* 65: 957–8.

121. Olovnikov I., Chan K., Sachidanandam R., Newman D.K. and Aravin A.A. (2013) Bacterial Argonaute samples the transcriptome to identify foreign DNA. *Molecular Cell* 51: 594–605.

122. Swarts D.C. et al. (2014) DNA-guided DNA interference by a prokaryotic Argonaute. *Nature* 507: 258–61.

123. Swarts D.C. et al. (2017) Autonomous generation and loading of DNA guides by bacterial Argonaute. *Molecular Cell* 65: 985–98.

124. Kaya E. et al. (2016) A bacterial Argonaute with non-canonical guide RNA specificity. *Proceedings of the National Academy of Sciences USA* 113: 4057–62.

125. Drinnenberg I.A., Fink G.R. and Bartel D.P. (2011) Compatibility with killer explains the rise of RNAi-deficient fungi. *Science* 333: 1592.

126. Becker B. and Schmitt M.J. (2017) Yeast killer toxin K28: Biology and unique strategy of host cell intoxication and killing. *Toxins* 9: 333.

127. Wassarman K.M., Repoila F., Rosenow C., Storz G. and Gottesman S. (2001) Identification of novel small RNAs using comparative genomics and microarrays. *Genes & Development* 15: 1637–51.

128. Møller T. et al. (2002) Hfq: A bacterial Sm-like protein that mediates RNA-RNA interaction. *Molecular Cell* 9: 23–30.

129. Zhang A., Wassarman K.M., Ortega J., Steven A.C. and Storz G. (2002) The Sm-like Hfq protein increases OxyS RNA interaction with target mRNAs. *Molecular Cell* 9: 11–22.

130. Schumacher M.A., Pearson R.F., Møller T., Valentin-Hansen P. and Brennan R.G. (2002) Structures of the pleiotropic translational regulator Hfq and an Hfq-RNA complex: A bacterial Sm-like protein. *EMBO Journal* 21: 3546–56.

131. Geissmann T.A. and Touati D. (2004) Hfq, a new chaperoning role: Binding to messenger RNA determines access for small RNA regulator. *EMBO Journal* 23: 396–405.

132. Lenz D.H. et al. (2004) The small RNA chaperone Hfq and multiple small RNAs control quorum sensing in *Vibrio harveyi* and *Vibrio cholerae*. *Cell* 118: 69–82.

133. Jose A.M. and Hunter C.P. (2007) Transport of sequence-specific RNA interference information between cells. *Annual Review of Genetics* 41: 305–30.

134. Winston W.M., Molodowitch C. and Hunter C.P. (2002) Systemic RNAi in *C. elegans* requires the putative transmembrane protein SID-1. *Science* 295: 2456–9.

135. Winston W.M., Sutherlin M., Wright A.J., Feinberg E.H. and Hunter C.P. (2007) *Caenorhabditis elegans* SID-2 is required for environmental RNA interference. *Proceedings of the National Academy of Sciences USA* 104: 10565–70.

136. Nguyen T.A. et al. (2017) SIDT2 transports extracellular dsRNA into the cytoplasm for innate immune recognition. *Immunity* 47: 498–509.

137. Nguyen T.A. et al. (2019) SIDT1 localizes to endolysosomes and mediates double-stranded RNA transport into the cytoplasm. *Journal of Immunology* 202: 3483–92.

138. Molnar A. et al. (2010) Small silencing RNAs in plants are mobile and direct epigenetic modification in recipient cells. *Science* 328: 872–5.

139. Dickson E. and Robertson H.D. (1976) Potential regulatory roles for RNA in cellular development. *Cancer Research* 36: 3387–93.

140. Benner S.A. (1988) Extracellular 'communicator RNA'. *FEBS Letters* 233: 225–8.

141. Valadi H. et al. (2007) Exosome-mediated transfer of mRNAs and microRNAs is a novel mechanism of genetic exchange between cells. *Nature Cell Biology* 9: 654–9.

142. Dinger M.E., Mercer T.R. and Mattick J.S. (2008) RNAs as extracellular signaling molecules. *Journal of Molecular Endocrinology* 40: 151–9.

143. Wahlgren J. et al. (2012) Plasma exosomes can deliver exogenous short interfering RNA to monocytes and lymphocytes. *Nucleic Acids Research* 40: e130.

144. O'Brien K., Breyne K., Ughetto S., Laurent L.C. and Breakefield X.O. (2020) RNA delivery by extracellular vesicles in mammalian cells and its applications. *Nature Reviews Molecular Cell Biology* 21: 585–606.

145. Mittelbrunn M. et al. (2011) Unidirectional transfer of microRNA-loaded exosomes from T cells to antigen-presenting cells. *Nature Communications* 2: 282.

146. Montecalvo A. et al. (2012) Mechanism of transfer of functional microRNAs between mouse dendritic cells via exosomes. *Blood* 119: 756–66.

147. Okoye I.S. et al. (2014) MicroRNA-containing T-regulatory-cell-derived exosomes suppress pathogenic T helper 1 cells. *Immunity* 41: 89–103.

148. Thomou T. et al. (2017) Adipose-derived circulating miRNAs regulate gene expression in other tissues. *Nature* 542: 450–5.

149. Turchinovich A., Drapkina O. and Tonevitsky A. (2019) Transcriptome of extracellular vesicles: State-of-the-art. *Frontiers in Immunology* 10: 202.

150. Reggiardo R.E. et al. (2020) Epigenomic reprogramming of repetitive noncoding RNAs and IFN-stimulated genes by mutant KRAS. *bioRxiv*: 2020.11.04.367771.

151. Zhang Y. et al. (2017) Hypothalamic stem cells control ageing speed partly through exosomal miRNAs. *Nature* 548: 52–7.

152. Skog J. et al. (2008) Glioblastoma microvesicles transport RNA and proteins that promote tumour growth and provide diagnostic biomarkers. *Nature Cell Biology* 10: 1470–6.

153. Azmi A.S., Bao B. and Sarkar F.H. (2013) Exosomes in cancer development, metastasis, and drug resistance: A comprehensive review. *Cancer and Metastasis Reviews* 32: 623–42.

154. Rana S., Malinowska K. and Zöller M. (2013) Exosomal tumor microRNA modulates premetastatic organ cells. *Neoplasia* 15: 281–95.

155. Albanese M. et al. (2021) MicroRNAs are minor constituents of extracellular vesicles that are rarely delivered to target cells. *PLOS Genetics* 17: e1009951.

156. Garcia-Martin R. et al. (2022) MicroRNA sequence codes for small extracellular vesicle release and cellular retention. *Nature* 601: 446–51.

157. Mette M.F., van der Winden J., Matzke M.A. and Matzke A.J.M. (1999) Production of aberrant promoter transcripts contributes to methylation and silencing of unlinked homologous promoters in trans. *EMBO Journal* 18: 241–8.

158. Morel J., Mourrain P., Beclin C. and Vaucheret H. (2000) DNA methylation and chromatin structure affect transcriptional and post-transcriptional transgene silencing in *Arabidopsis*. *Current Biology* 10: 1591–4.

159. Matzke M.A. and Mosher R.A. (2014) RNA-directed DNA methylation: An epigenetic pathway of increasing complexity. *Nature Reviews Genetics* 15: 394–408.

160. Erdmann R.M. and Picard C.L. (2020) RNA-directed DNA methylation. *PLOS Genetics* 16: e1009034.

161. Wassenegger M. (2000) RNA-directed DNA methylation. *Plant Molecular Biology* 43: 203–20.

162. Pal-Bhadra M. et al. (2004) Heterochromatic silencing and HP1 localization in *Drosophila* are dependent on the RNAi machinery. *Science* 303: 669–72.

163. Fukagawa T. et al. (2004) Dicer is essential for formation of the heterochromatin structure in vertebrate cells. *Nature Cell Biology* 6: 784–91.

164. Morris K.V., Chan S.W., Jacobsen S.E. and Looney D.J. (2004) Small interfering RNA-induced transcriptional gene silencing in human cells. *Science* 305: 1289–92.

165. Castel S.E. and Martienssen R.A. (2013) RNA interference in the nucleus: Roles for small RNAs in transcription, epigenetics and beyond. *Nature Reviews Genetics* 14: 100–12.

166. Janowski B.A. et al. (2006) Involvement of AGO1 and AGO2 in mammalian transcriptional silencing. *Nature Structural & Molecular Biology* 13: 787–92.

167. Kim D.H., Villeneuve L.M., Morris K.V. and Rossi J.J. (2006) Argonaute-1 directs siRNA-mediated transcriptional gene silencing in human cells. *Nature Structural & Molecular Biology* 13: 793–7.

168. Vagin V.V. et al. (2006) A distinct small RNA pathway silences selfish genetic elements in the germline. *Science* 313: 320–4.

169. Gutbrod M.J. and Martienssen R.A. (2020) Conserved chromosomal functions of RNA interference. *Nature Reviews Genetics* 21: 311–31.

170. Volpe T.A. et al. (2002) Regulation of heterochromatic silencing and histone H3 lysine-9 methylation by RNAi. *Science* 297: 1833–7.

171. Hall I.M. et al. (2002) Establishment and maintenance of a heterochromatin domain. *Science* 297: 2232–7.

172. Lippman Z. et al. (2004) Role of transposable elements in heterochromatin and epigenetic control. *Nature* 430: 471–6.

173. Lippman Z. and Martienssen R. (2004) The role of RNA interference in heterochromatic silencing. *Nature* 431: 364–70.

174. Verdel A. et al. (2004) RNAi-mediated targeting of heterochromatin by the RITS complex. *Science* 303: 672–6.

175. Martienssen R.A., Zaratiegui M. and Goto D.B. (2005) RNA interference and heterochromatin in the fission yeast *Schizosaccharomyces pombe*. *Trends in Genetics* 21: 450–6.

176. Buhler M., Verdel A. and Moazed D. (2006) Tethering RITS to a nascent transcript initiates RNAi- and heterochromatin-dependent gene silencing. *Cell* 125: 873–86.

177. Buhler M., Haas W., Gygi S.P. and Moazed D. (2007) RNAi-dependent and -independent RNA turnover mechanisms contribute to heterochromatic gene silencing. *Cell* 129: 707–21.

178. Buhler M. and Moazed D. (2007) Transcription and RNAi in heterochromatic gene silencing. *Nature Structural & Molecular Biology* 14: 1041–8.

179. Slotkin R.K. and Martienssen R. (2007) Transposable elements and the epigenetic regulation of the genome. *Nature Reviews Genetics* 8: 272–85.

180. Bayne E.H. et al. (2008) Splicing factors facilitate RNAi-directed silencing in fission yeast. *Science* 322: 602.

181. Shimada Y., Mohn F. and Bühler M. (2016) The RNA-induced transcriptional silencing complex targets chromatin exclusively via interacting with nascent transcripts. *Genes & Development* 30: 2571–80.

182. Reinhart B.J. and Bartel D.P. (2002) Small RNAs correspond to centromere heterochromatic repeats. *Science* 297: 1831.

183. Volpe T. et al. (2003) RNA interference is required for normal centromere function in fission yeast. *Chromosome Research* 11: 137–46.

184. Hall I.M., Noma K. and Grewal S.I. (2003) RNA interference machinery regulates chromosome dynamics during mitosis and meiosis in fission yeast. *Proceedings of the National Academy of Sciences USA* 100: 193–8.

185. Kanellopoulou C. et al. (2005) Dicer-deficient mouse embryonic stem cells are defective in differentiation and centromeric silencing. *Genes & Development* 19: 489–501.

186. Schramke V. and Allshire R. (2003) Hairpin RNAs and retrotransposon LTRs effect RNAi and chromatin-based gene silencing. *Science* 301: 1069–74.

187. Mochizuki K., Fine N.A., Fujisawa T. and Gorovsky M.A. (2002) Analysis of a piwi-related gene implicates small RNAs in genome rearrangement in *Tetrahymena*. *Cell* 110: 689–99.

188. Mochizuki K. and Gorovsky M.A. (2004) Small RNAs in genome rearrangement in *Tetrahymena*. *Current Opinion in Genetics and Development* 14: 181–7.

189. Nowacki M. et al. (2007) RNA-mediated epigenetic programming of a genome-rearrangement pathway. *Nature* 451: 153–8.

190. Nowacki M., Shetty K. and Landweber L.F. (2011) RNA-mediated epigenetic programming of genome rearrangements. *Annual Review of Genomics and Human Genetics* 12: 367–89.

191. Swiezewski S. et al. (2007) Small RNA-mediated chromatin silencing directed to the 3′ region of the *Arabidopsis* gene encoding the developmental regulator, FLC. *Proceedings of the National Academy of Sciences USA* 104: 3633–8.

192. Hu H. and Gatti R.A. (2011) MicroRNAs: New players in the DNA damage response. *Journal of Molecular Cell Biology* 3: 151–8.

193. Bhattacharjee S., Roche B. and Martienssen R.A. (2019) RNA-induced initiation of transcriptional silencing (RITS) complex structure and function. *RNA Biology* 16: 1133–46.

194. Collins J., Saari B. and Anderson P. (1987) Activation of a transposable element in the germ line but not the soma of *Caenorhabditis elegans*. *Nature* 328: 726–8.

195. Tabara H. et al. (1999) The rde-1 gene, RNA interference, and transposon silencing in *C. elegans*. *Cell* 99: 123–32.

196. Vastenhouw N.L. and Plasterk R.H.A. (2004) RNAi protects the *Caenorhabditis elegans* germline against transposition. *Trends in Genetics* 20: 314–9.

197. Yang N. and Kazazian H.H. (2006) L1 retrotransposition is suppressed by endogenously encoded small interfering RNAs in human cultured cells. *Nature Structural & Molecular Biology* 13: 763–71.

198. Ghildiyal M. et al. (2008) Endogenous siRNAs derived from transposons and mRNAs in *Drosophila* somatic cells. *Science* 320: 1077–81.

199. Bernstein E. et al. (2003) Dicer is essential for mouse development. *Nature Genetics* 35: 215–7.

200. Schuettengruber B., Chourrout D., Vervoort M., Leblanc B. and Cavalli G. (2007) Genome regulation by polycomb and trithorax proteins. *Cell* 128: 735–45.

201. Onodera Y. et al. (2005) Plant nuclear RNA polymerase IV mediates siRNA and DNA methylation-dependent heterochromatin formation. *Cell* 120: 613–22.

202. Pontier D. et al. (2005) Reinforcement of silencing at transposons and highly repeated sequences requires the concerted action of two distinct RNA polymerases IV in *Arabidopsis*. *Genes & Development* 19: 2030–40.

203. Wierzbicki A.T., Haag J.R. and Pikaard C.S. (2008) Noncoding transcription by RNA polymerase Pol IVb/Pol V mediates transcriptional silencing of overlapping and adjacent genes. *Cell* 135: 635–48.

204. Matzke M., Kanno T., Daxinger L., Huettel B. and Matzke A. (2009) RNA-mediated chromatin-based silencing in plants. *Current Opinion in Cell Biology* 21: 367–76.

205. Law J.A. and Jacobsen S.E. (2010) Establishing, maintaining and modifying DNA methylation patterns in plants and animals. *Nature Reviews Genetics* 11: 204–20.

206. Carbonell A. and Carrington J.C. (2015) Antiviral roles of plant ARGONAUTES. *Current Opinion in Plant Biology* 27: 111–7.

207. Rymen B., Ferrafiat L. and Blevins T. (2020) Non-coding RNA polymerases that silence transposable elements and reprogram gene expression in plants. *Transcription* 11: 172–91.

208. Rechavi O., Minevich G. and Hobert O. (2011) Transgenerational inheritance of an acquired small RNA-based antiviral response in C. elegans. *Cell* 147: 1248–56.

209. Ashe A. et al. (2012) piRNAs can trigger a multigenerational epigenetic memory in the germline of C. elegans. *Cell* 150: 88–99.

210. Duempelmann L., Skribbe M. and Bühler M. (2020) Small RNAs in the transgenerational inheritance of epigenetic information. *Trends in Genetics* 36: 203–14.

211. Shukla A. et al. (2020) poly(UG)-tailed RNAs in genome protection and epigenetic inheritance. *Nature* 582: 283–8.

212. Burton N.O., Burkhart K.B. and Kennedy S. (2011) Nuclear RNAi maintains heritable gene silencing in *Caenorhabditis elegans*. *Proceedings of the National Academy of Sciences USA* 108: 19683–8.

213. Kiani J. et al. (2013) RNA–mediated epigenetic heredity requires the cytosine methyltransferase Dnmt2. *PLOS Genetics* 9: e1003498.

214. Wheway G. et al. (2015) An siRNA-based functional genomics screen for the identification of regulators of ciliogenesis and ciliopathy genes. *Nature Cell Biology* 17: 1074–87.

215. Kok K.-H., Lei T. and Jin D.-Y. (2009) siRNA and shRNA screens advance key understanding of host factors required for HIV-1 replication. *Retrovirology* 6: 78.

216. Mendes-Pereira A.M. et al. (2012) Genome-wide functional screen identifies a compendium of genes affecting sensitivity to tamoxifen. *Proceedings of the National Academy of Sciences USA* 109: 2730–5.

217. Setten R.L., Rossi J.J. and Han S.-P. (2019) The current state and future directions of RNAi-based therapeutics. *Nature Reviews Drug Discovery* 18: 421–46.

218. Lykken E.A., Shyng C., Edwards R.J., Rozenberg A. and Gray S.J. (2018) Recent progress and considerations for AAV gene therapies targeting the central nervous system. *Journal of Neurodevelopmental Disorders* 10: 16.

219. Lorenzer C., Dirin M., Winkler A.-M., Baumann V. and Winkler J. (2015) Going beyond the liver: Progress and challenges of targeted delivery of siRNA therapeutics. *Journal of Controlled Release* 203: 1–15.

220. Daya S. and Berns K.I. (2008) Gene therapy using adeno-associated virus vectors. *Clinical Microbiology Reviews* 21: 583–93.

221. Lundstrom K. (2018) Viral vectors in gene therapy. *Diseases* 6: 42.

222. Kaczmarek J.C., Kowalski P.S. and Anderson D.G. (2017) Advances in the delivery of RNA therapeutics: From concept to clinical reality. *Genome Medicine* 9: 60.

223. Tatiparti K., Sau S., Kashaw S.K. and Iyer A.K. (2017) siRNA delivery strategies: A comprehensive review of recent developments. *Nanomaterials* 7: 77.

224. Dong Y., Siegwart D.J. and Anderson D.G. (2019) Strategies, design, and chemistry in siRNA delivery systems. *Advanced Drug Delivery Reviews* 144: 133–47.

225. Paunovska K., Loughrey D. and Dahlman J.E. (2022) Drug delivery systems for RNA therapeutics. *Nature Reviews Genetics.* epub ahead of print: https://www.nature.com/articles/s41576-021-00439-4.

226. Guo Q., Liu Q., Smith N.A., Liang G. and Wang M.-B. (2016) RNA silencing in plants: Mechanisms, technologies and applications in horticultural crops. *Current Genomics* 17: 476–89.

227. Dalakouras A. et al. (2020) Genetically modified organism-free RNA Interference: Exogenous application of RNA molecules in plants. *Plant Physiology* 182: 38–50.

228. Naso M.F., Tomkowicz B., Perry 3rd W.L., and Strohl W.R. (2017) Adeno-associated virus (AAV) as a vector for gene therapy. *BioDrugs: Clinical Immunotherapeutics, Biopharmaceuticals and Gene Therapy* 31: 317–34.

229. Goswami R. et al. (2019) Gene therapy leaves a vicious cycle. *Frontiers in Oncology* 9: 297.

230. Wilton S.D. et al. (2007) Antisense oligonucleotide-induced exon skipping across the human Dystrophin gene transcript. *Molecular Therapy* 15: 1288–96.

231. Dolgin E. (2019) News Feature: Gene therapy successes point to better therapies. *Proceedings of the National Academy of Sciences USA* 116: 23866–70.

232. Nguyen Q. and Yokota T. (2019) Antisense oligonucleotides for the treatment of cardiomyopathy in Duchenne muscular dystrophy. *American Journal of Translational Research* 11: 1202–18.

233. Russell S. *et al.* (2017) Efficacy and safety of voretigene neparvovec (AAV2-hRPE65v2) in patients with RPE65-mediated inherited retinal dystrophy: A randomised, controlled, open-label, phase 3 trial. *Lancet* 390: 849–60.

234. Lee R.C., Feinbaum R.L. and Ambros V. (1993) The *C. elegans* heterochronic *gene lin-4* encodes small RNAs with antisense complementarity to *lin-14. Cell* 75: 843–54.

235. Reinhart B.J. et al. (2000) The 21-nucleotide let-7 RNA regulates developmental timing in *Caenorhabditis elegans. Nature* 403: 901–6.

236. Pasquinelli A.E. et al. (2000) Conservation of the sequence and temporal expression of let-7 heterochronic regulatory RNA. *Nature* 408: 86–9.

237. Lagos-Quintana M., Rauhut R., Lendeckel W. and Tuschl T. (2001) Identification of novel genes coding for small expressed RNAs. *Science* 294: 853–8.

238. Slack F.J. et al. (2000) The *lin-41 RBCC* gene acts in the *C. elegans* heterochronic pathway between the let-7 regulatory RNA and the LIN-29 transcription factor. *Molecular Cell* 5: 659–69.

239. Grishok A. et al. (2001) Genes and mechanisms related to RNA interference regulate expression of the small temporal RNAs that control *C. elegans* developmental timing. *Cell* 106: 23–34.

240. Ambros V. (2008) The evolution of our thinking about microRNAs. *Nature Medicine* 14: 1036–40.

241. Lau N.C., Lim L.P., Weinstein E.G. and Bartel D.P. (2001) An abundant class of tiny RNAs with probable regulatory roles in *Caenorhabditis elegans. Science* 294: 858–62.

242. Lee R.C. and Ambros V. (2001) An extensive class of small RNAs in *Caenorhabditis elegans. Science* 294: 862–4.

243. Grad Y. et al. (2003) Computational and experimental identification of *C. elegans* microRNAs. *Molecular Cell* 11: 1253–63.

244. Lagos-Quintana M. et al. (2002) Identification of tissue-specific microRNAs from mouse. *Current Biology* 12: 735–9.

245. Brennecke J. and Cohen S.M. (2003) Towards a complete description of the microRNA complement of animal genomes. *Genome Biology* 4: 228.

246. Lagos-Quintana M., Rauhut R., Meyer J., Borkhardt A. and Tuschl T. (2003) New microRNAs from mouse and human. *RNA* 9: 175–9.

247. Bentwich I. et al. (2005) Identification of hundreds of conserved and nonconserved human microRNAs. *Nature Genetics* 37: 766–70.

248. Berezikov E. et al. (2006) Diversity of microRNAs in human and chimpanzee brain. *Nature Genetics* 38: 1375–7.

249. Berezikov E. et al. (2006) Many novel mammalian microRNA candidates identified by extensive cloning and RAKE analysis. *Genome Research* 16: 1289–98.

250. Pasquinelli A.E. (2002) MicroRNAs: Deviants no longer. *Trends in Genetics* 18: 171–3.

251. Ruvkun G. (2001) Glimpses of a tiny RNA world. *Science* 294: 797–9.

252. Clurman B.E. and Hayward W.S. (1989) Multiple proto-oncogene activations in avian leukosis virus-induced lymphomas: Evidence for stage-specific events. *Molecular and Cellular Biology* 9: 2657–64.

253. Tam W., Ben-Yehuda D. and Hayward W.S. (1997) *bic,* a novel gene activated by proviral insertions in avian leukosis virus-induced lymphomas, is likely to function through its noncoding RNA. *Molecular and Cellular Biology* 17: 1490–502.

254. Eis P.S. et al. (2005) Accumulation of miR-155 and BIC RNA in human B cell lymphomas. *Proceedings of the National Academy of Sciences USA* 102: 3627–32.

255. Nguyen L.X.T. et al. (2021) Cytoplasmic DROSHA and non-canonical mechanisms of MiR-155 biogenesis in FLT3-ITD acute myeloid leukemia. *Leukemia* 35: 2285–98.

256. Cai X. and Cullen B.R. (2007) The imprinted H19 non-coding RNA is a primary microRNA precursor. *RNA* 13: 313–6.

257. Keniry A. et al. (2012) The H19 lincRNA is a developmental reservoir of miR-675 that suppresses growth and Igf1r. *Nature Cell Biology* 14: 659–65.

258. Du Q. et al. (2019) MIR205HG Is a long noncoding RNA that regulates growth hormone and prolactin production in the anterior pituitary. *Developmental Cell* 49: 618–31.

259. Llave C., Kasschau K.D., Rector M.A. and Carrington J.C. (2002) Endogenous and silencing-associated small RNAs in plants. *Plant Cell* 14: 1605–19.

260. Rodriguez A., Griffiths-Jones S., Ashurst J.L. and Bradley A. (2004) Identification of mammalian microRNA host genes and transcription units. *Genome Research* 14: 1902–10.

261. Lee Y. et al. (2004) MicroRNA genes are transcribed by RNA polymerase II. *EMBO Journal* 23: 4051–60.

262. Borchert G.M., Lanier W. and Davidson B.L. (2006) RNA polymerase III transcribes human microRNAs. *Nature Structural & Molecular Biology* 13: 1097–101.

263. Kim Y.K. and Kim V.N. (2007) Processing of intronic microRNAs. *EMBO Journal* 26: 775–83.

264. Dieci G., Fiorino G., Castelnuovo M., Teichmann M. and Pagano A. (2007) The expanding RNA polymerase III transcriptome. *Trends in Genetics* 23: 614–22.

265. Babiarz J.E., Ruby J.G., Wang Y., Bartel D.P. and Blelloch R. (2008) Mouse ES cells express endogenous shRNAs, siRNAs, and other Microprocessor-independent, Dicer-dependent small RNAs. *Genes & Development* 22: 2773–85.

266. Lee Y. et al. (2003) The nuclear RNase III Drosha initiates microRNA processing. *Nature* 425: 415–9.

267. Han J. et al. (2004) The Drosha-DGCR8 complex in primary microRNA processing. *Genes & Development* 18: 3016–27.

268. Denli A.M., Tops B.B.J., Plasterk R.H.A., Ketting R.F. and Hannon G.J. (2004) Processing of primary microRNAs by the Microprocessor complex. *Nature* 432: 231–5.

269. Gregory R.I. et al. (2004) The Microprocessor complex mediates the genesis of microRNAs. *Nature* 432: 235–40.

270. Luhur A., Chawla G., Wu Y.-C., Li J. and Sokol N.S. (2014) Drosha-independent DGCR8/Pasha pathway regulates neuronal morphogenesis. *Proceedings of the National Academy of Sciences USA* 111: 1421–6.

271. Ruby J.G., Jan C.H. and Bartel D.P. (2007) Intronic microRNA precursors that bypass Drosha processing. *Nature* 448: 83–6.

272. Okamura K., Hagen J.W., Duan H., Tyler D.M. and Lai E.C. (2007) The mirtron pathway generates microRNA-class regulatory RNAs in *Drosophila*. *Cell* 130: 89–100.

273. Berezikov E., Chung W.J., Willis J., Cuppen E. and Lai E.C. (2007) Mammalian mirtron genes. *Molecular Cell* 28: 328–36.

274. Carrington J.C. and Ambros V. (2003) Role of microRNAs in plant and animal development. *Science* 301: 336–8.

275. Jinek M. and Doudna J.A. (2009) A three-dimensional view of the molecular machinery of RNA interference. *Nature* 457: 405–12.

276. Wilson R.C. and Doudna J.A. (2013) Molecular mechanisms of RNA interference. *Annual Review of Biophysics* 42: 217–39.

277. Houbaviy H.B., Murray M.F. and Sharp P.A. (2003) Embryonic stem cell-specific microRNAs. *Developmental Cell* 5: 351–8.

278. Suh M.-R. et al. (2004) Human embryonic stem cells express a unique set of microRNAs. *Developmental Biology* 270: 488–98.

279. Wheeler G., Ntounia-Fousara S., Granda B., Rathjen T. and Dalmay T. (2006) Identification of new central nervous system specific mouse microRNAs. *FEBS Letters* 580: 2195–200.

280. Xu S., Witmer P.D., Lumayag S., Kovacs B. and Valle D. (2007) MicroRNA (miRNA) transcriptome of mouse retina and identification of a sensory organ-specific miRNA cluster. *Journal of Biological Chemistry* 282: 25053–66.

281. Axtell M.J. and Bartel D.P. (2005) Antiquity of microRNAs and their targets in land plants. *Plant Cell* 17: 1658–73.

282. Peterson K.J., Dietrich M.R. and McPeek M.A. (2009) MicroRNAs and metazoan macroevolution: Insights into canalization, complexity, and the Cambrian explosion. *BioEssays* 31: 736–47.

283. Hake S. (2003) MicroRNAs: A role in plant development. *Current Biology* 13: R851–2.

284. DeVeale B., Swindlehurst-Chan J. and Blelloch R. (2021) The roles of microRNAs in mouse development. *Nature Reviews Genetics* 22: 307–23.

285. Chakraborty M. et al. (2020) MicroRNAs organize intrinsic variation into stem cell states. *Proceedings of the National Academy of Sciences USA* 117: 6942–50.

286. Brennecke J., Hipfner D.R., Stark A., Russell R.B. and Cohen S.M. (2003) *bantam* encodes a developmentally regulated microRNA that controls cell proliferation and regulates the proapoptotic gene *hid* in *Drosophila*. *Cell* 113: 25–36.

287. Gregory P.A. et al. (2008) The miR-200 family and miR-205 regulate epithelial to mesenchymal transition by targeting ZEB1 and SIP1. *Nature Cell Biology* 10: 593–601.

288. Giraldez A.J. et al. (2005) MicroRNAs regulate brain morphogenesis in zebrafish. *Science* 308: 833–8.

289. Chen C.Z., Li L., Lodish H.F. and Bartel D.P. (2004) MicroRNAs modulate hematopoietic lineage differentiation. *Science* 303: 83–6.

290. Johnston R.J. and Hobert O. (2003) A microRNA controlling left/right neuronal asymmetry in *Caenorhabditis elegans*. *Nature* 426: 845–9.

291. Schratt G.M. et al. (2006) A brain-specific microRNA regulates dendritic spine development. *Nature* 439: 283–9.

292. O'Rourke J.R. et al. (2007) Essential role for Dicer during skeletal muscle development. *Developmental Biology* 311: 359–68.

293. Baehrecke E.H. (2003) miRNAs: Micro managers of programmed cell death. *Current Biology* 13: R473–5.

294. Taganov K.D., Boldin M.P., Chang K.-J. and Baltimore D. (2006) NF-kappaB-dependent induction of microRNA miR-146, an inhibitor targeted to signaling proteins of innate immune responses. *Proceedings of the National Academy of Sciences USA* 103: 12481–6.

295. Boehm M. and Slack F.J. (2006) MicroRNA control of lifespan and metabolism. *Cell Cycle* 5: 837–40.

296. Palatnik J.F. et al. (2003) Control of leaf morphogenesis by microRNAs. *Nature* 425: 257–63.

297. Mallory A.C. et al. (2004) MicroRNA control of PHABULOSA in leaf development: Importance of pairing to the microRNA 5′ region. *EMBO Journal* 23: 3356–64.

298. Aukerman M.J. and Sakai H. (2003) Regulation of flowering time and floral organ identity by a microRNA and its *APETALA2*-like target genes. *Plant Cell* 15: 2730–41.

299. Blow M.J. et al. (2006) RNA editing of human microRNAs. *Genome Biology* 7: R27.

300. Fernandez-Valverde S.L., Taft R.J. and Mattick J.S. (2010) Dynamic isomiR regulation in *Drosophila* development. *RNA* 16: 1881–8.

301. Calin G.A. et al. (2005) A microRNA signature associated with prognosis and progression in chronic lymphocytic leukemia. *New England Journal of Medicine* 353: 1793–801.

302. Hammond S.M. (2006) MicroRNAs as oncogenes. *Current Opinion in Genetics and Development* 16: 4–9.

303. Vannini I., Fanini F. and Fabbri M. (2018) Emerging roles of microRNAs in cancer. *Current Opinion in Genetics and Development* 48: 128–33.

304. Esquela-Kerscher A. and Slack F.J. (2006) Oncomirs - microRNAs with a role in cancer. *Nature Reviews Cancer* 6: 259–69.

305. Reynolds A. et al. (2006) Induction of the interferon response by siRNA is cell type- and duplex length-dependent. *RNA* 12: 988–93.

306. Zhao T. et al. (2007) A complex system of small RNAs in the unicellular green alga *Chlamydomonas reinhardtii*. *Genes & Development* 21: 1190–203.

307. Hertel J. et al. (2006) The expansion of the metazoan microRNA repertoire. *BMC Genomics* 7: 25.

308. Prochnik S.E., Rokhsar D.S. and Aboobaker A.A. (2007) Evidence for a microRNA expansion in the bilaterian ancestor. *Development Genes and Evolution* 217: 73–7.

309. John B. et al. (2004) Human microRNA targets. *PLOS Biology* 2: e363.

310. Lewis B.P., Burge C.B. and Bartel D.P. (2005) Conserved seed pairing, often flanked by adenosines, indicates that thousands of human genes are microRNA targets. *Cell* 120: 15–20.

311. Baek D. et al. (2008) The impact of microRNAs on protein output. *Nature* 455: 64–71.

312. Lim L.P. et al. (2005) Microarray analysis shows that some microRNAs downregulate large numbers of target mRNAs. *Nature* 433: 769–73.

313. Friedman R.C., Farh K.K., Burge C.B. and Bartel D.P. (2009) Most mammalian mRNAs are conserved targets of microRNAs. *Genome Research* 19: 92–105.

314. Didiano D. and Hobert O. (2006) Perfect seed pairing is not a generally reliable predictor for miRNA-target interactions. *Nature Structural & Molecular Biology* 13: 849–51.

315. Grimson A. et al. (2007) MicroRNA targeting specificity in mammals: Determinants beyond seed pairing. *Molecular Cell* 27: 91–105.

316. McGeary S.E. et al. (2019) The biochemical basis of microRNA targeting efficacy. *Science* 366: eaav1741.

317. Alberti C. and Cochella L. (2017) A framework for understanding the roles of miRNAs in animal development. *Development* 144: 2548–59.

318. Sarshad A.A. et al. (2018) Argonaute-miRNA complexes silence target mRNAs in the nucleus of mammalian stem cells. *Molecular Cell* 71: 1040–50.

319. Shuaib M. et al. (2019) Nuclear AGO1 regulates gene expression by affecting chromatin architecture in human cells. *Cell Systems* 9: 446–58.

320. Robb G.B., Brown K.M., Khurana J. and Rana T.M. (2005) Specific and potent RNAi in the nucleus of human cells. *Nature Structural & Molecular Biology* 12: 133–7.

321. Cesana M. et al. (2011) A long noncoding RNA controls muscle differentiation by functioning as a competing endogenous RNA. *Cell* 147: 358–69.

322. Wu H.J., Wang Z.M., Wang M. and Wang X.J. (2013) Wide-spread long non-coding RNAs (lncRNAs) as endogenous target mimics (eTMs) for microRNAs in plants. *Plant Physiology* 161: 1875–84.

323. Memczak S. et al. (2013) Circular RNAs are a large class of animal RNAs with regulatory potency. *Nature* 495: 333–8.

324. Creasey K.M. et al. (2014) miRNAs trigger widespread epigenetically activated siRNAs from transposons in *Arabidopsis*. *Nature* 508: 411–5.

325. Borges F. et al. (2018) Transposon-derived small RNAs triggered by miR845 mediate genome dosage response in Arabidopsis. *Nature Genetics* 50: 186–92.

326. Bao N., Lye K.-W. and Barton M.K. (2004) MicroRNA binding sites in Arabidopsis class III HD-ZIP mRNAs are required for methylation of the template chromosome. *Developmental Cell* 7: 653–62.

327. Sinkkonen L. et al. (2008) MicroRNAs control de novo DNA methylation through regulation of transcriptional repressors in mouse embryonic stem cells. *Nature Structural & Molecular Biology* 15: 259–67.

328. Wu H. et al. (2020) Plant 22-nt siRNAs mediate translational repression and stress adaptation. *Nature* 581: 89–93.

329. Tam O.H. et al. (2008) Pseudogene-derived small interfering RNAs regulate gene expression in mouse oocytes. *Nature* 453: 534–8.

330. Watanabe T. et al. (2008) Endogenous siRNAs from naturally formed dsRNAs regulate transcripts in mouse oocytes. *Nature* 453: 539–43.

331. Sasidharan R. and Gerstein M. (2008) Protein fossils live on as RNA. *Nature* 453: 729–31.

332. Okamura K. and Lai E.C. (2008) Endogenous small interfering RNAs in animals. *Nature Reviews Molecular Cell Biology* 9: 673–8.

333. Aravin A.A. et al. (2001) Double-stranded RNA-mediated silencing of genomic tandem repeats and transposable elements in the *D. melanogaster* germline. *Current Biology* 11: 1017–27.

334. Ketting R.F., Haverkamp T.H., van Luenen H.G. and Plasterk R.H. (1999) Mut-7 of *C. elegans*, required for transposon silencing and RNA interference, is a homolog of Werner syndrome helicase and RNaseD. *Cell* 99: 133–41.

335. Kawamura Y. et al. (2008) *Drosophila* endogenous small RNAs bind to Argonaute 2 in somatic cells. *Nature* 453: 793–7.

336. Czech B. et al. (2008) An endogenous small interfering RNA pathway in *Drosophila*. *Nature* 453: 798–802.

337. Okamura K. et al. (2008) The *Drosophila* hairpin RNA pathway generates endogenous short interfering RNAs. *Nature* 453: 803–6.

338. Grivna S.T., Beyret E., Wang Z. and Lin H. (2006) A novel class of small RNAs in mouse spermatogenic cells. *Genes & Development* 20: 1709–14.

339. Watanabe T. et al. (2006) Identification and characterization of two novel classes of small RNAs in the mouse germline: Retrotransposon-derived siRNAs in oocytes and germline small RNAs in testes. *Genes & Development* 20: 1732–43.

340. Girard A., Sachidanandam R., Hannon G.J. and Carmell M.A. (2006) A germline-specific class of small RNAs binds mammalian Piwi proteins. *Nature* 442: 199–202.

341. Aravin A. et al. (2006) A novel class of small RNAs bind to MILI protein in mouse testes. *Nature* 442: 203–7.

342. Lau N.C. et al. (2006) Characterization of the piRNA complex from rat testes. *Science* 313: 363–7.

343. Aravin A.A. et al. (2003) The small RNA profile during *Drosophila melanogaster* development. *Developmental Cell* 5: 337–50.

344. Lin H. and Spradling A.C. (1997) A novel group of *pumilio* mutations affects the asymmetric division of germline stem cells in the Drosophila ovary. *Development* 124: 2463–76.

345. Kuramochi-Miyagawa S. et al. (2001) Two mouse piwi-related genes: miwi and mili. *Mechanisms of Development* 108: 121–33.

346. Ernst C., Odom D.T. and Kutter C. (2017) The emergence of piRNAs against transposon invasion to preserve mammalian genome integrity. *Nature Communications* 8: 1411.

347. Ji L. and Chen X. (2012) Regulation of small RNA stability: Methylation and beyond. *Cell Research* 22: 624–36.

348. Kirino Y. and Mourelatos Z. (2007) Mouse Piwi-interacting RNAs are 2′-O-methylated at their 3′ termini. *Nature Structural & Molecular Biology* 14: 347–8.

349. Simon B. et al. (2011) Recognition of 2′-O-methylated 3′-end of piRNA by the PAZ domain of a Piwi protein. *Structure* 19: 172–80.

350. Tian Y., Simanshu D.K., Ma J.-B. and Patel D.J. (2011) Structural basis for piRNA 2′-O-methylated 3′-end recognition by Piwi PAZ (Piwi/Argonaute/Zwille) domains. *Proceedings of the National Academy of Sciences USA* 108: 903–10.

351. Lakshmi S.S. and Agrawal S. (2008) piRNABank: A web resource on classified and clustered Piwi-interacting RNAs. *Nucleic Acids Research* 36: D173–7.

352. Özata D.M., Gainetdinov I., Zoch A., O'Carroll D. and Zamore P.D. (2019) PIWI-interacting RNAs: Small RNAs with big functions. *Nature Reviews Genetics* 20: 89–108.

353. Lau N.C. et al. (2009) Abundant primary piRNAs, endo-siRNAs, and microRNAs in a *Drosophila* ovary cell line. *Genome Research* 19: 1776–85.

354. Malone C.D. et al. (2009) Specialized piRNA pathways act in germline and somatic tissues of the *Drosophila* ovary. *Cell* 137: 522–35.

355. Kabayama Y. et al. (2017) Roles of MIWI, MILI and PLD6 in small RNA regulation in mouse growing oocytes. *Nucleic Acids Research* 45: 5387–98.

356. Taborska E. et al. (2019) Restricted and non-essential redundancy of RNAi and piRNA pathways in mouse oocytes. *PLOS Genetics* 15: e1008261.

357. Rojas-Ríos P. and Simonelig M. (2018) piRNAs and PIWI proteins: Regulators of gene expression in development and stem cells. *Development* 145: dev161786.

358. Halbach R. et al. (2020) A satellite repeat-derived piRNA controls embryonic development of *Aedes*. *Nature* 580: 274–7.

359. Li C. et al. (2009) Collapse of germline piRNAs in the absence of Argonaute3 reveals somatic piRNAs in flies. *Cell* 137: 509–21.

360. Rajasethupathy P. et al. (2012) A role for neuronal piRNAs in the epigenetic control of memory-related synaptic plasticity. *Cell* 149: 693–707.

361. Moazed D. (2012) A piRNA to remember. *Cell* 149: 512–4.

362. Ghosheh Y. et al. (2016) Characterization of piRNAs across postnatal development in mouse brain. *Scientific Reports* 6: 25039.

363. Gasperini C. et al. (2020) The piRNA pathway sustains adult neurogenesis by repressing protein synthesis. *bioRxiv:* 2020.09.15.297739.

364. Cheng Y. et al. (2019) Emerging roles of piRNAs in cancer: Challenges and prospects. *Aging* 11: 9932–46.

365. Soper S.F.C. et al. (2008) Mouse Maelstrom, a component of nuage, is essential for spermatogenesis and transposon repression in meiosis. *Developmental Cell* 15: 285–97.

366. Nott T.J. et al. (2015) Phase transition of a disordered nuage protein generates environmentally responsive membraneless organelles. *Molecular Cell* 57: 936–47.

367. Aravin A.A., Sachidanandam R., Girard A., Fejes-Toth K. and Hannon G.J. (2007) Developmentally regulated piRNA clusters implicate MILI in transposon control. *Science* 316: 744–7.

368. Aravin A.A., Hannon G.J. and Brennecke J. (2007) The Piwi-piRNA pathway provides an adaptive defense in the transposon arms race. *Science* 318: 761–4.

369. Kuramochi-Miyagawa S. et al. (2008) DNA methylation of retrotransposon genes is regulated by Piwi family members MILI and MIWI2 in murine fetal testes. *Genes & Development* 22: 908–17.

370. Jehn J. et al. (2018) PIWI genes and piRNAs are ubiquitously expressed in mollusks and show patterns of lineage-specific adaptation. *Communications Biology* 1: 137.

371. Zhang S., Pointer B. and Kelleher E. (2020) Rapid evolution of piRNA-mediated silencing of an invading transposable element was driven by abundant de novo mutations. *Genome Research* 30: 566–75.

372. Watanabe T. et al. (2011) Role for piRNAs and noncoding RNA in de novo DNA methylation of the imprinted mouse *Rasgrf1* locus. *Science* 332: 848–52.

373. Reddy H.M. et al. (2021) Y chromosomal noncoding RNAs regulate autosomal gene expression via piRNAs in mouse testis. *BMC Biology* 19: 198.

374. Pantano L. et al. (2015) The small RNA content of human sperm reveals pseudogene-derived piRNAs complementary to protein-coding genes. *RNA* 21: 1085–95.

375. Robine N. et al. (2009) A broadly conserved pathway generates 3′UTR-directed primary piRNAs. *Current Biology* 19: 2066–76.

376. Saito K. et al. (2009) A regulatory circuit for piwi by the large Maf gene *traffic jam* in *Drosophila*. *Nature* 461: 1296–9.

377. Rigoutsos I. (2010) Short RNAs: How big is this iceberg? *Current Biology* 20: R110–3.

378. Guida V. et al. (2016) Production of small noncoding RNAs from the *flamenco* locus Is regulated by the *gypsy* retrotransposon of *Drosophila melanogaster*. *Genetics* 204: 631–44.

379. Sarkar A., Volff J.-N. and Vaury C. (2017) piRNAs and their diverse roles: A transposable element-driven tactic for gene regulation? *FASEB Journal* 31: 436–46.

380. Ohtani H. and Iwasaki Y.W. (2021) Rewiring of chromatin state and gene expression by transposable elements. *Development, Growth & Differentiation* 63: 262–73.

381. Iwasaki Y.W. et al. (2021) Piwi–piRNA complexes induce stepwise changes in nuclear architecture at target loci. *EMBO Journal* 40: e108345.

382. Zheng K. and Wang P.J. (2012) Blockade of pachytene piRNA biogenesis reveals a novel requirement for maintaining post-meiotic germline genome integrity. *PLOS Genetics* 8: e1003038.

383. Quénerch'du E., Anand A. and Kai T. (2016) The piRNA pathway is developmentally regulated during spermatogenesis in *Drosophila*. *RNA* 22: 1044–54.

384. Gan H. et al. (2011) piRNA profiling during specific stages of mouse spermatogenesis. *RNA* 17: 1191–203.

385. Czech B. et al. (2018) piRNA-guided genome defense: From biogenesis to silencing. *Annual Review of Genetics* 52: 131–57.

386. Werner A., Piatek M.J. and Mattick J.S. (2015) Transpositional shuffling and quality control in male germ cells to enhance evolution of complex organisms. *Annals of the New York Academy of Sciences* 1341: 156–63.

387. Werner A. et al. (2021) Widespread formation of double-stranded RNAs in testis. *Genome Research* 31: 1174–86.

388. Goh W.S.S. et al. (2015) piRNA-directed cleavage of meiotic transcripts regulates spermatogenesis. *Genes & Development* 29: 1032–44.

389. Wu P.-H. et al. (2020) The evolutionarily conserved piRNA-producing locus pi6 is required for male mouse fertility. *Nature Genetics* 52: 728–39.

390. Barucci G. et al. (2020) Small-RNA-mediated transgenerational silencing of histone genes impairs fertility in piRNA mutants. *Nature Cell Biology* 22: 235–45.

391. Lambert M., Benmoussa A. and Provost P. (2019) Small non-coding RNAs derived from eukaryotic ribosomal RNA. *Noncoding RNA* 5: 16.

392. Cherlin T. et al. (2020) Ribosomal RNA fragmentation into short RNAs (rRFs) is modulated in a sex- and population of origin-specific manner. *BMC Biology* 18: 38.

393. Guan L. and Grigoriev A. (2021) Computational meta-analysis of ribosomal RNA fragments: Potential targets and interaction mechanisms. *Nucleic Acids Research* 49: 4085–103.

394. Lee H.C. et al. (2009) qiRNA is a new type of small interfering RNA induced by DNA damage. *Nature* 459: 274–7.

395. Francia S. et al. (2012) Site-specific DICER and DROSHA RNA products control the DNA-damage response. *Nature* 488: 231–5.

396. Michelini F. et al. (2017) Damage-induced lncRNAs control the DNA damage response through interaction with DDRNAs at individual double-strand breaks. *Nature Cell Biology* 19: 1400–11.

397. Pessina F. et al. (2019) Functional transcription promoters at DNA double-strand breaks mediate RNA-driven phase separation of damage-response factors. *Nature Cell Biology* 21: 1286–99.

398. Wahba L., Hansen L. and Fire A.Z. (2021) An essential role for the piRNA pathway in regulating the ribosomal RNA pool in *C. elegans*. *Developmental Cell* 56: 2295–312.

399. Kawaji H. et al. (2008) Hidden layers of human small RNAs. *BMC Genomics* 9: 157.

400. Cole C. et al. (2009) Filtering of deep sequencing data reveals the existence of abundant Dicer-dependent small RNAs derived from tRNAs. *RNA* 15: 2147–60.

401. Lee Y.S., Shibata Y., Malhotra A. and Dutta A. (2009) A novel class of small RNAs: tRNA-derived RNA fragments (tRFs). *Genes & Development* 23: 2639–49.

402. Anderson P. and Ivanov P. (2014) tRNA fragments in human health and disease. *FEBS Letters* 588: 4297–304.

403. Pederson T. (2010) Regulatory RNAs derived from transfer RNA? *RNA* 16: 1865–9.

404. Bracken C.P. et al. (2011) Global analysis of the mammalian RNA degradome reveals widespread miRNA-dependent and miRNA-independent endonucleolytic cleavage. *Nucleic Acids Research* 39: 5658–68.

405. Krishna S. et al. (2019) Dynamic expression of tRNA-derived small RNAs define cellular states. *EMBO Reports* 20: e47789.

406. Torres A.G., Reina O., Stephan-Otto Attolini C. and Ribas de Pouplana L. (2019) Differential expression of human tRNA genes drives the abundance of tRNA-derived fragments. *Proceedings of the National Academy of Sciences USA* 116: 8451–6.

407. Yeung M.L. et al. (2009) Pyrosequencing of small non-coding RNAs in HIV-1 infected cells: Evidence for the processing of a viral-cellular double-stranded RNA hybrid. *Nucleic Acids Research* 37: 6575–86.

408. Haussecker D. et al. (2010) Human tRNA-derived small RNAs in the global regulation of RNA silencing. *RNA* 16: 673–95.

409. Li Z. et al. (2012) Extensive terminal and asymmetric processing of small RNAs from rRNAs, snoRNAs, snRNAs, and tRNAs. *Nucleic Acids Research* 40: 6787–99.

410. Maute R.L. et al. (2013) tRNA-derived microRNA modulates proliferation and the DNA damage response and is down-regulated in B cell lymphoma. *Proceedings of the National Academy of Sciences USA* 110: 1404–9.

411. Kuscu C. et al. (2018) tRNA fragments (tRFs) guide Ago to regulate gene expression post-transcriptionally in a Dicer-independent manner. *RNA* 24: 1093–105.

412. Schaefer M. et al. (2010) RNA methylation by Dnmt2 protects transfer RNAs against stress-induced cleavage. *Genes & Development* 24: 1590–5.

413. Blanco S. et al. (2014) Aberrant methylation of tRNAs links cellular stress to neuro-developmental disorders. *EMBO Journal* 33: 2020–39.

414. He C. et al. (2021) TET2 chemically modifies tRNAs and regulates tRNA fragment levels. *Nature Structural & Molecular Biology* 28: 62–70.

415. Thompson D.M., Lu C., Green P.J. and Parker R. (2008) tRNA cleavage is a conserved response to oxidative stress in eukaryotes. *RNA* 14: 2095–103.

416. Yamasaki S., Ivanov P., Hu G.-F. and Anderson P. (2009) Angiogenin cleaves tRNA and promotes stress-induced translational repression. *Journal of Cell Biology* 185: 35–42.

417. Ivanov P., Emara M.M., Villen J., Gygi S.P. and Anderson P. (2011) Angiogenin-Induced tRNA fragments inhibit translation initiation. *Molecular Cell* 43: 613–23.

418. Sobala A. and Hutvagner G. (2013) Small RNAs derived from the 5′ end of tRNA can inhibit protein translation in human cells. *RNA Biology* 10: 553–63.

419. Vitali P. and Kiss T. (2019) Cooperative 2′-O-methylation of the wobble cytidine of human elongator tRNAMet(CAT) by a nucleolar and a Cajal body-specific box C/D RNP. *Genes & Development* 33: 741–6.

420. Sharma U. et al. (2016) Biogenesis and function of tRNA fragments during sperm maturation and fertilization in mammals. *Science* 351: 391–6.

421. Schorn A.J., Gutbrod M.J., LeBlanc C. and Martienssen R. (2017) LTR-retrotransposon control by tRNA-derived small RNAs. *Cell* 170: 61–71.

422. Hsieh L.-C., Lin S.-I., Kuo H.-F. and Chiou T.-J. (2010) Abundance of tRNA-derived small RNAs in phosphate-starved *Arabidopsis* roots. *Plant Signaling & Behavior* 5: 537–9.

423. Ren B., Wang X., Duan J. and Ma J. (2019) Rhizobial tRNA-derived small RNAs are signal molecules regulating plant nodulation. *Science* 365: 919–22.

424. Baglio S.R. et al. (2015) Human bone marrow- and adipose-mesenchymal stem cells secrete exosomes enriched in distinctive miRNA and tRNA species. *Stem Cell Research & Therapy* 6: 127.

425. Chiou N.-T., Kageyama R. and Ansel K.M. (2018) Selective export into extracellular vesicles and function of tRNA fragments during T cell activation. *Cell Reports* 25: 3356–70.

426. Peng H. et al. (2012) A novel class of tRNA-derived small RNAs extremely enriched in mature mouse sperm. *Cell Research* 22: 1609–12.

427. Vojtech L. et al. (2014) Exosomes in human semen carry a distinctive repertoire of small non-coding RNAs with potential regulatory functions. *Nucleic Acids Research* 42: 7290–304.

428. Chen Q. et al. (2016) Sperm tsRNAs contribute to inter-generational inheritance of an acquired metabolic disorder. *Science* 351: 397–400.

429. Gapp K. and Miska E.A. (2016) tRNA fragments: Novel players in intergenerational inheritance. *Cell Research* 26: 395–6.

430. Zhang Y. et al. (2018) Dnmt2 mediates intergenerational transmission of paternally acquired metabolic disorders through sperm small non-coding RNAs. *Nature Cell Biology* 20: 535–40.

431. Boskovic A., Bing X.Y., Kaymak E. and Rando O.J. (2020) Control of noncoding RNA production and histone levels by a 5′ tRNA fragment. *Genes & Development* 34: 118–31.

432. Wilusz J.E., Freier S.M. and Spector D.L. (2008) 3′ end processing of a long nuclear-retained noncoding RNA yields a tRNA-like cytoplasmic RNA. *Cell* 135: 919–32.

433. Lu X. et al. (2020) The tRNA-like small noncoding RNA mascRNA promotes global protein translation. *EMBO Reports* 21: e49684.

434. Zhang B. et al. (2017) Identification and characterization of a class of MALAT1-like genomic loci. *Cell Reports* 19: 1723–38.

435. Ender C. et al. (2008) A human snoRNA with microRNA-like functions. *Molecular Cell* 32: 519–28.

436. Taft R.J. et al. (2009) Small RNAs derived from snoR-NAs. *RNA* 15: 1233–40.

437. Saraiya A.A. and Wang C.C. (2008) snoRNA, a novel precursor of microRNA in *Giardia lamblia*. *PLOS Pathogens* 4: e1000224.

438. Scott M.S., Avolio F., Ono M., Lamond A.I. and Barton G.J. (2009) Human miRNA precursors with box H/ACA snoRNA features. *PLOS Computational Biology* 5: e1000507.

439. Patterson D.G. et al. (2017) Human snoRNA-93 is processed into a microRNA-like RNA that promotes breast cancer cell invasion. *Breast Cancer* 3: 25.

440. Hussain S. et al. (2013) NSun2-mediated cytosine-5 methylation of vault noncoding RNA determines its processing into regulatory small RNAs. *Cell Reports* 4: 255–61.

441. Hansen T.B. et al. (2016) Argonaute-associated short introns are a novel class of gene regulators. *Nature Communications* 7: 11538.

442. Taft R.J. et al. (2009) Tiny RNAs associated with transcription start sites in animals. *Nature Genetics* 41: 572–8.

443. Taft R.J., Kaplan C.D., Simons C. and Mattick J.S. (2009) Evolution, biogenesis and function of promoter-associated RNAs. *Cell Cycle* 8: 2332–8.

444. Taft R.J. et al. (2010) Nuclear-localized tiny RNAs are associated with transcription initiation and splice sites in metazoans. *Nature Structural & Molecular Biology* 17: 1030–4.

445. Timmons L. and Fire A. (1998) Specific interference by ingested dsRNA. *Nature* 395: 854.

446. Moriano-Gutierrez S. et al. (2020) The noncoding small RNA SsrA is released by *Vibrio fischeri* and modulates critical host responses. *PLOS Biology* 18: e3000934.

447. Kaletsky R. et al. (2020) *C. elegans* interprets bacterial non-coding RNAs to learn pathogenic avoidance. *Nature* 586: 445–51.

448. Moore R.S. et al. (2021) The role of the Cer1 transposon in horizontal transfer of transgenerational memory. *Cell* 184: 4697–712.

449. Pagliuso A. et al. (2019) An RNA-binding protein secreted by a bacterial pathogen modulates RIG-I signaling. *Cell Host & Microbe* 26: 823–35.

450. Wong-Bajracharya J. et al. (2022) The ectomycorrhizal fungus *Pisolithus microcarpus* encodes a microRNA involved in cross-kingdom gene silencing during symbiosis. *Proceedings of the National Academy of Sciences USA* 119: e2103527119.

451. Porquier A. et al. (2021) Retrotransposons as pathogenicity factors of the plant pathogenic fungus *Botrytis cinerea*. *Genome Biology* 22: 225.

452. Jayasinghe W.H., Kim H., Nakada Y. and Masuta C. (2021) A plant virus satellite RNA directly accelerates wing formation in its insect vector for spread. *Nature Communications* 12: 7087.

453. Sarkies P. and Miska E.A. (2013) Is there social RNA? *Science* 341: 467–8.

454. Chen X. and Rechavi O. (2021) Plant and animal small RNA communications between cells and organisms. *Nature Reviews Molecular Cell Biology* 23: 185–203.

455. Buck A.H. et al. (2014) Exosomes secreted by nematode parasites transfer small RNAs to mammalian cells and modulate innate immunity. *Nature Communications* 5: 5488.

456. Cai Q. et al. (2018) Plants send small RNAs in extracellular vesicles to fungal pathogen to silence virulence genes. *Science* 360: 1126–9.

457. Betti F. et al. (2021) Exogenous miRNAs induce post-transcriptional gene silencing in plants. *Nature Plants* 7: 1379–88.

458. Shahid S. et al. (2018) MicroRNAs from the parasitic plant *Cuscuta campestris* target host messenger RNAs. *Nature* 553: 82–5.

459. Maori E. et al. (2019) A transmissible RNA pathway in honey bees. *Cell Reports* 27: 1949–59.

460. Maori E. et al. (2019) A secreted RNA binding protein forms RNA-stabilizing granules in the honeybee royal jelly. *Molecular Cell* 74: 598–608.

461. Hunter W. et al. (2010) Large-scale field application of RNAi technology reducing Israeli acute paralysis virus disease in honey bees (Apis mellifera, Hymenoptera: Apidae). *PLOS Pathogens* 6: e1001160.

462. Lampson B.C. et al. (1989) Reverse transcriptase in a clinical strain of *Escherichia coli*: Production of branched RNA-linked msDNA. *Science* 243: 1033–8.

463. Lim D. and Maas W.K. (1989) Reverse transcriptase-dependent synthesis of a covalently linked, branched DNA-RNA compound in *E. coli* B. *Cell* 56: 891–904.

464. Temin H.M. (1989) Retrons in bacteria. *Nature* 339: 254–5.

465. Millman A. et al. (2020) Bacterial retrons function in anti-phage defense. *Cell* 183: 1551–61.

466. van der Oost J., Westra E.R., Jackson R.N. and Wiedenheft B. (2014) Unravelling the structural and mechanistic basis of CRISPR–Cas systems. *Nature Reviews Microbiology* 12: 479–92.

467. Marraffini L.A. (2015) CRISPR-Cas immunity in prokaryotes. *Nature* 526: 55–61.

468. Mojica F.J.M. and Rodriguez-Valera F. (2016) The discovery of CRISPR in archaea and bacteria. *FEBS Journal* 283: 3162–9.

469. Lander E.S. (2016) The heroes of CRISPR. *Cell* 164: 18–28.

470. Ishino Y., Krupovic M. and Forterre P. (2018) History of CRISPR-Cas from encounter with a mysterious repeated sequence to genome editing technology. *Journal of Bacteriology* 200: e00580.

471. Ishino Y., Shinagawa H., Makino K., Amemura M. and Nakata A. (1987) Nucleotide sequence of the *iap* gene, responsible for alkaline phosphatase isozyme conversion in *Escherichia coli*, and identification of the gene product. *Journal of Bacteriology* 169: 5429–33.

472. Mojica F.J.M., Ferrer C., Juez G. and Rodríguez-Valera F. (1995) Long stretches of short tandem repeats are present in the largest replicons of the *Archaea Haloferax mediterranei* and *Haloferax volcanii* and could be involved in replicon partitioning. *Molecular Microbiology* 17: 85–93.

473. Mojica F.J.M., Díez-Villaseñor C., Soria E. and Juez G. (2000) Biological significance of a family of regularly spaced repeats in the genomes of Archaea, Bacteria and mitochondria. *Molecular Microbiology* 36: 244–6.

474. Jansen R., Van Embden J.D.A., Gaastra W. and Schouls L.M. (2002) Identification of genes that are associated with DNA repeats in prokaryotes. *Molecular Microbiology* 43: 1565–75.

475. Gasiunas G., Barrangou R., Horvath P. and Siksnys V. (2012) Cas9–crRNA ribonucleoprotein complex mediates specific DNA cleavage for adaptive immunity in bacteria. *Proceedings of the National Academy of Sciences USA* 109: E2579–86.

476. Ledford H. (2016) The unsung heroes of CRISPR. *Nature* 535: 342–4.

477. Mojica F.J.M., Díez-Villaseñor C.S., García-Martínez J. and Soria E. (2005) Intervening sequences of regularly spaced prokaryotic repeats derive from foreign genetic elements. *Journal of Molecular Evolution* 60: 174–82.

478. Pourcel C., Salvignol G. and Vergnaud G. (2005) CRISPR elements in *Yersinia pestis* acquire new repeats by preferential uptake of bacteriophage DNA, and provide additional tools for evolutionary studies. *Microbiology* 151: 653–63.

479. Bolotin A., Quinquis B., Sorokin A. and Ehrlich S.D. (2005) Clustered regularly interspaced short palindrome repeats (CRISPRs) have spacers of extrachromosomal origin. *Microbiology* 151: 2551–61.

480. Makarova K.S., Grishin N.V., Shabalina S.A., Wolf Y.I. and Koonin E.V. (2006) A putative RNA-interference-based immune system in prokaryotes: Computational analysis of the predicted enzymatic machinery, functional analogies with eukaryotic RNAi, and hypothetical mechanisms of action. *Biology Direct* 1: 7.

481. Barrangou R. et al. (2007) CRISPR provides acquired resistance against viruses in prokaryotes. *Science* 315: 1709–12.

482. Brouns S.J.J. et al. (2008) Small CRISPR RNAs guide antiviral defense in prokaryotes. *Science* 321: 960–4.

483. Horvath P. et al. (2008) Diversity, activity, and evolution of CRISPR loci in *Streptococcus thermophilus*. *Journal of Bacteriology* 190: 1401–12.

484. Deveau H. et al. (2008) Phage response to CRISPR-encoded resistance in *Streptococcus thermophilus*. *Journal of Bacteriology* 190: 1390–400.

485. Jinek M. et al. (2012) A programmable dual-RNA–guided DNA endonuclease in adaptive bacterial immunity. *Science* 337: 816–21.

486. Jiang F. and Doudna J.A. (2017) CRISPR–Cas9 structures and mechanisms. *Annual Review of Biophysics* 46: 505–29.

487. Marraffini L.A. and Sontheimer E.J. (2008) CRISPR interference limits horizontal gene transfer in Staphylococci by targeting DNA. *Science* 322: 1843–5.

488. Garneau J.E. et al. (2010) The CRISPR/Cas bacterial immune system cleaves bacteriophage and plasmid DNA. *Nature* 468: 67–71.

489. Altae-Tran H. et al. (2021) The widespread IS200/IS605 transposon family encodes diverse programmable RNA-guided endonucleases. *Science* 374: 57–65.

490. Saito M. et al. (2021) Dual modes of CRISPR-associated transposon homing. *Cell* 184: 2441–53.

491. Mangold M. et al. (2004) Synthesis of group A streptococcal virulence factors is controlled by a regulatory RNA molecule. *Molecular Microbiology* 53: 1515–27.

492. Deltcheva E. et al. (2011) CRISPR RNA maturation by trans-encoded small RNA and host factor RNase III. *Nature* 471: 602–7.

493. Workman R.E. et al. (2021) A natural single-guide RNA repurposes Cas9 to autoregulate CRISPR-Cas expression. *Cell* 184: 675–88.

494. Rees H.A. and Liu D.R. (2018) Base editing: Precision chemistry on the genome and transcriptome of living cells. *Nature Reviews Genetics* 19: 770–88.

495. Pickar-Oliver A. and Gersbach C.A. (2019) The next generation of CRISPR–Cas technologies and applications. *Nature Reviews Molecular Cell Biology* 20: 490–507.

496. Haft D.H., Selengut J., Mongodin E.F. and Nelson K.E. (2005) A guild of 45 CRISPR-associated (Cas) protein families and multiple CRISPR/Cas subtypes exist in prokaryotic genomes. *PLOS Computational Biology* 1: e60.

497. Harrington L.B. et al. (2018) Programmed DNA destruction by miniature CRISPR-Cas14 enzymes. *Science* 362: 839–42.

498. Sapranauskas R. et al. (2011) The *Streptococcus thermophilus* CRISPR/Cas system provides immunity in *Escherichia coli*. *Nucleic Acids Research* 39: 9275–82.

499. Koonin E.V. and Makarova K.S. (2022) Evolutionary plasticity and functional versatility of CRISPR systems. *PLOS Biology* 20: e3001481.

500. Cong L. et al. (2013) Multiplex genome engineering using CRISPR/Cas systems. *Science* 339: 819–23.

501. Mali P. et al. (2013) RNA-guided human genome engineering via Cas9. *Science* 339: 823–6.

502. Cui X. and Davis G. (2007) Mobile group II intron targeting: Applications in prokaryotes and perspectives in eukaryotes. *Frontiers in Bioscience* 12: 4972–85.

503. Lambowitz A.M. and Zimmerly S. (2011) Group II introns: Mobile ribozymes that invade DNA. *Cold Spring Harbor Perspectives in Biology* 3: a003616.

504. Belfort M. and Bonocora R.P. (2014) Homing endonucleases: From genetic anomalies to programmable genomic clippers, in D. Edgell (ed.) *Homing Endonucleases: Methods and Protocols* (Humana Press, Totowa, NJ).

505. Burt A. (2003) Site-specific selfish genes as tools for the control and genetic engineering of natural populations. *Proceedings of the Royal Society B: Biological Sciences* 270: 921–8.

506. Wood A.J. et al. (2011) Targeted genome editing across species using ZFNs and TALENs. *Science* 333: 307.

507. Kim E. et al. (2012) Precision genome engineering with programmable DNA-nicking enzymes. *Genome Research* 22: 1327–33.

508. Gaj T., Gersbach C.A. and Barbas III C.F., (2013) ZFN, TALEN, and CRISPR/Cas-based methods for genome engineering. *Trends in Biotechnology* 31: 397–405.

509. Chen B. et al. (2013) Dynamic imaging of genomic loci in living human cells by an optimized CRISPR/Cas system. *Cell* 155: 1479–91.

510. Zetsche B. et al. (2015) Cpf1 Is a single RNA-guided endonuclease of a class 2 CRISPR-Cas System. *Cell* 163: 759–71.

511. Abudayyeh O.O. et al. (2017) RNA targeting with CRISPR-Cas13. *Nature* 550: 280–4.

512. Wang H. et al. (2019) CRISPR-mediated live imaging of genome editing and transcription. *Science* 365: 1301–5.

513. Li Y., Li S., Wang J. and Liu G. (2019) CRISPR/Cas systems towards next-generation biosensing. *Trends in Biotechnology* 37: 730–43.

514. Smargon A.A., Shi Y.J. and Yeo G.W. (2020) RNA-targeting CRISPR systems from metagenomic discovery to transcriptomic engineering. *Nature Cell Biology* 22: 143–50.

515. Liu H. *et al.* (2020) High-throughput CRISPR/Cas9 mutagenesis streamlines trait gene identification in maize. *Plant Cell* 32: 1397–413.

516. Donohoue P.D., Barrangou R. and May A.P. (2018) Advances in industrial biotechnology using CRISPR-Cas systems. *Trends in Biotechnology* 36: 134–46.

517. Zhang S. et al. (2020) Recent advances of CRISPR/Cas9-based genetic engineering and transcriptional regulation in industrial biology. *Frontiers in Bioengineering and Biotechnology* 7: 459.

518. Jaganathan D., Ramasamy K., Sellamuthu G., Jayabalan S. and Venkataraman G. (2018) CRISPR for crop improvement: An update review. *Frontiers in Plant Science* 9: 985.

519. Lin Q. et al. (2020) Prime genome editing in rice and wheat. *Nature Biotechnology* 38: 582–5.

520. Knott G.J. and Doudna J.A. (2018) CRISPR-Cas guides the future of genetic engineering. *Science* 361: 866–9.

521. Liu X.S. et al. (2016) Editing DNA methylation in the mammalian genome. *Cell* 167: 233–47.

522. McDonald J.I. et al. (2016) Reprogrammable CRISPR/Cas9-based system for inducing site-specific DNA methylation. *Biology Open* 5: 866–74.

523. Klann T.S. et al. (2017) CRISPR–Cas9 epigenome editing enables high-throughput screening for functional regulatory elements in the human genome. *Nature Biotechnology* 35: 561–8.

524. Lei Y., Huang Y.-H. and Goodell M.A. (2018) DNA methylation and de-methylation using hybrid site-targeting proteins. *Genome Biology* 19: 187.

525. Kang J.G., Park J.S., Ko J.-H. and Kim Y.-S. (2019) Regulation of gene expression by altered promoter methylation using a CRISPR/Cas9-mediated epigenetic editing system. *Scientific Reports* 9: 11960.

526. Devesa-Guerra I. et al. (2020) DNA methylation editing by CRISPR-guided excision of 5-methylcytosine. *Journal of Molecular Biology* 432: 2204–16.

527. Gilbert L.A. et al. (2013) CRISPR-mediated modular RNA-guided regulation of transcription in eukaryotes. *Cell* 154: 442–51.

528. Perez-Pinera P. et al. (2013) RNA-guided gene activation by CRISPR-Cas9–based transcription factors. *Nature Methods* 10: 973–6.

529. Konermann S. et al. (2015) Genome-scale transcriptional activation by an engineered CRISPR-Cas9 complex. *Nature* 517: 583–8.

530. Chavez A. et al. (2015) Highly efficient Cas9-mediated transcriptional programming. *Nature Methods* 12: 326–8.

531. Yeo N.C. et al. (2018) An enhanced CRISPR repressor for targeted mammalian gene regulation. *Nature Methods* 15: 611–6.

532. Komor A.C., Kim Y.B., Packer M.S., Zuris J.A. and Liu D.R. (2016) Programmable editing of a target base in genomic DNA without double-stranded DNA cleavage. *Nature* 533: 420–4.

533. Gaudelli N.M. et al. (2017) Programmable base editing of A•T to G•C in genomic DNA without DNA cleavage. *Nature* 551: 464–71.

534. Anzalone A.V. et al. (2019) Search-and-replace genome editing without double-strand breaks or donor DNA. *Nature* 576: 149–57.

535. Anzalone A.V. et al. (2021) Programmable deletion, replacement, integration and inversion of large DNA sequences with twin prime editing. *Nature Biotechnology* epub ahead of print: https://www.nature.com/articles/s41587-021-01133-w.

536. Staahl B.T. et al. (2017) Efficient genome editing in the mouse brain by local delivery of engineered Cas9 ribonucleoprotein complexes. *Nature Biotechnology* 35: 431–4.

537. Yeh W.-H. et al. (2020) In vivo base editing restores sensory transduction and transiently improves auditory function in a mouse model of recessive deafness. *Science Translational Medicine* 12: eaay9101.

538. Gillmore J.D. et al. (2021) CRISPR-Cas9 In vivo gene editing for Transthyretin Amyloidosis. *New England Journal of Medicine* 385: 493–502.

539. Özcan A. et al. (2021) Programmable RNA targeting with the single-protein CRISPR effector Cas7–11. *Nature* 597: 720–5.

540. Montiel-Gonzalez M.F., Vallecillo-Viejo I., Yudowski G.A. and Rosenthal J.J.C. (2013) Correction of mutations within the cystic fibrosis transmembrane conductance regulator by site-directed RNA editing. *Proceedings of the National Academy of Sciences USA* 110: 18285–90.

541. Sinnamon J.R. et al. (2017) Site-directed RNA repair of endogenous Mecp2 RNA in neurons. *Proceedings of the National Academy of Sciences USA* 114: E9395–402.

542. Vogel P. et al. (2018) Efficient and precise editing of endogenous transcripts with SNAP-tagged ADARs. *Nature Methods* 15: 535–8.

543. Cox D.B.T. et al. (2017) RNA editing with CRISPR-Cas13. *Science* 358: 1019–27.

544. Gootenberg J.S. et al. (2018) Multiplexed and portable nucleic acid detection platform with Cas13, Cas12a, and Csm6. *Science* 360: 439–44.

545. Myhrvold C. et al. (2018) Field-deployable viral diagnostics using CRISPR-Cas13. *Science* 360: 444–8.

546. Jiao C. et al. (2021) Noncanonical crRNAs derived from host transcripts enable multiplexable RNA detection by Cas9. *Science* 372: 941–8.

547. Curtis C.F. (1968) Possible use of translocations to fix desirable genes in insect pest populations. *Nature* 218: 368–9.

548. Curtis C.F. and Hill W.G. (1971) Theoretical studies on the use of translocations for the control of tsetse flies and other disease vectors. *Theoretical Population Biology* 2: 71–90.
549. Yan Y. and Finnigan G.C. (2018) Development of a multi-locus CRISPR gene drive system in budding yeast. *Scientific Reports* 8: 17277.
550. Yan Y. and Finnigan G.C. (2019) Analysis of CRISPR gene drive design in budding yeast. *Access Microbiology* 1: e000059.
551. Chan Y.-S., Naujoks D.A., Huen D.S. and Russell S. (2011) Insect population control by homing endonuclease-based gene drive: An evaluation in *Drosophila melanogaster*. *Genetics* 188: 33–44.
552. Windbichler N. et al. (2011) A synthetic homing endonuclease-based gene drive system in the human malaria mosquito. *Nature* 473: 212–5.
553. Gantz V.M. et al. (2015) Highly efficient Cas9-mediated gene drive for population modification of the malaria vector mosquito Anopheles stephensi. *Proceedings of the National Academy of Sciences USA* 112: E6736–43.
554. Hammond A. et al. (2016) A CRISPR-Cas9 gene drive system targeting female reproduction in the malaria mosquito vector *Anopheles gambiae*. *Nature Biotechnology* 34: 78–83.
555. Esvelt K.M., Smidler A.L., Catteruccia F. and Church G.M. (2014) Concerning RNA-guided gene drives for the alteration of wild populations. *eLife* 3: e03401.
556. Champer J., Buchman A. and Akbari O.S. (2016) Cheating evolution: Engineering gene drives to manipulate the fate of wild populations. *Nature Reviews Genetics* 17: 146–59.
557. DiCarlo J.E., Chavez A., Dietz S.L., Esvelt K.M. and Church G.M. (2015) Safeguarding CRISPR-Cas9 gene drives in yeast. *Nature Biotechnology* 33: 1250–5.
558. Roggenkamp E. et al. (2018) Tuning CRISPR-Cas9 in *Saccharomyces cerevisiae*. *G3 (Genes, Genomes, Genetics)* 8: 999–1018.
559. Noble C. et al. (2019) Daisy-chain gene drives for the alteration of local populations. *Proceedings of the National Academy of Sciences USA* 116: 8275–82.
560. Xu X.S. et al. (2020) Active genetic neutralizing elements for halting or deleting gene drives. *Molecular Cell* 80: 246–62.
561. Wei T., Cheng Q., Min Y.-L., Olson E.N. and Siegwart D.J. (2020) Systemic nanoparticle delivery of CRISPR-Cas9 ribonucleoproteins for effective tissue specific genome editing. *Nature Communications* 11: 3232.
562. Rosenblum D. et al. (2020) CRISPR-Cas9 genome editing using targeted lipid nanoparticles for cancer therapy. *Science Advances* 6: eabc9450.
563. Hou X., Zaks T., Langer R. and Dong Y. (2021) Lipid nanoparticles for mRNA delivery. *Nature Reviews Materials* 6: 1078–94.

CHAPTER 13

1. Storz G. (2002) An expanding universe of noncoding RNAs. *Science* 296: 1260–3.
2. Storz G., Altuvia S. and Wassarman K.M. (2005) An abundance of RNA regulators. *Annual Review of Biochemistry* 74: 199–217.
3. Croft L. et al. (2000) ISIS, the intron information system, reveals the high frequency of alternative splicing in the human genome. *Nature Genetics* 24: 340–1.
4. Deveson I.W. et al. (2018) Universal alternative splicing of noncoding exons. *Cell Systems* 6: 245–55.
5. Graves P.R. and Haystead T.A.J. (2002) Molecular biologist's guide to proteomics. *Microbiology and Molecular Biology Reviews* 66: 39–63.
6. Ahmad Y. and Lamond A.I. (2014) A perspective on proteomics in cell biology. *Trends in Cell Biology* 24: 257–64.
7. Adams M.D., Soares M.B., Kerlavage A.R., Fields C. and Venter J.C. (1993) Rapid cDNA sequencing (expressed sequence tags) from a directionally cloned human infant brain cDNA library. *Nature Genetics* 4: 373–80.
8. Fields C., Adams M.D., White O. and Venter J.C. (1994) How many genes in the human genome? *Nature Genetics* 7: 345–6.
9. Adesnik M. and Darnell J.E. (1972) Biogenesis and characterization of histone messenger RNA in HeLa cells. *Journal of Molecular Biology* 67: 397–406.
10. Milcarek C., Price R. and Penman S. (1974) The metabolism of a poly(A) minus mRNA fraction in HeLa cells. *Cell* 3: 1–10.
11. Katinakis P.K., Slater A. and Burdon R.H. (1980) Non-polyadenylated mRNAs from eukaryotes. *FEBS Letters* 116: 1–7.
12. Salditt-Georgieff M., Harpold M.M., Wilson M.C. and Darnell Jr. J.E., (1981) Large heterogeneous nuclear ribonucleic acid has three times as many 5′ caps as polyadenylic acid segments, and most caps do not enter polyribosomes. *Molecular Cell Biology* 1: 179–87.
13. Reis E.M. et al. (2004) Antisense intronic non-coding RNA levels correlate to the degree of tumor differentiation in prostate cancer. *Oncogene* 23: 6684–92.
14. Carninci P. et al. (2005) The transcriptional landscape of the mammalian genome. *Science* 309: 1559–63.
15. Roest Crollius H. et al. (2000) Estimate of human gene number provided by genome-wide analysis using *Tetraodon nigroviridis* DNA sequence. *Nature Genetics* 25: 235–8.
16. Strausberg R.L., Feingold E.A., Klausner R.D. and Collins F.S. (1999) The mammalian gene collection. *Science* 286: 455–7.
17. Strausberg R.L. et al. (2002) Generation and initial analysis of more than 15,000 full-length human and mouse cDNA sequences. *Proceedings of the National Academy of Sciences USA* 99: 16899–903.
18. Dias Neto E. et al. (2000) Shotgun sequencing of the human transcriptome with ORF expressed sequence tags. *Proceedings of the National Academy of Sciences USA* 97: 3491–6.
19. de Souza S.J. et al. (2000) Identification of human chromosome 22 transcribed sequences with ORF expressed sequence tags. *Proceedings of the National Academy of Sciences USA* 97: 12690–3.
20. Camargo A.A. et al. (2001) The contribution of 700,000 ORF sequence tags to the definition of the human transcriptome. *Proceedings of the National Academy of Sciences USA* 98: 12103–8.
21. Brentani H. et al. (2003) The generation and utilization of a cancer-oriented representation of the human transcriptome by using expressed sequence tags. *Proceedings of the National Academy of Sciences USA* 100: 13418–23.

22. Deveson I.W., Hardwick S.A., Mercer T.R. and Mattick J.S. (2017) The dimensions, dynamics, and relevance of the mammalian noncoding transcriptome. *Trends in Genetics* 33: 464–78.

23. Bishop J.O., Morton J.G., Rosbash M. and Richardson M. (1974) Three abundance classes in HeLa cell messenger RNA. *Nature* 250: 199–204.

24. Coupar B.E., Davies J.A. and Chesterton C.J. (1978) Quantification of hepatic transcribing RNA polymerase molecules, polyribonucleotide elongation rates and messenger RNA complexity in fed and fasted rats. *European Journal of Biochemistry* 84: 611–23.

25. Su A.I. et al. (2004) A gene atlas of the mouse and human protein-encoding transcriptomes. *Proceedings of the National Academy of Sciences USA* 101: 6062–7.

26. Bonaldo M.F., Lennon G. and Soares M.B. (1996) Normalization and subtraction: Two approaches to facilitate gene discovery. *Genome Research* 6: 791–806.

27. Carninci P. et al. (2000) Normalization and subtraction of cap-trapper-selected cDNAs to prepare full-length cDNA libraries for rapid discovery of new genes. *Genome Research* 10: 1617–30.

28. Cheng J. et al. (2005) Transcriptional maps of 10 human chromosomes at 5-nucleotide resolution. *Science* 308: 1149–54.

29. Carninci P. et al. (1996) High-efficiency full-length cDNA cloning by biotinylated CAP trapper. *Genomics* 37: 327–36.

30. Shiraki T. et al. (2003) Cap analysis gene expression for high-throughput analysis of transcriptional starting point and identification of promoter usage. *Proceedings of the National Academy of Sciences USA* 100: 15776–81.

31. Hirozane-Kishikawa T. et al. (2003) Subtraction of cap-trapped full-length cDNA libraries to select rare transcripts. *Biotechniques* 35: 510–8.

32. Carninci P. et al. (2003) Targeting a complex transcriptome: The construction of the mouse full-length cDNA encyclopedia. *Genome Research* 13: 1273–89.

33. de Hoon M., Shin J.W. and Carninci P. (2015) Paradigm shifts in genomics through the FANTOM projects. *Mammalian Genome* 26: 391–402.

34. Abugessaisa I. et al. (2021) FANTOM enters 20th year: Expansion of transcriptomic atlases and functional annotation of non-coding RNAs. *Nucleic Acids Research* 49: D892–8.

35. Okazaki Y. et al. (2002) Analysis of the mouse transcriptome based on functional annotation of 60,770 full-length cDNAs. *Nature* 420: 563–73.

36. Numata K. et al. (2003) Identification of putative noncoding RNAs among the RIKEN mouse full-length cDNA collection. *Genome Research* 13: 1301–6.

37. Ravasi T. et al. (2006) Experimental validation of the regulated expression of large numbers of non-coding RNAs from the mouse genome. *Genome Research* 16: 11–9.

38. Kiyosawa H., Yamanaka I., Osato N., Kondo S. and Hayashizaki Y. (2003) Antisense transcripts with FANTOM2 clone set and their implications for gene regulation. *Genome Research* 13: 1324–34.

39. Katayama S. *et al.* (2005) Antisense transcription in the mammalian transcriptome. *Science* 309: 1564–6.

40. Dahary D., Elroy-Stein O. and Sorek R. (2005) Naturally occurring antisense: Transcriptional leakage or real overlap? *Genome Research* 15: 364–8.

41. Shendure J. and Church G.M. (2002) Computational discovery of sense-antisense transcription in the human and mouse genomes. *Genome Biology* 3: 0044.

42. Yelin R. et al. (2003) Widespread occurrence of antisense transcription in the human genome. *Nature Biotechnology* 21: 379–86.

43. Trinklein N.D. et al. (2004) An abundance of bidirectional promoters in the human genome. *Genome Research* 14: 62–6.

44. Werner A. and Berdal A. (2005) Natural antisense transcripts: Sound or silence? *Physiological Genomics* 23: 125–31.

45. Engstrom P.G. et al. (2006) Complex loci in human and mouse genomes. *PLOS Genetics* 2: e47.

46. Nakaya H.I. et al. (2007) Genome mapping and expression analyses of human intronic noncoding RNAs reveal tissue-specific patterns and enrichment in genes related to regulation of transcription. *Genome Biology* 8: R43.

47. Xu Z. et al. (2009) Bidirectional promoters generate pervasive transcription in yeast. *Nature* 457: 1033–7.

48. Ozsolak F. et al. (2010) Comprehensive polyadenylation site maps in yeast and human reveal pervasive alternative polyadenylation. *Cell* 143: 1018–29.

49. Brown J.B. et al. (2014) Diversity and dynamics of the *Drosophila* transcriptome. *Nature* 512: 393–9.

50. Krzyczmonik K., Wroblewska-Swiniarska A. and Swiezewski S. (2017) Developmental transitions in *Arabidopsis* are regulated by antisense RNAs resulting from bidirectionally transcribed genes. *RNA Biology* 14: 838–42.

51. Zhao X. et al. (2018) Global identification of *Arabidopsis* lncRNAs reveals the regulation of MAF4 by a natural antisense RNA. *Nature Communications* 9: 5056.

52. Louro R., El-Jundi T., Nakaya H.I., Reis E.M. and Verjovski-Almeida S. (2008) Conserved tissue expression signatures of intronic noncoding RNAs transcribed from human and mouse loci. *Genomics* 92: 18–25.

53. Johnson J.M., Edwards S., Shoemaker D. and Schadt E.E. (2005) Dark matter in the genome: Evidence of widespread transcription detected by microarray tiling experiments. *Trends in Genetics* 21: 93–102.

54. Kapranov P. et al. (2002) Large-scale transcriptional activity in chromosomes 21 and 22. *Science* 296: 916–9.

55. Rinn J.L. et al. (2003) The transcriptional activity of human chromosome 22. *Genes & Development* 17: 529–40.

56. Bertone P. et al. (2004) Global identification of human transcribed sequences with genome tiling arrays. *Science* 306: 2242–6.

57. Kapranov P. et al. (2005) Examples of the complex architecture of the human transcriptome revealed by RACE and high-density tiling arrays. *Genome Research* 15: 987–97.

58. Efroni S. et al. (2008) Global transcription in pluripotent embryonic stem cells. *Cell Stem Cell* 2: 437–47.

59. St Laurent G. et al. (2012) Intronic RNAs constitute the major fraction of the non-coding RNA in mammalian cells. *BMC Genomics* 13: 504.

60. Espinoza C.A., Allen T.A., Hieb A.R., Kugel J.F. and Goodrich J.A. (2004) B2 RNA binds directly to RNA polymerase II to repress transcript synthesis. *Nature Structural & Molecular Biology* 11: 822–9.

61. Espinoza C.A., Goodrich J.A. and Kugel J.F. (2007) Characterization of the structure, function, and mechanism of B2 RNA, an ncRNA repressor of RNA polymerase II transcription. *RNA* 13: 583–96.

62. Mariner P.D. et al. (2008) Human Alu RNA is a modular transacting repressor of mRNA transcription during heat shock. *Molecular Cell* 29: 499–509.

63. Parrott A.M. and Mathews M.B. (2007) Novel rapidly evolving hominid RNAs bind nuclear factor 90 and display tissue-restricted distribution. *Nucleic Acids Research* 35: 6249–58.

64. Parrott A.M. et al. (2011) The evolution and expression of the snaR family of small non-coding RNAs. *Nucleic Acids Research* 39: 1485–500.

65. Conti A. et al. (2015) Identification of RNA polymerase III-transcribed Alu loci by computational screening of RNA-Seq data. *Nucleic Acids Research* 43: 817–35.

66. Stribling D. et al. (2021) A noncanonical microRNA derived from the snaR-A noncoding RNA targets a metastasis inhibitor. *RNA* 27: 694–709.

67. Van Bortle K. et al. (2021) A cancer-associated RNA polymerase III identity drives robust transcription and expression of SNAR-A noncoding RNA. *bioRxiv*: 2021.04.21.440535.

68. Lunyak V.V. et al. (2007) Developmentally regulated activation of a SINE B2 repeat as a domain boundary in organogenesis. *Science* 317: 248–51.

69. Dieci G., Fiorino G., Castelnuovo M., Teichmann M. and Pagano A. (2007) The expanding RNA polymerase III transcriptome. *Trends in Genetics* 23: 614–22.

70. Pagano A. et al. (2007) New small nuclear RNA gene-like transcriptional units as sources of regulatory transcripts. *PLOS Genetics* 3: e1.

71. White R.J. (2011) Transcription by RNA polymerase III: More complex than we thought. *Nature Reviews Genetics* 12: 459–63.

72. Djebali S. et al. (2012) Landscape of transcription in human cells. *Nature* 489: 101–8.

73. Fejes-Toth K. et al. (2009) Post-transcriptional processing generates a diversity of 5′-modified long and short RNAs. *Nature* 457: 1028–32.

74. Mercer T.R. et al. (2010) Regulated post-transcriptional RNA cleavage diversifies the eukaryotic transcriptome. *Genome Research* 20: 1639–50.

75. Mercer T.R. et al. (2011) Expression of distinct RNAs from 3′ untranslated regions. *Nucleic Acids Research* 39: 2393–403.

76. Kapranov P. et al. (2010) The majority of total nuclear-encoded non-ribosomal RNA in a human cell is 'dark matter' un-annotated RNA. *BMC Biology* 8: 149.

77. Xu A.G. et al. (2010) Intergenic and repeat transcription in human, chimpanzee and macaque brains measured by RNA-Seq. *PLOS Computational Biology* 6: e1000843.

78. Yamada K. et al. (2003) Empirical analysis of transcriptional activity in the *Arabidopsis* genome. *Science* 302: 842–6.

79. Imanishi T. et al. (2004) Integrative annotation of 21,037 human genes validated by full-length cDNA clones. *PLOS Biology* 2: 856–75.

80. Ota T. et al. (2004) Complete sequencing and characterization of 21,243 full-length human cDNAs. *Nature Genetics* 36: 40–5.

81. Seki M. et al. (2004) RIKEN Arabidopsis full-length (RAFL) cDNA and its applications for expression profiling under abiotic stress conditions. *Journal of Experimental Botany* 55: 213–23.

82. Chen J. et al. (2002) Identifying novel transcripts and novel genes in the human genome by using novel SAGE tags. *Proceedings of the National Academy of Sciences USA* 99: 12257–62.

83. Saha S. et al. (2002) Using the transcriptome to annotate the genome. *Nature Biotechnology* 20: 508–12.

84. Meyers B.C. et al. (2004) Analysis of the transcriptional complexity of *Arabidopsis thaliana* by massively parallel signature sequencing. *Nature Biotechnology* 22: 1006–11.

85. Jongeneel C.V. et al. (2005) An atlas of human gene expression from massively parallel signature sequencing. *Genome Research* 15: 1007–14.

86. Schadt E.E. et al. (2004) A comprehensive transcript index of the human genome generated using microarrays and computational approaches. *Genome Biology* 5: R73.

87. David L. et al. (2006) A high-resolution map of transcription in the yeast genome. *Proceedings of the National Academy of Sciences USA* 103: 5320–5.

88. He H. et al. (2007) Mapping the *C. elegans* noncoding transcriptome with a whole-genome tiling microarray. *Genome Research* 17: 1471–7.

89. Rinn J.L. et al. (2007) Functional demarcation of active and silent chromatin domains in human *HOX* loci by non-coding RNAs. *Cell* 129: 1311–23.

90. Stolc V. et al. (2004) A gene expression map for the euchromatic genome of *Drosophila melanogaster*. *Science* 306: 655–60.

91. Manak J.R. et al. (2006) Biological function of unannotated transcription during the early development of *Drosophila melanogaster*. *Nature Genetics* 38: 1151–8.

92. Willingham A.T. et al. (2006) Transcriptional landscape of the human and fly genomes: Nonlinear and multifunctional modular model of transcriptomes. *Cold Spring Harbor Symposia on Quantitative Biology* 71: 101–10.

93. Roy S. et al. (2010) Identification of functional elements and regulatory circuits by *Drosophila* modENCODE. *Science* 330: 1787–97.

94. Cherbas L. et al. (2011) The transcriptional diversity of 25 *Drosophila* cell lines. *Genome Research* 21: 301–14.

95. Graveley B.R. et al. (2011) The developmental transcriptome of *Drosophila melanogaster*. *Nature* 471: 473–9.

96. Birney E. et al. (2007) Identification and analysis of functional elements in 1% of the human genome by the ENCODE pilot project. *Nature* 447: 799–816.

97. Mercer T.R., Dinger M.E. and Mattick J.S. (2009) Long noncoding RNAs: Insights into function. *Nature Reviews Genetics* 10: 155–9.

98. Sasaki Y.T. et al. (2007) Identification and characterization of human non-coding RNAs with tissue-specific expression. *Biochemical and Biophysical Research Communications* 357: 991–6.

99. Amaral P.P. and Mattick J.S. (2008) Noncoding RNA in development. *Mammalian Genome* 19: 454–92.

100. Furuno M. et al. (2006) Clusters of internally primed transcripts reveal novel long noncoding RNAs. *PLOS Genetics* 2: e37.

101. Hackermüller J. et al. (2014) Cell cycle, oncogenic and tumor suppressor pathways regulate numerous long and macro non-protein-coding RNAs. *Genome Biology* 15: R48.

102. Lazorthes S. et al. (2015) A vlincRNA participates in senescence maintenance by relieving H2AZ-mediated repression at the INK4 locus. *Nature Communications* 6: 5971.

103. Sleutels F., Zwart R. and Barlow D.P. (2002) The noncoding Air RNA is required for silencing autosomal imprinted genes. *Nature* 415: 810–3.

104. Braidotti G. et al. (2004) The Air noncoding RNA: An imprinted cis-silencing transcript. *Cold Spring Harbor Symposia on Quantitative Biology* 69: 55–66.

105. Lorenzi L. et al. (2021) The RNA Atlas expands the catalog of human non-coding RNAs. *Nature Biotechnology* 39: 1453–65.

106. Mendoza-Vargas A. et al. (2009) Genome-wide identification of transcription start sites, promoters and transcription factor binding sites in *E. coli*. *PLOS ONE* 4: e7526.

107. Zhang G. et al. (2007) Antisense transcription in the human cytomegalovirus transcriptome. *Journal of Virology* 81: 11267–81.

108. Manghera M., Magnusson A. and Douville R.N. (2017) The sense behind retroviral anti-sense transcription. *Virology Journal* 14: 9.

109. Liu G., Mattick J.S. and Taft R.J. (2013) A meta-analysis of the genomic and transcriptomic composition of complex life. *Cell Cycle* 12: 2061–72.

110. Denoeud F. et al. (2007) Prominent use of distal 5′ transcription start sites and discovery of a large number of additional exons in ENCODE regions. *Genome Research* 17: 746–59.

111. Kapranov P. et al. (2007) RNA maps reveal new RNA classes and a possible function for pervasive transcription. *Science* 316: 1484–8.

112. St Laurent G. et al. (2013) VlincRNAs controlled by retroviral elements are a hallmark of pluripotency and cancer. *Genome Biology* 14: R73.

113. Faulkner G.J. et al. (2009) The regulated retrotransposon transcriptome of mammalian cells. *Nature Genetics* 41: 563–71.

114. Yeo G., Holste D., Kreiman G. and Burge C.B. (2004) Variation in alternative splicing across human tissues. *Genome Biology* 5: R74.

115. Yeo G.W., Van Nostrand E., Holste D., Poggio T. and Burge C.B. (2005) Identification and analysis of alternative splicing events conserved in human and mouse. *Proceedings of the National Academy of Sciences USA* 102: 2850–5.

116. Wang E.T. et al. (2008) Alternative isoform regulation in human tissue transcriptomes. *Nature* 456: 470–6.

117. Pan Q., Shai O., Lee L.J., Frey B.J. and Blencowe B.J. (2008) Deep surveying of alternative splicing complexity in the human transcriptome by high-throughput sequencing. *Nature Genetics* 40: 1413–5.

118. Raj B. and Blencowe B.J. (2015) Alternative splicing in the mammalian nervous system: Recent insights into mechanisms and functional roles. *Neuron* 87: 14–27.

119. Mercer T.R. et al. (2015) Genome-wide discovery of human splicing branchpoints. *Genome Research* 25: 290–303.

120. Tilgner H. et al. (2012) Deep sequencing of subcellular RNA fractions shows splicing to be predominantly co-transcriptional in the human genome but inefficient for lncRNAs. *Genome Research* 22: 1616–25.

121. Mele M. et al. (2017) Chromatin environment, transcriptional regulation, and splicing distinguish lincRNAs and mRNAs. *Genome Research* 27: 27–37.

122. Mukherjee N. et al. (2017) Integrative classification of human coding and noncoding genes through RNA metabolism profiles. *Nature Structural & Molecular Biology* 24: 86–96.

123. Yin Y. et al. (2020) U1 snRNP regulates chromatin retention of noncoding RNAs. *Nature* 580: 147–50.

124. Kelley D. and Rinn J. (2012) Transposable elements reveal a stem cell-specific class of long noncoding RNAs. *Genome Biology* 13: R107.

125. Kapusta A. et al. (2013) Transposable elements are major contributors to the origin, diversification, and regulation of vertebrate long noncoding RNAs. *PLOS Genetics* 9: e1003470.

126. Johnson R. and Guigo R. (2014) The RIDL hypothesis: Transposable elements as functional domains of long noncoding RNAs. *RNA* 20: 959–76.

127. Fort A. et al. (2014) Deep transcriptome profiling of mammalian stem cells supports a regulatory role for retrotransposons in pluripotency maintenance. *Nature Genetics* 46: 558–66.

128. Sasidharan R. and Gerstein M. (2008) Protein fossils live on as RNA. *Nature* 453: 729–31.

129. Li W., Yang W. and Wang X.-J. (2013) Pseudogenes: Pseudo or real functional elements? *Journal of Genetics and Genomics* 40: 171–7.

130. Cheetham S.W., Faulkner G.J. and Dinger M.E. (2020) Overcoming challenges and dogmas to understand the functions of pseudogenes. *Nature Reviews Genetics* 21: 191–201.

131. Ma Y. et al. (2021) Genome-wide analysis of pseudogenes reveals HBBP1's human-specific essentiality in erythropoiesis and implication in β-thalassemia. *Developmental Cell* 56: 478–93.

132. Sanchez-Herrero E. and Akam M. (1989) Spatially ordered transcription of regulatory DNA in the *bithorax* complex of *Drosophila*. *Development* 107: 321–9.

133. Ashe H.L., Monks J., Wijgerde M., Fraser P. and Proudfoot N.J. (1997) Intergenic transcription and transinduction of the human beta-globin locus. *Genes & Development* 11: 2494–509.

134. Feng J. et al. (2006) The Evf-2 noncoding RNA is transcribed from the Dlx-5/6 ultraconserved region and functions as a Dlx-2 transcriptional coactivator. *Genes & Development* 20: 1470–84.

135. Calin G.A. et al. (2007) Ultraconserved regions encoding ncRNAs are altered in human leukemias and carcinomas. *Cancer Cell* 12: 215–29.

136. Amaral P.P. et al. (2009) Complex architecture and regulated expression of the Sox2ot locus during vertebrate development. *RNA* 15: 2013–27.

137. Kim T.-K. et al. (2010) Widespread transcription at neuronal activity-regulated enhancers. *Nature* 465: 182–7.

138. De Santa F. et al. (2010) A large fraction of extragenic RNA pol II transcription sites overlap enhancers. *PLOS Biology* 8: e1000384.

139. Wang D. et al. (2011) Reprogramming transcription by distinct classes of enhancers functionally defined by eRNA. *Nature* 474: 390–4.

140. The Encode Project Consortium (2012) An integrated encyclopedia of DNA elements in the human genome. *Nature* 489: 57–74.

141. Kim T.-K., Hemberg M. and Gray J.M. (2015) Enhancer RNAs: A class of long noncoding RNAs synthesized at enhancers. *Cold Spring Harbor Perspectives in Biology* 7: a018622.

142. Arnold P.R., Wells A.D. and Li X.C. (2020) Diversity and emerging roles of enhancer RNA in regulation of gene expression and cell fate. *Frontiers in Cell and Developmental Biology* 7: 377.

143. Jacquier A. (2009) The complex eukaryotic transcriptome: Unexpected pervasive transcription and novel small RNAs. *Nature Reviews Genetics* 10: 833–44.

144. Davis C.A. and Ares Jr. M., (2006) Accumulation of unstable promoter-associated transcripts upon loss of the nuclear exosome subunit Rrp6p in *Saccharomyces cerevisiae*. *Proceedings of the National Academy of Sciences USA* 103: 3262–7.

145. Preker P. et al. (2008) RNA exosome depletion reveals transcription upstream of active human promoters. *Science* 322: 1851–4.

146. Taft R.J. et al. (2009) Tiny RNAs associated with transcription start sites in animals. *Nature Genetics* 41: 572–8.

147. Seila A.C. et al. (2008) Divergent transcription from active promoters. *Science* 322: 1849–51.

148. Schmitz K.M., Mayer C., Postepska A. and Grummt I. (2010) Interaction of noncoding RNA with the rDNA promoter mediates recruitment of DNMT3b and silencing of rRNA genes. *Genes & Development* 24: 2264–9.

149. Mayer C., Schmitz K.M., Li J., Grummt I. and Santoro R. (2006) Intergenic transcripts regulate the epigenetic state of rRNA genes. *Molecular Cell* 22: 351–61.

150. Mayer C., Neubert M. and Grummt I. (2008) The structure of NoRC-associated RNA is crucial for targeting the chromatin remodelling complex NoRC to the nucleolus. *EMBO Reports* 9: 774–80.

151. Santoro R., Schmitz K.M., Sandoval J. and Grummt I. (2010) Intergenic transcripts originating from a subclass of ribosomal DNA repeats silence ribosomal RNA genes in trans. *EMBO Reports* 11: 52–8.

152. Memczak S. et al. (2013) Circular RNAs are a large class of animal RNAs with regulatory potency. *Nature* 495: 333–8.

153. Hansen T.B. et al. (2013) Natural RNA circles function as efficient microRNA sponges. *Nature* 495: 384–8.

154. Kristensen L.S. et al. (2019) The biogenesis, biology and characterization of circular RNAs. *Nature Reviews Genetics* 20: 675–91.

155. Yin Q.-F. et al. (2012) Long noncoding RNAs with snoRNA ends. *Molecular Cell* 48: 219–30.

156. Ner-Gaon H. et al. (2004) Intron retention is a major phenomenon in alternative splicing in Arabidopsis. *Plant Journal* 39: 877–85.

157. Wong J.J.-L. et al. (2013) Orchestrated intron retention regulates normal granulocyte differentiation. *Cell* 154: 583–95.

158. Braunschweig U. et al. (2014) Widespread intron retention in mammals functionally tunes transcriptomes. *Genome Research* 24: 1774–86.

159. Pimentel H. et al. (2015) A dynamic intron retention program enriched in RNA processing genes regulates gene expression during terminal erythropoiesis. *Nucleic Acids Research* 44: 838–51.

160. Schmitz U. et al. (2017) Intron retention enhances gene regulatory complexity in vertebrates. *Genome Biology* 18: 216.

161. Olthof A.M., Hyatt K.C. and Kanadia R.N. (2019) Minor intron splicing revisited: Identification of new minor intron-containing genes and tissue-dependent retention and alternative splicing of minor introns. *BMC Genomics* 20: 686.

162. Zeng C. and Hamada M. (2020) RNA-seq analysis reveals localization-associated alternative splicing across 13 cell lines. *Genes* 11: 820.

163. Dumbović G. et al. (2021) Nuclear compartmentalization of TERT mRNA and TUG1 lncRNA is driven by intron retention. *Nature Communications* 12: 3308.

164. Flynn R.A. et al. (2021) Small RNAs are modified with N-glycans and displayed on the surface of living cells. *Cell* 184: 3109.

165. Huang N. et al. (2020) Natural display of nuclear-encoded RNA on the cell surface and its impact on cell interaction. *Genome Biology* 21: 225.

166. Sharma C.M. et al. (2010) The primary transcriptome of the major human pathogen *Helicobacter pylori*. *Nature* 464: 250–5.

167. Puerta-Fernandez E., Barrick J.E., Roth A. and Breaker R.R. (2006) Identification of a large noncoding RNA in extremophilic eubacteria. *Proceedings of the National Academy of Sciences USA* 103: 19490–5.

168. Ko J.-H. and Altman S. (2007) OLE RNA, an RNA motif that is highly conserved in several extremophilic bacteria, is a substrate for and can be regulated by RNase P RNA. *Proceedings of the National Academy of Sciences USA* 104: 7815–20.

169. Weinberg Z., Perreault J., Meyer M.M. and Breaker R.R. (2009) Exceptional structured noncoding RNAs revealed by bacterial metagenome analysis. *Nature* 462: 656–9.

170. Harris K.A. and Breaker R.R. (2018) Large noncoding RNAs in bacteria. *Microbiology Spectrum* 6. doi:10.1128/microbiolspec.RWR-0005-2017.

171. Frith M.C., Pheasant M. and Mattick J.S. (2005) The amazing complexity of the human transcriptome. *European Journal of Human Genetics* 13: 894–7.

172. Mattick J.S. and Makunin I.V. (2006) Non-coding RNA. *Human Molecular Genetics* 15: R17–29.

173. Kapranov P., Willingham A.T. and Gingeras T.R. (2007) Genome-wide transcription and the implications for genomic organization. *Nature Reviews Genetics* 8: 413–23.

174. Gingeras T.R. (2007) Origin of phenotypes: Genes and transcripts. *Genome Research* 17: 682–90.

175. Perez C.A.G. et al. (2021) Sense-overlapping lncRNA as a decoy of translational repressor protein for dimorphic gene expression. *PLOS Genetics* 17: e1009683.

176. Morris K.V. and Mattick J.S. (2014) The rise of regulatory RNA. *Nature Reviews Genetics* 15: 423–37.

177. Harrow J. et al. (2012) GENCODE: The reference human genome annotation for The ENCODE Project. *Genome Research* 22: 1760–74.

178. Mattick J.S. (2003) Challenging the dogma: The hidden layer of non-protein-coding RNAs in complex organisms. *BioEssays* 25: 930–9.

179. Dinger M.E., Amaral P.P., Mercer T.R. and Mattick J.S. (2009) Pervasive transcription of the eukaryotic genome: Functional indices and conceptual implications. *Briefings in Functional Genomics and Proteomics* 8: 407–23.

180. Amaral P.P., Dinger M.E., Mercer T.R. and Mattick J.S. (2008) The eukaryotic genome as an RNA machine. *Science* 319: 1787–9.

181. Dinger M.E., Pang K.C., Mercer T.R. and Mattick J.S. (2008) Differentiating protein-coding and noncoding RNA: Challenges and ambiguities. *PLOS Computational Biology* 4: e1000176.

182. Galindo M.I., Pueyo J.I., Fouix S., Bishop S.A. and Couso J.P. (2007) Peptides encoded by short ORFs control development and define a new eukaryotic gene family. *PLOS Biology* 5: e106.

183. Gultyaev A.P. and Roussis A. (2007) Identification of conserved secondary structures and expansion segments in enod40 reveals new enod40 homologues in plants. *Nucleic Acids Research* 35: 3144–52.

184. Hanyu-Nakamura K., Sonobe-Nojima H., Tanigawa A., Lasko P. and Nakamura A. (2008) *Drosophila* Pgc protein inhibits P-TEFb recruitment to chromatin in primordial germ cells. *Nature* 451: 730–3.

185. Cabili M.N. et al. (2011) Integrative annotation of human large intergenic noncoding RNAs reveals global properties and specific subclasses. *Genes & Development* 25: 1915–27.

186. Gascoigne D.K. et al. (2012) Pinstripe: A suite of programs for integrating transcriptomic and proteomic datasets identifies novel proteins and improves differentiation of protein-coding and non-coding genes. *Bioinformatics* 28: 3042–50.

187. Frith M.C. et al. (2006) The abundance of short proteins in the mammalian proteome. *PLOS Genetics* 2: e52.

188. Magny E.G. et al. (2013) Conserved regulation of cardiac calcium uptake by peptides encoded in small open reading frames. *Science* 341: 1116–20.

189. Lauressergues D. et al. (2015) Primary transcripts of microRNAs encode regulatory peptides. *Nature* 520: 90–3.

190. Anderson D.M. et al. (2015) A micropeptide encoded by a putative long noncoding RNA regulates muscle performance. *Cell* 160: 595–606.

191. Matsumoto A. et al. (2017) mTORC1 and muscle regeneration are regulated by the LINC00961-encoded SPAR polypeptide. *Nature* 541: 228–32.

192. Stein C.S. et al. (2018) Mitoregulin: A lncRNA-encoded microprotein that supports mitochondrial supercomplexes and respiratory efficiency. *Cell Reports* 23: 3710–20.

193. Ransohoff J.D., Wei Y. and Khavari P.A. (2018) The functions and unique features of long intergenic non-coding RNA. *Nature Reviews Molecular Cell Biology* 19: 143–57.

194. Lewandowski J.P. et al. (2020) The *Tug1* lncRNA locus is essential for male fertility. *Genome Biology* 21: 237.

195. Chen J. et al. (2020) Pervasive functional translation of noncanonical human open reading frames. *Science* 367: 1140–6.

196. Eckhart L., Lachner J., Tschachler E. and Rice R.H. (2020) TINCR is not a non-coding RNA but encodes a protein component of cornified epidermal keratinocytes. *Experimental Dermatology* 29: 376–9.

197. Nita A. et al. (2021) A ubiquitin-like protein encoded by the "noncoding" RNA TINCR promotes keratinocyte proliferation and wound healing. *PLOS Genetics* 17: e1009686.

198. Guttman M., Russell P., Ingolia N.T., Weissman J.S. and Lander E.S. (2013) Ribosome profiling provides evidence that large noncoding RNAs do not encode proteins. *Cell* 154: 240–51.

199. Röhrig H., Schmidt J., Miklashevichs E., Schell J. and John M. (2002) Soybean *ENOD40* encodes two peptides that bind to sucrose synthase. *Proceedings of the National Academy of Sciences USA* 99: 1915–20.

200. Dinger M.E., Gascoigne D.K. and Mattick J.S. (2011) The evolution of RNAs with multiple functions. *Biochimie* 93: 2013–8.

201. Candeias M.M. et al. (2008) p53 mRNA controls p53 activity by managing Mdm2 functions. *Nature Cell Biology* 10: 1098–105.

202. Cooper C. et al. (2011) Steroid receptor RNA activator bi-faceted genetic system: Heads or Tails? *Biochimie* 93: 1973–80.

203. Kumari P. and Sampath K. (2015) cncRNAs: Bi-functional RNAs with protein coding and non-coding functions. *Seminars in Cell & Developmental Biology* 47–48: 40–51.

204. Gilot D. et al. (2017) A non-coding function of TYRP1 mRNA promotes melanoma growth. *Nature Cell Biology* 19: 1348–57.

205. Ivanyi-Nagy R. et al. (2018) The RNA interactome of human telomerase RNA reveals a coding-independent role for a histone mRNA in telomere homeostasis. *eLife* 7: e40037.

206. Crerar H. et al. (2019) Regulation of NGF signaling by an axonal untranslated mRNA. *Neuron* 102: 553–63.

207. Gaertner B. et al. (2020) A human ESC-based screen identifies a role for the translated lncRNA LINC00261 in pancreatic endocrine differentiation. *eLife* 9: e58659.

208. Prasad A., Sharma N. and Prasad M. (2021) Noncoding but coding: Pri-miRNA into the action. *Trends in Plant Science* 26: 204–6.

209. Broadwell L.J. et al. (2021) Myosin 7b is a regulatory long noncoding RNA (lncMYH7b) in the human heart. *Journal of Biological Chemistry* 296: 100694.

210. Boraas L. et al. (2021) Non-coding function for mRNAs in focal adhesion architecture and mechanotransduction. *bioRxiv*: 2021.10.04.463097.

211. Lecuyer E. et al. (2007) Global analysis of mRNA localization reveals a prominent role in organizing cellular architecture and function. *Cell* 131: 174–87.

212. Ryder P.V. and Lerit D.A. (2018) RNA localization regulates diverse and dynamic cellular processes. *Traffic* 19: 496–502.

213. Grelet S. et al. (2017) A regulated PNUTS mRNA to lncRNA splice switch mediates EMT and tumour progression. *Nature Cell Biology* 19: 1105–15.

214. Hube F. et al. (2006) Alternative splicing of the first intron of the steroid receptor RNA activator (SRA) participates in the generation of coding and noncoding RNA isoforms in breast cancer cell lines. *DNA and Cell Biology* 25: 418–28.

215. Leygue E. (2007) Steroid receptor RNA activator (SRA1): Unusual bifaceted gene products with suspected relevance to breast cancer. *Nuclear Receptor Signaling* 5: e006.

216. Hashimoto K. et al. (2009) A liver X receptor (LXR)-β alternative splicing variant (LXRBSV) acts as an RNA co-activator of LXR-β. *Biochemical and Biophysical Research Communications* 390: 1260–5.

217. Ulveling D., Francastel C. and Hube F. (2011) When one is better than two: RNA with dual functions. *Biochimie* 93: 633–44.

218. Ulveling D., Francastel C. and Hube F. (2011) Identification of potentially new bifunctional RNA based on genome-wide data-mining of alternative splicing events. *Biochimie* 93: 2024–7.

219. Williamson L. et al. (2017) UV irradiation induces a non-coding RNA that functionally opposes the protein encoded by the same gene. *Cell* 168: 843–55.

220. Gautron A. et al. (2021) Human TYRP1: Two functions for a single gene? *Pigment Cell & Melanoma Research* 34: 836–52.

221. Qu S. et al. (2021) PD-L1 lncRNA splice isoform promotes lung adenocarcinoma progression via enhancing c-Myc activity. *Genome Biology* 22: 104.

222. Gonzàlez-Porta M., Frankish A., Rung J., Harrow J. and Brazma A. (2013) Transcriptome analysis of human tissues and cell lines reveals one dominant transcript per gene. *Genome Biology* 14: R70.

223. Duret L., Chureau C., Samain S., Weissenbach J. and Avner P. (2006) The *Xist* RNA gene evolved in eutherians by pseudogenization of a protein-coding gene. *Science* 312: 1653–5.

224. Elisaphenko E.A. et al. (2008) A dual origin of the *Xist* gene from a protein-coding gene and a set of transposable elements. *PLOS ONE* 3: e2521.

225. Hezroni H. et al. (2017) A subset of conserved mammalian long non-coding RNAs are fossils of ancestral protein-coding genes. *Genome Biology* 18: 162.

226. Beck-Engeser G.B. et al. (2008) Pvt1-encoded microRNAs in oncogenesis. *Retrovirology* 5: 4.

227. He D. et al. (2021) miRNA-independent function of long noncoding pri-miRNA loci. *Proceedings of the National Academy of Sciences USA* 118: e2017562118.

228. Immarigeon C. et al. (2021) Identification of a micropeptide and multiple secondary cell genes that modulate *Drosophila* male reproductive success. *Proceedings of the National Academy of Sciences USA* 118: e2001897118.

229. Montigny A. et al. (2021) *Drosophila* primary microRNA-8 encodes a microRNA-encoded peptide acting in parallel of miR-8. *Genome Biology* 22: 118.

230. Yang T. et al. (2017) lncRNA PVT1 and its splicing variant function as competing endogenous RNA to regulate clear cell renal cell carcinoma progression. *Oncotarget* 8: 85353–67.

231. Holdt L.M. and Teupser D. (2018) Long noncoding RNA ANRIL: Lnc-ing genetic variation at the chromosome 9p21 locus to molecular mechanisms of atherosclerosis. *Frontiers in Cardiovascular Medicine* 5: 145.

232. Hubberten M. et al. (2019) Linear isoforms of the long noncoding RNA CDKN2B-AS1 regulate the c-myc-enhancer binding factor RBMS1. *European Journal of Human Genetics* 27: 80–9.

233. Ghetti M., Vannini I., Storlazzi C.T., Martinelli G. and Simonetti G. (2020) Linear and circular PVT1 in hematological malignancies and immune response: Two faces of the same coin. *Molecular Cancer* 19: 69.

234. Shuai T. et al. (2021) lncRNA TRMP-S directs dual mechanisms to regulate p27-mediated cellular senescence. *Molecular Therapy - Nucleic Acids* 24: 971–85.

235. Wanowska E., Kubiak M., Makałowska I. and Szcześniak M.W. (2021) A chromatin-associated splicing isoform of OIP5-AS1 acts in cis to regulate the OIP5 oncogene. *RNA Biology* 18: 1834–45.

236. Dennis C. (2002) The brave new world of RNA. *Nature* 418: 122–4.

237. Jarvis K. and Robertson M. (2011) The noncoding universe. *BMC Biology* 9: 52.

238. Babak T., Blencowe B.J. and Hughes T.R. (2005) A systematic search for new mammalian noncoding RNAs indicates little conserved intergenic transcription. *BMC Genomics* 6: 104.

239. van Bakel H., Nislow C., Blencowe B.J. and Hughes T.R. (2010) Most "dark matter" transcripts are associated with known genes. *PLOS Biology* 8: e1000371.

240. Palazzo A.F. and Lee E.S. (2015) Non-coding RNA: What is functional and what is junk? *Frontiers in Genetics* 6: 2.

241. Lloréns-Rico V. et al. (2016) Bacterial antisense RNAs are mainly the product of transcriptional noise. *Science Advances* 2: e1501363.

242. Elowitz M.B., Levine A.J., Siggia E.D. and Swain P.S. (2002) Stochastic gene expression in a single cell. *Science* 297: 1183–6.

243. Raser J.M. and O'Shea E.K. (2005) Noise in gene expression: Origins, consequences, and control. *Science* 309: 2010–3.

244. Blake W.J., M. K.A., Cantor C.R. and Collins J.J. (2003) Noise in eukaryotic gene expression. *Nature* 422: 633–7.

245. Arias A.M. and Hayward P. (2006) Filtering transcriptional noise during development: concepts and mechanisms. *Nature Reviews Genetics* 7: 34–44.

246. Ramsey S. et al. (2006) Transcriptional noise and cellular heterogeneity in mammalian macrophages. *Philosophical Transactions of the Royal Society B: Biological Sciences* 361: 495–506.

247. Struhl K. (2007) Transcriptional noise and the fidelity of initiation by RNA polymerase II. *Nature Structural & Molecular Biology* 14: 103–5.

248. Clark M.B. et al. (2011) The reality of pervasive transcription. *PLOS Biology* 9: e1000625.

249. van Bakel H., Nislow C., Blencowe B.J. and Hughes T.R. (2011) Response to "the reality of pervasive transcription". *PLOS Biology* 9: e1001102.

250. Graur D. et al. (2013) On the immortality of television sets: "function" in the human genome according to the evolution-free gospel of encode. *Genome Biology and Evolution* 5: 578–90.

251. Mattick J.S. and Dinger M.E. (2013) The extent of functionality in the human genome. *HUGO Journal* 7: 2.

252. Mercer T.R. et al. (2012) Targeted RNA sequencing reveals the deep complexity of the human transcriptome. *Nature Biotechnology* 30: 99–104.

253. Clark M.B. et al. (2015) Quantitative gene profiling of long noncoding RNAs with targeted RNA sequencing. *Nature Methods* 12: 339–42.

254. Gloss B.S. and Dinger M.E. (2016) The specificity of long noncoding RNA expression. *Biochimica et Biophysica Acta* 1859: 16–22.

255. Xu Q. et al. (2017) Systematic comparison of lncRNAs with protein coding mRNAs in population expression and their response to environmental change. *BMC Plant Biology* 17: 42.

256. Sharova L.V. et al. (2009) Database for mRNA half-life of 19 977 genes obtained by DNA microarray analysis of pluripotent and differentiating mouse embryonic stem cells. *DNA Research* 16: 45–58.

257. Clark M. et al. (2012) Genome-wide analysis of long noncoding RNA stability. *Genome Research* 21: 885–98.

258. Field A.R. et al. (2019) Structurally conserved primate lncRNAs are transiently expressed during human cortical differentiation and influence cell-type-specific genes. *Stem Cell Reports* 12: 245–57.

259. Houseley J., Rubbi L., Grunstein M., Tollervey D. and Vogelauer M. (2008) A ncRNA modulates histone modification and mRNA induction in the yeast GAL gene cluster. *Molecular Cell* 32: 685–95.

260. Wang K.C. et al. (2011) A long noncoding RNA maintains active chromatin to coordinate homeotic gene expression. *Nature* 472: 120–4.

261. Li Z. et al. (2013) The long noncoding RNA *THRIL* regulates TNFα expression through its interaction with hnRNPL. *Proceedings of the National Academy of Sciences USA* 111: 1002–7.

262. Li M.A. et al. (2017) A lncRNA fine tunes the dynamics of a cell state transition involving Lin28, let-7 and de novo DNA methylation. *eLife* 6: e23468.

263. Xing Y.-H. et al. (2017) SLERT regulates DDX21 rings associated with Pol I transcription. *Cell* 169: 664–78.

264. Seiler J. et al. (2017) The lncRNA VELUCT strongly regulates viability of lung cancer cells despite its extremely low abundance. *Nucleic Acids Research* 45: 5458–69.

265. Hao Q. et al. (2020) The S-phase-induced lncRNA SUNO1 promotes cell proliferation by controlling YAP1/Hippo signaling pathway. *eLife* 9: e55102.

266. Rowland T.J., Dumbović G., Hass E.P., Rinn J.L. and Cech T.R. (2019) Single-cell imaging reveals unexpected heterogeneity of telomerase reverse transcriptase expression across human cancer cell lines. *Proceedings of the National Academy of Sciences USA* 116: 18488–97.

267. Mitchell P., Petfalski E., Shevchenko A., Mann M. and Tollervey D. (1997) The exosome: A conserved eukaryotic RNA processing complex containing multiple 3'-->5' exoribonucleases. *Cell* 91: 457–66.

268. Mitchell P. and Tollervey D. (2000) Musing on the structural organization of the exosome complex. *Nature Structural Biology* 7: 843–6.

269. Houseley J., LaCava J. and Tollervey D. (2006) RNA-quality control by the exosome. *Nature Reviews Molecular Cell Biology* 7: 529–39.

270. Losson R. and Lacroute F. (1979) Interference of nonsense mutations with eukaryotic messenger RNA stability. *Proceedings of the National Academy of Sciences USA* 76: 5134–7.

271. Nagy E. and Maquat L.E. (1998) A rule for termination-codon position within intron-containing genes: When nonsense affects RNA abundance. *Trends in Biochemical Sciences* 23: 198–9.

272. Conti E. and Izaurralde E. (2005) Nonsense-mediated mRNA decay: Molecular insights and mechanistic variations across species. *Current Opinion in Cell Biology* 17: 316–25.

273. Behm-Ansmant I. et al. (2007) mRNA quality control: An ancient machinery recognizes and degrades mRNAs with nonsense codons. *FEBS Letters* 581: 2845–53.

274. Isken O. and Maquat L.E. (2007) Quality control of eukaryotic mRNA: Safeguarding cells from abnormal mRNA function. *Genes & Development* 21: 1833–56.

275. Kurihara Y. et al. (2009) Genome-wide suppression of aberrant mRNA-like noncoding RNAs by NMD in Arabidopsis. *Proceedings of the National Academy of Sciences USA* 106: 2453–8.

276. Boehm V. and Gehring N.H. (2016) Exon junction complexes: Supervising the gene expression assembly line. *Trends in Genetics* 32: 724–35.

277. Isken O. and Maquat L.E. (2008) The multiple lives of NMD factors: Balancing roles in gene and genome regulation. *Nature Reviews Genetics* 9: 699–712.

278. Neu-Yilik G. and Kulozik A.E. (2008) NMD: Multitasking between mRNA surveillance and modulation of gene expression. *Advances in Genetics* 62: 185–243.

279. Karam R., Wengrod J., Gardner L.B. and Wilkinson M.F. (2013) Regulation of nonsense-mediated mRNA decay: Implications for physiology and disease. *Biochimica et Biophysica Acta* 1829: 624–33.

280. Andjus S., Morillon A. and Wery M. (2021) From yeast to mammals, the nonsense-mediated mRNA decay as a master regulator of long non-coding RNAs functional trajectory. *Non-Coding RNA* 7: 44.

281. Karam R. et al. (2015) The unfolded protein response is shaped by the NMD pathway. *EMBO Reports* 16: 599–609.

282. Huang L. and Wilkinson M.F. (2012) Regulation of nonsense-mediated mRNA decay. *WIREs RNA* 3: 807–28.

283. Li T. et al. (2015) Smg6/Est1 licenses embryonic stem cell differentiation via nonsense-mediated mRNA decay. *EMBO Journal* 34: 1630–47.

284. Lou C.-H. et al. (2016) Nonsense-mediated RNA decay influences human embryonic stem cell fate. *Stem Cell Reports* 6: 844–57.

285. Jaffrey S.R. and Wilkinson M.F. (2018) Nonsense-mediated RNA decay in the brain: Emerging modulator of neural development and disease. *Nature Reviews Neuroscience* 19: 715–28.

286. Wyers F. et al. (2005) Cryptic pol II transcripts are degraded by a nuclear quality control pathway involving a new poly(A) polymerase. *Cell* 121: 725–37.

287. Thompson D.M. and Parker R. (2007) Cytoplasmic decay of intergenic transcripts in *Saccharomyces cerevisiae*. *Molecular and Cellular Biology* 27: 92–101.

288. Chekanova J.A. et al. (2007) Genome-wide high-resolution mapping of exosome substrates reveals hidden features in the *Arabidopsis* transcriptome. *Cell* 131: 1340–53.

289. Camblong J., Iglesias N., Fickentscher C., Dieppois G. and Stutz F. (2007) Antisense RNA stabilization induces transcriptional gene silencing via histone deacetylation in *S. cerevisiae*. *Cell* 131: 706–17.

290. Houseley J., Kotovic K., El Hage A. and Tollervey D. (2007) Trf4 targets ncRNAs from telomeric and rDNA spacer regions and functions in rDNA copy number control. *EMBO Journal* 26: 4996–5006.

291. Uhler J.P., Hertel C. and Svejstrup J.Q. (2007) A role for noncoding transcription in activation of the yeast PHO5 gene. *Proceedings of the National Academy of Sciences USA* 104: 8011–6.

292. Vasiljeva L., Kim M., Terzi N., Soares L.M. and Buratowski S. (2008) Transcription termination and RNA degradation contribute to silencing of RNA polymerase II transcription within heterochromatin. *Molecular Cell* 29: 313–23.

293. Berretta J., Pinskaya M. and Morillon A. (2008) A cryptic unstable transcript mediates transcriptional trans-silencing of the Ty1 retrotransposon in *S. cerevisiae*. *Genes & Development* 22: 615–26.

294. Toesca I., Nery C.R., Fernandez C.F., Sayani S. and Chanfreau G.F. (2011) Cryptic transcription mediates repression of subtelomeric metal homeostasis genes. *PLOS Genetics* 7: e1002163.

295. Yin S. et al. (2020) The cryptic unstable transcripts are associated with developmentally regulated gene expression in blood-stage *Plasmodium falciparum*. *RNA Biology* 17: 828–42.

296. Balarezo-Cisneros L.N. et al. (2021) Functional and transcriptional profiling of non-coding RNAs in yeast reveal context-dependent phenotypes and in trans effects on the protein regulatory network. *PLOS Genetics* 17: e1008761.

297. Andersson R. et al. (2014) An atlas of active enhancers across human cell types and tissues. *Nature* 507: 455–61.

298. Pefanis E. et al. (2015) RNA exosome-regulated long non-coding RNA transcription controls super-enhancer activity. *Cell* 161: 774–89.

299. Sone M. et al. (2007) The mRNA-like noncoding RNA Gomafu constitutes a novel nuclear domain in a subset of neurons. *Journal of Cell Science* 120: 2498–506.

300. Mercer T.R., Dinger M.E., Sunkin S.M., Mehler M.F. and Mattick J.S. (2008) Specific expression of long noncoding RNAs in the mouse brain. *Proceedings of the National Academy of Sciences USA* 105: 716–2.

301. Goff L.A. et al. (2015) Spatiotemporal expression and transcriptional perturbations by long noncoding RNAs in the mouse brain. *Proceedings of the National Academy of Sciences USA* 112: 6855–62.

302. Bocchi V.D. et al. (2021) The coding and long noncoding single-cell atlas of the developing human fetal striatum. *Science* 372: eabf5759.

303. Pauli A. et al. (2012) Systematic identification of long noncoding RNAs expressed during zebrafish embryogenesis. *Genome Research* 22: 577–91.

304. Inagaki S. et al. (2005) Identification and expression analysis of putative mRNA-like non-coding RNA in *Drosophila*. *Genes to Cells* 10: 1163–73.

305. Tupy J.L. et al. (2005) Identification of putative noncoding polyadenylated transcripts in *Drosophila melanogaster*. *Proceedings of the National Academy of Sciences USA* 102: 5495–500.

306. Sawata M. et al. (2002) Identification and punctate nuclear localization of a novel noncoding RNA, Ks-1, from the honeybee brain. *RNA* 8: 772–85.

307. Sawata M., Takeuchi H. and Kubo T. (2004) Identification and analysis of the minimal promoter activity of a novel noncoding nuclear RNA gene, *AncR-1*, from the honeybee (*Apis mellifera* L.). *RNA* 10: 1047–58.

308. Wang M. et al. (2021) Sex-specific development in haplodiploid honeybee is controlled by the female-embryo-specific activation of thousands of intronic lncRNAs. *Frontiers in Cell and Developmental Biology* 9: 690167.

309. Gribnau J., Diderich K., Pruzina S., Calzolari R. and Fraser P. (2000) Intergenic transcription and developmental remodeling of chromatin subdomains in the human beta-globin locus. *Molecular Cell* 5: 377–86.

310. Chen H., Du G., Song X. and Li L. (2017) Non-coding transcripts from enhancers: New insights into enhancer activity and gene expression regulation. *Genomics, Proteomics & Bioinformatics* 15: 201–7.

311. Ivaldi M.S. et al. (2018) Fetal γ-globin genes are regulated by the *BGLT3* long noncoding RNA locus. *Blood* 132: 1963–73.

312. Rogan D.F. et al. (2004) Analysis of intergenic transcription in the human IL-4/IL-13 gene cluster. *Proceedings of the National Academy of Sciences USA* 101: 2446–51.

313. Jones E.A. and Flavell R.A. (2005) Distal enhancer elements transcribe intergenic RNA in the IL-10 family gene cluster. *Journal of Immunology* 175: 7437–46.

314. Islam S. et al. (2011) Characterization of the single-cell transcriptional landscape by highly multiplex RNA-seq. *Genome Research* 21: 1160–7.

315. Kim Daniel H. et al. (2015) Single-cell transcriptome analysis reveals dynamic changes in lncRNA expression during reprogramming. *Cell Stem Cell* 16: 88–101.

316. Liu S.J. et al. (2016) Single-cell analysis of long non-coding RNAs in the developing human neocortex. *Genome Biology* 17: 67.

317. Lencer E., Prekeris R. and Artinger K.B. (2021) Single-cell RNA analysis identifies pre-migratory neural crest cells expressing markers of differentiated derivatives. *eLife* 10: e66078.

318. Isakova A., Neff N. and Quake S.R. (2021) Single-cell quantification of a broad RNA spectrum reveals unique noncoding patterns associated with cell types and states. *Proceedings of the National Academy of Sciences USA* 118: e2113568118.

319. Brena C., Chipman A.D., Minelli A. and Akam M. (2006) Expression of trunk Hox genes in the centipede *Strigamia maritima*: Sense and anti-sense transcripts. *Evolution & Development* 8: 252–65.

320. Deveson I.W. et al. (2016) Representing genetic variation with synthetic DNA standards. *Nature Methods* 13: 784–91.

321. Hardwick S.A. et al. (2016) Spliced synthetic genes as internal controls in RNA sequencing experiments. *Nature Methods* 13: 792–8.

322. Wong T., Deveson I.W., Hardwick S.A. and Mercer T.R. (2017) ANAQUIN: A software toolkit for the analysis of spike-in controls for next generation sequencing. *Bioinformatics* 33: 1723–4.

323. Blackburn J. et al. (2019) Use of synthetic DNA spike-in controls (sequins) for human genome sequencing. *Nature Protocols* 14: 2119–51.

324. Deveson I.W. et al. (2019) Chiral DNA sequences as commutable controls for clinical genomics. *Nature Communications* 10: 1342.

325. Hardwick S.A. et al. (2018) Synthetic microbe communities provide internal reference standards for metagenome sequencing and analysis. *Nature Communications* 9: 3096.

326. Mercer T.R. et al. (2014) Targeted sequencing for gene discovery and quantification using RNA CaptureSeq. *Nature Protocols* 9: 989–1009.

327. Bussotti G. et al. (2016) Improved definition of the mouse transcriptome via targeted RNA sequencing. *Genome Research* 26: 705–16.

328. Lagarde J. et al. (2017) High-throughput annotation of full-length long noncoding RNAs with capture long-read sequencing. *Nature Genetics* 49: 1731–40.

329. Bartonicek N. et al. (2017) Intergenic disease-associated regions are abundant in novel transcripts. *Genome Biology* 18: 241.

330. Hardwick S.A. et al. (2019) Targeted, high-resolution RNA sequencing of non-coding genomic regions associated with neuropsychiatric functions. *Frontiers in Genetics* 10: 309.

331. de Goede O.M. et al. (2021) Population-scale tissue transcriptomics maps long non-coding RNAs to complex disease. *Cell* 184: 2633–48.

332. Hon C.-C. et al. (2017) An atlas of human long non-coding RNAs with accurate 5′ ends. *Nature* 543: 199–204.

333. Nasser J. et al. (2021) Genome-wide enhancer maps link risk variants to disease genes. *Nature* 593: 238–43.

334. Eze U.C., Bhaduri A., Haeussler M., Nowakowski T.J. and Kriegstein A.R. (2021) Single-cell atlas of early human brain development highlights heterogeneity of human neuroepithelial cells and early radial glia. *Nature Neuroscience* 24: 584–94.

335. Orom U.A. et al. (2010) Long noncoding RNAs with enhancer-like function in human cells. *Cell* 143: 46–58.

336. Derrien T. et al. (2012) The GENCODE v7 catalog of human long noncoding RNAs: Analysis of their gene structure, evolution, and expression. *Genome Research* 22: 1775–89.

337. Unfried J.P. et al. (2019) Identification of coding and long noncoding RNAs differentially expressed in tumors and preferentially expressed in healthy tissues. *Cancer Research* 79: 5167–80.

338. Lein E.S. et al. (2007) Genome-wide atlas of gene expression in the adult mouse brain. *Nature* 445: 168–76.

339. Sunkin S.M. and Hohmann J.G. (2007) Insights from spatially mapped gene expression in the mouse brain. *Human Molecular Genetics* 16: R209–19.

340. Gaiti F. et al. (2015) Dynamic and widespread lncRNA expression in a sponge and the origin of animal complexity. *Molecular Biology and Evolution* 32: 2367–82.

341. Dinger M.E. et al. (2008) Long noncoding RNAs in mouse embryonic stem cell pluripotency and differentiation. *Genome Research* 18: 1433–45.

342. Sheik Mohamed J., Gaughwin P.M., Lim B., Robson P. and Lipovich L. (2010) Conserved long noncoding RNAs transcriptionally regulated by Oct4 and Nanog modulate pluripotency in mouse embryonic stem cells. *RNA* 16: 324–37.

343. Kohtz J.D. and Fishell G. (2004) Developmental regulation of EVF-1, a novel non-coding RNA transcribed upstream of the mouse Dlx6 gene. *Gene Expression Patterns* 4: 407–12.

344. Johnson R. et al. (2009) Regulation of neural macroRNAs by the transcriptional repressor REST. *RNA* 15: 85–96.

345. Mercer T.R. et al. (2010) Long noncoding RNAs in neuronal-glial fate specification and oligodendrocyte lineage maturation. *BMC Neuroscience* 11: 14.

346. Sunwoo H. et al. (2009) MEN epsilon/beta nuclear-retained non-coding RNAs are up-regulated upon muscle differentiation and are essential components of paraspeckles. *Genome Research* 19: 347–59.

347. Askarian-Amiri M.E. et al. (2011) SNORD-host RNA Zfas1 is a regulator of mammary development and a potential marker for breast cancer. *RNA* 17: 878–91.

348. Pang K.C. et al. (2009) Genome-wide identification of long noncoding RNAs in CD8+ T cells. *Journal of Immunology* 182: 7738–48.

349. Zhang X. et al. (2009) A myelopoiesis-associated regulatory intergenic noncoding RNA transcript within the human HOXA cluster. *Blood* 113: 2526–34.

350. Dinger M.E. et al. (2009) NRED: A database of long noncoding RNA expression. *Nucleic Acids Research* 37: D122–6.

351. Warren W.C. et al. (2010) The genome of a songbird. *Nature* 464: 757–62.

352. Amaral P.P., Dinger M.E. and Mattick J.S. (2013) Non-coding RNAs in homeostasis, disease and stress responses: An evolutionary perspective. *Briefings in Functional Genomics* 12: 254–78.

353. Bhatia G., Sharma S., Upadhyay S.K. and Singh K. (2019) Long non-coding RNAs coordinate developmental transitions and other key biological processes in grapevine. *Scientific Reports* 9: 3552.

354. Yu Y., Zhang Y., Chen X. and Chen Y. (2019) Plant noncoding RNAs: Hidden players in development and stress responses. *Annual Review of Cell and Developmental Biology* 35: 407–31.

355. Kim W., Miguel-Rojas C., Wang J., Townsend J.P. and Trail F. (2018) Developmental dynamics of long noncoding RNA expression during sexual fruiting body formation in *Fusarium graminearum*. *mBio* 9: e01292.

356. Tiedge H., Fremeau Jr. R.T., Weinstock P.H., Arancio O. and Brosius J. (1991) Dendritic location of neural BC1 RNA. *Proceedings of the National Academy of Sciences USA* 88: 2093–7.

357. Shimada T., Yamashita A. and Yamamoto M. (2003) The fission yeast meiotic regulator Mei2p forms a dot structure in the horse-tail nucleus in association with the *sme2* locus on chromosome II. *Molecular Biology of the Cell* 14: 2461–9.

358. Hutchinson J.N. et al. (2007) A screen for nuclear transcripts identifies two linked noncoding RNAs associated with SC35 splicing domains. *BMC Genomics* 8: 39.

359. Clemson C.M. et al. (2009) An architectural role for a nuclear noncoding RNA: NEAT1 RNA is essential for the structure of paraspeckles. *Molecular Cell* 33: 717–26.

360. Clark M.B. and Mattick J.S. (2011) Long noncoding RNAs in cell biology. *Seminars in Cell & Developmental Biology* 22: 366–76.

361. Carrieri C. et al. (2012) Long non-coding antisense RNA controls Uchl1 translation through an embedded SINEB2 repeat. *Nature* 491: 454–7.

362. Cabili M.N. et al. (2015) Localization and abundance analysis of human lncRNAs at single-cell and single-molecule resolution. *Genome Biology* 16: 20.

363. Wilk R., Hu J., Blotsky D. and Krause H.M. (2016) Diverse and pervasive subcellular distributions for both coding and long noncoding RNAs. *Genes & Development* 30: 594–609.

364. Chen L.-L. (2016) Linking long noncoding RNA localization and function. *Trends in Biochemical Sciences* 41: 761–72.

365. Mas-Ponte D. et al. (2017) LncATLAS database for subcellular localization of long noncoding RNAs. *RNA* 23: 1080–7.

366. Wen X. et al. (2018) lncSLdb: A resource for long noncoding RNA subcellular localization. *Database* 2018: bay085.

367. Benoit Bouvrette L.P. et al. (2018) CeFra-seq reveals broad asymmetric mRNA and noncoding RNA distribution profiles in *Drosophila* and human cells. *RNA* 24: 98–113.

368. Palazzo A.F. and Lee E.S. (2018) Sequence determinants for nuclear retention and cytoplasmic export of mRNAs and lncRNAs. *Frontiers in Genetics* 9: 440.

369. Zhao Y. et al. (2019) Aberrant shuttling of long noncoding RNAs during the mitochondria-nuclear crosstalk in hepatocellular carcinoma cells. *American Journal of Cancer Research* 9: 999–1008.

370. Miao H. et al. (2019) A long noncoding RNA distributed in both nucleus and cytoplasm operates in the PYCARD-regulated apoptosis by coordinating the epigenetic and translational regulation. *PLOS Genetics* 15: e1008144.

371. Guo C.-J. et al. (2020) Distinct processing of lncRNAs contributes to non-conserved functions in stem cells. *Cell* 181: 621–36.

372. Bridges M.C., Daulagala A.C. and Kourtidis A. (2021) LNCcation: lncRNA localization and function. *Journal of Cell Biology* 220: e202009045.

373. Tani H. et al. (2012) Genome-wide determination of RNA stability reveals hundreds of short-lived noncoding transcripts in mammals. *Genome Research* 22: 947–56.

374. Shi K., Liu T., Fu H., Li W. and Zheng X. (2021) Genome-wide analysis of lncRNA stability in human. *PLOS Computational Biology* 17: e1008918.

375. Mattick J.S. (2009) The gene377. Cawley Stic signatures of noncoding RNAs. *PLOS Genetics* 5: e1000459.

376. Martone R. et al. (2003) Distribution of NF-kappaB-binding sites across human chromosome 22. *Proceedings of the National Academy of Sciences USA* 100: 12247–52.

377. Cawley S. et al. (2004) Unbiased mapping of transcription factor binding sites along human chromosomes 21 and 22 points to widespread regulation of noncoding RNAs. *Cell* 116: 499–509.

378. Euskirchen G. et al. (2004) CREB binds to multiple loci on human chromosome 22. *Molecular Cell Biology* 24: 3804–14.

379. Kim T.H. et al. (2005) A high-resolution map of active promoters in the human genome. *Nature* 436: 876–80.

380. Odom D.T. et al. (2006) Core transcriptional regulatory circuitry in human hepatocytes. *Molecular Systems Biology* 2: 2006.0017.

381. Louro R. et al. (2007) Androgen responsive intronic noncoding RNAs. *BMC Biology* 5: 4.

382. Louro R., Smirnova A.S. and Verjovski-Almeida S. (2009) Long intronic noncoding RNA transcription: Expression noise or expression choice? *Genomics* 93: 291–8.

383. Lopez F., Granjeaud S., Ara T., Ghattas B. and Gautheret D. (2006) The disparate nature of "intergenic" polyadenylation sites. *RNA* 12: 1794–801.

384. Ponjavic J., Ponting C.P. and Lunter G. (2007) Functionality or transcriptional noise? Evidence for selection within long noncoding RNAs. *Genome Research* 17: 556–65.

385. Guttman M. et al. (2009) Chromatin signature reveals over a thousand highly conserved large non-coding RNAs in mammals. *Nature* 458: 223–7.

386. Smith M.A., Gesell T., Stadler P.F. and Mattick J.S. (2013) Widespread purifying selection on RNA structure in mammals. *Nucleic Acids Research* 41: 8220–36.

387. Hiller M. et al. (2009) Conserved introns reveal novel transcripts in *Drosophila melanogaster. Genome Research* 19: 1289–300.

388. Will S., Yu M. and Berger B. (2013) Structure-based whole genome realignment reveals many novel non-coding RNAs. *Genome Research* 6: 1018–27.

389. Nitsche A., Rose D., Fasold M., Reiche K. and Stadler P.F. (2015) Comparison of splice sites reveals that long noncoding RNAs are evolutionarily well conserved. *RNA* 21: 801–12.

390. Vancura A. et al. (2021) Cancer LncRNA Census 2 (CLC2): An enhanced resource reveals clinical features of cancer lncRNAs. *NAR Cancer* 3: zcab013.

391. Chodroff R.A. et al. (2010) Long noncoding RNA genes: Conservation of sequence and brain expression among diverse amniotes. *Genome Biology* 11: R72.

392. Hezroni H. et al. (2015) Principles of long noncoding RNA evolution derived from direct comparison of transcriptomes in 17 species. *Cell Reports* 11: 1110–22.

393. Amaral P.P. et al. (2018) Genomic positional conservation identifies topological anchor point RNAs linked to developmental loci. *Genome Biology* 19: 32.

394. Washietl S., Hofacker I.L., Lukasser M., Huttenhofer A. and Stadler P.F. (2005) Mapping of conserved RNA secondary structures predicts thousands of functional noncoding RNAs in the human genome. *Nature Biotechnology* 23: 1383–90.

395. Torarinsson E., Sawera M., Havgaard J.H., Fredholm M. and Gorodkin J. (2006) Thousands of corresponding human and mouse genomic regions unalignable in primary sequence contain common RNA structure. *Genome Research* 16: 885–9.

396. Torarinsson E. et al. (2008) Comparative genomics beyond sequence-based alignments: RNA structures in the ENCODE regions. *Genome Research* 18: 242–51.

397. Vandivier L.E., Anderson S.J., Foley S.W. and Gregory B.D. (2016) The conservation and function of RNA secondary structure in plants. *Annual Review of Plant Biology* 67: 463–88.

398. Delli Ponti R., Armaos A., Marti S. and Tartaglia G.G. (2018) A method for RNA structure prediction shows evidence for structure in lncRNAs. *Frontiers in Molecular Biosciences* 5: 111.

399. Steigele S., Huber W., Stocsits C., Stadler P.F. and Nieselt K. (2007) Comparative analysis of structured RNAs in *S. cerevisiae* indicates a multitude of different functions. *BMC Biology* 5: 25.

400. Jung C.H., Makunin I.V. and Mattick J.S. (2010) Identification of conserved Drosophila-specific euchromatin-restricted non-coding sequence motifs. *Genomics* 96: 154–66.

401. Ross C.J. et al. (2021) Uncovering deeply conserved motif combinations in rapidly evolving noncoding sequences. *Genome Biology* 22: 29.

402. Pang K.C., Frith M.C. and Mattick J.S. (2006) Rapid evolution of noncoding RNAs: Lack of conservation does not mean lack of function. *Trends in Genetics* 22: 1–5.

403. Quinn J.J. et al. (2016) Rapid evolutionary turnover underlies conserved lncRNA-genome interactions. *Genes & Development* 30: 191–207.

404. Karner H. et al. (2020) Functional conservation of lncRNA *JPX* despite sequence and structural divergence. *Journal of Molecular Biology* 432: 283–300.

405. Oh H.J. and Lee J.T. (2020) Long noncoding RNA functionality beyond sequence: The *Jpx* model. *Journal of Molecular Biology* 432: 301–4.

406. Lyle R. et al. (2000) The imprinted antisense RNA at the *Igf2r* locus overlaps but does not imprint *Mas1*. *Nature Genetics* 25: 19–21.

407. Millar J.K. et al. (2000) Disruption of two novel genes by a translocation co-segregating with schizophrenia. *Human Molecular Genetics* 9: 1415–23.

408. Liu A.Y., Torchia B.S., Migeon B.R. and Siliciano R.F. (1997) The human *NTT* gene: Identification of a novel 17-kb noncoding nuclear RNA expressed in activated CD4+ T cells. *Genomics* 39: 171–84.

409. Takeda K. et al. (1998) Identification of a novel bone morphogenetic protein-responsive gene that may function as a noncoding RNA. *Journal of Biological Chemistry* 273: 17079–85.

410. Michel U., Kallmann B., Rieckmann P. and Isbrandt D. (2002) UM 9(5)h and UM 9(5)p, human and porcine noncoding transcripts with preferential expression in the cerebellum. *RNA* 8: 1538–47.

411. Ulitsky I., Shkumatava A., Jan C.H., Sive H. and Bartel D.P. (2011) Conserved function of lincRNAs in vertebrate embryonic development despite rapid sequence evolution. *Cell* 147: 1537–50.

412. Pheasant M. and Mattick J.S. (2007) Raising the estimate of functional human sequences. *Genome Research* 17: 1245–53.

413. Smith N.G., Brandstrom M. and Ellegren H. (2004) Evidence for turnover of functional noncoding DNA in mammalian genome evolution. *Genomics* 84: 806–13.

414. Kutter C. et al. (2012) Rapid turnover of long noncoding RNAs and the evolution of gene expression. *PLOS Genetics* 8: e1002841.

415. Dai H. et al. (2008) The evolution of courtship behaviors through the origination of a new gene in *Drosophila*. *Proceedings of the National Academy of Sciences USA* 105: 7478–83.

416. Paralkar V.R. et al. (2014) Lineage and species-specific long noncoding RNAs during erythro-megakaryocytic development. *Blood* 123: 1927–37.

417. de Almeida R.A., Fraczek M.G., Parker S., Delneri D. and O'Keefe R.T. (2016) Non-coding RNAs and disease: The classical ncRNAs make a comeback. *Biochemical Society Transactions* 44: 1073–8.

418. Sulisalo T. et al. (1993) Cartilage-hair hypoplasia gene assigned to chromosome 9 by linkage analysis. *Nature Genetics* 3: 338–41.

419. Ridanpää M. et al. (2001) Mutations in the RNA component of RNase MRP cause a pleiotropic human disease, cartilage-hair hypoplasia. *Cell* 104: 195–203.

420. Rogler L.E. et al. (2014) Small RNAs derived from lncRNA RNase MRP have gene-silencing activity relevant to human cartilage-hair hypoplasia. *Human Molecular Genetics* 23: 368–82.

421. Huang W. et al. (2015) DDX5 and its associated lncRNA Rmrp modulate TH17 cell effector functions. *Nature* 528: 517–22.

422. Chen Y. et al. (2021) Inactivation of the tumor suppressor p53 by long noncoding RNA RMRP. *Proceedings of the National Academy of Sciences USA* 118: e2026813118.

423. Murdock D.R. et al. (2021) Transcriptome-directed analysis for Mendelian disease diagnosis overcomes limitations of conventional genomic testing. *Journal of Clinical Investigation* 131: e141500.

424. Stenson P.D. et al. (2017) The human gene mutation database: Towards a comprehensive repository of inherited mutation data for medical research, genetic diagnosis and next-generation sequencing studies. *Human Genetics* 136: 665–77.

425. Broadbent H.M. et al. (2007) Susceptibility to coronary artery disease and diabetes is encoded by distinct, tightly linked, SNPs in the ANRIL locus on chromosome 9p. *Human Molecular Genetics* 17: 806–14.

426. Pasmant E., Sabbagh A., Vidaud M. and Bièche I. (2010) ANRIL, a long, noncoding RNA, is an unexpected major hotspot in GWAS. *FASEB Journal* 25: 444–8.

427. Tan J.Y. et al. (2017) Cis-acting complex-trait-associated lincRNA expression correlates with modulation of chromosomal architecture. *Cell Reports* 18: 2280–8.

428. Zhang D.-D. et al. (2017) Long noncoding RNA LINC00305 promotes inflammation by activating the AHRR-NF-κB pathway in human monocytes. *Scientific Reports* 7: 46204.

429. Cho H. et al. (2019) Long noncoding RNA *ANRIL* regulates endothelial cell activities associated with coronary artery disease by up-regulating *CLIP1*, *EZR*, and *LYVE1* genes. *Journal of Biological Chemistry* 294: 3881–98.

430. Allou L. et al. (2021) Non-coding deletions identify *Maenli* lncRNA as a limb-specific En1 regulator. *Nature* 592: 93–8.

431. Seifuddin F. et al. (2020) lncRNAKB, a knowledgebase of tissue-specific functional annotation and trait association of long noncoding RNA. *Scientific Data* 7: 326.

432. Asgari Y., Heng J.I.T., Lovell N., Forrest A.R.R. and Alinejad-Rokny H. (2020) Evidence for enhancer noncoding RNAs (enhancer-ncRNAs) with gene regulatory functions relevant to neurodevelopmental disorders. *bioRxiv*: 2020.05.16.087395v2.

433. Kelley R.L. et al. (1999) Epigenetic spreading of the *Drosophila* dosage compensation complex from *roX* RNA genes into flanking chromatin. *Cell* 98: 513–22.

434. Meller V.H. and Rattner B.P. (2002) The *roX* genes encode redundant *male-specific lethal* transcripts required for targeting of the MSL complex. *EMBO Journal* 21: 1084–91.

435. Georges M., Charlier C. and Cockett N. (2003) The callipyge locus: Evidence for the trans interaction of reciprocally imprinted genes. *Trends in Genetics* 19: 248–52.

436. Davis E. et al. (2005) RNAi-mediated allelic *trans*-Interaction at the imprinted *Rtl1/Peg11* locus. *Current Biology* 15: 743–9.

437. Takeda H. et al. (2006) The callipyge mutation enhances bidirectional long-range *DLK1-GTL2* intergenic transcription in *cis*. *Proceedings of the National Academy of Sciences USA* 103: 8119–24.

438. Mikovic J. et al. (2018) MicroRNA and long non-coding RNA regulation in skeletal muscle from growth to old age shows striking dysregulation of the *callipyge* locus. *Frontiers in Genetics* 9: 548.

439. Van Laere A.S. et al. (2003) A regulatory mutation in IGF2 causes a major QTL effect on muscle growth in the pig. *Nature* 425: 832–6.

440. Li J. et al. (2021) The crest phenotype in domestic chicken is caused by a 197 bp duplication in the intron of HOXC10. *G3 (Genes, Genomes, Genetics)* 11: jkaa048.

441. Alberti C. and Cochella L. (2017) A framework for understanding the roles of miRNAs in animal development. *Development* 144: 2548–59.

442. Miska E.A. et al. (2007) Most *Caenorhabditis elegans* microRNAs are individually not essential for development or viability. *PLOS Genetics* 3: e215.

443. Soshnev A.A. et al. (2011) A conserved long noncoding RNA affects sleep behavior in *Drosophila*. *Genetics* 189: 455–68.

444. Jenny A. et al. (2006) A translation-independent role of oskar RNA in early *Drosophila* oogenesis. *Development* 133: 2827–33.

445. Kenny A., Morgan M.B. and Macdonald P.M. (2021) Different roles for the adjoining and structurally similar A-rich and poly(A) domains of oskar mRNA: Only the A-rich domain is required for oskar noncoding RNA function, which includes MTOC positioning. *Developmental Biology* 476: 117–27.

446. Taft R.J., Pang K.C., Mercer T.R., Dinger M. and Mattick J.S. (2009) Non-coding RNAs: Regulators of disease. *Journal of Pathology* 220: 126–39.

447. Wapinski O. and Chang H.Y. (2011) Long noncoding RNAs and human disease. *Trends in Cell Biology* 21: 354–61.

448. Wan P., Su W. and Zhuo Y. (2017) The role of long non-coding RNAs in neurodegenerative diseases. *Molecular Neurobiology* 54: 2012–21.

449. Wanowska E., Kubiak M.R., Rosikiewicz W., Makałowska I. and Szcześniak M.W. (2018) Natural antisense transcripts in diseases: From modes of action to targeted therapies. *Wiley Interdisciplinary Reviews RNA* 9: e1461.

450. Aznaourova M., Schmerer N., Schmeck B. and Schulte L.N. (2020) Disease-causing mutations and rearrangements in long non-coding RNA gene loci. *Frontiers in Genetics* 11: 527484.

451. Rom A. et al. (2019) Regulation of CHD2 expression by the *Chaserr* long noncoding RNA gene is essential for viability. *Nature Communications* 10: 5092.

452. Sutherland H.F. et al. (1996) Identification of a novel transcript disrupted by a balanced translocation associated with DiGeorge syndrome. *American Journal of Human Genetics* 59: 23–31.

453. Abe Y. et al. (2014) Xq26.1–26.2 gain identified on array comparative genomic hybridization in bilateral periventricular nodular heterotopia with overlying polymicrogyria. *Developmental Medicine & Child Neurology* 56: 1221–4.

454. Ang C.E. et al. (2019) The novel lncRNA lnc-NR2F1 is pro-neurogenic and mutated in human neurodevelopmental disorders. *eLife* 8: e41770.

455. Long H.K. et al. (2020) Loss of extreme long-range enhancers in human neural crest drives a craniofacial disorder. *Cell Stem Cell* 27: 765–83.

456. Maass P.G. et al. (2012) A misplaced lncRNA causes brachydactyly in humans. *Journal of Clinical Investigation* 122: 3990–4002.

457. Shepherdson J.L., Zheng H., Amarillo I.E., McAlinden A. and Shinawi M. (2021) Delineation of the 1q24.3 microdeletion syndrome provides further evidence for the potential role of non-coding RNAs in regulating the skeletal phenotype. *Bone* 142: 115705.

458. Yu T.-T. et al. (2021) Deletion at an 1q24 locus reveals a critical role of long noncoding RNA DNM3OS in skeletal development. *Cell & Bioscience* 11: 47.

459. Meng L. et al. (2015) Towards a therapy for Angelman syndrome by targeting a long non-coding RNA. *Nature* 518: 409–12.

460. Rougeulle C., Cardoso C., Fontes M., Colleaux L. and Lalande M. (1998) An imprinted antisense RNA overlaps *UBE3A* and a second maternally expressed transcript. *Nature Genetics* 19: 15–6.

461. Jong M.T. et al. (1999) A novel imprinted gene, encoding a RING zinc-finger protein, and overlapping antisense transcript in the Prader-Willi syndrome critical region. *Human Molecular Genetics* 8: 783–93.

462. Royo H. et al. (2007) Bsr, a nuclear-retained RNA with monoallelic expression. *Molecular Biology of the Cell* 18: 2817–27.

463. Vitali P., Royo H., Marty V., Bortolin-Cavaille M.L. and Cavaille J. (2010) Long nuclear-retained non-coding RNAs and allele-specific higher-order chromatin organization at imprinted snoRNA gene arrays. *Journal of Cell Science* 123: 70–83.

464. Polesskaya O.O., Haroutunian V., Davis K.L., Hernandez I. and Sokolov B.P. (2003) Novel putative nonprotein-coding RNA gene from 11q14 displays decreased expression in brains of patients with schizophrenia. *Journal of Neuroscience Research* 74: 111–22.

465. Barry G. et al. (2014) The long non-coding RNA *Gomafu* is acutely regulated in response to neuronal activation and involved in schizophrenia-associated alternative splicing. *Molecular Psychiatry* 19: 486–94.

466. Ip J.Y. et al. (2016) Gomafu lncRNA knockout mice exhibit mild hyperactivity with enhanced responsiveness to the psychostimulant methamphetamine. *Scientific Reports* 6: 27204.

467. Stamou M. et al. (2020) A balanced translocation in Kallmann syndrome implicates a long noncoding RNA, *RMST*, as a GnRH neuronal regulator. *Journal of Clinical Endocrinology & Metabolism* 105: e231–44.

468. Chillambhi S. et al. (2010) Deletion of the noncoding *GNAS* Antisense transcript causes pseudohypoparathyroidism type ib and biparental defects of *GNAS* methylation in cis. *Journal of Clinical Endocrinology & Metabolism* 95: 3993–4002.

469. Bohnsack J.P., Teppen T., Kyzar E.J., Dzitoyeva S. and Pandey S.C. (2019) The lncRNA BDNF-AS is an epigenetic regulator in the human amygdala in early on*set al*cohol use disorders. *Translational Psychiatry* 9: 34.

470. Ruan X. et al. (2021) Identification of human long noncoding RNAs associated with nonalcoholic fatty liver disease and metabolic homeostasis. *Journal of Clinical investigation* 131: e136336.

471. Zemmour D., Pratama A., Loughhead S.M., Mathis D. and Benoist C. (2017) Flicr, a long noncoding RNA, modulates Foxp3 expression and autoimmunity. *Proceedings of the National Academy of Sciences USA* 114: E3472–80.

472. Zhang F., Liu G., Wei C., Gao C. and Hao J. (2017) Linc-MAF-4 regulates Th1/Th2 differentiation and is associated with the pathogenesis of multiple sclerosis by targeting MAF. *FASEB Journal* 31: 519–25.

473. Shirasawa S. et al. (2004) SNPs in the promoter of a B cell-specific antisense transcript, SAS-ZFAT, determine susceptibility to autoimmune thyroid disease. *Human Molecular Genetics* 13: 2221–31.

474. Wang J. et al. (2016) Upregulation of long noncoding RNA TMEVPG1 enhances T helper type 1 cell response in patients with Sjögren syndrome. *Immunologic Research* 64: 489–96.

475. Castellanos-Rubio A. et al. (2016) A long noncoding RNA associated with susceptibility to celiac disease. *Science* 352: 91–5.

476. Kotzin J.J. et al. (2016) The long non-coding RNA Morrbid regulates Bim and short-lived myeloid cell lifespan. *Nature* 537: 239–43.

477. Sonkoly E. et al. (2005) Identification and characterization of a novel, psoriasis susceptibility-related noncoding RNA gene, PRINS. *Journal of Biological Chemistry* 280: 24159–67.

478. Széll M., Danis J., Bata-Csörgő Z. and Kemény L. (2016) PRINS, a primate-specific long non-coding RNA, plays a role in the keratinocyte stress response and psoriasis pathogenesis. *European Journal of Physiology* 468: 935–43.

479. Padua D. et al. (2016) A long noncoding RNA signature for ulcerative colitis identifies IFNG-AS1 as an enhancer of inflammation. *American Journal of Physiology-Gastrointestinal and Liver Physiology* 311: G446–G57.

480. Hu Y.-W. et al. (2019) Long noncoding RNA NEXN-AS1 mitigates atherosclerosis by regulating the actin-binding protein NEXN. *Journal of Clinical Investigation* 129: 1115–28.

481. Ou M., Li X., Zhao S., Cui S. and Tu J. (2020) Long noncoding RNA CDKN2B-AS1 contributes to atherosclerotic plaque formation by forming RNA-DNA triplex in the CDKN2B promoter. *EBioMedicine* 55: 102694.

482. Han P. et al. (2014) A long noncoding RNA protects the heart from pathological hypertrophy. *Nature* 514: 102–6.

483. Faghihi M.A. et al. (2008) Expression of a noncoding RNA is elevated in Alzheimer's disease and drives rapid feed-forward regulation of beta-secretase. *Nature Medicine* 14: 723–30.

484. Mutsuddi M., Marshall C.M., Benzow K.A., Koob M.D. and Rebay I. (2004) The spinocerebellar ataxia 8 noncoding RNA causes neurodegeneration and associates with staufen in *Drosophila*. *Current Biology* 14: 302–8.

485. Sopher B.L. et al. (2011) CTCF regulates ataxin-7 expression through promotion of a convergently transcribed, antisense noncoding RNA. *Neuron* 70: 1071–84.

486. Ishii N. et al. (2006) Identification of a novel non-coding RNA, MIAT, that confers risk of myocardial infarction. *Journal of Human Genetics* 51: 1087–99.

487. Tufarelli C. et al. (2003) Transcription of antisense RNA leading to gene silencing and methylation as a novel cause of human genetic disease. *Nature Genetics* 34: 157–65.

488. Giannopoulou E. et al. (2012) A single nucleotide polymorphism in the *HbbP1* gene in the human β-globin locus is associated with a mild β-thalassemia disease phenotype. *Hemoglobin* 36: 433–45.

489. Li Y. et al. (2021) A noncoding RNA modulator potentiates phenylalanine metabolism in mice. *Science* 373: 662–73.

490. He F. et al. (2019) Integrative analysis of somatic mutations in non-coding regions altering RNA secondary structures in cancer genomes. *Scientific Reports* 9: 8205.

491. Huarte M. (2015) The emerging role of lncRNAs in cancer. *Nature Medicine* 21: 1253–61.

492. Wang Z. et al. (2018) lncRNA epigenetic landscape analysis identifies EPIC1 as an oncogenic lncRNA that interacts with MYC and promotes cell-cycle progression in cancer. *Cancer Cell* 33: 706–20.

493. Jiang M.-C., Ni J.-J., Cui W.-Y., Wang B.-Y. and Zhuo W. (2019) Emerging roles of lncRNA in cancer and therapeutic opportunities. *American Journal of Cancer Research* 9: 1354–66.

494. Carlevaro-Fita J. et al. (2020) Cancer LncRNA census reveals evidence for deep functional conservation of long noncoding RNAs in tumorigenesis. *Communications Biology* 3: 56.

495. Guo F., Li L., Yang W., Hu J.-F. and Cui J. (2021) Long noncoding RNA: A resident staff of genomic instability regulation in tumorigenesis. *Cancer Letters* 503: 103–9.

496. Liu S.J., Dang H.X., Lim D.A., Feng F.Y. and Maher C.A. (2021) Long noncoding RNAs in cancer metastasis. *Nature Reviews Cancer* 21: 446–60.

497. Landskron L. et al. (2018) The asymmetrically segregating lncRNA cherub is required for transforming stem cells into malignant cells. *eLife* 7: e31347.

498. Young T.L., Matsuda T. and Cepko C.L. (2005) The noncoding RNA taurine upregulated gene 1 is required for differentiation of the murine retina. *Current Biology* 15: 501–12.

499. Ji P. et al. (2003) MALAT-1, a novel noncoding RNA, and thymosin beta4 predict metastasis and survival in early-stage non-small cell lung cancer. *Oncogene* 22: 6087–97.

500. Berteaux N. et al. (2005) H19 mRNA-like noncoding RNA promotes breast cancer cell proliferation through positive control by E2F1. *Journal of Biological Chemistry* 280: 29625–36.

501. Reis E.M. et al. (2005) Large-scale transcriptome analyses reveal new genetic marker candidates of head, neck, and thyroid cancer. *Cancer Research* 65: 1693–9.

502. Lottin S. et al. (2005) The human H19 gene is frequently overexpressed in myometrium and stroma during pathological endometrial proliferative events. *European Journal of Cancer* 41: 168–77.

503. Barsyte-Lovejoy D. et al. (2006) The c-Myc oncogene directly induces the H19 noncoding RNA by allele-specific binding to potentiate tumorigenesis. *Cancer Research* 66: 5330–7.

504. Matouk I.J. et al. (2007) The H19 non-coding RNA is essential for human tumor growth. *PLOS ONE* 2: e845.

505. Perez D.S. et al. (2008) Long, abundantly expressed non-coding transcripts are altered in cancer. *Human Molecular Genetics* 17: 642–55.

506. Yoshimizu T. et al. (2008) The H19 locus acts in vivo as a tumor suppressor. *Proceedings of the National Academy of Sciences USA* 105: 12417–22.

507. Luo M. et al. (2013) Long non-coding RNA H19 increases bladder cancer metastasis by associating with EZH2 and inhibiting E-cadherin expression. *Cancer Letters* 333: 213–21.

508. Tseng Y.-Y. et al. (2014) *PVT1* dependence in cancer with *MYC* copy-number increase. *Nature* 512: 82–6.

509. Prensner J.R. et al. (2014) *PCAT-1*: A long noncoding RNA, regulates BRCA2 and controls homologous recombination in cancer. *Cancer Research* 74: 1651–60.

510. Iyer M.K. et al. (2015) The landscape of long noncoding RNAs in the human transcriptome. *Nature Genetics* 47: 199–208.

511. Adriaens C. et al. (2016) p53 induces formation of *NEAT1* lncRNA-containing paraspeckles that modulate replication stress response and chemosensitivity. *Nature Medicine* 22: 861–8.

512. Liu Z. et al. (2018) Long non-coding RNA *MIAT* promotes growth and metastasis of colorectal cancer cells through regulation of miR-132/Derlin-1 pathway. *Cancer Cell International* 18: 59.

513. Alipoor F.J., Asadi M.H. and Torkzadeh-Mahani M. (2018) *MIAT* lncRNA is overexpressed in breast cancer and its inhibition triggers senescence and G1 arrest in MCF7 cell line. *Journal of Cellular Biochemistry* 119: 6470–81.

514. Shin V.Y. et al. (2019) Long non-coding RNA *NEAT1* confers oncogenic role in triple-negative breast cancer through modulating chemoresistance and cancer stemness. *Cell Death & Disease* 10: 270.

515. Dong P. et al. (2019) Long noncoding RNA *NEAT1* drives aggressive endometrial cancer progression via miR-361-regulated networks involving STAT3 and tumor microenvironment-related genes. *Journal of Experimental & Clinical Cancer Research* 38: 295.

516. Chiu H.-S. et al. (2018) Pan-cancer analysis of lncRNA regulation supports their targeting of cancer genes in each tumor context. *Cell Reports* 23: 297–312.

517. Thapar R. et al. (2020) Mechanism of efficient double-strand break repair by a long non-coding RNA. *Nucleic Acids Research* 48: 10953–72.

518. Tolomeo D., Agostini A., Visci G., Traversa D. and Tiziana Storlazzi C. (2021) PVT1: A long non-coding RNA recurrently involved in neoplasia-associated fusion transcripts. *Gene* 779: 145497.

519. Zhang J. et al. (2019) ALKBH5 promotes invasion and metastasis of gastric cancer by decreasing methylation of the lncRNA NEAT1. *Journal of Physiology and Biochemistry* 75: 379–89.

520. Saitou M., Sugimoto J., Hatakeyama T., Russo G. and Isobe M. (2000) Identification of the TCL6 genes within the breakpoint cluster region on chromosome 14q32 in T-cell leukemia. *Oncogene* 19: 2796–802.

521. Tanaka R. et al. (2000) Intronic U50 small-nucleolar-RNA (snoRNA) host gene of no protein-coding potential is mapped at the chromosome breakpoint t(3;6)(q27;q15) of human B-cell lymphoma. *Genes to Cells* 5: 277–87.

522. Karreth F.A. et al. (2015) The *BRAF* pseudogene functions as a competitive endogenous RNA and induces lymphoma in vivo. *Cell* 161: 319–32.

523. Fatima R., Choudhury S.R., Divya T. R., Bhaduri U. and Rao M.R.S. (2019) A novel enhancer RNA, *Hmrhl*, positively regulates its host gene, *phkb*, in chronic myelogenous leukemia. *Noncoding RNA Research* 4: 96–108.

524. Rahman S. and Mansour M.R. (2019) The role of non-coding mutations in blood cancers. *Disease Models & Mechanisms* 12: dmm041988.

525. Napoli S. et al. (2021) Characterization of GECPAR, a noncoding RNA that regulates the transcriptional program of diffuse large B cell lymphoma. *Haematologica.* epub ahead of print: https://haematologica.org/article/view/haematol.2020.267096.

526. Leucci E. et al. (2016) Melanoma addiction to the long non-coding RNA SAMMSON. *Nature* 531: 518–22.

527. Hanniford D. et al. (2020) Epigenetic silencing of *CDR1as* drives IGF2BP3-mediated melanoma invasion and metastasis. *Cancer Cell* 37: 55–70.

528. Chen S., Zhou L. and Wang Y. (2020) ALKBH5-mediated m⁶A demethylation of lncRNA PVT1 plays an oncogenic role in osteosarcoma. *Cancer Cell International* 20: 34.

529. Du T. et al. (2016) Decreased expression of long non-coding RNA *WT1-AS* promotes cell proliferation and invasion in gastric cancer. *Biochimica et Biophysica Acta* 1862: 12–9.

530. Zhang S., Guan Y., Liu X., Ju M. and Zhang Q. (2019) Long non-coding RNA *DLEU1* exerts an oncogenic function in non-small cell lung cancer. *Biomedicine & Pharmacotherapy* 109: 985–90.

531. Zhang Y. et al. (2016) Long noncoding RNA *LINP1* regulates repair of DNA double-strand breaks in triple-negative breast cancer. *Nature Structural & Molecular Biology* 23: 522–30.

532. Marjaneh M.M. et al. (2020) Non-coding RNAs underlie genetic predisposition to breast cancer. *Genome Biology* 21: 7.

533. Salameh A. et al. (2015) PRUNE2 is a human prostate cancer suppressor regulated by the intronic long noncoding RNA PCA3. *Proceedings of the National Academy of Sciences USA* 112: 8403–8.

534. Guo H. et al. (2016) Modulation of long noncoding RNAs by risk SNPs underlying genetic predispositions to prostate cancer. *Nature Genetics* 48: 1142–50.

535. Hua J.T. et al. (2018) Risk SNP-mediated promoter-enhancer switching drives prostate cancer through lncRNA *PCAT19*. *Cell* 174: 564–75.

536. Xiong T. et al. (2020) LncRNA *NRON* promotes the proliferation, metastasis and EMT process in bladder cancer. *Journal of Cancer* 11: 1751–60.

537. Li C. et al. (2017) A ROR1–HER3–lncRNA signalling axis modulates the Hippo–YAP pathway to regulate bone metastasis. *Nature Cell Biology* 19: 106–19.

538. Schmitt A.M. and Chang H.Y. (2016) Long noncoding RNAs in cancer pathways. *Cancer Cell* 29: 452–63.

539. Slack F.J. and Chinnaiyan A.M. (2019) The role of non-coding RNAs in oncology. *Cell* 179: 1033–55.

540. Andergassen D. and Rinn J.L. (2021) From genotype to phenotype: Genetics of mammalian long non-coding RNAs in vivo. *Nature Reviews Genetics* 23: 229–43.

541. Ashburner M. et al. (1999) An exploration of the sequence of a 2.9-Mb region of the genome of *Drosophila melanogaster*: The *Adh* region. *Genetics* 153: 179–219.

542. The *C. elegans* Sequencing Consortium (1998) Genome sequence of the nematode *C. elegans*: A platform for investigating biology. *Science* 282: 2012–8.

543. Lek M. et al. (2016) Analysis of protein-coding genetic variation in 60,706 humans. *Nature* 536: 285–91.

544. Hayden E.C. (2016) A radical revision of human genetics: Why many 'deadly' gene mutations are turning out to be harmless. *Nature* 538: 154–7.

545. Lewejohann L. et al. (2004) Role of a neuronal small non-messenger RNA: Behavioural alterations in BC1 RNA-deleted mice. *Behavioural Brain Research* 154: 273–89.

546. Briz V. et al. (2017) The non-coding RNA BC1 regulates experience-dependent structural plasticity and learning. *Nature Communications* 8: 293.

547. Nakagawa S., Naganuma T., Shioi G. and Hirose T. (2011) Paraspeckles are subpopulation-specific nuclear bodies that are not essential in mice. *Journal of Cell Biology* 193: 31–9.

548. Nakagawa S. et al. (2012) *Malat1* is not an essential component of nuclear speckles in mice. *RNA* 18: 1487–99.

549. Nakagawa S. et al. (2014) The lncRNA *Neat1* is required for corpus luteum formation and the establishment of pregnancy in a subpopulation of mice. *Development* 141: 4618–27.

550. Kukharsky M.S. et al. (2020) Long non-coding RNA *Neat1* regulates adaptive behavioural response to stress in mice. *Translational Psychiatry* 10: 171.

551. Gutschner T., Hämmerle M. and Diederichs S. (2013) MALAT1 — a paradigm for long noncoding RNA function in cancer. *Journal of Molecular Medicine* 91: 791–801.

552. Zhang X., Hamblin M.H. and Yin K.-J. (2017) The long noncoding RNA Malat1: Its physiological and pathophysiological functions. *RNA Biology* 14: 1705–14.

553. Ji Q. et al. (2019) MALAT1 regulates the transcriptional and translational levels of proto-oncogene RUNX2 in colorectal cancer metastasis. *Cell Death & Disease* 10: 378.

554. Mehta S.L., Kim T. and Vemuganti R. (2015) Long noncoding RNA FosDT promotes ischemic brain injury by interacting with REST-associated chromatin-modifying proteins. *Journal of Neuroscience* 35: 16443–9.

555. Mehta S.L. et al. (2021) Long noncoding RNA *Fos Downstream Transcript* is developmentally dispensable but vital for shaping the poststroke functional outcome. *Stroke* 52: 2381–92.

556. Ramos A.D. et al. (2015) The long noncoding RNA Pnky regulates neuronal differentiation of embryonic and postnatal neural stem cells. *Cell Stem Cell* 16: 439–47.

557. Andersen R.E. et al. (2019) The long noncoding RNA Pnky is a trans-acting regulator of cortical development in vivo. *Developmental Cell* 49: 632–42.

558. Sauvageau M. et al. (2013) Multiple knockout mouse models reveal lincRNAs are required for life and brain development. *eLife* 2: e01749.

559. Lai K.-M.V. et al. (2015) Diverse phenotypes and specific transcription patterns in twenty mouse lines with ablated lincRNAs. *PLOS ONE* 10: e0125522.

560. Ramilowski J.A. et al. (2020) Functional annotation of human long noncoding RNAs via molecular phenotyping. *Genome Research* 30: 1060–72.

561. Cao H. et al. (2021) Very long intergenic non-coding (vlinc) RNAs directly regulate multiple genes in cis and trans. *BMC Biology* 19: 108.

562. Han X. et al. (2018) Mouse knockout models reveal largely dispensable but context-dependent functions of lncRNAs during development. *Journal of Molecular Cell Biology* 10: 175–8.

563. Willingham A.T. et al. (2005) A strategy for probing the function of noncoding RNAs finds a repressor of NFAT. *Science* 309: 1570–3.

564. Joung J. et al. (2017) Genome-scale activation screen identifies a lncRNA locus regulating a gene neighbourhood. *Nature* 548: 343–6.

565. Liu S. et al. (2020) Wnt-regulated lncRNA discovery enhanced by *in vivo* identification and CRISPRi functional validation. *Genome Medicine* 12: 89.

566. Liu S.J. et al. (2020) CRISPRi-based radiation modifier screen identifies long non-coding RNA therapeutic targets in glioma. *Genome Biology* 21: 83.

567. Zhao W. et al. (2016) The long noncoding RNA *Sprightly* regulates cell proliferation in primary human melanocytes. *Journal of Investigative Dermatology* 136: 819–28.

568. Wen K. et al. (2016) Critical roles of long noncoding RNAs in *Drosophila* spermatogenesis. *Genome Research* 26: 1233–44.

569. Esposito R. et al. (2021) Multi-hallmark long noncoding RNA maps reveal non-small cell lung cancer vulnerabilities. *bioRxiv*: 2021.10.19.464956.

570. Liu Y. et al. (2018) Genome-wide screening for functional long noncoding RNAs in human cells by Cas9 targeting of splice sites. *Nature Biotechnology* 36: 1203–10.

571. Liu S.J. et al. (2017) CRISPRi-based genome-scale identification of functional long noncoding RNA loci in human cells. *Science* 355: eaah7111.

572. Rodriguez-Lopez M. et al. (2022) Functional profiling of long intergenic non-coding RNAs in fission yeast. *eLife* 11: e76000.

573. Wilkinson M.F. (2019) Genetic paradox explained by nonsense. *Nature* 568: 179–80.

574. Hong J.-W., Hendrix D.A. and Levine M.S. (2008) Shadow enhancers as a source of evolutionary novelty. *Science* 321: 1314.

575. Cannavò E. et al. (2016) Shadow enhancers are pervasive features of developmental regulatory networks. *Current Biology* 26: 38–51.

576. Osterwalder M. et al. (2018) Enhancer redundancy provides phenotypic robustness in mammalian development. *Nature* 554: 239–43.

577. Waymack R., Fletcher A., Enciso G. and Wunderlich Z. (2020) Shadow enhancers can suppress input transcription factor noise through distinct regulatory logic. *eLife* 9: e59351.

578. Kvon E.Z., Waymack R., Gad M. and Wunderlich Z. (2021) Enhancer redundancy in development and disease. *Nature Reviews Genetics* 22: 324–36.

579. Hirotsune S. et al. (2003) An expressed pseudogene regulates the messenger-RNA stability of its homologous coding gene. *Nature* 423: 91–6.

580. Zhang J. et al. (2006) NANOGP8 is a retrogene expressed in cancers. *FEBS Journal* 273: 1723–30.

581. Tam O.H. et al. (2008) Pseudogene-derived small interfering RNAs regulate gene expression in mouse oocytes. *Nature* 453: 534–8.

582. Poliseno L. et al. (2010) A coding-independent function of gene and pseudogene mRNAs regulates tumour biology. *Nature* 465: 1033–8.

583. Johnsson P. et al. (2013) A pseudogene long-noncoding-RNA network regulates PTEN transcription and translation in human cells. *Nature Structural & Molecular Biology* 20: 440–6.

584. Chiang J.J. et al. (2018) Viral unmasking of cellular 5S rRNA pseudogene transcripts induces RIG-I-mediated immunity. *Nature Immunology* 19: 53–62.

585. Capel B. et al. (1993) Circular transcripts of the testis-determining gene Sry in adult mouse testis. *Cell* 73: 1019–30.

586. Gruhl F., Janich P., Kaessmann H. and Gatfield D. (2021) Circular RNA repertoires are associated with evolutionarily young transposable elements. *eLife* 10: e67991.

587. Salzman J., Gawad C., Wang P.L., Lacayo N. and Brown P.O. (2012) Circular RNAs Are the predominant transcript isoform from hundreds of human genes in diverse cell types. *PLOS ONE* 7: e30733.

588. Jeck W.R. et al. (2013) Circular RNAs are abundant, conserved, and associated with ALU repeats. *RNA* 19: 426.

589. Patop I.L., Wüst S. and Kadener S. (2019) Past, present, and future of circRNAs. *EMBO Journal* 38: e100836.

590. Li Z. et al. (2015) Exon-intron circular RNAs regulate transcription in the nucleus. *Nature Structural & Molecular Biology* 22: 256–64.

591. Zhang Y. et al. (2013) Circular intronic long noncoding RNAs. *Molecular Cell* 51: 792–806.

592. Piwecka M. et al. (2017) Loss of a mammalian circular RNA locus causes miRNA deregulation and affects brain function. *Science* 357: eaam8526.

593. Suenkel C., Cavalli D., Massalini S., Calegari F. and Rajewsky N. (2020) A highly conserved circular RNA is required to keep neural cells in a progenitor state in the mammalian brain. *Cell Reports* 30: 2170–9.

594. Hafez A.K. et al. (2022) A bidirectional competitive interaction between circHomer1 and Homer1b within the orbitofrontal cortex regulates reversal learning. *Cell Reports* 38: 110282.

595. Weigelt C.M. et al. (2020) An insulin-sensitive circular RNA that regulates lifespan in *Drosophila*. *Molecular Cell* 79: 268–79.

596. Liu Y. et al. (2020) Back-spliced RNA from retrotransposon binds to centromere and regulates centromeric chromatin loops in maize. *PLOS Biology* 18: e3000582.

597. Statello L., Guo C.-J., Chen L.-L. and Huarte M. (2021) Gene regulation by long non-coding RNAs and its biological functions. *Nature Reviews Molecular Cell Biology* 22: 96–118.

598. Harrow J. et al. (2006) GENCODE: Producing a reference annotation for ENCODE. *Genome Biology* 7: S4.

599. Frankish A. et al. (2021) GENCODE 2021. *Nucleic Acids Research* 49: D916–23.

600. The RNAcentral Consortium (2021) RNAcentral 2021: Secondary structure integration, improved sequence search and new member databases. *Nucleic Acids Research* 49: D212–20.

601. Uszczynska-Ratajczak B., Lagarde J., Frankish A., Guigó R. and Johnson R. (2018) Towards a complete map of the human long non-coding RNA transcriptome. *Nature Reviews Genetics* 19: 535–48.

602. Zimmer-Bensch G. (2019) Emerging roles of long noncoding RNAs as drivers of brain evolution. *Cells* 8: 1399.

603. Glinsky G. et al. (2018) Single cell expression analysis of primate-specific retroviruses-derived HPAT lincRNAs in viable human blastocysts identifies embryonic cells co-expressing genetic markers of multiple lineages. *Heliyon* 4: e00667.

604. Yang D. et al. (2018) N6-methyladenosine modification of lincRNA 1281 is critically required for mESC differentiation potential. *Nucleic Acids Research* 46: 3906–20.

605. Jandura A. and Krause H.M. (2017) The new RNA world: Growing evidence for long noncoding RNA functionality. *Trends in Genetics* 33: 665–76.

606. Rinn J.L. and Chang H.Y. (2020) Long noncoding RNAs: Molecular modalities to organismal functions. *Annual Review of Biochemistry* 89: 283–308.

607. Flynn R.A. and Chang H.Y. (2014) Long noncoding RNAs in cell-fate programming and reprogramming. *Cell Stem Cell* 14: 752–61.

608. Fatica A. and Bozzoni I. (2014) Long non-coding RNAs: New players in cell differentiation and development. *Nature Reviews Genetics* 15: 7–21.

609. Xu M. et al. (2019) Long noncoding RNA SMRG regulates *Drosophila* macrochaetes by antagonizing scute through E(spl)mβ. *RNA Biology* 16: 42–53.

610. Zhang X., X.U. Yn, Chen B. and Kang L. (2020) Long noncoding RNA PAHAL modulates locust behavioural plasticity through the feedback regulation of dopamine biosynthesis. *PLOS Genetics* 16: e1008771.

611. Tang Y. et al. (2021) The long noncoding RNA FRILAIR regulates strawberry fruit ripening by functioning as a noncanonical target mimic. *PLOS Genetics* 17: e1009461.

612. Nowacki M., Shetty K. and Landweber L.F. (2011) RNA-mediated epigenetic programming of genome rearrangements. *Annual Review of Genomics and Human Genetics* 12: 367–89.

613. Andric V. et al. (2021) A scaffold lncRNA shapes the mitosis to meiosis switch. *Nature Communications* 12: 770.

614. Ding D.Q. et al. (2012) Meiosis-specific noncoding RNA mediates robust pairing of homologous chromosomes in meiosis. *Science* 336: 732–6.

615. Ding D.-Q. et al. (2019) Chromosome-associated RNA–protein complexes promote pairing of homologous chromosomes during meiosis in *Schizosaccharomyces pombe*. *Nature Communications* 10: 5598.

616. Parra-Rivero O., Pardo-Medina J., Gutiérrez G., Limón M.C. and Avalos J. (2020) A novel lncRNA as a positive regulator of carotenoid biosynthesis in *Fusarium*. *Scientific Reports* 10: 678.

617. Loewer S. et al. (2010) Large intergenic non-coding RNA-RoR modulates reprogramming of human induced pluripotent stem cells. *Nature Genetics* 42: 1113–7.

618. Guttman M. et al. (2011) lincRNAs act in the circuitry controlling pluripotency and differentiation. *Nature* 477: 295–300.

619. Klattenhoff C.A. et al. (2013) Braveheart, a long noncoding RNA required for cardiovascular lineage commitment. *Cell* 152: 570–83.

620. Wang Y. et al. (2013) Endogenous miRNA sponge *lincRNA-RoR* regulates *Oct4, Nanog,* and *Sox2* in human embryonic stem cell self-renewal. *Developmental Cell* 25: 69–80.

621. Lin N. et al. (2014) An evolutionarily conserved long non-coding RNA TUNA controls pluripotency and neural lineage commitment. *Molecular Cell* 53: 1005–19.

622. Yin Y. et al. (2015) Opposing roles for the lncRNA Haunt and its genomic locus in regulating HOXA gene activation during embryonic stem cell differentiation. *Cell Stem Cell* 16: 504–16.

623. Durruthy-Durruthy J. et al. (2016) The primate-specific noncoding RNA HPAT5 regulates pluripotency during human preimplantation development and nuclear reprogramming. *Nature Genetics* 48: 44–52.

624. Percharde M. et al. (2018) A LINE1-nucleolin partnership regulates early development and ESC identity. *Cell* 174: 391–405.

625. Zhu P. et al. (2018) LncGata6 maintains stemness of intestinal stem cells and promotes intestinal tumorigenesis. *Nature Cell Biology* 20: 1134–44.

626. Yan P. et al. (2020) LncRNA *Platr22* promotes super-enhancer activity and stem cell pluripotency. *Journal of Molecular Cell Biology* 13: 295–313.

627. Pal D. et al. (2021) LncRNA *Mrhl* orchestrates differentiation programs in mouse embryonic stem cells through chromatin mediated regulation. *Stem Cell Research* 53: 102250.

628. Wang H., Yang Y., Liu J. and Qian L. (2021) Direct cell reprogramming: Approaches, mechanisms and progress. *Nature Reviews Molecular Cell Biology* 22: 410–24.

629. Cipriano A. et al. (2021) Epigenetic regulation of Wnt7b expression by the cis-acting long noncoding RNA Lnc-Rewind in muscle stem cells. *eLife* 10: e54782.

630. Guo Q. et al. (2014) BRAF-activated long non-coding RNA contributes to colorectal cancer migration by inducing epithelial-mesenchymal transition. *Oncology Letters* 8: 869–75.

631. Dill T.L., Carroll A., Pinheiro A., Gao J. and Naya F.J. (2021) The long noncoding RNA Meg3 regulates myoblast plasticity and muscle regeneration through epithelial-mesenchymal transition. *Development* 148: dev194027.

632. Luo S. et al. (2016) Divergent lncRNAs regulate gene expression and lineage differentiation in pluripotent cells. *Cell Stem Cell* 18: 637–52.

633. Daneshvar K. et al. (2016) *DIGIT* Is a conserved long noncoding RNA that regulates GSC expression to control definitive endoderm differentiation of embryonic stem cells. *Cell Reports* 17: 353–65.

634. Alexanian M. et al. (2017) A transcribed enhancer dictates mesendoderm specification in pluripotency. *Nature Communications* 8: 1806.

635. Yoneda R., Ueda N., Uranishi K., Hirasaki M. and Kurokawa R. (2020) Long noncoding RNA *pncRNA-D* reduces cyclin D1 gene expression and arrests cell cycle through RNA m6A modification. *Journal of Biological Chemistry* 295: 5626–39.

636. Pruunsild P., Kazantseva A., Aid T., Palm K. and Timmusk T. (2007) Dissecting the human BDNF locus: Bidirectional transcription, complex splicing, and multiple promoters. *Genomics* 90: 397–406.

637. Modarresi F. et al. (2012) Inhibition of natural antisense transcripts in vivo results in gene-specific transcriptional upregulation. *Nature Biotechnology* 30: 453–9.

638. Wilson K.D. et al. (2020) Endogenous retrovirus-derived lncRNA BANCR promotes cardiomyocyte migration in humans and non-human primates. *Developmental Cell* 54: 694–709.

639. Wang X. et al. (2021) Mutual dependency between lncRNA LETN and protein NPM1 in controlling the nucleolar structure and functions sustaining cell proliferation. *Cell Research* 31: 664–83.

640. Ducoli L. et al. (2021) LETR1 is a lymphatic endothelial-specific lncRNA governing cell proliferation and migration through KLF4 and SEMA3C. *Nature Communications* 12: 925.

641. Huang D. et al. (2018) NKILA lncRNA promotes tumor immune evasion by sensitizing T cells to activation-induced cell death. *Nature Immunology* 19: 1112–25.

642. Briggs J.A., Wolvetang E.J., Mattick J.S., Rinn J.L. and Barry G. (2015) Mechanisms of long non-coding RNAs in mammalian nervous system development, plasticity, disease, and evolution. *Neuron* 88: 861–77.

643. Onoguchi M., Hirabayashi Y., Koseki H. and Gotoh Y. (2012) A noncoding RNA regulates the *neurogenin1* gene locus during mouse neocortical development. *Proceedings of the National Academy of Sciences USA* 109: 16939–44.

644. Cajigas I. et al. (2018) The *Evf2* ultraconserved enhancer lncRNA functionally and spatially organizes megabase distant genes in the developing forebrain. *Molecular Cell* 71: 956–72.

645. Cajigas I. et al. (2021) Sox2-*Evf2* lncRNA mechanisms of chromosome topological control in developing forebrain. *Development* 148: dev197202.

646. Rapicavoli N.A., Poth E.M. and Blackshaw S. (2010) The long noncoding RNA RNCR2 directs mouse retinal cell specification. *BMC Developmental Biology* 10: 49.

647. Krol J. et al. (2015) A network comprising short and long noncoding RNAs and RNA helicase controls mouse retina architecture. *Nature Communications* 6: 7305.

648. Bond A.M. et al. (2009) Balanced gene regulation by an embryonic brain ncRNA is critical for adult hippocampal GABA circuitry. *Nature Neuroscience* 12: 1020–7.

649. Rea J. et al. (2020) *HOTAIRM1* regulates neuronal differentiation by modulating NEUROGENIN 2 and the downstream neurogenic cascade. *Cell Death & Disease* 11: 527.

650. Hollensen A.K. et al. (2020) circZNF827 nucleates a transcription inhibitory complex to balance neuronal differentiation. *eLife* 9: e58478.

651. Perry R.B.-T., Hezroni H., Goldrich M.J. and Ulitsky I. (2018) Regulation of neuroregeneration by long noncoding RNAs. *Molecular Cell* 72: 553–67.

652. Grelet S. et al. (2022) TGFβ-induced expression of long noncoding lincRNA Platr18 controls breast cancer axonogenesis. *Life Science Alliance* 5: e202101261.

653. Andreassi C. et al. (2021) Cytoplasmic cleavage of IMPA1 3′ UTR is necessary for maintaining axon integrity. *Cell Reports* 34: 108778.

654. Martinez-Moreno M. et al. (2017) Regulation of peripheral myelination through transcriptional buffering of Egr2 by an antisense long non-coding RNA. *Cell Reports* 20: 1950–63.

655. Bernard D. et al. (2010) A long nuclear-retained non-coding RNA regulates synaptogenesis by modulating gene expression. *EMBO Journal* 29: 3082–93.

656. You X. et al. (2015) Neural circular RNAs are derived from synaptic genes and regulated by development and plasticity. *Nature Neuroscience* 18: 603–10.

657. Raveendra B.L. et al. (2018) Long noncoding RNA GM12371 acts as a transcriptional regulator of synapse function. *Proceedings of the National Academy of Sciences USA* 115: E10197–205.

658. Ma M. et al. (2020) A novel pathway regulates social hierarchy via lncRNA AtLAS and postsynaptic synapsin IIb. *Cell Research* 30: 105–18.

659. Wang F. et al. (2021) The long noncoding RNA *Synage* regulates synapse stability and neuronal function in the cerebellum. *Cell Death & Differentiation* 28: 2634–50.

660. Grinman E. et al. (2021) Activity-regulated synaptic targeting of lncRNA ADEPTR mediates structural plasticity by localizing Sptn1 and AnkB in dendrites. *Science Advances* 7: eabf0605.

661. Li D. et al. (2018) Activity dependent LoNA regulates translation by coordinating rRNA transcription and methylation. *Nature Communications* 9: 1726.

662. Li X. et al. (2021) On the discovery of ADRAM, an experience-dependent long noncoding RNA that drives fear extinction through a direct interaction with the chaperone protein 14-3-3. *bioRxiv*: 2021.08.01.454607.

663. Issler O. et al. (2020) Sex-specific role for the long non-coding RNA LINC00473 in depression. *Neuron* 106: 912–26.

664. Shore A.N. et al. (2012) Pregnancy-induced noncoding RNA (PINC) associates with polycomb repressive complex 2 and regulates mammary epithelial differentiation. *PLOS Genetics* 8: e1002840.

665. Stafford D.A., Dichmann D.S., Chang J.K. and Harland R.M. (2017) Deletion of the sclerotome-enriched lncRNA PEAT augments ribosomal protein expression. *Proceedings of the National Academy of Sciences USA* 114: 101–6.

666. Grote P. et al. (2013) The tissue-specific lncRNA Fendrr is an essential regulator of heart and body wall development in the mouse. *Developmental Cell* 24: 206–14.

667. Anderson K.M. et al. (2016) Transcription of the non-coding RNA upperhand controls *Hand2* expression and heart development. *Nature* 539: 433–6.

668. Han X. et al. (2019) The lncRNA *Hand2os1/Uph* locus orchestrates heart development through regulation of precise expression of *Hand2*. *Development* 146: dev176198.

669. Ritter N. et al. (2019) The lncRNA locus Handsdown regulates cardiac gene programs and is essential for early mouse development. *Developmental Cell* 50: 644–57.

670. Szafranski P. et al. (2013) Small noncoding differentially methylated copy-number variants, including lncRNA genes, cause a lethal lung developmental disorder. *Genome Research* 23: 23–33.

671. Delpretti S. et al. (2013) Multiple enhancers regulate *Hoxd* genes and the Hotdog lncRNA during cecum budding. *Cell Reports* 5: 137–50.

672. Cesana M. et al. (2011) A long noncoding RNA controls muscle differentiation by functioning as a competing endogenous RNA. *Cell* 147: 358–69.

673. Ounzain S. et al. (2015) CARMEN, a human super enhancer-associated long noncoding RNA controlling cardiac specification, differentiation and homeostasis. *Journal of Molecular and Cellular Cardiology* 89: 98–112.

674. Wang Z. et al. (2016) The long noncoding RNA Chaer defines an epigenetic checkpoint in cardiac hypertrophy. *Nature Medicine* 22: 1131–9.

675. Liu J., Li Y., Lin B., Sheng Y. and Yang L. (2017) HBL1 Is a human long noncoding RNA that modulates cardiomyocyte development from pluripotent stem cells by counteracting MIR1. *Developmental Cell* 42: 333–48.

676. Micheletti R. et al. (2017) The long noncoding RNA *Wisper* controls cardiac fibrosis and remodeling. *Science Translational Medicine* 9: eaai9118.

677. Ballarino M. et al. (2018) Deficiency in the nuclear long noncoding RNA Charme causes myogenic defects and heart remodeling in mice. *EMBO Journal* 37: e99697.

678. Alessio E. et al. (2019) Single cell analysis reveals the involvement of the long non-coding RNA Pvt1 in the modulation of muscle atrophy and mitochondrial network. *Nucleic Acids Research* 47: 1653–70.

679. Kang X., Zhao Y., Van Arsdell G., Nelson S.F. and Touma M. (2020) Ppp1r1b-lncRNA inhibits PRC2 at myogenic regulatory genes to promote cardiac and skeletal muscle development in mouse and human. *RNA* 26: 481–91.

680. Schutt C. et al. (2020) Linc-MYH configures INO80 to regulate muscle stem cell numbers and skeletal muscle hypertrophy. *EMBO Journal* 39: e105098.

681. Dong K. et al. (2021) *CARMN* Is an evolutionarily conserved smooth muscle cell-specific lncRNA that maintains contractile phenotype by binding myocardin. *Circulation* 144: 1856–75.

682. Anderson D.M. et al. (2021) A myocardin-adjacent lncRNA balances SRF-dependent gene transcription in the heart. *Genes & Development* 35: 835–40.

683. Sato M. et al. (2021) The lncRNA Caren antagonizes heart failure by inactivating DNA damage response and activating mitochondrial biogenesis. *Nature Communications* 12: 2529.

684. Vacante F. et al. (2021) CARMN loss regulates smooth muscle cells and accelerates atherosclerosis in mice. *Circulation Research* 128: 1258–75.

685. Lu L. et al. (2013) Genome-wide survey by ChIP-seq reveals YY1 regulation of lincRNAs in skeletal myogenesis. *EMBO journal* 32: 2575–88.

686. Legnini I., Morlando M., Mangiavacchi A., Fatica A. and Bozzoni I. (2014) A feedforward regulatory loop between HuR and the long noncoding RNA linc-MD1 controls early phases of myogenesis. *Molecular Cell* 53: 506–14.

687. Raz V., Riaz M., Tatum Z., Kielbasa S.M. and t Hoen P.A.C. (2018) The distinct transcriptomes of slow and fast adult muscles are delineated by noncoding RNAs. *FASEB Journal* 32: 1579–90.

688. Dou M. et al. (2020) The long noncoding RNA MyHC IIA/X-AS contributes to skeletal muscle myogenesis and maintains the fast fiber phenotype. *Journal of Biological Chemistry* 295: 4937–49.

689. Yu J.-A. et al. (2021) LncRNA-FKBP1C regulates muscle fiber type switching by affecting the stability of MYH1B. *Cell Death Discovery* 7: 73.

690. Cai B. et al. (2021) Long noncoding RNA *SMUL* suppresses *SMURF2* production-mediated muscle atrophy via nonsense-mediated mRNA decay. *Molecular Therapy - Nucleic Acids* 23: 512–26.

691. Wang Y. (2018) Long noncoding RNA lncHand2 promotes liver repopulation via c-Met signaling. *Journal of Hepatology* 69: 861–72.

692. Wang Y. et al. (2019) LncRNA HAND2-AS1 promotes liver cancer stem cell self-renewal via BMP signaling. *EMBO Journal* 38: e101110.

693. Hennessy E.J. et al. (2019) The long noncoding RNA CHROME regulates cholesterol homeostasis in primates. *Nature Metabolism* 1: 98–110.

694. Leisegang M.S. et al. (2017) Long noncoding RNA *MANTIS* facilitates endothelial angiogenic function. *Circulation* 136: 65–79.

695. Mushimiyimana I. et al. (2021) Genomic landscapes of noncoding RNAs regulating *VEGFA* and *VEGFC* expression in endothelial cells. *Molecular and Cellular Biology* 41: e00594-20.

696. Chignon A. et al. (2022) Genome-wide chromatin contacts of super-enhancer-associated lncRNA identify LINC01013 as a regulator of fibrosis in the aortic valve. *PLOS Genetics* 18: e1010010.

697. Lyu Q. et al. (2019) *SENCR* stabilizes vascular endothelial cell adherens junctions through interaction with CKAP4. *Proceedings of the National Academy of Sciences USA* 116: 546–55.

698. Sehgal P. et al. (2021) LncRNA VEAL2 regulates PRKCB2 to modulate endothelial permeability in diabetic retinopathy. *EMBO Journal* 2021: e107134.

699. Lewandowski J.P. et al. (2019) The *Firre* locus produces a trans-acting RNA molecule that functions in hematopoiesis. *Nature Communications* 10: 5137.

700. Ranzani V. et al. (2015) The long intergenic noncoding RNA landscape of human lymphocytes highlights the regulation of T cell differentiation by linc-MAF-4. *Nature Immunology* 16: 318–25.

701. Kretz M. et al. (2012) Suppression of progenitor differentiation requires the long noncoding RNA ANCR. *Genes & development* 26: 338–43.

702. Kretz M. et al. (2013) Control of somatic tissue differentiation by the long non-coding RNA TINCR. *Nature* 493: 231–5.

703. Cai P. et al. (2019) A genome-wide long noncoding RNA CRISPRi screen identifies PRANCR as a novel regulator of epidermal homeostasis. *Genome Research* 30: 22–34.

704. Arriaga-Canon C. et al. (2014) A long non-coding RNA promotes full activation of adult gene expression in the chicken α-globin domain. *Epigenetics* 9: 173–81.

705. Werner M.S. et al. (2017) Chromatin-enriched lncRNAs can act as cell-type specific activators of proximal gene transcription. *Nature Structural & Molecular Biology* 24: 596–603.

706. Alvarez-Dominguez J.R., Knoll M., Gromatzky A.A. and Lodish H.F. (2017) The super-enhancer-derived *alncRNA-EC7/Bloodlinc* potentiates red blood cell development in trans. *Cell Reports* 19: 2503–14.

707. Morrison T.A. et al. (2018) A long noncoding RNA from the *HBS1L-MYB* intergenic region on chr6q23 regulates human fetal hemoglobin expression. *Blood Cells, Molecules, and Diseases* 69: 1–9.

708. Jiang M. et al. (2018) Self-recognition of an inducible host lncRNA by RIG-I feedback restricts innate immune response. *Cell* 173: 906–19.

709. Zhou B. et al. (2019) Endogenous retrovirus-derived long noncoding RNA enhances innate immune responses via derepressing RELA expression. *mBio* 10: e00937-19.

710. Liu W. et al. (2020) LncRNA Malat1 inhibition of TDP43 cleavage suppresses IRF3-initiated antiviral innate immunity. *Proceedings of the National Academy of Sciences USA* 117: 23695–706.

711. Hewitson J.P. et al. (2020) Malat1 suppresses immunity to infection through promoting expression of Maf and IL-10 in Th cells. *Journal of Immunology* 204: 2949–60.

712. Flores-Concha M. and Oñate Á.A. (2020) Long non-coding RNAs in the regulation of the immune response and trained immunity. *Frontiers in Genetics* 11: 718.

713. Kotzin J.J. et al. (2019) The long noncoding RNA Morrbid regulates CD8 T cells in response to viral infection. *Proceedings of the National Academy of Sciences USA* 116: 11916–25.

714. Lai C. et al. (2021) Long noncoding RNA AVAN promotes antiviral innate immunity by interacting with TRIM25 and enhancing the transcription of FOXO3a. *Cell Death & Differentiation* 28: 2900–15.

715. Peltier D. et al. (2021) RNA-seq of human T cells after hematopoietic stem cell transplantation identifies Linc00402 as a regulator of T cell alloimmunity. *Science Translational Medicine* 13: eaaz0316.

716. Sui B. et al. (2020) A novel antiviral lncRNA, EDAL, shields a T309 O-GlcNAcylation site to promote EZH2 lysosomal degradation. *Genome Biology* 21: 228.

717. Vigneau S., Rohrlich P.S., Brahic M. and Bureau J.F. (2003) *Tmevpg1*, a candidate gene for the control of Theiler's virus persistence, could be implicated in the regulation of gamma interferon. *Journal of Virology* 77: 5632–8.

718. Collier S.P., Collins P.L., Williams C.L., Boothby M.R. and Aune T.M. (2012) Cutting edge: Influence of *Tmevpg1*, a long intergenic noncoding RNA, on the expression of *Ifng* by Th1 cells. *Journal of Immunology* 189: 2084–8.

719. Gomez J.A. et al. (2013) The *NeST* long ncRNA controls microbial susceptibility and epigenetic activation of the interferon-gamma locus. *Cell* 152: 743–54.

720. Stein N. et al. (2019) IFNG-AS1 enhances interferon gamma production in human natural killer cells. *iScience* 11: 466–73.

721. Vollmers A.C. et al. (2021) A conserved long noncoding RNA, GAPLINC, modulates the immune response during endotoxic shock. *Proceedings of the National Academy of Sciences USA* 118: e2016648118.

722. Rothschild G. et al. (2020) Noncoding RNA transcription alters chromosomal topology to promote isotype-specific class switch recombination. *Science Immunology* 5: eaay5864.

723. Nair L. et al. (2021) Mechanism of noncoding RNA-associated N6-methyladenosine recognition by an RNA processing complex during IgH DNA recombination. *Molecular Cell* 81: 3949–64.

724. Zhao X. et al. (2013) A long noncoding RNA contributes to neuropathic pain by silencing Kcna2 in primary afferent neurons. *Nature Neuroscience* 16: 1024–31.

725. Atianand M.K. et al. (2016) A long noncoding RNA lincRNA-EPS acts as a transcriptional brake to restrain inflammation. *Cell* 165: 1672–85.

726. Hu G. et al. (2016) LincRNA-Cox2 promotes late inflammatory gene transcription in macrophages through modulating SWI/SNF-mediated chromatin remodeling. *Journal of Immunology* 196: 2799–808.

727. Liu X. et al. (2019) A long noncoding RNA, Antisense IL-7, promotes inflammatory gene transcription through facilitating histone acetylation and switch/sucrose nonfermentable chromatin remodeling. *Journal of Immunology* 203: 1548–59.

728. Miyata K. et al. (2021) Pericentromeric noncoding RNA changes DNA binding of CTCF and inflammatory gene expression in senescence and cancer. *Proceedings of the National Academy of Sciences USA* 118: e2025647118.

729. Du Q. et al. (2019) MIR205HG Is a long noncoding RNA that regulates growth hormone and prolactin production in the anterior pituitary. *Developmental Cell* 49: 618–31.

730. Liang L. et al. (2021) The long noncoding RNA HOTAIRM1 controlled by AML1 enhances glucocorticoid resistance by activating RHOA/ROCK1 pathway through suppressing ARHGAP18. *Cell Death & Disease* 12: 702.

731. Heinen T.J.A.J., Staubach F., Häming D. and Tautz D. (2009) Emergence of a new gene from an intergenic region. *Current Biology* 19: 1527–31.

732. Nakajima R., Sato T., Ogawa T., Okano H. and Noce T. (2017) A noncoding RNA containing a SINE-B1 motif associates with meiotic metaphase chromatin and has an indispensable function during spermatogenesis. *PLOS ONE* 12: e0179585.

733. Bouska M.J. and Bai H. (2021) Long noncoding RNA regulation of spermatogenesis via the spectrin cytoskeleton in *Drosophila. G3 (Genes, Genomes, Genetics)* 11: jkab080.

734. Schmitt A.M. et al. (2016) An inducible long noncoding RNA amplifies DNA damage signaling. *Nature Genetics* 48: 1370–6.

735. Michelini F. et al. (2017) Damage-induced lncRNAs control the DNA damage response through interaction with DDRNAs at individual double-strand breaks. *Nature Cell Biology* 19: 1400–11.

736. Tran K.-V. et al. (2020) Human thermogenic adipocyte regulation by the long noncoding RNA LINC00473. *Nature Metabolism* 2: 397–412.

737. Scheele C. et al. (2007) The human PINK1 locus is regulated in vivo by a non-coding natural antisense RNA during modulation of mitochondrial function. *BMC Genomics* 8: 74.

738. Lanz R.B. et al. (1999) A steroid receptor coactivator, SRA, functions as an RNA and is present in an SRC-1 complex. *Cell* 97: 17–27.

739. Lanz R.B., Razani B., Goldberg A.D. and O'Malley B.W. (2002) Distinct RNA motifs are important for coactivation of steroid hormone receptors by steroid receptor RNA activator (SRA). *Proceedings of the National Academy of Sciences USA* 99: 16081–6.

740. Kino T., Hurt D.E., Ichijo T., Nader N. and Chrousos G.P. (2010) Noncoding RNA gas5 is a growth arrest- and starvation-associated repressor of the glucocorticoid receptor. *Science Signaling* 3: ra8.

741. Campalans A., Kondorosi A. and Crespi M. (2004) Enod40, a short open reading frame-containing mRNA, induces cytoplasmic localization of a nuclear RNA binding protein in *Medicago truncatula. Plant Cell* 16: 1047–59.

742. Khvorova A., Kwak Y.-G., Tamkun M., Majerfeld I. and Yarus M. (1999) RNAs that bind and change the permeability of phospholipid membranes. *Proceedings of the National Academy of Sciences USA* 96: 10649–54.

743. McClintock M.A. et al. (2018) RNA-directed activation of cytoplasmic dynein-1 in reconstituted transport RNPs. *eLife* 7: e36312.

744. Lin A. et al. (2017) The *LINK-A* lncRNA interacts with PtdIns(3,4,5)P3 to hyperactivate AKT and confer resistance to AKT inhibitors. *Nature Cell Biology* 19: 238–51.

745. Sang L. et al. (2018) LncRNA *CamK-A* regulates Ca2+-signaling-mediated tumor microenvironment remodeling. *Molecular Cell* 72: 71–83.

746. Ma Y., Zhang J., Wen L. and Lin A. (2018) Membrane-lipid associated lncRNA: A new regulator in cancer signaling. *Cancer Letters* 419: 27–9.

747. Zheng X. et al. (2017) LncRNA wires up Hippo and Hedgehog signaling to reprogramme glucose metabolism. *EMBO Journal* 36: 3325–35.

748. Zucchelli S. et al. (2015) SINEUPs: A new class of natural and synthetic antisense long non-coding RNAs that activate translation. *RNA Biology* 12: 771–9.

749. Wang H. et al. (2005) Dendritic BC1 RNA in translational control mechanisms. *Journal of Cell Biology* 171: 811–21.

750. Deforges J. et al. (2019) Control of cognate sense mRNA translation by cis-natural antisense RNAs. *Plant Physiology* 180: 305–22.

751. Munroe S.H. and Lazar M.A. (1991) Inhibition of c-erbA mRNA splicing by a naturally occurring antisense RNA. *Journal of Biological Chemistry* 266: 22083–6.

752. Yan M.D. et al. (2005) Identification and characterization of a novel gene Saf transcribed from the opposite strand of Fas. *Human Molecular Genetics* 14: 1465–74.

753. Kishore S. and Stamm S. (2006) The snoRNA HBII-52 regulates alternative splicing of the serotonin receptor 2C. *Science* 311: 230–2.

754. Beltran M. et al. (2008) A natural antisense transcript regulates Zeb2/Sip1 gene expression during Snail1-induced epithelial-mesenchymal transition. *Genes & Development* 22: 756–69.

755. Tripathi V. et al. (2010) The nuclear-retained noncoding RNA *MALAT1* regulates alternative splicing by modulating SR splicing factor phosphorylation. *Molecular Cell* 39: 925–38.

756. Morrissy A.S., Griffith M. and Marra M.A. (2011) Extensive relationship between antisense transcription and alternative splicing in the human genome. *Genome Research* 21: 1203–12.

757. Gonzalez I. et al. (2015) A lncRNA regulates alternative splicing via establishment of a splicing-specific chromatin signature. *Nature Structural & Molecular Biology* 22: 370–6.

758. Yap K. et al. (2018) A short tandem repeat-enriched RNA assembles a nuclear compartment to control alternative splicing and promote cell survival. *Molecular Cell* 72: 525–40.

759. Romero-Barrios N., Legascue M.F., Benhamed M., Ariel F. and Crespi M. (2018) Splicing regulation by long noncoding RNAs. *Nucleic Acids Research* 46: 2169–84.

760. Pisignano G. and Ladomery M. (2021) Epigenetic regulation of alternative splicing: How lncRNAs tailor the message. *Non-Coding RNA* 7: 21.

761. Otten A.B.C. et al. (2021) The noncoding RNA PRANCR regulates splicing of Fibronectin-1 to control keratinocyte proliferation and migration. *bioRxiv*: 2021.06.22.449364.

762. Simone R. et al. (2021) MIR-NATs repress MAPT translation and aid proteostasis in neurodegeneration. *Nature* 594: 117–23.

763. Peters N.T., Rohrbach J.A., Zalewski B.A., Byrkett C.M. and Vaughn J.C. (2003) RNA editing and regulation of *Drosophila 4f-rnp* expression by sas-10 antisense readthrough mRNA transcripts. *RNA* 9: 698–710.

764. Prasanth K.V. et al. (2005) Regulating gene expression through RNA nuclear retention. *Cell* 123: 249–63.

765. Werner A. (2005) Natural antisense transcripts. *RNA Biology* 2: 53–62.

766. Li K. et al. (2010) A noncoding antisense RNA in *tie-1* locus regulates tie-1 function *in vivo*. *Blood* 115: 133–9.

767. Gong C. and Maquat L.E. (2011) lncRNAs transactivate STAU1-mediated mRNA decay by duplexing with 3′ UTRs via Alu elements. *Nature* 470: 284–8.

768. Goodrich J.A. and Kugel J.F. (2006) Non-coding-RNA regulators of RNA polymerase II transcription. *Nature Reviews Molecular Cell Biology* 7: 612–6.

769. Studniarek C., Egloff S. and Murphy S. (2021) Noncoding RNAs set the stage for RNA Polymerase II transcription. *Trends in Genetics* 37: 279–91.

770. Yang Z., Zhu Q., Luo K. and Zhou Q. (2001) The 7SK small nuclear RNA inhibits the CDK9/cyclin T1 kinase to control transcription. *Nature* 414: 317–22.

771. Nguyen V.T., Kiss T., Michels A.A. and Bensaude O. (2001) 7SK small nuclear RNA binds to and inhibits the activity of CDK9/cyclin T complexes. *Nature* 414: 322–5.

772. Kohoutek J. (2009) P-TEFb- the final frontier. *Cell Division* 4: 19.

773. Lenasi T. and Barboric M. (2010) P-TEFb stimulates transcription elongation and pre-mRNA splicing through multilateral mechanisms. *RNA Biology* 7: 145–50.

774. Peterlin B.M., Brogie J.E. and Price D.H. (2012) 7SK snRNA: A noncoding RNA that plays a major role in regulating eukaryotic transcription. *Wiley Interdisciplinary Reviews RNA* 3: 92–103.

775. Castelo-Branco G. et al. (2013) The non-coding snRNA 7SK controls transcriptional termination, poising, and bidirectionality in embryonic stem cells. *Genome Biology* 14: R98.

776. Quaresma A.J.C., Bugai A. and Barboric M. (2016) Cracking the control of RNA polymerase II elongation by 7SK snRNP and P-TEFb. *Nucleic Acids Research* 44: 7527–39.

777. Flynn R.A. et al. (2016) 7SK-BAF axis controls pervasive transcription at enhancers. *Nature Structural & Molecular Biology* 23: 231–8.

778. Studniarek C. et al. (2021) The 7SK/P-TEFb snRNP controls ultraviolet radiation-induced transcriptional reprogramming. *Cell Reports* 35: 108965.

779. Egloff S., Studniarek C. and Kiss T. (2018) 7SK small nuclear RNA, a multifunctional transcriptional regulatory RNA with gene-specific features. *Transcription* 9: 95–101.

780. Briese M. and Sendtner M. (2021) Keeping the balance: The noncoding RNA 7SK as a master regulator for neuron development and function. *Bioessays* 43: e2100092.

781. Wang X. et al. (2008) Induced ncRNAs allosterically modify RNA-binding proteins in cis to inhibit transcription. *Nature* 454: 126–30.

782. Martianov I., Ramadass A., Serra Barros A., Chow N. and Akoulitchev A. (2007) Repression of the human dihydrofolate reductase gene by a non-coding interfering transcript. *Nature* 445: 666–70.

783. Li Y., Syed J. and Sugiyama H. (2016) RNA-DNA triplex formation by long noncoding RNAs. *Cell Chemical Biology* 23: 1325–33.

784. Imamura T. et al. (2004) Non-coding RNA directed DNA demethylation of *Sphk1* CpG island. *Biochemical and Biophysical Research Communications* 322: 593–600.

785. Postepska-Igielska A. et al. (2015) LncRNA *Khps1* regulates expression of the proto-oncogene *SPHK1* via triplex-mediated changes in chromatin structure. *Molecular Cell* 60: 626–36.

786. Fan S. et al. (2020) lncRNA *CISAL* Inhibits *BRCA1* transcription by forming a tertiary structure at its promoter. *iScience* 23: 100835.

787. Azzalin C.M., Reichenbach P., Khoriauli L., Giulotto E. and Lingner J. (2007) Telomeric repeat containing RNA and RNA surveillance factors at mammalian chromosome ends. *Science* 318: 798–801.

788. Schoeftner S. and Blasco M.A. (2008) Developmentally regulated transcription of mammalian telomeres by DNA-dependent RNA polymerase II. *Nature Cell Biology* 10: 228–36.

789. Xu Y., Kimura T. and Komiyama M. (2008) Human telomere RNA and DNA form an intermolecular G-quadruplex. *Nucleic Acids Symposium Series (Oxford)* 52: 169–70.

790. Montero J.J. et al. (2018) TERRA recruitment of polycomb to telomeres is essential for histone trymethylation marks at telomeric heterochromatin. *Nature Communications* 9: 1548.

791. Chu H.-P. et al. (2017) TERRA RNA antagonizes ATRX and protects telomeres. *Cell* 170: 86–101.

792. Marión R.M. et al. (2019) TERRA regulate the transcriptional landscape of pluripotent cells through TRF1-dependent recruitment of PRC2. *eLife* 8: e44656.

793. Silva B., Arora R., Bione S. and Azzalin C.M. (2021) TERRA transcription destabilizes telomere integrity to initiate break-induced replication in human ALT cells. *Nature Communications* 12: 3760.

794. Graf M. et al. (2017) Telomere length determines *TERRA* and R-loop regulation through the cell cycle. *Cell* 170: 72–85.

795. Kwapisz M. and Morillon A. (2020) Subtelomeric transcription and its regulation. *Journal of Molecular Biology* 432: 4199–219.

796. Mattick J.S., Amaral P.P., Dinger M.E., Mercer T.R. and Mehler M.F. (2009) RNA regulation of epigenetic processes. *BioEssays* 31: 51–9.

797. Koziol M.J. and Rinn J.L. (2010) RNA traffic control of chromatin complexes. *Current Opinion in Genetics and Development* 20: 142–8.

798. Mattick J.S. and Gagen M.J. (2001) The evolution of controlled multitasked gene networks: The role of introns and other noncoding RNAs in the development of complex organisms. *Molecular Biology and Evolution* 18: 1611–30.

799. Lee J.T. (2012) Epigenetic regulation by long noncoding RNAs. *Science* 338: 1435–9.

800. Mercer T.R. and Mattick J.S. (2013) Structure and function of long noncoding RNAs in epigenetic regulation. *Nature Structural & Molecular Biology* 20: 300–7.

801. Bernstein E. et al. (2006) Mouse polycomb proteins bind differentially to methylated histone H3 and RNA and are enriched in facultative heterochromatin. *Molecular Cell Biology* 26: 2560–9.

802. Yu W. et al. (2008) Epigenetic silencing of tumour suppressor gene p15 by its antisense RNA. *Nature* 451: 202–6.

803. Yap K.L. et al. (2010) Molecular interplay of the noncoding RNA ANRIL and methylated histone H3 lysine 27 by polycomb CBX7 in transcriptional silencing of INK4a. *Molecular Cell* 38: 662–74.

804. Kotake Y. et al. (2011) Long non-coding RNA ANRIL is required for the PRC2 recruitment to and silencing of p15(INK4B) tumor suppressor gene. *Oncogene* 30: 1956–62.

805. Alfeghaly C. et al. (2021) Implication of repeat insertion domains in the trans-activity of the long non-coding RNA ANRIL. *Nucleic Acids Research* 49: 4954–70.

806. Bae E., Calhoun V.C., Levine M., Lewis E.B. and Drewell R.A. (2002) Characterization of the intergenic RNA profile at *abdominal-A* and *abdominal-B* in the *Drosophila bithorax* complex. *Proceedings of the National Academy of Sciences USA* 99: 16847–52.

807. Bender W. and Fitzgerald D.P. (2002) Transcription activates repressed domains in the *Drosophila bithorax* complex. *Development* 129: 4923–30.

808. Drewell R.A., Bae E., Burr J. and Lewis E.B. (2002) Transcription defines the embryonic domains of cis-regulatory activity at the *Drosophila bithorax* complex. *Proceedings of the National Academy of Sciences USA* 99: 16853–8.

809. Hogga I. and Karch F. (2002) Transcription through the iab-7 cis-regulatory domain of the bithorax complex interferes with maintenance of Polycomb-mediated silencing. *Development* 129: 4915–22.

810. Rank G., Prestel M. and Paro R. (2002) Transcription through intergenic chromosomal memory elements of the Drosophila bithorax complex correlates with an epigenetic switch. *Molecular and Cellular Biology* 22: 8026–34.

811. Schmitt S., Prestel M. and Paro R. (2005) Intergenic transcription through a polycomb group response element counteracts silencing. *Genes & Development* 19: 697–708.

812. Lempradl A. and Ringrose L. (2008) How does noncoding transcription regulate Hox genes? *BioEssays* 30: 110–21.

813. Khalil A.M. et al. (2009) Many human large intergenic noncoding RNAs associate with chromatin-modifying complexes and affect gene expression. *Proceedings of the National Academy of Sciences USA* 106: 11667–72.

814. Zhao J. et al. (2010) Genome-wide identification of polycomb-associated RNAs by RIP-seq. *Molecular Cell* 40: 939–53.

815. Davidovich C., Zheng L., Goodrich K.J. and Cech T.R. (2013) Promiscuous RNA binding by Polycomb repressive complex 2. *Nature Structural & Molecular Biology* 20: 1250–7.

816. Tsai M.C. et al. (2010) Long noncoding RNA as modular scaffold of histone modification complexes. *Science* 329: 689–93.

817. Gupta R.A. et al. (2010) Long non-coding RNA HOTAIR reprograms chromatin state to promote cancer metastasis. *Nature* 464: 1071–6.

818. Spitale R.C., Tsai M.C. and Chang H.Y. (2011) RNA templating the epigenome: Long noncoding RNAs as molecular scaffolds. *Epigenetics* 6: 539–43.

819. Guttman M. and Rinn J.L. (2012) Modular regulatory principles of large non-coding RNAs. *Nature* 482: 339–46.

820. Wierzbicki A.T., Blevins T. and Swiezewski S. (2021) Long noncoding RNAs in plants. *Annual Review of Plant Biology* 72: 245–71.

821. Ietswaart R., Wu Z. and Dean C. (2012) Flowering time control: Another window to the connection between antisense RNA and chromatin. *Trends in Genetics* 28: 445–53.

822. Swiezewski S., Liu F., Magusin A. and Dean C. (2009) Cold-induced silencing by long antisense transcripts of an *Arabidopsis* Polycomb target. *Nature* 462: 799–802.

823. Heo J.B. and Sung S. (2011) Vernalization-mediated epigenetic silencing by a long intronic noncoding RNA. *Science* 331: 76–9.

824. Sun Q., Csorba T., Skourti-Stathaki K., Proudfoot N.J. and Dean C. (2013) R-Loop stabilization represses antisense transcription at the *Arabidopsis* FLC locus. *Science* 340: 619–21.

825. Csorba T., Questa J.I., Sun Q. and Dean C. (2014) Antisense COOLAIR mediates the coordinated switching of chromatin states at FLC during vernalization. *Proceedings of the National Academy of Sciences USA* 111: 16160–5.

826. Zhu D., Rosa S. and Dean C. (2015) Nuclear organization changes and the epigenetic silencing of FLC during vernalization. *Journal of Molecular Biology* 427: 659–69.

827. Hawkes E.J. et al. (2016) COOLAIR antisense RNAs form evolutionarily conserved elaborate secondary structures. *Cell Reports* 16: 3087–96.

828. Kim D.-H., Xi Y. and Sung S. (2017) Modular function of long noncoding RNA, COLDAIR, in the vernalization response. *PLOS Genetics* 13: e1006939.

829. Kim D.H. and Sung S. (2017) Vernalization-triggered intragenic chromatin loop formation by long noncoding RNAs. *Developmental Cell* 40: 302–12.

830. Yang H. et al. (2017) Distinct phases of Polycomb silencing to hold epigenetic memory of cold in *Arabidopsis*. *Science* 357: 1142–5.

831. Whittaker C. and Dean C. (2017) The FLC locus: A platform for discoveries in epigenetics and adaptation. *Annual Review of Cell and Developmental Biology* 33: 555–75.

832. Tian Y. et al. (2019) PRC2 recruitment and H3K27me3 deposition at *FLC* require FCA binding of *COOLAIR*. *Science Advances* 5: eaau7246.

833. Fang X. et al. (2019) Arabidopsis FLL2 promotes liquid–liquid phase separation of polyadenylation complexes. *Nature* 569: 265–9.

834. Xu C. et al. (2021) R-loop resolution promotes co-transcriptional chromatin silencing. *Nature Communications* 12: 1790.

835. Zhu P., Lister C. and Dean C. (2021) Cold-induced Arabidopsis FRIGIDA nuclear condensates for FLC repression. *Nature* 599: 657–61.

836. Rosa S., Duncan S. and Dean C. (2016) Mutually exclusive sense–antisense transcription at FLC facilitates environmentally induced gene repression. *Nature Communications* 7: 13031.

837. Li P., Tao Z. and Dean C. (2015) Phenotypic evolution through variation in splicing of the noncoding RNA COOLAIR. *Genes & Development* 29: 696–701.

838. Berry S. and Dean C. (2015) Environmental perception and epigenetic memory: Mechanistic insight through *FLC*. *Plant Journal* 83: 133–48.

839. Jin Y., Ivanov M., Dittrich A.N., Nelson A.D.L. and Marquardt S. (2021) A *trans*-acting long non-coding RNA represses flowering in *Arabidopsis*. *bioRxiv*: 2021.11.15.468639.

840. Chen Y. et al. (2021) Hovlinc is a recently evolved class of ribozyme found in human lncRNA. *Nature Chemical Biology* 17: 601–7.

841. Shen S. et al. (2021) circPDE4B prevents articular cartilage degeneration and promotes repair by acting as a scaffold for RIC8A and MID1. *Annals of the Rheumatic Diseases* 80: 1209–19.

842. Banks I.R., Zhang Y., Wiggins B.E., Heck G.R. and Ivashuta S. (2012) RNA decoys: An emerging component of plant regulatory networks? *Plant Signaling & Behavior* 7: 1188–93.

843. Thomson D.W. and Dinger M.E. (2016) Endogenous microRNA sponges: Evidence and controversy. *Nature Reviews Genetics* 17: 272–83.

844. Nowak R. (1994) Mining treasures from 'junk DNA'. *Science* 263: 608–10.

845. Pennisi E. (2010) Shining a light on the genome's 'dark matter'. *Science* 330: 1614.

846. Lee H., Zhang Z. and Krause H.M. (2019) Long noncoding RNAs and repetitive elements: Junk or intimate evolutionary partners? *Trends in Genetics* 35: 892–902.

847. Scherer S.W. et al. (2003) Human chromosome 7: DNA sequence and biology. *Science* 300: 767–72.

848. Duan Y., Zhao M., Jiang M., Li Z. and Ni C. (2020) LINC02476 promotes the malignant phenotype of hepatocellular carcinoma by sponging *miR-497* and increasing *HMGA2* expression. *OncoTargets and Therapy* 13: 2701–10.

849. Cech T.R. and Steitz J.A. (2014) The noncoding RNA revolution - trashing old rules to forge new ones. *Cell* 157: 77–94.

850. Stent G.S. (1972) Prematurity and uniqueness in scientific discovery. *Scientific American* 227: 84–93.

CHAPTER 14

1. Waddington C.H. (1942) Canalization of development and the inheritance of acquired characters. *Nature* 150: 563–5.

2. Waddington C.H. (1957) *The Strategy of the Genes: A Discussion of Some Aspects of Theoretical Biology* (George Allen and Unwin, Crows Nest).

3. Waddington C.H. (1966) *Principles of Development and Differentiation* (Macmillan, London).

4. Robertson A. (1977) Conrad Hal Waddington, 8 November 1905–26 September 1975. *Biographical Memoirs of Fellows of the Royal Society* 23: 575–622.

5. Allen M. (2015) Compelled by the diagram: Thinking through C. H. Waddington's epigenetic landscape. *Contemporaneity: Historical Presence in Visual Culture* 4: 119–42.

6. Saksouk N., Simboeck E. and Déjardin J. (2015) Constitutive heterochromatin formation and transcription in mammals. *Epigenetics & Chromatin* 8: 3.

7. Żylicz J.J. and Heard E. (2020) Molecular mechanisms of facultative heterochromatin formation: An X-chromosome perspective. *Annual Review of Biochemistry* 89: 255–82.

8. Dixon J.R. et al. (2015) Chromatin architecture reorganization during stem cell differentiation. *Nature* 518: 331–6.

9. Maeda R.K. and Karch F. (2015) The open for business model of the *bithorax* complex in *Drosophila*. *Chromosoma* 124: 293–307.

10. Olins A.L. and Olins D.E. (1974) Spheroid chromatin units (ν bodies). *Science* 183: 330–2.

11. Kornberg R.D. and Thomas J.O. (1974) Chromatin structure; oligomers of the histones. *Science* 184: 865–8.

12. Kornberg R.D. (1974) Chromatin structure: A repeating unit of histones and DNA. *Science* 184: 868–71.

13. Woodcock C.L.F., Safer J.P. and Stanchfield J.E. (1976) Structural repeating units in chromatin: I. Evidence for their general occurrence. *Experimental Cell Research* 97: 101–10.

14. Olins D.E. and Olins A.L. (2003) Chromatin history: Our view from the bridge. *Nature Reviews Molecular Cell Biology* 4: 809–14.

15. Allfrey V.G. and Mirsky A.E. (1964) Structural modifications of histones and their possible role in the regulation of RNA synthesis. *Science* 144: 559.

16. Allfrey V.G., Faulkner R. and Mirsky A.E. (1964) Acetylation and methylation of histones and their possible role in the regulation of RNA synthesis. *Proceedings of the National Academy of Sciences USA* 51: 786–94.

17. Allfrey V.G., Pogo B.G., Littau V.C., Gershey E.L. and Mirsky A.E. (1968) Histone acetylation in insect chromosomes. *Science* 159: 314–6.

18. Gaspar-Maia A., Alajem A., Meshorer E. and Ramalho-Santos M. (2011) Open chromatin in pluripotency and reprogramming. *Nature Reviews Molecular Cell Biology* 12: 36–47.

19. Finch J.T. and Klug A. (1976) Solenoidal model for superstructure in chromatin. *Proceedings of the National Academy of Sciences USA* 73: 1897–901.

20. Tremethick D.J. (2007) Higher-order structures of chromatin: The elusive 30 nm fiber. *Cell* 128: 651–4.

21. Grigoryev S.A. and Woodcock C.L. (2012) Chromatin organization — the 30nm fiber. *Experimental Cell Research* 318: 1448–55.

22. Krietenstein N. and Rando O.J. (2020) Mesoscale organization of the chromatin fiber. *Current Opinion in Genetics and Development* 61: 32–6.

23. Paulson J.R. and Laemmli U.K. (1977) The structure of histone-depleted metaphase chromosomes. *Cell* 12: 817–28.

24. Marston A.L. and Amon A. (2004) Meiosis: Cell-cycle controls shuffle and deal. *Nature Reviews Molecular Cell Biology* 5: 983–97.

25. McIntosh J.R. (2016) Mitosis. *Cold Spring Harbor Perspectives in Biology* 8: a023218.

26. Fitz-James M.H. et al. (2020) Large domains of heterochromatin direct the formation of short mitotic chromosome loops. *eLife* 9: e57212.

27. Bannister A.J. and Kouzarides T. (2011) Regulation of chromatin by histone modifications. *Cell Research* 21: 381–95.

28. Liu J., Ali M. and Zhou Q. (2020) Establishment and evolution of heterochromatin. *Annals of the New York Academy of Sciences* 1476: 59–77.

29. Pederson T. and Bhorjee J.S. (1979) Evidence for a role of RNA in eukaryotic chromosome structure. Metabolically stable, small nuclear RNA species are covalently linked to chromosomal DNA in HeLa cells. *Journal of Molecular Biology* 128: 451–80.

30. Schubert T. et al. (2012) Df31 protein and snoRNAs maintain accessible higher-order structures of chromatin. *Molecular Cell* 48: 434–44.

31. Schwartz U. et al. (2019) Characterizing the nuclease accessibility of DNA in human cells to map higher order structures of chromatin. *Nucleic Acids Research* 47: 1239–54.

32. Caspersson T., Zech L. and Johansson C. (1970) Differential binding of alkylating fluorochromes in human chromosomes. *Experimental Cell Research* 60: 315–9.

33. Korenberg J.R. and Rykowski M.C. (1988) Human genome organization: Alu, LINES, and the molecular structure of metaphase chromosome bands. *Cell* 53: 391–400.

34. Bernardi G. (1989) The isochore organization of the human genome. *Annual Review of Genetics* 23: 637–59.

35. Craig J.M. and Bickmore W.A. (1993) Genes and genomes: Chromosome bands – flavours to savour. *BioEssays* 15: 349–54.

36. Craig J.M. and Bickmore W.A. (1994) The distribution of CpG islands in mammalian chromosomes. *Nature Genetics* 7: 376–82.

37. Versteeg R. et al. (2003) The human transcriptome map reveals extremes in gene density, intron length, GC content, and repeat pattern for domains of highly and weakly expressed genes. *Genome Research* 13: 1998–2004.

38. Zhimulev I.F., Belyaeva E.S., Vatolina T.Y. and Demakov S.A. (2012) Banding patterns in *Drosophila melanogaster* polytene chromosomes correlate with DNA-binding protein occupancy. *BioEssays* 34: 498–508.

39. Bernardi G. (2015) Chromosome architecture and genome organization. *PLOS ONE* 10: e0143739.

40. Bickmore W.A. (2019) Patterns in the genome. *Heredity* 123: 50–7.

41. Lu J.Y. et al. (2021) Homotypic clustering of L1 and B1/Alu repeats compartmentalizes the 3D genome. *Cell Research* 31: 613–30.

42. Lu J.Y. et al. (2020) Genomic repeats categorize genes with distinct functions for orchestrated regulation. *Cell Reports* 30: 3296–311.

43. Croft J.A. et al. (1999) Differences in the localization and morphology of chromosomes in the human nucleus. *Journal of Cell Biology* 145: 1119–31.

44. Cremer T. and Cremer C. (2001) Chromosome territories, nuclear architecture and gene regulation in mammalian cells. *Nature Reviews Genetics* 2: 292–301.

45. Tanabe H. et al. (2002) Evolutionary conservation of chromosome territory arrangements in cell nuclei from higher primates. *Proceedings of the National Academy of Sciences USA* 99: 4424–9.

46. Bolzer A. et al. (2005) Three-dimensional maps of all chromosomes in human male fibroblast nuclei and prometaphase rosettes. *PLOS Biology* 3: e157.

47. Cremer T. and Cremer M. (2010) Chromosome territories. *Cold Spring Harbor Perspectives in Biology* 2: a003889.

48. Stevens T.J. et al. (2017) 3D structures of individual mammalian genomes studied by single-cell Hi-C. *Nature* 544: 59–64.

49. Rabi C. (1885) On cell division. *Morphological Yearbook* 10: 214–330.

50. Boveri T. (1909) Die Blastomerenkerne von *Ascaris megalocephala* und die Theorie der Chromosomenindividualität. *Archiv für Zellforschung* 3: 181–268.

51. Comings D.E. (1980) Arrangement of chromatin in the nucleus. *Human Genetics* 53: 131–43.

52. Albiez H. et al. (2006) Chromatin domains and the interchromatin compartment form structurally defined and functionally interacting nuclear networks. *Chromosome Research* 14: 707–33.

53. Küpper K. et al. (2007) Radial chromatin positioning is shaped by local gene density, not by gene expression. *Chromosoma* 116: 285–306.

54. Harr J.C., Gonzalez-Sandoval A. and Gasser S.M. (2016) Histones and histone modifications in perinuclear chromatin anchoring: From yeast to man. *EMBO Reports* 17: 139–55.

55. Hu B. et al. (2019) Plant lamin-like proteins mediate chromatin tethering at the nuclear periphery. *Genome Biology* 20: 87.

56. Girelli G. et al. (2020) GPSeq reveals the radial organization of chromatin in the cell nucleus. *Nature Biotechnology* 38: 1184–93.

57. Lieberman-Aiden E. et al. (2009) Comprehensive mapping of long-range interactions reveals folding principles of the human genome. *Science* 326: 289–93.

58. Dong P. et al. (2017) 3D chromatin architecture of large plant genomes determined by local A/B compartments. *Molecular Plant* 10: 1497–509.

59. Dekker J., Rippe K., Dekker M. and Kleckner N. (2002) Capturing chromosome conformation. *Science* 295: 1306–11.

60. Dixon J.R. et al. (2012) Topological domains in mammalian genomes identified by analysis of chromatin interactions. *Nature* 485: 376–80.

61. Nora E.P., Dekker J. and Heard E. (2013) Segmental folding of chromosomes: A basis for structural and regulatory chromosomal neighborhoods? *BioEssays* 35: 818–28.

62. Gibcus J.H. and Dekker J. (2013) The hierarchy of the 3D genome. *Molecular Cell* 49: 773–82.

63. Pombo A. and Dillon N. (2015) Three-dimensional genome architecture: Players and mechanisms. *Nature Reviews Molecular Cell Biology* 16: 245–57.

64. Lupiáñez D.G. et al. (2015) Disruptions of topological chromatin domains cause pathogenic rewiring of gene-enhancer interactions. *Cell* 161: 1012–25.

65. Dekker J. and Heard E. (2015) Structural and functional diversity of topologically associating domains. *FEBS Letters* 589: 2877–84.

66. Dekker J. and Mirny L. (2016) The 3D genome as moderator of chromosomal communication. *Cell* 164: 1110–21.

67. Fortin J.-P. and Hansen K.D. (2015) Reconstructing A/B compartments as revealed by Hi-C using long-range correlations in epigenetic data. *Genome Biology* 16: 180.

68. Rowley M.J. et al. (2017) Evolutionarily conserved principles predict 3D chromatin organization. *Molecular Cell* 67: 837–52.

69. Falk M. et al. (2019) Heterochromatin drives compartmentalization of inverted and conventional nuclei. *Nature* 570: 395–9.

70. Krietenstein N. et al. (2020) Ultrastructural details of mammalian chromosome architecture. *Molecular Cell* 78: 554–65.

71. Grubert F. et al. (2020) Landscape of cohesin-mediated chromatin loops in the human genome. *Nature* 583: 737–43.

72. Chadwick B.P. (2008) DXZ4 chromatin adopts an opposing conformation to that of the surrounding chromosome and acquires a novel inactive X-specific role involving CTCF and antisense transcripts. *Genome Research* 18: 1259–69.

73. Giorgetti L. et al. (2016) Structural organization of the inactive X chromosome in the mouse. *Nature* 535: 575–9.

74. Bonora G. et al. (2018) Orientation-dependent Dxz4 contacts shape the 3D structure of the inactive X chromosome. *Nature Communications* 9: 1445.

75. Andergassen D. et al. (2019) In vivo Firre and Dxz4 deletion elucidates roles for autosomal gene regulation. *eLife* 8: e47214.

76. Nora E.P. et al. (2012) Spatial partitioning of the regulatory landscape of the X-inactivation centre. *Nature* 485: 381–5.

77. van Bemmel J.G. et al. (2019) The bipartite TAD organization of the X-inactivation center ensures opposing developmental regulation of Tsix and Xist. *Nature Genetics* 51: 1024–34.

78. Polymenidou M. (2018) The RNA face of phase separation. *Science* 360: 859–60.

79. Sabari B.R. et al. (2018) Coactivator condensation at super-enhancers links phase separation and gene control. *Science* 361: eaar3958.

80. Berry J., Weber S.C., Vaidya N., Haataja M. and Brangwynne C.P. (2015) RNA transcription modulates phase transition-driven nuclear body assembly. *Proceedings of the National Academy of Sciences USA* 112: E5237–45.

81. Sawyer I.A. and Dundr M. (2017) Chromatin loops and causality loops: The influence of RNA upon spatial nuclear architecture. *Chromosoma* 126: 541–57.

82. Strom A.R. et al. (2017) Phase separation drives heterochromatin domain formation. *Nature* 547: 241–5.

83. Zenk F. et al. (2021) HP1 drives de novo 3D genome reorganization in early *Drosophila* embryos. *Nature* 593: 613–30.

84. Bonev B. et al. (2017) Multiscale 3D genome rewiring during mouse neural development. *Cell* 171: 557–72.

85. Ulianov S.V. et al. (2016) Active chromatin and transcription play a key role in chromosome partitioning into topologically associating domains. *Genome Research* 26: 70–84.

86. Peters J.-M., Tedeschi A. and Schmitz J. (2008) The cohesin complex and its roles in chromosome biology. *Genes & Development* 22: 3089–114.

87. Kim S., Yu N.-K. and Kaang B.-K. (2015) CTCF as a multifunctional protein in genome regulation and gene expression. *Experimental & Molecular Medicine* 47: e166.

88. Sanborn A.L. et al. (2015) Chromatin extrusion explains key features of loop and domain formation in wild-type and engineered genomes. *Proceedings of the National Academy of Sciences USA* 112: E6456–65.

89. Wutz G. et al. (2017) Topologically associating domains and chromatin loops depend on cohesin and are regulated by CTCF, WAPL, and PDS5 proteins. *EMBO Journal* 36: 3573–99.

90. Vian L. et al. (2018) The energetics and physiological impact of cohesin extrusion. *Cell* 173: 1165–78.

91. Pugacheva E.M. et al. (2020) CTCF mediates chromatin looping via N-terminal domain-dependent cohesin retention. *Proceedings of the National Academy of Sciences USA* 117: 2020–31.

92. Ryu J.-K. et al. (2021) Bridging-induced phase separation induced by cohesin SMC protein complexes. *Science Advances* 7: eabe5905.

93. Diehl A.G., Ouyang N. and Boyle A.P. (2020) Transposable elements contribute to cell and species-specific chromatin looping and gene regulation in mammalian genomes. *Nature Communications* 11: 1796.

94. Nanavaty V. et al. (2020) DNA methylation regulates alternative polyadenylation via CTCF and the cohesin complex. *Molecular Cell* 78: 752–64.

95. Mukhopadhyay R. et al. (2004) The binding sites for the chromatin insulator protein CTCF map to DNA methylation-free domains genome-wide. *Genome Research* 14: 1594–602.

96. Wang H. et al. (2012) Widespread plasticity in CTCF occupancy linked to DNA methylation. *Genome Research* 22: 1680–8.

97. Maurano M.T. et al. (2015) Role of DNA methylation in modulating transcription factor occupancy. *Cell Reports* 12: 1184–95.

98. Hashimoto H. et al. (2017) Structural basis for the versatile and methylation-dependent binding of CTCF to DNA. *Molecular Cell* 66: 711–20.

99. Wiehle L. et al. (2019) DNA (de)methylation in embryonic stem cells controls CTCF-dependent chromatin boundaries. *Genome Research* 29: 750–61.

100. Damaschke N.A. et al. (2020) CTCF loss mediates unique DNA hypermethylation landscapes in human cancers. *Clinical Epigenetics* 12: 80.

101. Guelen L. et al. (2008) Domain organization of human chromosomes revealed by mapping of nuclear lamina interactions. *Nature* 453: 948–51.

102. Dobrzynska A., Gonzalo S., Shanahan C. and Askjaer P. (2016) The nuclear lamina in health and disease. *Nucleus* 7: 233–48.

103. Costantini M., Cammarano R. and Bernardi G. (2009) The evolution of isochore patterns in vertebrate genomes. *BMC Genomics* 10: 146.

104. Bernardi G. (2018) The formation of chromatin domains involves a primary step based on the 3-D structure of DNA. *Scientific Reports* 8: 17821.

105. Yunis J.J. (1981) Mid-prophase human chromosomes. The attainment of 2000 bands. *Human Genetics* 56: 293–8.

106. Drouin R. and Richer C.-L. (1989) High-resolution R-banding at the 1250-band level. II. Schematic representation and nomenclature of human RBG-banded chromosomes. *Genome* 32: 425–39.

107. Eagen K.P., Hartl T.A. and Kornberg R.D. (2015) Stable chromosome condensation revealed by chromosome conformation capture. *Cell* 163: 934–46.

108. Macgregor H.C. (2012) Chromomeres revisited. *Chromosome Research* 20: 911–24.

109. Eagen K.P. (2018) Principles of chromosome architecture revealed by Hi-C. *Trends in Biochemical Sciences* 43: 469–78.

110. Beagan J.A. and Phillips-Cremins J.E. (2020) On the existence and functionality of topologically associating domains. *Nature Genetics* 52: 8–16.

111. Pope B.D. et al. (2014) Topologically associating domains are stable units of replication-timing regulation. *Nature* 515: 402–5.

112. Dixon J.R., Gorkin D.U. and Ren B. (2016) Chromatin domains: The unit of chromosome organization. *Molecular Cell* 62: 668–80.

113. Harmston N. et al. (2017) Topologically associating domains are ancient features that coincide with Metazoan clusters of extreme noncoding conservation. *Nature Communications* 8: 441.

114. Yu M. and Ren B. (2017) The three-dimensional organization of mammalian genomes. *Annual Review of Cell and Developmental Biology* 33: 265–89.

115. Winick-Ng W. et al. (2021) Cell-type specialization is encoded by specific chromatin topologies. *Nature* 599: 684–91.

116. Symmons O. et al. (2014) Functional and topological characteristics of mammalian regulatory domains. *Genome Research* 24: 390–400.

117. Zhang Y. et al. (2019) Transcriptionally active HERV-H retrotransposons demarcate topologically associating domains in human pluripotent stem cells. *Nature Genetics* 51: 1380–8.

118. Shuaib M. et al. (2019) Nuclear AGO1 regulates gene expression by affecting chromatin architecture in human cells. *Cell Systems* 9: 446–58.

119. Lonfat N. and Duboule D. (2015) Structure, function and evolution of topologically associating domains (TADs) at HOX loci. *FEBS Letters* 589: 2869–76.

120. Le Dily F. and Beato M. (2015) TADs as modular and dynamic units for gene regulation by hormones. *FEBS Letters* 589: 2885–92.

121. Le Dily F. and Beato M. (2018) Signaling by steroid hormones in the 3D nuclear space. *International Journal of Molecular Sciences* 19: 306.

122. Beagan J.A. et al. (2020) Three-dimensional genome restructuring across timescales of activity-induced neuronal gene expression. *Nature Neuroscience* 23: 707–17.

123. Arnould C. et al. (2021) Loop extrusion as a mechanism for formation of DNA damage repair foci. *Nature* 590: 660–5.

124. Krijger P.H.L. and de Laat W. (2016) Regulation of disease-associated gene expression in the 3D genome. *Nature Reviews Molecular Cell Biology* 17: 771–82.

125. Kaiser V.B. and Semple C.A. (2017) When TADs go bad: Chromatin structure and nuclear organisation in human disease. *F1000Research* 6: 314.

126. Weber B., Zicola J., Oka R. and Stam M. (2016) Plant enhancers: A call for discovery. *Trends in Plant Science* 21: 974–87.

127. Hnisz D. et al. (2013) Super-enhancers in the control of cell identity and disease. *Cell* 155: 934–47.

128. Shlyueva D., Stampfel G. and Stark A. (2014) Transcriptional enhancers: From properties to genome-wide predictions. *Nature Reviews Genetics* 15: 272–86.

129. Fulco C.P. et al. (2019) Activity-by-contact model of enhancer–promoter regulation from thousands of CRISPR perturbations. *Nature Genetics* 51: 1664–9.

130. Arnold P.R., Wells A.D. and Li X.C. (2020) Diversity and emerging roles of enhancer RNA in regulation of gene expression and cell fate. *Frontiers in Cell and Developmental Biology* 7: 377.

131. Souaid C., Bloyer S. and Noordermeer D. (2018) Promoter–enhancer looping and regulatory neighborhoods: Gene regulation in the framework of topologically associating domains, in C. Lavelle and J.-M. Victor (eds.) *Nuclear Architecture and Dynamics* (Academic Press, Cambridge, MA).

132. Lim B. and Levine M.S. (2021) Enhancer-promoter communication: Hubs or loops? *Current Opinion in Genetics & Development* 67: 5–9.

133. Zhu I., Song W., Ovcharenko I. and Landsman D. (2021) A model of active transcription hubs that unifies the roles of active promoters and enhancers. *Nucleic Acids Research* 49: 4493–505.

134. Banerji J., Olson L. and Schaffner W. (1983) A lymphocyte-specific cellular enhancer is located downstream of the joining region in immunoglobulin heavy chain genes. *Cell* 33: 729–40.

135. Kong S., Bohl D., Li C. and Tuan D. (1997) Transcription of the HS2 enhancer toward a cis-linked gene is independent of the orientation, position, and distance of the enhancer relative to the gene. *Molecular and Cellular Biology* 17: 3955–65.

136. Smith E. and Shilatifard A. (2014) Enhancer biology and enhanceropathies. *Nature Structural & Molecular Biology* 21: 210–9.

137. Murakawa Y. et al. (2016) Enhanced identification of transcriptional enhancers provides mechanistic insights into diseases. *Trends in Genetics* 32: 76–88.

138. Buffry A.D., Mendes C.C. and McGregor A.P. (2016) The functionality and evolution of eukaryotic transcriptional enhancers. *Advances in Genetics* 96: 143–206.

139. Rickels R. and Shilatifard A. (2018) Enhancer logic and mechanics in development and disease. *Trends in Cell Biology* 28: 608–30.

140. Schoenfelder S. and Fraser P. (2019) Long-range enhancer–promoter contacts in gene expression control. *Nature Reviews Genetics* 20: 437–55.

141. Hogness D.S. et al. (1985) Regulation and products of the Ubx domain of the bithorax complex. *Cold Spring Harbor Symposia on Quantitative Biology* 50: 181–94.

142. Akam M.E., Martinez-Arias A., Weinzierl R. and Wilde C.D. (1985) Function and expression of ultrabithorax in the *Drosophila* embryo. *Cold Spring Harbor Symposia on Quantitative Biology* 50: 195–200.

143. McCall K., O'Connor M.B. and Bender W. (1994) Enhancer traps in the *Drosophila bithorax* complex mark parasegmental domains. *Genetics* 138: 387–99.

144. Banerji J., Rusconi S. and Schaffner W. (1981) Expression of a β-globin gene is enhanced by remote SV40 DNA sequences. *Cell* 27: 299–308.

145. Deniz Ö et al. (2020) Endogenous retroviruses are a source of enhancers with oncogenic potential in acute myeloid leukaemia. *Nature Communications* 11: 3506.

146. Li Q., Peterson K.R., Fang X. and Stamatoyannopoulos G. (2002) Locus control regions. *Blood* 100: 3077–86.

147. Gillies S.D., Morrison S.L., Oi V.T. and Tonegawa S. (1983) A tissue-specific transcription enhancer element is located in the major intron of a rearranged immunoglobulin heavy chain gene. *Cell* 33: 717–28.

148. Maeda R.K. and Karch F. (2006) The ABC of the *BX-C*: The *bithorax* complex explained. *Development* 133: 1413–22.

149. Park B.K. et al. (2004) Intergenic enhancers with distinct activities regulate Dlx gene expression in the mesenchyme of the branchial arches. *Developmental Biology* 268: 532–45.

150. Miyagi S. et al. (2006) The Sox2 regulatory region 2 functions as a neural stem cell-specific enhancer in the telencephalon. *Journal of Biological Chemistry* 281: 13374–81.

151. Perry M.W., Boettiger A.N. and Levine M. (2011) Multiple enhancers ensure precision of gap gene-expression patterns in the *Drosophila* embryo. *Proceedings of the National Academy of Sciences USA* 108: 13570–5.

152. Stathopoulos A., Van Drenth M., Erives A., Markstein M. and Levine M. (2002) Whole-genome analysis of dorsal-ventral patterning in the *Drosophila* embryo. *Cell* 111: 687–701.

153. Ntini E. and Marsico A. (2019) Functional impacts of non-coding RNA processing on enhancer activity and target gene expression. *Journal of Molecular Cell Biology* 11: 868–79.

154. Halfon M.S. (2019) Studying transcriptional enhancers: The founder fallacy, validation creep, and other biases. *Trends in Genetics* 35: 93–103.

155. Zehnder T., Benner P. and Vingron M. (2019) Predicting enhancers in mammalian genomes using supervised hidden Markov models. *BMC Bioinformatics* 20: 157.

156. O'Kane C.J. and Gehring W.J. (1987) Detection in situ of genomic regulatory elements in *Drosophila*. *Proceedings of the National Academy of Sciences USA* 84: 9123–7.

157. Galloni M., Gyurkovics H., Schedl P. and Karch F. (1993) The bluetail transposon: Evidence for independent cis-regulatory domains and domain boundaries in the bithorax complex. *EMBO Journal* 12: 1087–97.

158. Springer P.S. (2000) Gene traps: Tools for plant development and genomics. *Plant Cell* 12: 1007–20.

159. Trinh L.A. and Fraser S.E. (2013) Enhancer and gene traps for molecular imaging and genetic analysis in zebrafish. *Development, Growth & Differentiation* 55: 434–45.

160. Ogryzko V.V., Schiltz R.L., Russanova V., Howard B.H. and Nakatani Y. (1996) The transcriptional coactivators p300 and CBP are histone acetyltransferases. *Cell* 87: 953–9.

161. Yin J.-W. and Wang G. (2014) The Mediator complex: A master coordinator of transcription and cell lineage development. *Development* 141: 977–87.

162. Allen B.L. and Taatjes D.J. (2015) The Mediator complex: A central integrator of transcription. *Nature Reviews Molecular Cell Biology* 16: 155–66.

163. Soutourina J. (2018) Transcription regulation by the Mediator complex. *Nature Reviews Molecular Cell Biology* 19: 262–74.

164. Lai F. et al. (2013) Activating RNAs associate with Mediator to enhance chromatin architecture and transcription. *Nature* 494: 497–501.

165. Yang Y. et al. (2016) Enhancer RNA-driven looping enhances the transcription of the long noncoding RNA DHRS4-AS1, a controller of the DHRS4 gene cluster. *Scientific Reports* 6: 20961.

166. Bose D.A. et al. (2017) RNA binding to CBP stimulates histone acetylation and transcription. *Cell* 168: 135–49.

167. Heintzman N.D. et al. (2007) Distinct and predictive chromatin signatures of transcriptional promoters and enhancers in the human genome. *Nature Genetics* 39: 311–8.

168. Heintzman N.D. et al. (2009) Histone modifications at human enhancers reflect global cell-type-specific gene expression. *Nature* 459: 108–12.

169. Henriques T. et al. (2018) Widespread transcriptional pausing and elongation control at enhancers. *Genes & Development* 32: 26–41.

170. Shen Y. et al. (2012) A map of the cis-regulatory sequences in the mouse genome. *Nature* 488: 116–20.

171. Pradeepa M.M. et al. (2016) Histone H3 globular domain acetylation identifies a new class of enhancers. *Nature Genetics* 48: 681–6.

172. De Santa F. et al. (2010) A large fraction of extragenic RNA pol II transcription sites overlap enhancers. *PLOS Biology* 8: e1000384.

173. Wang D. et al. (2011) Reprogramming transcription by distinct classes of enhancers functionally defined by eRNA. *Nature* 474: 390–4.

174. Wu H. et al. (2014) Tissue-specific RNA expression marks distant-acting developmental enhancers. *PLOS Genetics* 10: e1004610.

175. Kim T.-K., Hemberg M. and Gray J.M. (2015) Enhancer RNAs: A class of long noncoding RNAs synthesized at enhancers. *Cold Spring Harbor Perspectives in Biology* 7: a018622.

176. Kim Y.W., Lee S., Yun J. and Kim A. (2015) Chromatin looping and eRNA transcription precede the transcriptional activation of gene in the beta-globin locus. *Bioscience Reports* 35: e00179.

177. Arner E. et al. (2015) Transcribed enhancers lead waves of coordinated transcription in transitioning mammalian cells. *Science* 347: 1010–4.

178. Chen H., Du G., Song X. and Li L. (2017) Non-coding transcripts from enhancers: New insights into enhancer activity and gene expression regulation. *Genomics, Proteomics & Bioinformatics* 15: 201–7.

179. Sartorelli V. and Lauberth S.M. (2020) Enhancer RNAs are an important regulatory layer of the epigenome. *Nature Structural & Molecular Biology* 27: 521–8.

180. Heidari N. et al. (2014) Genome-wide map of regulatory interactions in the human genome. *Genome Research* 24: 1905–17.

181. Rubinstein M. and de Souza F.S.J. (2013) Evolution of transcriptional enhancers and animal diversity. *Philosophical Transactions of the Royal Society B: Biological Sciences* 368: 20130017.

182. Sebé-Pedrós A. et al. (2016) The dynamic regulatory genome of *Capsaspora* and the origin of animal multicellularity. *Cell* 165: 1224–37.

183. Closser M. et al. (2021) An expansion of the non-coding genome and its regulatory potential underlies vertebrate neuronal diversity. *Neuron* 110: 70–85.

184. Glinsky G. and Barakat T.S. (2019) The evolution of Great Apes has shaped the functional enhancers' landscape in human embryonic stem cells. *Stem Cell Research* 37: 101456.

185. Aldea D. et al. (2021) Repeated mutation of a developmental enhancer contributed to human thermoregulatory evolution. *Proceedings of the National Academy of Sciences USA* 118: e2021722118.

186. Prabhakar S. et al. (2008) Human-specific gain of function in a developmental enhancer. *Science* 321: 1346–50.

187. Woltering J.M. and Duboule D. (2010) The origin of digits: Expression patterns versus regulatory mechanisms. *Developmental Cell* 18: 526–32.

188. Whyte W.A. et al. (2013) Master transcription factors and mediator establish super-enhancers at key cell identity genes. *Cell* 153: 307–19.

189. Parker S.C.J. et al. (2013) Chromatin stretch enhancer states drive cell-specific gene regulation and harbor human disease risk variants. *Proceedings of the National Academy of Sciences USA* 110: 17921–6.

190. Pott S. and Lieb J.D. (2015) What are super-enhancers? *Nature Genetics* 47: 8–12.

191. Wang X., Cairns M.J. and Yan J. (2019) Super-enhancers in transcriptional regulation and genome organization. *Nucleic Acids Research* 47: 11481–96.

192. Chen H. and Liang H. (2020) A high-resolution map of human enhancer RNA loci characterizes super-enhancer activities in cancer. *Cancer Cell* 38: 701–15.

193. Li S. and Ovcharenko I. (2020) Enhancer jungles establish robust tissue-specific regulatory control in the human genome. *Genomics* 112: 2261–70.

194. Herz H.-M. (2016) Enhancer deregulation in cancer and other diseases. *BioEssays* 38: 1003–15.

195. Nott A. et al. (2019) Brain cell type-specific enhancer-promoter interactome maps and disease-risk association. *Science* 366: 1134–9.

196. Nasser J. et al. (2021) Genome-wide enhancer maps link risk variants to disease genes. *Nature* 593: 238–43.

197. Dong X. et al. (2018) Enhancers active in dopamine neurons are a primary link between genetic variation and neuropsychiatric disease. *Nature Neuroscience* 21: 1482–92.

198. Ptashne M. (1988) How eukaryotic transcriptional activators work. *Nature* 335: 683–9.

199. Jindal G.A. and Farley E.K. (2021) Enhancer grammar in development, evolution, and disease: Dependencies and interplay. *Developmental Cell* 56: 575–87.

200. Larke M.S.C. et al. (2021) Enhancers predominantly regulate gene expression during differentiation via transcription initiation. *Molecular Cell* 81: 983–97.

201. Hansen A.S., Cattoglio C., Darzacq X. and Tjian R. (2018) Recent evidence that TADs and chromatin loops are dynamic structures. *Nucleus* 9: 20–32.

202. Postika N. et al. (2018) Boundaries mediate long-distance interactions between enhancers and promoters in the *Drosophila bithorax* complex. *PLOS Genetics* 14: e1007702.

203. Furlong E.E.M. and Levine M. (2018) Developmental enhancers and chromosome topology. *Science* 361: 1341–5.

204. Ghavi-Helm Y. et al. (2019) Highly rearranged chromosomes reveal uncoupling between genome topology and gene expression. *Nature Genetics* 51: 1272–82.

205. Carullo N.V.N. et al. (2020) Enhancer RNAs predict enhancer–gene regulatory links and are critical for enhancer function in neuronal systems. *Nucleic Acids Research* 48: 9550–70.

206. Benabdallah N.S. et al. (2019) Decreased enhancer-promoter proximity accompanying enhancer activation. *Molecular Cell* 76: 473–84.

207. Kim T.-K. et al. (2010) Widespread transcription at neuronal activity-regulated enhancers. *Nature* 465: 182–7.

208. Li W., Notani D. and Rosenfeld M.G. (2016) Enhancers as non-coding RNA transcription units: Recent insights and future perspectives. *Nature Reviews Genetics* 17: 207–23.

209. Lewis M.W., Li S. and Franco H.L. (2019) Transcriptional control by enhancers and enhancer RNAs. *Transcription* 10: 171–86.

210. Azofeifa J.G. et al. (2018) Enhancer RNA profiling predicts transcription factor activity. *Genome Research* 28: 334–44.

211. Grossman S.R. et al. (2018) Positional specificity of different transcription factor classes within enhancers. *Proceedings of the National Academy of Sciences USA* 115: E7222–30.

212. Hon C.-C. et al. (2017) An atlas of human long non-coding RNAs with accurate 5′ ends. *Nature* 543: 199–204.

213. Soibam B. (2017) Super-lncRNAs: Identification of lncRNAs that target super-enhancers via RNA:DNA:DNA triplex formation. *RNA* 23: 1729–42.

214. Mishra K. and Kanduri C. (2019) Understanding long non-coding RNA and chromatin interactions: What we know so far. *Noncoding RNA* 5: 54.

215. Cai Z. et al. (2020) RIC-seq for global in situ profiling of RNA–RNA spatial interactions. *Nature* 582: 432–7.

216. Thurman R.E. et al. (2012) The accessible chromatin landscape of the human genome. *Nature* 489: 75–82.

217. Dunham I. et al. (2012) An integrated encyclopedia of DNA elements in the human genome. *Nature* 489: 57–74.

218. Zhu J. et al. (2013) Genome-wide chromatin state transitions associated with developmental and environmental cues. *Cell* 152: 642–54.

219. Andersson R. et al. (2014) An atlas of active enhancers across human cell types and tissues. *Nature* 507: 455–61.

220. Woodcock C.L. and Ghosh R.P. (2010) Chromatin higher-order structure and dynamics. *Cold Spring Harbor Perspectives in Biology* 2: a000596.

221. Alberts B. et al. (2002) Chromosomal DNA and Its packaging in the chromatin fiber, in *Molecular Biology of the Cell*, 4th edition (Garland Science, New York).

222. McGhee J.D. and Felsenfeld G. (1980) Nucleosome structure. *Annual Review of Biochemistry* 49: 1115–56.

223. Luger K., Mäder A.W., Richmond R.K., Sargent D.F. and Richmond T.J. (1997) Crystal structure of the nucleosome core particle at 2.8 Å resolution. *Nature* 389: 251–60.

224. Luger K. and Richmond T.J. (1998) The histone tails of the nucleosome. *Current Opinion in Genetics & Development* 8: 140–6.

225. Kornberg R.D. and Lorch Y. (1999) Twenty-five years of the nucleosome, fundamental particle of the eukaryote chromosome. *Cell* 98: 285–94.

226. Zhao Y.-Q., Jordan K. and Lunyak V. (2013) Epigenetics components of aging in the central nervous system. *Neurotherapeutics* 10: 647–63.

227. Mattiroli F. et al. (2017) Structure of histone-based chromatin in Archaea. *Science* 357: 609–12.

228. Attar N. et al. (2020) The histone H3-H4 tetramer is a copper reductase enzyme. *Science* 369: 59–64.

229. Anbar A.D. (2008) Elements and evolution. *Science* 322: 1481–3.

230. Jamrich M., Greenleaf A.L. and Bautz E.K. (1977) Localization of RNA polymerase in polytene chromosomes of *Drosophila melanogaster*. *Proceedings of the National Academy of Sciences USA* 74: 2079–83.

231. Cutter A.R. and Hayes J.J. (2015) A brief review of nucleosome structure. *FEBS Letters* 589: 2914–22.

232. Kasinsky H.E., Lewis J.D., Dacks J.B. and Ausló J. (2001) Origin of H1 linker histones. *FASEB Journal* 15: 34–42.

233. Venkatesh S. and Workman J.L. (2015) Histone exchange, chromatin structure and the regulation of transcription. *Nature Reviews Molecular Cell Biology* 16: 178–89.

234. Lai W.K.M. and Pugh B.F. (2017) Understanding nucleosome dynamics and their links to gene expression and DNA replication. *Nature Reviews Molecular Cell Biology* 18: 548–62.

235. Zhou K., Gaullier G. and Luger K. (2019) Nucleosome structure and dynamics are coming of age. *Nature Structural & Molecular Biology* 26: 3–13.

236. Carone B.R. et al. (2014) High-resolution mapping of chromatin packaging in mouse embryonic stem cells and sperm. *Developmental Cell* 30: 11–22.

237. Jin C. and Felsenfeld G. (2007) Nucleosome stability mediated by histone variants H3.3 and H2A.Z. *Genes & Development* 21: 1519–29.

238. Jin C. et al. (2009) H3.3/H2A.Z double variant-containing nucleosomes mark 'nucleosome-free regions' of active promoters and other regulatory regions. *Nature Genetics* 41: 941–5.

239. He H.H. et al. (2010) Nucleosome dynamics define transcriptional enhancers. *Nature Genetics* 42: 343–7.

240. Xi Y., Yao J., Chen R., Li W. and He X. (2011) Nucleosome fragility reveals novel functional states of chromatin and poises genes for activation. *Genome Research* 21: 718–24.

241. Voong L.N. et al. (2016) Insights into nucleosome organization in mouse embryonic stem cells through chemical mapping. *Cell* 167: 1555–70.

242. Jeffers T.E. and Lieb J.D. (2017) Nucleosome fragility is associated with future transcriptional response to developmental cues and stress in *C. elegans*. *Genome Research* 27: 75–86.

243. Oruba A., Saccani S. and van Essen D. (2020) Role of cell-type specific nucleosome positioning in inducible activation of mammalian promoters. *Nature Communications* 11: 1075.

244. van Daal A. and Elgin S.C. (1992) A histone variant, H2AvD, is essential in *Drosophila melanogaster*. *Molecular Biology of the Cell* 3: 593–602.

245. Clarkson M.J., Wells J.R.E., Gibson F., Saint R. and Tremethick D.J. (1999) Regions of variant histone His2AvD required for *Drosophila* development. *Nature* 399: 694–7.

246. Faast R. et al. (2001) Histone variant H2A.Z is required for early mammalian development. *Current Biology* 11: 1183–7.

247. Ridgway P., Brown K.D., Rangasamy D., Svensson U. and Tremethick D.J. (2004) Unique residues on the H2A.Z containing nucleosome surface are important for *Xenopus laevis* development. *Journal of Biological Chemistry* 279: 43815–20.

248. March-Díaz R. et al. (2008) Histone H2A.Z and homologues of components of the SWR1 complex are required to control immunity in *Arabidopsis*. *Plant Journal* 53: 475–87.

249. Stefanelli G. et al. (2018) Learning and age-related changes in genome-wide H2A.Z binding in the mouse hippocampus. *Cell Reports* 22: 1124–31.

250. Bönisch C. et al. (2012) H2A.Z.2.2 is an alternatively spliced histone H2A.Z variant that causes severe nucleosome destabilization. *Nucleic Acids Research* 40: 5951–64.

251. Giaimo B.D., Ferrante F., Herchenröther A., Hake S.B. and Borggrefe T. (2019) The histone variant H2A.Z in gene regulation. *Epigenetics & Chromatin* 12: 37.

252. Greenberg R.S., Long H.K., Swigut T. and Wysocka J. (2019) Single amino acid change underlies distinct roles of H2A.Z subtypes in human syndrome. *Cell* 178: 1421–36.

253. Vardabasso C. et al. (2015) Histone variant H2A.Z.2 mediates proliferation and drug sensitivity of malignant melanoma. *Molecular Cell* 59: 75–88.

254. Paull T.T. et al. (2000) A critical role for histone H2AX in recruitment of repair factors to nuclear foci after DNA damage. *Current Biology* 10: 886–95.

255. Scully R. and Xie A. (2013) Double strand break repair functions of histone H2AX. *Mutation Research* 750: 5–14.

256. Chakravarthy S. et al. (2005) Structural characterization of the histone variant macroH2A. *Molecular and Cellular biology* 25: 7616–24.

257. Ladurner A.G. (2003) Inactivating chromosomes: A macro domain that minimizes transcription. *Molecular Cell* 12: 1–3.

258. Gaspar-Maia A. et al. (2013) MacroH2A histone variants act as a barrier upon reprogramming towards pluripotency. *Nature Communications* 4: 1565.

259. Douet J. et al. (2017) MacroH2A histone variants maintain nuclear organization and heterochromatin architecture. *Journal of Cell Science* 130: 1570–82.

260. Jiang X., Soboleva T.A. and Tremethick D.J. (2020) Short histone H2A variants: Small in stature but not in function. *Cells* 9: 867.

261. Soboleva T.A. et al. (2012) A unique H2A histone variant occupies the transcriptional start site of active genes. *Nature Structural & Molecular Biology* 19: 25–30.

262. Soboleva T.A. et al. (2017) A new link between transcriptional initiation and pre-mRNA splicing: The RNA binding histone variant H2A.B. *PLOS Genetics* 13: e1006633.

263. Anuar N.D. et al. (2019) Gene editing of the multi-copy H2A.B gene and its importance for fertility. *Genome Biology* 20: 23.

264. Molaro A. et al. (2020) Biparental contributions of the H2A.B histone variant control embryonic development in mice. *PLOS Biology* 18: e3001001.

265. Bao Y. et al. (2004) Nucleosomes containing the histone variant H2A.Bbd organize only 118 base pairs of DNA. *EMBO Journal* 23: 3314–24.

266. Hoghoughi N. et al. (2020) RNA-guided genomic localization of H2A.L.2 histone variant. *Cells* 9: 474.

267. Dahm R. (2010) From discovering to understanding. Friedrich Miescher's attempts to uncover the function of DNA. *EMBO Reports* 11: 153–60.

268. Tagami H., Ray-Gallet D., Almouzni G. and Nakatani Y. (2004) Histone H3.1 and H3.3 complexes mediate nucleosome assembly pathways dependent or independent of DNA synthesis. *Cell* 116: 51–61.

269. Mito Y., Henikoff J.G. and Henikoff S. (2005) Genome-scale profiling of histone H3.3 replacement patterns. *Nature Genetics* 37: 1090–7.

270. Elsaesser S.J., Goldberg A.D. and Allis C.D. (2010) New functions for an old variant: No substitute for histone H3.3. *Current Opinion in Genetics & Development* 20: 110–7.

271. Goldberg A.D. et al. (2010) Distinct factors control histone variant H3.3 localization at specific genomic regions. *Cell* 140: 678–91.

272. Szenker E., Ray-Gallet D. and Almouzni G. (2011) The double face of the histone variant H3.3. *Cell Research* 21: 421–34.

273. van der Heijden G.W. et al. (2007) Chromosome-wide nucleosome replacement and H3.3 incorporation during mammalian meiotic sex chromosome inactivation. *Nature Genetics* 39: 251–8.

274. Ghanim G.E. et al. (2021) Structure of human telomerase holoenzyme with bound telomeric DNA. *Nature* 593: 449–53.

275. Hödl M. and Basler K. (2009) Transcription in the absence of histone H3.3. *Current Biology* 19: 1221–6.

276. Cox S.G. et al. (2012) An essential role of variant histone H3.3 for ectomesenchyme potential of the cranial neural crest. *PLOS Genetics* 8: e1002938.

277. Szenker E., Lacoste N. and Almouzni G. (2012) A developmental requirement for HIRA-dependent H3.3 deposition revealed at gastrulation in *Xenopus*. *Cell Reports* 1: 730–40.

278. Maze I. et al. (2015) Critical role of histone turnover in neuronal transcription and plasticity. *Neuron* 87: 77–94.

279. Bryant L. et al. (2020) Histone H3.3 beyond cancer: Germline mutations in *Histone 3* Family 3A and 3B cause a previously unidentified neurodegenerative disorder in 46 patients. *Science Advances* 6: eabc9207.

280. Schwartzentruber J. et al. (2012) Driver mutations in histone H3.3 and chromatin remodelling genes in paediatric glioblastoma. *Nature* 482: 226–31.

281. Wu G. et al. (2012) Somatic histone H3 alterations in pediatric diffuse intrinsic pontine gliomas and non-brainstem glioblastomas. *Nature Genetics* 44: 251–3.

282. Muhire B.M., Booker M.A. and Tolstorukov M.Y. (2019) Non-neutral evolution of H3.3-encoding genes occurs without alterations in protein sequence. *Scientific Reports* 9: 8472.

283. Couldrey C., Carlton M.B.L., Nolan P.M., Colledge W.H. and Evans M.J. (1999) A retroviral gene trap insertion into the histone 3.3A gene causes partial neonatal lethality, stunted growth, neuromuscular deficits and male subfertility in transgenic mice. *Human Molecular Genetics* 8: 2489–95.

284. Santenard A. et al. (2010) Heterochromatin formation in the mouse embryo requires critical residues of the histone variant H3.3. *Nature Cell Biology* 12: 853–62.

285. Jang C.-W., Shibata Y., Starmer J., Yee D. and Magnuson T. (2015) Histone H3.3 maintains genome integrity during mammalian development. *Genes & Development* 29: 1377–92.

286. Sharma A.B., Dimitrov S., Hamiche A. and Van Dyck E. (2018) Centromeric and ectopic assembly of CENP-A chromatin in health and cancer: Old marks and new tracks. *Nucleic Acids Research* 47: 1051–69.

287. McKinley K.L. and Cheeseman I.M. (2016) The molecular basis for centromere identity and function. *Nature Reviews Molecular Cell Biology* 17: 16–29.

288. Borg M. et al. (2020) Targeted reprogramming of H3K27me3 resets epigenetic memory in plant paternal chromatin. *Nature Cell Biology* 22: 621–9.

289. Hajkova P. et al. (2008) Chromatin dynamics during epigenetic reprogramming in the mouse germ line. *Nature* 452: 877–81.

290. Grosschedl R., Giese K. and Pagel J. (1994) HMG domain proteins: Architectural elements in the assembly of nucleoprotein structures. *Trends in Genetics* 10: 94–100.

291. Bianchi M.E. and Agresti A. (2005) HMG proteins: Dynamic players in gene regulation and differentiation. *Current Opinion in Genetics and Development* 15: 496–506.

292. Mallik R., Kundu A. and Chaudhuri S. (2018) High mobility group proteins: The multifaceted regulators of chromatin dynamics. *Nucleus* 61: 213–26.

293. Dodonova S.O., Zhu F., Dienemann C., Taipale J. and Cramer P. (2020) Nucleosome-bound SOX2 and SOX11 structures elucidate pioneer factor function. *Nature* 580: 669–72.

294. Friedman J.R. and Kaestner K.H. (2006) The Foxa family of transcription factors in development and metabolism. *Cellular and Molecular Life Sciences* 63: 2317–28.

295. Sekiya T., Muthurajan U.M., Luger K., Tulin A.V. and Zaret K.S. (2009) Nucleosome-binding affinity as a primary determinant of the nuclear mobility of the pioneer transcription factor FoxA. *Genes & Development* 23: 804–9.

296. Zaret K.S. and Carroll J.S. (2011) Pioneer transcription factors: Establishing competence for gene expression. *Genes & Development* 25: 2227–41.

297. Fournier M. et al. (2016) FOXA and master transcription factors recruit Mediator and Cohesin to the core transcriptional regulatory circuitry of cancer cells. *Scientific Reports* 6: 34962.

298. Iwafuchi-Doi M. et al. (2016) The pioneer transcription factor FoxA maintains an accessible nucleosome configuration at enhancers for tissue-specific gene activation. *Molecular Cell* 62: 79–91.

299. Gurard-Levin Z.A., Quivy J.-P. and Almouzni G. (2014) Histone chaperones: Assisting histone traffic and nucleosome dynamics. *Annual Review of Biochemistry* 83: 487–517.

300. Hammond C.M., Strømme C.B., Huang H., Patel D.J. and Groth A. (2017) Histone chaperone networks shaping chromatin function. *Nature Reviews Molecular Cell Biology* 18: 141–58.

301. Clapier C.R., Iwasa J., Cairns B.R. and Peterson C.L. (2017) Mechanisms of action and regulation of ATP-dependent chromatin-remodelling complexes. *Nature Reviews Molecular Cell Biology* 18: 407–22.

302. Grunstein M. (1990) Nucleosomes: Regulators of transcription. *Trends in Genetics* 6: 395–400.

303. Lee D.Y., Hayes J.J., Pruss D. and Wolffe A.P. (1993) A positive role for histone acetylation in transcription factor access to nucleosomal DNA. *Cell* 72: 73–84.

304. Guarente L. (1995) Transcriptional coactivators in yeast and beyond. *Trends in Biochemical Sciences* 20: 517–21.

305. Durrin L.K., Mann R.K., Kayne P.S. and Grunstein M. (1991) Yeast histone H4 N-terminal sequence is required for promoter activation in vivo. *Cell* 65: 1023–31.

306. Brownell J.E. et al. (1996) Tetrahymena histone acetyltransferase A: A homolog to yeast Gcn5p linking histone acetylation to gene activation. *Cell* 84: 843–51.

307. Candau R., Zhou J.X., Allis C.D. and Berger S.L. (1997) Histone acetyltransferase activity and interaction with ADA2 are critical for GCN5 function in vivo. *EMBO Journal* 16: 555–65.

308. Taunton J., Hassig C.A. and Schreiber S.L. (1996) A mammalian histone deacetylase related to the yeast transcriptional regulator Rpd3p. *Science* 272: 408–11.

309. Bannister A.J. and Kouzarides T. (1996) The CBP co-activator is a histone acetyltransferase. *Nature* 384: 641–3.

310. Grant P.A. et al. (1997) Yeast Gcn5 functions in two multisubunit complexes to acetylate nucleosomal histones: Characterization of an Ada complex and the SAGA (Spt/Ada) complex. *Genes & Development* 11: 1640–50.

311. Tan M. et al. (2011) Identification of 67 histone marks and histone lysine crotonylation as a new type of histone modification. *Cell* 146: 1016–28.

312. Arnaudo A.M. and Garcia B.A. (2013) Proteomic characterization of novel histone post-translational modifications. *Epigenetics & Chromatin* 6: 24.

313. Peng Z., Mizianty M.J., Xue B., Kurgan L. and Uversky V.N. (2012) More than just tails: Intrinsic disorder in histone proteins. *Molecular BioSystems* 8: 1886–901.

314. Watson M. and Stott K. (2019) Disordered domains in chromatin-binding proteins. *Essays in Biochemistry* 63: 147–56.

315. Cuthbert G.L. et al. (2004) Histone deimination antagonizes arginine methylation. *Cell* 118: 545–53.

316. Christophorou M.A. et al. (2014) Citrullination regulates pluripotency and histone H1 binding to chromatin. *Nature* 507: 104–8.

317. Li P. et al. (2010) PAD4 is essential for antibacterial innate immunity mediated by neutrophil extracellular traps. *Journal of Experimental Medicine* 207: 1853–62.

318. Brinkmann V. and Zychlinsky A. (2012) Neutrophil extracellular traps: Is immunity the second function of chromatin? *Journal of Cell Biology* 198: 773–83.

319. Apel F. et al. (2021) The cytosolic DNA sensor cGAS recognizes neutrophil extracellular traps. *Science Signaling* 14: eaax7942.

320. Kouzarides T. (2007) Chromatin modifications and their function. *Cell* 128: 693–705.

321. Suganuma T. and Workman J.L. (2011) Signals and combinatorial functions of histone modifications. *Annual Review of Biochemistry* 80: 473–99.

322. Wang Y. et al. (2009) Histone hypercitrullination mediates chromatin decondensation and neutrophil extracellular trap formation. *Journal of Cell Biology* 184: 205–13.

323. Kebede A.F. et al. (2017) Histone propionylation is a mark of active chromatin. *Nature Structural & Molecular Biology* 24: 1048–56.

324. Wan J., Liu H., Chu J. and Zhang H. (2019) Functions and mechanisms of lysine crotonylation. *Journal of Cellular and Molecular Medicine* 23: 7163–9.

325. Qin B. et al. (2019) UFL1 promotes histone H4 ufmylation and ATM activation. *Nature Communications* 10: 1242.

326. Qin B. et al. (2020) STK38 promotes ATM activation by acting as a reader of histone H4 ufmylation. *Science Advances* 6: eaax8214.

327. Zhang D. et al. (2019) Metabolic regulation of gene expression by histone lactylation. *Nature* 574: 575–80.

328. Lepack A.E. et al. (2020) Dopaminylation of histone H3 in ventral tegmental area regulates cocaine seeking. *Science* 368: 197–201.

329. Farrelly L.A. et al. (2019) Histone serotonylation is a permissive modification that enhances TFIID binding to H3K4me3. *Nature* 567: 535–9.

330. Mews P. et al. (2019) Alcohol metabolism contributes to brain histone acetylation. *Nature* 574: 717–21.

331. Renthal W. and Nestler E.J. (2009) Histone acetylation in drug addiction. *Seminars in Cell & Developmental Biology* 20: 387–94.

332. Hyun K., Jeon J., Park K. and Kim J. (2017) Writing, erasing and reading histone lysine methylations. *Experimental & Molecular Medicine* 49: e324.

333. Lavarone E., Barbieri C.M. and Pasini D. (2019) Dissecting the role of H3K27 acetylation and methylation in PRC2 mediated control of cellular identity. *Nature Communications* 10: 1679.

334. Cooper S. et al. (2014) Targeting polycomb to pericentric heterochromatin in embryonic stem cells reveals a role for H2AK119u1 in PRC2 recruitment. *Cell Reports* 7: 1456–70.

335. Cao Q. et al. (2014) The central role of EED in the orchestration of polycomb group complexes. *Nature Communications* 5: 3127.

336. van der Vlag J. and Otte A.P. (1999) Transcriptional repression mediated by the human polycomb-group protein EED involves histone deacetylation. *Nature Genetics* 23: 474–8.

337. Herz H.-M., Garruss A. and Shilatifard A. (2013) SET for life: Biochemical activities and biological functions of SET domain-containing proteins. *Trends in Biochemical Sciences* 38: 621–39.

338. Ma R.-G., Zhang Y., Sun T.-T and Cheng B. (2014) Epigenetic regulation by polycomb group complexes: Docus on roles of CBX proteins. *Journal of Zhejiang University. Science B* 15: 412–28.

339. Gil J. and O'Loghlen A. (2014) PRC1 complex diversity: Where is it taking us? *Trends in Cell Biology* 24: 632–41.

340. Chittock E.C., Latwiel S., Miller T.C.R. and Müller C.W. (2017) Molecular architecture of polycomb repressive complexes. *Biochemical Society Transactions* 45: 193–205.

341. Ren X. and Kerppola T.K. (2011) REST interacts with Cbx proteins and regulates polycomb repressive complex 1 occupancy at RE1 elements. *Molecular and Cellular Biology* 31: 2100–10.

342. Plys A.J. et al. (2019) Phase separation of Polycomb-repressive complex 1 is governed by a charged disordered region of CBX2. *Genes & Development* 33: 1–15.

343. Conaway R.C. and Conaway J.W. (2011) Origins and activity of the mediator complex. *Seminars in Cell & Developmental Biology* 22: 729–34.

344. Heger P., Marin B., Bartkuhn M., Schierenberg E. and Wiehe T. (2012) The chromatin insulator CTCF and the emergence of metazoan diversity. *Proceedings of the National Academy of Sciences USA* 109: 17507–12.

345. Tchasovnikarova I.A. et al. (2015) Epigenetic silencing by the HUSH complex mediates position-effect variegation in human cells. *Science* 348: 1481–5.

346. Emerson R.O. and Thomas J.H. (2009) Adaptive evolution in zinc finger transcription factors. *PLOS Genetics* 5: e1000325.

347. Turelli P. et al. (2020) Primate-restricted KRAB zinc finger proteins and target retrotransposons control gene expression in human neurons. *Science Advances* 6: eaba3200.

348. Bannister A.J. et al. (2001) Selective recognition of methylated lysine 9 on histone H3 by the HP1 chromo domain. *Nature* 410: 120–4.

349. Ruthenburg A.J., Allis C.D. and Wysocka J. (2007) Methylation of lysine 4 on histone H3: Intricacy of writing and reading a single epigenetic mark. *Molecular Cell* 25: 15–30.

350. Bonasio R., Lecona E. and Reinberg D. (2010) MBT domain proteins in development and disease. *Seminars in Cell & Developmental Biology* 21: 221–30.

351. Yun M., Wu J., Workman J.L. and Li B. (2011) Readers of histone modifications. *Cell Research* 21: 564–78.

352. Sanchez R. and Zhou M.-M. (2011) The PHD finger: A versatile epigenome reader. *Trends in Biochemical Sciences* 36: 364–72.

353. Liu W. et al. (2013) Brd4 and JMJD6-associated anti-pause enhancers in regulation of transcriptional pause release. *Cell* 155: 1581–95.

354. Shelton S.B. et al. (2018) Crosstalk between the RNA methylation and histone-binding activities of MePCE regulates P-TEFb activation on chromatin. *Cell Reports* 22: 1374–83.

355. Mita M.M. and Mita A.C. (2020) Bromodomain inhibitors a decade later: A promise unfulfilled? *British Journal of Cancer* 123: 1713–4.

356. Kouzarides T. (2000) Acetylation: A regulatory modification to rival phosphorylation? *EMBO Journal* 19: 1176–9.

357. Zucconi B.E. et al. (2019) Combination targeting of the bromodomain and acetyltransferase active site of p300/CBP. *Biochemistry* 58: 2133–43.

358. Maurer-Stroh S. et al. (2003) The Tudor domain 'Royal Family': Tudor, plant Agenet, chromo, PWWP and MBT domains. *Trends in Biochemical Sciences* 28: 69–74.

359. Hard R. et al. (2018) Deciphering and engineering chromodomain-methyllysine peptide recognition. *Science Advances* 4: eaau1447.

360. Chi P., Allis C.D. and Wang G.G. (2010) Covalent histone modifications - miswritten, misinterpreted and mis-erased in human cancers. *Nature Reviews Cancer* 10: 457–69.

361. Audia J.E. and Campbell R.M. (2016) Histone modifications and cancer. *Cold Spring Harbor Perspectives in Biology* 8: a019521.

362. Reddy T.E. (2017) The functional genome: Epigenetics and epigenomics, in G. S. Ginsburg, H. F. Willard, J. Strickler and M. Stuart McKinney (eds.) *Genomic and Precision Medicine* (3rd Edition) (Academic Press, Cambridge, MA).

363. Nacev B.A. et al. (2019) The expanding landscape of 'oncohistone' mutations in human cancers. *Nature* 567: 473–8.

364. Zhao Z. and Shilatifard A. (2019) Epigenetic modifications of histones in cancer. *Genome Biology* 20: 245.

365. Winters A.C. and Bernt K.M. (2017) MLL-rearranged leukemias an update on science and clinical approaches. *Frontiers in Pediatrics* 5: 4.

366. Wang G.G. et al. (2009) Haematopoietic malignancies caused by dysregulation of a chromatin-binding PHD finger. *Nature* 459: 847–51.

367. Williams S.R. et al. (2010) Haploinsufficiency of HDAC4 causes Brachydactyly Mental Retardation Syndrome, with Brachydactyly Type E, developmental delays, and behavioral problems. *American Journal of Human Genetics* 87: 219–28.

368. Park S.-Y. and Kim J.-S. (2020) A short guide to histone deacetylases including recent progress on class II enzymes. *Experimental & Molecular Medicine* 52: 204–12.

369. Abascal F. et al. (2020) Perspectives on ENCODE. *Nature* 583: 693–8.

370. Strahl B.D. and Allis C.D. (2000) The language of covalent histone modifications. *Nature* 403: 41–5.

371. Allis C.D. (2015) "Modifying" my career toward chromatin biology. *Journal of Biological Chemistry* 290: 15904–8.

372. Ptashne M. (1967) Isolation of the lambda phage repressor. *Proceedings of the National Academy of Sciences USA* 57: 306–13.

373. Ptashne M. (1967) Specific binding of the lambda phage repressor to lambda DNA. *Nature* 214: 232–4.

374. Brent R. and Ptashne M. (1985) A eukaryotic transcriptional activator bearing the DNA specificity of a prokaryotic repressor. *Cell* 43: 729–36.

375. Ptashne M. (2013) Epigenetics: Core misconcept. *Proceedings of the National Academy of Sciences USA* 110: 7101–3.

376. Saha S., Ansari A.Z., Jarrell K.A. and Ptashne M. (2003) RNA sequences that work as transcriptional activating regions. *Nucleic Acids Research* 31: 1565–70.

377. Goldberg A.D., Allis C.D. and Bernstein E. (2007) Epigenetics: A landscape takes shape. *Cell* 128: 635–8.

378. Allis C.D. et al. (2007) New nomenclature for chromatin-modifying enzymes. *Cell* 131: 633–6.

379. Turner B.M. (1993) Decoding the nucleosome. *Cell* 75: 5–8.

380. Allis C.D., Caparros M.-L., Jenuwein T. and Reinberg D. (2015) *Epigenetics* (Cold Spring Harbor Laboratory Press, Cold Spring Harbor, NY).

381. Allis C.D. and Jenuwein T. (2016) The molecular hallmarks of epigenetic control. *Nature Reviews Genetics* 17: 487–500.

382. Barski A. et al. (2007) High-resolution profiling of histone methylations in the human genome. *Cell* 129: 823–37.

383. The ENCODE Consortium (2007) Identification and analysis of functional elements in 1% of the human genome by the ENCODE pilot project. *Nature* 447: 799–816.

384. Wang Z. et al. (2008) Combinatorial patterns of histone acetylations and methylations in the human genome. *Nature Genetics* 40: 897–903.

385. The Encode Project Consortium (2012) An integrated encyclopedia of DNA elements in the human genome. *Nature* 489: 57–74.

386. Kundaje A. et al. (2015) Integrative analysis of 111 reference human epigenomes. *Nature* 518: 317–30.

387. Gardner K.E., Allis C.D. and Strahl B.D. (2011) OPERating ON chromatin, a colorful language where context matters. *Journal of Molecular Biology* 409: 36–46.

388. Henikoff S. and Shilatifard A. (2011) Histone modification: Cause or cog? *Trends in Genetics* 27: 389–96.

389. Valencia-Sánchez M.I. et al. (2021) Regulation of the Dot1 histone H3K79 methyltransferase by histone H4K16 acetylation. *Science* 371: eabc6663.

390. Pokholok D.K. et al. (2005) Genome-wide map of nucleosome acetylation and methylation in yeast. *Cell* 122: 517–27.

391. Basilicata M.F. et al. (2018) De novo mutations in MSL3 cause an X-linked syndrome marked by impaired histone H4 lysine 16 acetylation. *Nature Genetics* 50: 1442–51.

392. Monserrat J. et al. (2021) Disruption of the MSL complex inhibits tumour maintenance by exacerbating chromosomal instability. *Nature Cell Biology* 23: 401–12.

393. Motazedian A. and Dawson M.A. (2021) MSL pushes genomic instability over the edge. *Nature Cell Biology* 23: 295–6.

394. Radzisheuskaya A. et al. (2021) Complex-dependent histone acetyltransferase activity of KAT8 determines its role in transcription and cellular homeostasis. *Molecular Cell* 81: 1749–65.

395. Creyghton M.P. et al. (2010) Histone H3K27ac separates active from poised enhancers and predicts developmental state. *Proceedings of the National Academy of Sciences USA* 107: 21931–6.

396. Zhang T., Zhang Z., Dong Q., Xiong J. and Zhu B. (2020) Histone H3K27 acetylation is dispensable for enhancer activity in mouse embryonic stem cells. *Genome Biology* 21: 45.

397. Trojer P. and Reinberg D. (2007) Facultative heterochromatin: Is there a distinctive molecular signature? *Molecular Cell* 28: 1–13.

398. Khan A.A., Lee A.J. and Roh T.-Y. (2015) Polycomb group protein-mediated histone modifications during cell differentiation. *Epigenomics* 7: 75–84.

399. Gibson W.T. et al. (2012) Mutations in EZH2 cause weaver syndrome. *American Journal of Human Genetics* 90: 110–8.

400. Zenk F. et al. (2017) Germ line–inherited H3K27me3 restricts enhancer function during maternal-to-zygotic transition. *Science* 357: 212–6.

401. Guo Y., Zhao S. and Wang G.G. (2021) Polycomb gene silencing mechanisms: PRC2 chromatin targeting, H3K27me3 'readout', and phase separation-based compaction. *Trends in Genetics* 37: 547–65.

402. Cai Y. et al. (2021) H3K27me3-rich genomic regions can function as silencers to repress gene expression via chromatin interactions. *Nature Communications* 12: 719.

403. Ahmad K. and Henikoff S. (2021) The H3.3K27M oncohistone antagonizes reprogramming in *Drosophila*. *PLOS Genetics* 17: e1009225.

404. Pengelly A.R., Copur Ö., Jäckle H., Herzig A. and Müller J. (2013) A histone mutant reproduces the phenotype caused by loss of histone-modifying factor Polycomb. *Science* 339: 698–9.

405. Pease N.A. et al. (2021) Tunable, division-independent control of gene activation timing by a polycomb switch. *Cell Reports* 34: 108888.

406. Heurteau A. et al. (2020) Insulator-based loops mediate the spreading of H3K27me3 over distant micro-domains repressing euchromatin genes. *Genome Biology* 21: 193.

407. Kwasnieski J.C., Fiore C., Chaudhari H.G. and Cohen B.A. (2014) High-throughput functional testing of ENCODE segmentation predictions. *Genome Research* 24: 1595–602.

408. Bonn S. et al. (2012) Tissue-specific analysis of chromatin state identifies temporal signatures of enhancer activity during embryonic development. *Nature Genetics* 44: 148–56.

409. Taylor G.C.A., Eskeland R., Hekimoglu-Balkan B., Pradeepa M.M. and Bickmore W.A. (2013) H4K16 acetylation marks active genes and enhancers of embryonic stem cells, but does not alter chromatin compaction. *Genome Research* 23: 2053–65.

410. Osmala M. and Lähdesmäki H. (2020) Enhancer prediction in the human genome by probabilistic modelling of the chromatin feature patterns. *BMC Bioinformatics* 21: 317.

411. Santos-Rosa H. et al. (2002) Active genes are tri-methylated at K4 of histone H3. *Nature* 419: 407–11.

412. Adelman K. and Lis J.T. (2012) Promoter-proximal pausing of RNA polymerase II: Emerging roles in metazoans. *Nature Reviews Genetics* 13: 720–31.

413. Jonkers I. and Lis J.T. (2015) Getting up to speed with transcription elongation by RNA polymerase II. *Nature Reviews Molecular Cell Biology* 16: 167–77.

414. Saldi T., Riemondy K., Erickson B. and Bentley D.L. (2021) Alternative RNA structures formed during transcription depend on elongation rate and modify RNA processing. *Molecular Cell* 81: 1789–801.

415. Schneider R. et al. (2004) Histone H3 lysine 4 methylation patterns in higher eukaryotic genes. *Nature Cell Biology* 6: 73–7.

416. Guenther M.G., Levine S.S., Boyer L.A., Jaenisch R. and Young R.A. (2007) A chromatin landmark and transcription initiation at most promoters in human cells. *Cell* 130: 77–88.

417. Mikkelsen T.S. et al. (2007) Genome-wide maps of chromatin state in pluripotent and lineage-committed cells. *Nature* 448: 553–60.

418. Vermeulen M. et al. (2007) Selective anchoring of TFIID to nucleosomes by trimethylation of histone H3 lysine 4. *Cell* 131: 58–69.

419. Lauberth S.M. et al. (2013) H3K4me3 interactions with TAF3 regulate preinitiation complex assembly and selective gene activation. *Cell* 152: 1021–36.

420. Howe F.S., Fischl H., Murray S.C. and Mellor J. (2017) Is H3K4me3 instructive for transcription activation? *BioEssays* 39: e201600095.

421. Mercer T.R. et al. (2011) Expression of distinct RNAs from 3′ untranslated regions. *Nucleic Acids Research* 39: 2393–403.

422. Bernstein B.E. et al. (2006) A bivalent chromatin structure marks key developmental genes in embryonic stem cells. *Cell* 125: 315–26.

423. Vastenhouw N.L. et al. (2010) Chromatin signature of embryonic pluripotency is established during genome activation. *Nature* 464: 922–6.

424. Chovanec P. et al. (2021) Widespread reorganisation of pluripotent factor binding and gene regulatory interactions between human pluripotent states. *Nature Communications* 12: 2098.

425. Mei H. et al. (2021) H2AK119ub1 guides maternal inheritance and zygotic deposition of H3K27me3 in mouse embryos. *Nature Genetics* 53: 539–50.

426. Chen Z., Djekidel M.N. and Zhang Y. (2021) Distinct dynamics and functions of H2AK119ub1 and H3K27me3 in mouse preimplantation embryos. *Nature Genetics* 53: 551–63.

427. Armache A. et al. (2020) Histone H3.3 phosphorylation amplifies stimulation-induced transcription. *Nature* 583: 852–7.

428. Lachner M., O'Carroll D., Rea S., Mechtler K. and Jenuwein T. (2001) Methylation of histone H3 lysine 9 creates a binding site for HP1 proteins. *Nature* 410: 116–20.

429. Cubeñas-Potts C. and Matunis M.J. (2013) SUMO: A multifaceted modifier of chromatin structure and function. *Developmental Cell* 24: 1–12.

430. Kolasinska-Zwierz P. et al. (2009) Differential chromatin marking of introns and expressed exons by H3K36me3. *Nature Genetics* 41: 376–81.

431. Leung C.S. et al. (2019) H3K36 methylation and the chromodomain protein Eaf3 are required for proper cotranscriptional spliceosome assembly. *Cell Reports* 27: 3760–9.

432. Hilton I.B. et al. (2015) Epigenome editing by a CRISPR-Cas9-based acetyltransferase activates genes from promoters and enhancers. *Nature Biotechnology* 33: 510–7.

433. Kearns N.A. et al. (2015) Functional annotation of native enhancers with a Cas9-histone demethylase fusion. *Nature Methods* 12: 401–3.

434. Cano-Rodriguez D. et al. (2016) Writing of H3K4Me3 overcomes epigenetic silencing in a sustained but context-dependent manner. *Nature Communications* 7: 12284.

435. Labrie V. et al. (2016) Lactase nonpersistence is directed by DNA-variation-dependent epigenetic aging. *Nature Structural & Molecular Biology* 23: 566–73.

436. Chen L.-F. et al. (2019) Enhancer histone acetylation modulates transcriptional bursting dynamics of neuronal activity-inducible genes. *Cell Reports* 26: 1174–88.

437. Li J. et al. (2021) Programmable human histone phosphorylation and gene activation using a CRISPR/Cas9-based chromatin kinase. *Nature Communications* 12: 896.

438. Liscovitch-Brauer N. et al. (2021) Profiling the genetic determinants of chromatin accessibility with scalable single-cell CRISPR screens. *Nature Biotechnology* 39: 1270–7.

439. Cheung P. et al. (2018) Single-cell chromatin modification profiling reveals increased epigenetic variations with aging. *Cell* 173: 1385–97.

440. Schwartz S., Meshorer E. and Ast G. (2009) Chromatin organization marks exon-intron structure. *Nature Structural & Molecular Biology* 16: 990–5.

441. Tilgner H. et al. (2009) Nucleosome positioning as a determinant of exon recognition. *Nature Structural & Molecular Biology* 16: 996–1001.

442. Andersson R., Enroth S., Rada-Iglesias A., Wadelius C. and Komorowski J. (2009) Nucleosomes are well positioned in exons and carry characteristic histone modifications. *Genome Research* 19: 1732–41.

443. Nahkuri S., Taft R.J. and Mattick J.S. (2009) Nucleosomes are preferentially positioned at exons in somatic and sperm cells. *Cell Cycle* 8: 3420–4.

444. Naftelberg S., Schor I.E., Ast G. and Kornblihtt A.R. (2015) Regulation of alternative splicing through coupling with transcription and chromatin structure. *Annual Review of Biochemistry* 84: 165–98.

445. Mercer T.R. et al. (2013) DNase I-hypersensitive exons colocalize with promoters and distal regulatory elements. *Nature Genetics* 45: 852–9.

446. Schor I.E., Rascovan N., Pelisch F., Alló M. and Kornblihtt A.R. (2009) Neuronal cell depolarization induces intragenic chromatin modifications affecting NCAM alternative splicing. *Proceedings of the National Academy of Sciences USA* 106: 4325–30.

447. Luco R.F. et al. (2010) Regulation of alternative splicing by histone modifications. *Science* 327: 996–1000.

448. Luco R.F. and Misteli T. (2011) More than a splicing code: Integrating the role of RNA, chromatin and non-coding RNA in alternative splicing regulation. *Current Opinion in Genetics and Development* 21: 366–72.

449. Schor I.E., Fiszbein A., Petrillo E. and Kornblihtt A.R. (2013) Intragenic epigenetic changes modulate NCAM alternative splicing in neuronal differentiation. *EMBO Journal* 32: 2264–74.

450. Xu Y., Zhao W., Olson S.D., Prabhakara K.S. and Zhou X. (2018) Alternative splicing links histone modifications to stem cell fate decision. *Genome Biology* 19: 133.

451. Rahhal R. and Seto E. (2019) Emerging roles of histone modifications and HDACs in RNA splicing. *Nucleic Acids Research* 47: 4911–26.

452. Hu Q., Greene C.S. and Heller E.A. (2020) Specific histone modifications associate with alternative exon selection during mammalian development. *Nucleic Acids Research* 48: 4709–24.

453. Agirre E., Oldfield A.J., Bellora N., Segelle A. and Luco R.F. (2021) Splicing-associated chromatin signatures: A combinatorial and position-dependent role for histone marks in splicing definition. *Nature Communications* 12: 682.

454. Alló M. et al. (2009) Control of alternative splicing through siRNA-mediated transcriptional gene silencing. *Nature Structural & Molecular Biology* 16: 717–24.

455. Guang S. et al. (2010) Small regulatory RNAs inhibit RNA polymerase II during the elongation phase of transcription. *Nature* 465: 1097–101.

456. Taft R.J. et al. (2010) Nuclear-localized tiny RNAs are associated with transcription initiation and splice sites in metazoans. *Nature Structural & Molecular Biology* 17: 1030–4.

457. Fan J., Krautkramer K.A., Feldman J.L. and Denu J.M. (2015) Metabolic regulation of histone post-translational modifications. *ACS Chemical Biology* 10: 95–108.

458. Berger S.L. and Sassone-Corsi P. (2016) Metabolic signaling to chromatin. *Cold Spring Harbor Perspectives in Biology* 8: a019463.

459. Etchegaray J.-P. and Mostoslavsky R. (2016) Interplay between metabolism and epigenetics: A nuclear adaptation to environmental changes. *Molecular Cell* 62: 695–711.

460. Costello K.R. and Schones D.E. (2018) Chromatin modifications in metabolic disease: Potential mediators of long-term disease risk. *WIREs Systems Biology and Medicine* 10: e1416.

461. Musselman C.A. and Kutateladze T.G. (2021) Characterization of functional disordered regions within chromatin-associated proteins. *iScience* 24: 102070.

462. Hatos A. et al. (2020) DisProt: Intrinsic protein disorder annotation in 2020. *Nucleic Acids Research* 48: D269–76.

463. Samata M. et al. (2020) Intergenerationally maintained histone H4 lysine 16 acetylation is instructive for future gene activation. *Cell* 182: 127–44.

464. Richard Albert J. and Greenberg M.V.C. (2021) The Polycomb landscape in mouse development. *Nature Genetics* 53: 427–9.

465. Kang H. et al. (2020) Dynamic regulation of histone modifications and long-range chromosomal interactions during postmitotic transcriptional reactivation. *Genes & Development* 34: 913–30.

466. Pelham-Webb B. et al. (2021) H3K27ac bookmarking promotes rapid post-mitotic activation of the pluripotent stem cell program without impacting 3D chromatin reorganization. *Molecular Cell* 81: 1732–48.

467. Escobar T.M. et al. (2019) Active and repressed chromatin domains exhibit distinct nucleosome segregation during DNA replication. *Cell* 179: 953–63.

468. Stewart-Morgan K.R., Petryk N. and Groth A. (2020) Chromatin replication and epigenetic cell memory. *Nature Cell Biology* 22: 361–71.

469. Escobar T.M., Loyola A. and Reinberg D. (2021) Parental nucleosome segregation and the inheritance of cellular identity. *Nature Reviews Genetics* 22: 379–92.

470. Klein K.N. et al. (2021) Replication timing maintains the global epigenetic state in human cells. *Science* 372: 371–8.

471. Saxton D.S. and Rine J. (2019) Epigenetic memory independent of symmetric histone inheritance. *eLife* 8: e51421.

472. Francis N.J., Follmer N.E., Simon M.D., Aghia G. and Butler J.D. (2009) Polycomb proteins remain bound to chromatin and DNA during DNA replication in vitro. *Cell* 137: 110–22.

473. Budhavarapu V.N., Chavez M. and Tyler J.K. (2013) How is epigenetic information maintained through DNA replication? *Epigenetics & Chromatin* 6: 32.

474. Métivier R. et al. (2003) Estrogen receptor-α directs ordered, cyclical, and combinatorial recruitment of cofactors on a natural target promoter. *Cell* 115: 751–63.

475. Métivier R., Reid G. and Gannon F. (2006) Transcription in four dimensions: Nuclear receptor-directed initiation of gene expression. *EMBO Reports* 7: 161–7.

476. Margueron R. and Reinberg D. (2010) Chromatin structure and the inheritance of epigenetic information. *Nature Reviews Genetics* 11: 285–96.

477. Moazed D. (2011) Mechanisms for the inheritance of chromatin states. *Cell* 146: 510–8.

478. Korlach J. and Turner S.W. (2012) Going beyond five bases in DNA sequencing. *Current Opinion in Structural Biology* 22: 251–61.

479. Sood A.J., Viner C. and Hoffman M.M. (2019) DNAmod: The DNA modification database. *Journal of Cheminformatics* 11: 30.

480. Chen C. et al. (2017) Convergence of DNA methylation and phosphorothioation epigenetics in bacterial genomes. *Proceedings of the National Academy of Sciences USA* 114: 4501–6.

481. Wion D. and Casadesús J. (2006) N6-methyl-adenine: An epigenetic signal for DNA–protein interactions. *Nature Reviews Microbiology* 4: 183–92.

482. Casadesús J. and Low D. (2006) Epigenetic gene regulation in the bacterial world. *Microbiology and Molecular Biology Reviews* 70: 830–56.

483. Blow M.J. et al. (2016) The epigenomic landscape of prokaryotes. *PLOS Genetics* 12: e1005854.

484. Adhikari S. and Curtis P.D. (2016) DNA methyltransferases and epigenetic regulation in bacteria. *FEMS Microbiology Reviews* 40: 575–91.

485. Hattman S. (2005) DNA-[adenine] methylation in lower eukaryotes. *Biochemistry (Moscow)* 70: 550–8.

486. Ma C. et al. (2019) N6-methyldeoxyadenine is a transgenerational epigenetic signal for mitochondrial stress adaptation. *Nature Cell Biology* 21: 319–27.

487. Fu Y. et al. (2015) N6-methyldeoxyadenosine marks active transcription start sites in *Chlamydomonas*. *Cell* 161: 879–92.

488. Zhang G. et al. (2015) N^6-methyladenine DNA modification in *Drosophila*. *Cell* 161: 893–906.

489. Liu J. et al. (2016) Abundant DNA 6mA methylation during early embryogenesis of zebrafish and pig. *Nature Communications* 7: 13052.

490. Meyer K.D. and Jaffrey S.R. (2016) Expanding the diversity of DNA base modifications with N^6-methyldeoxyadenosine. *Genome Biology* 17: 5.

491. Wu T.P. et al. (2016) DNA methylation on N6-adenine in mammalian embryonic stem cells. *Nature* 532: 329–33.

492. Li X. et al. (2019) The DNA modification N6-methyl-2'-deoxyadenosine (m6dA) drives activity-induced gene expression and is required for fear extinction. *Nature Neuroscience* 22: 534–44.

493. O'Brown Z.K. et al. (2019) Sources of artifact in measurements of 6mA and 4mC abundance in eukaryotic genomic DNA. *BMC Genomics* 20: 445.

494. Douvlataniotis K., Bensberg M., Lentini A., Gylemo B. and Nestor C.E. (2020) No evidence for DNA N6-methyladenine in mammals. *Science Advances* 6: eaay3335.

495. Fedoroff N.V. (1989) About maize transposable elements and development. *Cell* 56: 181–91.

496. Smith Z.D. and Meissner A. (2013) DNA methylation: Roles in mammalian development. *Nature Reviews Genetics* 14: 204–20.

497. Calarco J.P. et al. (2012) Reprogramming of DNA methylation in pollen guides epigenetic inheritance via small RNA. *Cell* 151: 194–205.

498. Law J.A. and Jacobsen S.E. (2010) Establishing, maintaining and modifying DNA methylation patterns in plants and animals. *Nature Reviews Genetics* 11: 204–20.

499. Zhang H., Lang Z. and Zhu J.-K. (2018) Dynamics and function of DNA methylation in plants. *Nature Reviews Molecular Cell Biology* 19: 489–506.

500. Chandler V.L. and Walbot V. (1986) DNA modification of a maize transposable element correlates with loss of activity. *Proceedings of the National Academy of Sciences USA* 83: 1767–71.

501. Chomet P.S., Wessler S. and Dellaporta S.L. (1987) Inactivation of the maize transposable element Activator (Ac) is associated with its DNA modification. *EMBO Journal* 6: 295–302.

502. Regev A., Lamb M.J. and Jablonka E. (1998) The role of DNA methylation in invertebrates: Developmental regulation or genome defense? *Molecular Biology and Evolution* 15: 880–91.

503. Dunwell T.L. and Pfeifer G.P. (2014) Drosophila genomic methylation: New evidence and new questions. *Epigenomics* 6: 459–61.

504. Tweedie S., Charlton J., Clark V. and Bird A. (1997) Methylation of genomes and genes at the invertebrate-vertebrate boundary. *Molecular and Cellular Biology* 17: 1469–75.

505. Riggs A.D. (1975) X inactivation, differentiation, and DNA methylation. *Cytogenetics and Cell Genetics* 14: 9–25.

506. Holliday R. and Pugh J.E. (1975) DNA modification mechanisms and gene activity during development. *Science* 187: 226–32.

507. Jiang L. et al. (2013) Sperm, but not oocyte, DNA methylome is inherited by zebrafish early embryos. *Cell* 153: 773–84.

508. Potok M.E., Nix D.A., Parnell T.J. and Cairns B.R. (2013) Reprogramming the maternal zebrafish genome after fertilization to match the paternal methylation pattern. *Cell* 153: 759–72.

509. Macleod D., Clark V.H. and Bird A. (1999) Absence of genome-wide changes in DNA methylation during development of the zebrafish. *Nature Genetics* 23: 139–40.

510. Ortega-Recalde O., Day R.C., Gemmell N.J. and Hore T.A. (2019) Zebrafish preserve global germline DNA methylation while sex-linked rDNA is amplified and demethylated during feminisation. *Nature Communications* 10: 3053.

511. Skvortsova K. et al. (2019) Retention of paternal DNA methylome in the developing zebrafish germline. *Nature Communications* 10: 3054.

512. Veenstra G.J.C. and Wolffe A.P. (2001) Constitutive genomic methylation during embryonic development of *Xenopus. Biochimica et Biophysica Acta* 1521: 39–44.

513. Lee Heather J., Hore Timothy A. and Reik W. (2014) Reprogramming the methylome: Erasing memory and creating diversity. *Cell Stem Cell* 14: 710–9.

514. Messerschmidt D.M., Knowles B.B. and Solter D. (2014) DNA methylation dynamics during epigenetic reprogramming in the germline and preimplantation embryos. *Genes & Development* 28: 812–28.

515. Miyoshi N. et al. (2016) Erasure of DNA methylation, genomic imprints, and epimutations in a primordial germ-cell model derived from mouse pluripotent stem cells. *Proceedings of the National Academy of Sciences USA* 113: 9545–50.

516. Zeng Y. and Chen T. (2019) DNA methylation reprogramming during mammalian development. *Genes* 10: 257.

517. Hackett J.A. and Surani M.A. (2013) DNA methylation dynamics during the mammalian life cycle. *Philosophical Transactions of the Royal Society B: Biological Sciences* 368: 20110328.

518. Li C. et al. (2018) DNA methylation reprogramming of functional elements during mammalian embryonic development. *Cell Discovery* 4: 41.

519. Jones P.A. and Taylor S.M. (1980) Cellular differentiation, cytidine analogs and DNA methylation. *Cell* 20: 85–93.

520. Carr B.I., Reilly J.G., Smith S.S., Winberg C. and Riggs A. (1984) The tumorigenicity of 5-azacytidine in the male Fischer rat. *Carcinogenesis* 5: 1583–90.

521. Riggs A.D. (2002) X chromosome inactivation, differentiation, and DNA methylation revisited, with a tribute to Susumu Ohno. *Cytogenetic and Genome Research* 99: 17–24.

522. Hansen R.S. (2003) X inactivation-specific methylation of LINE-1 elements by DNMT3B: Implications for the Lyon repeat hypothesis. *Human Molecular Genetics* 12: 2559–67.

523. Bird A., Taggart M., Frommer M., Miller O.J. and Macleod D. (1985) A fraction of the mouse genome that is derived from islands of nonmethylated, CpG-rich DNA. *Cell* 40: 91–9.

524. Wutz A. et al. (1997) Imprinted expression of the *Igf2r* gene depends on an intronic CpG island. *Nature* 389: 745–9.

525. Smilinich N.J. et al. (1999) A maternally methylated CpG island in KvLQT1 is associated with an antisense paternal transcript and loss of imprinting in Beckwith-Wiedemann syndrome. *Proceedings of the National Academy of Sciences USA* 96: 8064–9.

526. Bird A. (2002) DNA methylation patterns and epigenetic memory. *Genes & Development* 16: 6–21.

527. Deaton A.M. and Bird A. (2011) CpG islands and the regulation of transcription. *Genes & Development* 25: 1010–22.

528. Deniz Ö., Frost J.M. and Branco M.R. (2019) Regulation of transposable elements by DNA modifications. *Nature Reviews Genetics* 20: 417–31.

529. Saxonov S., Berg P. and Brutlag D.L. (2006) A genome-wide analysis of CpG dinucleotides in the human genome distinguishes two distinct classes of promoters. *Proceedings of the National Academy of Sciences USA* 103: 1412–7.

530. Fuks F. et al. (2003) The methyl-CpG-binding protein MeCP2 links DNA methylation to histone methylation. *Journal of Biological Chemistry* 278: 4035–40.

531. Rose N.R. and Klose R.J. (2014) Understanding the relationship between DNA methylation and histone lysine methylation. *Biochimica et Biophysica Acta* 1839: 1362–72.

532. Bird A.P. (1986) CpG-rich islands and the function of DNA methylation. *Nature* 321: 209–13.

533. Gardiner-Garden M. and Frommer M. (1987) CpG islands in vertebrate genomes. *Journal of Molecular Biology* 196: 261–82.

534. Brinkman A.B. et al. (2012) Sequential ChIP-bisulfite sequencing enables direct genome-scale investigation of chromatin and DNA methylation cross-talk. *Genome Research* 22: 1128–38.

535. Vavouri T. and Lehner B. (2012) Human genes with CpG island promoters have a distinct transcription-associated chromatin organization. *Genome Biology* 13: R110.

536. Jeziorska D.M. et al. (2017) DNA methylation of intragenic CpG islands depends on their transcriptional activity during differentiation and disease. *Proceedings of the National Academy of Sciences USA* 114: E7526–35.

537. Weinberg D.N. et al. (2019) The histone mark H3K36me2 recruits DNMT3A and shapes the intergenic DNA methylation landscape. *Nature* 573: 281–6.

538. Zaghi M., Broccoli V. and Sessa A. (2020) H3K36 methylation in neural development and associated diseases. *Frontiers in Genetics* 10: 1291.

539. Huff J.T. and Zilberman D. (2014) Dnmt1-independent CG methylation contributes to nucleosome positioning in diverse eukaryotes. *Cell* 156: 1286–97.

540. Lev Maor G., Yearim A. and Ast G. (2015) The alternative role of DNA methylation in splicing regulation. *Trends in Genetics* 31: 274–80.

541. Jadhav U. et al. (2019) Extensive recovery of embryonic enhancer and gene memory stored in hypomethylated enhancer DNA. *Molecular Cell* 74: 542–54.

542. Reizel Y. et al. (2021) FoxA-dependent demethylation of DNA initiates epigenetic memory of cellular identity. *Developmental Cell* 56: 602–12.

543. Li E., Bestor T.H. and Jaenisch R. (1992) Targeted mutation of the DNA methyltransferase gene results in embryonic lethality. *Cell* 69: 915–26.

544. Okano M., Bell D.W., Haber D.A. and Li E. (1999) DNA methyltransferases Dnmt3a and Dnmt3b are essential for de novo methylation and mammalian development. *Cell* 99: 247–57.

545. Cui D. and Xu X. (2018) DNA methyltransferases, DNA methylation, and age-associated cognitive function. *International Journal of Molecular Sciences* 19: 1315.

546. Nishiyama A. et al. (2013) Uhrf1-dependent H3K23 ubiq-uitylation couples maintenance DNA methylation and replication. *Nature* 502: 249–53.
547. Métivier R. et al. (2008) Cyclical DNA methylation of a transcriptionally active promoter. *Nature* 452: 45–50.
548. Kangaspeska S. et al. (2008) Transient cyclical methylation of promoter DNA. *Nature* 452: 112–5.
549. Miller C.A. and Sweatt J.D. (2007) Covalent modification of DNA regulates memory formation. *Neuron* 53: 857–69.
550. Ip J.P.K., Mellios N. and Sur M. (2018) Rett syndrome: Insights into genetic, molecular and circuit mechanisms. *Nature Reviews Neuroscience* 19: 368–82.
551. de Mendoza A. et al. (2021) The emergence of the brain non-CpG methylation system in vertebrates. *Nature Ecology & Evolution* 5: 369–78.
552. Goll M.G. et al. (2006) Methylation of tRNA^{Asp} by the DNA methyltransferase homolog Dnmt2. *Science* 311: 395–8.
553. Rai K. et al. (2007) Dnmt2 functions in the cytoplasm to promote liver, brain, and retina development in zebrafish. *Genes & Development* 21: 261–6.
554. Jeltsch A. et al. (2017) Mechanism and biological role of Dnmt2 in nucleic acid methylation. *RNA Biology* 14: 1108–23.
555. Tahiliani M. et al. (2009) Conversion of 5-methylcytosine to 5-hydroxymethylcytosine in mammalian DNA by MLL partner TET1. *Science* 324: 930–5.
556. Ito S. et al. (2011) Tet proteins can convert 5-methylcytosine to 5-formylcytosine and 5-carboxylcytosine. *Science* 333: 1300–3.
557. Ito S. et al. (2010) Role of Tet proteins in 5mC to 5hmC conversion, ES-cell self-renewal and inner cell mass specification. *Nature* 466: 1129–33.
558. Yang J., Bashkenova N., Zang R., Huang X. and Wang J. (2020) The roles of TET family proteins in development and stem cells. *Development* 147: dev183129.
559. Lu F., Liu Y., Jiang L., Yamaguchi S. and Zhang Y. (2014) Role of Tet proteins in enhancer activity and telomere elongation. *Genes & Development* 28: 2103–19.
560. Costa Y. et al. (2013) NANOG-dependent function of TET1 and TET2 in establishment of pluripotency. *Nature* 495: 370–4.
561. Dixon G. et al. (2021) QSER1 protects DNA methylation valleys from de novo methylation. *Science* 372: eabd0875.
562. Gu T. and Goodell M.A. (2021) The push and pull of DNA methylation. *Science* 372: 128–9.
563. Kriaucionis S. and Heintz N. (2009) The nuclear DNA base 5-hydroxymethylcytosine is present in Purkinje neurons and the brain. *Science* 324: 929–30.
564. Antunes C. et al. (2021) Tet3 ablation in adult brain neurons increases anxiety-like behavior and regulates cognitive function in mice. *Molecular Psychiatry* 26: 1445–57.
565. Li X. et al. (2014) Neocortical Tet3-mediated accumulation of 5-hydroxymethylcytosine promotes rapid behavioral adaptation. *Proceedings of the National Academy of Sciences USA* 111: 7120–5.
566. Yu H. et al. (2015) Tet3 regulates synaptic transmission and homeostatic plasticity via DNA oxidation and repair. *Nature Neuroscience* 18: 836–43.
567. Frommer M. et al. (1992) A genomic sequencing protocol that yields a positive display of 5-methylcytosine residues in individual DNA strands. *Proceedings of the National Academy of Sciences USA* 89: 1827–31.
568. Clark S.J., Harrison J., Paul C.L. and Frommer M. (1994) High sensitivity mapping of methylated cytosines. *Nucleic acids research* 22: 2990–7.
569. Simpson J.T. et al. (2017) Detecting DNA cytosine methylation using nanopore sequencing. *Nature Methods* 14: 407–10.
570. Rand A.C. et al. (2017) Mapping DNA methylation with high-throughput nanopore sequencing. *Nature methods* 14: 411–3.
571. Swallow D.M. and Troelsen J.T. (2016) Escape from epigenetic silencing of lactase expression is triggered by a single-nucleotide change. *Nature Structural & Molecular Biology* 23: 505–7.
572. Horvath S. (2013) DNA methylation age of human tissues and cell types. *Genome Biology* 14: R115.
573. Wagner W. (2017) Epigenetic aging clocks in mice and men. *Genome Biology* 18: 107.
574. Boix C.A., James B.T., Park Y.P., Meuleman W. and Kellis M. (2021) Regulatory genomic circuitry of human disease loci by integrative epigenomics. *Nature* 590: 300–7.
575. Lin X. et al. (2020) Genome-wide analysis of aberrant methylation of enhancer DNA in human osteoarthritis. *BMC Medical Genomics* 13: 1.
576. Bartosovic M., Kabbe M. and Castelo-Branco G. (2021) Single-cell CUT&Tag profiles histone modifications and transcription factors in complex tissues. *Nature Biotechnology* 39: 825–35.
577. Wu S.J. et al. (2021) Single-cell CUT&Tag analysis of chromatin modifications in differentiation and tumor progression. *Nature Biotechnology* 39: 819–24.
578. Näär A.M., Lemon B.D. and Tjian R. (2001) Transcriptional coactivator complexes. *Annual Review of Biochemistry* 70: 475–501.
579. Iwafuchi-Doi M. and Zaret K.S. (2014) Pioneer transcription factors in cell reprogramming. *Genes & Development* 28: 2679–92.
580. Soufi A. et al. (2015) Pioneer transcription factors target partial DNA motifs on nucleosomes to initiate reprogramming. *Cell* 161: 555–68.
581. Iwafuchi-Doi M. (2019) The mechanistic basis for chromatin regulation by pioneer transcription factors. *WIREs Systems Biology and Medicine* 11: e1427.

CHAPTER 15

1. Maturana H.R. and Varela F.J. (1972) *Autopoiesis and Cognition: The Realization of the Living* (D. Reidel Publishing Company, Dordrecht).
2. Ohtomo Y., Kakegawa T., Ishida A., Nagase T. and Rosing M.T. (2014) Evidence for biogenic graphite in early Archaean Isua metasedimentary rocks. *Nature Geoscience* 7: 25–8.
3. Hoyle F. and Wickramasinghe N.C. (1986) The case for life as a cosmic phenomenon. *Nature* 322: 509–11.
4. Pelletier J.F. et al. (2021) Genetic requirements for cell division in a genomically minimal cell. *Cell* 184: 2430–40.
5. Etchegaray E., Naville M., Volff J.-N. and Haftek-Terreau Z. (2021) Transposable element-derived sequences in vertebrate development. *Mobile DNA* 12: 1.
6. Cosby R.L. et al. (2021) Recurrent evolution of vertebrate transcription factors by transposase capture. *Science* 371: eabc6405.

7. Wang K. et al. (2021) African lungfish genome sheds light on the vertebrate water-to-land transition. *Cell* 184: 1362–76.

8. van de Lagemaat L.N., Landry J.-R., Mager D.L. and Medstrand P. (2003) Transposable elements in mammals promote regulatory variation and diversification of genes with specialized functions. *Trends in Genetics* 19: 530–6.

9. Roller M. et al. (2021) LINE retrotransposons characterize mammalian tissue-specific and evolutionarily dynamic regulatory regions. *Genome Biology* 22: 62.

10. Jacques P.-É., Jeyakani J. and Bourque G. (2013) The majority of primate-specific regulatory sequences are derived from transposable elements. *PLOS Genetics* 9: e1003504.

11. Trizzino M. et al. (2017) Transposable elements are the primary source of novelty in primate gene regulation. *Genome Research* 27: 1623–33.

12. Bianconi E. et al. (2013) An estimation of the number of cells in the human body. *Annals of Human Biology* 40: 463–71.

13. Sender R., Fuchs S. and Milo R. (2016) Revised estimates for the number of human and bacteria cells in the body. *PLOS Biology* 14: e1002533.

14. Rivron N.C. et al. (2018) Blastocyst-like structures generated solely from stem cells. *Nature* 557: 106–11.

15. Sozen B. et al. (2018) Self-assembly of embryonic and two extra-embryonic stem cell types into gastrulating embryo-like structures. *Nature Cell Biology* 20: 979–89.

16. Li R. et al. (2019) Generation of blastocyst-like structures from mouse embryonic and adult cell cultures. *Cell* 179: 687–702.

17. Yu L. et al. (2021) Blastocyst-like structures generated from human pluripotent stem cells. *Nature* 591: 620–6.

18. Liu X. et al. (2021) Modelling human blastocysts by reprogramming fibroblasts into iBlastoids. *Nature* 591: 627–32.

19. Clevers H. (2016) Modeling development and disease with organoids. *Cell* 165: 1586–97.

20. Kim J., Koo B.-K. and Knoblich J.A. (2020) Human organoids: Model systems for human biology and medicine. *Nature Reviews Molecular Cell Biology* 21: 571–84.

21. Matthews K.R.W. et al. (2021) Rethinking human embryo research policies. *Hastings Center Report* 51: 47–51.

22. Miller A.J. et al. (2019) Generation of lung organoids from human pluripotent stem cells in vitro. *Nature Protocols* 14: 518–40.

23. Prior N., Inacio P. and Huch M. (2019) Liver organoids: From basic research to therapeutic applications. *Gut* 68: 2228–37.

24. Pleguezuelos-Manzano C. et al. (2020) Establishment and culture of human intestinal organoids derived from adult stem cells. *Current Protocols in Immunology* 130: e106.

25. Sato T. et al. (2011) Long-term expansion of epithelial organoids from human colon, adenoma, adenocarcinoma, and barrett's epithelium. *Gastroenterology* 141: 1762–72.

26. Nishinakamura R. (2019) Human kidney organoids: Progress and remaining challenges. *Nature Reviews Nephrology* 15: 613–24.

27. O'Hara-Wright M. and Gonzalez-Cordero A. (2020) Retinal organoids: A window into human retinal development. *Development* 147: dev189746.

28. Lancaster M.A. et al. (2013) Cerebral organoids model human brain development and microcephaly. *Nature* 501: 373–9.

29. Luo C. et al. (2016) Cerebral organoids recapitulate epigenomic signatures of the human fetal brain. *Cell Reports* 17: 3369–84.

30. Lancaster M.A. et al. (2017) Guided self-organization and cortical plate formation in human brain organoids. *Nature Biotechnology* 35: 659–66.

31. Sachs N., Tsukamoto Y., Kujala P., Peters P.J. and Clevers H. (2017) Intestinal epithelial organoids fuse to form self-organizing tubes in floating collagen gels. *Development* 144: 1107–12.

32. Giandomenico S.L. et al. (2019) Cerebral organoids at the air–liquid interface generate diverse nerve tracts with functional output. *Nature Neuroscience* 22: 669–79.

33. Pellegrini L. et al. (2020) Human CNS barrier-forming organoids with cerebrospinal fluid production. *Science* 369: eaaz5626.

34. Benito-Kwiecinski S. et al. (2021) An early cell shape transition drives evolutionary expansion of the human forebrain. *Cell* 184: 2084–102.

35. Korpal M., Lee E.S., Hu G. and Kang Y. (2008) The miR-200 family inhibits epithelial-mesenchymal transition and cancer cell migration by direct targeting of E-cadherin transcriptional repressors ZEB1 and ZEB2. *Journal of Biological Chemistry* 283: 14910–4.

36. Beltran M. et al. (2008) A natural antisense transcript regulates Zeb2/Sip1 gene expression during Snail1-induced epithelial-mesenchymal transition. *Genes & Development* 22: 756–69.

37. Bernardes de Jesus B. et al. (2018) Silencing of the lncRNA *Zeb2-NAT* facilitates reprogramming of aged fibroblasts and safeguards stem cell pluripotency. *Nature Communications* 9: 94.

38. Chen X. et al. (2016) Upregulation of long noncoding RNA HOTTIP promotes metastasis of esophageal squamous cell carcinoma via induction of EMT. *Oncotarget* 7: 84480–5.

39. Xu Q. et al. (2016) Long non-coding RNA regulation of epithelial–mesenchymal transition in cancer metastasis. *Cell Death & Disease* 7: e2254.

40. Shen X. et al. (2020) The long noncoding RNA TUG1 is required for TGF-β/TWIST1/EMT-mediated metastasis in colorectal cancer cells. *Cell Death & Disease* 11: 65.

41. Xiong T. et al. (2020) LncRNA *NRON* promotes the proliferation, metastasis and EMT process in bladder cancer. *Journal of Cancer* 11: 1751–60.

42. Eze U.C., Bhaduri A., Haeussler M., Nowakowski T.J. and Kriegstein A.R. (2021) Single-cell atlas of early human brain development highlights heterogeneity of human neuroepithelial cells and early radial glia. *Nature Neuroscience* 24: 584–94.

43. Heintzman N.D. et al. (2009) Histone modifications at human enhancers reflect global cell-type-specific gene expression. *Nature* 459: 108–12.

44. Waddington C.H. (1957) *The Strategy of the Genes: A Discussion of Some Aspects of Theoretical Biology* (George Allen and Unwin, Crows Nest).

45. Reilly M.B., Cros C., Varol E., Yemini E. and Hobert O. (2020) Unique homeobox codes delineate all the neuron classes of C. elegans. *Nature* 584: 595–601.

46. Rinn J.L., Bondre C., Gladstone H.B., Brown P.O. and Chang H.Y. (2006) Anatomic demarcation by positional variation in fibroblast gene expression programs. *PLOS Genetics* 2: e119.

47. Rinn J.L. et al. (2008) A dermal HOX transcriptional program regulates site-specific epidermal fate. *Genes & Development* 22: 303–7.
48. Frank-Bertoncelj M. et al. (2017) Epigenetically-driven anatomical diversity of synovial fibroblasts guides joint-specific fibroblast functions. *Nature Communications* 8: 14852.
49. Marioni J.C. and Arendt D. (2017) How single-cell genomics is changing evolutionary and developmental biology. *Annual Review of Cell and Developmental Biology* 33: 537–53.
50. Arendt D., Bertucci P.Y., Achim K. and Musser J.M. (2019) Evolution of neuronal types and families. *Current Opinion in Neurobiology* 56: 144–52.
51. Vinogradov A.E. and Anatskaya O.V. (2007) Organismal complexity, cell differentiation and gene expression: Human over mouse. *Nucleic Acids Research* 35: 6350–6.
52. Shapiro E., Biezuner T. and Linnarsson S. (2013) Single-cell sequencing-based technologies will revolutionize whole-organism science. *Nature Reviews Genetics* 14: 618–30.
53. Stuart T. and Satija R. (2019) Integrative single-cell analysis. *Nature Reviews Genetics* 20: 257–72.
54. Kashima Y. et al. (2020) Single-cell sequencing techniques from individual to multiomics analyses. *Experimental & Molecular Medicine* 52: 1419–27.
55. Sagar and Grün D. (2020) Deciphering cell fate decision by integrated single-cell sequencing analysis. *Annual Review of Biomedical Data Science* 3: 1–22.
56. Cardona-Alberich A., Tourbez M., Pearce S.F. and Sibley C.R. (2021) Elucidating the cellular dynamics of the brain with single-cell RNA sequencing. *RNA Biology* 18: 1063–84.
57. Sladitschek H.L. et al. (2020) MorphoSeq: Full single-cell transcriptome dynamics up to gastrulation in a chordate. *Cell* 181: 922–35.
58. Lencer E., Prekeris R. and Artinger K.B. (2021) Single-cell RNA analysis identifies pre-migratory neural crest cells expressing markers of differentiated derivatives. *eLife* 10: e66078.
59. Fang R. et al. (2021) Comprehensive analysis of single cell ATAC-seq data with SnapATAC. *Nature Communications* 12: 1337.
60. Kelley D. and Rinn J. (2012) Transposable elements reveal a stem cell-specific class of long noncoding RNAs. *Genome Biology* 13: R107.
61. Kelley D.R., Hendrickson D.G., Tenen D. and Rinn J.L. (2014) Transposable elements modulate human RNA abundance and splicing via specific RNA-protein interactions. *Genome Biology* 15: 537.
62. Kim Daniel H. et al. (2015) Single-cell transcriptome analysis reveals dynamic changes in lncRNA expression during reprogramming. *Cell Stem Cell* 16: 88–101.
63. Ma Q. and Chang H.Y. (2016) Single-cell profiling of lncRNAs in the developing human brain. *Genome Biology* 17: 68.
64. Liu S.J. et al. (2016) Single-cell analysis of long non-coding RNAs in the developing human neocortex. *Genome Biology* 17: 67.
65. Tasic B. et al. (2018) Shared and distinct transcriptomic cell types across neocortical areas. *Nature* 563: 72–8.
66. Alessio E. et al. (2019) Single cell analysis reveals the involvement of the long non-coding RNA Pvt1 in the modulation of muscle atrophy and mitochondrial network. *Nucleic Acids Research* 47: 1653–70.
67. Bakken T.E. et al. (2021) Comparative cellular analysis of motor cortex in human, marmoset and mouse. *Nature* 598: 111–9.
68. Bocchi V.D. et al. (2021) The coding and long noncoding single-cell atlas of the developing human fetal striatum. *Science* 372: eabf5759.
69. Manji S.S. et al. (2006) Molecular characterization and expression of maternally expressed gene 3 (Meg3/Gtl2) RNA in the mouse inner ear. *Journal of Neuroscience Research* 83: 181–90.
70. Moradi Chameh H. et al. (2021) Diversity amongst human cortical pyramidal neurons revealed via their sag currents and frequency preferences. *Nature Communications* 12: 2497.
71. Bray S.J. (2016) Notch signalling in context. *Nature Reviews Molecular Cell Biology* 17: 722–35.
72. Oh E.C. and Katsanis N. (2012) Cilia in vertebrate development and disease. *Development* 139: 443–8.
73. Guignard L. et al. (2020) Contact area–dependent cell communication and the morphological invariance of ascidian embryogenesis. *Science* 369: eaar5663.
74. Priya R. et al. (2020) Tension heterogeneity directs form and fate to pattern the myocardial wall. *Nature* 588: 130–4.
75. Levin M. (2021) Bioelectric signaling: Reprogrammable circuits underlying embryogenesis, regeneration, and cancer. *Cell* 184: 1971–89.
76. Zepp J.A. et al. (2021) Genomic, epigenomic, and biophysical cues controlling the emergence of the lung alveolus. *Science* 371: eabc3172.
77. Mallo M. and Alonso C.R. (2013) The regulation of Hox gene expression during animal development. *Development* 140: 3951–63.
78. Wolpert L. (1969) Positional information and the spatial pattern of cellular differentiation. *Journal of Theoretical Biology* 25: 1–47.
79. Sulston J.E. and Horvitz H.R. (1977) Post-embryonic cell lineages of the nematode, *Caenorhabditis elegans*. *Developmental Biology* 56: 110–56.
80. Sulston J.E., Schierenberg E., White J.G. and Thomson J.N. (1983) The embryonic cell lineage of the nematode *Caenorhabditis elegans*. *Developmental Biology* 100: 64–119.
81. Kemphues K.J. (2016) Horvitz and Sulston on *Caenorhabditis elegans* cell lineage mutants. *Genetics* 203: 1485–7.
82. Hobert O. (2010) Neurogenesis in the nematode *Caenorhabditis elegans*, in *Wormbook: The Online Review of C. elegans biology*. http://www.wormbook.org.
83. Jonsson H. et al. (2021) Differences between germline genomes of monozygotic twins. *Nature Genetics* 53: 27–34.
84. Witherspoon D.J. et al. (2007) Genetic similarities within and between human populations. *Genetics* 176: 351–9.
85. Huynh J.-R. and St Johnston D. (2004) The origin of asymmetry: Early polarisation of the *Drosophila* germline cyst and oocyte. *Current Biology* 14: R438–49.
86. Chen Q., Shi J., Tao Y. and Zernicka-Goetz M. (2018) Tracing the origin of heterogeneity and symmetry breaking in the early mammalian embryo. *Nature Communications* 9: 1819.
87. Jankele R., Jelier R. and Gönczy P. (2021) Physically asymmetric division of the *C. elegans* zygote ensures invariably successful embryogenesis. *eLife* 10: e61714.

88. Sunchu B. and Cabernard C. (2020) Principles and mechanisms of asymmetric cell division. *Development* 147: dev167650.

89. Cardelli L., Csikász-Nagy A., Dalchau N., Tribastone M. and Tschaikowski M. (2016) Noise reduction in complex biological switches. *Scientific Reports* 6: 20214.

90. Klosin A. et al. (2020) Phase separation provides a mechanism to reduce noise in cells. *Science* 367: 464–8.

91. Exelby K. et al. (2021) Precision of tissue patterning is controlled by dynamical properties of gene regulatory networks. *Development* 148: dev197566.

92. Block F.E. (1987) Analog and digital computer theory. *International Journal of Clinical Monitoring and Computing* 4: 47–51.

93. Wong E.S. et al. (2015) Decoupling of evolutionary changes in transcription factor binding and gene expression in mammals. *Genome Research* 25: 167–78.

94. Berthelot C., Villar D., Horvath J.E., Odom D.T. and Flicek P. (2018) Complexity and conservation of regulatory landscapes underlie evolutionary resilience of mammalian gene expression. *Nature Ecology & Evolution* 2: 152–63.

95. Peterson K.J., Dietrich M.R. and McPeek M.A. (2009) MicroRNAs and metazoan macroevolution: Insights into canalization, complexity, and the Cambrian explosion. *BioEssays* 31: 736–47.

96. Ebert M.S. and Sharp P.A. (2012) Roles for microRNAs in conferring robustness to biological processes. *Cell* 149: 515–24.

97. Plavskin Y. et al. (2016) Ancient trans-acting siRNAs confer robustness and sensitivity onto the auxin response. *Developmental Cell* 36: 276–89.

98. Hong J.-W., Hendrix D.A. and Levine M.S. (2008) Shadow enhancers as a source of evolutionary novelty. *Science* 321: 1314.

99. Cannavò E. et al. (2016) Shadow enhancers are pervasive features of developmental regulatory networks. *Current Biology* 26: 38–51.

100. Osterwalder M. et al. (2018) Enhancer redundancy provides phenotypic robustness in mammalian development. *Nature* 554: 239–43.

101. Waymack R., Fletcher A., Enciso G. and Wunderlich Z. (2020) Shadow enhancers can suppress input transcription factor noise through distinct regulatory logic. *eLife* 9: e59351.

102. Kvon E.Z., Waymack R., Gad M. and Wunderlich Z. (2021) Enhancer redundancy in development and disease. *Nature Reviews Genetics* 22: 324–36.

103. Sultana T., Zamborlini A., Cristofari G. and Lesage P. (2017) Integration site selection by retroviruses and transposable elements in eukaryotes. *Nature Reviews Genetics* 18: 292–308.

104. Barth N.K.H., Li L. and Taher L. (2020) Independent transposon exaptation is a widespread mechanism of redundant enhancer evolution in the mammalian genome. *Genome Biology and Evolution* 12: 1–17.

105. McKinley K.L. and Cheeseman I.M. (2016) The molecular basis for centromere identity and function. *Nature Reviews Molecular Cell Biology* 17: 16–29.

106. Trivedi P. et al. (2019) The inner centromere is a biomolecular condensate scaffolded by the chromosomal passenger complex. *Nature Structural & Molecular Biology* 21: 1127–37.

107. Hartley G. and O'Neill R.J. (2019) Centromere repeats: Hidden gems of the genome. *Genes* 10: 223.

108. Zackroff R.V., Rosenfeld A.C. and Weisenberg R.C. (1976) Effects of RNase and RNA on *in vitro* aster assembly. *Journal of Supramolecular Structure* 5: 577–89.

109. Topp C.N., Zhong C.X. and Dawe R.K. (2004) Centromere-encoded RNAs are integral components of the maize kinetochore. *Proceedings of the National Academy of Sciences USA* 101: 15986–91.

110. Folco H.D., Pidoux A.L., Urano T. and Allshire R.C. (2008) Heterochromatin and RNAi are required to establish CENP-A chromatin at centromeres. *Science* 319: 94–7.

111. Wong L.H. et al. (2007) Centromere RNA is a key component for the assembly of nucleoproteins at the nucleolus and centromere. *Genome Research* 17: 1146–60.

112. Li F., Sonbuchner L., Kyes S.A., Epp C. and Deitsch K.W. (2008) Nuclear non-coding RNAs are transcribed from the centromeres of *Plasmodium falciparum* and are associated with centromeric chromatin. *Journal of Biological Chemistry* 283: 5692–8.

113. Ferri F., Bouzinba-Segard H., Velasco G., Hubé F. and Francastel C. (2009) Non-coding murine centromeric transcripts associate with and potentiate Aurora B kinase. *Nucleic Acids Research* 37: 5071–80.

114. Malik H.S. and Henikoff S. (2009) Major evolutionary transitions in centromere complexity. *Cell* 138: 1067–82.

115. Alliegro M.C. (2011) The centrosome and spindle as a ribonucleoprotein complex. *Chromosome Research* 19: 367–76.

116. Zwicker D., Decker M., Jaensch S., Hyman A.A. and Jülicher F. (2014) Centrosomes are autocatalytic droplets of pericentriolar material organized by centrioles. *Proceedings of the National Academy of Sciences USA* 111: E2636–45.

117. Rošić S., Köhler F. and Erhardt S. (2014) Repetitive centromeric satellite RNA is essential for kinetochore formation and cell division. *Journal of Cell Biology* 207: 335–49.

118. Quénet D. and Dalal Y. (2014) A long non-coding RNA is required for targeting centromeric protein A to the human centromere. *eLife* 3: e26016.

119. Rošić S. and Erhardt S. (2016) No longer a nuisance: Long non-coding RNAs join CENP-A in epigenetic centromere regulation. *Cellular and Molecular Life Sciences* 73: 1387–98.

120. Mutazono M. et al. (2017) The intron in centromeric non-coding RNA facilitates RNAi-mediated formation of heterochromatin. *PLOS Genetics* 13: e1006606.

121. Smurova K. and De Wulf P. (2018) Centromere and pericentromere transcription: Roles and regulation … in sickness and in health. *Frontiers in Genetics* 9: 674.

122. Ling Y.H. and Yuen K.W.Y. (2019) Point centromere activity requires an optimal level of centromeric noncoding RNA. *Proceedings of the National Academy of Sciences USA* 116: 6270–9.

123. Bergalet J. et al. (2020) Inter-dependent centrosomal co-localization of the cen and ik2 cis-natural antisense mRNAs in *Drosophila*. *Cell Reports* 30: 3339–52.

124. Habermann K. and Lange B.M.H. (2012) New insights into subcomplex assembly and modifications of centrosomal proteins. *Cell Division* 7: 17.

125. Chavali P.L., Pütz M. and Gergely F. (2014) Small organelle, big responsibility: The role of centrosomes in development and disease. *Philosophical Transactions of the Royal Society B: Biological Sciences* 369: 20130468.

126. Conduit P.T., Wainman A. and Raff J.W. (2015) Centrosome function and assembly in animal cells. *Nature Reviews Molecular Cell Biology* 16: 611–24.

127. Joukov V. and De Nicolo A. (2019) The centrosome and the primary cilium: The yin and yang of a hybrid organelle. *Cells* 8: 701.

128. Métivier R. et al. (2003) Estrogen receptor-α directs ordered, cyclical, and combinatorial recruitment of cofactors on a natural target promoter. *Cell* 115: 751–63.

129. Niklas K.J. (2014) The evolutionary-developmental origins of multicellularity. *American Journal of Botany* 101: 6–25.

130. Weintraub H. et al. (1991) Muscle-specific transcriptional activation by MyoD. *Genes & Development* 5: 1377–86.

131. Loewer S. et al. (2010) Large intergenic non-coding RNA-RoR modulates reprogramming of human induced pluripotent stem cells. *Nature Genetics* 42: 1113–7.

132. Takahashi K. and Yamanaka S. (2006) Induction of pluripotent stem cells from mouse embryonic and adult fibroblast cultures by defined factors. *Cell* 126: 663–76.

133. Nakagawa M. et al. (2008) Generation of induced pluripotent stem cells without Myc from mouse and human fibroblasts. *Nature Biotechnology* 26: 101–6.

134. Zalc A. et al. (2021) Reactivation of the pluripotency program precedes formation of the cranial neural crest. *Science* 371: eabb4776.

135. Odom D.T. et al. (2006) Core transcriptional regulatory circuitry in human hepatocytes. *Molecular Systems Biology* 2: 2006.0017.

136. Zaret K.S. and Carroll J.S. (2011) Pioneer transcription factors: Establishing competence for gene expression. *Genes & Development* 25: 2227–41.

137. Iwafuchi-Doi M. and Zaret K.S. (2014) Pioneer transcription factors in cell reprogramming. *Genes & Development* 28: 2679–92.

138. Soufi A. et al. (2015) Pioneer transcription factors target partial DNA motifs on nucleosomes to initiate reprogramming. *Cell* 161: 555–68.

139. Iwafuchi-Doi M. (2019) The mechanistic basis for chromatin regulation by pioneer transcription factors. *WIREs Systems Biology and Medicine* 11: e1427.

140. Whyte W.A. et al. (2013) Master transcription factors and mediator establish super-enhancers at key cell identity genes. *Cell* 153: 307–19.

141. Amaral P.P. et al. (2018) Genomic positional conservation identifies topological anchor point RNAs linked to developmental loci. *Genome Biology* 19: 32.

142. Winick-Ng W. et al. (2021) Cell-type specialization is encoded by specific chromatin topologies. *Nature* 599: 684–91.

143. Claes P. et al. (2018) Genome-wide mapping of global-to-local genetic effects on human facial shape. *Nature Genetics* 50: 414–23.

144. Long H.K. et al. (2020) Loss of extreme long-range enhancers in human neural crest drives a craniofacial disorder. *Cell Stem Cell* 27: 765–83.

145. White J.D. et al. (2021) Insights into the genetic architecture of the human face. *Nature Genetics* 53: 45–53.

146. Naqvi S. et al. (2021) Shared heritability of human face and brain shape. *Nature Genetics* 53: 830–9.

147. Liu C. et al. (2021) Genome scans of facial features in East Africans and cross-population comparisons reveal novel associations. *PLOS Genetics* 17: e1009695.

148. Fedoroff N.V. (2012) Transposable elements, epigenetics, and genome evolution. *Science* 338: 758–67.

149. Otto S.P. and Whitton J. (2000) Polyploid incidence and evolution. *Annual Review of Genetics* 34: 401–37.

150. Adams K.L. and Wendel J.F. (2005) Polyploidy and genome evolution in plants. *Current Opinion in Plant Biology* 8: 135–41.

151. Bumgarner S.L. et al. (2012) Single-cell analysis reveals that noncoding RNAs contribute to clonal heterogeneity by modulating transcription factor recruitment. *Molecular Cell* 45: 470–82.

152. van Werven F.J. et al. (2012) Transcription of two long noncoding RNAs mediates mating-type control of gametogenesis in budding yeast. *Cell* 150: 1170–81.

153. Hiriart E. and Verdel A. (2013) Long noncoding RNA-based chromatin control of germ cell differentiation: A yeast perspective. *Chromosome Research* 21: 653–63.

154. Moretto F., Wood N.E., Kelly G., Doncic A. and van Werven F.J. (2018) A regulatory circuit of two lncRNAs and a master regulator directs cell fate in yeast. *Nature Communications* 9: 780.

155. Moretto F. et al. (2021) Transcription levels of a noncoding RNA orchestrate opposing regulatory and cell fate outcomes in yeast. *Cell Reports* 34: 108643.

156. Chacko N. et al. (2015) The lncRNA RZE1 controls cryptococcal morphological transition. *PLOS Genetics* 11: e1005692.

157. Kim W., Miguel-Rojas C., Wang J., Townsend J.P. and Trail F. (2018) Developmental dynamics of long noncoding RNA expression during sexual fruiting body formation in *Fusarium graminearum*. *mBio* 9: e01292.

158. Avesson L. et al. (2011) Abundant class of non-coding RNA regulates development in the social amoeba *Dictyostelium discoideum*. *RNA Biology* 8: 1094–104.

159. Kjellin J. et al. (2021) Abundantly expressed class of noncoding RNAs conserved through the multicellular evolution of dictyostelid social amoebas. *Genome Research* 31: 436–47.

160. Nowacki M. et al. (2007) RNA-mediated epigenetic programming of a genome-rearrangement pathway. *Nature* 451: 153–8.

161. Lindblad K.A., Bracht J.R., Williams A.E. and Landweber L.F. (2017) Thousands of RNA-cached copies of whole chromosomes are present in the ciliate *Oxytricha* during development. *RNA* 23: 1200–8.

162. Allen S.E. and Nowacki M. (2020) Roles of noncoding RNAs in ciliate genome architecture. *Journal of Molecular Biology* 432: 4186–98.

163. Barcons-Simon A., Cordon-Obras C., Guizetti J., Bryant J.M. and Scherf A. (2020) CRISPR interference of a clonally variant GC-rich noncoding RNA family leads to general repression of *var* genes in *Plasmodium falciparum*. *mBio* 11: e03054–19.

164. van Noort V. and Huynen M.A. (2006) Combinatorial gene regulation in *Plasmodium falciparum*. *Trends in Genetics* 22: 73–8.

165. Sebé-Pedrós A. et al. (2016) The dynamic regulatory genome of *Capsaspora* and the origin of animal multicellularity. *Cell* 165: 1224–37.

166. Schwaiger M. et al. (2014) Evolutionary conservation of the eumetazoan gene regulatory landscape. *Genome Research* 24: 639–50.

167. Gaiti F. et al. (2017) Landscape of histone modifications in a sponge reveals the origin of animal cis-regulatory complexity. *eLife* 6: e22194.

168. Wong E.S. et al. (2020) Deep conservation of the enhancer regulatory code in animals. *Science* 370: eaax8137.

169. Olsen P.H. and Ambros V. (1999) The *lin-4* regulatory RNA controls developmental timing in *Caenorhabditis elegans* by blocking LIN-14 protein synthesis after the initiation of translation. *Developmental Biology* 216: 671–80.

170. Reinhart B.J. et al. (2000) The 21-nucleotide let-7 RNA regulates developmental timing in *Caenorhabditis elegans*. *Nature* 403: 901–6.

171. Slack F.J. et al. (2000) The *lin-41* RBCC gene acts in the *C. elegans* heterochronic pathway between the let-7 regulatory RNA and the LIN-29 transcription factor. *Molecular Cell* 5: 659–69.

172. Abbott A.L. et al. (2005) The *let-7* microRNA family members *mir-48*, *mir-84*, and *mir-241* function together to regulate developmental timing in *Caenorhabditis elegans*. *Developmental Cell* 9: 403–14.

173. Tsialikas J., Romens M.A., Abbott A. and Moss E.G. (2017) Stage-specific timing of the microRNA regulation of *lin-28* by the heterochronic gene *lin-14* in *Caenorhabditis elegans*. *Genetics* 205: 251–62.

174. Lawson H.N., Wexler L.R., Wnuk H.K. and Portman D.S. (2020) Dynamic, non-binary specification of sexual state in the *C. elegans* nervous system. *Current Biology* 30: 3617–23.

175. Jukam D., Shariati S.A.M. and Skotheim J.M. (2017) Zygotic genome activation in vertebrates. *Developmental Cell* 42: 316–32.

176. Vastenhouw N.L., Cao W.X. and Lipshitz H.D. (2019) The maternal-to-zygotic transition revisited. *Development* 146: dev161471.

177. Cuykendall T.N. and Houston D.W. (2010) Identification of germ plasm-associated transcripts by microarray analysis of *Xenopus* vegetal cortex RNA. *Developmental Dynamics* 239: 1838–48.

178. D'Orazio F.M. et al. (2021) Germ cell differentiation requires Tdrd7-dependent chromatin and transcriptome reprogramming marked by germ plasm relocalization. *Developmental Cell* 56: 641–56.

179. Sharma U. (2019) Paternal contributions to offspring health: Role of sperm small RNAs in intergenerational transmission of epigenetic information. *Frontiers in Cell and Developmental Biology* 7: 215.

180. Zhang Y., Shi J., Rassoulzadegan M., Tuorto F. and Chen Q. (2019) Sperm RNA code programmes the metabolic health of offspring. *Nature Reviews Endocrinology* 15: 489–98.

181. Piasecka B., Lichocki P., Moretti S., Bergmann S. and Robinson-Rechavi M. (2013) The hourglass and the early conservation models—co-existing patterns of developmental constraints in vertebrates. *PLOS Genetics* 9: e1003476.

182. Irie N. and Kuratani S. (2014) The developmental hourglass model: A predictor of the basic body plan? *Development* 141: 4649–55.

183. Atallah J. and Lott S.E. (2018) Evolution of maternal and zygotic mRNA complements in the early *Drosophila* embryo. *PLOS Genetics* 14: e1007838.

184. Kalinka A.T. et al. (2010) Gene expression divergence recapitulates the developmental hourglass model. *Nature* 468: 811–4.

185. Ninova M., Ronshaugen M. and Griffiths-Jones S. (2014) Conserved temporal patterns of microRNA expression in *Drosophila* support a developmental hourglass model. *Genome Biology and Evolution* 6: 2459–67.

186. Kauffman S. (1969) Homeostasis and differentiation in random genetic control networks. *Nature* 224: 177–8.

187. Glass L. and Kauffman S.A. (1973) The logical analysis of continuous, non-linear biochemical control networks. *Journal of Theoretical Biology* 39: 103–29.

188. Kauffman S. (1974) The large scale structure and dynamics of gene control circuits: An ensemble approach. *Journal of Theoretical Biology* 44: 167–90.

189. Kauffman S. (2003) Understanding genetic regulatory networks. *International Journal of Astrobiology* 2: 131–9.

190. Kauffman S. (2004) A proposal for using the ensemble approach to understand genetic regulatory networks. *Journal of Theoretical Biology* 230: 581–90.

191. Reményi A., Schöler H.R. and Wilmanns M. (2004) Combinatorial control of gene expression. *Nature Structural & Molecular Biology* 11: 812–5.

192. Levine M. and Davidson E.H. (2005) Gene regulatory networks for development. *Proceedings of the National Academy of Sciences USA* 102: 4936–42.

193. Yuh C.H., Bolouri H. and Davidson E.H. (1998) Genomic cis-regulatory logic: Experimental and computational analysis of a sea urchin gene. *Science* 279: 1896–902.

194. Howard M.L. and Davidson E.H. (2004) cis-Regulatory control circuits in development. *Developmental Biology* 271: 109–18.

195. Davidson E.H. and Erwin D.H. (2006) Gene regulatory networks and the evolution of animal body plans. *Science* 311: 796–800.

196. Peter I.S. and Davidson E.H. (2011) Evolution of gene regulatory networks controlling body plan development. *Cell* 144: 970–85.

197. Albert R. and Othmer H.G. (2003) The topology of the regulatory interactions predicts the expression pattern of the segment polarity genes in *Drosophila melanogaster*. *Journal of Theoretical Biology* 223: 1–18.

198. Espinosa-Soto C., Padilla-Longoria P. and Alvarez-Buylla E.R. (2004) A gene regulatory network model for cell-fate determination during *Arabidopsis thaliana* flower development that is robust and recovers experimental gene expression profiles. *Plant Cell* 16: 2923.

199. Li F., Long T., Lu Y., Ouyang Q. and Tang C. (2004) The yeast cell-cycle network is robustly designed. *Proceedings of the National Academy of Sciences USA* 101: 4781–6.

200. Bornholdt S. (2008) Boolean network models of cellular regulation: Prospects and limitations. *Journal of The Royal Society Interface* 5: S85–94.

201. Payankaulam S., Li L.M. and Arnosti D.N. (2010) Transcriptional repression: Conserved and evolved features. *Current Biology* 20: R764–71.

202. Bintu L. et al. (2005) Transcriptional regulation by the numbers: Models. *Current Opinion in Genetics and Development* 15: 116–24.

203. van Hijum S.A.F.T., Medema M.H. and Kuipers O.P. (2009) Mechanisms and evolution of control logic in prokaryotic transcriptional regulation. *Microbiology and Molecular Biology Reviews* 73: 481–509.

204. Ferraris L. et al. (2011) Combinatorial binding of transcription factors in the pluripotency control regions of the genome. *Genome Research* 21: 1055–64.

205. Luna-Zurita L. et al. (2016) Complex interdependence regulates heterotypic transcription factor distribution and coordinates cardiogenesis. *Cell* 164: 999–1014.

206. Vierstra J. et al. (2020) Global reference mapping of human transcription factor footprints. *Nature* 583: 729–36.

207. Moore J.E. et al. (2020) Expanded encyclopaedias of DNA elements in the human and mouse genomes. *Nature* 583: 699–710.

208. Thurman R.E. et al. (2012) The accessible chromatin landscape of the human genome. *Nature* 489: 75–82.

209. Pliner H.A. et al. (2018) Cicero predicts cis-regulatory DNA interactions from single-cell chromatin accessibility data. *Molecular Cell* 71: 858–71.

210. Cusanovich D.A. et al. (2018) A single-cell atlas of *in vivo* mammalian chromatin accessibility. *Cell* 174: 1309–24.

211. Ptashne M. (1967) Specific binding of the lambda phage repressor to lambda DNA. *Nature* 214: 232–4.

212. Ptashne M. (1988) How eukaryotic transcriptional activators work. *Nature* 335: 683–9.

213. Hasty J., Pradines J., Dolnik M. and Collins J.J. (2000) Noise-based switches and amplifiers for gene expression. *Proceedings of the National Academy of Sciences USA* 97: 2075–80.

214. Isaacs F.J., Hasty J., Cantor C.R. and Collins J.J. (2003) Prediction and measurement of an autoregulatory genetic module. *Proceedings of the National Academy of Sciences USA* 100: 7714–9.

215. Niklas K.J., Dunker A.K. and Yruela I. (2018) The evolutionary origins of cell type diversification and the role of intrinsically disordered proteins. *Journal of Experimental Botany* 69: 1437–46.

216. Sigler P.B. (1988) Acid blobs and negative noodles. *Nature* 333: 210–2.

217. Niklas K.J., Bondos S.E., Dunker A.K. and Newman S.A. (2015) Rethinking gene regulatory networks in light of alternative splicing, intrinsically disordered protein domains, and post-translational modifications. *Frontiers in Cell and Developmental Biology* 3: 8.

218. Cumberworth A., Lamour G., Babu M.M. and Gsponer J. (2013) Promiscuity as a functional trait: Intrinsically disordered regions as central players of interactomes. *Biochemical Journal* 454: 361–9.

219. Protter D.S.W. et al. (2018) Intrinsically disordered regions can contribute promiscuous interactions to RNP granule assembly. *Cell Reports* 22: 1401–12.

220. Wang J. et al. (2012) Sequence features and chromatin structure around the genomic regions bound by 119 human transcription factors. *Genome Research* 22: 1798–812.

221. Smith C.W. and Valcarcel J. (2000) Alternative pre-mRNA splicing: The logic of combinatorial control. *Trends in Biochemical Sciences* 25: 381–8.

222. Levine M. and Tjian R. (2003) Transcription regulation and animal diversity. *Nature* 424: 147–51.

223. Arnone M.I. and Davidson E.H. (1997) The hardwiring of development: Organization and function of genomic regulatory systems. *Development* 124: 1851–64.

224. Davidson E.H. (2006) *The Regulatory Genome: Gene Regulatory Networks In Development And Evolution* (Academic Press, Cambridge, MA).

225. Carroll S.B. (2008) Evo-devo and an expanding evolutionary synthesis: A genetic theory of morphological evolution. *Cell* 134: 25–36.

226. Peter I.S. and Davidson E.H. (2011) A gene regulatory network controlling the embryonic specification of endoderm. *Nature* 474: 635–9.

227. Bhattacharjee S., Renganaath K., Mehrotra R. and Mehrotra S. (2013) Combinatorial control of gene expression. *BioMed Research International* 2013: 407263.

228. Peter I.S. and Davidson E.H. (2016) Implications of developmental gene regulatory networks inside and outside developmental biology. *Current Topics in Developmental Biology* 117: 237–51.

229. Scholes C., DePace A.H. and Sánchez Á. (2017) Combinatorial gene regulation through kinetic control of the transcription cycle. *Cell Systems* 4: 97–108.

230. Peter I.S. and Davidson E.H. (2017) Assessing regulatory information in developmental gene regulatory networks. *Proceedings of the National Academy of Sciences USA* 114: 5862–9.

231. Azpeitia E. et al. (2017) The combination of the functionalities of feedback circuits is determinant for the attractors' number and size in pathway-like Boolean networks. *Scientific Reports* 7: 42023.

232. Kauffman S.A. (1969) Metabolic stability and epigenesis in randomly constructed genetic nets. *Journal of Theoretical Biology* 22: 437–67.

233. Samuelsson B. and Troein C. (2003) Superpolynomial growth in the number of attractors in Kauffman networks. *Physical Review Letters* 90: 098701.

234. Buchler N.E., Gerland U. and Hwa T. (2003) On schemes of combinatorial transcription logic. *Proceedings of the National Academy of Sciences USA* 100: 5136–41.

235. Akhlaghpour H. (2022) An RNA-based theory of natural universal computation. *Journal of Theoretical Biology* 536: 110984.

236. Eddington A.S. (1956) The constants of nature, in J.R. Newman (ed.) *The World of Mathematics 2* (Simon & Schuster, New York).

237. Stampfel G. et al. (2015) Transcriptional regulators form diverse groups with context-dependent regulatory functions. *Nature* 528: 147–51.

238. Jolma A. et al. (2015) DNA-dependent formation of transcription factor pairs alters their binding specificity. *Nature* 527: 384–8.

239. Chovanec P. et al. (2021) Widespread reorganisation of pluripotent factor binding and gene regulatory interactions between human pluripotent states. *Nature Communications* 12: 2098.

240. Bright A.R. *et al.* (2021) Combinatorial transcription factor activities on open chromatin induce embryonic heterogeneity in vertebrates. *EMBO Journal* 40: e104913.

241. Thomas H.F. et al. (2021) Temporal dissection of an enhancer cluster reveals distinct temporal and functional contributions of individual elements. *Molecular Cell* 81: 969–82.

242. Soldatov R. et al. (2019) Spatiotemporal structure of cell fate decisions in murine neural crest. *Science* 364: eaas9536.

243. Croft L.J., Lercher M.J., Gagen M.J. and Mattick J.S. (2003) Is prokaryotic complexity limited by accelerated growth in regulatory overhead? *Genome Biology Preprint Depository* http://genomebiology.com/qc/2003/5/1/p2.

244. van Nimwegen E. (2003) Scaling laws in the functional content of genomes. *Trends in Genetics* 19: 479–84.

245. Mattick J.S. and Gagen M.J. (2005) Accelerating networks. *Science* 307: 856–8.

246. Molina N. and van Nimwegen E. (2008) Universal patterns of purifying selection at noncoding positions in bacteria. *Genome Research* 18: 148–60.

247. Charoensawan V., Wilson D. and Teichmann S.A. (2010) Genomic repertoires of DNA-binding transcription factors across the tree of life. *Nucleic Acids Research* 38: 7364–77.

248. Gagen M.J. and Mattick J.S. (2005) Inherent size constraints on prokaryote gene networks due to "accelerating" growth. *Theory in Bioscience* 123: 381–411.

249. Mattick J.S. (2011) The central role of RNA in human development and cognition. *FEBS Letters* 585: 1600–16.

250. Liu G., Mattick J.S. and Taft R.J. (2013) A meta-analysis of the genomic and transcriptomic composition of complex life. *Cell Cycle* 12: 2061–72.

251. Kapusta A. and Feschotte C. (2014) Volatile evolution of long noncoding RNA repertoires: Mechanisms and biological implications. *Trends in Genetics* 30: 439–52.

252. Alberti C. and Cochella L. (2017) A framework for understanding the roles of miRNAs in animal development. *Development* 144: 2548–59.

CHAPTER 16

1. Paul J. and Duerksen J.D. (1975) Chromatin-associated RNA content of heterochromatin and euchromatin. *Molecular and Cellular Biochemistry* 9: 9–16.

2. Pederson T. and Bhorjee J.S. (1979) Evidence for a role of RNA in eukaryotic chromosome structure. Metabolically stable, small nuclear RNA species are covalently linked to chromosomal DNA in HeLa cells. *Journal of Molecular Biology* 128: 451–80.

3. Bynum J.W. and Volkin E. (1980) Chromatin-associated RNA: Differential extraction and characterization. *Biochimica et Biophysica Acta* 607: 304–18.

4. Nickerson J.A., Krochmalnic G., Wan K.M. and Penman S. (1989) Chromatin architecture and nuclear RNA. *Proceedings of the National Academy of Sciences USA* 86: 177–81.

5. Bernstein E. and Allis C.D. (2005) RNA meets chromatin. *Genes & Development* 19: 1635–55.

6. Chakalova L., Debrand E., Mitchell J.A., Osborne C.S. and Fraser P. (2005) Replication and transcription: Shaping the landscape of the genome. *Nature Reviews Genetics* 6: 669–77.

7. Rodriguez-Campos A. and Azorin F. (2007) RNA Is an integral component of chromatin that contributes to its structural organization. *PLOS ONE* 2: e1182.

8. Mondal T., Rasmussen M., Pandey G.K., Isaksson A. and Kanduri C. (2010) Characterization of the RNA content of chromatin. *Genome Research* 20: 899–907.

9. Schubert T. et al. (2012) Df31 protein and snoRNAs maintain accessible higher-order structures of chromatin. *Molecular Cell* 48: 434–44.

10. Rinn J. and Guttman M. (2014) RNA and dynamic nuclear organization. *Science* 345: 1240–1.

11. Sridhar B. et al. (2017) Systematic mapping of RNA-chromatin interactions *in vivo*. *Current Biology* 27: 602–9.

12. Li X. et al. (2017) GRID-seq reveals the global RNA-chromatin interactome. *Nature Biotechnology* 35: 940–50.

13. Bell J.C. et al. (2018) Chromatin-associated RNA sequencing (ChAR-seq) maps genome-wide RNA-to-DNA contacts. *eLife* 7: e27024.

14. Subhash S. et al. (2018) H3K4me2 and WDR5 enriched chromatin interacting long non-coding RNAs maintain transcriptionally competent chromatin at divergent transcriptional units. *Nucleic Acids Research* 46: 9384–400.

15. Schwartz U. et al. (2019) Characterizing the nuclease accessibility of DNA in human cells to map higher order structures of chromatin. *Nucleic Acids Research* 47: 1239–54.

16. Li X. and Fu X.-D. (2019) Chromatin-associated RNAs as facilitators of functional genomic interactions. *Nature Reviews Genetics* 20: 503–19.

17. Mishra K. and Kanduri C. (2019) Understanding long noncoding RNA and chromatin interactions: What we know so far. *Noncoding RNA* 5: 54.

18. Thakur J. and Henikoff S. (2020) Architectural RNA in chromatin organization. *Biochemical Society Transactions* 48: 1967–78.

19. Chu C., Qu K., Zhong F.L., Artandi S.E. and Chang H.Y. (2011) Genomic maps of long noncoding RNA occupancy reveal principles of RNA-chromatin interactions. *Molecular Cell* 44: 667–78.

20. Werner M.S. and Ruthenburg A.J. (2015) Nuclear fractionation reveals thousands of chromatin-tethered noncoding rnas adjacent to active genes. *Cell Reports* 12: 1089–98.

21. Werner M.S. et al. (2017) Chromatin-enriched lncRNAs can act as cell-type specific activators of proximal gene transcription. *Nature Structural & Molecular Biology* 24: 596–603.

22. Li Z. et al. (2015) Exon-intron circular RNAs regulate transcription in the nucleus. *Nature Structural & Molecular Biology* 22: 256–64.

23. Conn V.M. et al. (2017) A circRNA from SEPALLATA3 regulates splicing of its cognate mRNA through R-loop formation. *Nature Plants* 3: 17053.

24. Studniarek C., Egloff S. and Murphy S. (2021) Noncoding RNAs set the stage for RNA polymerase II transcription. *Trends in Genetics* 37: 279–91.

25. Briese M. and Sendtner M. (2021) Keeping the balance: The noncoding RNA 7SK as a master regulator for neuron development and function. *Bioessays* 43: e2100092.

26. Vilborg A., Passarelli M.C., Yario T.A., Tycowski K.T. and Steitz J.A. (2015) Widespread inducible transcription downstream of human genes. *Molecular Cell* 59: 449–61.

27. Shevtsov S.P. and Dundr M. (2011) Nucleation of nuclear bodies by RNA. *Nature Cell Biology* 13: 167–73.

28. Sharma P. and Beato M. (2018) Long non-coding RNAs are driver to maintain the chromatin active regions at divergent transcriptional units. *Non-coding RNA Investigation* 2: 50.

29. Xiao R. et al. (2019) Pervasive chromatin-RNA binding protein interactions enable RNA-based regulation of transcription. *Cell* 178: 107–21.
30. Trigiante G., Blanes Ruiz N. and Cerase A. (2021) Emerging roles of repetitive and repeat-containing RNA in nuclear and chromatin organization and gene expression. *Frontiers in Cell and Developmental Biology* 9: 2730.
31. Hall L.L. et al. (2014) Stable C0T-1 repeat RNA Is abundant and is associated with euchromatic interphase chromosomes. *Cell* 156: 907–19.
32. Lu J.Y. et al. (2021) Homotypic clustering of L1 and B1/Alu repeats compartmentalizes the 3D genome. *Cell Research* 31: 613–30.
33. Fadloun A. et al. (2013) Chromatin signatures and retrotransposon profiling in mouse embryos reveal regulation of LINE-1 by RNA. *Nature Structural & Molecular Biology* 20: 332–8.
34. Percharde M. et al. (2018) A LINE1-nucleolin partnership regulates early development and ESC identity. *Cell* 174: 391–405.
35. Ballarino M. et al. (2018) Deficiency in the nuclear long noncoding RNA Charme causes myogenic defects and heart remodeling in mice. *EMBO Journal* 37: e99697.
36. Abdalla M.O.A. et al. (2019) The Eleanor ncRNAs activate the topological domain of the ESR1 locus to balance against apoptosis. *Nature Communications* 10: 3778.
37. Oh H.J. et al. (2021) Jpx RNA regulates CTCF anchor site selection and formation of chromosome loops. *Cell* 184: P6157–73.
38. Li L. et al. (2021) Global profiling of RNA–chromatin interactions reveals co-regulatory gene expression networks in Arabidopsis. *Nature Plants* 7: 1364–78.
39. Lei E.P. and Corces V.G. (2006) RNA interference machinery influences the nuclear organization of a chromatin insulator. *Nature Genetics* 38: 936–41.
40. Shuaib M. et al. (2019) Nuclear AGO1 regulates gene expression by affecting chromatin architecture in human cells. *Cell Systems* 9: 446–58.
41. Kim S., Yu N.-K. and Kaang B.-K. (2015) CTCF as a multifunctional protein in genome regulation and gene expression. *Experimental & Molecular Medicine* 47: e166.
42. Diehl A.G., Ouyang N. and Boyle A.P. (2020) Transposable elements contribute to cell and species-specific chromatin looping and gene regulation in mammalian genomes. *Nature Communications* 11: 1796.
43. Zhang Y. et al. (2019) Transcriptionally active HERV-H retrotransposons demarcate topologically associating domains in human pluripotent stem cells. *Nature Genetics* 51: 1380–8.
44. Lai F. et al. (2013) Activating RNAs associate with Mediator to enhance chromatin architecture and transcription. *Nature* 494: 497–501.
45. Luo S. et al. (2016) Divergent lncRNAs regulate gene expression and lineage differentiation in pluripotent cells. *Cell Stem Cell* 18: 637–52.
46. Chueh A.C., Northrop E.L., Brettingham-Moore K.H., Choo K.H. and Wong L.H. (2009) LINE retrotransposon RNA is an essential structural and functional epigenetic component of a core neocentromeric chromatin. *PLOS Genetics* 5: e1000354.
47. Du Y., Topp C.N. and Dawe R.K. (2010) DNA binding of centromere protein C (CENPC) Is stabilized by single-stranded RNA. *PLOS Genetics* 6: e1000835.
48. Rošić S., Köhler F. and Erhardt S. (2014) Repetitive centromeric satellite RNA is essential for kinetochore formation and cell division. *Journal of Cell Biology* 207: 335–49.
49. Corless S., Höcker S. and Erhardt S. (2020) Centromeric RNA and its function at and beyond centromeric chromatin. *Journal of Molecular Biology* 432: 4257–69.
50. Bierhoff H., Postepska-Igielska A. and Grummt I. (2014) Noisy silence: Non-coding RNA and heterochromatin formation at repetitive elements. *Epigenetics* 9: 53–61.
51. Hall I.M., Noma K. and Grewal S.I. (2003) RNA interference machinery regulates chromosome dynamics during mitosis and meiosis in fission yeast. *Proceedings of the National Academy of Sciences USA* 100: 193–8.
52. Pal-Bhadra M. et al. (2004) Heterochromatic silencing and HP1 localization in *Drosophila* are dependent on the RNAi machinery. *Science* 303: 669–72.
53. Fukagawa T. et al. (2004) Dicer is essential for formation of the heterochromatin structure in vertebrate cells. *Nature Cell Biology* 6: 784–91.
54. Zofall M. and Grewal S.I. (2006) RNAi-mediated heterochromatin assembly in fission yeast. *Cold Spring Harbor Symposia on Quantitative Biology* 71: 487–96.
55. Brower-Toland B. et al. (2007) Drosophila PIWI associates with chromatin and interacts directly with HP1a. *Genes & Development* 21: 2300–11.
56. Muchardt C. et al. (2002) Coordinated methyl and RNA binding is required for heterochromatin localization of mammalian HP1. *EMBO Reports* 3: 975–81.
57. Maison C. et al. (2002) Higher-order structure in pericentric heterochromatin involves a distinct pattern of histone modification and an RNA component. *Nature Genetics* 30: 329–34.
58. Camacho O.V. et al. (2017) Major satellite repeat RNA stabilize heterochromatin retention of Suv39h enzymes by RNA-nucleosome association and RNA:DNA hybrid formation. *eLife* 6: e25293.
59. Johnson W.L. et al. (2017) RNA-dependent stabilization of SUV39H1 at constitutive heterochromatin. *eLife* 6: e25299.
60. Bierhoff H. et al. (2014) Quiescence-induced lncRNAs trigger H4K20 trimethylation and transcriptional silencing. *Molecular Cell* 54: 675–82.
61. Azzalin C.M., Reichenbach P., Khoriauli L., Giulotto E. and Lingner J. (2007) Telomeric repeat containing RNA and RNA surveillance factors at mammalian chromosome ends. *Science* 318: 798–801.
62. Montero J.J. et al. (2018) TERRA recruitment of polycomb to telomeres is essential for histone trymethylation marks at telomeric heterochromatin. *Nature Communications* 9: 1548.
63. Ding D.Q. et al. (2012) Meiosis-specific noncoding RNA mediates robust pairing of homologous chromosomes in meiosis. *Science* 336: 732–6.
64. Ding D.-Q. et al. (2019) Chromosome-associated RNA–protein complexes promote pairing of homologous chromosomes during Cmeiosis in *Schizosaccharomyces pombe*. *Nature Communications* 10: 5598.

65. Hendrickson D.G., Kelley D.R., Tenen D., Bernstein B. and Rinn J.L. (2016) Widespread RNA binding by chromatin-associated proteins. *Genome Biology* 17: 28.

66. Guo F., Li L.L., Yang W., Hu J.-F. and Cui J. (2021) Long noncoding RNA: A resident staff of genomic instability regulation in tumorigenesis. *Cancer Letters* 503: 103–9.

67. Zaret K.S. and Carroll J.S. (2011) Pioneer transcription factors: Establishing competence for gene expression. *Genes & Development* 25: 2227–41.

68. Iwafuchi-Doi M. and Zaret K.S. (2014) Pioneer transcription factors in cell reprogramming. *Genes & Development* 28: 2679–92.

69. Soufi A. et al. (2015) Pioneer transcription factors target partial DNA motifs on nucleosomes to initiate reprogramming. *Cell* 161: 555–68.

70. Mayran A. and Drouin J. (2018) Pioneer transcription factors shape the epigenetic landscape. *Journal of Biological Chemistry* 293: 13795–804.

71. Iwafuchi-Doi M. (2019) The mechanistic basis for chromatin regulation by pioneer transcription factors. *WIREs Systems Biology and Medicine* 11: e1427.

72. Takahashi K. and Yamanaka S. (2006) Induction of pluripotent stem cells from mouse embryonic and adult fibroblast cultures by defined factors. *Cell* 126: 663–76.

73. Loh Y.H. et al. (2006) The Oct4 and Nanog transcription network regulates pluripotency in mouse embryonic stem cells. *Nature Genetics* 38: 431–40.

74. Masui S. et al. (2007) Pluripotency governed by Sox2 via regulation of Oct3/4 expression in mouse embryonic stem cells. *Nature Cell Biology* 9: 625–35.

75. Yu J. et al. (2007) Induced pluripotent stem cell lines derived from human somatic cells. *Science* 318: 1917–20.

76. Zalc A. et al. (2021) Reactivation of the pluripotency program precedes formation of the cranial neural crest. *Science* 371: eabb4776.

77. Wang Z., Oron E., Nelson B., Razis S. and Ivanova N. (2012) Distinct lineage specification roles for NANOG, OCT4, and SOX2 in human embryonic stem cells. *Cell Stem Cell* 10: 440–54.

78. Dunwell T.L. and Holland P.W.H. (2017) A sister of *NANOG* regulates genes expressed in pre-implantation human development. *Open Biology* 7: 170027.

79. Catron K.M., Iler N. and Abate C. (1993) Nucleotides flanking a conserved TAAT core dictate the DNA binding specificity of three murine homeodomain proteins. *Molecular and Cellular Biology* 13: 2354–65.

80. Bürglin T.R. and Affolter M. (2016) Homeodomain proteins: An update. *Chromosoma* 125: 497–521.

81. Jauch R., Ng C.K.L., Saikatendu K.S., Stevens R.C. and Kolatkar P.R. (2008) Crystal structure and DNA binding of the homeodomain of the stem cell transcription factor Nanog. *Journal of Molecular Biology* 376: 758–70.

82. Fang X. et al. (2011) Genome-wide analysis of OCT4 binding sites in glioblastoma cancer cells. *Journal of Zhejiang University. Science B* 12: 812–9.

83. Ferraris L. et al. (2011) Combinatorial binding of transcription factors in the pluripotency control regions of the genome. *Genome Research* 21: 1055–64.

84. Tantin D. (2013) Oct transcription factors in development and stem cells: Insights and mechanisms. *Development* 140: 2857–66.

85. Rodda D.J. et al. (2005) Transcriptional regulation of nanog by OCT4 and SOX2. *Journal of Biological Chemistry* 280: 24731–7.

86. Boyer L.A. et al. (2005) Core transcriptional regulatory circuitry in human embryonic stem cells. *Cell* 122: 947–56.

87. Wang J. et al. (2006) A protein interaction network for pluripotency of embryonic stem cells. *Nature* 444: 364–8.

88. Zhou Q., Chipperfield H., Melton D.A. and Wong W.H. (2007) A gene regulatory network in mouse embryonic stem cells. *Proceedings of the National Academy of Sciences USA* 104: 16438–43.

89. Kashyap V. et al. (2009) Regulation of stem cell pluripotency and differentiation involves a mutual regulatory circuit of the NANOG, OCT4, and SOX2 pluripotency transcription factors with polycomb repressive complexes and stem cell microRNAs. *Stem Cells and Development* 18: 1093–108.

90. Costa Y. et al. (2013) NANOG-dependent function of TET1 and TET2 in establishment of pluripotency. *Nature* 495: 370–4.

91. Tomita S. et al. (2015) A cluster of noncoding RNAs activates the ESR1 locus during breast cancer adaptation. *Nature Communications* 6: 6966.

92. Setten R.L., Chomchan P., Epps E.W., Burnett J.C. and Rossi J.J. (2021) CRED9: A differentially expressed elncRNA regulates expression of transcription factor CEBPA. *RNA* 27: 891–906.

93. Lin H., Shabbir A., Molnar M. and Lee T. (2007) Stem cell regulatory function mediated by expression of a novel mouse Oct4 pseudogene. *Biochemical and Biophysical Research Communications* 355: 111–6.

94. Hawkins P.G. and Morris K.V. (2010) Transcriptional regulation of Oct4 by a long non-coding RNA antisense to Oct4-pseudogene 5. *Transcription* 1: 165–75.

95. Wang Y. et al. (2013) Endogenous miRNA sponge *lincRNA-RoR* regulates *Oct4, Nanog,* and *Sox2* in human embryonic stem cell self-renewal. *Developmental Cell* 25: 69–80.

96. Rosa A. and Ballarino M. (2016) Long noncoding RNA regulation of pluripotency. *Stem Cells International* 2016: 1797692.

97. Scarola M. et al. (2015) Epigenetic silencing of Oct4 by a complex containing SUV39H1 and Oct4 pseudogene lncRNA. *Nature Communications* 6: 7631.

98. Scarola M. et al. (2020) FUS-dependent loading of SUV39H1 to OCT4 pseudogene-lncRNA programs a silencing complex with OCT4 promoter specificity. *Communications Biology* 3: 632.

99. Sheik Mohamed J., Gaughwin P.M., Lim B., Robson P. and Lipovich L. (2010) Conserved long noncoding RNAs transcriptionally regulated by Oct4 and Nanog modulate pluripotency in mouse embryonic stem cells. *RNA* 16: 324–37.

100. Booth H.A.F. and Holland P.W.H. (2004) Eleven daughters of NANOG. *Genomics* 84: 229–38.

101. Fairbanks D.J. and Maughan P.J. (2006) Evolution of the NANOG pseudogene family in the human and chimpanzee genomes. *BMC Evolutionary Biology* 6: 12.

102. Poursani E.M., Mohammad Soltani B. and Mowla S.J. (2016) Differential expression of *OCT4* pseudogenes in pluripotent and tumor cell lines. *Cell Journal* 18: 28–36.

103. Mehravar M., Ghaemimanesh F. and Poursani E.M. (2021) An overview on the complexity of OCT4: At the level of DNA, RNA and protein. *Stem Cell Reviews and Reports* 17: 1121–36.

104. Zhang J. et al. (2006) NANOGP8 is a retrogene expressed in cancers. *FEBS Journal* 273: 1723–30.

105. Svingen T. and Tonissen K.F. (2006) Hox transcription factors and their elusive mammalian gene targets. *Heredity* 97: 88–96.

106. Sharov A.A. et al. (2008) Identification of Pou5f1, Sox2, and Nanog downstream target genes with statistical confidence by applying a novel algorithm to time course microarray and genome-wide chromatin immunoprecipitation data. *BMC Genomics* 9: 269.

107. Hou L., Srivastava Y. and Jauch R. (2017) Molecular basis for the genome engagement by Sox proteins. *Seminars in Cell & Developmental Biology* 63: 2–12.

108. Li H. et al. (2019) The spatial binding model of the pioneer factor Oct4 with its target genes during cell reprogramming. *Computational and Structural Biotechnology Journal* 17: 1226–33.

109. Luo Z., Rhie S.K. and Farnham P.J. (2019) The enigmatic HOX genes: Can we crack their code? *Cancers* 11: 323.

110. Sherwood R.I. et al. (2014) Discovery of directional and nondirectional pioneer transcription factors by modeling DNase profile magnitude and shape. *Nature Biotechnology* 32: 171–8.

111. Chovanec P. et al. (2021) Widespread reorganisation of pluripotent factor binding and gene regulatory interactions between human pluripotent states. *Nature Communications* 12: 2098.

112. Pavlopoulos A. and Akam M. (2011) Hox gene *Ultrabithorax* regulates distinct sets of target genes at successive stages of *Drosophila* haltere morphogenesis. *Proceedings of the National Academy of Sciences USA* 108: 2855–60.

113. Dubnau J. and Struhl G. (1996) RNA recognition and translational regulation by a homeodomain protein. *Nature* 379: 694–9.

114. Rivera-Pomar R., Niessing D., Schmidt-Ott U., Gehring W.J. and Jacklé H. (1996) RNA binding and translational suppression by bicoid. *Nature* 379: 746–9.

115. Degani N., Lubelsky Y., Perry R.B.-T., Ainbinder E. and Ulitsky I. (2021) Highly conserved and cis-acting lncRNAs produced from paralogous regions in the center of *HOXA* and *HOXB* clusters in the endoderm lineage. *PLOS Genetics* 17: e1009681.

116. Grosschedl R., Giese K. and Pagel J. (1994) HMG domain proteins: Architectural elements in the assembly of nucleoprotein structures. *Trends in Genetics* 10: 94–100.

117. Hock R., Furusawa T., Ueda T. and Bustin M. (2007) HMG chromosomal proteins in development and disease. *Trends in Cell Biology* 17: 72–9.

118. Dodonova S.O., Zhu F., Dienemann C., Taipale J. and Cramer P. (2020) Nucleosome-bound SOX2 and SOX11 structures elucidate pioneer factor function. *Nature* 580: 669–72.

119. Bianchi M.E. and Agresti A. (2005) HMG proteins: Dynamic players in gene regulation and differentiation. *Current Opinion in Genetics and Development* 15: 496–506.

120. Štros M., Launholt D. and Grasser K.D. (2007) The HMG-box: A versatile protein domain occurring in a wide variety of DNA-binding proteins. *Cellular and Molecular Life Sciences* 64: 2590.

121. Holmes Z.E. et al. (2020) The Sox2 transcription factor binds RNA. *Nature Communications* 11: 1805.

122. Cajigas I. et al. (2021) Sox2-*Evf2* lncRNA mechanisms of chromosome topological control in developing forebrain. *Development* 148: dev197202.

123. Genzor P. and Bortvin A. (2015) A unique HMG-box domain of mouse Maelstrom binds structured RNA but not double stranded DNA. *PLOS ONE* 10: e0120268.

124. Veretnik S. and Gribskov M. (1999) RNA binding domain of HDV antigen is homologous to the HMG box of SRY. *Archives of Virology* 144: 1139–58.

125. Ng S.-Y., Johnson R. and Stanton L.W. (2012) Human long non-coding RNAs promote pluripotency and neuronal differentiation by association with chromatin modifiers and transcription factors. *EMBO Journal* 31: 522–33.

126. Ng S.-Y., Bogu G.K., Soh B.S. and Stanton L.W. (2013) The long noncoding RNA RMST interacts with SOX2 to regulate neurogenesis. *Molecular Cell* 51: 349–59.

127. Samudyata et al. (2019) Interaction of Sox2 with RNA binding proteins in mouse embryonic stem cells. *Experimental Cell Research* 381: 129–38.

128. Hou L. et al. (2020) Concurrent binding to DNA and RNA facilitates the pluripotency reprogramming activity of Sox2. *Nucleic Acids Research* 48: 3869–87.

129. Zhao Z., Dammert M.A., Grummt I. and Bierhoff H. (2016) lncRNA-Induced nucleosome repositioning reinforces transcriptional repression of RNA genes upon hypotonic stress. *Cell Reports* 14: 1876–82.

130. Wang X., Cairns M.J. and Yan J. (2019) Super-enhancers in transcriptional regulation and genome organization. *Nucleic Acids Research* 47: 11481–96.

131. Kelley D. and Rinn J. (2012) Transposable elements reveal a stem cell-specific class of long noncoding RNAs. *Genome Biology* 13: R107.

132. Lu X. et al. (2014) The retrovirus HERVH is a long non-coding RNA required for human embryonic stem cell identity. *Nature Structural & Molecular Biology* 21: 423–5.

133. Weintraub H. et al. (1991) Muscle-specific transcriptional activation by MyoD. *Genes & Development* 5: 1377–86.

134. Caretti G. et al. (2006) The RNA helicases p68/p72 and the noncoding RNA SRA are coregulators of MyoD and skeletal muscle differentiation. *Developmental Cell* 11: 547–60.

135. Dong A. et al. (2020) A long noncoding RNA, *LncMyoD*, modulates chromatin accessibility to regulate muscle stem cell myogenic lineage progression. *Proceedings of the National Academy of Sciences USA* 117: 32464–75.

136. Yu X. et al. (2017) Long non-coding RNA Linc-RAM enhances myogenic differentiation by interacting with MyoD. *Nature Communications* 8: 14016.

137. Pandorf C.E. et al. (2006) Dynamics of myosin heavy chain gene regulation in slow skeletal muscle: Role of natural antisense RNA. *Journal of Biological Chemistry* 281: 38330–42.

138. Haddad F. et al. (2008) Intergenic transcription and developmental regulation of cardiac myosin heavy chain genes. *American Journal of Physiology Heart and Circulation Physiology* 294: H29–40.

139. Dou M. et al. (2020) The long noncoding RNA MyHC IIA/X-AS contributes to skeletal muscle myogenesis and maintains the fast fiber phenotype. *Journal of Biological Chemistry* 295: 4937–49.

140. Bose D.A. et al. (2017) RNA binding to CBP stimulates histone acetylation and transcription. *Cell* 168: 135–49.

141. Becker P.B. and Hörz W. (2002) ATP-dependent nucleosome remodeling. *Annual Review of Biochemistry* 71: 247–73.

142. Clapier C.R., Iwasa J., Cairns B.R. and Peterson C.L. (2017) Mechanisms of action and regulation of ATP-dependent chromatin-remodelling complexes. *Nature Reviews Molecular Cell Biology* 18: 407–22.

143. Prensner J.R. et al. (2013) The long noncoding RNA SChLAP1 promotes aggressive prostate cancer and antagonizes the SWI/SNF complex. *Nature Genetics* 45: 1392–8.

144. Han P. et al. (2014) A long noncoding RNA protects the heart from pathological hypertrophy. *Nature* 514: 102–6.

145. Cajigas I. et al. (2015) Evf2 *lncRNA/BRG1/DLX1* interactions reveal RNA-dependent inhibition of chromatin remodeling. *Development* 142: 2641–52.

146. Hu G. et al. (2016) LincRNA-Cox2 promotes late inflammatory gene transcription in macrophages through modulating SWI/SNF-mediated chromatin remodeling. *Journal of Immunology* 196: 2799–808.

147. Atianand M.K. et al. (2016) A long noncoding RNA lincRNA-EPS acts as a transcriptional brake to restrain inflammation. *Cell* 165: 1672–85.

148. Tang Y. et al. (2017) Linking long non-coding RNAs and SWI/SNF complexes to chromatin remodeling in cancer. *Molecular Cancer* 16: 42.

149. Liu B. et al. (2017) Long noncoding RNA *lncKdm2b* is required for ILC3 maintenance by initiation of Zfp292 expression. *Nature Immunology* 18: 499–508.

150. Leisegang M.S. et al. (2017) Long noncoding RNA *MANTIS* facilitates endothelial angiogenic function. *Circulation* 136: 65–79.

151. Lino Cardenas C.L. et al. (2018) An HDAC9-MALAT1-BRG1 complex mediates smooth muscle dysfunction in thoracic aortic aneurysm. *Nature Communications* 9: 1009.

152. Wang Y. et al. (2018) Long noncoding RNA lncHand2 promotes liver repopulation via c-Met signaling. *Journal of Hepatology* 69: 861–72.

153. Wang Y. et al. (2019) LncRNA HAND2-AS1 promotes liver cancer stem cell self-renewal via BMP signaling. *EMBO Journal* 38: e101110.

154. Hu Y.-W. et al. (2019) Long noncoding RNA NEXN-AS1 mitigates atherosclerosis by regulating the actin-binding protein NEXN. *Journal of Clinical investigation* 129: 1115–28.

155. Liu X. et al. (2019) A long noncoding RNA, Antisense IL-7, promotes inflammatory gene transcription through facilitating histone acetylation and switch/sucrose nonfermentable chromatin remodeling. *Journal of Immunology* 203: 1548–59.

156. Jégu T. et al. (2019) Xist RNA antagonizes the SWI/SNF chromatin remodeler BRG1 on the inactive X chromosome. *Nature Structural & Molecular Biology* 26: 96–109.

157. Grossi E. et al. (2020) A lncRNA-SWI/SNF complex crosstalk controls transcriptional activation at specific promoter regions. *Nature Communications* 11: 936.

158. Schutt C. et al. (2020) Linc-MYH configures INO80 to regulate muscle stem cell numbers and skeletal muscle hypertrophy. *EMBO Journal* 39: e105098.

159. Patty B.J. and Hainer S.J. (2020) Non-coding RNAs and nucleosome remodeling complexes: An intricate regulatory relationship. *Biology* 9: 213.

160. Ducoli L. et al. (2021) LETR1 is a lymphatic endothelial-specific lncRNA governing cell proliferation and migration through KLF4 and SEMA3C. *Nature Communications* 12: 925.

161. Santos-Zavaleta A. et al. (2019) RegulonDB v 10.5: Tackling challenges to unify classic and high throughput knowledge of gene regulation in *E. coli* K-12. *Nucleic Acids Research* 47: D212–20.

162. Iuchi S. (2001) Three classes of C2H2 zinc finger proteins. *Cellular and Molecular Life Sciences CMLS* 58: 625–35.

163. Badis G. et al. (2009) Diversity and complexity in DNA recognition by transcription factors. *Science* 324: 1720–3.

164. Mendoza-Vargas A. et al. (2009) Genome-wide identification of transcription start sites, promoters and transcription factor binding sites in *E. coli*. *PLOS ONE* 4: e7526.

165. Jolma A. et al. (2013) DNA-binding specificities of human transcription factors. *Cell* 152: 327–39.

166. Castellanos M., Mothi N. and Muñoz V. (2020) Eukaryotic transcription factors can track and control their target genes using DNA antennas. *Nature Communications* 11: 540.

167. Negre N. et al. (2011) A cis-regulatory map of the *Drosophila* genome. *Nature* 471: 527–31.

168. Lee B.K. et al. (2012) Cell-type specific and combinatorial usage of diverse transcription factors revealed by genome-wide binding studies in multiple human cells. *Genome Research* 22: 9–24.

169. Wang J. et al. (2012) Sequence features and chromatin structure around the genomic regions bound by 119 human transcription factors. *Genome Research* 22: 1798–812.

170. Kheradpour P. and Kellis M. (2014) Systematic discovery and characterization of regulatory motifs in ENCODE TF binding experiments. *Nucleic Acids Research* 42: 2976–87.

171. Najafabadi H.S. et al. (2015) C2H2 zinc finger proteins greatly expand the human regulatory lexicon. *Nature Biotechnology* 33: 555–62.

172. Cassiday L.A. and Maher 3rd L.J. (2002) Having it both ways: Transcription factors that bind DNA and RNA. *Nucleic Acids Research* 30: 4118–26.

173. Burdach J., O'Connell M.R., Mackay J.P. and Crossley M. (2012) Two-timing zinc finger transcription factors liaising with RNA. *Trends in Biochemical Sciences* 37: 199–205.

174. Pelham H.R. and Brown D.D. (1980) A specific transcription factor that can bind either the 5S RNA gene or 5S RNA. *Proceedings of the National Academy of Sciences USA* 77: 4170–4.

175. Honda B.M. and Roeder R.G. (1980) Association of a 5S gene transcription factor with 5S RNA and altered levels of the factor during cell differentiation. *Cell* 22: 119–26.

176. Han H. et al. (2017) Multilayered control of alternative splicing regulatory networks by transcription factors. *Molecular Cell* 65: 539–53.

177. Hastie N.D. (2017) Wilms' tumour 1 (WT1) in development, homeostasis and disease. *Development* 144: 2862–72.

178. Larsson S.H. et al. (1995) Subnuclear localization of WT1 in splicing or transcription factor domains is regulated by alternative splicing. *Cell* 81: 391–401.

179. Kennedy D., Ramsdale T., Mattick J. and Little M. (1996) An RNA recognition motif in Wilms' tumour protein (WT1) revealed by structural modelling. *Nature Genetics* 12: 329–32.

180. Naftelberg S., Schor I.E., Ast G. and Kornblihtt A.R. (2015) Regulation of alternative splicing through coupling with transcription and chromatin structure. *Annual Review of Biochemistry* 84: 165–98.

181. Klamt B. et al. (1998) Frasier syndrome is caused by defective alternative splicing of WT1 leading to an altered ratio of WT1 +/−KTS splice isoforms. *Human Molecular Genetics* 7: 709–14.

182. Caricasole A. et al. (1996) RNA binding by the Wilms tumor suppressor zinc finger proteins. *Proceedings of the National Academy of Sciences USA* 93: 7562–6.

183. Bardeesy N. and Pelletier J. (1998) Overlapping RNA and DNA binding domains of the *wt1* tumor suppressor gene product. *Nucleic Acids Research* 26: 1784–92.

184. Zhai G., Iskandar M., Barilla K. and Romaniuk P.J. (2001) Characterization of RNA aptamer binding by the Wilms' tumor suppressor protein WT1. *Biochemistry* 40: 2032–40.

185. Niksic M., Slight J., Sanford J.R., Caceres J.F. and Hastie N.D. (2004) The Wilms' tumour protein (WT1) shuttles between nucleus and cytoplasm and is present in functional polysomes. *Human Molecular Genetics* 13: 463–71.

186. O'Connor L., Gilmour J. and Bonifer C. (2016) The role of the ubiquitously expressed transcription factor SP1 in tissue-specific transcriptional regulation and in disease. *Yale Journal of Biology and Medicine* 89: 513–25.

187. Shi Y. and Berg J.M. (1995) Specific DNA-RNA hybrid binding by zinc finger proteins. *Science* 268: 282–4.

188. Gordon S., Akopyan G., Garban H. and Bonavida B. (2006) Transcription factor YY1: Structure, function, and therapeutic implications in cancer biology. *Oncogene* 25: 1125–42.

189. Sigova A.A. et al. (2015) Transcription factor trapping by RNA in gene regulatory elements. *Science* 350: 978–81.

190. Weintraub A.S. et al. (2017) YY1 is a structural regulator of enhancer-promoter loops. *Cell* 171: 1573–88.

191. Yao H. et al. (2010) Mediation of CTCF transcriptional insulation by DEAD-box RNA-binding protein p68 and steroid receptor RNA activator SRA. *Genes & Development* 24: 2543–55.

192. Saldana-Meyer R. et al. (2014) CTCF regulates the human p53 gene through direct interaction with its natural antisense transcript, Wrap53. *Genes & Development* 28: 723–34.

193. Kung J.T. et al. (2015) Locus-specific targeting to the X chromosome revealed by the RNA interactome of CTCF. *Molecular Cell* 57: 361–75.

194. Hansen A.S. et al. (2019) Distinct classes of chromatin loops revealed by deletion of an RNA-binding region in CTCF. *Molecular Cell* 76: 395–411.

195. Kuang S. and Wang L. (2020) Identification and analysis of consensus RNA motifs binding to the genome regulator CTCF. *NAR Genomics and Bioinformatics* 2: lqaa031.

196. Wang I.X. et al. (2018) Human proteins that interact with RNA/DNA hybrids. *Genome Research* 28: 1405–14.

197. Abakir A. et al. (2020) N^6-methyladenosine regulates the stability of RNA:DNA hybrids in human cells. *Nature Genetics* 52: 48–55.

198. Crossley M.P., Bocek M. and Cimprich K.A. (2019) R-loops as cellular regulators and genomic threats. *Molecular Cell* 73: 398–411.

199. Niehrs C. and Luke B. (2020) Regulatory R-loops as facilitators of gene expression and genome stability. *Nature Reviews Molecular Cell Biology* 21: 167–78.

200. Wahba L., Costantino L., Tan F.J., Zimmer A. and Koshland D. (2016) S1-DRIP-seq identifies high expression and polyA tracts as major contributors to R-loop formation. *Genes & Development* 30: 1327–38.

201. Sanz L.A. et al. (2016) Prevalent, dynamic, and conserved R-loop structures associate with specific epigenomic signatures in mammals. *Molecular Cell* 63: 167–78.

202. Wickramasinghe V.O. and Venkitaraman A.R. (2016) RNA processing and genome stability: Cause and consequence. *Molecular Cell* 61: 496–505.

203. Rinaldi C., Pizzul P., Longhese M.P. and Bonetti D. (2021) Sensing R-loop-associated DNA damage to safeguard genome stability. *Frontiers in Cell and Developmental Biology* 8: 1657.

204. Chen P.B., Chen H.V., Acharya D., Rando O.J. and Fazzio T.G. (2015) R loops regulate promoter-proximal chromatin architecture and cellular differentiation. *Nature Structural & Molecular Biology* 22: 999–1007.

205. Boque-Sastre R. et al. (2015) Head-to-head antisense transcription and R-loop formation promotes transcriptional activation. *Proceedings of the National Academy of Sciences USA* 112: 5785–90.

206. Nadel J. et al. (2015) RNA:DNA hybrids in the human genome have distinctive nucleotide characteristics, chromatin composition, and transcriptional relationships. *Epigenetics & Chromatin* 8: 46.

207. Shiromoto Y., Sakurai M., Minakuchi M., Ariyoshi K. and Nishikura K. (2021) ADAR1 RNA editing enzyme regulates R-loop formation and genome stability at telomeres in cancer cells. *Nature Communications* 12: 1654.

208. van Holde K. and Zlatanova J. (1994) Unusual DNA structures, chromatin and transcription. *BioEssays* 16: 59–68.

209. Gilbert D.E. and Feigon J. (1999) Multistranded DNA structures. *Current Opinion in Structural Biology* 9: 305–14.

210. Mirkin S.M. (2008) Discovery of alternative DNA structures: A heroic decade (1979–1989). *Frontiers in Bioscience* 13: 1064–71.

211. Zeraati M. et al. (2018) I-motif DNA structures are formed in the nuclei of human cells. *Nature Chemistry* 10: 631–7.

212. Herbert A. (2019) Z-DNA and Z-RNA in human disease. *Communications Biology* 2: 7.

213. Herbert A. (2020) Simple repeats as building blocks for genetic computers. *Trends in Genetics* 36: 739–50.

214. Bekhor I., Bonner J. and Dahmus G.K. (1969) Hybridization of chromosomal RNA to native DNA. *Proceedings of the National Academy of Sciences USA* 62: 271–7.

215. Lee J.S., Woodsworth M.L., Latimer L.J. and Morgan A.R. (1984) Poly(pyrimidine).poly(purine) synthetic DNAs containing 5-methylcytosine form stable triplexes at neutral pH. *Nucleic Acids Research* 12: 6603–14.

216. Goni J.R., de la Cruz X. and Orozco M. (2004) Triplex-forming oligonucleotide target sequences in the human genome. *Nucleic Acids Research* 32: 354–60.

217. Jalali S., Singh A., Maiti S. and Scaria V. (2017) Genome-wide computational analysis of potential long noncoding RNA mediated DNA:DNA:RNA triplexes in the human genome. *Journal of Translational Medicine* 15: 186.

218. Lee J.S., Burkholder G.D., Latimer L.J., Haug B.L. and Braun R.P. (1987) A monoclonal antibody to triplex DNA binds to eucaryotic chromosomes. *Nucleic Acids Research* 15: 1047–61.

219. Agazie Y.M., Burkholder G.D. and Lee J.S. (1996) Triplex DNA in the nucleus: Direct binding of triplex-specific antibodies and their effect on transcription, replication and cell growth. *Biochemical Journal* 316: 461–6.

220. Ohno M., Fukagawa T., Lee J.S. and Ikemura T. (2002) Triplex-forming DNAs in the human interphase nucleus visualized in situ by polypurine/polypyrimidine DNA probes and antitriplex antibodies. *Chromosoma* 111: 201–13.

221. Cetin N.S. et al. (2019) Isolation and genome-wide characterization of cellular DNA:RNA triplex structures. *Nucleic Acids Research* 47: 2306–21.

222. Farabella I., Di Stefano M., Soler-Vila P., Marti-Marimon M. and Marti-Renom M.A. (2021) Three-dimensional genome organization via triplex-forming RNAs. *Nature Structural & Molecular Biology* 28: 945–54.

223. Soibam B. and Zhamangaraeva A. (2021) LncRNA:DNA triplex-forming sites are positioned at specific areas of genome organization and are predictors for Topologically Associated Domains. *BMC Genomics* 22: 397.

224. Vasquez K.M. and Wilson J.H. (1998) Triplex-directed modification of genes and gene activity. *Trends in Biochemical Sciences* 23: 4–9.

225. Carbone G.M., McGuffie E.M., Collier A. and Catapano C.V. (2003) Selective inhibition of transcription of the Ets2 gene in prostate cancer cells by a triplex-forming oligonucleotide. *Nucleic Acids Research* 31: 833–43.

226. Re R.N., Cook J.L. and Giardina J.F. (2004) The inhibition of tumor growth by triplex-forming oligonucleotides. *Cancer Letters* 209: 51–3.

227. Song J., Intody Z., Li M. and Wilson J.H. (2004) Activation of gene expression by triplex-directed psoralen crosslinks. *Gene* 324: 183–90.

228. Jandura A. and Krause H.M. (2017) The new RNA world: Growing evidence for long noncoding RNA functionality. *Trends in Genetics* 33: 665–76.

229. Grote P. et al. (2013) The tissue-specific lncRNA Fendrr is an essential regulator of heart and body wall development in the mouse. *Developmental Cell* 24: 206–14.

230. O'Leary V.B. et al. (2015) PARTICLE, a triplex-forming long ncRNA, regulates locus-specific methylation in response to low-dose irradiation. *Cell Reports* 11: 474–85.

231. Mondal T. et al. (2015) MEG3 long noncoding RNA regulates the TGF-beta pathway genes through formation of RNA-DNA triplex structures. *Nature Communications* 6: 7743.

232. Postepska-Igielska A. et al. (2015) LncRNA *Khps1* regulates expression of the proto-oncogene *SPHK1* via triplex-mediated changes in chromatin structure. *Molecular Cell* 60: 626–36.

233. Blank-Giwojna A., Postepska-Igielska A. and Grummt I. (2019) lncRNA *KHPS1* activates a poised enhancer by triplex-dependent recruitment of epigenomic regulators. *Cell Reports* 26: 2904–15.

234. Kalwa M. et al. (2016) The lncRNA HOTAIR impacts on mesenchymal stem cells via triple helix formation. *Nucleic Acids Research* 44: 10631–43.

235. Cloutier S.C. et al. (2016) Regulated formation of lncRNA-DNA hybrids enables faster transcriptional induction and environmental adaptation. *Molecular Cell* 61: 393–404.

236. Soibam B. (2017) Super-lncRNAs: Identification of lncRNAs that target super-enhancers via RNA:DNA:DNA triplex formation. *RNA* 23: 1729–42.

237. Wang S. et al. (2018) LncRNA MIR100HG promotes cell proliferation in triple-negative breast cancer through triplex formation with p27 loci. *Cell Death & Disease* 9: 805.

238. Kuo C.-C. et al. (2019) Detection of RNA–DNA binding sites in long noncoding RNAs. *Nucleic Acids Research* 47: e32.

239. Arab K. et al. (2019) GADD45A binds R-loops and recruits TET1 to CpG island promoters. *Nature Genetics* 51: 217–23.

240. Ou M., Li X., Zhao S., Cui S. and Tu J. (2020) Long noncoding RNA CDKN2B-AS1 contributes to atherosclerotic plaque formation by forming RNA-DNA triplex in the CDKN2B promoter. *EBioMedicine* 55: 102694.

241. Toscano-Garibay J.D. and Aquino-Jarquin G. (2014) Transcriptional regulation mechanism mediated by miRNA–DNA•DNA triplex structure stabilized by Argonaute. *Biochimica et Biophysica Acta* 1839: 1079–83.

242. Ariel F. et al. (2020) R-loop mediated trans action of the *APOLO* long noncoding RNA. *Molecular Cell* 77: 1055–65.

243. Moison M. et al. (2021) The lncRNA APOLO interacts with the transcription factor WRKY42 to trigger root hair cell expansion in response to cold. *Molecular Plant* 14: 937–48.

244. Fonouni-Farde C. et al. (2021) Sequence-unrelated long noncoding RNAs converged to modulate the activity of conserved epigenetic machineries across kingdoms. *bioRxiv*: 433017.

245. Ladomery M. (1997) Multifunctional proteins suggest connections between transcriptional and post-transcriptional processes. *BioEssays* 19: 903–9.

246. Matsumoto K. and Wolffe A.P. (1998) Gene regulation by Y-box proteins: Coupling control of transcription and translation. *Trends in Cell Biology* 8: 318–23.

247. Medvedovic J., Ebert A., Tagoh H. and Busslinger M. (2011) Pax5: A master regulator of B cell development and leukemogenesis, in F.W. Alt (ed.) *Advances in Immunology* (Academic Press, Cambridge, MA).

248. Lee N., Moss W.N., Yario T.A. and Steitz J.A. (2015) EBV noncoding RNA binds nascent RNA to drive host PAX5 to viral DNA. *Cell* 160: 607–18.

249. Xu Y. et al. (2021) ERα is an RNA-binding protein sustaining tumor cell survival and drug resistance. *Cell* 184: 5215–29.

250. Oberosler P., Hloch P., Ramsperger U. and Stahl H. (1993) p53-catalyzed annealing of complementary single-stranded nucleic acids. *The EMBO Journal* 12: 2389–96.

251. Schmitt A.M. et al. (2016) An inducible long noncoding RNA amplifies DNA damage signaling. *Nature Genetics* 48: 1370–6.

252. Wang X. et al. (2008) Induced ncRNAs allosterically modify RNA-binding proteins in cis to inhibit transcription. *Nature* 454: 126–30.

253. Yoneda R. et al. (2016) The binding specificity of Translocated in LipoSarcoma/FUsed in Sarcoma with lncRNA transcribed from the promoter region of cyclin D1. *Cell & Bioscience* 6: 4.

254. Yoneda R., Ueda N., Uranishi K., Hirasaki M. and Kurokawa R. (2020) Long noncoding RNA *pncRNA-D* reduces cyclin D1 gene expression and arrests cell cycle through RNA m⁶A modification. *Journal of Biological Chemistry* 295: 5626–39.

255. Allen T.A., Von Kaenel S., Goodrich J.A. and Kugel J.F. (2004) The SINE-encoded mouse B2 RNA represses mRNA transcription in response to heat shock. *Nature Structural & Molecular Biology* 11: 816–21.

256. Espinoza C.A., Allen T.A., Hieb A.R., Kugel J.F. and Goodrich J.A. (2004) B2 RNA binds directly to RNA polymerase II to repress transcript synthesis. *Nature Structural & Molecular Biology* 11: 822–9.

257. Mariner P.D. et al. (2008) Human Alu RNA is a modular transacting repressor of mRNA transcription during heat shock. *Molecular Cell* 29: 499–509.

258. Yakovchuk P., Goodrich J.A. and Kugel J.F. (2009) B2 RNA and Alu RNA repress transcription by disrupting contacts between RNA polymerase II and promoter DNA within assembled complexes. *Proceedings of the National Academy of Sciences USA* 106: 5569–74.

259. Ferrigno O. et al. (2001) Transposable B2 SINE elements can provide mobile RNA polymerase II promoters. *Nature Genetics* 28: 77–81.

260. Groner A.C. et al. (2010) KRAB-zinc finger proteins and KAP1 can mediate long-range transcriptional repression through heterochromatin spreading. *PLOS Genetics* 6: e1000869.

261. Pontis J. et al. (2019) Hominoid-specific transposable elements and KZFPs facilitate human embryonic genome activation and control transcription in naive human ESCs. *Cell Stem Cell* 24: 724–35.

262. Turelli P. et al. (2020) Primate-restricted KRAB zinc finger proteins and target retrotransposons control gene expression in human neurons. *Science Advances* 6: eaba3200.

263. Birtle Z. and Ponting C.P. (2006) Meisetz and the birth of the KRAB motif. *Bioinformatics* 22: 2841–5.

264. Urrutia R. (2003) KRAB-containing zinc-finger repressor proteins. *Genome Biology* 4: 231.

265. Huntley S. et al. (2006) A comprehensive catalog of human KRAB-associated zinc finger genes: Insights into the evolutionary history of a large family of transcriptional repressors. *Genome Research* 16: 669–77.

266. Emerson R.O. and Thomas J.H. (2009) Adaptive evolution in zinc finger transcription factors. *PLOS Genetics* 5: e1000325.

267. Lupo A. et al. (2013) KRAB-zinc finger proteins: A repressor family displaying multiple biological functions. *Current Genomics* 14: 268–78.

268. Imbeault M., Helleboid P.-Y. and Trono D. (2017) KRAB zinc-finger proteins contribute to the evolution of gene regulatory networks. *Nature* 543: 550–4.

269. Yang P., Wang Y. and Macfarlan T.S. (2017) The role of KRAB-ZFPs in transposable element repression and mammalian evolution. *Trends in Genetics* 33: 871–81.

270. Cosby R.L. et al. (2021) Recurrent evolution of vertebrate transcription factors by transposase capture. *Science* 371: eabc6405.

271. Minezaki Y., Homma K., Kinjo A.R. and Nishikawa K. (2006) Human transcription factors contain a high fraction of intrinsically disordered regions essential for transcriptional regulation. *Journal of Molecular Biology* 359: 1137–49.

272. Liu J. et al. (2006) Intrinsic disorder in transcription factors. *Biochemistry* 45: 6873–88.

273. Guo X., Bulyk M.L. and Hartemink A.J. (2012) Intrinsic disorder within and flanking the DNA-binding domains of human transcription factors. *Pacific Symposium on Biocomputing. Pacific Symposium on Biocomputing* 2012: 104–15.

274. Staby L. et al. (2017) Eukaryotic transcription factors: Paradigms of protein intrinsic disorder. *Biochemical Journal* 474: 2509–32.

275. Brodsky S. et al. (2020) Intrinsically disordered regions direct transcription factor *in vivo* binding specificity. *Molecular Cell* 79: 459–71.

276. Mattick J.S., Taft R.J. and Faulkner G.J. (2010) A global view of genomic information--moving beyond the gene and the master regulator. *Trends in Genetics* 26: 21–8.

277. Erdmann R.M. and Picard C.L. (2020) RNA-directed DNA methylation. *PLOS Genetics* 16: e1009034.

278. Sigman M.J. et al. (2021) An siRNA-guided Argonaute protein directs RNA polymerase V to initiate DNA methylation. *Nature Plants* 7: 1461–74.

279. Morris K.V., Chan S.W., Jacobsen S.E. and Looney D.J. (2004) Small interfering RNA-induced transcriptional gene silencing in human cells. *Science* 305: 1289–92.

280. Castel S.E. and Martienssen R.A. (2013) RNA interference in the nucleus: Roles for small RNAs in transcription, epigenetics and beyond. *Nature Reviews Genetics* 14: 100–12.

281. Janowski B.A. et al. (2006) Involvement of AGO1 and AGO2 in mammalian transcriptional silencing. *Nature Structural & Molecular Biology* 13: 787–92.

282. Kim D.H., Villeneuve L.M., Morris K.V. and Rossi J.J. (2006) Argonaute-1 directs siRNA-mediated transcriptional gene silencing in human cells. *Nature Structural & Molecular Biology* 13: 793–7.

283. Vagin V.V. et al. (2006) A distinct small RNA pathway silences selfish genetic elements in the germline. *Science* 313: 320–4.

284. Benetti R. et al. (2008) A mammalian microRNA cluster controls DNA methylation and telomere recombination via Rbl2-dependent regulation of DNA methyltransferases. *Nature Structural & Molecular Biology* 15: 268–79.

285. Aravin A.A. et al. (2008) A piRNA pathway primed by individual transposons is linked to de novo DNA methylation in mice. *Molecular Cell* 31: 785–99.

286. Ting A.H., Schuebel K.E., Herman J.G. and Baylin S.B. (2005) Short double-stranded RNA induces transcriptional gene silencing in human cancer cells in the absence of DNA methylation. *Nature Genetics* 37: 906–10.

287. Jeffery L. and Nakielny S. (2004) Components of the DNA methylation system of chromatin control are RNA-binding proteins. *Journal of Biological Chemistry* 279: 49479–87.

288. Mohammad F., Mondal T., Guseva N., Pandey G.K. and Kanduri C. (2010) Kcnq1ot1 noncoding RNA mediates transcriptional gene silencing by interacting with Dnmt1. *Development* 137: 2493–9.

289. Di Ruscio A. et al. (2013) DNMT1-interacting RNAs block gene-specific DNA methylation. *Nature* 503: 371–6.

290. Merry C.R. et al. (2015) DNMT1-associated long non-coding RNAs regulate global gene expression and DNA methylation in colon cancer. *Human Molecular Genetics* 24: 6240–53.

291. Somasundaram S. et al. (2018) The DNMT1-associated lincRNA DACOR1 reprograms genome-wide DNA methylation in colon cancer. *Clinical Epigenetics* 10: 127.

292. Weiss A., Keshet I., Razin A. and Cedar H. (1996) DNA demethylation in vitro: Involvement of RNA. *Cell* 86: 709–18.

293. Jost J.-P., Frémont M., Siegmann M. and Hofsteenge J. (1997) The RNA moiety of chick embryo 5-methylcyto-sine-DNA glycosylase targets DNA demethylation. *Nucleic Acids Research* 25: 4545–50.

294. Imamura T. et al. (2004) Non-coding RNA directed DNA demethylation of *Sphk1* CpG island. *Biochemical and Biophysical Research Communications* 322: 593–600.

295. Sytnikova Y.A., Kubarenko A.V., Schäfer A., Weber A.N.R. and Niehrs C. (2011) Gadd45a Is an RNA binding protein and is localized in nuclear speckles. *PLOS ONE* 6: e14500.

296. He C. et al. (2016) High-resolution mapping of RNA-binding regions in the nuclear proteome of embryonic stem cells. *Molecular Cell* 64: 416–30.

297. He C. et al. (2021) TET2 chemically modifies tRNAs and regulates tRNA fragment levels. *Nature Structural & Molecular Biology* 28: 62–70.

298. Guallar D. et al. (2018) RNA-dependent chromatin targeting of TET2 for endogenous retrovirus control in pluripotent stem cells. *Nature Genetics* 50: 443–51.

299. Landers C.C. et al. (2021) Ectopic expression of pericentric HSATII RNA results in nuclear RNA accumulation, MeCP2 recruitment, and cell division defects. *Chromosoma* 130: 75–90.

300. O'Leary V.B. et al. (2017) Long non-coding RNA PARTICLE bridges histone and DNA methylation. *Scientific Reports* 7: 1790.

301. Tresaugues L. et al. (2006) Structural characterization of Set1 RNA recognition motifs and their role in histone H3 lysine 4 methylation. *Journal of Molecular Biology* 359: 1170–81.

302. Akhtar A., Zink D. and Becker P.B. (2000) Chromodomains are protein-RNA interaction modules. *Nature* 407: 405–9.

303. Yap K.L. et al. (2010) Molecular interplay of the noncoding RNA ANRIL and methylated histone H3 lysine 27 by polycomb CBX7 in transcriptional silencing of INK4a. *Molecular Cell* 38: 662–74.

304. Pek J.W., Anand A. and Kai T. (2012) Tudor domain proteins in development. *Development* 139: 2255–66.

305. Zhao J., Sun B.K., Erwin J.A., Song J.J. and Lee J.T. (2008) Polycomb proteins targeted by a short repeat RNA to the mouse X chromosome. *Science* 322: 750–6.

306. Zhao J. et al. (2010) Genome-wide identification of polycomb-associated RNAs by RIP-seq. *Molecular Cell* 40: 939–53.

307. Kanhere A. et al. (2010) Short RNAs are transcribed from repressed polycomb target genes and interact with polycomb repressive complex-2. *Molecular Cell* 38: 675–88.

308. Cifuentes-Rojas C., Hernandez A.J., Sarma K. and Lee J.T. (2014) Regulatory interactions between RNA and polycomb repressive complex 2. *Molecular Cell* 55: 171–85.

309. Kaneko S. et al. (2014) Interactions between JARID2 and noncoding RNAs regulate PRC2 recruitment to chromatin. *Molecular Cell* 53: 290–300.

310. Betancur J.G. and Tomari Y. (2015) Cryptic RNA-binding by PRC2 components EZH2 and SUZ12. *RNA Biology* 12: 959–65.

311. Tsai M.C. et al. (2010) Long noncoding RNA as modular scaffold of histone modification complexes. *Science* 329: 689–93.

312. Huang R.C. and Bonner J. (1965) Histone-bound RNA, a component of native nucleohistone. *Proceedings of the National Academy of Sciences USA* 54: 960–7.

313. Benjamin W., Levander O.A., Gellhorn A. and DeBellis R.H. (1966) An RNA-histone complex in mammalian cells: The isolation and characterization of a new RNA species. *Proceedings of the National Academy of Sciences USA* 55: 858–65.

314. Denisenko O., Shnyreva M., Suzuki H. and Bomsztyk K. (1998) Point mutations in the WD40 domain of Eed block its interaction with Ezh2. *Molecular and Cellular Biology* 18: 5634–42.

315. Zhang H. et al. (2004) The *C. elegans* Polycomb gene *SOP-2* encodes an RNA binding protein. *Molecular Cell* 14: 841–7.

316. Bernstein E. et al. (2006) Mouse polycomb proteins bind differentially to methylated histone H3 and RNA and are enriched in facultative heterochromatin. *Molecular Cell Biology* 26: 2560–9.

317. Schmitt S., Prestel M. and Paro R. (2005) Intergenic transcription through a polycomb group response element counteracts silencing. *Genes & Development* 19: 697–708.

318. Rinn J.L. et al. (2007) Functional demarcation of active and silent chromatin domains in human *HOX* loci by non-coding RNAs. *Cell* 129: 1311–23.

319. Sessa L. et al. (2007) Noncoding RNA synthesis and loss of polycomb group repression accompanies the colinear activation of the human HOXA cluster. *RNA* 13: 223–39.

320. Nagano T. et al. (2008) The Air noncoding RNA epigenetically silences transcription by targeting G9a to chromatin. *Science* 322: 1717–20.

321. Pandey R.R. et al. (2008) Kcnq1ot1 antisense noncoding RNA mediates lineage-specific transcriptional silencing through chromatin-level regulation. *Molecular Cell* 32: 232–46.

322. Kanduri C. (2016) Long noncoding RNAs: Lessons from genomic imprinting. *Biochimica et Biophysica Acta* 1859: 102–11.

323. Dinger M.E. et al. (2008) Long noncoding RNAs in mouse embryonic stem cell pluripotency and differentiation. *Genome Research* 18: 1433–45.

324. Wang K.C. et al. (2011) A long noncoding RNA maintains active chromatin to coordinate homeotic gene expression. *Nature* 472: 120–4.

325. Gomez J.A. et al. (2013) The *NeST* long ncRNA controls microbial susceptibility and epigenetic activation of the interferon-gamma locus. *Cell* 152: 743–54.

326. Yang Y.W. et al. (2014) Essential role of lncRNA binding for WDR5 maintenance of active chromatin and embryonic stem cell pluripotency. *eLife* 3: e02046.

327. He X. et al. (2015) An Lnc RNA (GAS5)/SnoRNA-derived piRNA induces activation of TRAIL gene by site-specifically recruiting MLL/COMPASS-like complexes. *Nucleic Acids Research* 43: 3712–25.

328. Deng C. et al. (2016) *HoxBlinc* RNA recruits Set1/MLL complexes to activate *Hox* gene expression patterns and mesoderm lineage development. *Cell Reports* 14: 103–14.

329. Wang X.Q.D. and Dostie J. (2017) Reciprocal regulation of chromatin state and architecture by HOTAIRM1 contributes to temporal collinear HOXA gene activation. *Nucleic Acids Research* 45: 1091–104.

330. Alexanian M. et al. (2017) A transcribed enhancer dictates mesendoderm specification in pluripotency. *Nature Communications* 8: 1806.

331. Su W. et al. (2017) Long noncoding RNA ZEB1-AS1 epigenetically regulates the expressions of ZEB1 and downstream molecules in prostate cancer. *Molecular Cancer* 16: 142.

332. Luo H. et al. (2019) HOTTIP lncRNA promotes hematopoietic stem cell self-renewal leading to AML-like disease in mice. *Cancer Cell* 36: 645–59.

333. Fanucchi S. et al. (2019) Immune genes are primed for robust transcription by proximal long noncoding RNAs located in nuclear compartments. *Nature Genetics* 51: 138–50.

334. Hu A. et al. (2021) Long non-coding RNA *ROR* recruits histone transmethylase MLL1 to up-regulate *TIMP3* expression and promote breast cancer progression. *Journal of Translational Medicine* 19: 95.

335. Wang Y. et al. (2018) Overexpressing lncRNA LAIR increases grain yield and regulates neighbouring gene cluster expression in rice. *Nature Communications* 9: 3516.

336. Khalil A.M. et al. (2009) Many human large intergenic noncoding RNAs associate with chromatin-modifying complexes and affect gene expression. *Proceedings of the National Academy of Sciences USA* 106: 11667–72.

337. Kotake Y. et al. (2011) Long non-coding RNA ANRIL is required for the PRC2 recruitment to and silencing of p15(INK4B) tumor suppressor gene. *Oncogene* 30: 1956–62.

338. Shore A.N. et al. (2012) Pregnancy-induced noncoding RNA (PINC) associates with polycomb repressive complex 2 and regulates mammary epithelial differentiation. *PLOS Genetics* 8: e1002840.

339. Davidovich C. and Cech T.R. (2015) The recruitment of chromatin modifiers by long noncoding RNAs: Lessons from PRC2. *RNA* 21: 2007–22.

340. Marchese F.P., Raimondi I. and Huarte M. (2017) The multidimensional mechanisms of long noncoding RNA function. *Genome Biology* 18: 206.

341. Dill T.L., Carroll A., Pinheiro A., Gao J. and Naya F.J. (2021) The long noncoding RNA Meg3 regulates myoblast plasticity and muscle regeneration through epithelial-mesenchymal transition. *Development* 148: dev194027.

342. Balas M.M. et al. (2021) Establishing RNA-RNA interactions remodels lncRNA structure and promotes PRC2 activity. *Science Advances* 7: eabc9191.

343. Long Y. et al. (2020) RNA is essential for PRC2 chromatin occupancy and function in human pluripotent stem cells. *Nature Genetics* 52: 931–8.

344. Wang X. et al. (2017) Targeting of polycomb repressive complex 2 to RNA by short repeats of consecutive guanines. *Molecular Cell* 65: 1056–67.

345. Beltran M. et al. (2019) G-tract RNA removes Polycomb repressive complex 2 from genes. *Nature Structural & Molecular Biology* 26: 899–909.

346. Zhang Q. et al. (2019) RNA exploits an exposed regulatory site to inhibit the enzymatic activity of PRC2. *Nature Structural & Molecular Biology* 26: 237–47.

347. Plys A.J. et al. (2019) Phase separation of Polycomb-repressive complex 1 is governed by a charged disordered region of CBX2. *Genes & Development* 33: 1–15.

348. Davidovich C., Zheng L., Goodrich K.J. and Cech T.R. (2013) Promiscuous RNA binding by Polycomb repressive complex 2. *Nature Structural & Molecular Biology* 20: 1250–7.

349. Davidovich C. et al. (2015) Toward a consensus on the binding specificity and promiscuity of PRC2 for RNA. *Molecular Cell* 57: 552–8.

350. Martinez A.M. and Cavalli G. (2006) The role of polycomb group proteins in cell cycle regulation during development. *Cell Cycle* 5: 1189–97.

351. Bantignies F. and Cavalli G. (2006) Cellular memory and dynamic regulation of polycomb group proteins. *Current Opinion in Cell Biology* 18: 275–83.

352. Boyer L.A. et al. (2006) Polycomb complexes repress developmental regulators in murine embryonic stem cells. *Nature* 441: 349–53.

353. Lee T.I. et al. (2006) Control of developmental regulators by polycomb in human embryonic stem cells. *Cell* 125: 301–13.

354. Poynter S.T. and Kadoch C. (2016) Polycomb and trithorax opposition in development and disease. *Wiley Interdisciplinary Reviews Developmental Biology* 5: 659–88.

355. Schuettengruber B., Bourbon H.-M., Di Croce L. and Cavalli G. (2017) Genome regulation by Polycomb and Trithorax: 70 years and counting. *Cell* 171: 34–57.

356. Herzog V.A. et al. (2014) A strand-specific switch in noncoding transcription switches the function of a Polycomb/Trithorax response element. *Nature Genetics* 46: 973–81.

357. Beltran M. et al. (2016) The interaction of PRC2 with RNA or chromatin is mutually antagonistic. *Genome Research* 26: 896–907.

358. Zovoilis A., Cifuentes-Rojas C., Chu H.-P., Hernandez A.J. and Lee J.T. (2016) Destabilization of B2 RNA by EZH2 activates the stress response. *Cell* 167: 1788–802.

359. Singh I. et al. (2018) MiCEE is a ncRNA-protein complex that mediates epigenetic silencing and nucleolar organization. *Nature Genetics* 50: 990–1001.

360. Kang X., Zhao Y., Van Arsdell G., Nelson S.F. and Touma M. (2020) Ppp1r1b-lncRNA inhibits PRC2 at myogenic regulatory genes to promote cardiac and skeletal muscle development in mouse and human. *RNA* 26: 481–91.

361. Rosenberg M. et al. (2021) Motif-driven interactions between RNA and PRC2 are rheostats that regulate transcription elongation. *Nature Structural & Molecular Biology* 28: 103–17.

362. Guo Y., Zhao S. and Wang G.G. (2021) Polycomb gene silencing mechanisms: PRC2 chromatin targeting, H3K27me3 'readout', and phase separation-based compaction. *Trends in Genetics* 37: 547–65.

363. Kraft K. et al. (2020) Polycomb-mediated genome architecture enables long-range spreading of H3K27 methylation. *bioRxiv*: 2020.07.27.223438.

364. Ringrose L. (2017) Noncoding RNAs in Polycomb and Trithorax regulation: A quantitative perspective. *Annual Review of Genetics* 51: 385–411.

365. Jantrapirom S. et al. (2021) Long noncoding RNA-dependent methylation of nonhistone proteins. *WIREs RNA* 12: e1661.

366. Butler A.A., Johnston D.R., Kaur S. and Lubin F.D. (2019) Long noncoding RNA NEAT1 mediates neuronal histone methylation and age-related memory impairment. *Science Signaling* 12: eaaw9277.

367. Budhavarapu V.N., Chavez M. and Tyler J.K. (2013) How is epigenetic information maintained through DNA replication? *Epigenetics & Chromatin* 6: 32.

368. Kanduri C., Whitehead J. and Mohammad F. (2009) The long and the short of it: RNA-directed chromatin asymmetry in mammalian X-chromosome inactivation. *FEBS Letters* 583: 857–64.

369. Gendrel A.-V. and Heard E. (2014) Noncoding RNAs and epigenetic mechanisms during X-chromosome inactivation. *Annual Review of Cell and Developmental Biology* 30: 561–80.

370. Brown C.J. *et al.* (1992) The human *XIST* gene: Analysis of a 17 kb inactive X-specific RNA that contains conserved repeats and is highly localized within the nucleus. *Cell* 71: 527–42.

371. Chu C. et al. (2015) Systematic discovery of Xist RNA binding proteins. *Cell* 161: 404–16.

372. Minajigi A. et al. (2015) Chromosomes. A comprehensive Xist interactome reveals cohesin repulsion and an RNA-directed chromosome conformation. *Science* 349: aab2276.

373. Smola M.J. et al. (2016) SHAPE reveals transcript-wide interactions, complex structural domains, and protein interactions across the Xist lncRNA in living cells. *Proceedings of the National Academy of Sciences USA* 113: 10322–7.

374. Lu Z. et al. (2016) RNA duplex map in living cells reveals higher-order transcriptome structure. *Cell* 165: 1267–79.

375. Sunwoo H., Wu J.Y. and Lee J.T. (2015) The Xist RNA-PRC2 complex at 20-nm resolution reveals a low Xist stoichiometry and suggests a hit-and-run mechanism in mouse cells. *Proceedings of the National Academy of Sciences USA* 112: E4216–25.

376. Markaki Y. et al. (2021) Xist nucleates local protein gradients to propagate silencing across the X chromosome. *Cell* 184: 6174–92.

377. Chaumeil J., Le Baccon P., Wutz A. and Heard E. (2006) A novel role for Xist RNA in the formation of a repressive nuclear compartment into which genes are recruited when silenced. *Genes & Development* 20: 2223–37.

378. Yamada N. et al. (2015) *Xist* exon 7 contributes to the stable localization of Xist RNA on the inactive X-chromosome. *PLOS Genetics* 11: e1005430.

379. Sunwoo H., Colognori D., Froberg J.E., Jeon Y. and Lee J.T. (2017) Repeat E anchors Xist RNA to the inactive X chromosomal compartment through CDKN1A-interacting protein (CIZ1). *Proceedings of the National Academy of Sciences USA* 114: 10654–9.

380. Yue M. et al. (2017) Xist RNA repeat E is essential for ASH2L recruitment to the inactive X and regulates histone modifications and escape gene expression. *PLOS Genetics* 13: e1006890.

381. Wang Y. et al. (2019) Identification of a Xist silencing domain by tiling CRISPR. *Scientific Reports* 9: 2408.

382. Wei G., Almeida M., Bowness J.S., Nesterova T.B. and Brockdorff N. (2021) Xist Repeats B and C, but not Repeat A, mediate de novo recruitment of the Polycomb system in X chromosome inactivation. *Developmental Cell* 56: 1234–5.

383. Csankovszki G., Nagy A. and Jaenisch R. (2001) Synergism of *Xist* RNA, DNA methylation, and histone hypoacetylation in maintaining X chromosome inactivation. *Journal of Cell Biology* 153: 773–84.

384. McHugh C.A. et al. (2015) The Xist lncRNA interacts directly with SHARP to silence transcription through HDAC3. *Nature* 521: 232–6.

385. Yu B. et al. (2021) B cell-specific XIST complex enforces X-inactivation and restrains atypical B cells. *Cell* 184: 1790–803.

386. Plath K. et al. (2004) Developmentally regulated alterations in polycomb repressive complex 1 proteins on the inactive X chromosome. *Journal of Cell Biology* 167: 1025–35.

387. Maenner S. et al. (2010) 2-D structure of the A region of Xist RNA and its implication for PRC2 association. *PLOS Biology* 8: e1000276.

388. Splinter E. et al. (2011) The inactive X chromosome adopts a unique three-dimensional conformation that is dependent on Xist RNA. *Genes & Development* 25: 1371–83.

389. Nora E.P. et al. (2012) Spatial partitioning of the regulatory landscape of the X-inactivation centre. *Nature* 485: 381–5.

390. Engreitz J.M. et al. (2013) The Xist lncRNA exploits three-dimensional genome architecture to spread across the X chromosome. *Science* 341: 1237973.

391. van Bemmel J.G. et al. (2019) The bipartite TAD organization of the X-inactivation center ensures opposing developmental regulation of Tsix and Xist. *Nature Genetics* 51: 1024–34.

392. Kriz A.J., Colognori D., Sunwoo H., Nabet B. and Lee J.T. (2021) Balancing cohesin eviction and retention prevents aberrant chromosomal interactions, Polycomb-mediated repression, and X-inactivation. *Molecular Cell* 81: 1970–87.

393. Cerase A. et al. (2019) Phase separation drives X-chromosome inactivation: A hypothesis. *Nature Structural & Molecular Biology* 26: 331–4.

394. Pandya-Jones A. et al. (2020) A protein assembly mediates Xist localization and gene silencing. *Nature* 587: 145–51.

395. Chow J.C. et al. (2010) LINE-1 activity in facultative heterochromatin formation during X chromosome inactivation. *Cell* 141: 956–69.

396. Tannan N.B. et al. (2014) DNA methylation profiling in X;autosome translocations supports a role for L1 repeats in the spread of X chromosome inactivation. *Human Molecular Genetics* 23: 1224–36.

397. Brockdorff N. (2018) Local tandem repeat expansion in Xist RNA as a model for the functionalisation of ncRNA. *Noncoding RNA* 4: 28.

398. Carter A.C. et al. (2020) Spen links RNA-mediated endogenous retrovirus silencing and X chromosome inactivation. *eLife* 9: e54508.

399. Loda A. and Heard E. (2019) Xist RNA in action: Past, present, and future. *PLOS Genetics* 15: e1008333.

400. Lyon M.F. (1998) X-Chromosome inactivation: A repeat hypothesis. *Cytogenetic and Genome Research* 80: 133–7.

401. Lyon M.F. (2000) LINE-1 elements and X chromosome inactivation: A function for "junk" DNA? *Proceedings of the National Academy of Sciences USA* 97: 6248–9.

402. Matsuno Y., Yamashita T., Wagatsuma M. and Yamakage H. (2019) Convergence in LINE-1 nucleotide variations can benefit redundantly forming triplexes with lncRNA in mammalian X-chromosome inactivation. *Mobile DNA* 10: 33.

403. Yildirim E. et al. (2013) Xist RNA is a potent suppressor of hematologic cancer in mice. *Cell* 152: 727–42.

404. Sado T., Wang Z., Sasaki H. and Li E. (2001) Regulation of imprinted X-chromosome inactivation in mice by Tsix. *Development* 128: 1275–86.

405. Spencer R.J. et al. (2011) A boundary element between *Tsix* and *Xist* binds the chromatin insulator Ctcf and contributes to initiation of X-chromosome inactivation. *Genetics* 189: 441–54.

406. Morey C. et al. (2004) The region 3′ to Xist mediates X chromosome counting and H3 Lys-4 dimethylation within the Xist gene. *EMBO Journal* 23: 594–604.

407. Tian D., Sun S. and Lee J.T. (2010) The long noncoding RNA, Jpx, is a molecular switch for X chromosome inactivation. *Cell* 143: 390–403.

408. Chureau C. et al. (2011) Ftx is a non-coding RNA which affects Xist expression and chromatin structure within the X-inactivation center region. *Human Molecular Genetics* 20: 705–18.

409. Sun S. et al. (2013) Jpx RNA activates Xist by evicting CTCF. *Cell* 153: 1537–51.

410. Karner H. et al. (2020) Functional conservation of lncRNA *JPX* despite sequence and structural divergence. *Journal of Molecular Biology* 432: 283–300.

411. Casanova M. et al. (2019) A primate-specific retroviral enhancer wires the XACT lncRNA into the core pluripotency network in humans. *Nature Communications* 10: 5652.

412. Anguera M.C. et al. (2011) *Tsx* produces a long noncoding RNA and has general functions in the germline, stem cells, and brain. *PLOS Genetics* 7: e1002248.

413. Yang F. et al. (2015) The lncRNA *Firre* anchors the inactive X chromosome to the nucleolus by binding CTCF and maintains H3K27me3 methylation. *Genome Biology* 16: 52.

414. Hacisuleyman E. et al. (2014) Topological organization of multichromosomal regions by the long intergenic noncoding RNA *Firre*. *Nature Structural & Molecular Biology* 21: 198–206.

415. Hacisuleyman E., Shukla C.J., Weiner C.L. and Rinn J.L. (2016) Function and evolution of local repeats in the *Firre* locus. *Nature Communications* 7: 11021.

416. Lewandowski J.P. et al. (2019) The *Firre* locus produces a trans-acting RNA molecule that functions in hematopoiesis. *Nature Communications* 10: 5137.

417. Ogawa Y., Sun B.K. and Lee J.T. (2008) Intersection of the RNA interference and X-inactivation pathways. *Science* 320: 1336–41.

418. Nesterova T.B. et al. (2008) Dicer regulates Xist promoter methylation in ES cells indirectly through transcriptional control of Dnmt3a. *Epigenetics & Chromatin* 1: 2.

419. Patil D.P. et al. (2016) m⁶A RNA methylation promotes XIST-mediated transcriptional repression. *Nature* 537: 369–73.

420. Irvine D.V. et al. (2006) Argonaute slicing is required for heterochromatic silencing and spreading. *Science* 313: 1134–7.

421. Li L.C. et al. (2006) Small dsRNAs induce transcriptional activation in human cells. *Proceedings of the National Academy of Sciences USA* 103: 17337–42.

422. Kelley R.L. (2004) Path to equality strewn with roX. *Developmental Biology* 269: 18–25.

423. Park S.-W., Kuroda M.I. and Park Y. (2008) Regulation of histone H4 Lys16 acetylation by predicted alternative secondary structures in roX noncoding RNAs. *Molecular and Cellular Biology* 28: 4952–62.

424. Valsecchi C.I.K. et al. (2020) RNA nucleation by MSL2 induces selective X chromosome compartmentalization. *Nature* 589: 137–42.

425. Ptashne M. (1988) How eukaryotic transcriptional activators work. *Nature* 335: 683–9.

426. Buecker C. and Wysocka J. (2012) Enhancers as information integration hubs in development: Lessons from genomics. *Trends in Genetics* 28: 276–84.

427. Smith E. and Shilatifard A. (2014) Enhancer biology and enhanceropathies. *Nature Structural & Molecular Biology* 21: 210–9.

428. Shlyueva D., Stampfel G. and Stark A. (2014) Transcriptional enhancers: From properties to genome-wide predictions. *Nature Reviews Genetics* 15: 272–86.

429. Krijger P.H.L. and de Laat W. (2016) Regulation of disease-associated gene expression in the 3D genome. *Nature Reviews Molecular Cell Biology* 17: 771–82.

430. Henriques T. et al. (2018) Widespread transcriptional pausing and elongation control at enhancers. *Genes & Development* 32: 26–41.

431. Souaid C., Bloyer S. and Noordermeer D. (2018) Promoter–enhancer looping and regulatory neighborhoods: Gene regulation in the framework of topologically associating domains, in C. Lavelle and J.-M. Victor (eds.) *Nuclear Architecture and Dynamics* (Academic Press, Cambridge, MA).

432. Gribnau J., Diderich K., Pruzina S., Calzolari R. and Fraser P. (2000) Intergenic transcription and developmental remodeling of chromatin subdomains in the human beta-globin locus. *Molecular Cell* 5: 377–86.

433. Masternak K., Peyraud N., Krawczyk M., Barras E. and Reith W. (2003) Chromatin remodeling and extragenic transcription at the MHC class II locus control region. *Nature Immunology* 4: 132–7.

434. Ling J. et al. (2004) HS2 enhancer function is blocked by a transcriptional terminator inserted between the enhancer and the promoter. *Journal of Biological Chemistry* 279: 51704–13.

435. Delpretti S. et al. (2013) Multiple enhancers regulate *Hoxd* genes and the Hotdog lncRNA during cecum budding. *Cell Reports* 5: 137–50.

436. Core L.J. et al. (2014) Analysis of nascent RNA identifies a unified architecture of initiation regions at mammalian promoters and enhancers. *Nature Genetics* 46: 1311–20.

437. Kim T.-K., Hemberg M. and Gray J.M. (2015) Enhancer RNAs: A class of long noncoding RNAs synthesized at enhancers. *Cold Spring Harbor Perspectives in Biology* 7: a018622.

438. Dong X. et al. (2018) Enhancers active in dopamine neurons are a primary link between genetic variation and neuropsychiatric disease. *Nature Neuroscience* 21: 1482–92.

439. Azofeifa J.G. et al. (2018) Enhancer RNA profiling predicts transcription factor activity. *Genome Research* 28: 334–44.

440. Grossman S.R. et al. (2018) Positional specificity of different transcription factor classes within enhancers. *Proceedings of the National Academy of Sciences USA* 115: E7222–30.

441. Arnold P.R., Wells A.D. and Li X.C. (2020) Diversity and emerging roles of enhancer RNA in regulation of gene expression and cell fate. *Frontiers in Cell and Developmental Biology* 7: 377.

442. Li W., Notani D. and Rosenfeld M.G. (2016) Enhancers as non-coding RNA transcription units: Recent insights and future perspectives. *Nature Reviews Genetics* 17: 207–23.

443. Arner E. et al. (2015) Transcribed enhancers lead waves of coordinated transcription in transitioning mammalian cells. *Science* 347: 1010–4.

444. Kim T.-K. et al. (2010) Widespread transcription at neuronal activity-regulated enhancers. *Nature* 465: 182–7.

445. Hnisz D. et al. (2013) Super-enhancers in the control of cell identity and disease. *Cell* 155: 934–47.

446. Hon C.-C. et al. (2017) An atlas of human long noncoding RNAs with accurate 5′ ends. *Nature* 543: 199–204.

447. Lewis M.W., Li S. and Franco H.L. (2019) Transcriptional control by enhancers and enhancer RNAs. *Transcription* 10: 171–86.

448. De Santa F. et al. (2010) A large fraction of extragenic RNA pol II transcription sites overlap enhancers. *PLOS Biology* 8: e1000384.

449. Wang D. et al. (2011) Reprogramming transcription by distinct classes of enhancers functionally defined by eRNA. *Nature* 474: 390–4.

450. Wu H. et al. (2014) Tissue-specific RNA expression marks distant-acting developmental enhancers. *PLOS Genetics* 10: e1004610.

451. Kim Y.W., Lee S., Yun J. and Kim A. (2015) Chromatin looping and eRNA transcription precede the transcriptional activation of gene in the beta-globin locus. *Bioscience Reports* 35: e00179.

452. Sartorelli V. and Lauberth S.M. (2020) Enhancer RNAs are an important regulatory layer of the epigenome. *Nature Structural & Molecular Biology* 27: 521–8.

453. Carullo N.V.N. et al. (2020) Enhancer RNAs predict enhancer–gene regulatory links and are critical for enhancer function in neuronal systems. *Nucleic Acids Research* 48: 9550–70.

454. Lin C.Y. et al. (2016) Active medulloblastoma enhancers reveal subgroup-specific cellular origins. *Nature* 530: 57–62.

455. Zhao Y. et al. (2016) Activation of P-TEFb by androgen receptor-regulated enhancer RNAs in castration-resistant prostate cancer. *Cell Reports* 15: 599–610.

456. Chen H. and Liang H. (2020) A high-resolution map of human enhancer RNA loci characterizes super-enhancer activities in cancer. *Cancer Cell* 38: 701–15.

457. Katsushima K. et al. (2021) The long noncoding RNA lnc-HLX-2–7 is oncogenic in Group 3 medulloblastomas. *Neuro-Oncology* 23: 572–85.

458. Thomas H.F. *et al.* (2021) Temporal dissection of an enhancer cluster reveals distinct temporal and functional contributions of individual elements. *Molecular Cell* 81: 969–82.

459. Melgar M.F., Collins F.S. and Sethupathy P. (2011) Discovery of active enhancers through bidirectional expression of short transcripts. *Genome Biology* 12: R113.

460. Andersson R. et al. (2014) An atlas of active enhancers across human cell types and tissues. *Nature* 507: 455–61.

461. Pnueli L., Rudnizky S., Yosefzon Y. and Melamed P. (2015) RNA transcribed from a distal enhancer is required for activating the chromatin at the promoter of the gonadotropin alpha-subunit gene. *Proceedings of the National Academy of Sciences USA* 112: 4369–74.

462. Kouno T. et al. (2019) C1 CAGE detects transcription start sites and enhancer activity at single-cell resolution. *Nature Communications* 10: 360.

463. Seila A.C. et al. (2008) Divergent transcription from active promoters. *Science* 322: 1849–51.

464. Young R.S., Kumar Y., Bickmore W.A. and Taylor M.S. (2017) Bidirectional transcription initiation marks accessible chromatin and is not specific to enhancers. *Genome Biology* 18: 242.

465. Mikhaylichenko O. et al. (2018) The degree of enhancer or promoter activity is reflected by the levels and directionality of eRNA transcription. *Genes & Development* 32: 42–57.

466. Pefanis E. et al. (2015) RNA exosome-regulated long noncoding RNA transcription controls super-enhancer activity. *Cell* 161: 774–89.

467. Tippens N.D., Vihervaara A. and Lis J.T. (2018) Enhancer transcription: What, where, when, and why? *Genes & Development* 32: 1–3.

468. Harman C.C.D. et al. (2021) An in vivo screen of noncoding loci reveals that *Daedalus* is a gatekeeper of an Ikaros-dependent checkpoint during haematopoiesis. *Proceedings of the National Academy of Sciences USA* 118: e1918062118.

469. Drewell R.A., Bae E., Burr J. and Lewis E.B. (2002) Transcription defines the embryonic domains of cisregulatory activity at the *Drosophila bithorax* complex. *Proceedings of the National Academy of Sciences USA* 99: 16853–8.

470. Orom U.A. et al. (2010) Long noncoding RNAs with enhancer-like function in human cells. *Cell* 143: 46–58.

471. Orom U.A. and Shiekhattar R. (2011) Noncoding RNAs and enhancers: Complications of a long-distance relationship. *Trends in Genetics* 27: 433–9.

472. Alvarez-Dominguez J.R., Knoll M., Gromatzky A.A. and Lodish H.F. (2017) The super-enhancer-derived *alncRNA-EC7/Bloodlinc* potentiates red blood cell development in trans. *Cell Reports* 19: 2503–14.

473. Cajigas I. et al. (2018) The *Evf2* ultraconserved enhancer lncRNA functionally and spatially organizes megabase distant genes in the developing forebrain. *Molecular Cell* 71: 956–72.

474. Tsai P.-F. et al. (2018) A muscle-specific enhancer RNA mediates cohesin recruitment and regulates transcription in trans. *Molecular Cell* 71: 129–41.

475. Morrison T.A. et al. (2018) A long noncoding RNA from the *HBS1L-MYB* intergenic region on chr6q23 regulates human fetal hemoglobin expression. *Blood Cells, Molecules, and Diseases* 69: 1–9.

476. Ntini E. and Marsico A. (2019) Functional impacts of noncoding RNA processing on enhancer activity and target gene expression. *Journal of Molecular Cell Biology* 11: 868–79.

477. Fatima R., Choudhury S.R., Divya T. R., Bhaduri U. and Rao M.R.S. (2019) A novel enhancer RNA, *Hmrhl*, positively regulates its host gene, *phkb*, in chronic myelogenous leukemia. *Noncoding RNA Research* 4: 96–108.

478. Yan P. et al. (2020) LncRNA *Platr22* promotes super-enhancer activity and stem cell pluripotency. *Journal of Molecular Cell Biology* 13: 295–313.

479. Allou L. et al. (2021) Non-coding deletions identify *Maenli* lncRNA as a limb-specific En1 regulator. *Nature* 592: 93–8.

480. Borsari B. et al. (2021) Enhancers with tissue-specific activity are enriched in intronic regions. *Genome Research* 31: 1325–36.

481. Gibcus J.H. and Dekker J. (2013) The hierarchy of the 3D genome. *Molecular Cell* 49: 773–82.

482. Isoda T. et al. (2017) Non-coding transcription instructs chromatin folding and compartmentalization to dictate enhancer-promoter communication and T cell fate. *Cell* 171: 103–19.

483. Yang Y. et al. (2016) Enhancer RNA-driven looping enhances the transcription of the long noncoding RNA DHRS4-AS1, a controller of the DHRS4 gene cluster. *Scientific Reports* 6: 20961.

484. Benabdallah N.S. et al. (2019) Decreased enhancer-promoter proximity accompanying enhancer activation. *Molecular Cell* 76: 473–84.

485. Crump N.T. et al. (2021) BET inhibition disrupts transcription but retains enhancer-promoter contact. *Nature Communications* 12: 223.

486. Kornienko A.E., Guenzl P.M., Barlow D.P. and Pauler F.M. (2013) Gene regulation by the act of long non-coding RNA transcription. *BMC Biology* 11: 59.

487. de Lara J.C.-F., Arzate-Mejía R.G. and Recillas-Targa F. (2019) Enhancer RNAs: Insights into their biological role. *Epigenetics Insights* 12: 1–7.

488. Levine M., Cattoglio C. and Tjian R. (2014) Looping back to leap forward: Transcription enters a new era. *Cell* 157: 13–25.

489. Halfon M.S. (2019) Studying transcriptional enhancers: The founder fallacy, validation creep, and other biases. *Trends in Genetics* 35: 93–103.

490. Engreitz J.M. et al. (2016) Local regulation of gene expression by lncRNA promoters, transcription and splicing. *Nature* 539: 452–5.

491. Paralkar Vikram R. et al. (2016) Unlinking an lncRNA from its associated cis element. *Molecular Cell* 62: 104–10.

492. Kioussis D. and Festenstein R. (1997) Locus control regions: Overcoming heterochromatin-induced gene inactivation in mammals. *Current Opinion in Genetics and Development* 7: 614–9.

493. Sipos L. et al. (1998) Transvection in the *Drosophila Abd-B* domain: Extensive upstream sequences are involved in anchoring distant cis-regulatory regions to the promoter. *Genetics* 149: 1031–50.

494. Melo C.A. et al. (2013) eRNAs are required for p53-dependent enhancer activity and gene transcription. *Molecular Cell* 49: 524–35.

495. Lam M.T.Y., Li W., Rosenfeld M.G. and Glass C.K. (2014) Enhancer RNAs and regulated transcriptional programs. *Trends in Biochemical Sciences* 39: 170–82.

496. Yin Y. et al. (2015) Opposing roles for the lncRNA Haunt and its genomic locus in regulating HOXA gene activation during embryonic stem cell differentiation. *Cell Stem Cell* 16: 504–16.

497. Stafford D.A., Dichmann D.S., Chang J.K. and Harland R.M. (2017) Deletion of the sclerotome-enriched lncRNA PEAT augments ribosomal protein expression. *Proceedings of the National Academy of Sciences USA* 114: 101–6.

498. Andergassen D. et al. (2019) The Airn lncRNA does not require any DNA elements within its locus to silence distant imprinted genes. *PLOS Genetics* 15: e1008268.

499. Bond A.M. et al. (2009) Balanced gene regulation by an embryonic brain ncRNA is critical for adult hippocampal GABA circuitry. *Nature Neuroscience* 12: 1020–7.

500. Aguilo F. et al. (2016) Deposition of 5-methylcytosine on enhancer RNAs enables the coactivator function of PGC-1a. *Cell Reports* 14: 479–92.

501. Cheng J.X. et al. (2018) RNA cytosine methylation and methyltransferases mediate chromatin organization and 5-azacytidine response and resistance in leukaemia. *Nature Communications* 9: 1163.

502. Tan J.Y., Biasini A., Young R.S. and Marques A.C. (2020) Splicing of enhancer-associated lincRNAs contributes to enhancer activity. *Life Science Alliance* 3: e202000663.

503. Maass P.G. et al. (2012) A misplaced lncRNA causes brachydactyly in humans. *Journal of Clinical Investigation* 122: 3990–4002.

504. Li W. et al. (2013) Functional roles of enhancer RNAs for oestrogen-dependent transcriptional activation. *Nature* 498: 516–20.

505. Xiang J.F. et al. (2014) Human colorectal cancer-specific CCAT1-L lncRNA regulates long-range chromatin interactions at the MYC locus. *Cell Research* 24: 513–31.

506. Sun J. et al. (2014) A novel antisense long noncoding RNA within the IGF1R gene locus is imprinted in hematopoietic malignancies. *Nucleic Acids Research* 42: 9588–601.

507. Paralkar V.R. et al. (2014) Lineage and species-specific long noncoding RNAs during erythro-megakaryocytic development. *Blood* 123: 1927–37.

508. Lu L. et al. (2013) Genome-wide survey by ChIP-seq reveals YY1 regulation of lincRNAs in skeletal myogenesis. *EMBO journal* 32: 2575–88.

509. Shii L., Song L., Maurer K., Zhang Z. and Sullivan K.E. (2017) SERPINB2 is regulated by dynamic interactions with pause-release proteins and enhancer RNAs. *Molecular Immunology* 88: 20–31.

510. Han X. et al. (2019) The lncRNA *Hand2os1/Uph* locus orchestrates heart development through regulation of precise expression of *Hand2*. *Development* 146: dev176198.

511. Andergassen D. and Rinn J.L. (2021) From genotype to phenotype: Genetics of mammalian long non-coding RNAs in vivo. *Nature Reviews Genetics* 23: 229–243.

512. Rahnamoun H. et al. (2018) RNAs interact with BRD4 to promote enhanced chromatin engagement and transcription activation. *Nature Structural & Molecular Biology* 25: 687–97.

513. Fang K. et al. (2020) Cis-acting lnc-eRNA SEELA directly binds histone H4 to promote histone recognition and leukemia progression. *Genome Biology* 21: 269.

514. Groff A.F., Barutcu A.R., Lewandowski J.P. and Rinn J.L. (2018) Enhancers in the Peril lincRNA locus regulate distant but not local genes. *Genome Biology* 19: 219.

515. Hu T. et al. (2017) Long non-coding RNAs transcribed by ERV-9 LTR retrotransposon act in cis to modulate long-range LTR enhancer function. *Nucleic Acids Research* 45: 4479–92.

516. Rothschild G. et al. (2020) Noncoding RNA transcription alters chromosomal topology to promote isotype-specific class switch recombination. *Science Immunology* 5: eaay5864.

517. Chen H., Du G., Song X. and Li L. (2017) Non-coding transcripts from enhancers: New insights into enhancer activity and gene expression regulation. *Genomics, Proteomics & Bioinformatics* 15: 201–7.

518. Cai Z. et al. (2020) RIC-seq for global in situ profiling of RNA–RNA spatial interactions. *Nature* 582: 432–7.

519. Lim B. and Levine M.S. (2021) Enhancer-promoter communication: Hubs or loops? *Current Opinion in Genetics & Development* 67: 5–9.

520. Zhu I., Song W., Ovcharenko I. and Landsman D. (2021) A model of active transcription hubs that unifies the roles of active promoters and enhancers. *Nucleic Acids Research* 49: 4493–505.

521. Mele M. and Rinn J.L. (2016) "Cat's cradling" the 3D genome by the act of lncRNA transcription. *Molecular Cell* 62: 657–64.

522. Morf J., Basu S. and Amaral P.P. (2020) RNA, genome output and input. *Frontiers in Genetics* 11: 1330.

523. Berry J., Weber S.C., Vaidya N., Haataja M. and Brangwynne C.P. (2015) RNA transcription modulates phase transition-driven nuclear body assembly. *Proceedings of the National Academy of Sciences USA* 112: E5237–45.

524. Creamer K.M., Kolpa H.J. and Lawrence J.B. (2021) Nascent RNA scaffolds contribute to chromosome territory architecture and counter chromatin compaction. *Molecular Cell* 81: 3509–25.

525. Cusanovich D.A. et al. (2018) A single-cell atlas of *in vivo* mammalian chromatin accessibility. *Cell* 174: 1309–24.

526. Fang R. et al. (2021) Comprehensive analysis of single cell ATAC-seq data with SnapATAC. *Nature Communications* 12: 1337.

527. Shen Y. et al. (2012) A map of the cis-regulatory sequences in the mouse genome. *Nature* 488: 116–20.

528. Thurman R.E. et al. (2012) The accessible chromatin landscape of the human genome. *Nature* 489: 75–82.

529. The Encode Project Consortium (2012) An integrated encyclopedia of DNA elements in the human genome. *Nature* 489: 57–74.

530. Zhu J. et al. (2013) Genome-wide chromatin state transitions associated with developmental and environmental cues. *Cell* 152: 642–54.

531. Heidari N. et al. (2014) Genome-wide map of regulatory interactions in the human genome. *Genome Research* 24: 1905–17.

532. Pott S. and Lieb J.D. (2015) What are super-enhancers? *Nature Genetics* 47: 8–12.

533. Li S. and Ovcharenko I. (2020) Enhancer jungles establish robust tissue-specific regulatory control in the human genome. *Genomics* 112: 2261–70.

534. Austenaa L.M.I. et al. (2021) A first exon termination checkpoint preferentially suppresses extragenic transcription. *Nature Structural & Molecular Biology* 28: 337–46.

535. Gil N. and Ulitsky I. (2021) Inefficient splicing curbs noncoding RNA transcription. *Nature Structural & Molecular Biology* 28: 327–8.

536. Laffleur B. et al. (2021) Noncoding RNA processing by DIS3 regulates chromosomal architecture and somatic hypermutation in B cells. *Nature Genetics* 53: 230–42.

537. Anderson P. and Kedersha N. (2009) RNA granules: Post-transcriptional and epigenetic modulators of gene expression. *Nature Reviews Molecular Cell Biology* 10: 430–6.

538. Buchan J.R. (2014) mRNP granules. *RNA Biology* 11: 1019–30.

539. Wright P.E. and Dyson H.J. (2015) Intrinsically disordered proteins in cellular signalling and regulation. *Nature Reviews Molecular Cell Biology* 16: 18–29.

540. Brangwynne C.P., Mitchison T.J. and Hyman A.A. (2011) Active liquid-like behavior of nucleoli determines their size and shape in *Xenopus laevis* oocytes. *Proceedings of the National Academy of Sciences USA* 108: 4334–9.

541. Hyman A.A., Weber C.A. and Jülicher F. (2014) Liquid-liquid phase separation in biology. *Annual Review of Cell and Developmental Biology* 30: 39–58.

542. Polymenidou M. (2018) The RNA face of phase separation. *Science* 360: 859–60.

543. Narlikar G.J. et al. (2021) Is transcriptional regulation just going through a phase? *Molecular Cell* 81: 1579–85.

544. Shin Y. and Brangwynne C.P. (2017) Liquid phase condensation in cell physiology and disease. *Science* 357: eaaf4382.

545. Yoshizawa T., Nozawa R.-S., Jia T.Z., Saio T. and Mori E. (2020) Biological phase separation: Cell biology meets biophysics. *Biophysical Reviews* 12: 519–39.

546. Lin Y., Protter D.S.W., Rosen M.K. and Parker R. (2015) Formation and maturation of phase-separated liquid droplets by RNA-binding proteins. *Molecular Cell* 60: 208–19.

547. Zhang H. et al. (2015) RNA controls polyQ protein phase transitions. *Molecular Cell* 60: 220–30.

548. Järvelin A.I., Noerenberg M., Davis I. and Castello A. (2016) The new (dis)order in RNA regulation. *Cell Communication and Signaling* 14: 9.

549. Fay M.M. and Anderson P.J. (2018) The role of RNA in biological phase separations. *Journal of Molecular Biology* 430: 4685–701.

550. Protter D.S.W. et al. (2018) Intrinsically disordered regions can contribute promiscuous interactions to RNP granule assembly. *Cell Reports* 22: 1401–12.

551. Hahn S. (2018) Phase separation, protein disorder, and enhancer function. *Cell* 175: 1723–5.

552. Sanders D.W. et al. (2020) Competing protein-RNA interaction networks control multiphase intracellular organization. *Cell* 181: 306–24.

553. Roden C. and Gladfelter A.S. (2021) RNA contributions to the form and function of biomolecular condensates. *Nature Reviews Molecular Cell Biology* 22: 183–95.

554. Uversky V.N. (2013) A decade and a half of protein intrinsic disorder: Biology still waits for physics. *Protein Science* 22: 693–724.

555. Uversky V.N. (2016) Dancing protein clouds: The strange biology and chaotic physics of intrinsically disordered proteins. *Journal of Biological Chemistry* 291: 6681–8.

556. Uversky V.N. (2019) Intrinsically disordered proteins and their "mysterious" (meta)physics. *Frontiers in Physics* 7: 10.

557. Kulkarni P. and Uversky V.N. (2018) Intrinsically disordered proteins: The dark horse of the dark proteome. *Proteomics* 18: 1800061.

558. Cumberworth A., Lamour G., Babu M.M. and Gsponer J. (2013) Promiscuity as a functional trait: Intrinsically disordered regions as central players of interactomes. *Biochemical Journal* 454: 361–9.

559. Niklas K.J., Bondos S.E., Dunker A.K. and Newman S.A. (2015) Rethinking gene regulatory networks in light of alternative splicing, intrinsically disordered protein domains, and post-translational modifications. *Frontiers in Cell and Developmental Biology* 3: 8.

560. Niklas K.J., Dunker A.K. and Yruela I. (2018) The evolutionary origins of cell type diversification and the role of intrinsically disordered proteins. *Journal of Experimental Botany* 69: 1437–46.

561. Macossay-Castillo M. et al. (2019) The balancing act of intrinsically disordered proteins: Enabling functional diversity while minimizing promiscuity. *Journal of Molecular Biology* 431: 1650–70.

562. Ozdilek B.A. et al. (2017) Intrinsically disordered RGG/RG domains mediate degenerate specificity in RNA binding. *Nucleic Acids Research* 45: 7984–96.

563. Balcerak A., Trebinska-Stryjewska A., Konopinski R., Wakula M. and Grzybowska E.A. (2019) RNA–protein interactions: Disorder, moonlighting and junk contribute to eukaryotic complexity. *Open Biology* 9: 190096.

564. Robertson N.O. et al. (2018) Disparate binding kinetics by an intrinsically disordered domain enables temporal regulation of transcriptional complex formation. *Proceedings of the National Academy of Sciences USA* 115: 4643–8.

565. Peeters E., Driessen R.P.C., Werner F. and Dame R.T. (2015) The interplay between nucleoid organization and transcription in archaeal genomes. *Nature Reviews Microbiology* 13: 333–41.

566. Monterroso B. et al. (2019) Bacterial FtsZ protein forms phase-separated condensates with its nucleoid-associated inhibitor SlmA. *EMBO Reports* 20: e45946.

567. Ladouceur A.-M. et al. (2020) Clusters of bacterial RNA polymerase are biomolecular condensates that assemble through liquid–liquid phase separation. *Proceedings of the National Academy of Sciences USA* 117: 18540–9.

568. Guilhas B. et al. (2020) ATP-driven separation of liquid phase condensates in bacteria. *Molecular Cell* 79: 293–303.

569. Niklas K.J. (2014) The evolutionary-developmental origins of multicellularity. *American Journal of Botany* 101: 6–25.

570. Korneta I. and Bujnicki J.M. (2012) Intrinsic disorder in the human spliceosomal proteome. *PLOS Computational Biology* 8: e1002641.

571. Tantos A., Han K.-H. and Tompa P. (2012) Intrinsic disorder in cell signaling and gene transcription. *Molecular and Cellular Endocrinology* 348: 457–65.

572. Chen W. and Moore M.J. (2014) The spliceosome: Disorder and dynamics defined. *Current Opinion in Structural Biology* 24: 141–9.

573. Lazar T. et al. (2016) Intrinsic protein disorder in histone lysine methylation. *Biology Direct* 11: 30.

574. Peng Z., Mizianty M.J., Xue B., Kurgan L. and Uversky V.N. (2012) More than just tails: Intrinsic disorder in histone proteins. *Molecular BioSystems* 8: 1886–901.

575. Watson M. and Stott K. (2019) Disordered domains in chromatin-binding proteins. *Essays in Biochemistry* 63: 147–56.

576. Quintero-Cadena P., Lenstra T.L. and Sternberg P.W. (2020) RNA Pol II length and disorder enable cooperative scaling of transcriptional bursting. *Molecular Cell* 79: 207–20.

577. Musselman C.A. and Kutateladze T.G. (2021) Characterization of functional disordered regions within chromatin-associated proteins. *iScience* 24: 102070.

578. Romero P.R. et al. (2006) Alternative splicing in concert with protein intrinsic disorder enables increased functional diversity in multicellular organisms. *Proceedings of the National Academy of Sciences USA* 103: 8390–5.

579. Buljan M. et al. (2012) Tissue-specific splicing of disordered segments that embed binding motifs rewires protein interaction networks. *Molecular Cell* 46: 871–83.

580. Ellis J.D. et al. (2012) Tissue-specific alternative splicing remodels protein-protein interaction networks. *Molecular Cell* 46: 884–92.

581. Barbosa-Morais N.L. et al. (2012) The evolutionary landscape of alternative splicing in vertebrate species. *Science* 338: 1587–93.

582. Gueroussov S. et al. (2017) Regulatory expansion in mammals of multivalent hnRNP assemblies that globally control alternative splicing. *Cell* 170: 324–39.

583. Weatheritt R.J., Davey N.E. and Gibson T.J. (2012) Linear motifs confer functional diversity onto splice variants. *Nucleic Acids Research* 40: 7123–31.

584. Weatheritt R.J. and Gibson T.J. (2012) Linear motifs: Lost in (pre)translation. *Trends in Biochemical Sciences* 37: 333–41.

585. Guillén-Boixet J. et al. (2020) RNA-Induced conformational switching and clustering of G3BP drive stress granule assembly by condensation. *Cell* 181: 346–61.

586. Meyer K. et al. (2018) Mutations in disordered regions can cause disease by creating dileucine motifs. *Cell* 175: 239–53.

587. Pechstein A. et al. (2020) Vesicle clustering in a living synapse depends on a synapsin region that mediates phase separation. *Cell Reports* 30: 2594–602.

588. Iakoucheva L.M. et al. (2004) The importance of intrinsic disorder for protein phosphorylation. *Nucleic Acids Research* 32: 1037–49.

589. Pejaver V. et al. (2014) The structural and functional signatures of proteins that undergo multiple events of post-translational modification. *Protein Science* 23: 1077–93.

590. Bah A. and Forman-Kay J.D. (2016) Modulation of intrinsically disordered protein function by post-translational modifications. *Journal of Biological Chemistry* 291: 6696–705.

591. Wei H.-M. et al. (2014) Arginine methylation of the cellular nucleic acid binding protein does not affect its subcellular localization but impedes RNA binding. *FEBS Letters* 588: 1542–8.

592. Nott T.J. et al. (2015) Phase transition of a disordered nuage protein generates environmentally responsive membraneless organelles. *Molecular Cell* 57: 936–47.

593. Castello A. et al. (2016) Comprehensive identification of RNA-binding domains in human cells. *Molecular Cell* 63: 696–710.

594. Chong P.A., Vernon R.M. and Forman-Kay J.D. (2018) RGG/RG motif regions in RNA binding and phase separation. *Journal of Molecular Biology* 430: 4650–65.

595. Wesseling H. et al. (2020) Tau PTM profiles identify patient heterogeneity and stages of Alzheimer's Disease. *Cell* 183: 1699–713.

596. Loughlin F.E. et al. (2021) Tandem RNA binding sites induce self-association of the stress granule marker protein TIA-1. *Nucleic Acids Research* 49: 2403–17.

597. Apicco D.J. et al. (2018) Reducing the RNA binding protein TIA1 protects against tau-mediated neurodegeneration in vivo. *Nature Neuroscience* 21: 72–80.

598. White M.R. et al. (2019) C9orf72 poly(PR) dipeptide repeats disturb biomolecular phase separation and disrupt nucleolar function. *Molecular Cell* 74: 713–28.

599. Kim T.H. et al. (2019) Phospho-dependent phase separation of FMRP and CAPRIN1 recapitulates regulation of translation and deadenylation. *Science* 365: 825–9.

600. Loganathan S., Lehmkuhl E.M., Eck R.J. and Zarnescu D.C. (2020) To be or not to be…toxic — is RNA association with TDP-43 complexes deleterious or protective in neurodegeneration? *Frontiers in Molecular Biosciences* 6: 154.

601. Yu H. et al. (2021) HSP70 chaperones RNA-free TDP-43 into anisotropic intranuclear liquid spherical shells. *Science* 371: eabb4309.

602. Bakthavachalu B. et al. (2018) RNP-granule assembly via ataxin-2 disordered domains is required for long-term memory and neurodegeneration. *Neuron* 98: 754–66.

603. Gonatopoulos-Pournatzis T. et al. (2020) Autism-misregulated eIF4G microexons control synaptic translation and higher order cognitive functions. *Molecular Cell* 77: 1176–92.

604. Gsponer J., Futschik M.E., Teichmann S.A. and Babu M.M. (2008) Tight regulation of unstructured proteins: From transcript synthesis to protein degradation. *Science* 322: 1365–8.

605. Vavouri T., Semple J.I., Garcia-Verdugo R. and Lehner B. (2009) Intrinsic protein disorder and interaction promiscuity are widely associated with dosage sensitivity. *Cell* 138: 198–208.

606. Jain A. and Vale R.D. (2017) RNA phase transitions in repeat expansion disorders. *Nature* 546: 243–7.

607. Ray S. et al. (2020) α-Synuclein aggregation nucleates through liquid–liquid phase separation. *Nature Chemistry* 12: 705–16.

608. Castello A., Fischer B., Hentze M.W. and Preiss T. (2013) RNA-binding proteins in Mendelian disease. *Trends in Genetics* 29: 318–27.

609. Strom A.R. et al. (2017) Phase separation drives heterochromatin domain formation. *Nature* 547: 241–5.

610. Hnisz D., Shrinivas K., Young R.A., Chakraborty A.K. and Sharp P.A. (2017) A phase separation model for transcriptional control. *Cell* 169: 13–23.

611. Garcia-Jove Navarro M. et al. (2019) RNA is a critical element for the sizing and the composition of phase-separated RNA–protein condensates. *Nature Communications* 10: 3230.

612. Frank L. and Rippe K. (2020) Repetitive RNAs as regulators of chromatin-associated subcompartment formation by phase separation. *Journal of Molecular Biology* 432: 4270–86.

613. Jacq A. et al. (2021) Direct RNA–RNA interaction between Neat1 and RNA targets, as a mechanism for RNAs paraspeckle retention. *RNA Biology* 18: 1–12.

614. Quinodoz S.A. et al. (2021) RNA promotes the formation of spatial compartments in the nucleus. *Cell* 184: 5775–90.

615. Courchaine E.M., Lu A. and Neugebauer K.M. (2016) Droplet organelles? *EMBO Journal* 35: 1603–12.

616. Parker M.W. et al. (2019) A new class of disordered elements controls DNA replication through initiator self-assembly. *eLife* 8: e48562.

617. Zhang H. et al. (2020) Nuclear body phase separation drives telomere clustering in ALT cancer cells. *Molecular Biology of the Cell* 31: 2048–56.

618. Zwicker D., Decker M., Jaensch S., Hyman A.A. and Jülicher F. (2014) Centrosomes are autocatalytic droplets of pericentriolar material organized by centrioles. *Proceedings of the National Academy of Sciences USA* 111: E2636–45.

619. Shimada T., Yamashita A. and Yamamoto M. (2003) The fission yeast meiotic regulator Mei2p forms a dot structure in the horse-tail nucleus in association with the *sme2* locus on chromosome II. *Molecular Biology of the Cell* 14: 2461–9.

620. Hiraoka Y. (2020) Phase separation drives pairing of homologous chromosomes. *Current Genetics* 66: 881–7.

621. Kistler K.E. et al. (2018) Phase transitioned nuclear Oskar promotes cell division of *Drosophila* primordial germ cells. *eLife* 7: e37949.

622. Trinkle-Mulcahy L. and Sleeman J.E. (2017) The Cajal body and the nucleolus: "In a relationship" or "It's complicated"? *RNA Biology* 14: 739–51.

623. Hur W. et al. (2020) CDK-regulated phase separation seeded by histone genes ensures precise growth and function of histone locus bodies. *Developmental Cell* 54: 379–94.

624. Yamazaki T. and Hirose T. (2021) Control of condensates dictates nucleolar architecture. *Science* 373: 486–7.

625. Alexander K.A. et al. (2021) p53 mediates target gene association with nuclear speckles for amplified RNA expression. *Molecular Cell* 81: 1666–81.

626. Liu S. et al. (2021) USP42 drives nuclear speckle mRNA splicing via directing dynamic phase separation to promote tumorigenesis. *Cell Death & Differentiation* 28: 2482–98.

627. Sone M. et al. (2007) The mRNA-like noncoding RNA Gomafu constitutes a novel nuclear domain in a subset of neurons. *Journal of Cell Science* 120: 2498–506.

628. Ishizuka A., Hasegawa Y., Ishida K., Yanaka K. and Nakagawa S. (2014) Formation of nuclear bodies by the lncRNA Gomafu-associating proteins Celf3 and SF1. *Genes to Cells* 19: 704–21.

629. Tripathi V. et al. (2010) The nuclear-retained noncoding RNA *MALAT1* regulates alternative splicing by modulating SR splicing factor phosphorylation. *Molecular Cell* 39: 925–38.

630. Sasaki Y.T., Ideue T., Sano M., Mituyama T. and Hirose T. (2009) MEN epsilon/beta noncoding RNAs are essential for structural integrity of nuclear paraspeckles. *Proceedings of the National Academy of Sciences USA* 106: 2525–30.

631. Fox A.H., Nakagawa S., Hirose T. and Bond C.S. (2018) Paraspeckles: Where long noncoding RNA meets phase separation. *Trends in Biochemical Sciences* 43: 124–35.

632. Fox A.H. et al. (2002) Paraspeckles: A novel nuclear domain. *Current Biology* 12: 13–25.

633. Yamazaki T. et al. (2018) Functional domains of NEAT1 architectural lncRNA induce paraspeckle assembly through phase separation. *Molecular Cell* 70: 1038–53.

634. Yamazaki T. et al. (2021) Paraspeckles are constructed as block copolymer micelles. *The EMBO Journal* 40: e107270.

635. Keenen M.M. et al. (2021) HP1 proteins compact DNA into mechanically and positionally stable phase separated domains. *eLife* 10: e64563.

636. Biamonti G. and Vourc'h C. (2010) Nuclear stress bodies. *Cold Spring Harbor Perspectives in Biology* 2: a000695.

637. Ninomiya K. et al. (2020) LncRNA-dependent nuclear stress bodies promote intron retention through SR protein phosphorylation. *EMBO Journal* 39: e102729.

638. Stortz M., Pecci A., Presman D.M. and Levi V. (2020) Unraveling the molecular interactions involved in phase separation of glucocorticoid receptor. *BMC Biology* 18: 59.

639. Chen H. et al. (2020) Liquid–liquid phase separation by SARS-CoV-2 nucleocapsid protein and RNA. *Cell Research* 30: 1143–5.

640. Emenecker R.J., Holehouse A.S. and Strader L.C. (2020) Emerging roles for phase separation in plants. *Developmental Cell* 55: 69–83.

641. Anderson P. and Kedersha N. (2006) RNA granules. *Journal of Cell Biology* 172: 803–8.

642. Mateu-Regué À., Nielsen F.C. and Christiansen J. (2020) Cytoplasmic mRNPs revisited: Singletons and condensates. *BioEssays* 42: 2000097.

643. Brangwynne C.P. et al. (2009) Germline P granules are liquid droplets that localize by controlled dissolution/condensation. *Science* 324: 1729–32.

644. Smith J. et al. (2016) Spatial patterning of P granules by RNA-induced phase separation of the intrinsically-disordered protein MEG-3. *eLife* 5: e21337.

645. Jin M. et al. (2017) Glycolytic enzymes coalesce in G bodies under hypoxic stress. *Cell Reports* 20: 895–908.

646. Molliex A. et al. (2015) Phase separation by low complexity domains promotes stress granule assembly and drives pathological fibrillization. *Cell* 163: 123–33.

647. Nakamura A., Amikura R., Mukai M., Kobayashi S. and Lasko P.F. (1996) Requirement for a noncoding RNA in *Drosophila* polar granules for germ cell establishment. *Science* 274: 2075–9.

648. Sankaranarayanan M. and Weil T.T. (2020) Granule regulation by phase separation during *Drosophila* oogenesis. *Emerging Topics in Life Sciences* 4: 355–64.

649. Weatheritt R.J., Gibson T.J. and Babu M.M. (2014) Asymmetric mRNA localization contributes to fidelity and sensitivity of spatially localized systems. *Nature Structural & Molecular Biology* 21: 833–9.

650. Chouaib R. et al. (2020) A dual protein-mRNA localization screen reveals compartmentalized translation and widespread co-translational RNA targeting. *Developmental Cell* 54: 773–91.

651. Chen X., Wu X., Wu H. and Zhang M. (2020) Phase separation at the synapse. *Nature Neuroscience* 23: 301–10.

652. Mao Y.S., Sunwoo H., Zhang B. and Spector D.L. (2011) Direct visualization of the co-transcriptional assembly of a nuclear body by noncoding RNAs. *Nature Cell Biology* 13: 95–101.

653. Henninger J.E. et al. (2021) RNA-mediated feedback control of transcriptional condensates. *Cell* 184: 207–25.

654. Luo J. et al. (2021) LncRNAs: Architectural scaffolds or more potential roles in phase separation. *Frontiers in Genetics* 12: 369.

655. Quinodoz S.A. and Guttman M. (2021) Essential roles for RNA in shaping nuclear organization. *Cold Spring Harbor Perspectives in Biology.* epub ahead of print: https://cshperspectives.cshlp.org/content/early/2021/08/16/cshperspect.a039719.long.

656. Klosin A. et al. (2020) Phase separation provides a mechanism to reduce noise in cells. *Science* 367: 464–8.

657. Ahn J.H. et al. (2021) Phase separation drives aberrant chromatin looping and cancer development. *Nature* 595: 591–5.

658. Sabari B.R. et al. (2018) Coactivator condensation at super-enhancers links phase separation and gene control. *Science* 361: eaar3958.

659. Nair S.J. et al. (2019) Phase separation of ligand-activated enhancers licenses cooperative chromosomal enhancer assembly. *Nature Structural & Molecular Biology* 26: 193–203.

660. Shrinivas K. et al. (2019) Enhancer features that drive formation of transcriptional condensates. *Molecular Cell* 75: 549–61.

661. Boija A. et al. (2018) Transcription factors activate genes through the phase-separation capacity of their activation domains. *Cell* 175: 1842–55.

662. Chong S. et al. (2018) Imaging dynamic and selective low-complexity domain interactions that control gene transcription. *Science* 361: eaar2555.

663. Cho W.-K. et al. (2018) Mediator and RNA polymerase II clusters associate in transcription-dependent condensates. *Science* 361: 412–5.

664. Shao W. et al. (2022) Phase separation of RNA-binding protein promotes polymerase binding and transcription. *Nature Chemical Biology* 18: 70–80.

665. Falk M. et al. (2019) Heterochromatin drives compartmentalization of inverted and conventional nuclei. *Nature* 570: 395–9.

666. Rawal C.C., Caridi C.P. and Chiolo I. (2019) Actin' between phase separated domains for heterochromatin repair. *DNA Repair* 81: 102646.

667. Hofmann J.W., Seeley W.W. and Huang E.J. (2019) RNA binding proteins and the pathogenesis of frontotemporal lobar degeneration. *Annual Review of Pathology: Mechanisms of Disease* 14: 469–95.

668. Williams J.F. et al. (2020) Phase separation enables heterochromatin domains to do mechanical work. *bioRxiv*: 2020.07.02.184127.

669. Novo C.L. *et al.* (2020) Satellite repeat transcripts modulate heterochromatin condensates and safeguard chromosome stability in mouse embryonic stem cells. *bioRxiv*: 2020.06.08.139642.

670. Hilbert L. et al. (2021) Transcription organizes euchromatin via microphase separation. *Nature Communications* 12: 1360.

671. Caudron-Herger M. et al. (2015) Alu element-containing RNAs maintain nucleolar structure and function. *EMBO Journal* 34: 2758–74.

672. Xing Y.-H. et al. (2017) SLERT regulates DDX21 rings associated with Pol I transcription. *Cell* 169: 664–78.

673. Wang X. et al. (2021) Mutual dependency between lncRNA LETN and protein NPM1 in controlling the nucleolar structure and functions sustaining cell proliferation. *Cell Research* 31: 664–83.

674. Wu M. et al. (2021) lncRNA SLERT controls phase separation of FC/DFCs to facilitate Pol I transcription. *Science* 373: 547–55.

675. Yap K. et al. (2018) A short tandem repeat-enriched RNA assembles a nuclear compartment to control alternative splicing and promote cell survival. *Molecular Cell* 72: 525–40.

676. Pessina F. et al. (2019) Functional transcription promoters at DNA double-strand breaks mediate RNA-driven phase separation of damage-response factors. *Nature Cell Biology* 21: 1286–99.

677. Pessina F. et al. (2021) DNA damage triggers a new phase in neurodegeneration. *Trends in Genetics* 37: 337–54.

678. Thapar R. et al. (2020) Mechanism of efficient double-strand break repair by a long non-coding RNA. *Nucleic Acids Research* 48: 10953–72.

679. Cabili M.N. et al. (2015) Localization and abundance analysis of human lncRNAs at single-cell and single-molecule resolution. *Genome Biology* 16: 20.

680. Goldstrohm A.C., Hall T.M.T. and McKenney K.M. (2018) Post-transcriptional regulatory functions of mammalian Pumilio proteins. *Trends in Genetics* 34: 972–90.

681. Tichon A. et al. (2016) A conserved abundant cytoplasmic long noncoding RNA modulates repression by Pumilio proteins in human cells. *Nature Communications* 7: 12209.

682. Lee S. et al. (2016) Noncoding RNA NORAD regulates genomic stability by sequestering PUMILIO proteins. *Cell* 164: 69–80.

683. Tichon A., Perry R.B.-T., Stojic L. and Ulitsky I. (2018) SAM68 is required for regulation of Pumilio by the NORAD long noncoding RNA. *Genes & Development* 32: 70–8.

684. Elguindy M.M. and Mendell J.T. (2021) NORAD-induced Pumilio phase separation is required for genome stability. *Nature* 595: 303–8.

685. Kopp F. et al. (2019) PUMILIO hyperactivity drives premature aging of Norad-deficient mice. *eLife* 8: e42650.

686. Brecht R.M. et al. (2020) Nucleolar localization of RAG1 modulates V(D)J recombination activity. *Proceedings of the National Academy of Sciences USA* 117: 4300–9.

687. Simone R. et al. (2021) MIR-NATs repress MAPT translation and aid proteostasis in neurodegeneration. *Nature* 594: 117–23.

688. Amé J.-C., Spenlehauer C. and de Murcia G. (2004) The PARP superfamily. *BioEssays* 26: 882–93.

689. Citarelli M., Teotia S. and Lamb R.S. (2010) Evolutionary history of the poly(ADP-ribose) polymerase gene family in eukaryotes. *BMC Evolutionary Biology* 10: 308.

690. Ray Chaudhuri A. and Nussenzweig A. (2017) The multifaceted roles of PARP1 in DNA repair and chromatin remodelling. *Nature Reviews Molecular Cell Biology* 18: 610–21.

691. Höfer K. et al. (2021) Viral ADP-ribosyltransferases attach RNA chains to host proteins. *bioRxiv*: 446905.

692. Guetg C., Scheifele F., Rosenthal F., Hottiger M.O. and Santoro R. (2012) Inheritance of silent rDNA chromatin is mediated by PARP1 via noncoding RNA. *Molecular Cell* 45: 790–800.

693. Qi H. et al. (2018) The long noncoding RNA lncPARP1 contributes to progression of hepatocellular carcinoma through up-regulation of PARP1. *Bioscience Reports* 38: BSR20180703.

694. Wang Y., Zhou P., Li P., Yang F. and Gao X.-Q. (2020) Long non-coding RNA H19 regulates proliferation and doxorubicin resistance in MCF-7 cells by targeting PARP1. *Bioengineered* 11: 536–46.

695. Stapleton K. et al. (2020) Novel long noncoding RNA, macrophage inflammation-suppressing transcript (*MIST*), regulates macrophage activation during obesity. *Arteriosclerosis, Thrombosis, and Vascular Biology* 40: 914–28.

696. Melikishvili M., Chariker J.H., Rouchka E.C. and Fondufe-Mittendorf Y.N. (2017) Transcriptome-wide identification of the RNA-binding landscape of the chromatin-associated protein PARP1 reveals functions in RNA biogenesis. *Cell Discovery* 3: 17043.

697. Ke Y., Zhang J., Lv X., Zeng X. and Ba X. (2019) Novel insights into PARPs in gene expression: Regulation of RNA metabolism. *Cellular and Molecular Life Sciences* 76: 3283–99.

698. Altmeyer M. et al. (2015) Liquid demixing of intrinsically disordered proteins is seeded by poly(ADP-ribose). *Nature Communications* 6: 8088.

699. Duan Y. et al. (2019) PARylation regulates stress granule dynamics, phase separation, and neurotoxicity of disease-related RNA-binding proteins. *Cell Research* 29: 233–47.

700. Murakami K., Oshimura M. and Kugoh H. (2007) Suggestive evidence for chromosomal localization of non-coding RNA from imprinted LIT1. *Journal of Human Genetics* 52: 926–33.

701. Royo H. et al. (2007) Bsr, a nuclear-retained RNA with monoallelic expression. *Molecular Biology of the Cell* 18: 2817–27.

702. Alessio E. et al. (2019) Single cell analysis reveals the involvement of the long non-coding RNA Pvt1 in the modulation of muscle atrophy and mitochondrial network. *Nucleic Acids Research* 47: 1653–70.

703. Jonkhout N. et al. (2021) Subcellular relocalization and nuclear redistribution of the RNA methyltransferases TRMT1 and TRMT1L upon neuronal activation. *RNA Biology* 18: 1905–19.

704. de Laat W. and Grosveld F. (2003) Spatial organization of gene expression: The active chromatin hub. *Chromosome Research* 11: 447–59.

705. Poudyal R.R., Pir Cakmak F., Keating C.D. and Bevilacqua P.C. (2018) Physical principles and extant biology reveal roles for RNA-containing membraneless compartments in origins of life chemistry. *Biochemistry* 57: 2509–19.

706. Strulson C.A., Molden R.C., Keating C.D. and Bevilacqua P.C. (2012) RNA catalysis through compartmentalization. *Nature Chemistry* 4: 941–6.

707. Drobot B. et al. (2018) Compartmentalised RNA catalysis in membrane-free coacervate protocells. *Nature Communications* 9: 3643.

708. Blanco C., Bayas M., Yan F. and Chen I.A. (2018) Analysis of evolutionarily independent protein-RNA complexes yields a criterion to evaluate the relevance of prebiotic scenarios. *Current Biology* 28: 526–37.

709. Son A., Horowitz S. and Seong B.L. (2021) Chaperna: Linking the ancient RNA and protein worlds. *RNA Biology* 18: 16–23.

710. Deveson I.W. et al. (2018) Universal alternative splicing of noncoding exons. *Cell Systems* 6: 245–55.

711. Furuno M. et al. (2006) Clusters of internally primed transcripts reveal novel long noncoding RNAs. *PLOS Genetics* 2: e37.

712. Koziol M.J. and Rinn J.L. (2010) RNA traffic control of chromatin complexes. *Current Opinion in Genetics and Development* 20: 142–8.

713. Guttman M. and Rinn J.L. (2012) Modular regulatory principles of large non-coding RNAs. *Nature* 482: 339–46.

714. Mercer T.R. and Mattick J.S. (2013) Structure and function of long noncoding RNAs in epigenetic regulation. *Nature Structural & Molecular Biology* 20: 300–7.

715. Ross C.J. et al. (2021) Uncovering deeply conserved motif combinations in rapidly evolving noncoding sequences. *Genome Biology* 22: 29.

716. Smith M.A., Gesell T., Stadler P.F. and Mattick J.S. (2013) Widespread purifying selection on RNA structure in mammals. *Nucleic Acids Research* 41: 8220–36.

717. Uroda T. et al. (2019) Conserved pseudoknots in lncRNA MEG3 are essential for stimulation of the p53 pathway. *Molecular Cell* 75: 982–95.

718. Smith M.A., Seemann S.E., Quek X.C. and Mattick J.S. (2017) DotAligner: Identification and clustering of RNA structure motifs. *Genome Biology* 18: 244.

719. Seemann S.E. et al. (2017) The identification and functional annotation of RNA structures conserved in vertebrates. *Genome Research* 27: 1371–83.

720. Miladi M. et al. (2017) RNAscClust: Clustering RNA sequences using structure conservation and graph based motifs. *Bioinformatics* 33: 2089–96.

721. Johnson R. and Guigo R. (2014) The RIDL hypothesis: Transposable elements as functional domains of long noncoding RNAs. *RNA* 20: 959–76.

722. Nesterova T.B. et al. (2001) Characterization of the genomic *Xist* locus in rodents reveals conservation of overall gene structure and tandem repeats but rapid evolution of unique sequence. *Genome Research* 11: 833–49.

723. Sprague D. et al. (2019) Nonlinear sequence similarity between the *Xist* and *Rsx* long noncoding RNAs suggests shared functions of tandem repeat domains. *RNA* 25: 1004–19.

724. Quinn J.J. et al. (2016) Rapid evolutionary turnover underlies conserved lncRNA-genome interactions. *Genes & Development* 30: 191–207.

725. Alfeghaly C. et al. (2021) Implication of repeat insertion domains in the trans-activity of the long non-coding RNA ANRIL. *Nucleic Acids Research* 49: 4954–70.

726. Kelley D.R., Hendrickson D.G., Tenen D. and Rinn J.L. (2014) Transposable elements modulate human RNA abundance and splicing via specific RNA-protein interactions. *Genome Biology* 15: 537.

727. Ilik I.A. et al. (2013) Tandem stem-loops in roX RNAs act together to mediate X chromosome dosage compensation in *Drosophila*. *Molecular Cell* 51: 156–73.

728. Kapusta A. et al. (2013) Transposable elements are major contributors to the origin, diversification, and regulation of vertebrate long noncoding RNAs. *PLOS Genetics* 9: e1003470.

729. McClintock M.A. et al. (2018) RNA-directed activation of cytoplasmic dynein-1 in reconstituted transport RNPs. *eLife* 7: e36312.

730. Somarowthu S. et al. (2015) HOTAIR forms an intricate and modular secondary structure. *Molecular Cell* 58: 353–61.

731. Spitale R.C. et al. (2015) Structural imprints in vivo decode RNA regulatory mechanisms. *Nature* 519: 486–90.

732. Chowdhury I.H. et al. (2017) Expression profiling of long noncoding RNA splice variants in human microvascular endothelial cells: Lipopolysaccharide effects *in vitro*. *Mediators of Inflammation* 2017: 3427461.

733. Kiegle E.A., Garden A., Lacchini E. and Kater M.M. (2018) A genomic view of alternative splicing of long noncoding RNAs during rice seed development reveals extensive splicing and lncRNAgene families. *Frontiers in Plant Science* 9: 115.

734. Ma X. et al. (2019) Overexpressed long noncoding RNA CRNDE with distinct alternatively spliced isoforms in multiple cancers. *Frontiers of Medicine* 13: 330–43.

735. Khan M.R., Wellinger R.J. and Laurent B. (2021) Exploring the alternative splicing of long noncoding RNAs. *Trends in Genetics* 37: 695–8.

736. Li P., Tao Z. and Dean C. (2015) Phenotypic evolution through variation in splicing of the noncoding RNA COOLAIR. *Genes & Development* 29: 696–701.

737. Wu H. et al. (2016) Unusual processing generates SPA LncRNAs that sequester multiple RNA binding proteins. *Molecular Cell* 64: 534–48.

738. Guo C.-J. et al. (2020) Distinct processing of lncRNAs contributes to non-conserved functions in stem cells. *Cell* 181: 621–36.

739. Rauzan B. et al. (2013) Kinetics and thermodynamics of DNA, RNA, and hybrid duplex formation. *Biochemistry* 52: 765–72.

740. Zhou J. et al. (2015) H19 lncRNA alters DNA methylation genome wide by regulating S-adenosylhomocysteine hydrolase. *Nature Communications* 6: 10221.

741. Mercer T.R. et al. (2013) DNase I-hypersensitive exons colocalize with promoters and distal regulatory elements. *Nature Genetics* 45: 852–9.

742. Torarinsson E. et al. (2008) Comparative genomics beyond sequence-based alignments: RNA structures in the ENCODE regions. *Genome Research* 18: 242–51.

743. Will S., Yu M. and Berger B. (2013) Structure-based whole genome realignment reveals many novel noncoding RNAs. *Genome Research* 23: 1018–27.

744. Miao Z. et al. (2015) RNA-Puzzles Round II: Assessment of RNA structure prediction programs applied to three large RNA structures. *RNA* 21: 1066–84.

745. Vandivier L.E., Anderson S.J., Foley S.W. and Gregory B.D. (2016) The conservation and function of RNA secondary structure in plants. *Annual Review of Plant Biology* 67: 463–88.

746. Zhao Q. et al. (2021) Review of machine learning methods for RNA secondary structure prediction. *PLOS Computational Biology* 17: e1009291.

747. Townshend R.J.L. et al. (2021) Geometric deep learning of RNA structure. *Science* 373: 1047–51.

748. Licatalosi D.D. et al. (2008) HITS-CLIP yields genome-wide insights into brain alternative RNA processing. *Nature* 456: 464–9.

749. Hafner M. et al. (2010) Transcriptome-wide Identification of RNA-binding protein and microRNA target sites by PAR-CLIP. *Cell* 141: 129–41.

750. König J. et al. (2010) iCLIP reveals the function of hnRNP particles in splicing at individual nucleotide resolution. *Nature Structural & Molecular Biology* 17: 909–15.

751. Zarnegar B.J. et al. (2016) irCLIP platform for efficient characterization of protein-RNA interactions. *Nature Methods* 13: 489–92.

752. McFadden E.J. and Hargrove A.E. (2016) Biochemical methods to investigate lncRNA and the influence of lncRNA: Protein complexes on chromatin. *Biochemistry* 55: 1615–30.

753. Bridges M.C., Daulagala A.C. and Kourtidis A. (2021) LNCcation: lncRNA localization and function. *Journal of Cell Biology* 220: e202009045.

754. Fan Y. et al. (2021) Genome-wide detection and quantitation of RNA distribution by ChIRC[13a]-seq. *Protocol Exchange.* Online publication: https://protocolexchange. researchsquare.com/article/pex–1416/v1.

755. Van Nostrand E.L. et al. (2020) A large-scale binding and functional map of human RNA-binding proteins. *Nature* 583: 711–9.

756. Kalvari I. et al. (2018) Rfam 13.0: Shifting to a genome-centric resource for non-coding RNA families. *Nucleic Acids Research* 46: D335–42.

757. El-Gebali S. et al. (2019) The Pfam protein families database in 2019. *Nucleic Acids Research* 47: D427–32.

758. Taft R.J., Pheasant M. and Mattick J.S. (2007) The relationship between non-protein-coding DNA and eukaryotic complexity. *BioEssays* 29: 288–99.

759. Liu G., Mattick J.S. and Taft R.J. (2013) A meta-analysis of the genomic and transcriptomic composition of complex life. *Cell Cycle* 12: 2061–72.

CHAPTER 17

1. D'Alessandro A. et al. (2016) AltitudeOmics: Red blood cell metabolic adaptation to high altitude hypoxia. *Journal of Proteome Research* 15: 3883–95.

2. Julian C.G. (2017) Epigenomics and human adaptation to high altitude. *Journal of Applied Physiology* 123: 1362–70.

3. Childebayeva A. et al. (2019) DNA methylation changes are associated with an incremental ascent to high altitude. *Frontiers in Genetics* 10: 1062.

4. Ling C. and Rönn T. (2019) Epigenetics in human obesity and type 2 diabetes. *Cell Metabolism* 29: 1028–44.

5. Khera A.V. et al. (2018) Genome-wide polygenic scores for common diseases identify individuals with risk equivalent to monogenic mutations. *Nature Genetics* 50: 1219–24.

6. Christensen B.C. et al. (2009) Aging and environmental exposures alter tissue-specific DNA methylation dependent upon CpG island context. *PLOS Genetics* 5: e1000602.

7. Day J.J. and Sweatt J.D. (2011) Cognitive neuroepigenetics: A role for epigenetic mechanisms in learning and memory. *Neurobiology of Learning and Memory* 96: 2–12.

8. Martin E.M. and Fry R.C. (2018) Environmental influences on the epigenome: Exposure- associated DNA methylation in human populations. *Annual Review of Public Health* 39: 309–33.

9. Kim S. and Kaang B.-K. (2017) Epigenetic regulation and chromatin remodeling in learning and memory. *Experimental & Molecular Medicine* 49: e281.

10. Collins B.E., Greer C.B., Coleman B.C. and Sweatt J.D. (2019) Histone H3 lysine K4 methylation and its role in learning and memory. *Epigenetics & Chromatin* 12: 7.

11. Mattick J.S., Amaral P.P., Dinger M.E., Mercer T.R. and Mehler M.F. (2009) RNA regulation of epigenetic processes. *BioEssays* 31: 51–9.

12. Mattick J.S. (2010) RNA as the substrate for epigenome-environment interactions. *BioEssays* 32: 548–52.

13. Dias B.G., Maddox S., Klengel T. and Ressler K.J. (2015) Epigenetic mechanisms underlying learning and the inheritance of learned behaviors. *Trends in Neuroscience* 38: 96–107.

14. Duempelmann L., Skribbe M. and Bühler M. (2020) Small RNAs in the transgenerational inheritance of epigenetic information. *Trends in Genetics* 36: 203–14.

15. Ramanathan A., Robb G.B. and Chan S.-H. (2016) mRNA capping: Biological functions and applications. *Nucleic Acids Research* 44: 7511–26.

16. Galloway A. and Cowling V.H. (2019) mRNA cap regulation in mammalian cell function and fate. *Biochimica et Biophysica Acta* 1862: 270–9.

17. Wang J. et al. (2019) Quantifying the RNA cap epitranscriptome reveals novel caps in cellular and viral RNA. *Nucleic Acids Research* 47: e130.

18. Bird J.G. et al. (2016) The mechanism of RNA 5′ capping with NAD+, NADH and desphospho-CoA. *Nature* 535: 444–7.

19. Jonkhout N. et al. (2017) The RNA modification landscape in human disease. *RNA* 23: 1754–69.

20. Roundtree I.A., Evans M.E., Pan T. and He C. (2017) Dynamic RNA modifications in gene expression regulation. *Cell* 169: 1187–200.

21. Boccaletto P. et al. (2018) MODOMICS: A database of RNA modification pathways. 2017 update. *Nucleic Acids Research* 46: D303–7.

22. Behrens A., Rodschinka G. and Nedialkova D.D. (2021) High-resolution quantitative profiling of tRNA abundance and modification status in eukaryotes by mim-tRNAseq. *Molecular Cell* 81: 1802–15.

23. Wolk S.K. et al. (2020) Modified nucleotides may have enhanced early RNA catalysis. *Proceedings of the National Academy of Sciences USA* 117: 8236–42.

24. Sloan K.E. et al. (2017) Tuning the ribosome: The influence of rRNA modification on eukaryotic ribosome biogenesis and function. *RNA Biology* 14: 1138–52.

25. Dominissini D. et al. (2012) Topology of the human and mouse m6A RNA methylomes revealed by m6A-seq. *Nature* 485: 201–6.

26. Meyer K.D. et al. (2012) Comprehensive analysis of mRNA methylation reveals enrichment in 3′ UTRs and near stop codons. *Cell* 149: 1635–46.

27. Squires J.E. et al. (2012) Widespread occurrence of 5-methylcytosine in human coding and non-coding RNA. *Nucleic Acids Research* 40: 5023–33.

28. Hussain S. et al. (2013) NSun2-mediated cytosine-5 methylation of vault noncoding RNA determines its processing into regulatory small RNAs. *Cell Reports* 4: 255–61.

29. Khoddami V. and Cairns B.R. (2013) Identification of direct targets and modified bases of RNA cytosine methyltransferases. *Nature Biotechnology* 31: 458–64.

30. Carlile T.M. et al. (2014) Pseudouridine profiling reveals regulated mRNA pseudouridylation in yeast and human cells. *Nature* 515: 143–6.

31. Schwartz S. et al. (2014) Transcriptome-wide mapping reveals widespread dynamic-regulated pseudouridylation of ncRNA and mRNA. *Cell* 159: 148–62.

32. Schwartz S. et al. (2014) Perturbation of m6A writers reveals two distinct classes of mRNA methylation at internal and 5′ sites. *Cell Reports* 8: 284–96.

33. Dominissini D. et al. (2016) The dynamic N(1)-methyladenosine methylome in eukaryotic messenger RNA. *Nature* 530: 441–6.

34. Alarcon C.R., Lee H., Goodarzi H., Halberg N. and Tavazoie S.F. (2015) N6-methyladenosine marks primary microRNAs for processing. *Nature* 519: 482–5.

35. Esteller M. and Pandolfi P.P. (2017) The epitranscriptome of noncoding RNAs in cancer. *Cancer Discovery* 7: 359–68.

36. Bohnsack K.E., Höbartner C. and Bohnsack M.T. (2019) Eukaryotic 5-methylcytosine (m5C) RNA methyltransferases: Mechanisms, cellular functions, and links to disease. *Genes* 10: 102.

37. Adachi H., De Zoysa M.D. and Yu Y.-T. (2019) Post-transcriptional pseudouridylation in mRNA as well as in some major types of noncoding RNAs. *Biochimica et Biophysica Acta* 1862: 230–9.

38. Jia G. et al. (2011) N6-methyladenosine in nuclear RNA is a major substrate of the obesity-associated FTO. *Nature Chemical Biology* 7: 885–7.

39. Zhao X. et al. (2014) FTO-dependent demethylation of N6-methyladenosine regulates mRNA splicing and is required for adipogenesis. *Cell Research* 24: 1403–19.

40. Liu F. et al. (2016) ALKBH1-mediated tRNA demethylation regulates translation. *Cell* 167: 816–28.

41. Saletore Y. et al. (2012) The birth of the Epitranscriptome: Deciphering the function of RNA modifications. *Genome Biology* 13: 175.

42. Novoa E.M., Mason C.E. and Mattick J.S. (2017) Charting the unknown epitranscriptome. *Nature Reviews Molecular Cell Biology* 18: 339–40.

43. Linder B. et al. (2015) Single-nucleotide-resolution mapping of m6A and m6Am throughout the transcriptome. *Nature Methods* 12: 767–72.

44. Batista P.J. et al. (2014) m(6)A RNA modification controls cell fate transition in mammalian embryonic stem cells. *Cell Stem Cell* 15: 707–19.

45. Li S. and Mason C.E. (2014) The pivotal regulatory landscape of RNA modifications. *Annual Review of Genomics and Human Genetics* 15: 127–50.

46. Trixl L. and Lusser A. (2019) The dynamic RNA modification 5-methylcytosine and its emerging role as an epitranscriptomic mark. *WIREs RNA* 10: e1510.

47. Mikutis S. et al. (2020) meCLICK-Seq, a substrate-hijacking and RNA degradation strategy for the study of RNA methylation. *ACS Central Science* 6: 2196–208.

48. Alarcón C.R. et al. (2015) HNRNPA2B1 Is a mediator of m6A-dependent nuclear RNA processing events. *Cell* 162: 1299–308.

49. Fustin J.-M. et al. (2013) RNA-methylation-dependent RNA processing controls the speed of the circadian clock. *Cell* 155: 793–806.

50. Zheng G. et al. (2013) ALKBH5 is a mammalian RNA demethylase that impacts RNA metabolism and mouse fertility. *Molecular Cell* 49: 18–29.

51. Wang X. et al. (2014) N6-methyladenosine-dependent regulation of messenger RNA stability. *Nature* 505: 117–20.

52. Wang Y. et al. (2014) N6-methyladenosine modification destabilizes developmental regulators in embryonic stem cells. *Nature Cell Biology* 16: 191–8.

53. Wang X. et al. (2015) N(6)-methyladenosine modulates messenger RNA translation efficiency. *Cell* 161: 1388–99.

54. Geula S. et al. (2015) m6A mRNA methylation facilitates resolution of naive pluripotency toward differentiation. *Science* 347: 1002–6.

55. Knuckles P. et al. (2017) RNA fate determination through cotranscriptional adenosine methylation and microprocessor binding. *Nature Structural & Molecular Biology* 24: 561–9.

56. Tang C. et al. (2018) ALKBH5-dependent m6A demethylation controls splicing and stability of long 3′-UTR mRNAs in male germ cells. *Proceedings of the National Academy of Sciences USA* 115: E325–33.

57. Kasowitz S.D. et al. (2018) Nuclear m6A reader YTHDC1 regulates alternative polyadenylation and splicing during mouse oocyte development. *PLOS Genetics* 14: e1007412.

58. Louloupi A., Ntini E., Conrad T. and Ørom U.A.V. (2018) Transient N-6-methyladenosine transcriptome sequencing reveals a regulatory role of m6A in splicing efficiency. *Cell Reports* 23: 3429–37.

59. Zaccara S., Ries R.J. and Jaffrey S.R. (2019) Reading, writing and erasing mRNA methylation. *Nature Reviews Molecular Cell Biology* 20: 608–24.

60. Liu N. et al. (2015) N6-methyladenosine-dependent RNA structural switches regulate RNA-protein interactions. *Nature* 518: 560–4.

61. Spitale R.C. et al. (2015) Structural imprints in vivo decode RNA regulatory mechanisms. *Nature* 519: 486–90.

62. Mendel M. et al. (2021) Splice site m6A methylation prevents binding of U2AF35 to inhibit RNA splicing. *Cell* 184: 3125–42.

63. Yang X. et al. (2019) m6A promotes R-loop formation to facilitate transcription termination. *Cell Research* 29: 1035–8.

64. Liu J. et al. (2020) N6-methyladenosine of chromosome-associated regulatory RNA regulates chromatin state and transcription. *Science* 367: 580–6.

65. Ries R.J. et al. (2019) m6A enhances the phase separation potential of mRNA. *Nature* 571: 424–8.

66. Abakir A. et al. (2020) N6-methyladenosine regulates the stability of RNA:DNA hybrids in human cells. *Nature Genetics* 52: 48–55.

67. Zhang C. et al. (2020) METTL3 and N6-methyladenosine promote homologous recombination-mediated repair of DSBs by modulating DNA-RNA hybrid accumulation. *Molecular Cell* 79: 425–42.

68. Wei J. and He C. (2021) Chromatin and transcriptional regulation by reversible RNA methylation. *Current Opinion in Cell Biology* 70: 109–15.

69. Kan R.L., Chen J. and Sallam T. (2021) Crosstalk between epitranscriptomic and epigenetic mechanisms in gene regulation. *Trends in Genetics* 38: 182–93.

70. Schwartz S. et al. (2013) High-resolution mapping reveals a conserved, widespread, dynamic mRNA methylation program in yeast meiosis. *Cell* 155: 1409–21.

71. Chelmicki T. et al. (2021) m⁶A RNA methylation regulates the fate of endogenous retroviruses. *Nature* 591: 312–6.

72. Xu W. et al. (2021) METTL3 regulates heterochromatin in mouse embryonic stem cells. *Nature* 591: 317–21.

73. Chen C. et al. (2021) Nuclear m⁶A reader Ythdc1 regulates the scaffold function of LINE1 RNA in mouse ESCs and early embryos. *Protein & Cell* 12: 455–74.

74. Duda K.J. et al. (2021) m6A RNA methylation of major satellite repeat transcripts facilitates chromatin association and RNA:DNA hybrid formation in mouse heterochromatin. *Nucleic Acids Research* 49: 5568–87.

75. Blanco S. et al. (2011) The RNA-methyltransferase Misu (NSun2) poises epidermal stem cells to differentiate. *PLOS Genetics* 7: e1002403.

76. Frye M. and Blanco S. (2016) Post-transcriptional modifications in development and stem cells. *Development* 143: 3871–81.

77. Frye M., Harada B.T., Behm M. and He C. (2018) RNA modifications modulate gene expression during development. *Science* 361: 1346–9.

78. Wang Y. et al. (2018) N⁶-methyladenosine RNA modification regulates embryonic neural stem cell self-renewal through histone modifications. *Nature Neuroscience* 21: 195–206.

79. Hongay C.F. and Orr-Weaver T.L. (2011) *Drosophila* Inducer of MEiosis 4 (IME4) is required for Notch signaling during oogenesis. *Proceedings of the National Academy of Sciences USA* 108: 14855–60.

80. Lin Z. and Tong M.-H. (2019) m⁶A mRNA modification regulates mammalian spermatogenesis. *Biochimica et Biophysica Acta* 1862: 403–11.

81. Sánchez-Vásquez E., Alata Jimenez N., Vázquez N.A. and Strobl-Mazzulla P.H. (2018) Emerging role of dynamic RNA modifications during animal development. *Mechanisms of Development* 154: 24–32.

82. Mendel M. et al. (2018) Methylation of structured RNA by the m⁶A writer METTL16 Is essential for mouse embryonic development. *Molecular Cell* 71: 986–1000.

83. Fustin J.-M. et al. (2018) Two *Ck1* transcripts regulated by m6A methylation code for two antagonistic kinases in the control of the circadian clock. *Proceedings of the National Academy of Sciences USA* 115: 5980–5.

84. Zhong X. et al. (2018) Circadian clock regulation of hepatic lipid metabolism by modulation of m⁶A mRNA methylation. *Cell Reports* 25: 1816–28.

85. Schaefer M. et al. (2010) RNA methylation by Dnmt2 protects transfer RNAs against stress-induced cleavage. *Genes & Development* 24: 1590–5.

86. Zhou J. et al. (2018) N6-methyladenosine guides mRNA alternative translation during integrated stress response. *Molecular Cell* 69: 636–47.

87. Alriquet M. et al. (2020) The protective role of m¹A during stress-induced granulation. *Journal of Molecular Cell Biology* 12: 870–80.

88. Vu L.P. et al. (2017) The N⁶-methyladenosine (m⁶A)-forming enzyme METTL3 controls myeloid differentiation of normal hematopoietic and leukemia cells. *Nature Medicine* 23: 1369–76.

89. Wang Y.-N., Yu C.-Y. and Jin H.-Z. (2020) RNA N⁶-methyladenosine modifications and the immune response. *Journal of Immunology Research* 2020: 6327614.

90. Wu C. et al. (2020) Interplay of m⁶A and H3K27 trimethylation restrains inflammation during bacterial infection. *Science Advances* 6: eaba0647.

91. Li S.-X., Yan W., Liu J.-P., Zhao Y.-J. and Chen L. (2021) Long noncoding RNA SNHG4 remits lipopolysaccharide-engendered inflammatory lung damage by inhibiting METTL3 – Mediated m6A level of STAT2 mRNA. *Molecular Immunology* 139: 10–22.

92. Nair L. et al. (2021) Mechanism of noncoding RNA-associated N6-methyladenosine recognition by an RNA processing complex during IgH DNA recombination. *Molecular Cell* 81: 3949–64.

93. Yoon K.J. et al. (2017) Temporal control of mammalian cortical neurogenesis by m⁶A methylation. *Cell* 171: 877–89.

94. Li L. et al. (2017) Fat mass and obesity-associated (FTO) protein regulates adult neurogenesis. *Human Molecular Genetics* 26: 2398–411.

95. Weng Y.-L. et al. (2018) Epitranscriptomic m⁶A regulation of axon regeneration in the adult mammalian nervous system. *Neuron* 97: 313–25.

96. Flores J.V. et al. (2017) Cytosine-5 RNA methylation regulates neural stem cell differentiation and motility. *Stem Cell Reports* 8: 112–24.

97. Wu R. et al. (2019) A novel m⁶A reader Prrc2a controls oligodendroglial specification and myelination. *Cell Research* 29: 23–41.

98. Hess M.E. et al. (2013) The fat mass and obesity associated gene (*Fto*) regulates activity of the dopaminergic midbrain circuitry. *Nature Neuroscience* 16: 1042–8.

99. Merkurjev D. et al. (2018) Synaptic N⁶-methyladenosine (m⁶A) epitranscriptome reveals functional partitioning of localized transcripts. *Nature Neuroscience* 21: 1004–14.

100. Hussain S. and Bashir Z.I. (2015) The epitranscriptome in modulating spatiotemporal RNA translation in neuronal post-synaptic function. *Frontiers in Cellular Neuroscience* 9: 420.

101. Yu J. et al. (2018) Dynamic m⁶A modification regulates local translation of mRNA in axons. *Nucleic Acids Research* 46: 1412–23.

102. Ma C. et al. (2018) RNA m⁶A methylation participates in regulation of postnatal development of the mouse cerebellum. *Genome Biology* 19: 68.

103. Wang C.-X. et al. (2018) METTL3-mediated m⁶A modification is required for cerebellar development. *PLOS Biology* 16: e2004880.

104. Widagdo J. et al. (2016) Experience-dependent accumulation of N-methyladenosine in the prefrontal cortex is associated with memory processes in mice. *Journal of Neuroscience* 36: 6771–7.

105. Walters B.J. et al. (2017) The role of the RNA demethylase FTO (fat mass and obesity-associated) and mRNA methylation in hippocampal memory formation. *Neuropsychopharmacology* 42: 1502–10.

106. Koranda J.L. et al. (2018) Mettl14 Is essential for epitranscriptomic regulation of striatal function and learning. *Neuron* 99: 283–92.

107. Engel M. et al. (2018) The role of m⁶A/m-RNA methylation in stress response regulation. *Neuron* 99: 389–403.

108. Zhang Z. et al. (2018) METTL3-mediated N⁶-methyladenosine mRNA modification enhances long-term memory consolidation. *Cell Research* 28: 1050–61.

109. Shi H. et al. (2018) m⁶A facilitates hippocampus-dependent learning and memory through YTHDF1. *Nature* 563: 249–53.

110. Xu Y. et al. (2020) Regulation of N6-methyladenosine in the differentiation of cancer stem cells and their fate. *Frontiers in Cell and Developmental Biology* 8: 950.

111. Haussmann I.U. et al. (2016) m⁶A potentiates Sxl alternative pre-mRNA splicing for robust *Drosophila* sex determination. *Nature* 540: 301–4.

112. Lence T. et al. (2016) m⁶A modulates neuronal functions and sex determination in *Drosophila*. *Nature* 540: 242–7.

113. Kan L. et al. (2017) The m⁶A pathway facilitates sex determination in *Drosophila*. *Nature Communications* 8: 15737.

114. Yu Q. et al. (2021) RNA demethylation increases the yield and biomass of rice and potato plants in field trials. *Nature Biotechnology* 39: 1581–8.

115. Xiao S. et al. (2019) The RNA N⁶-methyladenosine modification landscape of human fetal tissues. *Nature Cell Biology* 21: 651–61.

116. Zhou K.I. et al. (2016) N(6)-methyladenosine modification in a long noncoding RNA hairpin predisposes its conformation to protein binding. *Journal of Molecular Biology* 428: 822–33.

117. Wang X. et al. (2021) N6-methyladenosine modification of *MALAT1* promotes metastasis via reshaping nuclear speckles. *Developmental Cell* 56: 702–15.

118. Patil D.P. et al. (2016) m⁶A RNA methylation promotes XIST-mediated transcriptional repression. *Nature* 537: 369–73.

119. Yoneda R., Ueda N., Uranishi K., Hirasaki M. and Kurokawa R. (2020) Long noncoding RNA *pncRNA-D* reduces cyclin D1 gene expression and arrests cell cycle through RNA m⁶A modification. *Journal of Biological Chemistry* 295: 5626–39.

120. Xhemalce B., Robson S.C. and Kouzarides T. (2012) Human RNA methyltransferase BCDIN3D regulates microRNA processing. *Cell* 151: 278–88.

121. Blanco S. et al. (2014) Aberrant methylation of tRNAs links cellular stress to neuro-developmental disorders. *EMBO Journal* 33: 2020–39.

122. Shi H., Chai P., Jia R. and Fan X. (2020) Novel insight into the regulatory roles of diverse RNA modifications: Re-defining the bridge between transcription and translation. *Molecular Cancer* 19: 78.

123. Mauer J. et al. (2017) Reversible methylation of m⁶Aₘ in the 5′ cap controls mRNA stability. *Nature* 541: 371–5.

124. Cheng J.X. et al. (2018) RNA cytosine methylation and methyltransferases mediate chromatin organization and 5-azacytidine response and resistance in leukaemia. *Nature Communications* 9: 1163.

125. Schumann U. et al. (2020) Multiple links between 5-methylcytosine content of mRNA and translation. *BMC Biology* 18: 40.

126. Ali A.T., Idaghdour Y. and Hodgkinson A. (2020) Analysis of mitochondrial m1A/G RNA modification reveals links to nuclear genetic variants and associated disease processes. *Communications Biology* 3: 147.

127. Zhao X. et al. (2004) Regulation of nuclear receptor activity by a pseudouridine synthase through posttranscriptional modification of steroid receptor RNA activator. *Molecular Cell* 15: 549–58.

128. Liu W. et al. (2013) Brd4 and JMJD6-associated anti-pause enhancers in regulation of transcriptional pause release. *Cell* 155: 1581–95.

129. Zhao Y., Karijolich J., Glaunsinger B. and Zhou Q. (2016) Pseudouridylation of 7SK snRNA promotes 7SK snRNP formation to suppress HIV-1 transcription and escape from latency. *EMBO Reports* 17: 1441–51.

130. Barbieri I. and Kouzarides T. (2020) Role of RNA modifications in cancer. *Nature Reviews Cancer* 20: 303–22.

131. Chen S., Zhou L. and Wang Y. (2020) ALKBH5-mediated m⁶A demethylation of lncRNA PVT1 plays an oncogenic role in osteosarcoma. *Cancer Cell International* 20: 34.

132. Miano V., Codino A., Pandolfini L. and Barbieri I. (2021) The non-coding epitranscriptome in cancer. *Briefings in Functional Genomics* 20: 94–105.

133. Begik O. et al. (2020) Integrative analyses of the RNA modification machinery reveal tissue- and cancer-specific signatures. *Genome Biology* 21: 97.

134. Liu J. et al. (2014) A METTL3-METTL14 complex mediates mammalian nuclear RNA N6-adenosine methylation. *Nature Chemical Biology* 10: 93–5.

135. Meyer K.D. and Jaffrey S.R. (2017) Rethinking m⁶A readers, writers, and erasers. *Annual Review of Cell and Developmental Biology* 33: 319–42.

136. Pendleton K.E. et al. (2017) The U6 snRNA m⁶A methyltransferase METTL16 regulates SAM synthetase intron retention. *Cell* 169: 824–35.

137. Chi L. and Delgado-Olguín P. (2013) Expression of NOL1/NOP2/sun domain (Nsun) RNA methyltransferase family genes in early mouse embryogenesis. *Gene Expression Patterns* 13: 319–27.

138. Heissenberger C. et al. (2019) Loss of the ribosomal RNA methyltransferase NSUN5 impairs global protein synthesis and normal growth. *Nucleic Acids Research* 47: 11807–25.

139. Chen P., Zhang T., Yuan Z., Shen B. and Chen L. (2019) Expression of the RNA methyltransferase Nsun5 is essential for developing cerebral cortex. *Molecular Brain* 12: 74.

140. Haag S. et al. (2015) NSUN6 is a human RNA methyltransferase that catalyzes formation of m5C72 in specific tRNAs. *RNA (New York, N.Y.)* 21: 1532–43.

141. Selmi T. et al. (2020) Sequence- and structure-specific cytosine-5 mRNA methylation by NSUN6. *Nucleic Acids Research* 49: 1006–22.

142. Jonkhout N. et al. (2021) Subcellular relocalization and nuclear redistribution of the RNA methyltransferases TRMT1 and TRMT1L upon neuronal activation. *RNA Biology* 18: 1905–19.

143. Hauenschild R. et al. (2015) The reverse transcription signature of N-1-methyladenosine in RNA-Seq is sequence dependent. *Nucleic Acids Research* 43: 9950–64.

144. Brinkerhoff H., Kang A.S.W., Liu J., Aksimentiev A. and Dekker C. (2021) Multiple rereads of single proteins at single-amino acid resolution using nanopores. *Science* 374: 1509–13.

145. Bošković F. and Keyser U.F. (2021) Toward single-molecule proteomics. *Science* 374: 1443–4.

146. Garalde D.R. et al. (2018) Highly parallel direct RNA sequencing on an array of nanopores. *Nature Methods* 15: 201–6.

147. Smith A.M., Jain M., Mulroney L., Garalde D.R. and Akeson M. (2019) Reading canonical and modified nucleobases in 16S ribosomal RNA using nanopore native RNA sequencing. *PLOS ONE* 14: e0216709.

148. Liu H. et al. (2019) Accurate detection of m⁶A RNA modifications in native RNA sequences. *Nature Communications* 10: 4079.

149. Workman R.E. et al. (2019) Nanopore native RNA sequencing of a human poly(A) transcriptome. *Nature Methods* 16: 1297–305.

150. Leger A. et al. (2021) RNA modifications detection by comparative Nanopore direct RNA sequencing. *Nature Communications* 12: 7198.

151. Karikó K. et al. (2008) Incorporation of pseudouridine into mRNA yields superior nonimmunogenic vector with increased translational capacity and biological stability. *Molecular Therapy* 16: 1833–40.

152. Karikó K., Buckstein M., Ni H. and Weissman D. (2005) Suppression of RNA recognition by Toll-like receptors: The impact of nucleoside modification and the evolutionary origin of RNA. *Immunity* 23: 165–75.

153. Linares-Fernández S., Lacroix C., Exposito J.-Y. and Verrier B. (2020) Tailoring mRNA vaccine to balance innate/adaptive immune response. *Trends in Molecular Medicine* 26: 311–23.

154. Pardi N., Hogan M.J., Porter F.W. and Weissman D. (2018) mRNA vaccines — a new era in vaccinology. *Nature Reviews Drug Discovery* 17: 261–79.

155. Krienke C. et al. (2021) A noninflammatory mRNA vaccine for treatment of experimental autoimmune encephalomyelitis. *Science* 371: 145–53.

156. Wardell C.M. and Levings M.K. (2021) mRNA vaccines take on immune tolerance. *Nature Biotechnology* 39: 419–21.

157. Zangi L. et al. (2013) Modified mRNA directs the fate of heart progenitor cells and induces vascular regeneration after myocardial infarction. *Nature Biotechnology* 31: 898–907.

158. Collén A. et al. (2022) VEGFA mRNA for regenerative treatment of heart failure. *Nature Reviews Drug Discovery* 21: 79–80.

159. Hou X., Zaks T., Langer R. and Dong Y. (2021) Lipid nanoparticles for mRNA delivery. *Nature Reviews Materials* 6: 1078–94.

160. Winkle M., El-Daly S.M., Fabbri M. and Calin G.A. (2021) Noncoding RNA therapeutics — challenges and potential solutions. *Nature Reviews Drug Discovery* 20: 629–51.

161. Sletten A.C. et al. (2021) Loss of SNORA73 reprograms cellular metabolism and protects against steatohepatitis. *Nature Communications* 12: 5214.

162. Li Y. et al. (2021) A noncoding RNA modulator potentiates phenylalanine metabolism in mice. *Science* 373: 662–73.

163. Wei H. et al. (2021) Systematic analysis of purified astrocytes after SCI unveils Zeb2os function during astrogliosis. *Cell Reports* 34: 108721.

164. Benne R. et al. (1986) Major transcript of the frameshifted coxII gene from trypanosome mitochondria contains four nucleotides that are not encoded in the DNA. *Cell* 46: 819–26.

165. Blum B., Bakalara N. and Simpson L. (1990) A model for RNA editing in kinetoplastid mitochondria: "guide" RNA molecules transcribed from maxicircle DNA provide the edited information. *Cell* 60: 189–98.

166. Simpson L., Sbicego S. and Aphasizhev R. (2003) Uridine insertion/deletion RNA editing in trypanosome mitochondria: A complex business. *RNA* 9: 265–76.

167. Paz N. et al. (2007) Altered adenosine-to-inosine RNA editing in human cancer. *Genome Research* 17: 1586–95.

168. Nishikura K. (2015) A-to-I editing of coding and noncoding RNAs by ADARs. *Nature Reviews Molecular Cell Biology* 17: 83–96.

169. Paz-Yaacov N. et al. (2015) Elevated RNA editing activity is a major contributor to transcriptomic diversity in tumors. *Cell Reports* 13: 267–76.

170. Kung C.-P., Maggi L.B. and Weber J.D. (2018) The role of RNA editing in cancer development and metabolic disorders. *Frontiers in Endocrinology* 9: 762.

171. Ishizuka J.J. et al. (2019) Loss of ADAR1 in tumours overcomes resistance to immune checkpoint blockade. *Nature* 565: 43–8.

172. Bass B.L. and Weintraub H. (1988) An unwinding activity that covalently modifies its double-stranded RNA substrate. *Cell* 55: 1089–98.

173. Kimelman D. and Kirschner M.W. (1989) An antisense mRNA directs the covalent modification of the transcript encoding fibroblast growth factor in Xenopus oocytes. *Cell* 59: 687–96.

174. Bass B.L., Weintraub H., Cattaneo R. and Billeter M.A. (1989) Biased hypermutation of viral RNA genomes could be due to unwinding/modification of double-stranded RNA. *Cell* 56: 331.

175. Keegan L.P., Leroy A., Sproul D. and O'Connell M.A. (2004) Adenosine deaminases acting on RNA (ADARs): RNA-editing enzymes. *Genome Biology* 5: 209.

176. Valente L. and Nishikura K. (2005) ADAR gene family and A-to-I RNA editing: Diverse roles in posttranscriptional gene regulation. *Progress in Nucleic Acid Research and Molecular Biology* 79: 299–338.

177. Nishikura K. (2010) Functions and regulation of RNA editing by ADAR deaminases. *Annual Review of Biochemistry* 79: 321–49.

178. Farajollahi S. and Maas S. (2010) Molecular diversity through RNA editing: A balancing act. *Trends in Genetics* 26: 221–30.

179. Savva Y.A., Rieder L.E. and Reenan R.A. (2012) The ADAR protein family. *Genome Biology* 13: 252.

180. Walkley C.R. and Li J.B. (2017) Rewriting the transcriptome: Adenosine-to-inosine RNA editing by ADARs. *Genome Biology* 18: 205.

181. Tan M.H. et al. (2017) Dynamic landscape and regulation of RNA editing in mammals. *Nature* 550: 249–54.

182. Laurencikiene J., Kallman A.M., Fong N., Bentley D.L. and Ohman M. (2006) RNA editing and alternative splicing: The importance of co-transcriptional coordination. *EMBO Reports* 7: 303–7.

183. Rodriguez J., Menet J.S. and Rosbash M. (2012) Nascent-seq indicates widespread cotranscriptional RNA editing in *Drosophila*. *Molecular Cell* 47: 27–37.

184. Wong S.K., Sato S. and Lazinski D.W. (2001) Substrate recognition by ADAR1 and ADAR2. *RNA* 7: 846–58.

185. Uzonyi A. et al. (2021) Deciphering the principles of the RNA editing code via large-scale systematic probing. *Molecular Cell* 81: 2374–87.

186. Stefl R. and Allain F.H. (2005) A novel RNA pentaloop fold involved in targeting ADAR2. *RNA* 11: 592–7.

187. Bass B.L. (2002) RNA editing by adenosine deaminases that act on RNA. *Annual Review of Biochemistry* 71: 817–46.

188. Sapiro A.L. et al. (2019) Illuminating spatial A-to-I RNA editing signatures within the *Drosophila* brain. *Proceedings of the National Academy of Sciences USA* 116: 2318–27.

189. Anantharaman A. et al. (2017) ADAR2 regulates RNA stability by modifying access of decay-promoting RNA-binding proteins. *Nucleic Acids Research* 45: 4189–201.

190. Gong C. and Maquat L.E. (2011) lncRNAs transactivate STAU1-mediated mRNA decay by duplexing with 3′ UTRs via Alu elements. *Nature* 470: 284–8.

191. Wagner R.W., Smith J.E., Cooperman B.S. and Nishikura K. (1989) A double-stranded RNA unwinding activity introduces structural alterations by means of adenosine to inosine conversions in mammalian cells and Xenopus eggs. *Proceedings of the National Academy of Sciences USA* 86: 2647–51.

192. Sommer B., Köhler M., Sprengel R. and Seeburg P.H. (1991) RNA editing in brain controls a determinant of ion flow in glutamate-gated channels. *Cell* 67: 11–9.

193. Melcher T. *et al.* (1996) A mammalian RNA editing enzyme. *Nature* 379: 460–4.

194. Seeburg P.H. (2002) A-to-I editing: New and old sites, functions and speculations. *Neuron* 35: 17–20.

195. Hoopengardner B., Bhalla T., Staber C. and Reenan R. (2003) Nervous system targets of RNA editing identified by comparative genomics. *Science* 301: 832–6.

196. Schmauss C. (2003) Serotonin 2C receptors: Suicide, serotonin, and runaway RNA editing. *Neuroscientist* 9: 237–42.

197. Barlati S. and Barbon A. (2005) RNA editing: A molecular mechanism for the fine modulation of neuronal transmission. *Acta Neurochirurgica Supplementum* 93: 53–7.

198. Behm M., Wahlstedt H., Widmark A., Eriksson M. and Öhman M. (2017) Accumulation of nuclear ADAR2 regulates adenosine-to-inosine RNA editing during neuronal development. *Journal of Cell Science* 130: 745–53.

199. Yablonovitch A.L., Deng P., Jacobson D. and Li J.B. (2017) The evolution and adaptation of A-to-I RNA editing. *PLOS Genetics* 13: e1007064.

200. Mattick J.S. and Mehler M.F. (2008) RNA editing, DNA recoding and the evolution of human cognition. *Trends in Neuroscience* 31: 227–33.

201. Palladino M.J., Keegan L.P., O'Connell M.A. and Reenan R.A. (2000) dADAR, a *Drosophila* double-stranded RNA-specific adenosine deaminase is highly developmentally regulated and is itself a target for RNA editing. *RNA* 6: 1004–18.

202. Slavov D. and Gardiner K. (2002) Phylogenetic comparison of the pre-mRNA adenosine deaminase ADAR2 genes and transcripts: Conservation and diversity in editing site sequence and alternative splicing patterns. *Gene* 299: 83–94.

203. Dawson T.R., Sansam C.L. and Emeson R.B. (2004) Structure and sequence determinants required for the RNA editing of ADAR2 substrates. *Journal of Biological Chemistry* 279: 4941–51.

204. Xiang J.-F. et al. (2018) N6-methyladenosines modulate A-to-I RNA editing. *Molecular Cell* 69: 126–35.

205. Rueter S.M., Dawson T.R. and Emeson R.B. (1999) Regulation of alternative splicing by RNA editing. *Nature* 399: 75–80.

206. Zhang Z. and Carmichael G.G. (2001) The fate of dsRNA in the nucleus: A p54nrb-containing complex mediates the nuclear retention of promiscuously A-to-I edited RNAs. *Cell* 106: 465–76.

207. DeCerbo J. and Carmichael G.G. (2005) Retention and repression: Fates of hyperedited RNAs in the nucleus. *Current Opinion in Cell Biology* 17: 302–8.

208. Blow M.J. et al. (2006) RNA editing of human microRNAs. *Genome Biology* 7: R27.

209. Yang W. et al. (2006) Modulation of microRNA processing and expression through RNA editing by ADAR deaminases. *Nature Structural & Molecular Biology* 13: 13–21.

210. Kawahara Y., Zinshteyn B., Chendrimada T.P., Shiekhattar R. and Nishikura K. (2007) RNA editing of the microRNA-151 precursor blocks cleavage by the Dicer-TRBP complex. *EMBO Reports* 8: 763–9.

211. Kawahara Y. et al. (2007) Redirection of silencing targets by adenosine-to-inosine editing of miRNAs. *Science* 315: 1137–40.

212. Macbeth M.R. et al. (2005) Inositol hexakisphosphate is bound in the ADAR2 core and required for RNA editing. *Science* 309: 1534–9.

213. Valastro B. et al. (2001) Inositol hexakisphosphate-mediated regulation of glutamate receptors in rat brain sections. *Hippocampus* 11: 673–82.

214. Chalk A.M., Taylor S., Heraud-Farlow J.E. and Walkley C.R. (2019) The majority of A-to-I RNA editing is not required for mammalian homeostasis. *Genome Biology* 20: 268.

215. Tan B.Z., Huang H., Lam R. and Soong T.W. (2009) Dynamic regulation of RNA editing of ion channels and receptors in the mammalian nervous system. *Molecular Brain* 2: 13.

216. Herbert A., Lowenhaupt K., Spitzner J. and Rich A. (1995) Chicken double-stranded RNA adenosine deaminase has apparent specificity for Z-DNA. *Proceedings of the National Academy of Sciences USA* 92: 7550–4.

217. Herbert A. et al. (1997) A Z-DNA binding domain present in the human editing enzyme, double-stranded RNA adenosine deaminase. *Proceedings of the National Academy of Sciences USA* 94: 8421–6.

218. Herbert A. et al. (1998) The Zalpha domain from human ADAR1 binds to the Z-DNA conformer of many different sequences. *Nucleic Acids Research* 26: 3486–93.

219. Herbert A. and Rich A. (1999) Left-handed Z-DNA: Structure and function. *Genetica* 106: 37–47.

220. Herbert A. and Rich A. (2001) The role of binding domains for dsRNA and Z-DNA in the in vivo editing of minimal substrates by ADAR1. *Proceedings of the National Academy of Sciences USA* 98: 12132–7.

221. Mirkin S.M. (2008) Discovery of alternative DNA structures: A heroic decade (1979–1989). *Frontiers in Bioscience* 13: 1064–71.

222. Herbert A. (1996) RNA editing, introns and evolution. *Trends in Genetics* 12: 6–9.

223. Herbert A. (2019) Z-DNA and Z-RNA in human disease. *Communications Biology* 2: 7.

224. Kim J.I. et al. (2021) RNA editing at a limited number of sites is sufficient to prevent MDA5 activation in the mouse brain. *PLOS Genetics* 17: e1009516.

225. Liddicoat B.J. et al. (2015) RNA editing by ADAR1 prevents MDA5 sensing of endogenous dsRNA as nonself. *Science* 349: 1115–20.

226. Ahmad S. et al. (2018) Breaching self-tolerance to Alu duplex RNA underlies MDA5-mediated inflammation. *Cell* 172: 797–810.

227. Chung H. et al. (2018) Human ADAR1 prevents endogenous RNA from triggering translational shutdown. *Cell* 172: 811–24.

228. Terajima H. et al. (2021) N6-methyladenosine promotes induction of ADAR1-mediated A-to-I RNA editing to suppress aberrant antiviral innate immune responses. *PLOS Biology* 19: e3001292.

229. Marshall P.R. et al. (2020) Dynamic regulation of Z-DNA in the mouse prefrontal cortex by the RNA-editing enzyme Adar1 is required for fear extinction. *Nature Neuroscience* 23: 718–29.

230. McFadden M.J. and Horner S.M. (2021) N6-methyladenosine regulates host responses to viral infection. *Trends in Biochemical Sciences* 46: 366–77.

231. Hartner J.C. et al. (2004) Liver disintegration in the mouse embryocaused by deficiency in the RNA-editing enzyme ADAR1. *Journal of Biological Chemistry* 279: 4894–902.

232. Hartner J.C., Walkley C.R., Lu J. and Orkin S.H. (2009) ADAR1 is essential for the maintenance of hematopoiesis and suppression of interferon signaling. *Nature Immunology* 10: 109–15.

233. Rice G.I. et al. (2012) Mutations in ADAR1 cause Aicardi-Goutières syndrome associated with a type I interferon signature. *Nature Genetics* 44: 1243–8.

234. Crow Y.J. and Manel N. (2015) Aicardi–Goutières syndrome and the type I interferonopathies. *Nature Reviews Immunology* 15: 429–40.

235. Heraud-Farlow J.E. et al. (2017) Protein recoding by ADAR1-mediated RNA editing is not essential for normal development and homeostasis. *Genome Biology* 18: 166.

236. Tariq A. and Jantsch M. (2012) Transcript diversification in the nervous system: A to I RNA-editing in CNS function and disease development. *Frontiers in Neuroscience* 6: 99.

237. Sansam C.L., Wells K.S. and Emeson R.B. (2003) Modulation of RNA editing by functional nucleolar sequestration of ADAR2. *Proceedings of the National Academy of Sciences USA* 100: 14018–23.

238. Vitali P. et al. (2005) ADAR2-mediated editing of RNA substrates in the nucleolus is inhibited by C/D small nucleolar RNAs. *Journal of Cell Biology* 169: 745–53.

239. Tan T.Y. et al. (2020) Bi-allelic *ADARB1* variants associated with microcephaly, intellectual disability, and seizures. *American Journal of Human Genetics* 106: 467–83.

240. Maroofian R. et al. (2021) Biallelic variants in *ADARB1*, encoding a dsRNA-specific adenosine deaminase, cause a severe developmental and epileptic encephalopathy. *Journal of Medical Genetics* 58: 495–504.

241. Higuchi M. et al. (2000) Point mutation in an AMPA receptor gene rescues lethality in mice deficient in the RNA-editing enzyme ADAR2. *Nature* 406: 78–81.

242. Aruscavage P.J. and Bass B.L. (2000) A phylogenetic analysis reveals an unusual sequence conservation within introns involved in RNA editing. *RNA* 6: 257–69.

243. Maas S., Patt S., Schrey M. and Rich A. (2001) Underediting of glutamate receptor GluR-B mRNA in malignant gliomas. *Proceedings of the National Academy of Sciences USA* 98: 14687–92.

244. Seifert G., Zhou M. and Steinhäuser C. (1997) Analysis of AMPA receptor properties during postnatal development of mouse hippocampal astrocytes. *Journal of Neurophysiology* 78: 2916–23.

245. Jakowec M.W., Jackson-Lewis V., Chen X., Langston J.W. and Przedborski S. (1998) The postnatal development of AMPA receptor subunits in the basal ganglia of the rat. *Developmental Neuroscience* 20: 19–33.

246. Petralia R.S. et al. (1999) Selective acquisition of AMPA receptors over postnatal development suggests a molecular basis for silent synapses. *Nature Neuroscience* 2: 31–6.

247. Gordon S., Akopyan G., Garban H. and Bonavida B. (2006) Transcription factor YY1: Structure, function, and therapeutic implications in cancer biology. *Oncogene* 25: 1125–42.

248. Yoon Y.J., White S.L., Ni X., Gokin A.P. and Martin-Caraballo M. (2012) Downregulation of GluA2 AMPA receptor subunits reduces the dendritic arborization of developing spinal motoneurons. *PLOS ONE* 7: e49879.

249. Tzakis N. and Holahan M.R. (2020) Investigation of GluA1 and GluA2 AMPA receptor subtype distribution in the hippocampus and anterior cingulate cortex of Long Evans rats during development. *IBRO Reports* 8: 91–100.

250. Khan A. et al. (2020) Membrane and synaptic defects leading to neurodegeneration in Adar mutant Drosophila are rescued by increased autophagy. *BMC Biology* 18: 15.

251. Vallecillo-Viejo I.C. et al. (2020) Spatially regulated editing of genetic information within a neuron. *Nucleic Acids Research* 48: 3999–4012.

252. Horsch M. et al. (2011) Requirement of the RNA-editing enzyme ADAR2 for normal physiology in mice. *Journal of Biological Chemistry* 286: 18614–22.

253. Maas S. and Gommans W.M. (2009) Novel exon of mammalian ADAR2 extends open reading frame. *PLOS ONE* 4: e4225.

254. Gerber A., O'Connell M.A. and Keller W. (1997) Two forms of human double-stranded RNA-specific editase 1 (hRED1) generated by the insertion of an Alu cassette. *RNA* 3: 453–63.

255. Keegan L.P. et al. (2005) Tuning of RNA editing by ADAR is required in Drosophila. *EMBO Journal* 24: 2183–93.

256. Deng P. et al. (2020) Adar RNA editing-dependent and -independent effects are required for brain and innate immune functions in *Drosophila*. *Nature Communications* 11: 1580.

257. Palladino M.J., Keegan L.P., O'Connell M.A. and Reenan R.A. (2000) A-to-I pre-mRNA editing in *Drosophila* is primarily involved in adult nervous system function and integrity. *Cell* 102: 437–49.

258. Porath H.T. et al. (2019) RNA editing is abundant and correlates with task performance in a social bumblebee. *Nature Communications* 10: 1605.

259. Tonkin L.A. et al. (2002) RNA editing by ADARs is important for normal behavior in *Caenorhabditis elegans*. *EMBO Journal* 21: 6025–35.

260. Sebastiani P. et al. (2009) RNA editing genes associated with extreme old age in humans and with lifespan in *C. elegans*. *PLOS ONE* 4: e8210.

261. Goldstein B. et al. (2017) A-to-I RNA editing promotes developmental stage–specific gene and lncRNA expression. *Genome Research* 27: 462–70.

262. Melcher T. et al. (1996) RED2, a brain-specific member of the RNA-specific adenosine deaminase family. *Journal of Biological Chemistry* 271: 31795–8.

263. Chen C.-X. et al. (2000) A third member of the RNA-specific adenosine deaminase gene family, ADAR3, contains both single- and double-stranded RNA binding domains. *RNA* 6: 755–67.

264. Cho D.-S.C. et al. (2003) Requirement of dimerization for RNA editing activity of adenosine deaminases acting on RNA. *Journal of Biological Chemistry* 278: 17093–102.

265. Mladenova D. et al. (2018) Adar3 Is Involved in learning and memory in mice. *Frontiers in Neuroscience* 12: 243.

266. Patton D.E., Silva T. and Bezanilla F. (1997) RNA editing generates a diverse array of transcripts encoding squid Kv2 K+ channels with altered functional properties. *Neuron* 19: 711–22.

267. Alon S. et al. (2015) The majority of transcripts in the squid nervous system are extensively recoded by A-to-I RNA editing. *eLife* 4: e05198.

268. Albertin C.B. et al. (2015) The octopus genome and the evolution of cephalopod neural and morphological novelties. *Nature* 524: 220–4.

269. Liscovitch-Brauer N. et al. (2017) Trade-off between transcriptome plasticity and genome evolution in cephalopods. *Cell* 169: 191–202.

270. Levanon E.Y. et al. (2004) Systematic identification of abundant A-to-I editing sites in the human transcriptome. *Nature Biotechnology* 22: 1001–5.

271. Athanasiadis A., Rich A. and Maas S. (2004) Widespread A-to-I RNA editing of Alu-containing mRNAs in the human transcriptome. *PLOS Biology* 2: e391.

272. Kim D.D. et al. (2004) Widespread RNA editing of embedded alu elements in the human transcriptome. *Genome Research* 14: 1719–25.

273. Blow M., Futreal P.A., Wooster R. and Stratton M.R. (2004) A survey of RNA editing in human brain. *Genome Research* 14: 2379–87.

274. Porath H.T., Carmi S. and Levanon E.Y. (2014) A genome-wide map of hyper-edited RNA reveals numerous new sites. *Nature Communications* 5: 4726.

275. Solomon O. et al. (2017) RNA editing by ADAR1 leads to context-dependent transcriptome-wide changes in RNA secondary structure. *Nature Communications* 8: 1440.

276. Paz-Yaacov N. et al. (2010) Adenosine-to-inosine RNA editing shapes transcriptome diversity in primates. *Proceedings of the National Academy of Sciences USA* 107: 12174–9.

277. Britten R.J., Baron W.F., Stout D.B. and Davidson E.H. (1988) Sources and evolution of human Alu repeated sequences. *Proceedings of the National Academy of Sciences USA* 85: 4770–4.

278. Liu G.E., Alkan C., Jiang L., Zhao S. and Eichler E.E. (2009) Comparative analysis of Alu repeats in primate genomes. *Genome Research* 19: 876–85.

279. Bazak L. et al. (2014) A-to-I RNA editing occurs at over a hundred million genomic sites, located in a majority of human genes. *Genome Research* 24: 365–76.

280. Liu W.M., Chu W.M., Choudary P.V. and Schmid C.W. (1995) Cell stress and translational inhibitors transiently increase the abundance of mammalian SINE transcripts. *Nucleic Acids Research* 23: 1758–65.

281. Lev-Maor G., Sorek R., Shomron N. and Ast G. (2003) The birth of an alternatively spliced exon: 3′ splice-site selection in Alu exons. *Science* 300: 1288–91.

282. Jurka J. (2004) Evolutionary impact of human Alu repetitive elements. *Current Opinion in Genetics and Development* 14: 603–8.

283. Krull M., Brosius J. and Schmitz J. (2005) Alu-SINE exonization: En route to protein-coding function. *Molecular Biology and Evolution* 22: 1702–11.

284. Hasler J. and Strub K. (2006) Alu elements as regulators of gene expression. *Nucleic Acids Research* 34: 5491–7.

285. Britten R.J. (2010) Transposable element insertions have strongly affected human evolution. *Proceedings of the National Academy of Sciences USA* 107: 19945–8.

286. Jády B.E., Ketele A. and Kiss T. (2012) Human intron-encoded Alu RNAs are processed and packaged into Wdr79-associated nucleoplasmic box H/ACA RNPs. *Genes & Development* 26: 1897–910.

287. Lubelsky Y. and Ulitsky I. (2018) Sequences enriched in Alu repeats drive nuclear localization of long RNAs in human cells. *Nature* 555: 107–11.

288. Cheng Y. et al. (2020) Increased processing of SINE B2 ncRNAs unveils a novel type of transcriptome deregulation in amyloid beta neuropathology. *eLife* 9: e61265.

289. Hernandez A.J. et al. (2020) B2 and ALU retrotransposons are self-cleaving ribozymes whose activity is enhanced by EZH2. *Proceedings of the National Academy of Sciences USA* 117: 415–25.

290. Cheng Y. et al. (2021) Increased Alu RNA processing in Alzheimer brains is linked to gene expression changes. *EMBO Reports* 22: e52255.

291. Li W.Y., Reddy R., Henning D., Epstein P. and Busch H. (1982) Nucleotide sequence of 7S RNA. Homology to Alu DNA and La 4.5S RNA. *Journal of Biological Chemistry* 257: 5136–42.

292. Labuda D. and Zietkiewicz E. (1994) Evolution of secondary structure in the family of 7SL-like RNAs. *Journal of Molecular Evolution* 39: 506–18.

293. Navaratnam N. and Sarwar R. (2006) An overview of cytidine deaminases. *International Journal of Hematology* 83: 195–200.

294. Conticello S.G. (2008) The AID/APOBEC family of nucleic acid mutators. *Genome Biology* 9: 229.

295. Münk C., Willemsen A. and Bravo I.G. (2012) An ancient history of gene duplications, fusions and losses in the evolution of APOBEC3 mutators in mammals. *BMC Evolutionary Biology* 12: 71.

296. Salter J.D., Bennett R.P. and Smith H.C. (2016) The APOBEC protein family: United by structure, divergent in function. *Trends in Biochemical Sciences* 41: 578–94.

297. Chen S. et al. (1987) Apolipoprotein B-48 is the product of a messenger RNA with an organ-specific in-frame stop codon. *Science* 238: 363–6.

298. Powell L.M. et al. (1987) A novel form of tissue-specific RNA processing produces apolipoprotein-B48 in intestine. *Cell* 50: 831–40.

299. Davidson N.O. and Shelness G.S. (2000) Apolipoprotein B: mRNA editing, lipoprotein assembly, and presecretory degradation. *Annual Review of Nutrition* 20: 169–93.

300. Liu M.-C. et al. (2018) AID/APOBEC-like cytidine deaminases are ancient innate immune mediators in invertebrates. *Nature Communications* 9: 1948.

301. Schutsky E.K., Nabel C.S., Davis A.K.F., DeNizio J.E. and Kohli R.M. (2017) APOBEC3A efficiently deaminates methylated, but not TET-oxidized, cytosine bases in DNA. *Nucleic Acids Research* 45: 7655–65.

302. Chahwan R., Wontakal S.N. and Roa S. (2010) Crosstalk between genetic and epigenetic information through cytosine deamination. *Trends in Genetics* 26: 443–8.

303. Nelson V.R., Heaney J.D., Tesar P.J., Davidson N.O. and Nadeau J.H. (2012) Transgenerational epigenetic effects of the Apobec1 cytidine deaminase deficiency on testicular germ cell tumor susceptibility and embryonic viability. *Proceedings of the National Academy of Sciences USA* 109: E2766–73.

304. Severi F., Chicca A. and Conticello S.G. (2010) Analysis of reptilian APOBEC1 suggests that RNA editing may not be its ancestral function. *Molecular Biology and Evolution* 28: 1125–9.

305. Rogozin I.B., Basu M.K., Jordan I.K., Pavlov Y.I. and Koonin E.V. (2005) APOBEC4, a new member of the AID/APOBEC family of polynucleotide (deoxy)cytidine deaminases predicted by computational analysis. *Cell Cycle* 4: 1281–5.

306. Ito J., Gifford R.J. and Sato K. (2020) Retroviruses drive the rapid evolution of mammalian *APOBEC3* genes. *Proceedings of the National Academy of Sciences USA* 117: 610–8.

307. Sawyer S.L., Emerman M. and Malik H.S. (2004) Ancient adaptive evolution of the primate antiviral DNA-editing enzyme APOBEC3G. *PLOS Biology* 2: e275.

308. Zhang J. and Webb D.M. (2004) Rapid evolution of primate antiviral enzyme APOBEC3G. *Human Molecular Genetics* 13: 1785–91.

309. LaRue R.S. et al. (2009) Guidelines for naming nonprimate APOBEC3 genes and proteins. *Journal of Virology* 83: 494–7.

310. Krishnan A., Iyer L.M., Holland S.J., Boehm T. and Aravind L. (2018) Diversification of AID/APOBEC-like deaminases in metazoa: Multiplicity of clades and widespread roles in immunity. *Proceedings of the National Academy of Sciences USA* 115: E3201–10.

311. Koning F.A. (2009) Defining APOBEC3 expression patterns in human tissues and hematopoietic cell subsets. *Journal of Virology* 83: 9474–85.

312. Hill M.S. et al. (2006) APOBEC3G expression is restricted to neurons in the brains of pigtailed macaques. *AIDS Research and Human Retroviruses* 22: 541–50.

313. Harris R.S. and Dudley J.P. (2015) APOBECs and virus restriction. *Virology* 479: 131–45.

314. Chiu Y.-L. and Greene W.C. (2008) The APOBEC3 cytidine deaminases: An innate defensive network opposing exogenous retroviruses and endogenous retroelements. *Annual Review of Immunology* 26: 317–53.

315. Koyama T. et al. (2013) APOBEC3G oligomerization is associated with the inhibition of both Alu and Line-1 retrotransposition. *PLOS ONE* 8: e84228.

316. Mustafin R.N. and Khusnutdinova E.K. (2020) Involvement of transposable elements in neurogenesis. *Vavilov Journal of Genetics and Breeding* 24: 209–18.

317. Muotri A.R. et al. (2005) Somatic mosaicism in neuronal precursor cells mediated by L1 retrotransposition. *Nature* 435: 903–10.

318. Coufal N.G. et al. (2009) L1 retrotransposition in human neural progenitor cells. *Nature* 460: 1127–31.

319. Muotri A.R. et al. (2010) L1 retrotransposition in neurons is modulated by MeCP2. *Nature* 468: 443–6.

320. Salvador-Palomeque C. et al. (2019) Dynamic methylation of an L1 transduction family during reprogramming and neurodifferentiation. *Molecular and Cellular Biology* 39: e00499.

321. Sanchez-Luque F.J. et al. (2019) LINE-1 evasion of epigenetic repression in humans. *Molecular Cell* 75: 590–604.

322. Baillie J.K. et al. (2011) Somatic retrotransposition alters the genetic landscape of the human brain. *Nature* 479: 534–7.

323. Rowe H.M. et al. (2010) KAP1 controls endogenous retroviruses in embryonic stem cells. *Nature* 463: 237–40.

324. Thomas C.A. et al. (2017) Modeling of TREX1-dependent autoimmune disease using human stem cells highlights L1 accumulation as a source of neuroinflammation. *Cell Stem Cell* 21: 319–31.

325. Jönsson M.E., Garza R., Johansson P.A. and Jakobsson J. (2020) Transposable elements: A common feature of neurodevelopmental and neurodegenerative disorders. *Trends in Genetics* 36: 610–23.

326. Turelli P. et al. (2020) Primate-restricted KRAB zinc finger proteins and target retrotransposons control gene expression in human neurons. *Science Advances* 6: eaba3200.

327. Jönsson M.E. et al. (2021) Activation of endogenous retroviruses during brain development causes an inflammatory response. *EMBO Journal* 40: e106423.

328. Marchetto M.C.N. et al. (2013) Differential L1 regulation in pluripotent stem cells of humans and apes. *Nature* 503: 525–9.

329. Jönsson M.E. et al. (2019) Activation of neuronal genes via LINE-1 elements upon global DNA demethylation in human neural progenitors. *Nature Communications* 10: 3182.

330. Swanson L., Newman E., Araque A. and Dubinsky J.M.A., Katie (2017) *The Beautiful Brain: The Drawings of Santiago Ramón y Cajal* (Abrams, New York).

331. Tan L. et al. (2021) Changes in genome architecture and transcriptional dynamics progress independently of sensory experience during post-natal brain development. *Cell* 184: 741–58.

332. Edelman G.M. (1978) *The Mindful Brain: Cortical Organization and the Group-Selective Theory of Higher Brain Function* (MIT Press, Cambridge, MA).

333. Edelman G.M. (1988) *Topobiology: An Introduction to Molecular Embryology* (Basic Books, New York).

334. Mateos-Aparicio P. and Rodríguez-Moreno A. (2019) The impact of studying brain plasticity. *Frontiers in Cellular Neuroscience* 13: 66.

335. Niemi M.E.K. et al. (2018) Common genetic variants contribute to risk of rare severe neurodevelopmental disorders. *Nature* 562: 268–71.

336. Asgari Y., Heng J.I.T., Lovell N., Forrest A.R.R. and Alinejad-Rokny H. (2020) Evidence for enhancer noncoding RNAs (enhancer-ncRNAs) with gene regulatory functions relevant to neurodevelopmental disorders. *bioRxiv:* 2020.05.16.087395v2.

337. Sicot G. and Gomes-Pereira M. (2013) RNA toxicity in human disease and animal models: From the uncovering of a new mechanism to the development of promising therapies. *Biochimica et Biophysica Acta* 1832: 1390–409.

338. Riva P., Ratti A. and Venturin M. (2016) The long noncoding RNAs in neurodegenerative diseases: Novel mechanisms of pathogenesis. *Current Alzheimer Research* 13: 1219–31.

339. Wan P., Su W. and Zhuo Y. (2017) The role of long noncoding RNAs in neurodegenerative diseases. *Molecular Neurobiology* 54: 2012–21.

340. Prabhakar S., Noonan J.P., Paabo S. and Rubin E.M. (2006) Accelerated evolution of conserved noncoding sequences in humans. *Science* 314: 786.

341. Xu A.G. et al. (2010) Intergenic and repeat transcription in human, chimpanzee and macaque brains measured by RNA-Seq. *PLOS Computational Biology* 6: e1000843.

342. Liu S.J. et al. (2016) Single-cell analysis of long noncoding RNAs in the developing human neocortex. *Genome Biology* 17: 67.

343. Ma Q. and Chang H.Y. (2016) Single-cell profiling of lncRNAs in the developing human brain. *Genome Biology* 17: 68.

344. Eze U.C., Bhaduri A., Haeussler M., Nowakowski T.J. and Kriegstein A.R. (2021) Single-cell atlas of early human brain development highlights heterogeneity of human neuroepithelial cells and early radial glia. *Nature Neuroscience* 24: 584–94.

345. Bocchi V.D. et al. (2021) The coding and long noncoding single-cell atlas of the developing human fetal striatum. *Science* 372: eabf5759.

346. Briggs J.A., Wolvetang E.J., Mattick J.S., Rinn J.L. and Barry G. (2015) Mechanisms of long non-coding RNAs in mammalian nervous system development, plasticity, disease, and evolution. *Neuron* 88: 861–77.

347. Zwir I. et al. (2022) Evolution of genetic networks for human creativity. *Molecular Psychiatry* 27: 354–76.

348. Sas-Nowosielska H. and Magalska A. (2021) Long noncoding RNAs—crucial players organizing the landscape of the neuronal nucleus. *International Journal of Molecular Sciences* 22: 3478.

349. Trizzino M. et al. (2017) Transposable elements are the primary source of novelty in primate gene regulation. *Genome Research* 27: 1623–33.

350. Cosby R.L. et al. (2021) Recurrent evolution of vertebrate transcription factors by transposase capture. *Science* 371: eabc6405.

351. Mercer T.R. et al. (2011) Expression of distinct RNAs from 3′ untranslated regions. *Nucleic Acids Research* 39: 2393–403.

352. Kocabas A., Duarte T., Kumar S. and Hynes M.A. (2015) Widespread differential expression of coding region and 3′UTR sequences in neurons and other tissues. *Neuron* 88: 1149–56.

353. Andreassi C. et al. (2021) Cytoplasmic cleavage of IMPA1 3′ UTR is necessary for maintaining axon integrity. *Cell Reports* 34: 108778.

354. Smalheiser N.R. (2014) The RNA-centred view of the synapse: Non-coding RNAs and synaptic plasticity. *Philosophical Transactions of the Royal Society B: Biological Sciences* 369: 20130504.

355. Rogelj B. and Giese K.P. (2004) Expression and function of brain specific small RNAs. *Reviews in Neuroscience* 15: 185–98.

356. Rajasethupathy P. et al. (2012) A role for neuronal piRNAs in the epigenetic control of memory-related synaptic plasticity. *Cell* 149: 693–707.

357. Zuo L., Wang Z., Tan Y., Chen X. and Luo X. (2016) piRNAs and their functions in the brain. *International Journal of Human Genetics* 16: 53–60.

358. Gasperini C. et al. (2020) The piRNA pathway sustains adult neurogenesis by repressing protein synthesis. *bioRxiv:* 2020.09.15.297739.

359. Ashraf S.I. and Kunes S. (2006) A trace of silence: Memory and microRNA at the synapse. *Current Opinion in Neurobiology* 16: 535–9.

360. Sambandan S. et al. (2017) Activity-dependent spatially localized miRNA maturation in neuronal dendrites. *Science* 355: 634–7.

361. Zimmer-Bensch G. (2019) Emerging roles of long non-coding RNAs as drivers of brain evolution. *Cells* 8: 1399.

362. Ng S.-Y., Lin L., Soh B.S. and Stanton L.W. (2013) Long noncoding RNAs in development and disease of the central nervous system. *Trends in Genetics* 29: 461–8.

363. Clark B.S. and Blackshaw S. (2017) Understanding the role of lncRNAs in nervous system development. *Advances in Experimental Medicine and Biology* 1008: 253–82.

364. Korneev S.A. et al. (2013) Axonal trafficking of an antisense RNA transcribed from a pseudogene is regulated by classical conditioning. *Scientific Reports* 3: 1027.

365. Kleaveland B., Shi C.Y., Stefano J. and Bartel D.P. (2018) A network of noncoding regulatory RNAs acts in the mammalian brain. *Cell* 174: 350–62.

366. Maag J.L.V. et al. (2015) Dynamic expression of long noncoding RNAs and repeat elements in synaptic plasticity. *Frontiers in Neuroscience* 9: 351.

367. Bernard D. et al. (2010) A long nuclear-retained non-coding RNA regulates synaptogenesis by modulating gene expression. *EMBO Journal* 29: 3082–93.

368. Barry G. et al. (2014) The long non-coding RNA *Gomafu* is acutely regulated in response to neuronal activation and involved in schizophrenia-associated alternative splicing. *Molecular Psychiatry* 19: 486–94.

369. Ip J.Y. et al. (2016) Gomafu lncRNA knockout mice exhibit mild hyperactivity with enhanced responsiveness to the psychostimulant methamphetamine. *Scientific Reports* 6: 27204.

370. Barry G. et al. (2017) The long non-coding RNA *NEAT1* is responsive to neuronal activity and is associated with hyperexcitability states. *Scientific Reports* 7: 40127.

371. Butler A.A., Johnston D.R., Kaur S. and Lubin F.D. (2019) Long noncoding RNA NEAT1 mediates neuronal histone methylation and age-related memory impairment. *Science Signaling* 12: eaaw9277.

372. Kukharsky M.S. et al. (2020) Long non-coding RNA *Neat1* regulates adaptive behavioural response to stress in mice. *Translational Psychiatry* 10: 171.

373. Xu H. et al. (2020) Role of long noncoding RNA *Gas5* in cocaine action. *Biological Psychiatry* 88: 758–66.

374. Volff J.N. and Brosius J. (2007) Modern genomes with retro-look: Retrotransposed elements, retroposition and the origin of new genes. *Gene and Protein Evolution* 3: 175–90.

375. Centonze D. et al. (2007) The brain cytoplasmic RNA BC1 regulates dopamine D2 receptor-mediated transmission in the striatum. *Journal of Neuroscience* 27: 8885–92.

376. Zalfa F. et al. (2003) The Fragile X Syndrome protein FMRP associates with BC1 RNA and regulates the translation of specific mRNAs at synapses. *Cell* 112: 317–27.

377. Lewejohann L. et al. (2004) Role of a neuronal small non-messenger RNA: Behavioural alterations in BC1 RNA-deleted mice. *Behavioural Brain Research* 154: 273–89.

378. Chung A., Dahan N., Alarcon J.M. and Fenton A.A. (2017) Effects of regulatory BC1 RNA deletion on synaptic plasticity, learning, and memory. *Learning & Memory* 24: 646–9.

379. Labonté B. et al. (2021) Regulation of impulsive and aggressive behaviours by a novel lncRNA. *Molecular Psychiatry* 26: 3751–64.

380. Issler O. et al. (2020) Sex-specific role for the long non-coding RNA LINC00473 in depression. *Neuron* 106: 912–26.

381. Tan M.C. et al. (2017) The activity-induced long non-coding RNA Meg3 modulates AMPA receptor surface expression in primary cortical neurons. *Frontiers in Cellular Neuroscience* 11: 124.

382. Anguera M.C. et al. (2011) *Tsx* produces a long noncoding RNA and has general functions in the germline, stem cells, and brain. *PLOS Genetics* 7: e1002248.

383. Li D. et al. (2018) Activity dependent LoNA regulates translation by coordinating rRNA transcription and methylation. *Nature Communications* 9: 1726.

384. Ma M. et al. (2020) A novel pathway regulates social hierarchy via lncRNA AtLAS and postsynaptic synapsin IIb. *Cell Research* 30: 105–18.

385. Zhang X., X.U. Yn, Chen B. and Kang L. (2020) Long noncoding RNA PAHAL modulates locust behavioural plasticity through the feedback regulation of dopamine biosynthesis. *PLOS Genetics* 16: e1008771.

386. Liu F. et al. (2019) lncRNA profile of *Apis mellifera* and its possible role in behavioural transition from nurses to foragers. *BMC Genomics* 20: 393.

387. Levenson J.M. and Sweatt J.D. (2005) Epigenetic mechanisms in memory formation. *Nature Reviews Neuroscience* 6: 108–18.

388. Graff J. and Mansuy I.M. (2008) Epigenetic codes in cognition and behaviour. *Behavioural Brain Research* 192: 70–87.

389. Jarome T.J. and Lubin F.D. (2014) Epigenetic mechanisms of memory formation and reconsolidation. *Neurobiology of Learning and Memory* 115: 116–27.

390. Marshall P. and Bredy T.W. (2016) Cognitive neuroepigenetics: The next evolution in our understanding of the molecular mechanisms underlying learning and memory? *npj Science of Learning* 1: 16014.

391. Pachter J.S., de Vries H.E. and Fabry Z. (2003) The blood-brain barrier and its role in immune privilege in the central nervous system. *Journal of Neuropathology & Experimental Neurology* 62: 593–604.

392. Banks W.A. and Erickson M.A. (2010) The blood–brain barrier and immune function and dysfunction. *Neurobiology of Disease* 37: 26–32.

393. Majerova P. et al. (2019) Trafficking of immune cells across the blood-brain barrier is modulated by neurofibrillary pathology in tauopathies. *PLOS ONE* 14: e0217216.

394. Crockett A.M. et al. (2020) Immune activation of the blood brain barrier and implications for neuroinflammation in schizophrenia. *Journal of Immunology* 204(1 Supplement): 158.23.

395. Yoshihara Y., Oka S., Ikeda J. and Mori K. (1991) Immunoglobulin superfamily molecules in the nervous system. *Neuroscience Research* 10: 83–105.

396. Martin M.U. and Wesche H. (2002) Summary and comparison of the signaling mechanisms of the Toll/interleukin-1 receptor family. *Biochimica et Biophysica Acta* 1592: 265–80.

397. Rougon G. and Hobert O. (2003) New insights into the diversity and function of neuronal immunoglobulin superfamily molecules. *Annual Review of Neuroscience* 26: 207–38.

398. Steinman L. (2004) Elaborate interactions between the immune and nervous systems. *Nature Immunology* 5: 575–81.

399. Maness P.F. and Schachner M. (2007) Neural recognition molecules of the immunoglobulin superfamily: Signaling transducers of axon guidance and neuronal migration. *Nature Neuroscience* 10: 19–26.

400. Zinn K. and Özkan E. (2017) Neural immunoglobulin superfamily interaction networks. *Current Opinion in Neurobiology* 45: 99–105.

401. Morimoto K. and Nakajima K. (2019) Role of the immune system in the development of the central nervous system. *Frontiers in Neuroscience* 13: 916.

402. Anthoney N., Foldi I. and Hidalgo A. (2018) Toll and Toll-like receptor signalling in development. *Development* 145: dev156018.

403. Chen C.-Y., Shih Y.-C., Hung Y.-F. and Hsueh Y.-P. (2019) Beyond defense: Regulation of neuronal morphogenesis and brain functions via Toll-like receptors. *Journal of Biomedical Science* 26: 90.

404. Foldi I. et al. (2017) Three-tier regulation of cell number plasticity by neurotrophins and Tolls in *Drosophila*. *Journal of Cell Biology* 216: 1421–38.

405. Paemen L.R. et al. (1992) Glial localization of interleukin-1α in invertebrate ganglia. *Cellular and Molecular Neurobiology* 12: 463–72.

406. Galic M.A., Riazi K. and Pittman Q.J. (2012) Cytokines and brain excitability. *Frontiers in Neuroendocrinology* 33: 116–25.

407. Miller A.H., Haroon E., Raison C.L. and Felger J.C. (2013) Cytokine targets in the brain: Impact on neurotransmitters and neurocircuits. *Depression and Anxiety* 30: 297–306.

408. Chun J.J.M., Schatz D.G., Oettinger M.A., Jaenisch R. and Baltimore D. (1991) The recombination activating gene-1 (RAG-1) transcript is present in the murine central nervous system. *Cell* 64: 189–200.

409. Fugmann S.D., Lee A.I., Shockett P.E., Villey I.J. and Schatz D.G. (2000) The RAG proteins and V(D)J recombination: Complexes, ends, and transposition. *Annual Review of Immunology* 18: 495–527.

410. Jessen J.R., Jessen T.N., Vogel S.S. and Lin S. (2001) Concurrent expression of recombination activating genes 1 and 2 in zebrafish olfactory sensory neurons. *Genesis* 29: 156–62.

411. Álvarez-Lindo N., Baleriola J., de los Ríos V., Suárez T. and de la Rosa E.J. (2019) RAG-2 deficiency results in fewer phosphorylated histone H2AX foci, but increased retinal ganglion cell death and altered axonal growth. *Scientific Reports* 9: 18486.

412. Abarrategui I. and Krangel M.S. (2007) Noncoding transcription controls downstream promoters to regulate T-cell receptor alpha recombination. *EMBO Journal* 26: 4380–90.

413. Storici F., Bebenek K., Kunkel T.A., Gordenin D.A. and Resnick M.A. (2007) RNA-templated DNA repair. *Nature* 447: 338–41.

414. Nowacki M. et al. (2007) RNA-mediated epigenetic programming of a genome-rearrangement pathway. *Nature* 451: 153–8.

415. McGowan P.O., Hope T.A., Meck W.H., Kelsoe G. and Williams C.L. (2011) Impaired social recognition memory in recombination activating gene 1-deficient mice. *Brain Research* 1383: 187–95.

416. Matthews A.G. et al. (2007) RAG2 PHD finger couples histone H3 lysine 4 trimethylation with V(D)J recombination. *Nature* 450: 1106–10.

417. Lee M.-H. et al. (2018) Somatic APP gene recombination in Alzheimer's disease and normal neurons. *Nature* 563: 639–45.

418. Kaeser G. and Chun J. (2020) Brain cell somatic gene recombination and its phylogenetic foundations. *Journal of Biological Chemistry* 295: 12786–95.

419. Eyman M. et al. (2007) Local synthesis of axonal and presynaptic RNA in squid model systems. *European Journal of Neuroscience* 25: 341–50.

420. Sotelo J.R. et al. (2014) Glia to axon RNA transfer. *Developmental Neurobiology* 74: 292–302.

421. Kiebler M.A. and Bassell G.J. (2006) Neuronal RNA granules: Movers and makers. *Neuron* 51: 685–90.

422. Carson J.H. et al. (2008) Multiplexed RNA trafficking in oligodendrocytes and neurons. *Biochimica et Biophysica Acta* 1779: 453–8.

423. Garner C.C., Tucker R.P. and Matus A. (1988) Selective localization of messenger RNA for cytoskeletal protein MAP2 in dendrites. *Nature* 336: 674–7.

424. Burgin K.E. et al. (1990) In situ hybridization histochemistry of Ca2+/calmodulin-dependent protein kinase in developing rat brain. *Journal of Neuroscience* 10: 1788–98.

425. Kanai Y., Dohmae N. and Hirokawa N. (2004) Kinesin transports RNA: Isolation and characterization of an RNA-transporting granule. *Neuron* 43: 513–25.

426. Hirokawa N. (2006) mRNA transport in dendrites: RNA granules, motors, and tracks. *Journal of Neuroscience* 26: 7139–42.

427. Bramham C.R. and Wells D.G. (2007) Dendritic mRNA: Transport, translation and function. *Nature Reviews Neuroscience* 8: 776–89.

428. Yoon Y.J. et al. (2016) Glutamate-induced RNA localization and translation in neurons. *Proceedings of the National Academy of Sciences USA* 113: E6877–86.

429. Terenzio M., Schiavo G. and Fainzilber M. (2017) Compartmentalized signaling in neurons: From cell biology to neuroscience. *Neuron* 96: 667–79.

430. Madugalle S.U., Meyer K., Wang D.O. and Bredy T.W. (2020) RNA N^6-methyladenosine and the regulation of RNA localization and function in the brain. *Trends in Neurosciences* 43: 1011–23.

431. Raveendra B.L. et al. (2018) Long noncoding RNA GM12371 acts as a transcriptional regulator of synapse function. *Proceedings of the National Academy of Sciences USA* 115: E10197–205.

432. Grinman E. et al. (2021) Activity-regulated synaptic targeting of lncRNA ADEPTR mediates structural plasticity by localizing Sptn1 and AnkB in dendrites. *Science Advances* 7: eabf0605.

433. Wang F. et al. (2021) The long noncoding RNA *Synage* regulates synapse stability and neuronal function in the cerebellum. *Cell Death & Differentiation* 28: 2634–50.

434. Liau W.-S., Samaddar S., Banerjee S. and Bredy T.W. (2021) On the functional relevance of spatiotemporally-specific patterns of experience-dependent long noncoding RNA expression in the brain. *RNA Biology* 18: 1025–36.

435. McCurry C.L. et al. (2010) Loss of Arc renders the visual cortex impervious to the effects of sensory experience or deprivation. *Nature Neuroscience* 13: 450–7.

436. Guzowski J.F. et al. (2000) Inhibition of activity-dependent Arc protein expression in the rat hippocampus impairs the maintenance of long-term potentiation and the consolidation of long-term memory. *Journal of Neuroscience* 20: 3993–4001.

437. Plath N. et al. (2006) Arc/Arg3.1 Is essential for the consolidation of synaptic plasticity and memories. *Neuron* 52: 437–44.

438. Pastuzyn E.D. et al. (2018) The neuronal gene *Arc* encodes a repurposed retrotransposon Gag protein that mediates intercellular RNA transfer. *Cell* 172: 275–88.

439. Ashraf S.I., McLoon A.L., Sclarsic S.M. and Kunes S. (2006) Synaptic protein synthesis associated with memory is regulated by the RISC pathway in *Drosophila*. *Cell* 124: 191–205.

440. Moazed D. (2012) A piRNA to remember. *Cell* 149: 512–4.

441. Ghosheh Y. et al. (2016) Characterization of piRNAs across postnatal development in mouse brain. *Scientific Reports* 6: 25039.

442. Huang X. and Wong G. (2021) An old weapon with a new function: PIWI-interacting RNAs in neurodegenerative diseases. *Translational Neurodegeneration* 10: 9.

443. Perrat P.N. et al. (2013) Transposition-driven genomic heterogeneity in the *Drosophila* brain. *Science* 340: 91–5.

444. Nandi S. et al. (2016) Roles for small noncoding RNAs in silencing of retrotransposons in the mammalian brain. *Proceedings of the National Academy of Sciences USA* 113: 12697–702.

445. Knowles R.B. et al. (1996) Translocation of RNA granules in living neurons. *Journal of Neuroscience* 16: 7812–20.

446. Doyle M. and Kiebler M.A. (2011) Mechanisms of dendritic mRNA transport and its role in synaptic tagging. *EMBO Journal* 30: 3540–52.

447. Das S., Singer R.H. and Yoon Y.J. (2019) The travels of mRNAs in neurons: Do they know where they are going? *Current Opinion in Neurobiology* 57: 110–6.

448. Dubnau J. et al. (2003) The staufen/pumilio pathway is involved in *Drosophila* long-term memory. *Current Biology* 13: 286–96.

449. Lucas B.A. et al. (2018) Evidence for convergent evolution of SINE-directed Staufen-mediated mRNA decay. *Proceedings of the National Academy of Sciences USA* 115: 968–73.

450. Kim T.-K. et al. (2010) Widespread transcription at neuronal activity-regulated enhancers. *Nature* 465: 182–7.

451. Wu W. et al. (2021) Neuronal enhancers are hotspots for DNA single-strand break repair. *Nature* 593: 440–4.

452. Wei P.-C. et al. (2016) Long neural genes harbor recurrent DNA break clusters in neural stem/progenitor cells. *Cell* 164: 644–55.

453. Reid D.A. et al. (2021) Incorporation of a nucleoside analog maps genome repair sites in postmitotic human neurons. *Science* 372: 91–4.

454. Suberbielle E. et al. (2013) Physiologic brain activity causes DNA double-strand breaks in neurons, with exacerbation by amyloid-β. *Nature Neuroscience* 16: 613–21.

455. Stott R.T., Kritsky O. and Tsai L.-H. (2021) Profiling DNA break sites and transcriptional changes in response to contextual fear learning. *PLOS ONE* 16: e0249691.

456. Madabhushi R. et al. (2015) Activity-induced DNA breaks govern the expression of neuronal early-response genes. *Cell* 161: 1592–605.

457. Lou M.-M. et al. (2021) Long noncoding RNA BS-DRL1 modulates the DNA damage response and genome stability by interacting with HMGB1 in neurons. *Nature Communications* 12: 4075.

458. Li X. et al. (2019) The DNA repair-associated protein gadd45γ regulates the temporal coding of immediate early gene expression within the prelimbic prefrontal cortex and is required for the consolidation of associative fear memory. *Journal of Neuroscience* 39: 970–83.

459. Chow H. and Herrup K. (2015) Genomic integrity and the ageing brain. *Nature Reviews Neuroscience* 16: 672–84.

460. Meers C. et al. (2020) Genetic characterization of three distinct mechanisms supporting RNA-driven DNA repair and modification reveals major role of DNA polymerase ζ. *Molecular Cell* 79: 1037–50.

461. Chandramouly G. et al. (2021) Polθ reverse transcribes RNA and promotes RNA-templated DNA repair. *Science Advances* 7: eabf1771.

462. Chan K.Y., Li X., Ortega J., Gu L. and Li G.-M. (2021) DNA polymerase θ promotes CAG•CTG repeat expansions in Huntington's disease via insertion sequences of its catalytic domain. *Journal of Biological Chemistry* 297: 101144.

463. Grishok A., Tabara H. and Mello C.C. (2000) Genetic requirements for inheritance of RNAi in *C. elegans*. *Science* 287: 2494–7.

464. Kennerdell J.R. and Carthew R.W. (2000) Heritable gene silencing in *Drosophila* using double-stranded RNA. *Nature Biotechnology* 18: 896–8.

465. Vastenhouw N.L. et al. (2006) Gene expression: Long-term gene silencing by RNAi. *Nature* 442: 882.

466. Ashe A. et al. (2012) piRNAs can trigger a multigenerational epigenetic memory in the germline of *C. elegans*. *Cell* 150: 88–99.

467. Grentzinger T. et al. (2012) piRNA-mediated transgenerational inheritance of an acquired trait. *Genome Research* 22: 1877–88.

468. Shirayama M. et al. (2012) piRNAs initiate an epigenetic memory of nonself RNA in the *C. elegans* germline. *Cell* 150: 65–77.

469. Lewis A. et al. (2020) A family of Argonaute-interacting proteins gates nuclear RNAi. *Molecular Cell* 78: 862–75.

470. Jones L., Ratcliff F. and Baulcombe D.C. (2001) RNA-directed transcriptional gene silencing in plants can be inherited independently of the RNA trigger and requires Met1 for maintenance. *Current Biology* 11: 747–57.

471. Bateson W. and Pellew C. (1920) The genetics of "rogues" among culinary peas (*Pisum sativum*). *Proceedings of The Royal Society B: Biological Sciences* 91: 186–95.

472. Brink R.A. (1956) A genetic change associated with the R locus in maize which is directed and potentially reversible. *Genetics* 41: 872–89.

473. Brink R.A. (1958) Paramutation at the R locus in maize. *Cold Spring Harbor Symposia on Quantitative Biology* 23: 379–91.

474. Brink R.A., Styles E.D. and Axtell J.D. (1968) Paramutation: Directed genetic change. *Science* 159: 161–70.

475. Chandler V.L. (2007) Paramutation: From maize to mice. *Cell* 128: 641–5.

476. Pilu R. (2015) Paramutation phenomena in plants. *Seminars in Cell & Developmental Biology* 44: 2–10.

477. Hollick J.B. (2017) Paramutation and related phenomena in diverse species. *Nature Reviews Genetics* 18: 5–23.

478. Conine C.C. and Rando O.J. (2021) Soma-to-germline RNA communication. *Nature Reviews Genetics* 23: 73–88.

479. Bošković A. and Rando O.J. (2018) Transgenerational epigenetic inheritance. *Annual Review of Genetics* 52: 21–41.

480. Casier K., Boivin A., Carré C. and Teysset L. (2019) Environmentally-induced transgenerational epigenetic inheritance: Implication of PIWI interacting RNAs. *Cells* 8: 1108.

481. Zhao M. et al. (2021) The *mop1* mutation affects the recombination landscape in maize. *Proceedings of the National Academy of Sciences USA* 118: e2009475118.

482. Rassoulzadegan M. et al. (2006) RNA-mediated non-mendelian inheritance of an epigenetic change in the mouse. *Nature* 441: 469–74.

483. Pilu R. (2011) Paramutation: Just a curiosity or fine tuning of gene expression in the next generation? *Current Genomics* 12: 298–306.

484. Houri-Zeevi L., Korem Kohanim Y., Antonova O. and Rechavi O. (2020) Three rules explain transgenerational small RNA inheritance in *C. elegans*. *Cell* 182: 1186–97.

485. Herman H. et al. (2003) Trans allele methylation and paramutation-like effects in mice. *Nature Genetics* 34: 199–202.

486. Watanabe T. et al. (2011) Role for piRNAs and noncoding RNA in de novo DNA methylation of the imprinted mouse *Rasgrf1* locus. *Science* 332: 848–52.

487. Sapetschnig A., Sarkies P., Lehrbach N.J. and Miska E.A. (2015) Tertiary siRNAs mediate paramutation in *C. elegans*. *PLOS Genetics* 11: e1005078.

488. Kiani J. et al. (2013) RNA–mediated epigenetic heredity requires the cytosine methyltransferase Dnmt2. *PLOS Genetics* 9: e1003498.

489. Jeffreys A.J. (1987) Highly variable minisatellites and DNA fingerprints. *Biochemical Society Transactions* 15: 309–17.

490. Willems T. et al. (2014) The landscape of human STR variation. *Genome Research* 24: 1894–904.

491. Quilez J. et al. (2016) Polymorphic tandem repeats within gene promoters act as modifiers of gene expression and DNA methylation in humans. *Nucleic Acids Research* 44: 3750–62.

492. Gymrek M. (2017) A genomic view of short tandem repeats. *Current Opinion in Genetics and Development* 44: 9–16.

493. Hannan A.J. (2018) Tandem repeats mediating genetic plasticity in health and disease. *Nature Reviews Genetics* 19: 286–98.

494. Fondon III J.W., Hammock E.A.D., Hannan A.J. and King D.G. (2008) Simple sequence repeats: Genetic modulators of brain function and behavior. *Trends in Neurosciences* 31: 328–34.

495. Song J.H.T., Lowe C.B. and Kingsley D.M. (2018) Characterization of a human-specific tandem repeat associated with Bipolar Disorder and Schizophrenia. *American Journal of Human Genetics* 103: 421–30.

496. Fotsing S.F. et al. (2019) The impact of short tandem repeat variation on gene expression. *Nature Genetics* 51: 1652–9.

497. Trost B. et al. (2020) Genome-wide detection of tandem DNA repeats that are expanded in autism. *Nature* 586: 80–6.

498. Mitra I. et al. (2021) Patterns of de novo tandem repeat mutations and their role in autism. *Nature* 589: 246–50.

499. Hannan A.J. (2021) Repeat DNA expands our understanding of autism spectrum disorder. *Nature* 589: 200–2.

500. Xu T. et al. (2021) Polymorphic tandem DNA repeats activate the human telomerase reverse transcriptase gene. *Proceedings of the National Academy of Sciences USA* 118: e2019043118.

501. Stam M., Belele C., Dorweiler J.E. and Chandler V.L. (2002) Differential chromatin structure within a tandem array 100 kb upstream of the maize b1 locus is associated with paramutation. *Genes & Development* 16: 1906–18.

502. Hannan A.J. (2010) Tandem repeat polymorphisms: Modulators of disease susceptibility and candidates for 'missing heritability'. *Trends in Genetics* 26: 59–65.

503. Gymrek M. et al. (2016) Abundant contribution of short tandem repeats to gene expression variation in humans. *Nature Genetics* 48: 22–9.

504. Bennett S.T. et al. (1997) Insulin VNTR allele-specific effect in type 1 diabetes depends on identity of untransmitted paternal allele. *Nature Genetics* 17: 350–2.

505. Suzuki S., Miyabe E. and Inagaki S. (2018) Novel brain-expressed noncoding RNA, HSTR1, identified at a human-specific variable number tandem repeat locus with a human accelerated region. *Biochemical and Biophysical Research Communications* 503: 1478–83.

506. Hauser M.T., Aufsatz W., Jonak C. and Luschnig C. (2011) Transgenerational epigenetic inheritance in plants. *Biochimica et Biophysica Acta* 1809: 459–68.

507. Chatterjee N., Lin Y., Santillan B.A., Yotnda P. and Wilson J.H. (2015) Environmental stress induces trinucleotide repeat mutagenesis in human cells. *Proceedings of the National Academy of Sciences USA* 112: 3764–9.

508. Gouil Q. and Baulcombe D.C. (2018) Paramutation-like features of multiple natural epialleles in tomato. *BMC Genomics* 19: 203.

509. Anway M.D., Cupp A.S., Uzumcu M. and Skinner M.K. (2005) Epigenetic transgenerational actions of endocrine disruptors and male fertility. *Science* 308: 1466–9.

510. Franklin T.B. et al. (2010) Epigenetic transmission of the impact of early stress across generations. *Biological Psychiatry* 68: 408–15.

511. Crews D. et al. (2012) Epigenetic transgenerational inheritance of altered stress responses. *Proceedings of the National Academy of Sciences USA* 109: 9143–8.

512. Dias B.G. and Ressler K.J. (2014) Parental olfactory experience influences behavior and neural structure in subsequent generations. *Nature Neuroscience* 17: 89–96.

513. Liberman N., Wang S.Y. and Greer E.L. (2019) Transgenerational epigenetic inheritance: From phenomena to molecular mechanisms. *Current Opinion in Neurobiology* 59: 189–206.

514. Bozler J., Kacsoh B.Z. and Bosco G. (2019) Transgenerational inheritance of ethanol preference is caused by maternal NPF repression. *eLife* 8: e45391.

515. Pereira A.G., Gracida X., Kagias K. and Zhang Y. (2020) *C. elegans* aversive olfactory learning generates diverse intergenerational effects. *Journal of Neurogenetics* 34: 378–88.

516. Jung Y.H. et al. (2021) Recruitment of CTCF to an *Fto* enhancer is responsible for transgenerational inheritance of obesity. *bioRxiv*: 2020.11.20.391672.

517. Bodden C. et al. (2022) Intergenerational effects of a paternal Western diet during adolescence on offspring gut microbiota, stress reactivity, and social behavior. *FASEB Journal* 36: e21981.

518. Mohajer N., Joloya E.M., Seo J., Shioda T. and Blumberg B. (2021) Epigenetic transgenerational inheritance of the effects of obesogen exposure. *Frontiers in Endocrinology* 12: 1726.

519. Gapp K. et al. (2014) Implication of sperm RNAs in transgenerational inheritance of the effects of early trauma in mice. *Nature Neuroscience* 17: 667–9.

520. Gapp K. et al. (2020) Alterations in sperm long RNA contribute to the epigenetic inheritance of the effects of postnatal trauma. *Molecular Psychiatry* 25: 2162–74.

521. Carpenter B.L. et al. (2018) Mother–child transmission of epigenetic information by tunable polymorphic imprinting. *Proceedings of the National Academy of Sciences USA* 115: E11970–7.

522. Carpenter B.L. et al. (2021) Oocyte age and preconceptual alcohol use are highly correlated with epigenetic imprinting of a noncoding RNA (*nc886*). *Proceedings of the National Academy of Sciences USA* 118: e2026580118.

523. Smemo S. et al. (2014) Obesity-associated variants within FTO form long-range functional connections with IRX3. *Nature* 507: 371–5.

524. Ruud J. et al. (2019) The fat mass and obesity-associated protein (FTO) regulates locomotor responses to novelty via D2R medium spiny neurons. *Cell Reports* 27: 3182–98.

525. Grandjean V. et al. (2015) RNA-mediated paternal heredity of diet-induced obesity and metabolic disorders. *Scientific Reports* 5: 18193.

526. Sharma U. et al. (2016) Biogenesis and function of tRNA fragments during sperm maturation and fertilization in mammals. *Science* 351: 391–6.

527. Chen Q. et al. (2016) Sperm tsRNAs contribute to intergenerational inheritance of an acquired metabolic disorder. *Science* 351: 397–400.

528. Zhang Y. et al. (2018) Dnmt2 mediates intergenerational transmission of paternally acquired metabolic disorders through sperm small non-coding RNAs. *Nature Cell Biology* 20: 535–40.

529. Sarker G. et al. (2019) Maternal overnutrition programs hedonic and metabolic phenotypes across generations through sperm tsRNAs. *Proceedings of the National Academy of Sciences USA* 116: 10547–56.

530. Zhang Y., Shi J., Rassoulzadegan M., Tuorto F. and Chen Q. (2019) Sperm RNA code programmes the metabolic health of offspring. *Nature Reviews Endocrinology* 15: 489–98.

531. Skvortsova K., Iovino N. and Bogdanović O. (2018) Functions and mechanisms of epigenetic inheritance in animals. *Nature Reviews Molecular Cell Biology* 19: 774–90.

532. Bédécarrats A., Chen S., Pearce K., Cai D. and Glanzman D.L. (2018) RNA from trained *Aplysia* can induce an epigenetic engram for long-term sensitization in untrained Aplysia. *eNeuro* 5. doi: 10.1523/ENEURO.0038-18.2018.

533. van Steenwyk G. et al. (2020) Involvement of circulating factors in the transmission of paternal experiences through the germline. *EMBO Journal* 39: e104579.

534. Burton N.O. et al. (2020) Cysteine synthases CYSL-1 and CYSL-2 mediate *C. elegans* heritable adaptation to *P. vranovensis* infection. *Nature Communications* 11: 1741.

535. Silver M.J. et al. (2015) Independent genomewide screens identify the tumor suppressor *VTRNA2-1* as a human epiallele responsive to periconceptional environment. *Genome Biology* 16: 118.

536. Treppendahl M.B. et al. (2012) Allelic methylation levels of the noncoding VTRNA2-1 located on chromosome 5q31.1 predict outcome in AML. *Blood* 119: 206–16.

537. Romanelli V. et al. (2014) Variable maternal methylation overlapping the *nc886/vtRNA2-1* locus is locked between hypermethylated repeats and is frequently altered in cancer. *Epigenetics* 9: 783–90.

538. Katzmarski N. et al. (2021) Transmission of trained immunity and heterologous resistance to infections across generations. *Nature Immunology* 22: 1382–90.

539. Parrish N.F. *et al.* (2015) piRNAs derived from ancient viral processed pseudogenes as transgenerational sequence-specific immune memory in mammals. *RNA* 21: 1691–703.

540. Hettema J.M., Annas P., Neale M.C., Kendler K.S. and Fredrikson M. (2003) A twin study of the genetics of fear conditioning. *Archives of General Psychiatry* 60: 702–8.

541. Loken E.K., Hettema J.M., Aggen S.H. and Kendler K.S. (2014) The structure of genetic and environmental risk factors for fears and phobias. *Psychological Medicine* 44: 2375–84.

542. Grossniklaus U. et al. (2013) Transgenerational epigenetic inheritance: How important is it? *Nature Reviews Genetics* 14: 228–35.

543. Heard E. and Martienssen R.A. (2014) Transgenerational epigenetic inheritance: Myths and mechanisms. *Cell* 157: 95–109.

544. Horsthemke B. (2018) A critical view on transgenerational epigenetic inheritance in humans. *Nature Communications* 9: 2973.

545. Perez M.F. and Lehner B. (2019) Intergenerational and transgenerational epigenetic inheritance in animals. *Nature Cell Biology* 21: 143–51.

546. Bergmann M., Schindelmeiser J. and Greven H. (1984) The blood-testis barrier in vertebrates having different testicular organization. *Cell and Tissue Research* 238: 145–50.

547. Mruk D.D. and Cheng C.Y. (2015) The mammalian blood-testis barrier: Its biology and regulation. *Endocrine Reviews* 36: 564–91.

CHAPTER 18

1. Scherrer K. (2003) Historical review: The discovery of 'giant' RNA and RNA processing: 40 years of enigma. *Trends in Biochemical Sciences* 28: 566–71.

2. Niklas K.J. and Newman S.A. (2013) The origins of multicellular organisms. *Evolution & Development* 15: 41–52.

3. Niklas K.J. (2014) The evolutionary-developmental origins of multicellularity. *American Journal of Botany* 101: 6–25.

4. Darwin C. (1859) *On the Origin of Species by Means of Natural Selection* (John Murray, London).

5. Mattick J.S. (2009) Has evolution learnt how to learn? *EMBO Reports* 10: 665.

6. Barton R.A. and Venditti C. (2014) Rapid evolution of the cerebellum in humans and other great apes. *Current Biology* 24: 2440–4.

7. Gould S.J. (1986) Evolution and the triumph of homology, or why history matters. *American Scientist* 74: 60–9.

8. Gould S.J. (2004) The evolution of life on Earth. *Scientific American SA Special Editions* 14: 92–100.

9. Blount Z.D., Lenski R.E. and Losos J.B. (2018) Contingency and determinism in evolution: Replaying life's tape. *Science* 362: eaam5979.

10. Mattick J.S. (2007) A new paradigm for developmental biology. *Journal of Experimental Biology* 210: 1526–47.

11. Amaral P.P., Dinger M.E., Mercer T.R. and Mattick J.S. (2008) The eukaryotic genome as an RNA machine. *Science* 319: 1787–9.

12. Stoltzfus A. and Yampolsky L.Y. (2009) Climbing mount probable: Mutation as a cause of nonrandomness in evolution. *Journal of Heredity* 100: 637–47.

13. Baucom R.S. et al. (2009) Exceptional diversity, non-random distribution, and rapid evolution of retroelements in the Bw73 maize genome. *PLOS Genetics* 5: e1000732.

14. Martincorena I., Seshasayee A.S.N. and Luscombe N.M. (2012) Evidence of non-random mutation rates suggests an evolutionary risk management strategy. *Nature* 485: 95–8.

15. Martincorena I. and Luscombe N.M. (2013) Non-random mutation: The evolution of targeted hypermutation and hypomutation. *BioEssays* 35: 123–30.

16. Svensson E.I. and Berger D. (2019) The role of mutation bias in adaptive evolution. *Trends in Ecology & Evolution* 34: 422–34.

17. Storz J.F. et al. (2019) The role of mutation bias in adaptive molecular evolution: Insights from convergent changes in protein function. *Philosophical Transactions of the Royal Society B: Biological Sciences* 374: 20180238.

18. Monroe J.G. et al. (2022) Mutation bias reflects natural selection in *Arabidopsis thaliana*. *Nature* 602: 101–5.

19. Zhang J. (2022) Important genomic regions mutate less often than do other regions. *Nature*. News & Views: https://doi.org/10.1038/d41586-022-00017-6.

20. Downey R.G. and Fellows M.R. (1999) *Parameterized Complexity* (Springer, New York).

21. Neumann F. and Sutton A.M. (2020) Parameterized complexity analysis of randomized search heuristics, in Doerr B. and Neumann F. (eds.) *Theory of Evolutionary Computation: Recent Developments in Discrete Optimization* (Springer, Cham).

22. Dennett D. (1995) *Darwin's Dangerous Idea: Evolution and the Meanings of Life* (Simon Schuster, New York).

23. Werner A., Piatek M.J. and Mattick J.S. (2015) Transpositional shuffling and quality control in male germ cells to enhance evolution of complex organisms. *Annals of the New York Academy of Sciences* 1341: 156–63.

24. Werner A. et al. (2021) Widespread formation of double-stranded RNAs in testis. *Genome Research* 31: 1174–86.

25. Zhang W., Xie C., Ullrich K., Zhang Y.E. and Tautz D. (2021) The mutational load in natural populations is significantly affected by high primary rates of retroposition. *Proceedings of the National Academy of Sciences USA* 118: e2013043118.

26. Payne J.L. and Wagner A. (2019) The causes of evolvability and their evolution. *Nature Reviews Genetics* 20: 24–38.

27. Altenberg L. (1994) The evolution of evolvability in genetic programming, in K.E. Kinnear and P.J. Angeline (eds.) *Advances in Genetic Programming* (MIT Press, Cambridge, MA).

28. Wagner G.P. and Altenberg L. (1996) Perspective: Complex adaptations and the evolution of evolvability. *Evolution* 50: 967–76.

29. Eiben A.E. and Smith J. (2015) From evolutionary computation to the evolution of things. *Nature* 521: 476–82.

30. Melo D., Porto A., Cheverud J.M. and Marroig G. (2016) Modularity: Genes, development and evolution. *Annual Review of Ecology, Evolution, and Systematics* 47: 463–86.

31. Espinosa-Soto C. (2018) On the role of sparseness in the evolution of modularity in gene regulatory networks. *PLOS Computational Biology* 14: e1006172.

32. Dellinger A.S. et al. (2019) Modularity increases rate of floral evolution and adaptive success for functionally specialized pollination systems. *Communications Biology* 2: 453.

33. Clune J., Mouret J.-B. and Lipson H. (2013) The evolutionary origins of modularity. *Proceedings of the Royal Society B: Biological Sciences* 280: 1–11.

34. Diaz-de-la-Loza M-d-C., Loker R., Mann R.S. and Thompson B.J. (2020) Control of tissue morphogenesis by the HOX gene *Ultrabithorax*. *Development* 147: dev184564.

35. Kimura M. (1967) On the evolutionary adjustment of spontaneous mutation rates. *Genetical Research* 9: 23–34.

36. Tenaillon O., Taddei F., Radman M. and Matic I. (2001) Second-order selection in bacterial evolution: Selection acting on mutation and recombination rates in the course of adaptation. *Research in Microbiology* 152: 11–6.

37. Sultana T., Zamborlini A., Cristofari G. and Lesage P. (2017) Integration site selection by retroviruses and transposable elements in eukaryotes. *Nature Reviews Genetics* 18: 292–308.

38. Duret L. (2009) Mutation patterns in the human genome: More variable than expected. *PLOS Biology* 7: e1000028.

39. Hodgkinson A. and Eyre-Walker A. (2011) Variation in the mutation rate across mammalian genomes. *Nature Reviews Genetics* 12: 756–66.

40. Supek F. and Lehner B. (2015) Differential DNA mismatch repair underlies mutation rate variation across the human genome. *Nature* 521: 81–4.

41. Harris K. and Pritchard J.K. (2017) Rapid evolution of the human mutation spectrum. *eLife* 6: e24284.

42. Castellano D., Eyre-Walker A. and Munch K. (2019) Impact of mutation rate and selection at linked sites on DNA variation across the genomes of humans and other homininae. *Genome Biology and Evolution* 12: 3550–61.

43. Hawks J., Wang E.T., Cochran G.M., Harpending H.C. and Moyzis R.K. (2007) Recent acceleration of human adaptive evolution. *Proceedings of the National Academy of Sciences USA* 104: 20753–8.

44. Jablonka E. and Lamb M. (1994) *Epigenetic Inheritance and Evolution—The Lamarckian Dimension* (Oxford University Press, Oxford).

45. Jablonka E. and Lamb M.J. (1998) Epigenetic inheritance in evolution. *Journal of Evolutionary Biology* 11: 159–83.

46. Jablonka E. (2012) Epigenetic variations in heredity and evolution. *Clinical Pharmacology and Therapeutics* 92: 683–8.

47. Jablonka E. (2017) The evolutionary implications of epigenetic inheritance. *The Royal Society Interface Focus* 7: 20160135.

48. Skinner M.K. (2015) Environmental epigenetics and a unified theory of the molecular aspects of evolution: A neo-Lamarckian concept that facilitates neo-Darwinian evolution. *Genome Biology and Evolution* 7: 1296–302.

49. Drinnenberg I.A. et al. (2019) EvoChromo: Towards a synthesis of chromatin biology and evolution. *Development* 146: dev178962.

50. Ashe A., Colot V. and Oldroyd B.P. (2021) How does epigenetics influence the course of evolution? *Philosophical Transactions of the Royal Society B: Biological Sciences* 376: 20200111.

51. Melamed D. et al. (2022) De novo mutation rates at the single-mutation resolution in a human HBB gene-region associated with adaptation and genetic disease. *Genome Research* 32: 488–98.

52. Sharma S.V. et al. (2010) A chromatin-mediated reversible drug-tolerant state in cancer cell subpopulations. *Cell* 141: 69–80.

53. Planck M. (1949) *Scientific Autobiography and Other Papers* (Williams & Norgate, London).